Transgenic Animal Technology

Transgenic Animal Technology
A Laboratory Handbook

Third Edition

Edited by

Carl A. Pinkert

The University of Alabama
Tuscaloosa, AL, USA

AMSTERDAM • BOSTON • HEIDELBERG • LONDON • NEW YORK • OXFORD
PARIS • SAN DIEGO • SAN FRANCISCO • SINGAPORE • SYDNEY • TOKYO

Elsevier
32 Jamestown Road, London NW1 7BY
225 Wyman Street, Waltham, MA 02451, USA

Third edition 2014

Notices
Knowledge and best practice in this field are constantly changing. As new research and experience broaden our understanding, changes in research methods, professional practices, or medical treatment may become necessary.

Practitioners and researchers must always rely on their own experience and knowledge in evaluating and using any information, methods, compounds, or experiments described herein. In using such information or methods they should be mindful of their own safety and the safety of others, including parties for whom they have a professional responsibility.

To the fullest extent of the law, neither the Publisher nor the authors, contributors, or editors, assume any liability for any injury and/or damage to persons or property as a matter of products liability, negligence or otherwise, or from any use or operation of any methods, products, instructions, or ideas contained in the material herein.

British Library Cataloguing-in-Publication Data
A catalogue record for this book is available from the British Library

Library of Congress Cataloging-in-Publication Data
A catalog record for this book is available from the Library of Congress

ISBN: 978-0-12-410490-7

For information on all Elsevier publications
visit out website at store.elsevier.com

This book has been manufactured using Print On Demand technology. Each copy is produced to order and is limited to black ink. The online version of this book will show color figures where appropriate.

Contents

List of Contributors

Satoshi Akagi Animal Breeding and Reproduction Research Division, Institute of Livestock and Grassland Science, National Agriculture and Food Research Organization, Tsukuba, Ibaraki, Japan

Anna V. Anagnostopoulos Mouse Genome Informatics, The Jackson Laboratory, Bar Harbor, ME

Benjamin P. Beaton Division of Animal Sciences, University of Missouri, Columbia, MO

Cory F. Brayton Johns Hopkins University, Baltimore, MD

Steve Brown MRC Mammalian Genetics Unit, Harwell, Oxford, UK

Anthony W.S. Chan Yerkes National Primate Research Center, Department of Human Genetics, Department of Pediatrics, Emory University School of Medicine, Emory University, Atlanta, GA

Tom Doetschman University of Arizona, Tucson, AZ

Rex A. Dunham School of Fisheries, Aquaculture, and Aquatic Sciences, Auburn University, Auburn, AL

David A. Dunn Department of Biological Sciences, State University of New York at Oswego, Oswego, NY

Janan T. Eppig Mouse Genome Informatics, The Jackson Laboratory, Bar Harbor, ME

Almudena Fernández National Centre for Biotechnology (CNB-CSIC), Department of Molecular and Cellular Biology, Madrid, Spain

Tatiana Flisikowska Livestock Biotechnology, Technische Universität München, Freising, Germany

Vasiliy Galat Department of Pediatrics, Northwestern University Medical School and Developmental Biology Program, Ann & Robert H. Lurie Children's Hospital of Chicago Research Center, Chicago, IL

Robert A. Godke Embryo Biotechnology Laboratory, Department of Animal Science, Louisiana State University, Baton Rouge, LA

Philip Iannaccone Department of Pediatrics, Northwestern University Medical School and Developmental Biology Program, Ann & Robert H. Lurie Children's Hospital of Chicago Research Center, Chicago, IL

Michael H. Irwin Department of Pathobiology, College of Veterinary Medicine, Auburn University, Auburn, AL

Larry W. Johnson Department of Genetics, The University of Alabama at Birmingham, Birmingham, AL

Yoko Kato Laboratory of Animal Reproduction, College of Agriculture, Kinki University, Nakamachi, Nara, Japan

Teoan Kim Department of Physiology, Catholic University of Daegu School of Medicine, Daegu, Republic of Korea

Alexander Kind Livestock Biotechnology, Technische Universität München, Freising, Germany

Bon Chul Koo Department of Physiology, Catholic University of Daegu School of Medicine, Daegu, Republic of Korea

Mo Sun Kwon Department of Physiology, Catholic University of Daegu School of Medicine, Daegu, Republic of Korea

Daniel J. Ledbetter Department of Biology, University of Kentucky, Lexington, KY

Michael J. Martin Spring Point Project, Minneapolis, MN

Kazutsugu Matsukawa Multidisciplinary Science Cluster, Life and Environmental Medicine Science Unit, Kochi University, Nankoku, Kochi, Kerala, India

Colin McKerlie The Hospital for Sick Children and University of Toronto, Toronto, ON, Canada

Lluís Montoliu National Centre for Biotechnology (CNB-CSIC), Department of Molecular and Cellular Biology, Madrid, Spain

Paul E. Mozdziak Prestage Department of Poultry Science College of Agriculture and Life Sciences North Carolina State University Raleigh, NC USA

Akira Onishi Transgenic Pig Research Unit, Genetically Modified Organism Research Center, National Institute of Agrobiological Sciences, Tsukuba, Ibaraki, Japan

Paul A. Overbeek Department of Molecular and Cellular Biology, Baylor College of Medicine, Houston, TX

James N. Petitte Prestage Department of Poultry Science College of Agriculture and Life Sciences North Carolina State University Raleigh, NC USA

L. Philip Sanford University of Iowa, Iowa City, IA

Jorge A. Piedrahita Department of Molecular Biomedical Sciences, North Carolina State University, Raleigh, NC

Carl A. Pinkert Department of Biological Sciences, College of Arts and Sciences, The University of Alabama, Tuscaloosa, AL

Wendy K. Pogozelski Department of Chemistry, State University of New York at Geneseo, Geneseo, NY

H. Greg Polites Sanofi R + D Tucson, Sanofi Tucson Research Center, Oro Valley, AZ

Edmund B. Rucker III Department of Biology, University of Kentucky, Lexington, KY

Marina Sansinena Embryo Biotechnology Laboratory, Department of Animal Science, Louisiana State University, Baton Rouge, LA

Angelika Schnieke Livestock Biotechnology, Technische Universität München, Freising, Germany

Kumiko Takeda National Agricultural and Food Research Organization, NARO Institute of Livestock and Grassland Science, Tsukuba, Japan

James A. Thomson Plant Gene Expression Center, USDA-University of California, Berkeley, Albany, CA

Ian A. Trounce Center for Eye Research Australia, University of Melbourne, Department of Ophthalmology, Royal Victorian Eye and Ear Hospital, East Melbourne, Australia

Yukio Tsunoda Laboratory of Animal Reproduction, College of Agriculture, Kinki University, Nakamachi, Nara, Japan

Cristina Vicente-García National Centre for Biotechnology (CNB-CSIC), Department of Molecular and Cellular Biology, Madrid, Spain

Kevin D. Wells Division of Animal Sciences, National Swine Resource and Research Center, University of Missouri, Columbia, MO

Richard N. Winn Warnell School of Forest Resources, University of Georgia, Athens, GA

Curtis R. Youngs Department of Animal Science, Iowa State University, Ames, IA

Preface

In discussions regarding the third edition of "Transgenic Animal Technology: A Laboratory Handbook," we wrestled with the changing tide of technologies and a revision of the title for 2014. It has, after all, been over two decades since the original title was established. Was *transgenic* now a limiting term? Were the embryological, animal husbandry, reproductive biology, and molecular techniques something that should be highlighted in some bolder fashion? Yet it remained true that transgenic animal technologies and the ability to introduce and modify functional genes in animal models continue to represent powerful, dynamic, and evolving tools for dissecting complex biological processes. The questions to be addressed span the scientific spectrum from biomedical and biological applications to production agriculture. And yes, as transgenic methodologies continue to evolve, they have dramatically influenced a cross section of disciplines and are recognized as instrumental in expanding our understanding of gene expression, regulation, and function.

There are many general reviews on the topic of animal transgenesis and genetic engineering that are indeed very useful and timely. However, aside from the manuals devoted to mouse embryology, a single text illustrating the methodologies employed by leading laboratories in their respective disciplines had not previously been compiled prior to the first edition of this text. This third edition covers technical aspects of gene transfer in animals—from molecular methods to whole animal considerations across a host of species. Consequently, we kept the title and focus for this handbook, and it is envisioned as a bridge for researchers and a tool to facilitate training of students and technicians in the development and use of numerous transgenic animal model systems.

Clearly, much has changed from a technological perspective since the first two editions in 1994 and 2002. With this in mind, I would like to acknowledge all of the contributing authors both past and present, as well as those individuals in my laboratory who have assisted over the course of these three editions. In preparation of the three editions, a number of colleagues have graciously provided assistance in reviewing specific chapters and in some cases providing additional data for consideration prior to publication, for which we are all most grateful. I would also like to acknowledge all of my mentors, colleagues, and those who came through my laboratory over the years. Lastly, I am appreciative of the consideration and help provided by Halima Williams, Radhakrishnan Lakshmanan and their colleagues at Elsevier for facilitating publication of this third edition.

Carl A. Pinkert
Tuscaloosa, Alabama

Section One

Overview

1 Introduction to Transgenic Animal Technology

Carl A. Pinkert

Department of Biological Sciences, College of Arts and Sciences, The University of Alabama, Tuscaloosa, AL

I. Introduction

The last three decades witnessed a rapid advance of the application of genetic engineering techniques for increasingly complex organisms, from single-cell microbial and eukaryotic culture systems to multicellular whole-animal systems. The whole animal is generally recognized as an essential tool for biomedical and biological research, as well as for pharmaceutical development and toxicological/safety screening technologies. Moreover, an understanding of the developmental and tissue-specific regulation of gene expression is achieved only through *in vivo* whole-animal studies.

Today, particularly with gene editing technologies on the rise, transgenic animals (and animal biotechnologies) continue to embody one of the most potent and exciting research tools in the biological sciences. Transgenic animals represent unique models that are custom tailored to address specific biological questions. Hence, the ability to introduce functional genes into animals provides a very powerful tool for dissecting complex biological processes and systems. Gene transfer is of particular value in those animal species where long life cycles reduce the value of classical breeding practices for rapid genetic modification. For identification of interesting new models, genetic screening and characterization of chance mutations remain a long and arduous task. Furthermore, classical genetic monitoring cannot adequately engineer a specific genetic trait in a directed fashion.

II. Historical Background

In the early 1980s, only a handful of laboratories possessed the technology necessary to produce transgenic animals. With this in mind, this text is envisioned as a bridge to the development of various transgenic animal models. Hopefully, through the first two editions the curiosity and interests of researchers in diverse research fields were influenced productively. The gene transfer technology that is currently utilized across vertebrate species was pioneered using mouse and domestic animal models. Today, the mouse continues to serve as a starting point

Transgenic Animal Technology. DOI: http://dx.doi.org/10.1016/B978-0-12-410490-7.00001-3

for implementing a variety of gene transfer procedures and is the standard for optimizing experimental efficiencies for many species. Inherent species differences are frequently discounted by researchers who are planning studies with a more applicable species model. However, when one attempts to compare experimental results generated in mice to those obtained in other species, not surprisingly many differences become readily apparent. Therefore, an objective of this text will be to address the adaptation of relevant protocols.

When initiating work related to gene transfer, it is important to look at the rapid advancement of a technology that is still primitive by many standards. From an historical perspective, one readily contemplates potential technologies and methods that lie just ahead. Whereas modern recombinant DNA techniques are of primary importance, the techniques of early mammalian embryologists were crucial to the development of gene transfer technology. While we can look at well over three decades of transgenic animal production, the preliminary experiments leading to this text go back millennia to the first efforts to artificially regulate or synchronize embryo development. Amazingly, it has been more than a century since the first successful embryo transfer experiments, dating back to the efforts first published in the 1880s and to Heape's success in 1891 (Heape, 1891). By the time the studies by Hammond were reported in the late 1940s, culture systems were developed that sustained ova through several cleavage divisions. Such methods provided a means to systematically investigate and develop procedures for a variety of egg manipulations. These early studies led to experiments that ranged from mixing of mouse embryos and production of chimeric animals, to the transfer of inner cell mass cells and teratocarcinoma cells, to nuclear transfer and the first injections of nucleic acids into developing ova. Without the ability to culture or maintain ova *in vitro*, such manipulations or the requisite insights would not be possible (Brinster and Palmiter, 1986).

Gurdon (1977) transferred mRNA and DNA into *Xenopus* eggs and observed that the transferred nucleic acids could function in an appropriate manner. This was followed by a report by Brinster et al. (1980) of similar studies in a mammalian system, using fertilized mouse ova in initial experiments. Here, using rabbit globin mRNA, an appropriate translational product was obtained.

Major turning points in science continue to accelerate at an incredible pace. The technology available in 1994 had developed considerably and, as predicted, a number of areas from both the first and second editions of this text are clearly antiquated or obsolete today. It is amazing to look back at the major events related to genetic engineering of animals and how our ability to manipulate both the nuclear and mitochondrial genomes has come so far.

The production of transgenic mice has been hailed as a seminal event in the development of animal biotechnology. In reviewing the early events leading to the first genetically engineered mice, it is fascinating to note that the entire procedure for DNA microinjection was described nearly 50 years ago. While some progress seems extremely rapid, it is still difficult to believe that, following the first published report of a microinjection method in 1966 (Lin, 1966; Figures 1.1 and 1.2), it would be another 15 years before transgenic animals were created. The pioneering laboratories that reported success at gene transfer would

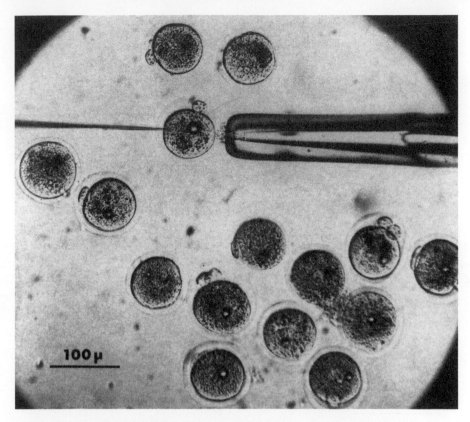

Figure 1.1 Microinjection of murine zygotes dating back to the 1960s. The initial procedures for DNA microinjection were outlined in 1966. Here, zygotes are being injected with oil droplets. The zygotes survived this mechanical trauma, from use of holding pipettes to insertion of an injection pipette.
Source: Reprinted with permission from Lin (1966).

not have been able to do so were it not for the recombinant DNA technologies necessary to develop protocols or document results (Gordon et al., 1980; Wagner et al., 1981a,b; Harbers et al., 1981; Brinster et al., 1981; Costantini and Lacy, 1981; Gordon and Ruddle, 1981). In gene transfer, animals carrying new genes (integrating foreign DNA segments into their genome) are referred to as *transgenic*, a term first coined by Gordon and Ruddle (1981). As such, transgenic animals were recognized as specific variants of species following the introduction and/or integration of a new gene or genes into the genome. As for many technologies, the definition of transgenic animals has taken on a broader meaning and perspective that is more inclusive and includes animals either integrating foreign DNA segments into their genome following gene transfer or resulting from the molecular manipulation of endogenous genomic DNA (Pinkert et al., 1995). Yet, as outlined by Beardmore (1997), this definition, too, required refinement as

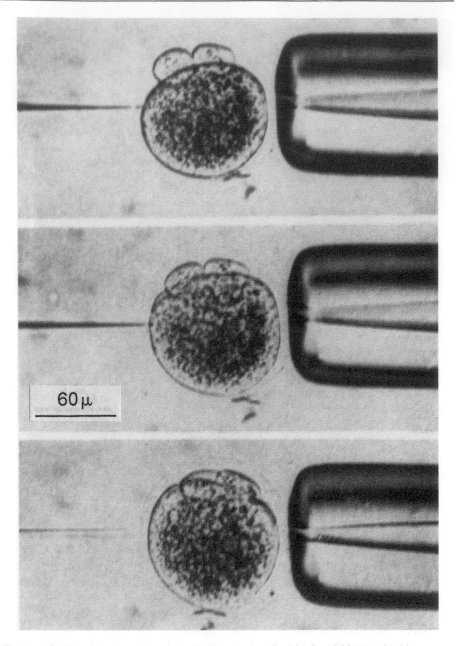

Figure 1.2 Microinjection of murine zygotes. As described in the 1966 paper by Lin, zygotes survived not only the mechanical trauma associated with the rudimentary injection procedures but also the injection of a bovine gg-globulin solution as well.
Source: Reprinted with permission from Lin (1966).

"state-of-the-art" technologies continue to evolve. And yes, we now include both the nuclear genome as well as the mitochondrial genome with the inclusion of transmitochondrial animal modeling first described in the mid-1990s (where we coined the term *mitomice* in 1997, see Wawrousek, 1998) (Dunn et al., 2012; Irwin et al., 2013).

Through the years, there have been literally thousands of excellent reviews that detail the production of transgenic animals, in addition to a journal, *Transgenic Research*, dedicated to this field. In the first edition, readers were referred to now classical reviews by Brinster and Palmiter (1986), Bürki (1986), Camper (1987), Cordaro (1989), First and Haseltine (1991), Grosveld and Kollias (1992), Hogan et al. (1986), Palmiter and Brinster (1986), Pattengale et al. (1989), Pinkert (1987), Pinkert et al. (1990), Pursel et al. (1989), Rusconi (1991), Scangos and Bieberich (1987), and Van Brunt (1988). However, to this day, in my opinion, the singular effort with the greatest influence on propelling this technology would not be among the initial reports just described. Rather, early work of Ralph Brinster and Richard Palmiter related to growth performance and the dramatic phenotype of growth hormone (GH) transgenic mice (Figure 1.3) subsequently influenced the emerging field in a most compelling manner for both basic and applied sciences (Palmiter et al., 1982, 1983).

III. Applications and Overview of Text

Scientists have envisioned many potential studies and applications if an animal genome could be readily modified. Therefore, the realization of the many technologies at hand today has opened new avenues of research promise. Production of transgenic mice marked the convergence of previous advances in the areas of recombinant DNA technology and manipulation and culture of animal cells and embryos. Transgenic mice provide a powerful model to explore the regulation of gene expression and of cellular and physiological processes.

The use of transgenic animals in biomedical, agricultural, biological, and biotechnological arenas requires the ability to target gene expression and to control the timing and level of expression of specific genes. Experimental designs have taken advantage of the ability to direct specific expression (including cell types, tissue, organ type, and a multiplicity of internal targets) and ubiquitous, whole-body expression *in vivo*.

From embryology to virology, the applications of transgenic mice provide models in many disciplines and research areas. Examples include the following:

- *Genetic bases of human and animal disease and the design and testing of strategies for therapy.* Many human diseases either do not exist in animals or are developed only by "higher" mammals, making models scarce and expensive. Many times, an animal model does not exist and the rationale for development is limited.
- *Disease resistance in humans and animals.* From a basic research and ethical standpoint, it would appear as a moral imperative that we develop models for enhancing characteristic well-being of all species.

Figure 1.3 Production of transgenic mice harboring a GH fusion transgene construct. Animals harboring and expressing the GH transgene grew at a rate two- to fourfold greater than control littermates, reaching a mature weight twice that of controls. This dramatic phenotype led the way for the exponential development of gene transfer technology. *Source*: Reprinted with permission from Palmiter et al. (1982).

- *Gene therapy*. Models for growth, immunological, neurological, reproductive, and hematological disorders have been developed. Circumvention and correction of genetic disorders are now possible to address using a variety of experimental methods.
- *Drug and product testing and screening*. Toxicological screening protocols are already in place that utilize transgenic animal systems. For preclinical drug development, from a fundamental research perspective, a whole-animal model for screening is essential for understanding disease etiology, investigating drug pharmacokinetics, and evaluating therapeutic efficacy. A comparable and validated need is crucial to product safety testing as well.

- *Novel product development through "molecular pharming."*[1] In domestic animals, biomedical proteins have been targeted to specific organs and body fluids with reasonable production efficiencies. Tissue plasminogen activator (TPA), human factor IX, and human α1 antitrypsin are a few products produced in transgenic animals in different stages of validation and commercialization.
- *Production agriculture.* Long term, it may become possible to produce animals with enhanced characteristics that will have profound influences on the food we eat, influences ranging from production efficiency to the inherent safety of our food supply.

The numerous strategies for producing genetically engineered animals extend from mechanistic (e.g., DNA microinjection, embryonic stem cell transfer, nuclear transfer, or retrovirus-mediated transfer) as well as molecular (cloning) techniques. As the chapters of this text unfold, it will be apparent that the technology has extended to a variety of animal species in addition to the mouse, including the production of transgenic rats, rabbits, swine, ruminants (sheep, goats, and cattle), poultry, and fish. Although genetically engineered amphibians, insects, nematodes, lower eukaryotes and prokaryotes, and members of the plant kingdom have been acknowledged in the literature, such models are beyond the scope of this text.

Advances in the understanding of promoter—enhancer sequences and external-transcription regulatory proteins involved in the control of gene expression continue to evolve using different model systems. In the systems explored in this text, gene transfer technology is a proven asset in science as a means of dissecting gene regulation and expression *in vivo*. However, the primary question that is addressed concerns the particular role of a single gene in development or in a given developmental pathway. With this caveat, considerations include the ramifications of gene activity from intracellular to inter- and extracellular events within a given tissue or cell-type milieu. Normally, gene function is influenced by cis-acting elements and trans-acting factors. For transferred genes, the cis- and trans-activators, in conjunction with the gene integration/insertion event within the host genome, influence regulation of both endogenous and transferred genes. Using genes that code for (or are composed of) reporter proteins (e.g., GH or lacZ constructs), analysis of transgenic animals revealed the importance of those three factors in determining the developmental timing, efficiency, and tissue distribution of gene expression. Additionally, transgenic animals have proved quite useful in unraveling *in vivo* artifacts of other model systems or techniques.

Although gene transfer technology continues to open new and unexplored biological frontiers, it also raises questions concerning regulatory and commercialization issues. It is not within the scope of this text, however, to fully address these issues. Suffice it to say that a number of issues exist and will continue to plague the development of many of the systems described herein. Major aspects

---

[1]In the first two editions, I resisted using the term "pharming" in light of the various nonpharmaceutical endpoints targeted in many research labs. However, the pharmaceutical industry has proven to be a driving force in effective utilization of genetically modified animals that benefit society at large.

of the regulation of this technology will focus on the following issues in the twenty-first century:

- Environmental impact following "release" of transgenic animals
- Public perceptions
- Ethical considerations
- Legislation
- Safety of transgenic foodstuffs
- Patent aspects and product uniformity/economics.

Contrary to the early prospects related to mainstreaming of this technology, there are numerous societal challenges regarding potential risks that are still ahead. The potential risks at hand in 1994 and 2002 are still with us, if not more complex, today. They still can be defined by scientific evidence and also in relation to public concern (whether perceived or real). Amazingly, the central questions still revolve around the proper safeguards to employ and the development of a coherent and unified regulation of the technology. Can new animal reservoirs of fatal human diseases be created? Can more virulent pathogens be artificially created? What is the environmental impact of the "release" of genetically engineered animals? Do the advantages of bioengineered products outweigh potential consequences of their use? These are but a few of the questions that researchers still cannot ignore and must approach. They are not alone, however, as the many regulatory hurdles that exist today will challenge not only scientists and policy makers, but sociologists, ethicists, and legal scholars as well. Fortuitously or not, such important and often controversial issues have provided continuing employment opportunities for many on either side of any discussion.

The chapters in this text outline the basic techniques that various laboratories currently use to genetically modify, develop, and characterize transgenic animals. The methods used to initiate experiments, develop vector systems, maintain animals (and the associated husbandry and experimental needs), and analyze and evaluate animals, with the requisite strategies to enhance experimental efficiency, are described at each step of experimentation. Discussion of all interlaboratory variations for each procedure is not feasible. As the authors have learned, the strategies associated with the production of transgenic animals are quite variable even between laboratories that utilize the same systems. Therefore, in some instances, alternatives to be published and commonly used techniques are presented. However, most of the techniques for extensions to other systems are unique and timely for new investigators. The overall efficiency of many procedures will vary, as will the cost−benefit ratios, but do not let the mechanics of experimentation outweigh the most important reason that one enters into these studies (unless you are a postdoctoral fellow looking for a niche and eventual job placement; again a critical concern to many embarking on their careers), which is the development and characterization of a biological model with specific utility.

With so many colleagues providing insights from their creative experimentation, our goal is to illustrate a number of variations or novel methods that differ from the standard protocols outlined in detail for the mouse. (Note some of the earlier references on embryology and micromanipulation of ova were cited earlier in addition to Brinster et al., 1985; Chen et al., 1986; Daniel, 1971; Dunn et al., 2005;

Hanahan, 1988; Hogan et al., 1994; Nagy et al., 2002; Rafferty, 1970) This text is organized in a manner to assist those interested in developing an understanding of the basic species differences in transgenic animal research.

For the novice or new trainee, as well as for the experienced researcher, this text should influence proficiency and ultimately help provide an increase in overall productivity. For those wishing to develop a transgenic animal research program, this text will provide an overview of the requirements needed for development of a comprehensive gene transfer program.

There is only one take-home message to readers beyond the development of a desired biological model. An appreciation for the effort involved in each step of experimentation is most important in order to see a project through, from its design and implementation to the validation of a defined animal model. From a personal standpoint, one cannot discount the equal importance of the many unrelated disciplines, from molecular to whole-animal biology, and the necessary training to ensure the overall success of transgenic animal technology.

References

Beardmore, J.A., 1997. Transgenics: autotransgenics and allotransgenics. Transgenic Res. 6, 107–108.

Brinster, R.L., Palmiter, R.D., 1986. Introduction of genes into the germ line of animals. Harvey Lect. 80, 1–38.

Brinster, R.L., Chen, H.Y., Trumbauer, M.E., Avarbock, M.R., 1980. Translation of globin messenger RNA by the mouse ovum. Nature (London) 282, 499–501.

Brinster, R.L., Chen, H.Y., Trumbauer, M.E., Senear, A.W., Warren, R., Palmiter, R.D., 1981. Somatic expression of herpes thymidine kinase in mice following injection of a fusion gene into eggs. Cell 27, 223–231.

Brinster, R.L., Chen, H.Y., Trumbauer, M.E., Yagle, M.K., Palmiter, R.D., 1985. Factors affecting the efficiency of introducing foreign DNA into mice by microinjecting eggs. Proc. Natl. Acad. Sci. USA 82, 4438–4442.

Bürki, K. (Ed.), 1986. Monographs in Developmental Biology, vol. 19. Karger, New York, NY.

Camper, S.A., 1987. Research applications of transgenic mice. BioTechniques 5, 638–650.

Chen, H.Y., Trumbauer, M.E., Ebert, K.M., Palmiter, R.D., Brinster, R.L., 1986. Developmental changes in the response of mouse eggs to injected genes. In: Bogorad, L. (Ed.), Molecular Developmental Biology. Alan R. Liss, New York, NY, pp. 149–159.

Cordaro, J.C., 1989. Transgenic mice as future tools in risk assessment. Risk Anal. 9, 157–168.

Costantini, F., Lacy, E., 1981. Introduction of a rabbit b-globin gene into the mouse germ line. Nature (London) 294, 92–94.

Daniel, J.C., 1971. Methods in Mammalian Embryology. Freeman, San Francisco, CA.

Dunn, D.A., Kooyman, D.L., Pinkert, C.A., 2005. Transgenic animals and their impact on the drug discovery industry. Drug Discov. Today 10, 757–767.

Dunn, D.A., Cannon, M.V., Irwin, M.H., Pinkert, C.A., 2012. Animal models of human mitochondrial DNA mutations. Biochim. Biophys. Acta 1820, 601–607.

First, N., Haseltine, F.P., 1991. Transgenic Animals. Butterworth-Heinemann, Stoneham, MA.

Gordon, J.W., Ruddle, F.H., 1981. Integration and stable germ line transmission of genes injected into mouse pronuclei. Science 214, 1244–1246.

Gordon, J.W., Scangos, G.A., Plotkin, D.J., Barbosa, J.A., Ruddle, F.H., 1980. Genetic transformation of mouse embryos by microinjection of purified DNA. Proc. Natl. Acad. Sci. USA 77, 7380−7384.

Grosveld, F., Kollias, G., 1992. Transgenic Animals. Academic Press, San Diego, CA.

Gurdon, J.B., 1977. Egg cytoplasm and gene control in development. Proc. R. Soc. Lond. B. 198, 211−247.

Hanahan, D., 1988. Dissecting multistep turmorigenesis in transgenic mice. Annu. Rev. Genet. 22, 479−519.

Harbers, K., Jahner, D., Jaenisch, R., 1981. Microinjection of cloned retroviral genomes into mouse zygotes; integration and expression in the animal. Nature (London) 293, 540−542.

Heape, W., 1891. Preliminary note on the transplantation and growth of mammalian ova within a uterine foster mother. Proc. R. Soc. Lond. 48, 457−458.

Hogan, B., Costantini, F., Lacy, E., 1986. Manipulating the Mouse Embryo: A Laboratory Manual. Cold Spring Harbor Laboratory, Cold Spring Harbor, NY.

Hogan, B., Beddington, R., Costantini, F., Lacy, E., 1994. Manipulating the Mouse Embryo: A Laboratory Manual. second ed. Cold Spring Harbor Press, Cold Spring Harbor, NY.

Irwin, M.H., Parameshwaran, K., Pinkert, C.A., 2013. Mouse models of mitochondrial complex I dysfunction. Int. J. Biochem. Cell Biol. 45, 34−40.

Lin, T.P., 1966. Microinjection of mouse eggs. Science 151, 333−337.

Nagy, A., Gertsenstein, M., Vintersten, K., Behringer, R., 2002. Manipulating the Mouse Embryo: A Laboratory Manual. third ed. Cold Spring Harbor Press, Cold Spring Harbor, NY.

Palmiter, R.D., Brinster, R.L., 1986. Germ-line transformation of mice. Annu. Rev. Genet. 20, 465−499.

Palmiter, R.D., Brinster, R.L., Hammer, R.E., Trumbauer, M.E., Rosenfeld, M.G., Birnberg, N.C., et al., 1982. Dramatic growth of mice that develop from eggs microinjected with metallothionein-growth hormone fusion genes. Nature (London) 300, 611−615.

Palmiter, R.D., Norstedt, G., Gelinas, R.E., Hammer, R.E., Brinster, R.L., 1983. Metallothionein−human GH fusion genes stimulate growth of mice. Science 222, 809−814.

Pattengale, P.K., Stewart, T.A., Leder, A., Sinn, E., Muller, W., Tepler, I., et al., 1989. Animal models of human disease. Am. J. Pathol. 135, 39−61.

Pinkert, C.A., 1987. Gene transfer and the production of transgenic livestock. Proc. U.S. Anim. Health Assoc. 91, 129−141.

Pinkert, C.A., Dyer, T.J., Kooyman, D.L., Kiehm, D.J., 1990. Characterization of transgenic livestock production. Domest. Anim. Endocrinol. 7, 1−18.

Pinkert, C.A., Irwin, M.H., Moffatt, R.J., 1995. Transgenic animal modeling. In: Meyers, R.A. (Ed.), Molecular Biology and Biotechnology. VCH Publishing, New York, NY, pp. 901−907.

Pursel, V.G., Pinkert, C.A., Miller, K.F., Bolt, D.J., Campbell, R.G., Palmiter, R.D., et al., 1989. Genetic engineering of livestock. Science 244, 1281−1288.

Rafferty, K.A., 1970. Methods in Experimental Embryology of the Mouse. Johns Hopkins Press, Baltimore, MD.

Rusconi, S., 1991. Transgenic regulation in laboratory animals. Experientia 47, 866−877.

Scangos, G., Bieberich, C., 1987. Gene transfer into mice. Adv. Genet. 25, 285−322.

Van Brunt, J., 1988. Molecular farming: transgenic animals as bioreactors. Biotechnology 6, 1149−1154.

Wagner, E.F., Stewart, T.A., Mintz, B., 1981a. The human b-globin gene and a functional viral thymidine kinase gene in developing mice. Proc. Natl. Acad. Sci. USA 78, 5016−5020.

Wagner, T.E., Hoppe, P.C., Jollick, J.D., Scholl, D.R., Hodinka, R.L., Gault, J.B., 1981b. Microinjection of a rabbit b-globin gene into zygotes and its subsequent expression in adult mice and their offspring. Proc. Natl. Acad. Sci. USA 78, 6376−6380.

Wawrousek, E., 1998. Meeting Review: IBC Conference on Engineered Animal Models, Washington, DC. Transgenic Res. 7, 141−145.

Section Two

Transgenic Animal Production Focusing on the Mouse Model

2 DNA Microinjection, Embryo Handling, and Germplasm Preservation

H. Greg Polites[1], Larry W. Johnson[2] and Carl A. Pinkert[3]

[1]Sanofi R + D Tucson, Sanofi Tucson Research Center, Oro Valley, AZ,
[2]Department of Genetics, The University of Alabama at Birmingham,
Birmingham, AL, [3]Department of Biological Sciences, College of
Arts and Sciences, The University of Alabama, Tuscaloosa, AL

I. Introduction

In the beginning, DNA microinjection was the most widely applied method for gene transfer in mammals. Then, with a variety of technologies building on DNA microinjection and related techniques, more options became available. While relative efficiencies are debated and new methodologies explored, DNA microinjection still has utility in a variety of applications. Additionally, and over the course of this chapter, it is our intention to identify important criteria for the selection of this methodology, principles and requirements of the technology, what expectations one might develop in the course of experimentation, and a number of ancillary or enabling technologies, including gamete cryopreservation for use after transgenic lineages are established. All animals utilized in these procedures or in the experiments that follow are cared for according to the NIH (National Institutes of Health) and OLAW (Office of Laboratory Animal Welfare, formerly the US Office for Protection from Research Risks) guidelines for appropriate husbandry under strict barrier adherence.

II. General Methods

A. DNA Preparation and Purification

1. DNA Construct/Fragment Structure

In comparison with other gene transfer methods (particularly in relation to retroviral packaging), it seemed that DNA microinjection was one of the only methods where fragment size was not at issue. However, fragment size constraints were originally imposed by plasmid (12 kb) and cosmid (45 kb) cloning vectors.

Transgenic Animal Technology. DOI: http://dx.doi.org/10.1016/B978-0-12-410490-7.00002-5

As related technology developed, the ability to clone larger *stable* constructs has been extended to yeast artificial chromosome (YAC), P1, and bacterial artificial chromosome (BAC) vectors (Sternberg, 1992; Schedl et al., 1993; Yang et al., 1997; Camper and Saunders, 2000), as well as to the transfer of large chromosomal fragments (megabase lengths; Richa and Lo, 1989).

Although genomic fragments can be rather large and difficult to isolate and clone, some studies demonstrate that homologous recombination can be effected using DNA injection into mouse zygotes. In the first example, using injection of a major histocompatibility complex (MHC) class II transgene, homologous recombination was effected in 1 of 500 mice incorporating the transgene (Brinster et al., 1989). In a second example (Pieper et al., 1992), three DNA fragments in the 30−40 kb range with 2.5−3.0 kb of overlapping ends were coinjected and homologously recombined before inserting into the mouse chromosome. Treating fragment ends or heterologous ends with phosphatase was not necessary to prevent concatemer formation. Apparently recombination was a faster event compared to DNA end ligation. As such, use of overlapping fragments may prove to be a convenient method to circumvent cloning and construction of large gene constructs.

2. Factors that Influence Transgenic Mouse Production

Many of the criteria for successful production of transgenic mice have been defined (Brinster et al., 1985; Brinster and Palmiter, 1986; Chen et al., 1986; Pinkert, 1994, 2002; Behringer et al., 2014; and many, many others since the mid-1980s). Briefly, to optimize experimental efficiencies the following considerations are important.

1. Linear DNA fragments integrate with greater efficiency than supercoiled DNA. The DNA fragment size/length does not generally affect integration frequency.
2. A low ionic strength microinjection buffer should be prepared (10 mM Tris pH 7.4 with 0.1−0.3 mM EDTA).
3. The efficiency of producing transgenic mice (DNA integration and development of microinjected eggs to term) appears most efficient at a DNA concentration between 1.0 and 3.0 ng/μL.
4. Linear DNA fragments with blunt ends have the lowest chromosomal integration frequency, whereas dissimilar ends are more efficient.
5. Injection of DNA into the male pronucleus is slightly more efficient than injection into the female pronucleus for producing transgenic mice.
6. Nuclear injection of foreign DNA is dramatically more efficient than cytoplasmic injection.

We used an elongation factor 1-alpha promoter (EF-1α; Kim et al., 1990) driving a lacZ reporter gene (EF-GAL) to confirm these findings. This construct allows the assessment of conditions affecting integration efficiency and minimizes the confounding losses during *in vivo* development. After microinjection of EF-GAL into pronuclear eggs, eggs were cultured to the hatched blastocyst stage and stained for β-galactosidase (β-gal) activity.

The EF-GAL construct was used to analyze the effect of DNA concentration and the type of DNA ends on foreign gene integration. Our results confirmed that integration increases rapidly with higher DNA concentrations; however, DNA is toxic to the egg, and viability diminishes as the DNA concentration increases. The integration frequency at 2 ng/μL was 22% and that at 6 ng/μL was 54%. In another series of experiments, we allowed the blastocysts to attach and the inner cell mass (ICM) to grow out. When they were stained, a surprisingly high degree of mosaicism in the positive-staining ICMs was observed.

In addition, we have used an elastase-EJ ras fusion construct (Quaife et al., 1987; Pinkert, 1990) for trainee preparation as well as foreign gene integration studies. The latter construct produces a visible phenotype (abdominal enlargement, pancreatic tumor formation, and ascites accumulation) at days 18−20 of gestation. While obviating the need for biochemical or molecular analyses, this construct affords the trainee an opportunity to evaluate the entire spectrum of procedures necessary to produce transgenic mice (described in greater detail in Chapter 3).

3. Preparation of DNA for Microinjection

a. Construction and Isolation of DNA

A given DNA fragment, after verification of correct cloning and confirmation of the nucleotide sequence, is purified through several steps to ensure that the DNA fragment does not contain nicks or strand breaks and is as pure as possible.

For these final steps, one should use the best reagents available. All solutions are made from tissue culture grade, 18 MΩ water. Commercially available water systems, in combination with media preparations described below, have been used since the early 1980s. Ultrafiltration or deionization followed by reverse osmosis and distillation provides water that is free of viruses and mycoplasmas. Good results can be obtained using water systems produced by most manufacturers (e.g., Milli-Q, Millipore, Bedford, MA; NANOpure, Barnstead, Dubuque, IA; Mega-Pure, Corning, Corning, NY). After preparation, solutions are also processed using a 0.45- or 0.2-μm filter.

Plasmid DNA can be replicated in any standard *Escherichia coli* host, but a DH5 methylase-defective strain (Invitrogen Life Technologies) that is both MCR⁻ and MRR⁻ will help maintain the methylation state of the gene construct during cloning. It also allows for transgene rescue to help in the analysis of flanking regions of the chromosomal insertion for toxicology applications or to localize insertional mutants (Short et al., 1989). Methods for basic cloning and assembly of gene constructs and methods for large-scale preparation of DNA are well documented (Green and Sambrook, 2012).

Before a given gene fragment has been linearized and is readied for microinjection, rather than isolation of supercoiled DNA from a cesium chloride (CsCl) gradient, plasmid DNA can be purified using a host of plasmid purification kits currently on the market (e.g., Qiagen or Promega). It is important that the fragment of interest is isolated from as many of the vector sequences as possible. Numerous studies point to the relative severity and/or influence of vector DNA sequences on the function of foreign genes after chromosomal integration (reviewed in Brinster and Palmiter, 1986; Rusconi, 1991). For the most part, we routinely minimize the amount of flanking vector sequences for injection fragments. Additionally, the construct should be designed so that comigrating bands of DNA are easily separated on agarose gels.

One should avoid any preparative steps or conditions that might introduce nicks or contaminants into the purified DNA. Overdigestion or partial digestion of DNA with restriction enzymes, use of excessive heat to dissolve dried DNA, vigorous vortexing or pipetting, use of unsterilized pipette tips or microcentrifuge tubes, exposure to short-wavelength ultraviolet light, exposure to varying temperatures and DNase activity, and exposure to equipment that has not been cleaned or sterilized are examples of standard problems that should be avoided or carefully controlled in preparing DNA for microinjection.

b.   DNA Purification

Once cut from a given vector, the DNA sample can be isolated by electrophoresis through agarose (SeaKem GTG grade; FMC, Rockland, ME). After adequate separation of the DNA bands in agarose, we generally purify fragments by either electroelution, followed by desalting and concentration by running the DNA on a column of DEAE-Sephacel, or glass bead adsorption (which we prefer; e.g., Qiaex II, Qiagen, Valencia, CA; Geneclean, Qbiogene, Inc., Carlsbad, CA; see also Green and Sambrook, 2012). Following purification, we generally measure the DNA in a 1-μL volume and then dilute to 2−3 ng/μL in microinjection buffer and determine the DNA concentration of the microinjection aliquot. At this point, the microinjection aliquot, now in a microcentrifuge tube sealed with Parafilm, may be stored at 4°C or frozen at −20°C.

There are numerous possible modifications to the purification protocol employed, and all are practical if they ensure that the final DNA preparation provides DNA that is (i) free of salts, organic solvents, or traces of agarose, (ii) in the correct and sterile buffer, and (iii) not degraded (e.g., sheared or nicked).

c.   Purification and Quantification of DNA

Routinely, we cut enough plasmid DNA containing the fragment of interest to yield 5−20 μg of insert and use a large preparative agarose gel to isolate a given DNA fragment. This allows for large losses common to long purification protocols and produces enough DNA for accurate quantification and dilution into the microinjection buffer. We usually try to obtain a final dilution ratio of 1:20 or greater and aliquot with a minimal volume of 300 μL in a 500-μL microcentrifuge tube, to prevent evaporation from changing the concentration.

Quantification of DNA can be performed using either a fluorometer with calf thymus standards or comparative size standards of known concentration on an agarose gel containing ethidium bromide. There are DNA-specific fluorometers and a host of accurate spectrophotometers that will provide accurate assessments (e.g., NanoDrop Spectrophotometers, Thermo Scientific; Qubit Fluorometer, Invitrogen Life Technologies). However, DNA comparisons on agarose gels allow a final quality control check on a given sample to verify both fragment size and purity.

With either method, precision is very important. Integration frequencies for foreign genes are concentration dependent (Brinster et al., 1985) but egg viability is inversely related to DNA concentration. At DNA concentrations of $1-2$ ng/μL injection buffer, viability averages 20% and still allows on average 30% liveborn transgenic mice. At higher concentrations (e.g., 6 ng/μL), integration frequencies can increase to 60%, but viability drops to 10% or less and reaches the point where there are insufficient uterine implantations for pregnancy maintenance.

B. Superovulation, Egg Culture, and Harvest

1. Mouse Strains for Microinjection and Transgenic Models

a. Strains

Several different hybrids are currently popular for DNA microinjection; the C57BL/6 × SJL (B6SJL) F1 hybrid has been shown to be efficient in the generation of transgenic mice (Brinster et al., 1985; Johnson et al., 1996; Vergara et al., 1997). The majority of the hybrids used for microinjection utilize the C57BL/6 inbred strain as one of the parental stocks due to favorable genetic and embryological characteristics. Hybrid mice are popular because hybrid vigor not only imparts desirable reproductive characteristics but also enhances the egg quality, hence leading to desirable microinjection characteristics. However, one should be concerned with the uniformity of the genetic background in which the transgene will be functioning, particularly in experimental designs where large populations or many generations of transgenic mice will be required.

Concepts related to the genetic background of mouse strains, including isogenicity (the degree to which individuals of a strain are genetically identical) and homogenicity (the degree to which individuals are homozygous at all genetic loci) are now influencing the characterization of transgenic models as lines are expanded. The initial selection of hybrid strains for gene transfer work focused on the efficiency of maintenance and reproduction, as well as on known embryology and responses to experimental manipulations. However, current applications necessitate transgenic mouse models produced in specific background strains. Early reports illustrated that transgene expression can be readily modulated or suppressed by "background" genetics of particular strains (Harris et al., 1988; Chisari et al., 1989; Engler et al., 1991). In these examples, transgenic mice backcrossed to particular strains show characteristic repression or enhancement of transgene expression. In experiments where other strains are utilized, losses in experimental efficiencies will be encountered. In some instances (e.g., using FVB or C57BL/6 strains), efficiency losses may be minimal, but conditions for manipulation and timing of specific biological end points (e.g., pronuclear egg formation, response to gonadotropins) using various strains must be identified (Johnson et al., 1996; Vergara et al., 1997). As with strains used in different fields (e.g., other inbreds or congenics), efficiency losses can range from a few percent to more than a 100-fold difference (using DBA/2 congenic mice as compared to the B6SJLF1 hybrid).

b. Genetic Variability and Transgene Expression

Significant variation in transgene expression among individuals, litters, or generations can doom a transgenic model at several stages. It can complicate the initial characterization of the model if the genetic background severely influences the transgene expression. Alternatively, as the model is bred through several generations, inbreeding can alter transgene expression slowly or bring out new recessive phenotypes unrelated to the transgene.

With hybrid strains, isogenicity and phenotypic uniformity are high in the F1 generation but rapidly drop as further generations are produced. Homozygosity in hybrids is low at the F1 generation and drops to very low in the F2 (Festing, 1979); however, inbred strains such as the C57BL/6 and FVB have very high characteristics for all these traits at any generation. Although efficiencies in inbred mice (related to superovulation, microinjection, and reproduction) are reduced, the utility associated with genetic characterization adds a dimension that may be required for particular projects. However, the ease related to production of transgenic mice using hybrid donors may warrant production of founder transgenics from hybrid stock, and *then* backcrossing to strains of choice.

2. Superovulation of Mice

a. Pregnant Mare Serum Gonadotropin and Human Chorionic Gonadotropin

For superovulation, we use PMSG (pregnant mare serum gonadotropin; Sigma-Aldrich) and HCG (human chorionic gonadotropin, e.g., Pregnyl, Baxter Pharmaceutical Solutions LLC, Bloomington, IN) both at 5.0—7.5 units i.p. per female mouse (3—8 weeks of age) using a 26-gauge needle (e.g., B-D, 305111) attached to a 1 or 3 cc syringe. Stocks are resuspended or diluted to 25 units/mL in phosphate-buffered saline (PBS) or water, then stored at 20°C (or lower temperatures) until thawed for use. These hormones have been stored in excess of 10 years at 70°C with no apparent loss of biological activity.

With a 12- to 14-h light cycle (06:00 to 18:00 or 07:00 to 21:00 light), we administer PMSG at noon followed by at 48 h with HCG for most strains (including a variety of hybrids as the B6SJLF1, as well as outbred and inbred strains of ICR or FVB mice) or at a 26- to 48-h interval for C57BL/6 mice. Natural variation in responsiveness to administration of exogenous hormones requires that the time and dosage be titrated for any new strain in the colony, supplier of hormones, or change in the colony conditions. A biological assay related to the quality and quantity of eggs as well as mating performance (plug formation) is evaluated following hormone batch preparation.

Excessive PMSG will lead to hormone refractoriness or will increase proportions of nonfertilized, crenated, or abnormal eggs. At a proper dosage, one can reasonably expect to obtain 20—30 eggs per female mouse (dependent on age and strain, in our experience, including B6SJL, C57BL/6 × DBA/2 [B6D2], and C57BL/6 × C3H [B6C3H] hybrids; C57BL/6 or BL/10 [and congenics], C3H or FVB inbreds; and outbred Swiss [Swiss-Webster, FVB, ICR, CD-1, and ND-4]).

Responsiveness of BALB/c and congenic strains developed with BALB/c mice is generally poorer with most uniform results obtained with 12-week-old donors as opposed to a greater uniformity and yield from 3- to 5-week-old donors using most other strains. Hybrid strains do show the lowest percentage of abnormal egg development, while inbred strains demonstrate higher proportions of abnormalities as a general rule. Normal eggs and some abnormal ova produced following superovulation with PMSG are illustrated in Figure 2.6C and include the following: (i) one-cell eggs with a degenerative cytoplasmic appearance (perivitelline space with multiple, fragmented polar bodies or devoid of polar bodies); (ii) highly fragmented, one-cell eggs with multiple unequal fragments in the perivitelline space; (iii) precociously matured, fragmented, two-cell eggs containing unequally divided blastomeres, with cytoplasmic fragmentation; and (iv) well-developed, one-cell, pronuclear-stage zygotes with a clear single (or double) polar body, two well-expanded pronuclei, and no cytoplasmic fragmentation.

b. Follicle Stimulating Hormone and Luteinizing Hormone Delivered with Osmotic Pumps

Use of FSH (follicle stimulating hormone; Sigma-Aldrich) delivered via an implantable osmotic minipump (Model 1007D; Alza Corp., Palo Alto, CA) and induction of ovulation with either HCG or LH (luteinizing hormone) can significantly improve egg quality and quantity in poorly responding strains. A method for mice was adapted from a protocol published by Leveille and Armstrong (1989) for superovulation in rats. For a 6-week-old, 20-g mouse (e.g., C57BL/6 strain), FSH is resuspended at 1 mg/mL and loaded into the pump, which is implanted subcutaneously following the manufacturer's recommendations. The female donors are anesthetized using Avertin before the pumps are implanted. The osmotic pumps supply a constant infusion of FSH, and the degree of superovulation can be regulated by changing the pump rate or the concentration of FSH. The pumps are implanted the morning of day -2 (07:00) and LH (Sigma-Aldrich; 0.1 mg/mL i.p.) or HCG (5.0 units, i.p.) is administered on day 0 at 13:00. The donors are then mated, and the pumps remain implanted in the females until the eggs are harvested the following day.

The FSH/pump protocol for superovulation is significantly more expensive and time-consuming than the PMSG/HCG regimen. The FSH osmotic pump method has been used successfully to induce superovulation in nonovulating (natural or PMSG/HCG-induced) transgenic females from a variety of lineages. In several attempts, we found that old, obese females that did not respond to PMSG can be successfully superovulated with the osmotic pump protocol. However, we adjusted the FSH concentration to a milligram per kilogram body weight level; for a 60-g female mouse, the FSH concentration is increased to 2 mg/mL.

3. Production of Eggs from Superovulated Females

a. Colony

The continuous production of superovulated females is dependent on a healthy colony with good management and environmental regulation. Additional factors are discussed in great detail in *The Mouse in Biomedical Research* (Foster et al., 1983) and include the number and training of animal care technicians handling the mice, feed, bedding, water, noise, and housing density. The reproductive performance of both donor females and stud males is the first characteristic to degenerate when husbandry conditions are suboptimal. It is imperative that all personnel are familiar with the daily operations and acceptable conditions in the mouse colony so that any aberrations or problems can be readily identified and rectified.

b. Influence of Litter Number

Interestingly, in setting up a breeding colony to produce B6SJLF1 donor stock, we found a strong correlation between the frequency of abnormal eggs and parity of the C57BL/6 female dams. Initial superovulation was attempted exclusively with donors derived from first parity females. A high frequency (35–45%) of abnormal eggs was obtained. As females derived from second and subsequent parity dams became available, the frequency of abnormal eggs declined to a fairly consistent range around 15%. This may relate to the superior reproductive performance from second through fourth parity females (particularly with inbred strains). However, variation in hormone treatments, breeding pairs, and especially colony environment are additional components to be carefully evaluated.

4. In Vitro *Culture of Superovulated Eggs*

a. Equipment

A dissecting stereozoom microscope (6.5–40×) with wide-field (16–20×) eye-pieces and long-working distance objective(s) offers excellent resolution for the harvest and manipulation of eggs (Figure 2.1). The binocular microscope is mounted on a mirrored transmitted light base for transillumination and egg contrast (e.g., Diagnostic Instruments, Inc., Sterling Heights, MI; Leeds, Minneapolis, MN).

There are several warming plates that precisely control the egg culture temperature at the laboratory bench to within 0.1°C (e.g., HT-400, Minitube of America, Verona, WI; 700 series, PMC Corp., San Diego, CA). Whether custom configured or commercially available, such plates will significantly enhance egg viability during manipulations (and with their adjustable range, readily function as do slide warmers to keep mice warm following surgery). In addition, the HT-400 can be equipped with a stage warming plate that is independently controlled to avoid temperature drops during egg manipulation under the microscope.

b. Atmosphere for Culturing Eggs

Eggs and medium are maintained in a 5% O_2, 5% CO_2, and 90% N_2 atmosphere (by volume), that is, controlled with a flowmeter (TSI, Shoreview, MN; Sierra Instruments, Monterey, CA). The mixture is passed through a gas humidifier, then into a small plastic incubator (e.g., one as inexpensive as using a pipette tip box lid with a hole cut to accommodate tubing for gas flow) covering the microdrop dishes containing eggs. Finally, equipment for basic surgical requirements can be obtained from several suppliers, as is well documented elsewhere.

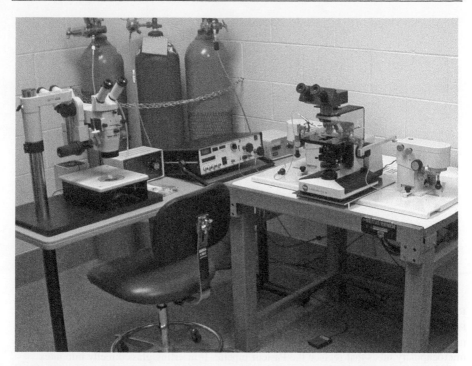

Figure 2.1 Microinjection station with binocular dual-head dissecting microscope (useful for both training and demonstration purposes) on left and microinjection microscope with manipulators on antivibration table on right. Custom-designed extension arms attached to manipulators and secondary microscope base plate allow for ergonomic placement of hands, which is helpful in preventing tendonitis following repeated/long sessions of microinjection using the injection rig as originally designed.

c. Glass Pipettes for Manipulating Eggs

Pipettes for manipulating eggs can be prepared from 9-in. borosilicate Pasteur pipettes or from 4-mm-diameter borosilicate glass tubing cut at 5 in. lengths. The taper of the Pasteur pipette or the center of the 5-in. tube is heated over a flame, and the tube is rapidly pulled apart when the glass becomes pliable. The smaller the area of tubing that is heated, the smaller the final tube diameter. The pulled end is pinched off to a length of 3 in., placed in a clean and/or sterile container, and sterilized. An alternative to pinching is to score the glass with the edge of a piece of 400-grit sandpaper, then bend the glass (with the scoring away from you) until it breaks. This usually results in a flat end with symmetrical opening. Using either method, with a little practice, one can learn how to pull pipettes/tubes of consistent diameter. Small diameters (i.e., slightly greater than one egg [$\sim$80–90 μm] in internal diameter) are appropriate for murine egg transfers; however, a larger inner diameter (ID; $\sim$200 μm) of the final taper facilitates a more rapid collection and transfer of eggs during the microinjection procedure.

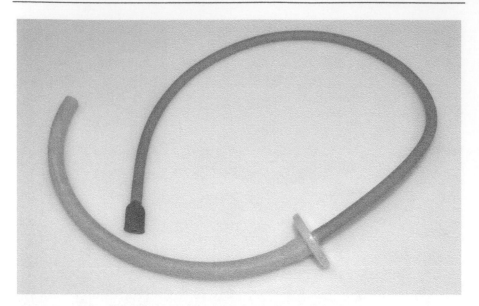

Figure 2.2 Mouth pipette assembly. Parts include: large-gauge latex tubing to attach to Pasteur pipette, 0.2–0.4 μm syringe filter (pore size may affect suction), small-gauge latex tubing, and mouthpiece. If desired, smaller gauge tubing can be used in conjunction with microcapillary tubes that replace the Pasteur pipette. If inserting glass directly into latex tubing, pipettes will become easier to remove from tubing with continued use.
Source: Adapted from Rafferty (1970).

Once the glass is prepared, the wide end of the glass pipette is attached with latex tubing to a capillary adapter (with cotton plug or filter; Figure 2.2). Using larger bore latex tubing, the adapter is then attached to a plastic mouthpiece. Use of this mouth-controlled pipetting device, with a little practice, greatly enhances control of egg manipulation and is superior to using handheld, micrometer-type devices (Rafferty, 1970).

d. Media

Two different egg culture media are used for manipulation and microinjection. First, a bicarbonate-buffered medium such as modified BMOC-3, requiring 5% O_2, 5% CO_2, 90% N_2 atmosphere may be used. Additionally, a hydroxyethyl piperazineethanesulfonic acid (HEPES)-buffered medium is used to maintain pH when eggs are removed from the controlled atmosphere (see recipes in Table 2.1). Egg viability is higher in the bicarbonate-buffered medium, which is the preferred medium for culture, whereas the HEPES-buffered medium is used only for extended periods when the triple gas or controlled atmosphere is unavailable.

The bicarbonate-buffered medium can be equilibrated with CO_2 and stored frozen for up to 6 months, with bovine serum albumin (BSA) added on thawing. The phenol red concentration can be reduced or omitted if feasible. Antibiotics are routinely

Table 2.1 Medium Recipes Suitable for the Culture of Mouse Ova (g/L)

Component	BMOC-3	M16	SECM	BMOC-3 + HEPES	M2
NaCl	5.200	5.533	5.540	5.200	5.533
KCl	0.356	0.356	0.356	0.356	0.356
$CaCl_2 \cdot 2H_2O$	0.252	0.252	–	0.252	0.252
Ca-lactatepentahydrate	–	–	0.527	–	–
KH_2PO_4	0.162	0.162	0.162	0.162	0.162
$MgSO_4 \cdot 7H_2O$	0.294	0.293	0.294	0.294	0.293
$NaHCO_3$	2.112	2.101	2.112	–	0.349
Sodium pyruvate	0.028	0.036	0.028	0.028	0.036
Sodium lactate	2.520	2.610	2.416	2.520	2.610
BSA	5.000	4.000	1.000	5.000	4.000
EDTA	0.037	–	–	0.037	–
Glucose	1.000	1.000	1.000	1.000	1.000
HEPES	–	–	–	5.950	4.969
Penicillin G potassium salt[a]	0.070	0.060	0.060	0.070	0.060
Streptomycin sulfate[a]	0.050	0.050	0.050	0.050	0.050
Phenol red[a]	0.010	0.010	0.005	–	0.010
Distilled water added to a final volume of 1 L.					

[a]Optional.

Note: BMOC-3 was defined in Brinster (1972) and Brinster et al. (1985). M2 and M16 were defined previously (original references: Quinn et al., 1982 and Whittingham, 1971, respectively). SECM was defined in Burki (1986) (original reference: Biggers et al., 1971).

omitted from medium preparations to minimize toxicity and to allow quick identification of possible contamination. All reagents should be cell culture tested and preferably of "hybridoma quality." Purchase of commercially available media (e.g., M2 and M16 media; Sigma-Aldrich) can simplify preparation and use. Preparing a given medium in large volumes (500 mL) allows for greater ease in accurate measurement of components while decreasing time-consuming preparation steps in the laboratory. Culture media have been stored at 4°C for over 3 years and provide egg manipulation-related efficiencies similar to those of freshly prepared aliquots (our only precondition was to exclude aliquots where contamination was grossly evident).

e. Water Sources
Water quality is a critical component and the importance of testing several sources to ensure quality is stressed. We have found that acceptable sources include in-house double processing systems (distilled then reverse osmosis-treated filter systems) and commercial cell culture suppliers that also screen for viral and mycoplasmal contamination.

f. Culture Dishes
Tissue culture dishes are routinely used for egg culture procedures. Small 35 mm tissue culture dishes (e.g., Corning 35 mm × 10 mm polystyrene, No. 25000) are filled with about 3 mL of medium and used for egg collection and "washing."

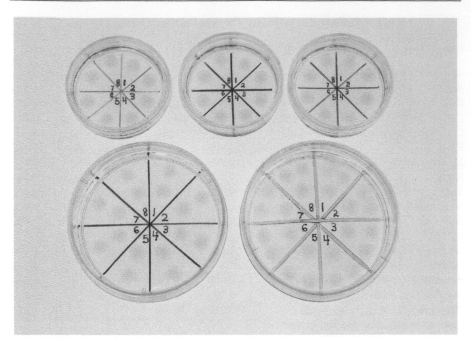

Figure 2.3 35 and 60 mm tissue culture dishes with 10−25 μL drops overlaid with silicone oil. Before medium and oil are added, dishes are labeled on the bottom surface accordingly.

Larger dishes (e.g., Corning 60 mm × 15 mm polystyrene, sterile tissue culture dishes, No. 25010) are used for holding large numbers of eggs before and after microinjection and for long-term culture. Small volumes of medium (20−50 μL in microdrops) are overlaid with silicone oil and placed in the dishes to maintain eggs during manipulation. The drops are best identified by marking quadrants/sections on the bottom of dishes *before* medium or oil is added to the dishes, similar to labeling quadrants on Petri dishes before pouring plates (Figure 2.3). Silicone oil (e.g., 200 Fluid; Dow Corning, Midland, MI) is used routinely and readily minimizes diffusion and evaporation of the microdrops. In our experience, washing or equilibrating oil has not proved to be necessary. However, to wash silicone oil, omit BSA from the medium, mix a 1:1 volume of oil and medium by shaking vigorously for 15 min, and allow overnight separation. Washed oil can then be separated from the top layer as needed. After microdrop dishes are prepared, they are placed in the temperature- and gas-controlled environment. A minimum of 10 min is required to equilibrate conditions before use.

5. Harvesting Superovulated Eggs

a. Mating
After female egg donors receive HCG, they are placed in cages with male mice. The next morning, the females are checked for copulatory plugs. Bred females are sacrificed for egg recovery.

b. Oviduct Dissection

Harvesting of the oviducts and eggs should be done within 8−9 h of the midpoint of the last dark cycle. This allows one to easily isolate the eggs as a mass from the oviduct, obviating the need to flush individual eggs. After donor females are euthanized, the oviducts are obtained through a midventral or dorsolateral approach. Either method is satisfactory, yet, personal preference will likely dictate the method of choice (note that the authors have mixed preferences here).

After cutting through the skin and body wall is complete, the fat pad by the ovary is located, and the uterotubal junction (where the uterus and oviduct join) is grasped with a pair of forceps. The uterus is then severed just below the forceps. A second cut between the oviduct and the ovary frees the oviduct, which is still grasped with forceps. The oviduct is then placed into a sterile Petri or tissue culture dish filled with bicarbonate buffer. The dissections from donor females continue as rapidly as possible to remove all the oviducts, but care is taken not to excessively manipulate oviducts, thus causing possible rupture and release or loss of egg masses.

c. Release of Eggs from the Oviduct

The oviducts are collected in a barrier hood, then brought to the microinjection laboratory and transferred to a clean dish. Some laboratories will collect or place oviducts into a hyaluronidase-containing medium at this point. We postpone hyaluronidase treatment until the eggs, contained in cumulus masses, are liberated from oviducts, to minimize enzyme exposure to eggs.

At this point, the swollen section (ampulla) of the oviduct should contain a large mass of eggs, and it can be observed using low-power magnification under a dissecting stereomicroscope (Figure 2.4). The oviduct or adjacent membrane/tissue, *but not the ampulla*, is first grasped with a pair of fine forceps or held in place with a small-gauge needle. The ampulla is then ruptured by tearing or pricking with another pair of forceps or a needle in proximity to the egg mass (Figure 2.5A−C). The intraoviductal pressure is such that the egg mass is expelled without further manipulation. Using a large-bore pipette (e.g., Pasteur pipette with bulb-suction filler or mouth pipette), the egg mass is then placed into a dish containing 3 mL of bicarbonate buffer supplemented with hyaluronidase (type IV-S, 100 units/mL; Sigma-Aldrich). The eggs remain until the cumulus cells surrounding the eggs are digested and the zonae pellucidae are free of any attached cells or debris (Figure 2.6). Practice is necessary to keep the exposure of eggs to hyaluronidase to a minimum ($\sim 3-6$ min in total).

To remove the eggs from the hyaluronidase-containing medium and for routine handling of eggs, a glass pipette (200-μm final ID) is first filled by capillary action with medium. Then two small air bubbles are drawn into the narrow bore to enhance fine control for egg manipulation (Figure 2.7).

The isolated eggs are carefully picked up, avoiding rough "vacuuming action" that may damage them. The eggs are placed in a new dish of hyaluronidase-free medium, minimizing the transfer of any remaining extraneous cells/debris. The eggs are then pipetted into a second wash and counted. Good quality eggs are then selected and placed in a dish containing microdrops of medium until readied for microinjection.

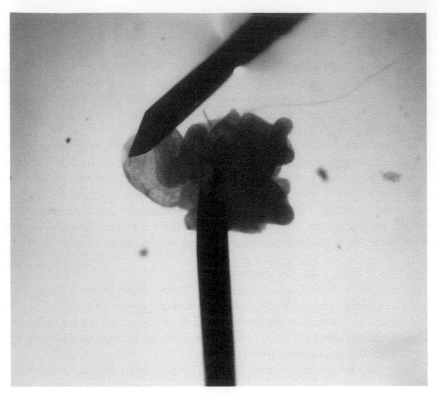

Figure 2.4 Oviduct collection. Low-power magnification of dissected oviduct; the uppermost needle is pointed to the ampulla where the cumulus mass is collected.

d. Egg Development and Microinjection Timing

During any manipulation with bicarbonate buffer, one must pay attention to the color of the medium and egg morphology to ensure pH balance. As the CO_2 diffuses, egg viability decreases rapidly. If eggs are harvested early before pronuclei are well formed, the microdrop dish can be stored in a modular incubator chamber (Billups Rothenberg, Del Mar, CA) and placed in a standard temperature-controlled incubator in order to maintain temperature, humidity, and atmospheric conditions relatively inexpensively. The optimal windows for microinjection range from the time when the male pronuclei at the periphery of the egg membrane are identifiable until the male and female pronuclei merge just prior to the first cleavage division. The time of this window varies between strains with outbred and hybrid strains demonstrating a more uniform and smaller time window compared to inbred mice. Typically, the injection window is 3—4 h, but it is affected by *in vitro* culturing methods and care during routine handling.

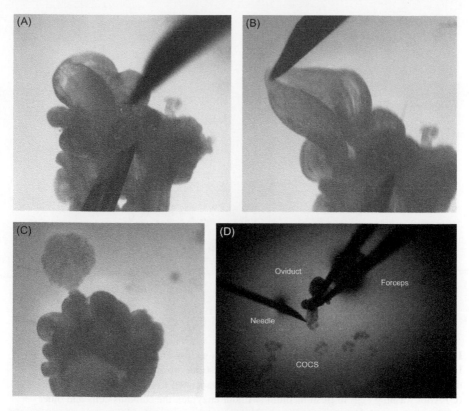

Figure 2.5 Cumulus mass collection. The sequence of events necessary to harvest cumulus masses is performed under high-power magnification. (A) The oviduct is positioned with two 25 ga. needles. (B) One needle is used to "prick" or cut the ampulla. (C) The cumulus mass exudes from the ampulla. (D) Labeled sections cumulus oocyte complexes (COCS).

e. Efficiency of Egg Culture

Media conditions and specific mouse strains were chosen to obviate the two-cell block in egg development. This allows one to test the quality of all media components and manipulations by culturing the eggs *in vitro* for several days. In optimal *in vitro* culturing conditions, we have obtained 95% or greater development of B6SJLF1 hybrid-derived zygotes to the blastocyst stage in 4 days, with 70% of the eggs capable of growing out ICMs within 3 days. Any suboptimal components will lower these efficiencies or delay development. It is important to run quality checks of media preparations periodically. Inbred and outbred strains of mice will exhibit lower developmental rates *in vitro*, and one should choose a representative strain with reasonable developmental uniformity for quality control checks of *in vitro* culturing conditions.

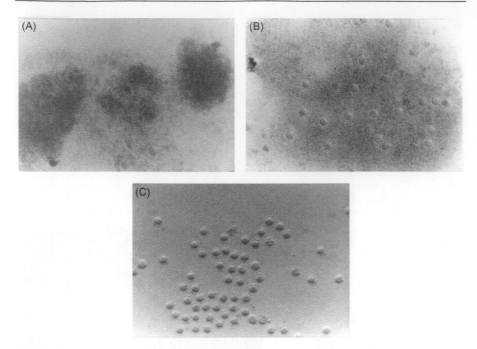

Figure 2.6 Egg harvest. (A) Three cumulus masses shortly after hyaluronidase treatment. Here, cumulus cells are just beginning to dissociate from the eggs. (B) After 5 min in medium containing hyaluronidase, eggs are well dissociated from cumulus cells and debris. (C) After washing eggs in fresh medium (without enzyme), the cumulus cells and other debris are removed. Note here that in obtaining eggs from an outbred donor, a proportion of degenerating or abnormal ova are obtained in addition to the pronuclear eggs.

C. Microinjection Needles and Slides

1. Types of Glass

In actual practice, either borosilicate or aluminosilicate glass is used to make microinjection needles. Additionally, thick-walled (ID:OD < 75%) or thin-walled (ID:OD > 75%) glass may be used for needle production. The ID:OD ratio we prefer is just under 75% (inner diameter 0.027 in., outer diameter 0.037 in.). This measurement is important because it determines whether the tip of the microinjection needle will be fused shut or left open after it is pulled (e.g., thick-walled glass will tend to shut at the tip). For further reference, an excellent introduction for micropipette pullers and microneedles was written by Flaming and Brown (1982). Two types of glass tubing configurations, as well, are used for microinjection: hollow capillary tubes (e.g., KG-33; Garner Glass, La Jolla, CA) and tubes with an internal filament (e.g., TW100F-4; World Precision Instruments, Sarasota, FL). The hollow microcapillaries require backfilling (from the tip back), which is performed after the needles are attached to oil-filled lines and an opening has been created at the tip.

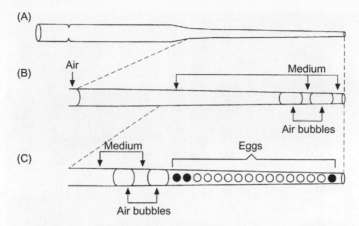

Figure 2.7 Egg handling pipette. After heating the tapered portion of a 9-in. Pasteur pipette to a molten state (orange color), the pipette can be pulled to an appropriate diameter (90–200 μm depending on particular needs). After cooling (~3 s), the two ends can again be pulled (without heat), breaking the glass evenly. Alternatively, microcapillary tubing may also be used. The diagrams illustrate the procedure from pulling a pipette (A) to loading the pipette (B and C represent enlargement of the distal end of the pipette). (B) Medium is drawn up into the pipette by capillary action. When the medium reaches equilibrium, the pipette is attached to the mouth pipette assembly and two air bubbles are drawn into the tip of the pipette. The addition of air bubbles provides a greater precision in egg movement. (C) Following the air bubbles, eggs are then drawn into the pipette. For egg transfers to recipient females, a tip diameter slightly larger than the eggs is used. As shown here, the smallest amount of medium is used as the eggs are drawn, side by side, into the pipette. For washing eggs or moving them from one treatment dish to another, a wider bore pipette (i.e., 200 μm diameter at tip) is more convenient. For ova transfer into oviducts, visualization of the placement of transferred ova into the ampulla may be beneficial. A technique first established in the Brinster laboratory involving the use of Sephadex beads (Bio-Rad ALG-1000) in the range of 75–100 μm proved to be most helpful. We have since found that visualization of two or three transferred Sephadex or Ficoll gradient beads as shown in C provides training assistance as well as assurance that any given group of ova was transferred appropriately.

The filament-containing tubes can be filled prior to attachment to hydraulic or pneumatic lines and before the tip is opened or cracked.

2. Preparation of Glass

Many protocols have been developed for pretreatment and washing of glass tubing to be used for forming microinjection needles, but we have bypassed such procedures without a loss of overall efficiency. We prefer to simplify the process of making transgenic animals, minimizing some procedures for the sake of expediency when possible.

Figure 2.8 Pipette puller. Uniform tapers and shapes of microneedles can be programmed using horizontal (depicted) or vertical pullers.

3. Pipette Pullers

Pipette pullers are available from several manufacturers and are of vertical design (e.g., Kopf, Tujunga, CA; Narishige, Greenvale, NY) or horizontal design (Sutter, San Raphael, CA; Figure 2.8). These units have distinct variations in filament and control options. In our experience, the Sutter pipette puller (P-87) has proved to be very reliable, delivering a high degree of uniformity and consistency. Its controls include five programmable options with the ability to cycle several combinations of parameters during the pulling of a single pipette. With a preset ramp program, one can evaluate the heating of any filament arriving at a "ramp" setting, allowing cross-referencing and use of different micropipettes and filaments.

4. Injection Pipette Shapes

We have used three basic shapes of pipettes/needles for microinjection that differ in two regions. The tip region is the final taper that directly enters the egg during microinjection. In turn, the shank is the taper from the tip back to the full diameter of the glass tube (Figure 2.9).

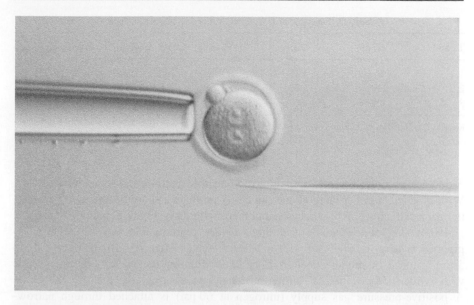

Figure 2.9 Injection and holding pipettes. With an egg in place (∼70 μm diameter) the relative size of the injection and holding pipettes becomes evident. The holding pipette may range from 15 to 50 μm in diameter depending on style (not polished to very polished), while the injection pipette is ∼3/4 μm in diameter. Generally, the injection pipette is aligned with the pronucleus before insertion into the egg.

The three basic shapes and applications are as follows. Type 1 injection pipettes have fast-tapering shanks and tips that produce a wide diameter bore at all sections of the micropipette. They can be used for large pronuclei at a late stage in development. Type 2 pipettes have shanks and tips with longer tapers and more gradual changes in tube diameter. The majority of eggs are injected with type 2 pipettes, which are fine enough to inject small pronuclei. Type 3 injection pipettes have extremely fine tips with very long shanks and can be used for injecting two-cell eggs or eggs with very small pronuclei just after fertilization.

It takes considerable time to optimize parameters for the pipette shape one desires, but the uniformity of each pipette from the Sutter unit is worthwhile (Table 2.2). One should understand the effects of altering individual parameters before trying to arrive at an optimal shape. In general, finer tips are achieved by increasing the heat or by increasing the pull. Decreasing the pressure, decreasing the time, or increasing the velocity increases the taper of the shank. The most difficult skill to develop in arriving at a desired pipette shape involves working with parameters that are interdependent but without equal influence on pipette shape.

5. Beveling Hollow Capillary Microinjection Needles

After the microinjection needle is pulled, one method for creating a sharp tip is to break the tip of the injection needle on the holding pipette. This method, while

Table 2.2 Examples of Sutter Settings for Injection Pipettes

Type	Heat	Pull	Velocity	Time	Pressure
1	765	175	60	250	800
2	750	170	60	250	800
3	780	180	70	250	800

Ramp = 729, box filament 3 mm × 3 mm, single cycle.

expedient, provides tips that are not as reproducible as those fashioned on a beveler. With practice of either method one can easily produce a sharp tip that will penetrate the zona pellucida and pronuclear membranes without lysing or damaging the egg.

If beveling of the needle is desired, a very reproducible way to create sharp pipettes is to grind the tips with a beveler (e.g., Sutter BV-10) using a 0.5- to 5-μm-grit-size diamond wheel. A fiber optic illuminator and a stereomicroscope with wide-field eyepieces allow adequate monitoring of the tip beveling. A positive-pressure gas supply (nitrogen at 60 psi) is attached through narrow-diameter tubing (e.g., Tygon, ID 1/32 in., OD 3/32 in., wall 1/32 in.) to the micro-injection needle. This pressure helps prevent grit from accumulating in the tip.

To bevel the microinjection needle, first attach it to the nitrogen source, then secure it in the beveler elevator. The pipette is positioned so that the grinding direction is parallel and rotating away from the pipette and not crosscutting the tip. Next, the needle tip is lowered onto the wheel with the coarse adjustment. The shadow of the pipette on the wheel is then brought into the viewing field of the binocular microscope, and the injection needle is carefully lowered with the coarse adjustment (then the fine control knob) until it just touches its shadow on the wheel.

The diameter of the tip is controlled by the length of time in contact with the wheel and by how far it is lowered onto the wheel. If the tip is ground too far, even with nitrogen pressure, it will accumulate grit within the pipette, making control of fluid flow difficult. The optimal bevel is observable on the microinjection scope at 120× magnification and takes 3–5 s to form with a minimal amount of pressure on the grinding wheel. Practice is needed to create uniform tips. Once the tips are beveled, the microinjection pipettes may be bent to a 30° angle similar to the holding pipette if so desired (Figure 2.10).

6.   Microforge Shaping of Pipettes

a.   Manufacturers

There are several microforge manufacturers (e.g., DeFonbrune/TPI, St. Louis, MO; Valiant Instruments, St. Louis, MO; Narishige, Greenvale, NY; Nikon, Melville, NY; Minitube of America, Verona, WI). The instruments have common features, including a heating element, cooling capability, glass holders, and stereomicroscope assembly (either attached or separate). As recommended, with the DeFonbrune-type unit (Figure 2.11) we use a 31-gauge heating filament made of

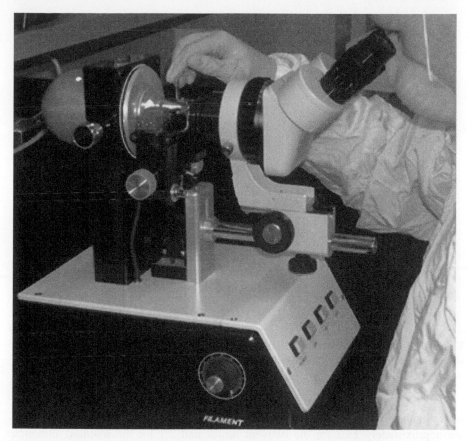

Figure 2.10 Microforge. Needles may be fire polished or bent to specification using the microforge mounted with either a compound or dissecting microscope head.

platinum and iridium. For magnification, we use either a stereozoom or compound microscope head with either type equipped with wide-field eyepieces. If desired, an eyepiece can be equipped with a micrometer/reticle to measure injection tip diameter and the overall dimensions of pipettes.

b. Heating Filament

The heating filament is bent to a U or V shape at the tip and held between two electrodes. A small glass bead is melted on the inside curve or tip of the filament. This allows uniform heating and will provide for uniform pipette breakage for production of holding pipettes as well as performing other shaping steps.

c. Holding Pipette

Either the Kopf or Sutter pipette pullers can be used to form holding pipettes. The Kopf puller is supplied with coiled platinum heating elements that can form much

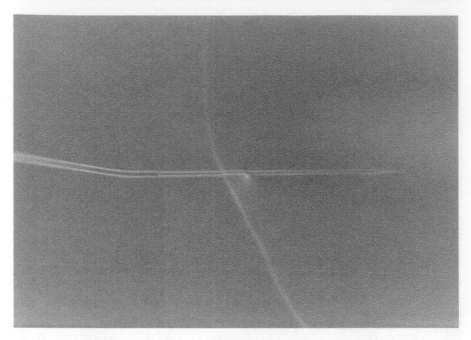

Figure 2.11 Microneedle with 30° bend. The bends in needles allow better access to cut-out slides and movements of needles parallel to the surface of slides or dishes.

wider diameter and longer tapering pipettes. For holding pipettes, the Kopf settings are approximately 13.2 for the filament heater and 2.1 for the solenoid, although these settings do require fine-tuning that is element dependent.

The holding pipette is formed by angling the pipette 30° off the horizontal plane above the glass bead on the microforge filament. The rheostat is turned on and adjusted until the bead just begins to glow (orange in color). The pipette is lowered to gently contact the bead and held for about 5 s; a minimal deformation of the pipette should be observed. The heat is then turned off, and, as the filament contracts on cooling, the fused bead and pipette break to form the tip of the pipette. The break will produce a flat face if the pipette angle was correct and the heat or fusing time was not excessive.

The holding pipette is then brought vertically over the filament glass bead, the electrode is turned on until the bead is again orange, and the pipette is lowered until the open tip just begins to melt. This polishes the glass tip, removing any sharp edges. This step, however, may be omitted. If an additional bend is desired so that the tip of the needle is in a plane horizontal to the injection slide, the pipette is again treated. The glass bead and filament are turned on until the bead is heated. Positioned near the side of the pipette, the filament is used to bend the pipette to a 30° angle (Figure 2.10). The pipette is then stored in a dust-free container.

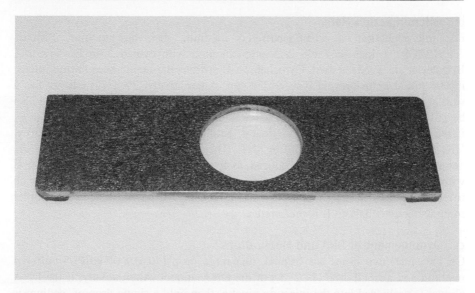

Figure 2.12 Cut-out slide. Custom slides use disposable cover slips for maintaining microdrops (with embryos and possibly DNA samples or ES cells) overlaid with oil. While it is possible to create cut-out slides from standard glass slides; metal slides with machined or welded risers (seen here) avoid inherent fragility/breakage in comparison to glass or plastic equivalents.

d. Injection Pipette

After several microinjection needles are prepared, they may be bent to a 30° angle should an elevated angle or straight approach to a microinjection slide or dish be desired. These needles may be stored in the same container as the holding pipettes, but sections of the container should be labeled for the types of needles stored to expedite selection.

7. Microinjection Slide for Eggs

a. Slide Designs

DNA and eggs can be held in depression slides or in cut-out slides (Figure 2.12); for cut-out slides, rather than a depression in the center, the equivalent area is cut out of a flat glass or metal slide. The cut-out slides make use of disposable glass cover slips for holding microdrops overlaid with oil. The cover slips are attached to the bottom of the cut-out slide with a ring of paraffin (outside the periphery of the hole). (*Note*: Use of 1−2 mm blocks/risers on the bottom of both ends of the slide keeps the cover slip from contacting surfaces.) This configuration offers an advantage to depression slides, because the cover slips are disposable and there is less light refraction. However, because a greater needle angle to manipulate eggs is required for working with dishes on a microscope, the needle tips must be bent (~30° angle, see Figure 2.10) prior to microinjection. Again, personal preference will dictate the method of choice.

b. Microinjection Media

For microinjection, a HEPES-buffered medium such as modified BMOC-3 + HEPES (recipes in Table 2.1) supplemented with cytochalasin B (5 μg/mL; Brinster et al., 1985) is used. Cytochalasin B stiffens the membranes during micro-injection and helps prevent lysis of the egg. Alternatively, 7% (v/v) ethanol has a similar effect (P. Hoppe, personal communication, 1991). Egg membranes can also be stiffened by lowering the slide temperature to 10°C with a stage cooler (e.g., Leica Microsystems, Wetzlar, Germany; Linkam Scientific Instruments, Surrey, UK; Physitemp Instruments, Clifton, NJ). This will not alter egg survival or the percentage of transgenic mice produced, as long as the eggs are returned to 37.5°C within approximately 45 min (unpublished data, 1992). The cooling effect takes up to 10 min before the egg membranes are noticeably stiffer; therefore, the simpler route using cytochalasin B is preferred.

c. Arrangement of DNA and Media Drops

On the slide, two drops are formed, one large drop (40 μL) of HEPES-buffered medium and a small drop (1−2 μL) of the DNA aliquot. Alternatively, if the DNA aliquot is preloaded into the injection pipette, then only a single drop of medium is needed. The large drop of medium is spread out from 1 to 1.5 cm so that the top surface is flattened to prevent refraction of light. The drops are then covered with silicone oil, making sure all the surfaces are submerged to prevent evaporation or optical distortion. If there is an optical distortion from making the depth of the drop too high, simply remove small amounts of medium until a clear image is obtained.

D. Microinjection Equipment

1. Microscopes

A number of microscopes and micromanipulators are marketed with different advantages and disadvantages. Rather than sell one model or another, we would rather highlight equipment representative of our experience.

Various microscope configurations (Figure 2.13) from upright to inverted styles (e.g., Laborlux and Labovert, Leica; Axioplan and Axiovert, Zeiss; Optiphot, Nikon; Diaphot, Olympus) afford excellent differential interference contrast (DIC; either Nomarski or Smith type). Magnification between 180× and 400× is commonly used, and a decision should be evaluated before purchase. Final DIC magnification may be limited by the microscope brand or configuration, and desired field diameter is generally dictated by personal experiences or past training. Usually, DIC is only required at the microinjection magnification (200×). Low-power magnification (50×), while providing a reasonably large working field for priming pipettes and sorting eggs, does not require DIC optics. Thus, the final configuration and necessary adjustments (e.g., light filament, lens, diaphragms, and focusing lenses) should be tested in advance of purchase, preferably when the manufacturer's representative and a few eggs are available at the same time. The representative can set an

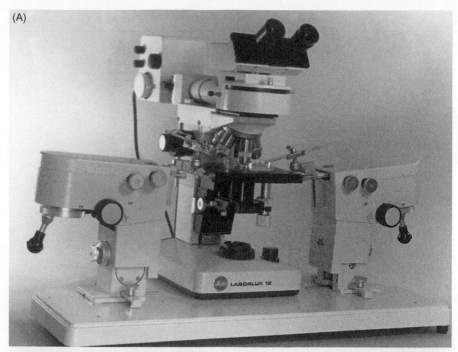

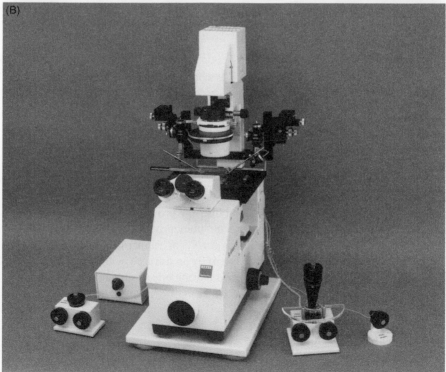

Figure 2.13 (A and C) A selection of inverted and upright DIC microscopes and micromanipulator assemblies used for DNA microinjection.
Source: Leitz courtesy of Leica Inc., Deerfield, IL; Nikon courtesy of Nikon Inc., Melville, NY; Narishige courtesy of Narishige USA, Greenvale, NY.

(C)

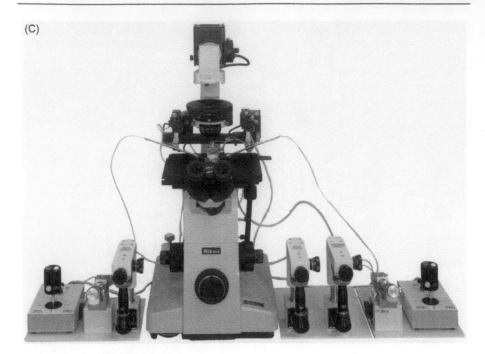

Figure 2.13 (Continued)

"optimal" DIC configuration, but this does not necessarily provide the best pronuclear images. Consult with the representative to ensure that all optical adjustments are demonstrated effectively and can be handled properly in his or her absence.

2. Micromanipulators

a. Micropipette Holders

A number of pipette holder/collar assemblies are provided by manipulator manufacturers. As an example, the Leica pipette holder consists of a collar assembly with two brass fittings that hold a fitted plastic or rubber/latex washer in place, thereby creating a tight junction. The assembly is covered with a metal, threaded cap that adjusts the assembly around the micropipette to create a leakproof seal. The tubing used as a washer should be of an appropriate diameter to match the outer diameter of the micropipette.

b. Micromanipulator Adjustments

The micromanipulators have several adjustable dials for regulating the sensitivity and tension on the joysticks. An authorized technician should be consulted to demonstrate the various adjustments that can be made on the micromanipulators. Not all the adjustments are obvious from the visible knobs and collars.

c. Antivibration Tables

When the microinjection needle is inserted into the egg, the slightest vibration will disrupt the egg, influencing lysis (immediately or delayed) and thereby making pronuclear placement difficult. To avoid frustration, one must use vibration-free tables. We have used a number of antivibration bases (Kinetic Systems, Roslindale, MA; Micro-G, Woburn, MA; Barry Wright, Burbank, CA), which support stainless steel-covered lead, composite, or steel casings on four nitrogen-driven pistons. In addition, and at lower cost, marble balance tables work well in most circumstances. Laboratory location and environment will play a role in defining specific needs.

3. Microinjection Systems

a. General

Systems for controlling microinjection of DNA and holding eggs are grouped into either air-driven systems (e.g., Narishige, Greenvale, NY; Nikon, Melville, NY; Carl Zeiss, Thornwood, NY; Eppendorf, Hauppauge, NY; Brinkman, Westbury, NY; Harvard Apparatus, Holliston, MA) or oil-driven hydraulic systems (Figure 2.14). The air-driven systems can also be attached to electronically controlled delivery systems (e.g., Eppendorf 5221; PLI-100, Medical Systems Corp., Great Neck, NY; Narishige IM-200) that allow preprogrammed regulation of fluid delivery volume, time, and pressure. Excellent results may be obtained with or without such systems, and the choice is dependent more on injector experience or bias. The technique, skill, and training of the injectionist are all factors that must ultimately be weighed in the choice of injection systems.

b. Oil-Driven Injection Assembly

The system we currently use is a custom hydraulic unit that is very inexpensive compared to other automated injector systems. The P-3 syringe assembly (Figure 2.14B) is a low-cost, low-maintenance microsyringe assembly that can be used with different syringes (from $100 \mu L$ to 2 mL) for a number of applications. We have found that the sensitivity and control of the P-3 system is greater than that of an oil-filled Eppendorf CellTram Vario system. The P-3, or other microsyringe assembly (e.g., Eppendorf, Huappauge, NY; Micro Instruments Ltd, Oxford, UK; Stolting, Chicago, IL), is attached to the injection pipette and holder via flexible tubing (e.g., PE-100; Clay Adams/BD, Parsippany, NJ). Additionally, a double Luer fitting tube (Bio-Rad, Richmond, CA, No. 732-8202) can be attached to a three-way Luer stopcock value (Bio-Rad, No. 732-8103) for greater ease of handling (e.g., purging air in lines). Ultimately, the micrometer assembly will allow precise positive and negative pressure as required.

The syringes are loaded with either silicone or mineral oil. The tubing, the microinjection needle holder, and the needle itself are filled with fluid to dampen control and allow continuous flow of DNA without repeated syringe adjustments.

The degree of control in the injection system can be reduced by using lower viscosity oil or greatly increased by higher viscosity oil (oils of 100- to 300-cSt

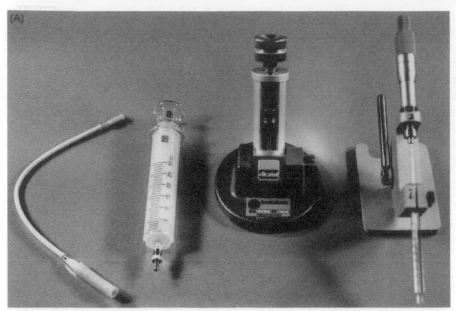

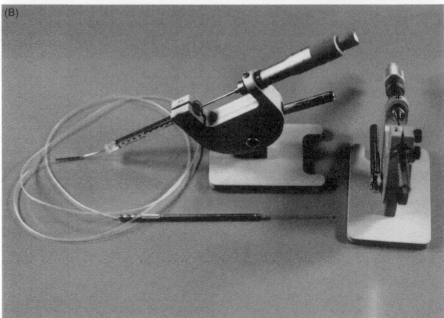

Figure 2.14 (A) Microsyringe assemblies (from left to right): mouth pipette control (length of tubing shortened here); glass syringe; and two, more elaborate, microsyringe assemblies. Whether air-driven or hydraulic systems are used, these assemblies provide a low-cost alternative to electronic systems, while providing different degrees of precision in delivery. (B) P-3 microsyringe assembly with tubing from micrometer-controlled Hamilton syringe to manipulator collar and microneedle. The manipulator collar is mounted on the micromanipulator and then the needle is aligned before use.

viscosities provide very fine control). One advantage in this system is that purging air bubbles and filling the system are readily performed using Luer tip syringes filled with the appropriate oil. The presence of air or any obstructions (broken pipettes, dirt, shaved pieces of tubing) should always be checked, and the system should be cleared before setting up to perform micromanipulations.

4. Oil-Driven Holding Pipettes

The holding pipette system we use is controlled by the same microsyringe assembly. The syringe is connected with PE-100 or Tygon tubing (e.g., ID 1/8 in., OD 1/4 in., and wall 1/16 in.) directly to the back of the pipette holder (Figure 2.14B). Dow Corning 200 fluid fills the entire system. Any air bubbles near the pipette holder or inside the barrel can be removed by holding the pipette holder vertically, removing the two brass collars, and gently tapping on the side of the pipette holder. The bubbles will slowly move up the tube and be visible once they emerge from the pipette holder. A few extra turns on the syringe will flush the bubble from the tube.

To recharge the holding pipette system with more oil, remove the brass collars, expel some fresh oil, and quickly plunge the tip into a beaker of oil. Fresh oil can then be pulled up by reversing the syringe pressure while making certain that air bubbles are not introduced.

5. Electronic Delivery Systems

Automatic microinjectors have been developed for controlled injection of cultured cells and have been applied to DNA microinjection techniques (e.g., Narishige, Nikon, Eppendorf). Similar controls on all models allow regulation of fluid delivery volume, time, and pressure. The pressure controls include separate regulators that allow a static pressure, an "injection" pressure for more rapid flow, and a maximum pressure for clearing the needle tip. In addition, some models provide suction delivery to facilitate holding the eggs. Advantages of automatic injectors over manually controlled micrometer syringe systems include the use of a foot pedal to activate the injection pressure and to allow rapid pressure changes on the fly, facilitating egg penetration and minimizing intracellular cytoplasmic disturbances (the period in which the needle is positioned within the cytoplasm). For novices, an important skill to develop is an ability to control the flow of DNA through the microinjection needle. If the flow is too great, the membranes of the egg will be pushed away from the tip as the microinjection needle enters the egg; hence, the needle tip will not successfully penetrate the pronucleus or deliver DNA properly. If the flow is too slow, the microinjection needle may easily penetrate the egg but the slow delivery of DNA will enhance the likelihood of egg lysis. With manual injection systems, a continuous flow rate between these two extremes is learned, but after much tribulation. An automated system can obviate one time-consuming stage of training and enhance overall injection efficiency.

It should be noted that not all electronically controlled systems have an easily accessible vacuum function to allow filling of the microinjection needle from the tip. Therefore, DNA is loaded from the back of the needles, as described for filament-containing tubes, or with the aide of microsyringes or capillary tubes. A nonmetallic syringe needle for filling micropipettes is available for such needs (MicroFil, World Precision Instruments).

E. Microinjection Procedure

1. Preparation of Injection and Holding Pipettes (Micrometer Syringe System)

a. Holding Pipette

The holding pipette can be inserted directly into the pipette holder, after the system has been checked for air bubbles. A few turns on the Hamilton syringe will push oil all the way to the tip of the pipette. The holding pipette is then oriented above the microscope lens so that the bend in the pipette is parallel to the slide. Finer adjustments are made by viewing under low power and adjusting the focus to ensure that the tip of the pipette is the lowest part of the holding needle.

b. Injection Pipette

The injection pipette is first filled with oil (the same as contained in the injection system) using a 4-in., 24-gauge needle and 5-mL Luer-Lok syringe. While filling, keep the injection needle vertical and fill from the tip back, making sure to trap only one air pocket in the tip of the pipette. Once the injection needle is filled and the injection system is cleared of air bubbles, turn the microsyringe micrometer to push some oil out the tip of the injection pipette holder. Insert the back of the injection needle into this oil drop and down into the pipette holder. The metal cap of the injection holder should be loose enough to avoid having to jam the injection needle through the brass collars or plastic/rubber/latex washer. This procedure is done with the injection needle held vertically. The microsyringe micrometer is then turned to increase the pressure on the air trapped in the tip of the injection needle, which will keep it in place until it is all pushed through the opening at the tip.

c. Aligning the Injection Needle

The injection needle and holder are then placed on the micromanipulator and adjusted to orient it parallel to the slide. As with the holding pipette, the injection pipette is first oriented so that it is perpendicular to the microscope. Next, under low power, adjust the angle of the injection pipette so that the full length of the bent tip section is in focus. This adjustment is sometimes time-consuming but is important because it ensures that the pipette enters the egg at a straight angle and avoids shearing the egg as one pushes the needle into the pronucleus. This adjustment will greatly increase the ease, speed, and precision of microinjection. It is also important to put the same degree of bend into each injection pipette to reduce the amount of time that is needed to realign each injection pipette in the micromanipulator.

The final adjustment is to check the tracking of the injection needle pipette under high power. With the tip of the needle in focus on one side of the field, move the needle directly across to the opposite side. If the tip goes out of focus, adjust the angle knob for the micromanipulator until the needle tip stays in focus across the entire field. This adjustment also increases the speed, accuracy, and precision of microinjection by having the injection pipette tip working plane overlap the focal plane. This saves time during injection of multiple eggs by eliminating the need to realign the injection tip and pronucleus for each egg.

d. Loading DNA into the Injection Needle

If the injection needle tip has been sufficiently beveled, then oil should be flowing out of the tip by the time orientation adjustments are complete. The size of the oil beads running down the pipette will indicate the diameter of the tip. The micrometer assembly is dialed back to reduce the flow, and the injection pipette is lowered into the DNA drop. The quality of the tip can be observed at this point and the needle replaced if any problems are found.

DNA is drawn up into the injection pipette as far as possible, and filling is allowed to continue while the eggs are being selected and prepared for microinjection (Figure 2.15). Once the injection pipette is loaded and the eggs are ready for microinjection, switch the stopcock valve to open the microsyringe micrometer and reverse oil flow to start the DNA/oil meniscus moving as slowly as possible down the pipette (observe under high power).

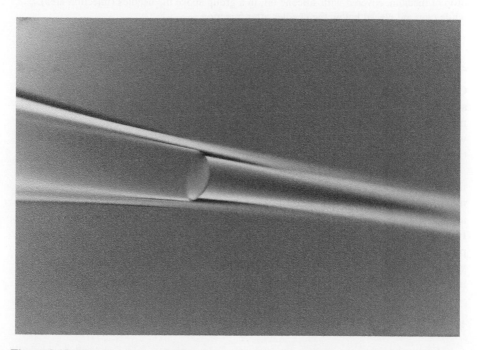

Figure 2.15 DNA meniscus in hollow fiber microinjection needle. The DNA flow rate can be readily monitored when using a hollow fiber injection needle. The meniscus can be used to determine or follow DNA flow during microinjection procedures.

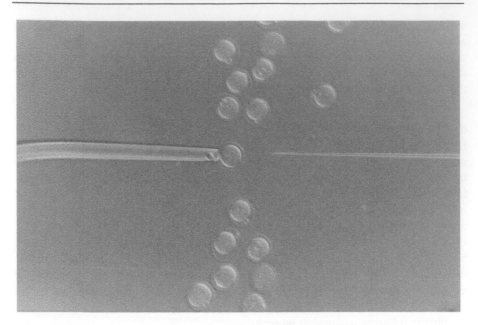

Figure 2.16 Egg organization during microinjection. Noninjected eggs are located in the drop of medium, covered with silicone oil, in a group above the needles (injection area), while eggs are placed below the needles after injection.

2.   Preparing Eggs for Microinjection

Once the DNA is readied in the injection pipette, the eggs to be injected are selected from the holding dish. The eggs are first transferred to a drop of injection medium (containing cytochalasin B) in a dish under oil. The dish is held on a warming plate for 5−7 min or until the egg membranes recover from "blebbing" (i.e., no longer exhibiting a scalloped outline).

Only eggs with visible pronuclei and normal appearance are selected and transferred to the drop of medium on the microinjection slide. There are many arrangements for the eggs on the slide, but we prefer to keep them organized in a line above and below the microinjection area (where needles are located; see Figure 2.16). The slide is placed on the microscope, the holding and injecting pipette are lowered into the drop of medium under low power, and the eggs are visualized. The high-magnification DIC objective is then moved into place, and microscope adjustments are made with the eggs in view.

3.   Steps for Efficient Manipulation of Eggs During Microinjection

a.   General

As with any precision skill, constant review should be focused on the efficiency of hand movements to reduce unnecessary work or strain on the operator.

A major facet of good microinjection skill is keeping the movements and adjustments of the microscope or manipulators simple and quick. The faster the egg can be returned to the correct temperature and medium (as well as *in vivo* reimplantation), the more likely that the eggs will survive the *in vitro* manipulations.

b. Micromanipulators

The joystick on the micromanipulators should be adjusted to provide adequate movement of the pipette with minimal pressure from the fingers. Pulling or pushing too hard on the joystick will slightly deflect the injection or holding pipette tip and reduce the accuracy of injecting the pronucleus. Keep in mind that such joysticks provide a three-dimensional movement of the micropipettes and do not afford the same accuracy as movement in two dimensions. In some instances, particularly in training, it is worthwhile to avoid use of three-dimensional movement.

c. Microinjection Needle

With the eggs on the slide, next adjust the focal plane under high magnification to be about 1.5 to 2 egg diameters above the cover slip. Bring the tip of the injection needle into focus. During the injection of eggs, the focus does not have to be readjusted nor does the height of the injection needle need to be altered. The flow of the DNA can be checked by waving the injection tip in the medium and looking for currents formed near the tip; dial back the flow rate if necessary. Fine micrometer adjustments are usually sufficient during the injection process to control the DNA flow rate (and are typically made once or twice during the injection of about 40–50 eggs). The rate of flow needed to effectively inject the pronucleus, without egg lysis, is determined empirically through trial and error; however, lysis can be minimized by slowing increasing pressure and paying close attention to the strength of the currents.

d. Controls and Egg Manipulation

During microinjection, eggs are pulled to the holding pipette by suction (negative pressure). The focus of the pronuclei is adjusted with the height control of the holding pipette. The positions of the egg and injection needle are kept as close as possible to the center of the viewing field by use of the joystick or manual two-dimensional control knobs.

The number of manipulations for each hand is kept to a minimum. One hand controls the injection needle position with the joystick (or control knobs) and the microsyringe assembly for suction control of the holding pipette. The other hand controls the position of the holding pipette and the focus of the egg with the *x*-axis control on the holding pipette. The first hand also occasionally moves the slide with the stage knob and also adjusts DNA flow. This arrangement allows a skilled operator to routinely microinject an average of 60–120 eggs per hour.

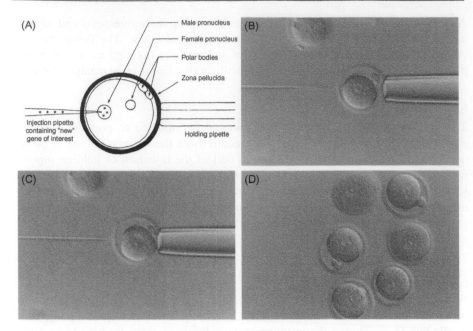

Figure 2.17 DNA microinjection. (A) Diagram of DNA microinjection into a pronuclear murine zygote (reproduced with permission from Pinkert, 1987). A pronuclear egg is held by a large-bore pipette using gentle suction. A small-bore injection pipette containing the DNA solution is inserted into the male pronucleus; the DNA solution is then slowly expelled into the pronucleus that expands approximately twofold. The diameter of the mouse egg is ~75 μm. (B) Male pronucleus in focus and injection needle aligned before DNA injection. Note diameter of pronucleus. (C) During injection, note the increase in diameter of the pronucleus and the uniform spherical periphery indicative of full expansion. (D) Within minutes of injection, some of the eggs will obviously lyse (at 11:00 and 3:00) and fill the zona pellucida.

4. Visualizing and Injecting the Egg Pronuclei

a. Orientation
The optimal orientation for microinjection of pronuclei is to have the male pronucleus closest to the injection needle (Figure 2.17). The height control of the holding pipette is raised and lowered until one is sure the center of the pronucleus is in focus. The injection needle tip should also be in focus prior to penetrating the egg.

b. Egg Membrane Penetration with the Injection Needle
Because the DNA aliquot is constantly running out the tip of the injection needle, one has to work efficiently to penetrate the zona pellucida membrane and pierce the pronuclear membrane to prevent the buffer and DNA from "pushing" the membranes (note cytoplasmic delivery in Figure 2.18). A sharp needle without debris stuck to the tip will have the best penetrating characteristics. The DNA flow may have to be finely adjusted if bubbling is still occurring with a good pipette.

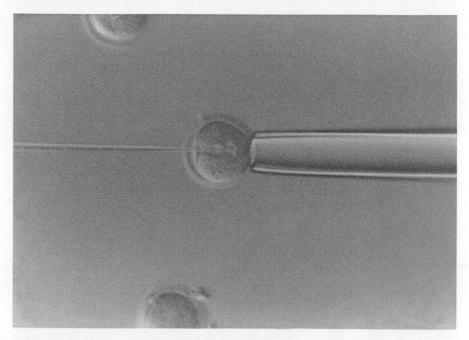

Figure 2.18 DNA microinjection into cytoplasm. Cytoplasmic injection is not desirable, but can happen if the injection needle is not aligned with the pronucleus or is too blunt for penetration, or if the flow rate is too great. It can result in the appearance of a cytoplasmic vacuole that may be maintained even after the needle is withdrawn.

c. Filling the Pronuclei with DNA Solution

The injection pipette should fill the pronucleus with DNA until the expansion of the pronucleus stops (an approximate 50–100% volume expansion), at which point the needle can be rapidly withdrawn to prevent overexpansion. One must simultaneously observe the swelling of the egg to ensure that not too much DNA is delivered. The average length of time inside the pronucleus ranges from 0.5 to 3.0 s. When first learning microinjection, one should try several different degrees of filling and record the results. Practice is the only way to understand the correct amount of DNA to deliver. The injection pipette does not always penetrate the pronucleus perfectly, and leaking of DNA out of the pronucleus or out the side of the pipette at the injection site will frequently occur. In this case, the pronucleus will not expand as rapidly, and the whole egg will enlarge. The egg will lyse if too much DNA (or solution volume) is delivered into the cytoplasm.

5. Criteria for Microinjection to Yield Transgenic Mice

a. Treatment After Microinjection

Regardless of the skill and speed of the injectionist, the eggs should be left on the slide no longer than 30 min. Once the injection slide is completed, the pipettes are raised,

Figure 2.19 Record keeping and animal care are paramount in transgenic experimentation. Animals are maintained in temperature-, humidity-, and photoperiod-controlled rooms in barrier caging that prevents the introduction of bacterial, viral, or other pathogens.

the slide carefully returned to the warming stage on the dissecting microscope, and the injected eggs transferred through two "wash" dishes of culture medium (without cytochalasin B) and then placed in an appropriately labeled microdrop dish.

b. Egg Lysis
Routinely 10–25% of eggs will lyse as a result of microinjection, but this figure is highly dependent on the skill of the injector, the egg background strain, the injection pipette shape, and the quality and concentration of the DNA preparation. If <10% of the microinjected eggs lyse, perhaps the volume of DNA delivered into the egg pronucleus was insufficient to generate transgenic mice efficiently. Again, this may depend on the skill of the injector, as lysis rates may approach 0% by skilled injectionists using appropriately quantified and high-purity DNA solutions.

c. Record Keeping
Accurate records during all of the microinjection-related procedures aid in the development of microinjection skills, help pinpoint sources of problems, and help maintain mouse identification (Figure 2.19). Representative data sheets that we use

are illustrated at the end of this chapter (see Appendix), whereas Chapter 3 details and explains specific needs and requirements.

d. *In Vitro* Culture of Microinjected Eggs

Zygotes can be routinely cultured to the two-cell or blastocyst stage, then transferred to the oviduct of day 1 pseudopregnant mice (Bronson and McLaren, 1970; Kooyman et al., 1990; Johnson et al., 1996). Using outbred (e.g., ICR) or hybrid (e.g., B6SJLF1) mice, typically 40–60% and 60–80% of injected eggs (60–80 and 75–100% of noninjected eggs), respectively, develop overnight to the two-cell stage in our experience. A high proportion of eggs going to the two-cell stage will usually proceed to the blastocyst stage. The significant decrease in egg viability following microinjection is due to the trauma associated with the ionic and physical environment of the zygote. Egg culture provides a reasonable assessment of microinjection proficiency. If survival frequencies approach noninjection efficiencies (culturing control, noninjected eggs simultaneously), then review of injection procedures is warranted. If survival frequencies drop way below controls, then experimental procedures should also be reevaluated. Inbred strains exhibit lower overall egg survivability and may not develop beyond the two-cell stage with the suggested media preparations outlined. As an aside, we have used progression to the two-cell stage as a useful gauge to indicate optimal BAC DNA concentrations for injection as a preinjection step prior to injecting the majority of eggs available for a particular injection session.

F. Egg Transfer

1. Ova Reimplantations

Procedures and factors influencing egg transfer have been described (Adams, 1982, and a number of subsequent reviews and laboratory manuals). Briefly, a pool of female mice are maintained and observed for visual evidence indicative of proestrus (vulval and vaginal appearance; Champlin et al., 1973). The day after mating to vasectomized males, those females exhibiting copulatory plugs are used as egg transfer recipients. (As a result of mating to vasectomized males, these females will have the ability to maintain the eggs through gestation, without uterine competition from naturally ovulated eggs. Because the mating to vasectomized males renders the females pseudopregnant, if fertilized eggs are not transferred, the females will not return to estrus for about 11 days.)

For transfer, recipients are anesthetized and placed in ventral recumbancy under a dissecting microscope. The surgical incision site is prepared (fur adjacent to the surgical site is removed) and then liberally swabbed with disinfectant (e.g., Betadine). Using sterile surgical instruments, a 5-mm transverse skin incision is made between 10 and 15 mm from the spine, posterior to the rib cage. By sliding the skin incision to a point located about one-third of the way between the dorsal and ventral midline, the ovarian fat pad will be located through the intact body wall. When the fat pad has been located, the body wall is gently grasped with blunt forceps, and a small incision is made in the body wall, taking care to avoid larger blood vessels. The ovarian fat pad will then be exteriorized (the fat pad is the land-mark used to locate the ovary and oviduct). The oviduct is maintained in place using paper sponges around the uterus, by placing a 4-cm-long 3-mm-wide sponge through the mesometrium, or by grasping the fat pad with a small surgical clamp (Behringer et al., 2014). The bursa surrounding the ovary is then either cut or torn to expose the ostium of the oviduct.

Once these steps are accomplished, eggs are obtained using mouth pipetting under a second dissecting microscope. Between 25 and 30 injected ova are then expelled into the ostium of the oviduct of the recipient and are blown through to the ampulla by gentle pressure through the pipette. (*Note*: The number of eggs transferred may vary based on injector skill or available microinjected eggs. One or two control eggs may be transferred to help ensure pregnancy maintenance if there is concern regarding proficiency or quantity of microinjected eggs, as just two implanted eggs have been shown to be as effective at maintaining pregnancy as larger numbers Johnson et al., 1996.) Alternatively, eggs may be deposited directly into the ampulla via a previously made hole or cut between the ostium and ampulla, thus avoiding trouble arising from anatomical variations often found in the ostium (Johnson et al., 1996). The reproductive tract is then carefully replaced into the abdomen, and the body wall is closed with one or two sutures followed by skin closure with wound clips. Surgery is performed with the animal resting on dry, soft padding. Body temperature is maintained during the procedure. Surgical depth of anesthesia is ensured by observation of respiratory rate, tail reflex, or foot-pinch reflex at frequent intervals.

2. Vasectomy

Male mice are vasectomized and used to mate proestrous stage females in order to induce pseudopregnancy. For this procedure, each male mouse is anesthetized. The incision site is prepared, and the abdomen is then thoroughly swabbed with Betadine. Using sterile instruments, a 10- to 15-mm midventral skin incision is fol-lowed by an incision through the body wall musculature. The fat pads by each tes-tis should be visible. The fat pad of one testis is grasped using blunt forceps to reveal and exteriorize the testis and vas deferens. The vas deferens can then be iso-lated and ligated or cauterized. Generally, the removal of 15−25 mm of the vas deferens is sufficient to ensure irreversible surgery (and limit worry of reanasto-mosis). The procedure is repeated for the second testis. Only one abdominal inci-sion is made for this procedure; testes are revealed and exteriorized through this single site. On completion of the vasectomy, the body wall is closed with sutures and the skin closed using wound clips or surgical adhesive. (*Note*: We generally place the sections of vas deferens aside following their removal to keep count of the number of sections removed. At least once in the last 15 years, a trainee indeed left one intact vas deferens but was able to identify the error while the mouse was still sedated before the wound clips were applied.)

Support care includes placement of the animal on clean soft padding during the procedure and warming anesthetized animals. Observation of respiratory rate and foot-pinch reflex of the mouse is routine throughout the procedure and postoperative period. If wound clips are used, they are removed after healing (i.e., in 7–10 days). At this time, the males are tested for mating ability. Copulatory plug formation and evidence of two infertile matings are required before the males are used in the vasectomized male battery. As a rule, we generally maintain twice the number of males that might be used for a given day of experiments. To arrive at this number, we estimate the number of recipients needed and assume that we will obtain copulatory plugs in approximately 50% of the paired proestrous females, selected on vulval and vaginal appearance. The vasectomized males are used until infirmities appear or there is no evidence of a successful mating with proestrous females after six consecutive pairings.

3. Postoperative Care

After completion of any surgical procedure, animals are kept at 30°C and carefully monitored until they regain reflex responsiveness. At that time, they are placed into clean cages and observed daily thereafter. As these surgeries are short in duration, are not invasive in nature, and aseptic techniques are employed, postoperative analgesia and antibiotic use are not favored.

G. Gamete Cryopreservation and In Vitro Fertilization

1. Introduction

Cryopreservation of preimplantation stage ova was first described in 1972, using dimethylsulfoxide (DMSO) as a cryoprotectant (Whittingham et al., 1972; Wilmut, 1972). Procedures for cryopreservation of murine eggs and most recently spermatozoa have been greatly simplified over the last few years. We have utilized a number of approaches and outline some straightforward techniques that have proved useful for preservation of gametes (Leibo and Oda, 1993; Sztein et al., 1997; Songsasen and Leibo, 1998). In conjunction with in vitro fertilization (IVF) protocols, the advantages of these enabling technologies provide a greater deal of flexibility in cost-effectively maintaining valuable (or potentially valuable) lineages.

The routine freezing of transgenic lineages provides several important advantages.

1. Should loss of transgene expression occur in latter generations of a given lineage, the original genetic background (and characterized transgene expression patterns) can be recovered?
2. For transgenic lines that may be in great demand, or where breeding spans a number of generations, genetic drift may be problematic. This is a crucial concern in the use of transgenic models to derive clinical correlations. As such, freezing ova within a given lineage at an early generation or at specific generational intervals can reduce the risks associated with genetic drift.
3. Disease or line contamination can unexpectedly destroy valuable transgenic animals. Preservation of a relatively small number of eggs can safeguard any given project.

4. In a catastrophic situation, the use of cryostorage can provide a means not only to safeguard research but also to influence rapid return to productivity.
5. Cryopreserving transgenic lineages can significantly reduce or eliminate the costs associated with housing and maintaining lines of mice.
6. Egg cryopreservation can be used in conjunction with rederivation of pathogen-free stocks of animals. In mouse experiments, embryo transfer of both frozen and nonfrozen eggs have been used to rederive important animals. Various rodent pathogens have been eliminated from colonies using embryo transfer. The classical method of rederiving pathogen-free stocks of mice involves the aseptic collection of fetuses by Cesarean section with transfer to pathogen-free foster mothers. However, while many infections can be vertically transmitted, the zona pellucida can provide a barrier to infection of pre-implantation ova (Reetz et al., 1988). Several groups have successfully obtained pathogen-free offspring from heavily contaminated stock using variations of embryo washing/transfer techniques (Reetz et al., 1988; Rouleau et al. (1993)). Similarly, we have successfully used these procedures to rederive mouse lines that were contaminated by indigenous murine pathogens.

2. Methods for Gamete Cryopreservation and IVF

a. Egg Cryopreservation

Procedures for cryopreservation of murine eggs have been greatly simplified over the last few years. We originally utilized two different approaches to freeze early stage eggs. A commercial vitrification kit was suitable for storage of eight-cell to early blastocyst stage ova. Using vitrification, survival rates >80% were reported for a number of mouse strains after vitrification and storage in liquid nitrogen (LN_2). While the reagents can be individually obtained and the procedure performed in the laboratory, we began using a one-step freezing method for cryopreservation of murine eggs as outlined by Leibo and Oda (1993). As described, for freezing medium, M2 (or BMOC-3 + HEPES) is supplemented with 2M ethylene glycol (Sigma E-9129) and 7.5% polyvinylpyrrolidone (PVP; Sigma P-2307). Eggs are incubated in the freezing medium for 5 min, drawn into a 0.5 mL sterile cryo-preservation straw (PETS, Canton, TX; 05-079-119-1; IMV Technologies USA; #005565), sealed and plunged into LN_2. To recover eggs, the straws are removed from the LN_2 tank and immediately thawed by submersion in a 37°C water bath. The contents of the straw are expelled into a dish of sterile culture medium and the eggs are washed twice before being readied for transfer. We have obtained a 50% or greater survival rate of two- to eight-cell eggs; however, our results with morula-stage eggs were significantly lower (Pinkert and Johnson, unpublished data). That said, we now use a JAX slow-cooling method for eggs (Ostermeier et al., 2008). Recently, The Center for Animal Resources and Development at Kumamoto University (CARD, Japan) published ova vitrification as an effective method of stockpiling eggs for use in IVF with fresh, cryopreserved, and cold-stored sperm (Nakagata et al., 2013).

a. Vitrification involves a brief exposure of the embryo to cryoprotectants followed by rapid cooling by submersion in liquid nitrogen. Cryopreservation via vitrification prevents ice crystal formation within the entire sample by creating a high concentration of a solute (such as DMSO) within cells. The sample maintains a glasslike state (*vitrum* means "glass") during storage, and the lack of water within the cell and the extremely low temperature (−190°C) shuts down virtually all metabolic processes.

b. Another popular method of cryopreservation is controlled rate cryopreservation. While the embryo itself is actually vitrified with this method, the process to achieve this is different. As with vitrification, controlled rate cryopreservation utilizes hypertonic concentrations of solutes (such as propylene glycol, glycerol, or ethylene glycol) to prevent ice crystal formation within the cell. However, unlike vitrification, there is enough water in the sample that ice crystals form elsewhere throughout the sample. In fact, ice crystal formation is purposely initiated at around the freezing point of the solution by touching the sample with forceps or another object cooled to the temperature of liquid nitrogen (a process termed "seeding"). While the sample crystallizes and water molecules are removed from the solution, the molarity of the solution and thus the cell increase. The temperature of the sample is slowly decreased to about −30°C, forcing continued removal of water from the solution and the cell. At this point, it is safe to rapidly cool the embryo and submerge the samples in liquid nitrogen for storage. The embryo remains in a state of suspended animation until thawed.

b. Sperm Cryopreservation

Great strides in sperm cryopreservation and accompanying IVF protocol efficiency and success rate were achieved in recent years. Contributions by multiple groups have culminated in a robust body of knowledge. When applied appropriately (e.g., being mindful of the methods used to preserve samples), this knowledge should prove effective in preserving and reanimating the majority of mouse strains or transgenic/knockout lines with high efficiency.

Arguably, the most effective and practical cryopreservation and reanimation protocols were published along with availability of cryopreservation kits (Nakagata et al., 2013). Such kits simplify and streamline procedures for those who do not perform these activities on a regular basis or who need a "gold standard" to ensure the success of critical projects.

Sperm samples are cryopreserved by suspending epididymal contents from one or two males (12 weeks or age or older) in a tall drop of ~120 µL of cryoprotectant (18% raffinose, 3% nonfat dry skim milk, and 100 mM L-glutamine) under oil. Then, during a 3-min incubation on a 37°C slide warmer, the dish is gently agitated in a circular motion to evenly distribute spermatozoa. Then each of about 10 10-µl samples is distributed into 0.25 cc cryopreservation straws. Before the sample is taken, a given volume (e.g., 100 µL) cryoprotectant is drawn into the straw. This provides ballast for the straw, keeping it submerged while it is stored in liquid nitrogen, and wets the polyvinyl alcohol plug at one end of the straw. Both ends of the straws are heat sealed and the straws immediately placed in the vapor phase of a storage Dewar using a custom-made holder floating on the liquid phase. After 10 min, the straws can be inserted into a liquid nitrogen-submerged goblet mounted on an aluminum cane and placed in the storage Dewar indefinitely.

Table 2.3 HTF Medium

Component	mg/100 mL	g/1000 mL
NaCl	593.8	5.938
KCL	35	0.35
$MgSO_4 \cdot 7H_2O$	4.9	0.049
KH_2PO_4	5.4	0.054
$CaCl_2 \cdot 2H_2O$	29.8	0.298
$NaHCO_3$	210	2.1
Glucose	50	0.5
Sodium pyruvate	3.7	0.037
Sodium lactate 70%	0.34 mL	3.4 mL

a. For IVF using cryopreserved sperm (following the above protocol), the sperm is thawed by removing the straw from the liquid nitrogen and holding it in air for 5 s, then plunging the straw into a 37°C water bath and leaving it for 10 min. The sample is then carefully transferred to a 120 μL drop of preincubation medium containing methyl-beta-cyclodextrin, which facilitates cholesterol efflux (Takeo et al., 2008), resulting in improved capacitation of cryopreserved sperm. Following a 30−45 min preincubation period, a 10-μl sample of sperm is introduced to a 90-μl drop of HTF medium supplemented with 1 mM reduced glutathione (Table 2.3; Quinn, 1995; Takeo and Nakagata, 2011) containing cumulus oocyte complexes (COCs). Tubes are gently agitated, allowing sperm to equilibrate in the HTF medium for 10 min at 37°C in a humidified environment (5% O_2, 5% CO_2, 90% N_2). The coculture of eggs and sperm lasts for approximately 3−4 h. At this point, only two-cell embryos are likely capable of normal development. These embryos are harvested for study or reimplantation to pseudopregnant recipient females.

b. After observing motile sperm, proceed to IVF protocol.

c. IVF Protocol

For IVF using fresh sperm, we currently use a technique outlined for studying strain differences in response to specific superovulation protocols (Vergara et al., 1997). This technique was also used to shorten the generational interval for a variety of breeding schemes. We have performed unilateral sperm recoveries (orchidectomy plus vasectomy) to test the fertility and germline penetrance of given mutations in genetically modified males and to develop large cohorts of offspring rapidly. In conjunction with marker-assisted selection protocols, IVF methodology can be used to identify and better characterize specific strain backgrounds (or identify animals with multiple germline modifications) and provide a useful tool in conjunction with the cryopreservation of fertilized eggs early in the characterization of novel lineages, as outlined here.

Table 2.4 IVF Medium (modified Tyrode solution; Whittingham, 1971)

Component	g/100 mL
NaCl	0.701
NaHCO$_3$	0.201
Glucose	0.100
KCl	0.0201
CaCl$_2$	0.026
NaHPO$_4$	0.004
Sodium pyruvate	0.0036
BSA	30 mg/mL (Sigma A7906)
MgSO$_4 \cdot$ 7H$_2$O	0.0294

1. PMSG is administered to females on day 1 and HCG on day 3 as noted for superovulation outlined in Section 2.I.B.
2. At 08:00 on day 4, an aliquot of spermatozoa is transferred to a fresh 35-mm tissue culture dish containing 0.5 mL IVF medium (Table 2.4) under oil, to give a final concentration of ~ 1 to 2×10^6 sperm/mL. The spermatozoa are then allowed to capacitate for 1 h at 37°C before oocytes are added to the incubation mixture.
3. At 09:30 on day 4, superovulated females are sacrificed and oviducts removed. If IVF includes the COC, oviducts are placed individually inside a fertilization dish (0.5 mL IVF or HTF medium containing spermatozoa under silicone oil), and cumulus masses are then expressed with forceps and a 26-gauge needle.
4. If IVF includes culture of cumulus-free oocytes (not necessary for most strains; Vergara et al., 1997), cumulus masses are enzymatically digested from oocytes in a 35-mm tissue culture dish containing BMOC-3 or M16 medium supplemented with hyaluronidase before oocytes are placed in the fertilization dish.
5. After all available COCs or cumulus-free oocytes are placed in the fertilization dish, the dish is incubated for ~ 6 h at 37°C. After incubation, eggs are removed from the IVF dish and rinsed twice in 2.0 mL of BMOC-3 or M16 medium, then placed in fresh medium in microdrops covered with silicone oil for overnight culture at 37°C. Following overnight culture (day 5), eggs are counted and the yield of two-cell ova is determined. These ova can then be cultured, cryopreserved, or transferred to pseudopregnant females as desired.

III. Summary

As summarized in Figure 2.20, the steps involved in producing and evaluating transgenic mice derived by DNA microinjection are straightforward, although mastering the steps can be very time-consuming. In general, it takes between 5 and 12 months of dedicated effort to become proficient at the outlined techniques. One

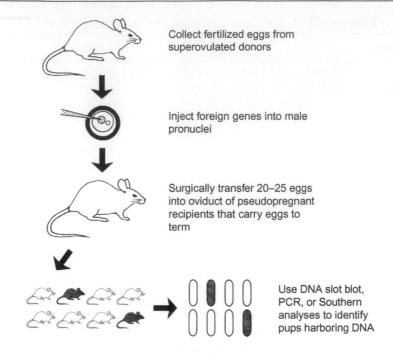

Collect fertilized eggs from superovulated donors

Inject foreign genes into male pronuclei

Surgically transfer 20–25 eggs into oviduct of pseudopregnant recipients that carry eggs to term

Use DNA slot blot, PCR, or Southern analyses to identify pups harboring DNA

• Perform tissue biopsies–analyze foreign DNA integration, mRNA transcription, and protein production

• Establish transgenic lines to study gene regulation in progeny

Figure 2.20 Gene transfer in mice. The methodology employed in the production and subsequent evaluation of transgenic mice.

must begin with embryo handling and practice embryo transfer skills to develop good coordination. These skills will ultimately maximize all gene transfer efforts. Then, microinjection practice will lead to the production of transgenic mice, but continued commitment will be necessary in order to develop proficiency at these techniques. Accordingly, the relative merit of other technologies for generation of founder transgenic mice, as outlined in succeeding chapters, may prove more expeditious. However, many of the associated technologies, including gamete preservation, represent important skill sets to master. We hope this chapter and the chapters that follow will illustrate the techniques necessary to create and maintain appropriate animal models both successfully and efficiently.

Acknowledgments

The authors gratefully acknowledge their mentors, colleagues, and trainees, whose invaluable assistance and many suggestions were most appreciated.

References

Adams, C.E., 1982. Factors affecting the success of egg transfer. In: Adams, C.E. (Ed.), Mammalian Egg Transfer. CRC Press, Boca Raton, FL, pp. 176–183.

Behringer, R., Gertsenstein, M., Nagy, K., Nagy, A., 2014. Manipulating the Mouse Embryo: A Laboratory Manual. fourth ed. Cold Spring Harbor Press, Cold Spring Harbor, NY.

Biggers, J.D., 1971. Metabolism of mouse embryos. J. Repord. Fertil. Suppl 14, 41–56.

Brinster, R.L., 1972. Culture of the mammalian embryo. In: Rothblat, G., Cristafalo, V. (Eds.), Nutrition and Metabolism of Cells in Culture, vol. II. Academic Press, New York, NY, pp. 251–286.

Brinster, R.L., Palmiter, R.D., 1986. Introduction of genes into the germ line of animals. Harvey Lectures, Series 80, pp. 1–38.

Brinster, R.L., Chen, H.Y., Trumbauer, M.E., Yagle, M.K., Palmiter, R.D., 1985. Factors affecting the efficiency of introducing foreign DNA into mice by microinjecting eggs. Proc. Natl. Acad. Sci. USA 82, 4438–4442.

Brinster, R.L., Braun, R.E., Lo, D., Avarbock, M.R., Oram, F., Palmiter, R.D., 1989. Targeted correction of a major histocompatibility class II Eα gene by DNA microinjected into mouse eggs. Proc. Natl. Acad. Sci. USA 86, 7087–7091.

Bronson, R.A., McLaren, A., 1970. Transfer to the mouse oviduct of eggs with and without the zona pellucida. J. Reprod. Fertil. 22, 129–137.

Burki, K., 1986. Experimental Embryology of the Mouse. Monographs in Developmental Biology, vol. 19. Karger, New York, NY.

Camper, S.A., Saunders, T.L., 2000. In: Accili, D. (Ed.), Genetic Manipulation of Receptor Expression and Function. Wiley-Liss, Inc., New York, NY, pp. 1–22.

Champlin, A.K., Dorr, D.L., Gates, A.H., 1973. Determining the stage of the estrous cycle in the mouse by the appearance of the vagina. Biol. Reprod. 8, 491–494.

Chen, H.Y., Trumbauer, M.E., Ebert, K.M., Palmiter, R.D., Brinster, R.L., 1986. Developmental changes in the response of mouse eggs to injected genes. In: Bogorad, L. (Ed.), Molecular Developmental Biology. Alan R. Liss, Inc., New York, NY.

Chisari, F.V., Klopchin, K., Moriyama, T., Pasquinelli, C., Dunsford, H.A., Sell, S., et al., 1989. Molecular pathogenesis of hepatocellular carcinoma in hepatitis B virus transgenic mice. Cell (Cambridge, MA). 59, 1145–1156.

Engler, P., Haasch, D., Pinkert, C.A., Doglio, L., Glymour, M., Brinster, R., et al., 1991. A strain-specific modifier on mouse chromosome 4 controls the methylation of independent transgene loci. Cell 65, 939–947.

Festing, M.F.W., 1979. Inbred Strains in Biomedical Research. Oxford University Press, New York, NY.

Flaming, D.G., Brown, K.T., 1982. Micropipette puller design: form of the heating filament and effects of filament width on tip length and diameter. J. Neurosci. Methods 6, 91–102.

Foster, H.L., Small, J.D., Fox, J.G., 1983. The Mouse in Biomedical Research. Academic Press, New York, NY.

Green, M.R., Sambrook, J., 2012. Molecular Cloning: A Laboratory Manual. fourth ed. Cold Spring Harbor Press, Cold Spring Harbor, NY.

Harris, A.W., Pinkert, C.A., Crawford, M., Langdon, W.Y., Brinster, R.L., Adams, J.M., 1988. The Eμ-*myc* transgenic mouse: a model for high incidence spontaneous lymphoma and leukemia of early B cells. J. Exp. Med. 167, 353–371.

Johnson, L.W., Moffatt, R.J., Bartol, F.F., Pinkert, C.A., 1996. Optimization of embryo transfer protocols for mice. Theriogenology 46, 1267–1276.

Kim, D.W., Uetsuki, T., Kaziro, Y., Yamaguchi, N., Sugano, S., 1990. Use of the human elongation factor 1α promoter as a versatile and efficient expression system. Gene 91, 217–223.

Kooyman, D.L., Baumgartner, A.P., Pinkert, C.A., 1990. The effect of asynchronous egg transfer in mice on fetal weight. J. Anim. Sci. 68 (Suppl. 1), 445.

Leibo, S.P., Oda, K., 1993. High survival of mouse zygotes and embryos cooled rapidly or slowly in ethylene glycol plus polyvinylpyrrolidone. Cryoletters 14, 133–144.

Leveille, M.-C., Armstrong, D.T., 1989. Preimplantation embryo development and serum steroid levels in immature rats induced to ovulate or superovulate with pregnant mares' serum gonadotropin injection or follicle-stimulating hormone infusions. Gamete Res. 23, 127–138.

Nakagata, N., Takeo, T., Fukumoto, K., Kondo, T., Haruguchi, Y., Takeshita, Y., et al., 2013. Applications of cryopreserved unfertilized mouse oocytes for *in vitro* fertilization. Cryobiology 67, 188–192.

Ostermeier, G.C., Wiles, M.V., Farley, J.S., Taft, R.A., 2008. Conserving, distributing and managing genetically modified mouse lines by sperm cryopreservation. PLoS One 30, e2792.

Pieper, F.R., deWit, I.C.M., Pronk, A.C.J., Kooiman, P.M., Strijker, R., Krimpenfort, P.K.A., et al., 1992. Efficient generation of functional transgenes by homologous recombination in murine zygotes. Nucleic Acids Res. 20, 1259–1264.

Pinkert, C.A., 1987. Gene transfer and the production of transgenic livestock. Proc. U.S. Anim. Health Assoc. 91, 129–141.

Pinkert, C.A., 1990. A rapid procedure to evaluate foreign DNA transfer into mammals. Biotechniques 9, 38–39.

Pinkert, C.A., 1994. Transgenic Animal Technology: A Laboratory Handbook. Academic Press, San Diego.

Pinkert, C.A., 2002. Transgenic Animal Technology: A Laboratory Handbook. second ed. Academic Press, San Diego.

Quaife, C.J., Pinkert, C.A., Ornitz, D.M., Palmiter, R.D., Brinster, R.L., 1987. Pancreatic neoplasia induced by *ras* expression in acinar cells of transgenic mice. Cell 48, 1023–1034.

Quinn, P., 1995. Enhanced results in mouse and human embryo culture using a modified human tubal fluid medium lacking glucose and phosphate. J. Assist. Reprod. Genet. 12, 97–105.

Quinn, P., Barros, C., Whittingham, D.G., 1982. Preservation of hamster oocytes to assay the fertilizing capacity of human spermatozoa. J. Reprod. Fertil. 66, 161–168.

Rafferty, K.A., 1970. Methods in Experimental Embryology of the Mouse. Johns Hopkins Press, Baltimore, MD.

Reetz, I.C., Wullenweber-Schmidt, M., Kraft, V., Hedrich, H.J., 1988. Rederivation of inbred strains of mice by means of embryo transfer. Lab. Anim. Sci. 38, 696–701.

Richa, J., Lo, C.W., 1989. Introduction of human DNA into mouse eggs by injection of dissected chromosome fragments. Nature 245, 175–177.

Rouleau, A.M., Kovacs, P.R., Kunz, H.W., Armstrong, D.T., 1993. Decontamination of rat embryos and transfer to specific pathogen-free recipients for the production of a breeding colony. Lab. Anim. Sci. 43, 611–615.

Rusconi, S., 1991. Transgenic regulation in laboratory animals. Experientia 47, 866–877.

Schedl, A., Montoliu, L., Kelsey, G., Schütz, G., 1993. A yeast artificial chromosome covering the tyrosinase gene confers copy number-dependent expression in transgenic mice. Nature. 362, 258–261.

Short, J.M., Blakeley, M., Sorge, J.A., Huse, W.D., Kohler, S.W., 1989. The effects of eukaryotic methylation on recovery of a lambda phage shuttle vector from transgenic mice. J. Cell. Biochem. (Suppl. 13b), 184. (Abstract).

Songsasen, N., Leibo, S.P., 1998. Live mice from cryopreserved embryos derived *in vitro* with cryopreserved ejaculated spermatozoa. Lab. Anim. Sci. 48, 275–281.

Sternberg, N.L., 1992. Cloning high molecular weight DNA fragments by the bacteriophage P1 system. Trends Genet. 8, 11–16.

Sztein, J.M., Farley, J., Young, A., Mobraaten, L.E., 1997. Motility of cryopreserved mouse spermatozoa affected by temperature of collection and rate of thawing. Cryobiology 35, 46–52.

Takeo, T., Nakagata, N., 2011. Reduced glutathione enhances fertility of frozen/thawed C57BL/6 mouse sperm after exposure to methyl-beta-cyclodextrin. Biol. Reprod. 85, 1066–1072.

Takeo, T., Hoshii, T., Kondo, Y., Toyodome, H., Arima, H., Yamamura, K.I., et al., 2008. Methyl-beta-cyclodextrin improves fertilizing ability of C57BL/6 mouse sperm after freezing and thawing by facilitating cholesterol efflux from the cells. Biol. Reprod. 78, 546–551.

Vergara, G.J., Irwin, M.H., Moffatt, R.J., Pinkert, C.A., 1997. *In vitro* fertilization in mice: strain differences in response to superovulation protocols and effect of cumulus cell removal. Theriogenology 47, 1245–1252.

Whittingham, D.G., 1971. Culture of mouse ova. J. Reprod. Fertil. (Suppl.) 14, 7–21.

Whittingham, D.G., Leibo, S.P., Mazur, P., 1972. Survival of mouse embryos frozen to −195°C and −269°C. Science 178, 411–414.

Wilmut, I., 1972. The effect of cooling rate, warming rate, cryoprotective agent and stage of development on survival of mouse embryos during freezing and thawing. Life Sci. 11, 1071–1079.

Yang, X.W., Model, P., Heintz, N., 1997. Homologous recombination-based modification in *E. coli* and germline transmission in transgenic mice of a bacterial artificial chromosome. Nat. Biotechnol. 15, 859–865.

Appendix

For record-keeping purposes, there are a number of database systems and programs available for experimentation and colony-related needs. The following figures highlight some of the records or windows that we have found to be of value in order to maintain a running status of projects.

TA/ESC COLONY SUMMARY

Date: _____   By:_____

COLONY TOTALS

	Room #	# of animals	# small cages	# large cages
Expt. Section				
Expt. Subtotal:				
Stock Section				
Stock Subtotal:				
Colony Total:				

• •

STOCK AVAILABLE

	# of animals	# of animals designated for use
S-W recipient ♀ [≤7wks/>7 wks]		
C57BL/6 ♀		
3-4 weeks		
5-8 weeks		
>8 weeks		
B6xSJL donor ♀		
3-4 weeks		
5-8 weeks		
>8 weeks		
11 or 12		

• •

WEANING

	Total	# to stock	# to mating
S-W ♀			
S-W ♂			
B6xSJL F1 ♀			
B6xSJL F1 ♂			
C57BL/6J ♀			
C57BL/6J ♂			
SJL/J ♀			
SJL/J ♂			
TOTAL WEANED:			

Figure A.1 Colony Summary Sheet. Additional pages are also updated weekly with identification of animals on a room-by-room basis.

INJECTION SCHEDULE from _____ to _____

Expt day/ Initials	# Mice/ Strain	Age (B/W/R)	Fri	Sat	Sun	Mon	Tues	Wed	Thurs

Figure A.2 Injection Schedule. A list is posted by the freezer with the aliquoted hormones in addition to individual cage cards identifying appropriate animals and treatments.

GENE LOG

Last Revised:

#	Name/Expt Description	Lab	Project	Rec'd	Date Injected	Eggs Trans	Trans Total	#Tg/ Total	Comments (# = transgenic animal log #)
1	El ras	CAP	DNA	5/1/91	practice/training	826	49	13/68	
2	AB1	PAW	ES cell	10/21/91	11/2/91 - 11/6/91	75	7	5/8-20	12 control eggs were used
3	MMTV-sIGF.3	CAP	DNA	10/22/91	12/19/91	259	12	16/58	
4									
5									
6									
7									
8									
9									
10									
11									

Figure A.3 Gene Log. All projects are kept on one central inventory.

MICROINJECTION

& NAME/EXP. DESCRIP.:_____

Notebook
Reference:_____

DATE:_____ STRAIN:_____ CLIENT:_____

INJECTOR:_____

OF DONORS OPENED:_____

OF EGGS OBTAINED:_____

OF EGGS GOOD:_____

OF EGGS GIVEN TO OTHER INJECTOR (NAME & #):_____

INJECTED	NOT INJ.	TOTAL	# ALIVE	START/FINISH	COMMENTS

EGGS FOR TRANSFER:_____

RECIPIENTS AVAILABLE:_____

START TRANSFER:_____

END TRANSFER:_____

Figure A.4 Microinjection Log. A detailed log is maintained for each day of injection.

TODAY'S DATE: ___/___/___

MATERNITY

EXPERIMENT (# & NAME):_____   REVIEWED: _____
DNA CONCENTRATION: _____

RECIPIENT D 1: ____/____/____   INJECTOR: _____
EGG STAGE (DAY): _____   TRANS DATE: ____/____/____   TRANSFER: _____
OPEN ON DAY*: _____   OPEN DATE: ____/____/____   OPENED: _____

ID #	[ep#]	Eggs transferred/side	L/C^z	Pups/ fetuses	RS	Comments
TOTALS						

_____/_____   FEMALES PREGNANT   _____ %

_____/_____   # IMPLANTATION   _____ %
                  SITES (PUP + RS)

_____/_____   PUPS (OR FETUSES)   _____ %

ADDITIONAL COMMENTS:
_____
_____
_____
_____
_____

^{z}Livebirth (L), Cesarian (C). If C, then indicate the side that the pup or resorption site located.
*Based on recipient

Figure A.5 Maternity Log. After the embryo transfers are completed, a copy of the day's transfers is placed into a central archive. While all cages are checked twice daily, the maternity logbook is also checked daily to be certain that no litters have gone unattended.

SAMPLE ANALYSIS

EXPERIMENT (#, name):____________ PAGE:___ OF ___ INVESTIGATOR:____________ ANALYSIS BY:____________ REVIEW:____

DATE SAMPLES READY:____________ DATE SAMPLES SENT:____________ DATE RESULTS RECEIVED:____________ DATE ANIMALS SHIPPED/SAC'D:____________

LOG #	ANIMAL ID	DAM	SIRE	DOB	DOD/DOS	TISSUE	DNA	RNA	METHOD	ANALYSIS/COMMENTS

Figure A.6 Sample Analysis. All tail samples or other biopsies are given a log number and maintained in a common notebook for the laboratory.

TRANSGENIC ANIMAL SUMMARY

EXPERIMENT (#, name): _____   Investigator: _____   Founder Transgenic Rate: ___ / ___   Date Revised: _____

ANIMAL ID	LINE #	STRAIN	DOB	DOD	DNA		mRNA		PROTEIN		PATH-OLOGY (Y/N)	PROGENY	COMMENTS
					LOG	RES	LOG	RES	LOG	RES			

Figure A.7 Transgenic Animal Summary. After founder transgenics are identified, a log is maintained by experiment (individual construct or gene and accession number) for founders (and subsequent offspring).

3 Factors Affecting Transgenic Animal Production

Paul A. Overbeek

Department of Molecular and Cellular Biology, Baylor College of Medicine, Houston, TX

I. Introduction

This chapter is intended to provide some general advice about generating transgenic mice. The major goal of the chapter is to help simplify some of the decisions that need to be made when a laboratory begins to do research with transgenic mice. What are good strains of mice to use? Should the transgenic mice be generated by microinjection, by retroviral or lentiviral infection, or by genetic manipulation of embryonic stem cells? If microinjection is chosen, what are good vectors to use for the initial training stages? What about record keeping for the mice?

The first half of the chapter provides information about factors to consider when getting started, including husbandry and record-keeping suggestions. The second half of the chapter discusses troubleshooting strategies and transgenic phenomenology. Most of the recommendations in this chapter are based on personal experience acquired while running a laboratory that generates transgenic mice by microinjection. The strategies suggested in this chapter have worked well for my laboratory, but they may need to be modified for other research settings. The information in this chapter represents a short introduction to inbred strains of mice, mouse husbandry, mouse breeding, record keeping, and transgenic phenomenology. More detailed information about these topics can be found in Behringer et al. (2014), Liu (2013), Festing (1990, 1992), Hetherington (1987), Otis and Foster (1983), Lang (1983), Rafferty (1970), Green (1975), Dickie (1975), and Les (1975).

II. Getting Started

A. Microinjection Versus Infection Versus Embryonic Stem Cells

The three general techniques for generating transgenic mice with newly integrated DNA are microinjection of one-cell-stage embryos, retroviral or lentiviral infection

Transgenic Animal Technology. DOI: http://dx.doi.org/10.1016/B978-0-12-410490-7.00003-7

of embryos, and genetic manipulation of embryonic stem (ES) cells. As a general rule, each of these techniques is used for a different purpose: microinjection of DNA is used when the major goal is to study expression of new genetic information; retroviral infection (of embryos or ES cells) is used for studies of cell lineage and for random insertional mutagenesis; and ES cells are used for site-directed mutagenesis by homologous recombination. In the past few years, new techniques have been developed for targeted manipulation of the genome in injected embryos, using zinc finger nucleases, TALE nucleases (TALENs), or the CRISPR/Cas9 endonuclease. These protocols typically involve the cytoplasmic injection of RNAs. They are discussed in detail in Chapter 8 by Dunn and Pinkert.

Microinjection is the most commonly used procedure, mainly because of its reliability. The technique requires a training period (often 3−6 months), but once the protocol is mastered, transgenic mice can be efficiently generated with almost any DNA construction, including very large DNAs like BACs. Transgenic DNAs introduced by microinjection have been found to be efficiently and reproducibly expressed in transgenic mice. Microinjection is the preferred strategy when the objective is to obtain expression of new genetic information (see Chapter 2).

Whereas microinjections are almost always performed on one-cell-stage embryos, retroviral infection can be done successfully at various stages of embryonic development, ranging from preimplantation to midgestation. As a result, retroviruses have served as useful markers for cell lineage studies (Soriano and Jacnisch, 1986). For random insertional mutagenesis studies, retroviruses have the valuable attribute of integrating into the host genome without deletions or rearrangements of the chromosomal DNA. One of the first insertional mutations characterized in transgenic mice was the inactivation of the $\alpha 1(I)$ collagen gene by retroviral integration in the *mov*13 family (Jaenisch et al., 1983; Schnieke et al., 1983; Harbers et al., 1984; Hartung et al., 1986). Retroviruses can also infect ES cells, allowing them to be used for promoter trap and gene trap experiments where there is a preselection for potentially interesting sites of integration (Friedrich and Soriano, 1991). In contrast to retroviruses, integration of microinjected DNA is often accompanied by major rearrangements or deletions in the host genomic DNA (Singh et al., 1991; Covarrubias et al., 1986, 1987; reviewed by Meisler, 1992). Such chromosomal changes can complicate the search for the coding sequences of insertionally inactivated genes. Retroviruses have also received attention because of their potential uses for gene therapy (see Chapter 5).

Integration of microinjected DNAs and of retroviruses typically does not occur by homologous recombination, so these procedures are not practical for site-directed mutagenesis. In ES cells, the frequency of integration by homologous recombination is high enough that the targeted integration events can be identified *in vitro* before the cells are used to generate transgenic mice. In those cases where a gene has been cloned, but the phenotype caused by mutation of the gene is not known, ES cells offer a strategy to generate the desired mutants. One technical hurdle to the use of ES cells is mastery of the tissue-culture system. The use of ES cells for targeted mutagenesis is described in detail in Chapter 4. Table 3.1 contains a summary of various features of the techniques for generating transgenic mice.

Table 3.1 Techniques for Generating Transgenic Mice

	Microinjection	Viral Infection of Embryos	Embryonic Stem Cells
DNA vector	Any cloned DNA, preferably linear, with vector sequences removed	Recombinant lentivirus or retrovirus	Cloned DNA or retroviruses
Introduction of DNA	Microinjection into pronucleus	Infection after removal of zona pellucid	Electroporation or retroviral infection
Embryonic stage	One-cell stage	One-cell stage or later	Totipotent ES cells
Embryo transfers	Oviduct	Oviduct or uterus	Into blastocoel, then into uterus
Genotype of founder mice	Usually nonmosaic	Mosaic	Chimeric
Screening of newborns	Dot blots, Southern blots, or PCR	Southern blots or PCR	Visual coat color markers plus PCR or Southern blots
Copy number at integration site	1–200	1	Can be varied by selection of method for introducing DNA
Percentage of potential founders that are transgenic	10–30%	5–40%	Up to 100%
Expression of the new DNA	Usually	Good (for lentiviral transgenes)	Enhancer trap, gene trap reporters
Integration	Random, nonhomologous, multicopy, single site	Integrase-mediated using long terminal repeats (LTRs)	Random plus targeted
Germ line transmission by founders	Usually	Usually	Occasionally a problem
Advantages	Straightforward procedure; Successful expression with many different constructs	Integration into open chromatin; Single-copy integration	Homologous recombination; Selection in vitro; Multiple independent insertions using retroviruses
Disadvantages	Physical damage of embryos during microinjections; Multiple copy integration; Lack of insertion by homologous recombination	High titers can be difficult to achieve	Difficult tissue-culture system

B.   Strains of Mice: Inbred, Outbred, and Hybrid

Inbred strains of mice are defined as strains that have been maintained by successive brother to sister matings over more than 20 generations (Green, 1975). Repetitive inbreeding removes genetic heterogeneity, so that mice of an inbred strain are considered to be genetically identical to each other. There are hundreds of different inbred strains of mice (Festing, 1992). Some of the more commonly used laboratory strains, along with their pigmentation, include the following: A (albino), BALB/c (albino), C3H (agouti), C57BL/6 (black), C57BL/10 (black), C57BR (brown), C58 (black), DBA (dilute brown), FVB (albino), NZB (black), NZW (white), SJL (albino), SWR (albino), and 129/Sv (usually albino or chinchilla). The full name for an inbred strain includes an abbreviation to designate the source of the mice. The abbreviation is placed after a slash (/) that follows the name of the inbred strain. For instance A/J mice would be strain A mice from the Jackson Laboratory, whereas FVB/N mice would be FVB mice from the National Institutes of Health (NIH). For those strains with a/ in the standard name, the abbreviation for the supplier is added to the end of the name (e.g., C57BL/6J mice from the Jackson Laboratory). For more detailed information about naming inbred strains of mice, see Festing (1992).

Fur pigmentation in inbred laboratory mice is controlled primarily by four different genetic loci: agouti (a), brown (b), albino (c), and dilute (d) (reviewed by Silvers, 1979). The albino locus encodes tyrosinase, the first enzyme in the pathway to melanin synthesis. When mice are homozygous (c/c) for mutations that inactivate the tyrosinase gene, the mice are albino regardless of the genotype at the other loci. Most of the common albino strains of laboratory mice have the same point mutation in the tyrosinase gene (Yokoyama et al., 1990), indicating that these strains are all derived from a common ancestor. When mice are homozygous or heterozygous for a nonmutated tyrosinase gene (i.e., C/C or C/c), then the color of pigmentation is determined by the condition of the other genes. If the mice have a wild-type agouti allele (either A/A or A/a), then the fur will contain both black and yellow bands of pigment and will be agouti (Figure 3.1). If both copies of agouti are mutated (a/a), then the hair becomes uniformly pigmented, and the mice are either black (B/B or B/b) or brown (b/b) (Figure 3.1). Mice that are homozygous for a mutation in the dilute gene (i.e., d/d) show a decreased intensity of pigmentation (not shown).

Hybrid mice are generated by mating mice from two different inbred strains. The mice from such a mating are termed F1 hybrid mice. They are genetically identical to one another but different from either inbred parent. When F1 mice are mated to one another, the offspring are referred to as F2 hybrids. F2 hybrid mice will be genetically different from one another (owing to meiotic recombination and random sorting of the chromosomes) and will contain different mixtures of the genetic variations that were present in the original inbred progenitors. Hybrid mice exhibit a phenomenon termed hybrid vigor. They show enhanced fertility, they respond better to superovulation regimens, and hybrid embryos can be grown efficiently from the one-cell to blastocyst stage *in vitro*. Outbred strains of mice are propagated by nonstandardized matings and therefore retain substantial genetic variability.

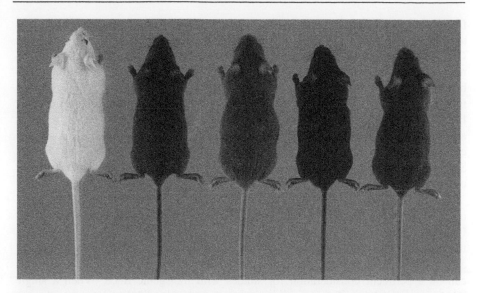

Figure 3.1 Pigmentation in laboratory mice. Some of the most common pigmentation phenotypes for laboratory mice are pictured. From left to right the mice are albino (genotype *c/c*), agouti (*A/a B/B C/c*), brown agouti (*A/a b/b C/c*), black (*a/a B/B C/c*), and brown (*a/a b/b C/c*). In agouti mice, the hairs of the fur show a subapical band of yellow melanin (pheomelanin) surrounded by regions of black melanin (eumelanin, not visible in the photograph). The pheomelanin band is missing in the black and brown mice that are mutated at the agouti locus.

Table 3.2 Mouse Strains Used for Transgenic Research

Strain	Typical Uses
Inbred Mice	
C57BL/6	Pronuclear microinjection, recipient blastocysts for embryonic stem (ES) cells
FVB	Pronuclear microinjection
129/SvEv	Generate ES cells
Outbred Mice	
ICR or CD-1	Pseudopregnant recipient females
Hybrid Mice	
B6SJL F1	Pronuclear microinjection
B6D2 F1	Vasectomized males

Inbred, outbred, and hybrid mice are used for transgenic research. Table 3.2 contains a short, subjective list of some of the frequently used strains. The inbred strains most commonly used for transgenic research include C57BL/6 (often referred to as "black 6"), FVB, and 129/SvEv mice. The C57BL/6 strain has been used for laboratory research for many years, so many known mouse mutations are available on the C57BL/6 background. Young C57BL/6 females superovulate well and C57BL/6 blastocysts have been found to be excellent recipients for genetically engineered ES cells (see Chapter 4). FVB embryos are often used for microinjections because they have large distinctive pronuclei (Figure 3.2) that are easy targets for microinjection (Taketo et al., 1991). Many of the ES cell lines have been derived from 129/SvEv mice. The most commonly used hybrid mice for transgenic research are B6SJL F1, derived by mating the inbred strains C57BL/6 and SJL. Brinster and colleagues (1985) have documented the advantages of the B6SJL mice. Outbred Swiss albino strains such as ICR or Swiss-Webster are often used as recipients for embryo transplantations.

Outbred mice are generally less expensive than inbred or hybrid mice. In general, outbred and hybrid mice have better fertility and larger litter sizes than inbred mice. However, the inbred FVB mice have fertility and embryo culture characteristics that nearly match the hybrid and outbred strains (Taketo et al., 1991). The biggest advantage of using inbred mice is the consistency of the genetic background. This can be a pertinent consideration for studies with transgenic mice. When a single DNA construct is used to generate transgenic mice, variability in expression or phenotype between different transgenic mice cannot be attributed to a variable genetic background if an inbred strain of mice is used.

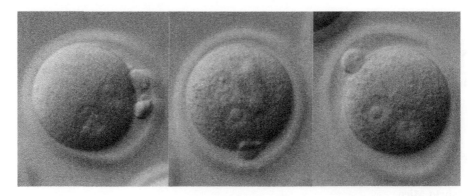

Figure 3.2 One-cell-stage FVB/N embryos. In each embryo, two pronuclei are readily visible. Each pronucleus contains one or more nucleoli (prominent circular organelles within each pronucleus). The two polar bodies are visible for the embryo at left.
The FVB/N embryos were cultured in the top of a plastic petri dish (Falcon 1006, Becton Dickinson Labware, Lincoln Park, NJ) and viewed using Hoffman optics.
Because the FVB/N pronuclei are large and distinctive, they are easy to inject. Moreover, the fact that FVB/N mice are inbred can simplify the subsequent interpretation of experimental results.
Source: This figure was originally published in Taketo et al. (1991) (magnification 400×).

C. Mouse Husbandry: Caging, Mating, Pregnancy, Record Keeping, etc.

1. Animal Welfare

All research with transgenic animals should be carried out as humanely as possible and in accordance with all federal and institutional policies for research with laboratory animals. In the United States, investigators and institutions are expected to comply with the Animal Welfare Act and the Guide for the Care and Use of Laboratory Animals (documents available from the Office of Laboratory Animal Welfare (OLAW), NIH, Bethesda, MD 20892). As part of this policy, each research institution is required to appoint an Institutional Animal Care and Use Committee, which is required to review and approve animal research proposals and protocols. Before initiating any research with transgenic mice, investigators must document their plans for humane care and research in order to obtain approval from the institutional review committee. In addition, investigators should always confer with the chief veterinarian for the facility where the animals will be housed. The veterinarian will be able to provide information about a variety of important issues, including space allocation for the transgenic mice, acceptable vendors for purchase of mice, responsibilities in the provision of daily and weekly care for the animals, animal husbandry charges, standard housing conditions for the animals, acceptable anesthetics, protocols for animal procedure rooms, treatment of sick animals, disposal of dead animals, and so forth. The chief veterinarian should be consulted concerning all of the husbandry recommendations made in this chapter to be certain that they conform to the standards of the animal facility.

2. Standard Housing Conditions

Laboratory mice are typically housed in polycarbonate cages that are equipped to provide food and water for the mice (see Figure 3.3; Lab Products, Inc., Maywood, NJ). Most facilities use two (or more) standard size cages: a small shoebox size of approximately 12 in. × 8 in. × 5 in. that can house four or five adult mice, or a nursing female with pups, and a larger cage (19 in. × 10 in. × 6 in.) that can house approximately twice as many mice.

Food for the mice is generally provided *ad libitum*. The food is placed either in a metal container that is positioned inside the cage or alternatively in the metal lid on top of the cage. Rodent chow is available in a number of different recipes from suppliers (e.g., Purina Mills, Inc., Richmond, IN). Most animal housing facilities purchase the rodent chow and provide the food as part of the standard animal husbandry services.

Water is also provided *ad libitum*, typically from water bottles equipped with rubber stoppers and sipper tubes (Figure 3.3). The water is often acidified to a pH of 2.5 by addition of HCl in order to slow the growth of bacteria in the water. Water bottles should be cleaned and replaced once per week.

Bedding for the mice is traditionally wood shavings, wood chips, or dried corn cob, and is typically provided by the animal husbandry service. Bedding should be replaced at least once per week. Pregnant females nearing delivery can be given a

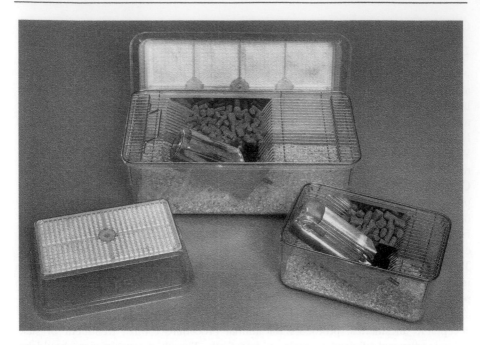

Figure 3.3 Mouse housing cages. Prototype small and large cages for housing laboratory mice are shown (Lab Products, Inc.). Microisolator tops for each cage size are also pictured. Cage card holders are not shown.

cotton block (e.g., Nestlets from Ancare Corporation, North Bellmore, NY) or paper towels to use to build a nest for the pups. Cages with nests should be checked regularly to be certain that the nests do not contact the sipper tubes on the water bottles. Such contact allows the water to drain from the water bottle, resulting in flooding of the animal cage. In general, we do not change the bedding in a cage with a female that is 18 days or more, pregnant, or is nursing pups that are less than 48 h old. It is best to leave females relatively undisturbed around the time of delivery.

Special microisolator caging systems have been designed to help prevent the spread of pathogens within a mouse colony. As one example, individual cages can be equipped with special filter tops (Figure 3.3) that prevent contamination by airborne debris from other cages in the room. When microisolator tops are used, the bedding should be changed more often, because the cage tops restrict evaporation and air circulation within the cages. Alternatively, the cages can be designed to dock with a ventillation system to provide regular air flow. Husbandry for the mice should be done inside a laminar flow hood to help prevent cross-contamination between the cages.

Housing conditions for mice vary according to the stringency of the procedures used to protect the mice from murine pathogens. The two most common conditions are referred to as "conventional" and "specific pathogen free (SPF)" (Lang, 1983; Otis and Foster, 1983). For SPF facilities, special precautions are employed to protect the mice from murine pathogens, particularly murine viruses. Personnel working with SPF mice are required to follow specific guidelines regarding personal hygiene and clothing when they enter the animal husbandry area. In addition, all experimental supplies and surgical equipment must be decontaminated before they are brought into an SPF area. Each SPF facility will have its own standard operating procedure, which should be adhered to conscientiously. Conventional housing facilities maintain less stringent hygiene standards, typically allowing mice to be returned to the facility after removal for surgical or experimental procedures.

3. Fertile Females and Embryos for Microinjection

Inbred, outbred, or hybrid one-cell-stage embryos can be used for microinjections. To generate inbred embryos, both mating partners must be of the same inbred strain (e.g., FVB × FVB), whereas hybrid embryos can be generated by mating mice of two different inbred strains. Hybrid embryos can also be generated by inbred × hybrid or by hybrid × hybrid matings. If either of the parents in the mating is outbred, the embryos are considered to be outbred.

The two most commonly used inbred strains are C57BL/6 and FVB. Although young C57BL/6 females (3−4 weeks old) superovulate well, often producing 30−40 embryos per female, there are some drawbacks to using the C57BL/6 strain to make transgenic mice, particularly for beginners. C57BL/6 embryos from superovulated females show a high percentage of unfertilized and abnormal embryos (10−40%), and the pronuclei can be small and difficult to identify. A later drawback is the fact that adult C57BL/6 females are not particularly fertile, with average litter sizes of only five or six mice. C57BL/6 females are sometimes poor mothers, so that newborn mortality can occur with a frustratingly high frequency. Both of these characteristics can make it difficult to generate extra offspring within a transgenic family in order to do further research.

In contrast, FVB/N embryos have large, well-defined pronuclei, making them easier to inject (Taketo et al., 1991). Superovulated FVB/N females yield an average of 15−25 embryos/female, they mate with high efficiencies, and they produce a high percentage (usually >90%) of fertilized healthy embryos. Moreover, FVB females are excellent mothers, they typically do not cannibalize dead newborns, and the average litter size is 9−10 pups. It should be noted that the FVB strain is not well known outside of transgenic mouse research. The FVB strain, particularly at the immunological loci, has not been as fully characterized as the C57BL/6 strain.

The other commonly used embryos for microinjection are hybrid F2 embryos generated from matings of the inbred strains C57BL/6 and SJL. B6SJL F2 mice have been used for many years by Brinster and colleagues (1985), who found that the hybrid embryos gave a substantially higher frequency of transgenic offspring than inbred C57BL/6 mice. The hybrid females superovulate well, and the embryos show the traditional hybrid vigor.

Embryos from outbred strains can also be used to generate transgenic mice. Because outbred mice are less expensive, the novice microinjector should consider using outbred embryos for the initial stages of training. Outbred females superovulate well, and one-cell-stage embryos can be cultured *in vitro* to the blastocyst stage in order to monitor percent survival. In contrast to outbred (and hybrid) embryos, embryos from inbred strains often stop development at the two-cell stage *in vitro* (referred to as the two-cell block). Embryos from a few inbred strains, such as FVB, will develop efficiently from the one-cell stage to the blastocyst stage in embryo culture medium such as M16 (Whittingham, 1971; Behringer et al., 2014) or KSOM (Erbach et al., 1994)

Female mice can be induced to superovulate by treatment with hormones. The two main advantages of superovulation are the increase in number of embryos and the synchronization of estrus. Without superovulation, female mice generally release 6−10 oocytes during each estrus. With superovulation the number of oocytes can be increased to as many as 40 per female (Rafferty, 1970). Superovulation often works best with younger females that have not yet started their own ovulatory cycle. For many strains of mice, females that are 3−5 weeks of age give the best yield of oocytes on superovulation. The young females will mate successfully, so that this strategy is the standard protocol to obtain embryos for microinjection. The regimen for hormone administration is described in Chapter 2.

In the absence of superovulation, female mice begin to ovulate and become fertile around 5−6 weeks of age. Adult female mice have an estrus cycle of 4−6 days, are receptive to mating only during estrus, and cycle regularly until 7−9 months of age for most strains. Female mice can be housed together in the same cage before mating, during pregnancy, and after newborns have been weaned. In general, pregnant females are placed in separate cages 1 or 2 days before their scheduled delivery. This helps prevent overcrowding, allows unambiguous identification of which pups belong to which mother, and eliminates the possibility that the pups from one mother will be attacked by another female in the cage.

In most research facilities, mice are housed in rooms with no external lighting. A timer-controlled lighting system is used to provide a consistent daily lights-on, lights-off cycle, which is typically 14 h of lighting, then 10 h of darkness (for example, lights on at 6:00, lights off at 20:00).

4. Fertile (Stud) Males

Male mice typically become fertile at 6−8 weeks of age. When males are housed together from weaning age, multiple males can generally be kept together in the same cage indefinitely. However, experienced males should not be housed together, since they will fight with one another, often until the most aggressive male has killed the other males in the cage. When males are ready to be used for matings, they should be placed in individual cages. In general it helps to give each male his own cage at least 24 h before setting up a mating in order to give the male time to establish the cage as his territory. Matings are initiated by adding one or more females to each male cage.

We typically use each fertile male for only one mating per week. A more frequent mating schedule leads to lower mating percentages and to a decrease in percent fertilization. Vasectomized males can be mated more frequently, since their ability to produce adequate sperm is not relevant. However, vasectomized males only occasionally mate more than twice a week, even if given the opportunity.

The fertility of the stud males typically begins to decline when the mice reach 9–10 months of age, at which time they should be replaced with younger males. Male mice remain fertile longer than female mice. In most strains of mice, experienced males remain fertile up to 14–18 months of age, and inexperienced males can often be successfully mated up to 10–12 months of age. Males that have stopped mating should be replaced.

The genetic background of the fertile males should be matched to that of the superovulated females. If the females are outbred, then the males can be outbred. If the females are inbred, then the males should be from the same strain in order to maintain the inbred background. Hybrid females are generally mated to either F1 or F2 hybrid males of the same genetic background.

We maintain a cage card with a mating record for each stud male. The mating record indicates whether a plug was found for each date when the male was set up for mating with a superovulated female. Males that do not produce plugs for four consecutive mating opportunities as well as males that mate less than 50% of the time are replaced. Overly aggressive males will fight with and injure females that are added to their cage. Males that show such behavior more than once are euthanized.

5. Vasectomized Males

The genetic background of the vasectomized males is usually not critical, since these males should not produce any direct descendents. To help recognize the occasional occurrence of an inadequate vasectomy (i.e., a fertile vasectomized male), coat color markers can be used. The vasectomized males can be selected so that inappropriate offspring would have a different coat color than the offspring from injected embryos (e.g., albino versus pigmented).

Vasectomized males should mate consistently and over a reasonable time span. Our favorite males are hybrid B6D2 F1 mice (generated by mating C57BL/6 to DBA/2 mice). The mice are inexpensive and readily available from commercial breeders. The males mate consistently up to at least 1 year of age, and they leave readily identifiable plugs. The males are docile and easy to work with. Males can be purchased already vasectomized (e.g., from Taconic).

6. Recipient Females

Once embryos have been genetically manipulated *in vitro*, they need to be transferred to pseudopregnant recipient females. Because the purpose of the recipient females is to carry the embryos to term and to nurse the newborns to weaning age (3 weeks), it is important to use females that are good mothers. The females should also be relatively docile so that they will allow inspection of the pups, and will accept pups from another mother. Outbred ICR females are inexpensive and make good recipient females.

One strategy to obtain pseudopregnant females is to set up random matings, placing one or two females per cage with vasectomized males. In a nonsynchronized population, 10−20% of the females will be in estrus on an average day, so that the total number of females to set up for matings should be 5−10 times the number that will be needed for the embryo transfers. Mated females are identified by inspection for copulation plugs (Rafferty, 1970). If embryo manipulations are done on consecutive days during the week, this system has the advantage that a large number of females can be set up on the first day, and on subsequent days only previously unmated females need to be checked for plugs.

An alternative strategy to obtain pseudopregnant recipient females is to set up new matings each day using females that are in estrus. Such females can be identified by inspection of the vagina (for swelling, redness, and a rippled folding (wrinkling) of the vaginal wall; see Behringer et al., 2014). When the females are prescreened for estrus, 50% or more of the females should mate overnight. The advantage of prescreening is that fewer females need to be checked for plugs each morning. The disadvantage is that new females need to be set up for matings every evening.

A third strategy to obtain pseudopregnant females is to perform superovulations to induce estrus in the recipient females, then to mate the females to vasectomized males. In our experience, naturally mated females have a substantially higher rate of pregnancy after embryo transfer than superovulated females, so we use superovulation only as a last resort.

Pseudopregnant females that are not used for embryo transfers will typically return to an estrus cycle 8−11 days after mating. The females can be saved and remated if desired.

7. Pregnancy

The gestational period in the laboratory mouse is 18−20 days. Females that are more than 11−12 days pregnant with a normal size litter of 7−10 pups can be recognized by visual inspection for abdominal enlargement.

Pregnant recipient females are generally given individual clean cages 1 or 2 days before delivery. If the females are provided with nesting materials such as paper towels or crushed cotton squares, they will generally build a nest in which to place the newborns after delivery. Females that do not build a nest often turn out to be poor mothers. Pregnant females that are housed with a male can be left with the male during delivery if additional offspring are desired. Female mice go into estrus postpartum and are receptive to mating on the day they deliver. As a result, female mice can be nursing one litter and pregnant with a second litter at the same time. The gestational time for the second litter will often be extended to 21−25 days, as a consequence of delayed implantation. The delayed implantation allows the first set of pups to be weaned at 21 days of age before the second set of newborns arrives.

Female mice often cannibalize fetuses that are dead at birth. If one wishes to look for prenatal or perinatal lethality of transgenic embryos, then a cesarean section (C-section) can be performed prior to the scheduled delivery. In those cases where one wishes to deliver live pups by C-section, the delivery should be done 24 h before the scheduled normal delivery, that is, at 18.5 days postfertilization. C-sections are also advisable for females that are pregnant with only one to three fetuses, since these fetuses can become oversized, impeding normal delivery. It is essential to have a foster mother to nurse the C-section pups if they are to be kept alive. The foster mother should have delivered her own litter within the previous 48 h, and ideally the natural pups of the foster mother should have a different coat color from the C-section pups so that they can be visually distinguished. In addition, the foster mother should have pups that are well fed, with milk visible in their stomachs.

The protocol used for the recovery of midgestation embryos can also be used for C-section delivery of live fetuses. The first step is to sacrifice the pregnant female humanely, then to open up the peritoneal cavity and externalize both horns of the uterus. By visual inspection, one can count the number of full-grown fetuses and also identify instances of postimplantation embryonic lethality. Embryos that die shortly after implantation can be recognized by the presence of degenerate (brownish) decidua. (Decidua are sites of cellular proliferation within the uterus induced by fetal implantation.) Fetuses that die at later stages can be identified by their small size and pale color.

To deliver live fetuses, use a pair of scissors to make an incision at the vaginal end of each uterine horn, then advance the scissors to open the antimesometrial side of the uterus up to the oviduct in order to expose the interior of each uterus. With a little practice, this procedure can be done so that each fetus remains encased within its own fetal yolk sac. To deliver each fetus, the yolk sac can be grasped with a pair of watchmaker's forceps and opened by tearing. The edges of the yolk sac can be peeled back around the fetus, so that the umbilical cord can be located and severed with a scissors. Each fetus can be cleaned up by rolling it back and forth on a paper towel. The fetuses can be placed on a paper towel on a slide warmer to help prevent hypothermia. The entire litter of newborns should be delivered as quickly as possible. Once the pups have been delivered, gently stimulate each fetus every 15–30 s to induce breathing. Forceps can be used to gently pinch the skin behind the neck, or to spread the legs apart, or to simply roll the newborns on the paper towel. Healthy fetuses will begin to make sporadic inhalations within 1–2 min of delivery, particularly on stimulation. Over the next 1–2 min the fetuses should begin to breathe more regularly. Once breathing becomes stable, the fetuses should become a healthy pinkish color as the blood becomes oxygenated. Fetuses that survive the C-section delivery should be allowed to stabilize, then transferred to a foster mother.

Our strategy for this transfer is to first remove the foster female from the inside of her cage and to place her on the cage top. Next a sufficient number of the natural pups are removed so that when the C-section newborns are added to the cage the total litter size will be five to eight pups. The foster mother is more likely to accept the new pups if some of her own pups are left in the cage. The C-section newborns are transferred to the recipient cage and placed with the remaining newborns of the foster female. The entire collection of pups is then buried under approximately 5 mm of the bedding that is already present in the cage of the foster female. This is done to transfer the odors of the cage to the C-section pups before the foster female is returned to the cage. After the appropriate information is entered on the cage card, the foster female is returned to her cage. One additional recommendation is to treat the nose of the foster female with an alcohol swab just prior to placing her back in her cage. The alcohol treatment will cause a short-term loss of olfactory discrimination. The behavior of the foster female should be monitored for the first few minutes after she is placed back in the cage. In most instances, the foster female will move around the inside of the cage sniffing at both the cage top and the bedding, as well as digging in the bedding at various locations in the cage. The pups in the cage will emerge from the bedding within 2–3 min if the female does not directly dig them out. When the foster female accepts the new pups, the female will begin to groom herself and the new pups. On the other hand, when the female does not accept the new pups, the usual response is to pick up the pups at the nape of the neck and to carry them rapidly around the inside of the cage as if looking for some way to dispose of them. If the newborns are not accepted by the first foster female, then it is necessary to repeat the procedure with a new foster mother.

If the natural pups cannot be distinguished from the C-section pups by coat color, then it may be helpful to physically mark one set of pups for later identification. One strategy is to use scissors to snip off a small toe on one of the feet. A small amount of bleeding accompanies toe removal, but the bleeding quickly stops and does not cause the foster female to reject the marked pups. The marked pups can be identified at weaning age by simple visual inspection for missing toes.

Milk production in lactating females is stimulated by nursing, and very small litters often do not provide adequate stimulation to maintain milk production. If the recipient females from 1 day's worth of injections all have small litters, it is often beneficial to consolidate the pups so that the litters average six to seven pups.

8.   Newborn Mice

New litters of mice should be monitored within 12–24 h after birth to check for newborns that have died and also to be certain that the newborns are being properly cared for and nursed. At the time of delivery, the mother will eat the placenta and fetal yolk sac associated with each newborn mouse. In addition, the mother will typically groom the newborns by licking in order to stimulate breathing. Within 2–4 h after birth, properly fed newborns will have milk in their stomachs, which can be seen through their skin. If the mice are not being fed 12 h after birth, or if they are found unattended and scattered around the cage, they should be transferred to a foster mother as described in the previous section.

Mice are typically old enough to wean by 21–23 days after birth. If the mother is pregnant with another litter, the first litter should be removed promptly at 21 days of age. If the older pups are not removed, newborn mice in the next litter will be trampled and will not survive. Adolescent mice should weigh 9—10 g or more at weaning. Mice that are smaller than this should be left with the mother, since they often have trouble surviving on their own. At weaning, the mice are sorted by sex and placed into new cages.

9. Mating of Transgenic Mice

Founder transgenic mice can be mated to nontransgenic partners to generate additional transgenic mice. Transgenic offspring of such matings will be hemizygous for the transmitted integration site(s). Interbreeding of mice from different transgenic families that carry the same construct is generally not done. Matings (intercrosses) between hemizygous mice within a given family can be set up in order to generate mice that are homozygous for a transgenic insert. In order to identify homozygous transgenic mice, quantitative Southern or dot-blot hybridizations are generally required. Homozygosity can be confirmed by mating (crossing) the mice to nontransgenic partners. All of the offspring will be transgenic if one of the parents is homozygous. In some families, viable homozygotes may be absent due to an insertional mutation that causes embryonic or perinatal lethality. When the homozygotes of both sexes are viable and fertile, they can be used for subsequent matings (incrossing) to maintain the transgenic family without the need for DNA screenings at each generation.

10. Record Keeping

There is no standardized record-keeping system for transgenic research of which I am aware. A good record-keeping system should provide critical information about every mouse in the research colony. For each mouse, it is important to know whether the mouse is transgenic and, if so, what DNA vector the mouse carries, who the parents and the offspring of the mouse are, when the mouse was born, and how the mouse has been characterized to date. I describe below some features of the record-keeping system that we use.

a. Identification Numbers for Each Mouse

Mice that belong to any one of the following three categories are assigned individual, unique identification numbers: recipient females, potential founder mice, and offspring within a transgenic family. Each category is allocated a series of numbers that can be preceded by a letter (e.g., R for recipient females, F for potential founders, and T for transgenic families). Within each category, new mice are assigned sequentially increasing numbers. (Logbooks are used to allocate the identification numbers.) In conjunction, we use ear tags that have the identification numbers engraved on them (National Brand and Tag Company, Newport, KY). When a potential transgenic mouse reaches weaning age, the appropriate ear tag is affixed. Ear tags are applied to recipient females at the time of embryo transplantation. Once the ear tags have been applied, individual mice are uniquely identified and can be readily located within the colony. For studies of embryonic development in transgenic families, the embryos are assigned identification numbers but from a series for which no corresponding ear tags are purchased.

A common alternative system for keeping records on transgenic mice is to keep track of mice within individual transgenic families. Each mating pair is assigned an identification number, and offspring are identified by ear punching (Behringer et al., 2014). Ear punching can be combined with toe clipping to mark up to 10,000 different mice. This system is less expensive than using ear tags, but mistakes in mouse husbandry are easier to recognize if the mice have ear tags.

b. Transgenic Family Identification Numbers

Investigators traditionally assign a family identification number (or letter) to each transgenic founder mouse and to all of its offspring. Most laboratories follow their own conventions in the assignment of these identifiers. However, the Committee on Transgenic Nomenclature of the Institute of Laboratory Animal Resources (ILAR) has adopted specific guidelines for naming transgenic families of mice. At a minimum, a laboratory-assigned number and laboratory identifier (three letters) should be given to each transgenic family. In this fashion each transgenic family becomes uniquely identified. The nomenclature rules should be adhered to by all transgenic research laboratories.

c. Cage Cards

Mouse cages are generally equipped with cage card holders for 3 in. × 5 in. index cards. Each card should carry information about the mice inside the cage. We use two different formats for the information: one for cages with mice that are all of the same sex and a different one for mating pairs or nursing females (Figure 3.4). For unisex cages, the following information is written on each cage card: sex of the mice, number of mice, identification numbers for each of the mice, date of birth, date of weaning (or date of receipt if the mice were received from a vendor), identification numbers for the parents, DNA vector, identification number of the transgenic family, information about recognizable phenotypes for any of the mice, and information about the results of screening for the transgenic DNA (see Figure 3.4). When a cage contains a mating pair, comparable information is entered on the cage card: identification numbers for the mating mice, date of birth of the mating pair, identification numbers of the parents of the mating pair, DNA vector, identification number of the transgenic family, date of birth and number of pups on delivery (including the number of stillborn mice), identification numbers assigned to the newborn mice, and comments about phenotypes of the newborn mice (see Figure 3.4).

To maintain some organization within the mouse housing area, mice that have similar genetic backgrounds are housed in proximity to each other. When more than one room is available to house the mice, each room can be allocated to different sets of mice. For example, one room could be specifically for the matings to generate embryos for injection and pseudopregnant females; a separate room could be used to house transgenic families. To simplify the recognition of various sets of mice, we have found that different colors of index cards are helpful. For example, we use one color card to identify FVB/N mice, a different color for outbred ICR mice, a third

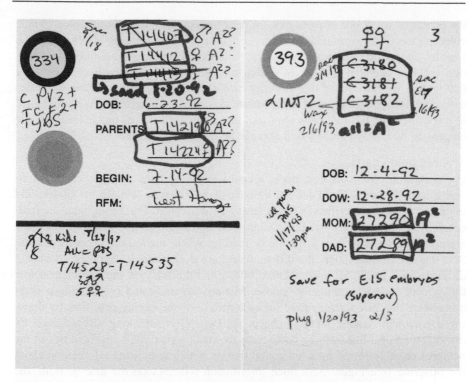

Figure 3.4 Cage cards. Information about mice within a cage is typically entered on a 3 in. × 5 in. index card that is held in a cage card holder on each cage. Prototype cage cards are shown for a mating pair (left) and a cage that contains only mice of the same sex (right). The identification numbers and sexes are entered at the top of each card, along with information about the date of birth (DOB), date of weaning (DOW), and parents. For mating pairs, the start date for the mating is entered (BEGIN), along with information about the reason for mating (RFM). When the offspring are born, the relevant information is recorded, including number of pups and date of birth. New identification numbers are assigned to the offspring and entered on the cage card. At weaning, the sexes of the offspring are recorded. To simplify the recognition of specific sets of related transgenic mice, each construction that is used for microinjection is assigned a code of one small colored dot within a different colored larger dot. These dots are adhesive and are placed in the upper left-hand corner of each card. Each founder mouse is assigned a family identification number, and the family number is written within the smaller dot (334 for the left card, 393 for the right card). The names of the transgenic constructions are written on the cards under the dots. When a mouse is sacrificed, a line is drawn through the corresponding identification number.

color for vasectomized B6D2 F1 males, and a fourth color for all of the cages that contain potential transgenic mice. In addition, we use a color-coding scheme to identify families of transgenic mice that were all derived from injections with the same DNA vector. Colored dots are added to each transgenic cage card (see Figure 3.4).

By placing a small dot of one color inside a larger dot of a different color, unique color patterns are generated and assigned to each microinjected DNA. When a founder transgenic mouse is identified, it is assigned a transgenic family identification number, and that number is written on the small dot on each cage card for that family. We have found that this cage card identification system greatly simplifies the search for specific cages of mice within an animal housing room. When the mice in a specific cage have all been removed, the cage card is saved and placed in a file (organized by family number) so that it can be retrieved if needed.

d. Pedigree Charts

For each family of transgenic mice, a continuously updated pedigree chart can be maintained (Figure 3.5). Males are designated by squares, females by circles, and mice that die or are sacrificed without sexing by diamonds. Transgenic mice are indicated by placing a diagonal line within their symbol. Putative homozygous mice are given an X within their symbol. For families where the founder mouse has more than one site of integration, the different sites are designated A, B, C, etc. (which is different from the ILAR nomenclature rules), and the relevant site is written below the symbol for each transgenic mouse. Matings are indicated by entering a short descending line below the symbol for each mouse of the mating pair, then by drawing a horizontal mating line to connect the two mice. Offspring are indicated by drawing a vertical line running down from the mating pair line to the next tier of the pedigree chart, followed by a horizontal line on which to indicate information about the offspring (see Figure 3.5). Such information will include the date of birth, the identification numbers, and the sexes of the mice, once determined. Mice that are deceased are indicated by drawing an X through the stem line that attaches to their sex symbol (see Figure 3.5). The pedigree charts are left in the mouse housing rooms and are updated whenever new information is available. These continuously updated pedigree charts provide a simple summary of the status of each transgenic family.

e. Newborn Records

Separate hard copy records about each litter of mice can also be maintained. Desirable information might include number of mice in the litter, identification number and sex of each mouse, parents, date of birth, date of weaning, transgenic vector, transgenic family identification number, phenotype of each mouse, results of genomic screening for each mouse, results of all assays for transgene expression, mating record for each mouse, and information about the date of death, cause of death, and status of any tissues that were saved for each animal. Investigators setting up a new transgenic laboratory will need to decide what type(s) of record-keeping system to establish, and then to establish the discipline necessary to maintain the records.

f. Computerized Record Keeping

In some laboratories, it may be preferable to employ a computerized record-keeping system, although such systems have not yet achieved widespread use. Ideally, a data management program would allow the relevant information for each cage of mice to be entered into the computer, and the computer would then print an appropriately labeled cage card. The computer could automatically assign identification numbers

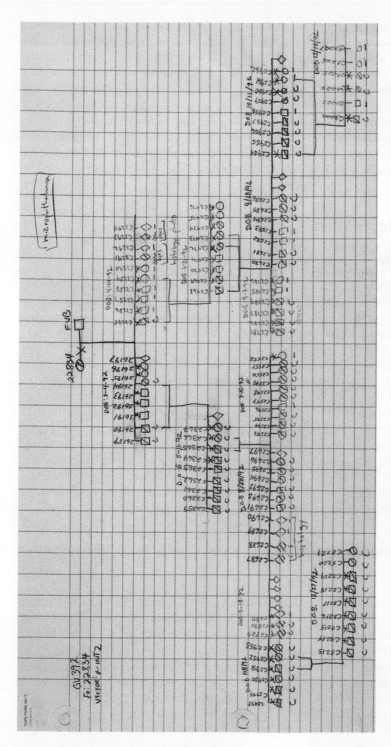

Figure 3.5 Pedigree chart. Pedigree charts are kept in the room where the mice are housed and are updated whenever new information becomes available. The charts provide a quick overview of the status and history of each transgenic family. Males are indicated by squares, females by circles, and mice that die (or are sacrificed) before sexing are indicated by diamonds. Identification numbers and dates of birth are written above the sex information for each mouse. Transgenic mice are indicated by placing a slash (/) within the sex symbol, and information about the phenotype or screening results is entered below the symbol (e.g., c for cataracts). Homozygous mice can be signified by placing an X inside the sex symbol. When mice are sacrificed, we indicate this by drawing an X through the vertical line that attaches the sex symbol to the horizontal litter line.

and could maintain the pedigree charts and newborn records for all of the transgenic families in the colony. Data would need to be entered into the computer for each recipient female at the time of embryo transfer, for each litter at the time of birth and also at the time of weaning, and for each mating pair. In addition, data concerning the results of assays on the mice and dates of sacrifice would need to be entered. The computer should be fully portable so that it can be brought into the mouse housing rooms whenever husbandry is performed. Commercial software programs that are designed for record keeping in a transgenic mouse colony and custom-designed programs for mouse record keeping do exist (for example, see http://bioinforx.com/lims/online-transgenic-mouse-colony-management-software-system/mlims).

III.   Troubleshooting

A.   Assaying for Successful Pronuclear Microinjection

One problem with learning to do microinjections is that it can be a long wait between the time the microinjections are done and the time that the results are known, particularly if one waits until the microinjected embryos have developed into weaning age mice before screening. By the time tail DNA is isolated and screened, it will be nearly 2 months after the time of the microinjections. If a novice is making a consistent mistake, a considerable amount of time and effort will be wasted.

At the outset, either the metallothionein−β-galactosidase (MT−β-gal) construct (Stevens et al., 1989) or alcohol dehydrogenase (ADH) under control of the Rous sarcoma virus promoter (RSV−ADH) (Nielsen and Pedersen, 1991) can be used to determine whether the microinjections are properly introducing DNA into the pronucleus. After correct microinjection, either MT−β-gal or RSV−ADH can be expressed in embryos at the two- to four-cell stage. As a result, embryos can be microinjected, incubated at 37°C for 24−48 h, then stained for β-gal or ADH activity (Stevens et al., 1989; Nielsen and Pedersen, 1991). When the microinjections are done properly, a significant proportion of the embryos will show histochemical staining.

B.   Visual Identification of Transgenic Mice

1.   Elastase−ras

The promoter for the elastase gene has been shown to be active in pancreatic acinar cells in transgenic mice (Swift et al., 1984; Ornitz et al., 1985). Transgenic mice that carry the elastase promoter linked to the *ras* oncogene develop pancreatic tumors (Quaife et al., 1987). These tumors are dominant and dramatic and can typically be recognized by simple visual inspection of the perinatal fetus. In a nontransgenic fetus, the whitish pancreas can be identified through the slightly transparent skin. Transgenic fetuses exhibit substantially enlarged pancreata (Quaife et al., 1987; Pinkert, 1990). Pancreatic neoplasia can be confirmed by surgery to allow visual inspection of the viscera. For the novice microinjector, elastase−*ras* permits rapid identification of successful generation of transgenic mice within 3 weeks after injection of the DNA (Pinkert, 1990). One of the advantages of this construct is the fact that it can be used with embryos of any genotype (in contrast to the tyrosinase minigene, which is described in the next section). One drawback to the elastase−*ras* construct is the fact that the pancreatic tumors are generally lethal at an early age for the transgenic mice (Quaife et al., 1987).

2. Tyrosinase Minigene

Classic albino strains of mice have a mutation in the gene encoding tyrosinase, the first enzyme in the pathway to melanin synthesis (Yokoyama et al., 1990). Microinjection of a tyrosinase minigene into embryos of an albino strain of mice can result in gene cure of the albino defect and the synthesis of pigment (Tanaka et al., 1990; Beermann et al., 1990; Yokoyama et al., 1990). The conversion from albinism to pigmentation is easy to recognize (Figure 3.6), so the tyrosinase minigene offers a number of advantages for microinjection training. The microinjections can be done using albino inbred strains such as FVB/N or BALB/c. Alternatively, inexpensive outbred albino strains such as ICR can be used. Pigmented mice have dark eyes that can be identified by simple visual inspection at birth. In fact, the pigment epithelial

Figure 3.6 Mice transgenic for a tyrosinase minigene. A nontransgenic FVB/N albino mouse is shown at left. The other three mice represent independent sites of integration (i.e., independent founder mice) for the tyrosine minigene *TyBS* (Yokoyama et al., 1990). The mouse at right is mosaic.
Source: This figure was originally published in a Dutch book by Schellekens (1993).

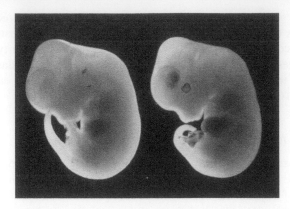

Figure 3.7 Visual identification of transgenic embryos. Most mice that are transgenic for the tyrosinase minigene can be identified by visual inspection for ocular pigmentation any time after embryonic day 11 (E11). A nontransgenic E12 embryo is shown at left, whereas a littermate embryo transgenic for the tyrosinase minigene is shown at right (magnification 10×).

cells of the retina begin to synthesize melanin by day 12 of embryonic development (Theiler, 1989), so that transgenic mice can typically be identified by visual inspection of the fetuses 2 weeks after microinjection (Figure 3.7). Another advantage of the tyrosinase minigene is the fact that it is not detrimental to the health of the transgenic mice. The tyrosinase minigene is not useful in strains of mice that are already pigmented.

C. Superovulation Problems

1. Poor or No Superovulation

After an overnight mating opportunity, 70% or more of superovulated females should have mated, on average. If the mating percentage is consistently below 50%, either the males or the females may be guilty. A good first step to take in diagnosing the problem is to check for oocytes in the oviducts of the females with no visible plug.

If the nonplugged females have ovulated a normal number of oocytes (10−20 side) and if the oocytes are mostly fertilized, then the problem is not in the superovulation regimen, but in the identification of copulation plugs. It may help to check for plugs earlier in the morning, before they have a chance to fall out. If the nonplugged females have ovulated, but the eggs are unfertilized, the fault may lie with the males rather than the females. Check to be sure that the males are not too old or mating too often. If the males are not the problem, then it may be the mating conditions. Check to be sure that the lights are set for the proper lights-on, lights-off cycle, and that the lights are truly turning off at night. Also check for problems with loud noises or other disturbances, and be certain that the temperature in the animal room is not elevated (80°F is too hot).

When the nonplugged females have not ovulated, the most likely explanation is a problem with the hormones, particularly the follicle stimulating hormone (FSH), which is usually provided by administration of pregnant mare's serum gonadotropin. If the stock solutions of hormones (50 units/mL) are more than 2 months old, make up new solutions and be sure to store them at 4°C. If the solutions were just recently prepared, check to be certain that they were prepared at the correct concentration. Another possible problem is superovulation of female mice too soon after their arrival in a new housing room. A minimum of 48 h is often required for acclimation.

If the mating percentage is fine, but there are very few embryos, it might be useful to isolate the embryos earlier in the day. In some cases, the cumulus cells disaggregate early and the embryos proceed down the oviduct, making them more difficult to identify and to isolate. The most conclusive test of poor superovulation is to save the females to see whether any of them become pregnant and whether they have large or small litters. If problems persist, try another strain of mice, or obtain the mice from a different vendor.

2. Unfertilized Eggs

If the females have copulation plugs, but the embryos are unfertilized, the problem lies with the males. In most cases, poor male fertility is caused by overuse of the males or by old age. Fertile males will typically show high fertilization rates if they are mated just once a week up to 9–10 months of age.

3. Malformed Eggs

The frequency of malformed eggs is generally a function of the strain of mice and the age of the females at superovulation. In our experience, 10–25% of the eggs from superovulated 4- to 5-week-old C57BL/6 females are immature or malformed, whereas analogous FVB females typically yield less then 5% defective eggs. The younger the females are at the time of superovulation, the higher the percentage of misshapen oocytes. If the percentage of defective eggs is over 25%, then the concentrations of the hormones used for superovulation should be rechecked. Poor health or abnormal stress of the superovulated females can also result in a high percentage of malformed eggs.

When almost all of the embryos appear unhealthy or abnormal, including the fertilized embryos, the most common cause is a problem with the osmolarity or pH of the mouse embryo medium. Alternatively, the CO_2 concentration in the 37°C incubator may be incorrect (it should be 5%, v/v), or the embryos may have been outside of the incubator for too long during embryo isolation or manipulation. (See Chapter 2 for information about embryo culture media and culture conditions.)

D. Colony Problems

1. Males Not Mating

Males not mating is a rare problem, most often caused by overbreeding of the males. If the females are in estrus, healthy adult males will mate at least 75% of the time.

2. No Recipient Females in Estrus

When female mice are housed together, they will often begin to cycle in synchrony. When this happens, there will be days with multiple recipient females in estrus and other days with very few recipients in estrus. One way to circumvent this problem is to superovulate recipient females when needed. The drawback to this strategy, as stated earlier, is that superovulated females tend to have a poor pregnancy rate when used as recipients. An alternative strategy is to maintain a larger stock of potential recipient females in the colony.

3. High Rates of Abortion in Recipients

Recipients that have as few as one healthy fetus will normally maintain pregnancy to full term. Pregnant females rarely miscarry or deliver before term (<1% of the time). Frequent miscarriage indicates that the females are either sick or unduly stressed.

4. Small Litter Sizes in Recipient Females

When recipient females consistently have small litters (only one or two pups), the problem is unlikely to be simply embryonic lethality of transgenic fetuses, and it is more likely to be a technical problem. The mouse embryo medium might be at fault, the microinjection procedure might be flawed, the concentration (or quality) of the microinjected DNA(s) might be incorrect, or there may be problems with the embryo transfer procedure. To identify the problem, a first step is to do embryo transfers using uninjected embryos. Approximately 75–100% of uninjected embryos should yield healthy newborns. If the percentage is substantially less than that, there is a serious problem with either the culture conditions or the embryo manipulation procedures. A second troubleshooting procedure is to culture injected and uninjected embryos overnight. Most (>90%) of the uninjected embryos should reach the two-cell stage after 24 h in culture. If the embryos do not divide, it suggests that there are serious deficiencies in the culture conditions. For embryos that survive microinjection, a majority (50–60%) should divide to the two-cell stage after overnight culture. If not, either the microinjection procedure is causing irreparable damage or the DNA solution is lethal. To test the microinjection procedure, do the injections with DNA buffer (10 mM Tris, 0.1 mM EDTA) alone. If the embryos survive mock injections but not injections of DNA, then the DNA should be repurified. The DNA must be free of contamination by phenol, chloroform, ethanol, etc., and the concentration should not exceed 5 µg/mL.

Another strategy to assess small litter sizes is to perform C-sections on recipient females at 19 days of gestation. When females are pregnant with one or more healthy fetuses, it will be possible to determine the total number of embryos that progressed to the implantation stage or beyond. Embryonic implantation induces proliferation of the wall of the uterus and formation of a deciduum. Even if the embryo ceases development shortly after implantation, the deciduum is not resorbed until after pregnancy is complete. Embryos that develop to midgestation or later can be recognized and their stage of development estimated. Contaminants in the microinjected DNA often cause early postimplantation lethality, resulting in multiple decidua per pregnant female. Recombinant DNAs that cause dominant embryonic lethality are very rare, so the absence of transgenic mice for a specific DNA construction is typically due to some factor other than prenatal lethality of transgenic fetuses. DNA purification for microinjection and removal of cloning vector sequences are discussed in Chapter 2.

5. Recipient Females Do Not Become Pregnant

If the recipient females do not become pregnant, do some test embryo transfers using uninjected embryos. If pregnancies still do not occur, then embryo transfers should be practiced until the embryos can be consistently transferred to the proper region of the oviduct and the recipient females consistently become pregnant. Lack of pregnancy can also be caused by health impairment, inappropriately prepared anesthetic, or stresses in the animal housing area. Much more likely factors are poor embryo culture conditions or improper reimplantation techniques. If the recipient females become pregnant when they receive uninjected embryos, but not when they receive injected embryos, then the problem lies with the microinjection technique (e.g., the microinjection needle is too wide or too sticky).

IV. Transgenic Phenomenology

Multiple independent transgenic families have been generated in my laboratory by microinjection of the tyrosinase minigene. They provide a visual demonstration of many of the common phenomena seen in transgenic mice.

A. Transgene Expression: Effects of Integration Site and Copy Number

Microinjected DNA usually integrates at only one site or a very limited number of different sites in individual embryos. The number of copies that integrate is highly variable, ranging from just one copy to tens or hundreds of copies. When multiple copies integrate, they are almost always found linked in a tandem head-to-tail array at the site of integration (Brinster et al., 1981). The molecular events responsible for this pattern of integration are not well understood.

There is often considerable variability in the level of transgene expression from one independent transgenic family to another. For example, with the tyrosinase minigene, pigmentation intensities have been found to range from nearly normal agouti to gray to brownish to tan to pigmented only in the ears (Figures 3.8 and 3.9). This variability is generally attributed to influences of the chromosomal sequences flanking the different sites of integration. For the tyrosinase minigene there is a modest, but not consistent, correlation between the copy number of the transgene and its level of expression (Table 3.3). Agouti pigmentation is present in some families that carry only two copies of the transgene, whereas other families

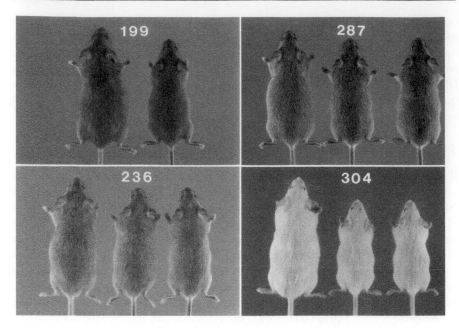

Figure 3.8 Variation in pigmentation. Different sites of integration of the tyrosinase minigene produce different levels of pigmentation, ranging from normal agouti (199) to gray (287) to brownish gray (236) to light tan (304). The color and intensity of pigmentation are consistent within each of the families. For each family, the mouse at left is a parent and representative pigmented offspring are shown at right.

that are pigmented only in their ears have much higher copy numbers (Table 3.3). The brown and tan families have consistently low copy numbers.

For some genes (such as the globins) regulatory sequences have been identified that will give levels of expression of transgenic DNA that are copy-number dependent and site-of-integration independent (Grosveld et al., 1987; Ryan et al., 1989). These regulatory sequences often contain DNase hypersensitive regions.

B. Genetic Mosaicism in Founder Mice

DNA synthesis in mouse embryos begins 12–14 h after fertilization, so replication of the genome is generally in progress by the time microinjections are performed. If the microinjected DNA integrates into an unreplicated location in the genome, then it will be duplicated along with the rest of the genome prior to cell division, and every cell in the developing embryo will receive a copy of the integrated transgenic DNA. However, if the integration site has replicated prior to integration, then the embryo will be a genetic mosaic composed of both transgenic and nontransgenic cells. When integration occurs after just one round of DNA replication, then the founder mouse will be approximately 50% mosaic, although

Figure 3.9 Pigmentation in the ears. In some of the tyrosinase transgenic families, pigmentation is readily visible only in the ears of the transgenic mice (left and center). A nontransgenic albino mouse is shown at right for comparison. (The right ear of the nontransgenic mouse has a scar where an ear tag was once located.)

Table 3.3 Comparison of Transgene Copy Number and Pigmentation Intensity for 78 Independent Integration Sites of the Tyrosinase Minigene

Pigmentation[a]	Number of Families	Copy Number (Range)[b]	Average Copy Number
Agouti	14	2–40	8.0
Gray	18	2–40	7.7
Himalayan	6	4–30	16.0
Brown	13	1–8	3.8
Tan	9	1–4	2.0
Ears only	5	2–30	9.6
Dark eyes only	2	2	2.0
Mottled	5	4–16	9.6
Variable	6	4–16	10.0

[a]Pigmentation intensities were classified subjectively. The agouti mice had nearly normal pigmentation intensity, the gray mice resembled a dilute black pigmentation, Himalayan mice were light gray with dark ears, brown mice showed a dilute brown phenotype, tan mice were very light brown, and some mice showed pigmentation exclusively in their ears or eyes. The mottled mice were partially pigmented, and the variable families showed variable intensities of pigmentation for a single integration site.
[b]Copy number was determined by Southern hybridizations. The intensity of the hybridization bands for the transgenic DNA was compared to the intensity for the endogenous tyrosinase gene.

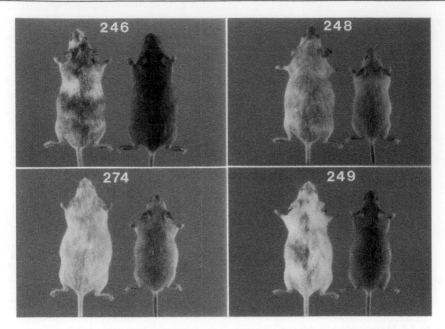

Figure 3.10 Mosaic founder mice. The tyrosinase minigene allows visual identification of mosaic founder mice, since the mosaic founder mice (at left in each panel) show a heterogeneous, mottled pattern of pigmentation, whereas their transgenic offspring (at right) show uniform pigmentation.

the percentage may vary depending on the rates of cell division and the relative contributions of the transgenic and nontransgenic cells to the various tissues of the developing mouse.

Nonmosaic transgenic mice with one site of integration should transmit the transgenic DNA in Mendelian fashion to about 50% of their offspring, whereas mosaic mice generally show a frequency of transmission of 25% or less. (*Note:* Transgenic founder mice that have more than one site of integration can produce litters where 75% or more of the offspring are transgenic, although the percent transmission for any one site of integration is expected to average 50% or less. See Section 3.IV.E for further discussion of multiple integration sites.) In some cases, the transgenic DNA will integrate after the two-cell stage, resulting in founder mice that are transgenic in substantially less than 50% of their cells. In most cases, mosaicism is recognized when the copy number of the transgenic insert is higher in the transgenic offspring than in the founder and when the percentage transmission is substantially below 50%. With the tyrosinase minigene, mosaic mice can be identified visually, since they show a mottled pigmentation pattern (Figure 3.10) that is reminiscent of chimeric mice (see Chapter 4). When the tyrosinase founder mice are mottled because of mosaicism, their transgenic offspring show a uniform pattern of pigmentation (Figure 3.10). The tyrosinase minigene allows visual identification of mosaic founder mice that have only a small

Figure 3.11 Mosaic expression. Some of the tyrosinase transgenic families show a persistent pattern of heterogeneous pigmentation even though the transgenic mice are genetically nonmosaic. The two mice in the top row are homozygous for the tyrosinase minigene and are the parents of the eight mice below (family OVE159). The transgenic mice in this family show variation in the intensity and specific pattern of pigmentation, but the mice are always mottled.

number of pigmented cells, and it also allows rapid identification of their rare transgenic offspring. Approximately 35% of the founder mice from our tyrosinase minigene injections were found to be genetic mosaics (data not shown). For five of the mottled founder mice, the transgenic offspring still exhibited a mosaic pattern of pigmentation (Figure 3.11), indicating that in these mice the mottled pigmentation was due to a mosaic pattern of transgene expression rather than to genetic mosaicism. These mice provide visual evidence that specific sites of integration not only can bias the level of transgene expression but may also influence the cell-by-cell pattern of expression.

C. Intrafamily Variation in Expression

In some transgenic families, the level of transgene expression may vary from mouse to mouse even though the mice are inbred and have the same site of integration (Figure 3.12). In our collection of tyrosinase transgenic families, over 90% of the integration sites gave stable patterns of expression. In these cases, all of the heterozygous transgenic mice in a given family had an identical, or nearly

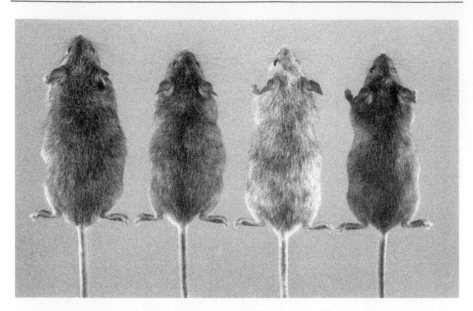

Figure 3.12 Inherited variability. Mice with the same site of integration can show variation in the intensity of pigmentation. In the family that is shown (OVE195), the parent at left produced the three offspring at right. Southern hybridizations showed an identical pattern of bands for all of the mice, indicating a single stable site of integration (data not shown). When the offspring mice were mated, the same variation in pigmentation was seen in the next generation. The lightly pigmented mouse had light and dark offspring, as did the more darkly pigmented mice.

identical, color, intensity, and pattern of pigmentation. The pigmentation did not vary from generation to generation, and the males showed the same pigmentation as the females. There were six transgenic families in which the intensity and color of pigmentation were not consistent (Figure 3.12). Southern hybridizations were done to look for multiple sites of integration in these families, but the hybridization patterns were stable and consistent within each family (data not shown). The mice in these six families show an innate variability in their level of transgene expression and demonstrate that mice of apparently identical genotype can exhibit variable phenotypes. The variation in expression is not correlated with the sex of the mice or of the transgenic parent, since the variability is seen between siblings.

The level of expression of transgenic DNA can also be influenced by genomic imprinting (Surani et al., 1990; Chaillet et al., 1991). No examples of genomic imprinting were seen in the families transgenic for the tyrosinase minigene.

D. Integration into a Sex Chromosome

Integration into either of the sex chromosomes can be identified by mating a (nonfounder) transgenic male to a nontransgenic female. If the transgenic DNA

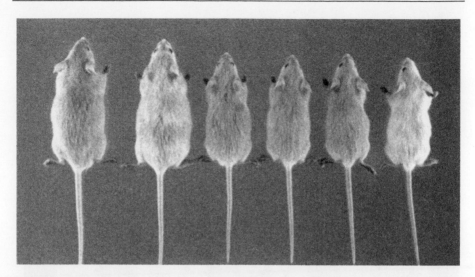

Figure 3.13 Inactivation of the X chromosome. A transgenic family with integration of the tyrosine minigene into the X chromosome is shown. The two mice at left are a heterozygous male and female, respectively, and are parents to the four mice at right (two males and two females). One of the females, second from the right, is homozygous for the tyrosinase minigene, whereas the other (far right) is heterozygous and mottled.

has integrated into a nonpseudoautosomal region of the X chromosome, then all the female offspring will be transgenic, whereas all the male offspring will be nontransgenic. In other words, the transgenic DNA will be transmitted along with the X chromosome to only the female offspring. The transmission pattern will be reversed for integration into the Y chromosome. Among the tyrosinase families, two integrations were found in the X chromosome and one in the Y chromosome. In the Y chromosome family, all the males and only the males were pigmented. The pigmentation was limited exclusively to the ears. In both instances, where the tyrosinase minigene integrated into the X chromosome, the transgene was subject to X chromosome inactivation (Figure 3.13). Heterozygous female mice showed a mottled pattern of pigmentation, whereas heterozygous males and homozygous females were uniformly pigmented.

E. Multiple Sites of Integration in Founder Mice

Founder transgenic mice occasionally have transgenic DNA integrated at more than one site in the genome. This phenomenon is typically recognized when the offspring are analyzed and found to contain two or more different copy numbers or hybridization patterns for the transgenic DNA. (Multiple sites of integration are difficult to identify by polymerase chain reaction (PCR) screening.) In general, different sites of integration are assumed to represent independent integration events.

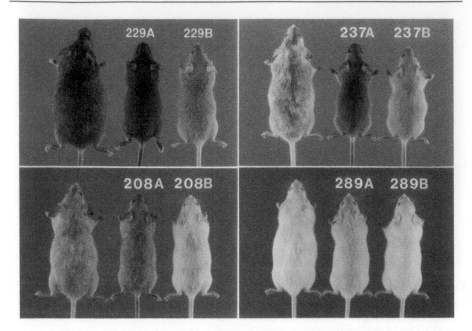

Figure 3.14 Founder mice with multiple integration sites. The founder mouse for each of these families (at left in each panel) produced offspring with two different integration sites and two distinct levels of pigmentation (termed A and B for each family).

The copy numbers are different, the levels of expression can be different, and there is typically independent transmission of each site to the F1 offspring. However, transgenic families have occasionally been discovered to have different insertions in the same region of the genome (Xiang et al., 1990).

The tyrosinase transgenic mice allow easy identification of transgenic founder mice with multiple sites of integration, because the founder mice produce offspring with two or more consistent colors or intensities of pigmentation (Figure 3.14). Approximately 15% of the tyrosinase founder mice had either two or three different integration sites that were recognized by visual inspection of the offspring and confirmed by Southern hybridizations (not shown).

F. Homozygous Versus Hemizygous

One of the major advantages of using the tyrosinase minigene to generate transgenic mice is the fact that homozygous mice in most families can be identified by simple visual inspection, since the homozygous mice have darker coat colors (Figure 3.15), reflecting the increased gene dosage. In addition, homozygotes can be verified by matings to nontransgenic partners, since homozygous mice will produce 100% pigmented offspring. The use of the tyrosinase minigene has allowed

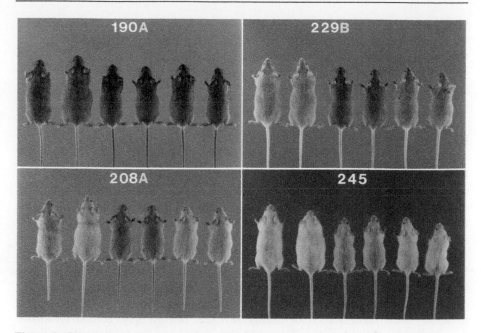

Figure 3.15 Visual identification of homozygotes. For most of the tyrosinase transgenic families, homozygous transgenic mice can be recognized by simple inspection for darker coat color, reflecting the twofold higher gene dosage. In each family shown, two heterozygous parents are shown at left, two homozygous offspring are shown in the middle, and two hemizygous offspring are shown on the right.

the identification of more than 100 unique and interesting insertional mutations (for example, see Bishop, et al., 2000; Caburet, et al., 2014; Hamilton, et al., 2011; Headon and Overbeek, 1999; Hicks, et al., 2012; Li, et al., 2010; Lu, et al., 1999; Moayedi, et al., 2014; Morgan, et al., 1998; Overbeek, et al., 2001; Wang, et al., 2011; Yang, et al., 2004; Yokoyama et al., 1993; for a more complete list search for Overbeek at http://www.informatics.jax.org/marker). Some tyrosinase families have insertional mutations that result in embryonic or neonatal lethality for the homozygous mice. These families have still been easy to maintain, since the transgenic mice in each generation can be visually identified.

G. Genetic Instability

In almost all instances, transgenic DNA is stably maintained once it has integrated into the genome. Even when there are multiple copies of the DNA integrated in a tandem head-to-tail array, the transgenic DNA is transmitted stably from one generation to the next without genomic rearrangements and without deletions. However, examples of transgenic families with genetic instability under selective pressure have been identified (Sandgren et al., 1991; Wilkie et al., 1991). The tyrosinase transgenic families have all displayed stable Southern hybridization patterns, and those families with stable pigmentation have maintained a uniform pigmentation intensity over more than 10 generations of mating. Because alterations of the integration site would be expected to change both tyrosinase expression and pigmentation, these results imply that genetic instability is very rare for these transgenic inserts.

V. Summary

In this chapter, I have tried to provide a simplified overview of factors that often need to be taken into consideration by laboratories that are just beginning to do research with transgenic mice. Overviews of different strains of mice, husbandry techniques, and troubleshooting protocols have been provided. In Section 3.III, I have reviewed some of the characteristics of transgenic mice generated by microinjection. In general, microinjected DNA appears to integrate randomly into the mouse genome. In most cases, the DNA integrates stably as a tandem head-to-tail repeat containing from 1 to hundreds of copies of the injected DNA. Multiple factors can influence the pattern of expression of the transgenic DNA. The tyrosinase minigene can be used to allow visual identification of transgenic mice. Interestingly, certain tyrosinase transgenic mice show that, even for a stable integration at a single site in the genome, there can be variations in the level and pattern of transgene expression.

Acknowledgments

I thank Lindsey Lampp for photographing the mice. Research using the tyrosinase minigene was supported by National Institutes of Health Grant HD 25340.

References

Beermann, F., Ruppert, S., Hummler, E., Bosch, F.X., Müller, G., Rüther, U., et al., 1990. Rescue of the albino phenotype by introduction of a functional tyrosinase gene into mice. EMBO J. 9, 2819−2826.

Behringer, R., Gertsenstein, M., Nagy, K., Nagy, A., 2014. Manipulating the Mouse Embryo: A Laboratory Manual. Fourth edition Cold Spring Harbor Laboratory Press, Cold Spring Harbor, NY.

Bishop, C.E., Whitworth, D.J., Qin, Y., Agoulnik, A.I., Agoulnik, I.U., Harrison, W.R., et al., 2000. A transgenic insertion upstream of sox9 is associated with dominant XX sex reversal in the mouse. Nat. Genet. 26, 490−494.

Brinster, R.L., Chen, H.Y., Trumbauer, M., 1981. Somatic expression of herpes thymidine kinase in mice following injection of a fusion gene into eggs. Cell 27, 223−231.

Brinster, R.L., Chen, H.Y., Trumbauer, M.E., Yagle, M.K., Palmiter, R.D., 1985. Factors affecting the efficiency of introducing foreign DNA into mice by microinjecting eggs. Proc. Natl. Acad. Sci. USA 82, 4438−4442.

Caburet, S., Arboleda, V.A., Llano, E., Overbeek, P.A., Barbero, J.L., Oka, K., et al., 2014. Mutant cohesin in premature ovarian failure. N. Engl. J Med. 370, 943–949.

Chaillet, J.R., Vogt, T.F., Beier, D.R., Leder, P., 1991. Parental-specific methylation of an imprinted transgene is established during gametogenesis and progressively changes during embryogenesis. Cell 66, 77–83.

Covarrubias, L., Nishida, Y., Mintz, B., 1986. Early postimplantation embryo lethality due to DNA rearrangements in a transgenic mouse strain. Proc. Natl. Acad. Sci. USA 83, 6020–6024.

Covarrubias, L., Nishida, Y., Terao, M., D'Eustachio, P., Mintz, B., 1987. Cellular DNA rearrangements and early developmental arrest caused by DNA insertion in transgenic mouse embryos. Mol. Cell. Biol. 7, 2243–2247.

Dickie, M.M., 1975. Keeping records. In: Green, E.L. (Ed.), Biology of the Laboratory Mouse, second ed. Dover, New York, NY, pp. 23–27.

Erbach, G.T., Lawitts, J.A., Papaioannou, V.E., Biggers, J.D., 1994. Differential growth of the mouse preimplantation embryo in chemically defined media. Biol. Reprod. 50, 1027–1033, Erratum in: Biol Reprod 1994 Aug;51(2):345.

Festing, M.F.W., 1990. Choice of an experimental animal and rodent genetics. In: Copp, A.J., Cockroft, D.L. (Eds.), Postimplantation Mammalian Embryos: A Practical Approach. Oxford University Press, London, pp. 205–219.

Festing, M.F.W., 1992. Origins and characteristics of inbred strains of mice, 14th listing. Mouse Genome 90, 231–352.

Friedrich, G., Soriano, P., 1991. Promoter traps in embryonic stem cells: a genetic screen to identify and mutate developmental genes in mice. Genes Dev. 5, 1513–1523.

Green, E.L., 1975. Breeding systems. In: Green, E.L. (Ed.), Biology of the Laboratory Mouse, second ed. Dover, New York, NY, pp. 11–22.

Grosveld, F., van Assendelft, G.B., Greaves, D.R., Kollias, G., 1987. Position-independent, high-level expression of the human β-globin gene in transgenic mice. Cell 51, 975–985.

Hamilton, S.M., Spencer, C.M., Harrison, W.R., Yuva-Paylor, L.A., Graham, D.F., Daza, R.A., et al., 2011. Multiple autism-like behaviors in a novel transgenic mouse model. Behav. Brain Res. 218, 29–41.

Harbers, K., Kuehn, M., Delius, H., Jaenisch, R., 1984. Insertion of retrovirus into the first intron of αl(I) collagen gene leads to embryonic lethal mutation in mice. Proc. Natl. Acad. Sci. USA 81, 1504–1508.

Hartung, S., Jacnisch, R., Breindl, M., 1986. Retrovirus insertion inactivates mouse αl(I) collagen gene by blocking initiation of transcription. Nature 320, 365–367.

Headon, D.J., Overbeek, P.A., 1999. Involvement of a novel Tnf receptor homologue in hair follicle induction. Nat. Genet. 22, 370–374.

Hetherington, C.M., 1987. Mouse husbandry. In: Monk, M. (Ed.), Mammalian Development: A Practical Approach. IRL Press, Oxford, pp. 1–12.

Hicks, A.N., Lorenzetti, D., Gilley, J., Lu, B., Andersson, K.E., Miligan, C., et al., 2012. Nicotinamide mononucleotide adenylyltransferase 2 (nmnat2) regulates axon integrity in the mouse embryo. PLoS One. 7 (10), e47869.

Jaenisch, R., Harbers, K., Schnieke, A., Chumakov, I., Jähner, D., Löhler, J., et al., 1983. Germline integration of Moloney murine leukemia virus at the Mov13 locus leads to recessive lethal mutation and early embryonic death. Cell 32, 209–216.

Lang, C.M., 1983. Design and management of research facilities. In: Foster, H.L., Small, J.D., Fox, J.G. (Eds.), The Mouse in Biomedical Research, vol. 3. Academic Press, New York, NY, pp. 38–50.

Les, E.P., 1975. Husbandry. In: Green, E.L. (Ed.), Biology of the Laboratory Mouse, second ed. Dover, New York, NY, pp. 29–37.

Li, W., Puertollano, R., Bonifacino, J.S., Overbeek, P.A., Everett, E.T., 2010. Disruption of the murine Ap2β1 gene causes nonsyndromic cleft palate. Cleft Palate Craniofac. J. 47, 566–573.

Liu, C., 2013. Strategies for designing transgenic DNA constructs. Methods Mol. Biol. 1027, 183–201.

Lu, W., Phillips, C.L., Killen, P.D., Hlaing, T., Harrison, W.R., Elder, F.F.B., et al., 1999. Insertional mutation of the collagen genes Col4a3 and Col4a4 in a mouse model of Alport syndrome. Genomics. 61, 113–124.

Meisler, M.H., 1992. Insertional mutation of "classical" and novel genes in transgenic mice. Trends Genet. 8, 341–344.

Moayedi, Y., Basch, M.L., Pacheco, N.L., Gao, S.S., Wang, R., Harrison, W., et al., 2014. The candidate splicing factor sfswap regulates growth and patterning of inner ear sensory organs. PLoS Genet. 10 (1), e1004055.

Morgan, D., Turnpenny, L., Goodship, J., Dai, W., Majumder, K., Matthews, L., et al., 1998. Inversin, a novel gene in the vertebrate left-right axis pathway, is partially deleted in the inv mouse. Nat. Genet. 20, 149–156.

Nielsen, L.L., Pedersen, R.A., 1991. Drosophila alcohol dehydrogenase: a novel reporter gene for use in mammalian embryos. J. Exp. Zool. 257, 128–133.

Ornitz, D.M., Palmiter, R.D., Hammer, R.E., Brinster, R.L., Swift, G.H., MacDonald, R.J., 1985. Specific expression of an elastase-human growth hormone fusion gene in pancreatic acinar cells of transgenic mice. Nature 313, 600–602.

Otis, A.P., Foster, H.L., 1983. Management and design of breeding facilities. In: Foster, H. L., Small, J.D., Fox, J.G. (Eds.), The Mouse in Biomedical Research, vol. 3. Academic Press, New York, NY, pp. 17–35.

Overbeek, P.A., Aguilar-Cordova, E., Hanten, G., Schaffner, D.L., Patel, P., Lebovitz, R.M., et al., 1991. Coinjection strategy for visual identification of transgenic mice. Transgenic Res. 1, 31–37.

Overbeek, P.A., Gorlov, I.P., Sutherland, R.W., Houston, H.B., Harrison, W.R., Boettger-Tong, H.L., et al., 2001. A transgenic insertion causing cryptorchidism in mice. Genesis. 30, 26–35.

Pinkert, C.A., 1990. A rapid procedure to evaluate foreign DNA transfer in mammals. Biotechniques 9, 38–39.

Quaife, C.J., Pinkert, C.A., Ornitz, D.M., Palmiter, R.D., Brinster, R.L., 1987. Pancreatic neoplasia induced by ras expression in acinar cells of transgenic mice. Cell 48, 1023–1034.

Rafferty Jr., K.A., 1970. Methods in Experimental Embryology of the Mouse. Johns Hopkins Press, Baltimore, MD.

Ryan, T.M., Behringer, R.R., Martin, N.C., Townes, T.M., Palmiter, R.D., Brinster, R.L., 1989. A single erythroid-specific DNase I superhypersensitive site activates high levels of human β-globin gene expression in transgenic mice. Genes Dev. 3, 314–323.

Sandgren, E.P., Palmiter, R.D., Heckel, J.L., Daugherty, C.C., Brinster, R.L., Degen, J.L., 1991. Complete hepatic regeneration after somatic deletion of an albumin-plasminogen activator transgene. Cell 66, 245–256.

Schellekens, H. (Ed.), 1993. De DNA-Makers, Architecten Van Het Leven. Natuur en Techniek, Maastricht, Netherlands.

Schnieke, A., Harbers, K., Jaenisch, R., 1983. Embryonic lethal mutation in mice induced by retroviral insertion into the αl(I) collagen gene. Nature 304, 315–320.

Silvers, W.K., 1979. The Coat Colors of Mice: A Model for Mammalian Gene Action and Interaction. Springer-Verlag, New York, NY.

Singh, G., Supp, D.M., Schreiner, C., McNeish, J., Merker, H.-J., Copeland, N.G., et al., 1991. *legless* insertional mutation: morphological, molecular, and genetic characterization. Genes Dev. 5, 2245–2255.

Soriano, P., Jacnisch, R., 1986. Retroviruses as probes for mammalian development: allocation of cells to the somatic and germ cell lineages. Cell 46, 19–29.

Stevens, M.E., Meneses, J.J., Pedersen, R.A., 1989. Expression of a mouse metallothionein-*Escherichia coli* β-galactosidase fusion gene (MT-βgal) in early mouse embryos. Exp. Cell Res. 183, 319–325.

Surani, M.A., Allen, N.D., Barton, S.C., Fundele, R., Howlett, S.K., Norris, M.L., et al., 1990. Developmental consequences of imprinting of parental chromosomes by DNA methylation. Philos. Trans R Soc Lond B Biol. Sci. 326, 313–327.

Swift, G.H., Hammer, R.E., MacDonald, R.J., Brinster, R.L., 1984. Tissue-specific expression of the rat pancreatic elastase I gene in transgenic mice. Cell 38, 639–646.

Taketo, M., Schroeder, A.C., Mobraaten, L.E., Gunning, K.B., Hanten, G., Fox, R.R., et al., 1991. FVB/N: an inbred mouse strain preferable for transgenic analyses. Proc. Natl. Acad. Sci. USA 88, 2065–2069.

Tanaka, S., Yamamoto, H., Takeuchi, S., Takeuchi, T., 1990. Melanization in albino mice transformed by introducing cloned mouse tyrosinase gene. Development. 108, 223–227.

Theiler, K., 1989. The House Mouse: Atlas of Embryonic Development. Springer-Verlag, New York, NY.

Wang, B., Harrison, W., Overbeek, P.A., Zheng, H., 2011. Transposon mutagenesis with coat color genotyping identifies an essential role for Skor2 in sonic hedgehog signaling and cerebellum development. Development. 138, 4487–4497.

Whittingham, D.G., 1971. Culture of mouse ova. J. Reprod. Fertil. 14, 7–21.

Wilkie, T.M., Braun, R.E., Ehrman, W.J., Palmitter, R.D., Hammer, R.E., 1991. Germ-line intrachromosomal recombination restores fertility in transgenic MyK-103 male mice. Genes Dev. 5, 38–48.

Xiang, X., Benson, K.F., Chada, K., 1990. Mini-mouse: disruption of the pygmy locus in a transgenic insertional mutant. Science 247, 967–969.

Yang, T., Liang, D., Koch, P.J., Hohl, D., Kheradmand, F., Overbeek, P.A., 2004. Epidermal detachment, desmosomal dissociation, and destabilization of corneodesmosin in Spink5-/- mice. Genes Dev. 18, 2354–2358.

Yokoyama, T., Silversides, D.W., Waymire, K.G., Kwon, B.S., Takeuchi, T., Overbeek, P.A., 1990. Conserved cysteine to serine mutation in tyrosinase is responsible for the classical albino mutation in laboratory mice. Nucleic Acids Res. 18, 7293–7298.

Yokoyama, T., Copeland, N.G., Jenkins, N.A., Montgomery, C.A., Elder, F.F.B., Overbeek, P.A., 1993. Reversal of left-right asymmetry: a situs inversus mutation. Science 260, 679–682.

4 Gene Targeting in Embryonic Stem Cells, I: History and Methodology

L. Philip Sanford[1] and Tom Doetschman[2]

[1]Transgenic and Genome Manipulation Core, University of Iowa, Iowa City, IA, [2]BIO5 Institute and Department Cellular & Molecular Medicine, University of Arizona, Tucson, AZ

I. Stem Cells of the Mouse Embryo

There have been extensive studies with embryonic stem cells from the mouse embryo, and these studies have led directly to a greater understanding of human genetic diseases, gene interactions, embryo development, and organogenesis both *in vivo* and *in vitro*. Pluripotential stem cells of the mouse can be classified into four major groups: embryonal carcinoma cells (ECCs), embryonic stem cells (ESCs), primordial germ cells (PGCs), and induced pluripotent stem cells (iPSCs). ECCs were first isolated from teratocarcinomas. ESCs were derived from the inner cell mass of 3.5 dpc mouse embryos. PGCs have been isolated from the posterior third of mouse embryos around 8.5 dpc. iPSCs can be produced from somatic cells via the coexpression of C-MYC, SOX2, OCT3/4, and KLF4 or NANOG.

These types of stem cells can be maintained in an undifferentiated state for many passages in appropriate culture conditions while retaining the pluripotency that makes them so valuable. It is this ability to be grown in culture in large numbers that makes them useful for the introduction of gene manipulations, the screening of very rare homologous recombination events resulting in precise manipulation of the genome, and finally for producing animal models with designed mutations of diverse types.

A. Establishment of Mouse ESCs in Culture

While the mystery has largely gone out of this process, a brief summary follows. ESC lines have been established from a wide variety of mouse strains and a growing number of mammalian species. Typically, these ESCs are derived from the outgrowth of the inner cell mass of blastocyst stage embryos. Mouse ESCs were originally produced from implantation-delayed blastocysts grown in embryonal carcinoma culture medium (Evans and Kaufman, 1981; Kaufman et al., 1983; Wobus et al., 1984), from nondelayed blastocysts grown in medium conditioned by ECCs (Martin, 1981), and from nondelayed blastocysts cultured in ECC medium (Robertson et al., 1983a;

Transgenic Animal Technology. DOI: http://dx.doi.org/10.1016/B978-0-12-410490-7.00004-9

Axelrod, 1984; Doetschman et al., 1985). ESCs were also established from PGCs (Resnick et al., 1992; Matsui et al., 1992; Stewart et al., 1994; Labosky et al., 1994). Prevention of differentiation during the establishment of these cell lines was accomplished by growth in conditioned medium from Buffalo Rat liver cells (Smith and Hooper, 1987), growth on STO cells (Evans and Kaufman, 1981; Martin, 1981) or in the presence of mouse embryonic fibroblasts (MEFs; Wobus et al., 1984; Doetschman et al., 1985), or in the absence of feeder cells but in the presence of leukemia inhibitory factor (LIF; Nichols et al., 1990; Pease et al., 1990). Under these conditions, undifferentiated mouse ESCs grow in tight colonies with refractory edges and no apparent cell borders. For excellent detailed descriptions of the process for deriving ESCs from mouse blastocysts, see Nagy and Vinterstein (2006) and Nagy and Nichols (2011). Mouse ESCs have also been cultivated in LIF and mitogen-activated protein kinase (MEK) inhibitors (Burdon et al., 1999), defined medium (Ying and Smith, 2003), knockout serum replacement (KOSR) medium plus LIF (Tanimoto et al., 2008), serum-free medium plus MEK and GSK3 inhibitors (2i) (Nichols et al., 2009; Kanda et al., 2012), or defined medium plus MEK, FGFR, and GSK3 inhibitors (3i) (Ying et al., 2008). The establishment of mouse ESC lines does not appear to be strain dependent (reviewed by Robertson et al., 1983b). ESCs have been established from embryos with null mutations (Magnuson et al., 1982; Martin et al., 1987; Conlon et al., 1991).

Many different strains of ESCs have been established, including strain 129 ESCs (Evans and Kaufman, 1981; Martin, 1981; Doetschman et al., 1985; Handyside et al., 1989), C57Bl/6 ESCs (Kontgen and Stewart, 1993; Nagy et al., 1993; Dinkel et al., 1999; Eggan et al., 2001; Cheng et al., 2004; George et al., 2007; Pettitt et al., 2009; Gertsenstein et al., 2010), Balb/C (Dinkel et al., 1999), C3H/Hej (Kitani et al., 1996), and many other induced pluripotential methodologies.

B.   Establishment of PGCs in Culture

Mouse PGCs arise from presumptive mesoderm cells located near the extra embryonic ectoderm prior to gastrulation at E6-6.5 (Lawson and Hage, 1994). A small number of alkaline phosphatase-positive PGCs are found in the extra embryonic mesoderm just posterior to the primitive streak and later in the primitive-streak mesoderm. At E8.5, the PGCs have increased to about 150 cells and occupy a region from the allantois to the hindgut endoderm (Ginsburg et al., 1990). At E10.5, PGCs have continued extensive expansion and have migrated to the genital ridges. With E8.5 embryos, embryonic germ (EG) cells are derived from the caudal third of the embryos (Resnick et al., 1992). PGCs have also been derived from E12.5 genital ridges (Matsui et al., 1992; Labosky et al., 1994). The cell lines established from these PGCs form cell aggregates visibly identical to ESCs (Figure 4.1A and B, respectively), and once established they are cultured in the same manner as ESCs (Shamblott et al., 1998). Supplements of Steel factor, LIF, and FGF2 have been used to establish PGC lines (Matsui et al., 1992; Labosky et al., 1994). However, these supplements are not required once the PGCs are established (Stewart et al., 1994). PGCs derived from

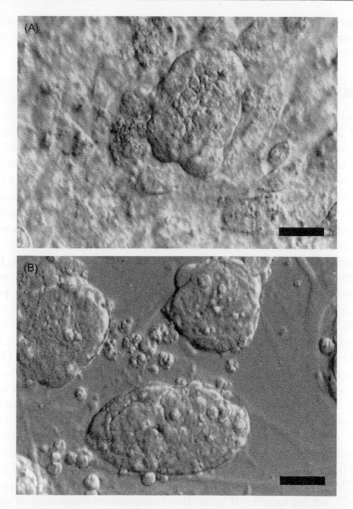

Figure 4.1 Visual comparison of human PGCs and mouse ESCs. (A) Human PGCs grown on mitotically inactivated STO feeder cells. (B) Mouse ES cells grown on mitotically inactivated MEFs. Bars represent 100 μM (Shamblott et al., 1998). Mouse PGCs are cultured similarly, with minor modifications (see text), and appear essentially the same as mouse ES cells. Their ability to form teratocarcinomas in severe combined immunodeficient (SCID) mice and to differentiate in culture (Figure 4.2) are also similar to mouse ESCs and PGCs.

E8.5-day embryos have colonized the germ line at frequencies similar to ESCs (Stewart et al., 1994; Labosky et al., 1994), but those from genital ridges have shown reduced germ line competence (Labosky et al., 1994). Both alleles of the insulin-like growth factor 2 receptor were found in the demethylated state in several, but not all, mouse embryonic germ cell lines (Labosky et al., 1994), and female PGCs were found to have two active X chromosomes (Stewart et al., 1994). In neither case could it be ruled out that the methylation state or X-activation state of the PGCs was affected by the culturing of the cells. Nonetheless, in neither case did these states affect the ability of PGCs to colonize the germ line.

C. Establishment of iPSCs in Culture

iPSCs were first reported by Takahashi and Yamanaka (2006). They screened a battery of 24 pluripotency-inducing candidate genes by transfection into MEFs. The cells were grown in DMEM on STO feeder cells plus conditioned medium with LIF and 10% FBS. They found that four genes were able to not only induce an embryonic gene *Fbx15* but also produce a cellular morphology and transcriptome similar to mouse ESCs. These genes were *c-Myc*, *Klf4*, *Sox2*, and *Oct3/4*. A year later it was shown that these iPSCs could be produced and maintained in the same conditions used for ESC culture (Hanna et al., 2007). Furthermore, they could be caused to differentiate into the hematopoietic progenitors similar to ESCs. Additionally, they demonstrated that these cells could be modified by ESC-type gene-targeting techniques, and that they could be used to correct a sickle-cell anemia mouse model by transplantation of gene-targeted hematopoietic progenitor cells into irradiated recipients. An interesting observation recently put forward (Yamanaka, 2012) is that both ESC and iPSCs are manmade in synthetic cell culture media. ESCs are not identical to inner cell mass cells and iPSCs are not identical to ESCs. Microarray studies have shown this to be the case (Takahashi and Yamanaka, 2006). These cell lines exhibit stem cell-like features such as pluripotency, the ability to form teratocarcinomas in immune-deficient mice, and the ability to make viable embryos. There does not appear to be an *in vivo* cell that is an exact sibling to these cells. Rather, they represent different elements in a continuum of stem cells that can be derived from animals (Yamanaka, 2012).

We would be remiss not to mention that not all cells that are derived from blastocyst outgrowths and show the ESC morphology will be useful for producing animals. Clones can become aneuploid, possess chromosome 8 or 11 trisomy and as such will not be capable of germ line transmission (Sugawara et al., 2006). As a rule female ESC lines have been avoided, because the use of very good male ESCs causes sex conversion, yielding a higher percentage of male chimeras, as reviewed by Fedorov et al. (1997). Additionally, the germ line of male mice can be much more rapidly interrogated for germ line transmission via harem breeding.

D. Differentiation Potential of ESCs, PGCs, and iPSCs

When grown in the absence of feeder layer cells or other inhibitors of differentiation, all three cell types will spontaneously differentiate in culture into identifiable embryonic structures. Initially, ESCs were found to differentiate into embryonic endoderm and ectoderm, yolk sac, nerve cells, epithelial cells, tubular structures, cartilage (Martin and Lock, 1983), heart muscle, and blood islets (Doetschman et al., 1985). Simple, complex, and cystic embryoid bodies with

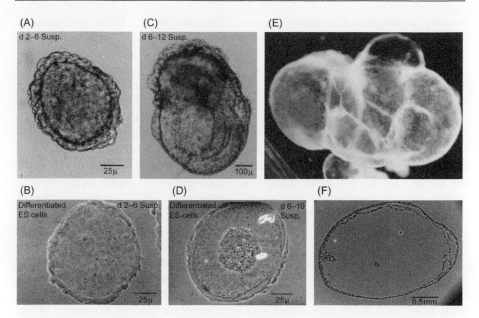

Figure 4.2 Embryoid bodies formed from differentiating ES cells. D3 ES cells were cultured in plastic Petri dishes in the absence of feeder cells, conditions under which ES cells spontaneously differentiate in suspension culture. (A) Example of simple embryoid body (EB) that can develop from 2 to 6 days of suspension culture with outer layer of embryonic endoderm (arrowhead). (B) Simple EB in cross section. (C) Example of complex EB that can develop between 6 and 12 days of suspension culture showing an additional inner layer(s) of embryonic endoderm (arrows). (D) Complex EB in cross section. (E) Cystic EB that can develop between 10 and 20 days of suspension culture showing multiple blood islands (red). (F) Cystic EB in cross section. (A–D, F, phase optics; E, dark-field optics, magnification 15×; B, D, F frozen sections.)

similarities to the pre-egg cylinder (day 5), egg cylinder (day 6), and yolk sac (day 9) stages of embryonic development, respectively, are shown in Figure 4.2A–F. When injected subcutaneously into syngeneic or immunosuppressed hosts, ESCs will form teratocarcinomas containing cartilage, secretory epithelium, glandular structures, keratinizing epithelium, melanin pigmentation, and muscle (Martin, 1981; Martin and Lock, 1983; Kaufman et al., 1983; Mann et al., 1990). When injected back into blastocysts, ESCs will colonize some extra embryonic and all embryonic tissues (Beddington and Robertson, 1989; Suemori et al., 1990; Lallemand and Brulet, 1990; Nagy et al., 1990) and all of the tissues tested in the adult (Gossler et al., 1986; Bradley and Robertson, 1986), including the germ line (Bradley et al., 1984). Tetraploid aggregation has been shown to produce entirely ESC-derived mice (Nagy et al., 1990).

The capacity of ESCs to differentiate into embryoid bodies and differentiated tissues with striking similarities to embryonic structures has provided the opportunity to characterize many aspects of developing tissues such as the growth factors involved in early postimplantation embryogenesis (Mummery et al., 1990), commitment of mesoderm to hematopoiesis (Robertson et al., 2000), or the biochemical steps involved in development of the heart (Metzger et al., 1994). Differentiating ESCs are also being used to dissect pathways of embryonic development through the use of pharmacologic or genetic manipulation. For example, manipulation of culture medium or the feeder layer can inhibit or enhance differentiation in particular directions: PI3 kinase inhibition restricts cardiomyocyte development (Klinz et al., 1999), retinoic acid treatment accelerates cardiomyocyte differentiation (Wobus et al., 1997), retinoic acid/db-cAMP treatment induces smooth muscle differentiation (Drab et al., 1997), BMP2 and BMP4 treatment induces chondrocyte differentiation (Kramer et al., 2000), and use of a stromal cell feeder layer plus 1-α,25-dihydroxyvitamin D_3 in stepwise fashion induces osteoclast differentiation (Yamane et al., 1997). Genetically altered differentiating ESCs have also been useful for dissecting genetic pathways of differentiation. For example, gene ablation for the LIF receptor attenuates adipocyte differentiation (Aubert et al., 1999) and genetic inactivation of β1 integrins retards myogenic and accelerates neuronal differentiation (Rohwedel et al., 1998). Cell ablation in embryoid bodies can also be used to isolate differentiated cell types. In one study, cardiomyocytes were isolated from G418-treated embryoid bodies made from ESCs stably transfected with cardiomyocyte-specific expression of the $G418^R$ gene (Klug et al., 1996).

It is beyond the scope of this chapter to exhaustively review all the tissues that mouse ESCs have the capacity to produce *in vitro*. A PubMed search for "tissues derived from ESCs" gives 1400-plus publications. Some recent reports include: the development of testicular cords (Pan et al., 2013), the production of neural precursors in defined medium (Peng et al., 2013), the formation of continuous stratified neuroepithelium by a laminin matrix (Nasu et al., 2012), tooth development (Marais et al., 2011), the formation of retina on three-dimensional structures (Colozza et al., 2012), hematogenic progenitor cells (Kim et al., 2013), differentiation into renal tubules (Morizane et al., 2009), the production of salivary gland cells (Kawakami et al., 2013), the generation of a pancreas (Matsunari et al., 2013), and many more.

In addition to using differentiating ESCs to characterize or alter differentiation, they can also be useful for gene discovery. Large-scale gene and enhancer trap screens have utilized LacZ marker gene expression in differentiating ESCs to reveal genes of potential importance for postimplantation embryogenesis (Gossler et al., 1989; Scherer et al., 1996; Salminen et al., 1998; Cannon et al., 1999). In addition, when gene ablation produces embryonic lethality that precludes investigation of a particular embryonic developmental process, differentiating ESCs can be employed to circumvent that lethality. For example, differentiating ESCs have demonstrated a role for *Gata6* in visceral endoderm (Morrisey et al., 1998) and a role for LIF receptor in adipocyte differentiation (Aubert et al., 1999).

One of the hallmarks of ESCs, PGCs, and iPSCs and their differentiated progeny is the maintenance of primary cell characteristics after cell culture. Since ESCs, PGCs, and iPSCs can reconstitute an animal through germ line chimerism, their differentiated progeny should also be able to contribute to the normal growth of tissue after transfer back into the animal. This has now been demonstrated for lymphocytes (Chen and Mok, 1995), cardiomyocytes (Klug et al., 1996), bone marrow (Kim et al., 2013), and neurons (Benninger et al., 2000).

There are a multitude of methods to produce mice from ESCs. There is the classical blastocyst injection of gene-targeted ESCs to produce chimeric mice followed by the breeding of the chimeras to produce founders (Bradley et al., 1984). Nagy et al. (1990) pioneered the production of conventional chimeras by diploid aggregation. In this method, gene-targeted ESCs were cocultivated with diploid embryos to produce blastocysts which were then transplanted into pseudopregnant female mice. This group went on to pioneer the tetraploid aggregation technique wherein two cell embryos are fused via an electric shock to make tetraploid embryos. These are cocultured with ESCs to produce blastocysts for implantation. The unique aspect of this method was that it can produce fully ESC-derived mice. A follow-up report by Jaenisch verified that only F1 ESCs could be used for this method (Eggan et al., 2001). Another method arose to produce completely ESC-derived mice (Poueymirou et al., 2007). This method involves the use of a laser to punch a hole on the zona pellucida of early eight-cell morulas into which ES cells were injected between the zona pellucida and aggregated cells.

With the availability of human ESC and PGC and iPSC cells (Thomson et al., 1998; Shamblott et al., 1998; Lowry and Quan, 2010), the potential of transplanting differentiated fetal cells for the purposes of regenerating damaged tissue (as in spinal cord injury, stroke, dementia, cardiomyopathy, or diabetes) has proven challenging. However, inroads are being made (Marais et al., 2011; Colozza et al., 2012; Borrell et al., 2012; Sasai, 2013; Pan et al., 2013; Matsunari et al., 2013; Peng et al., 2013).

II. Gene Targeting in ESCs

The vast majority of work being done with ESCs involves preplanned genetic modifications by homologous recombination between exogenous targeting DNA and an endogenous target gene. Multiple reviews on this topic provide considerable background (Baribault and Kemler, 1989; Capecchi, 1989; Mansour, 1990; Robertson, 1991; Koller and Smithies, 1992; Bradley et al., 1992; Soriano, 1995; Cheah and Behringer, 2000). Furthermore, countless genes have been modified in ESC repositories by the Knockout Mouse Project (KOMP) at U.C. Davis, Davis, CA (www.komp.org) and the International Knockout Mouse Consortium (IKMC) repository (www.knockoutmouse.org) in Munich, Germany (see also Chapter 24). Given the vast array of targeted mouse ESC clones available, one might ask why even consider gene targeting in mouse ESCs? There are a number of reasons why gene targeting will not be eliminated by the sizable gene targeting efforts cited above. First, not all genes have been targeted. Second, the largest number of targeted ESC clones in repositories are simple gene knockouts (KOs). Most of the conditional knockout (cKO) clones are at the IKMC and EUCOMM.

Those clones were targeted with the very well-designed KO first vector (Testa et al., 2004), which allows for the production of a KO or conditional KO (cKO) mouse. Both the KO and cKO actions of the IKMC clones are designed to produce null alleles. As gene discovery ensues, investigators often want to dissect their gene of interest by producing targeted alleles with single base pair polymorphisms (SNPs) that model known disease genes in the human population or with functional epitope deletions. These will be discussed in more detail below.

In this section, we will review some of the basic considerations that are to be taken into account when preparing to genetically engineer a gene through homologous recombination. Figure 4.3A and B shows the two general approaches which form the basis for most gene targeting schemes: replacement and insertion. The replacement or omega-type gene-targeting vector is very flexible. It can be used to produce insertions, deletions, or replacements of sequence such as changing the isoform of a gene. It can also be used to introduce site-specific recombinase sites, which are the hallmarks of conditional gene targeting.

A. Improving Gene-Targeting Efficiency

It has been reported that gene-targeting frequency is significantly higher when the target gene is actively expressed (Blackwell et al., 1986; Alt et al., 1986; Thomas and Rothstein, 1989; Nickoloff and Reynolds, 1990; Prado et al., 1997), suggesting the possibility that the increased accessibility to the chromosome DNA during transcription facilitates recombination.

Similarly, there are reports stating that gene targeting is highest when the cells are in M phase (Wong and Capecchi, 1987; Majumdar et al., 2003). These observations are the basis for many gene-targeting protocols recommending the feeding of ESC cultures 4 h before electroporation.

Lastly, the design of the gene-targeting vector, whether it be a replacement type or an insertion type, will affect the gene-targeting efficiency (Hasty et al., 1991b). Insertion-type vectors were shown to target the same gene at a nine-fold higher rate than replacement-type vectors, presumably because insertion-type gene-targeting vectors require only a single crossover. In spite of this advantage, only a small percentage of gene targetings have been done with insertion-type gene targeting.

Another approach to enhance the isolation of targeted clones is to decrease the survival of those ESCs with random integration of the gene-targeting vector. The most commonly used method is positive−negative selection (PNS) (Mansour et al., 1988). This scheme (Figure 4.3C) is similar to counter selection systems established in the 1980s for bacterial and yeast genetics, in which counter-selectable marker genes promote the death of the microorganism under appropriate culture conditions (reviewed by Reyrat et al., 1998). In the PNS scheme, it is assumed that nonhomologous integration events will insert the entire targeting vector and thus retain the negative selection gene, conferring sensitivity to either a drug (Mansour et al., 1988) or to the negative selection gene (Yu et al., 2000). In this case, the homologous recombination process will separate the negative selector gene DNA from the positive selection cassette. In most experiments, the neo^R gene is the positive

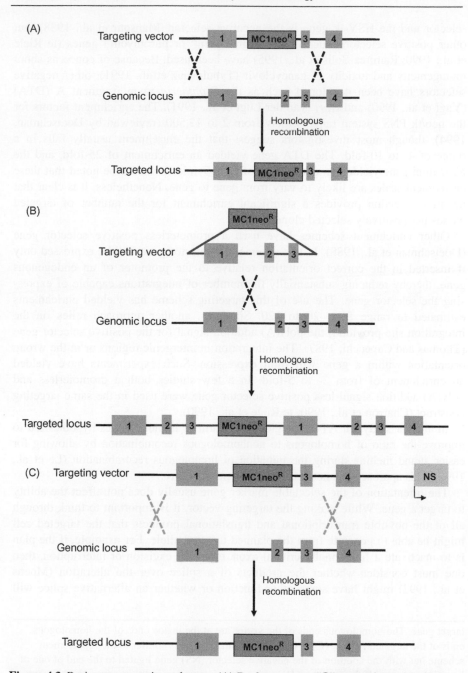

Figure 4.3 Basic gene-targeting schemes. (A) Replacement or "Ω"-type scheme in which the homologous arms of the targeting vector flank the marker gene (MC1neo^R). The homologous arms are oriented in the same direction as the genomic locus when the vector is linearized and their homologous ends are at the outside ends of the targeted region. (B) Insertional or "O"-type scheme in which the targeting vector is linearized within the

selector and the HSV-tk gene is the negative selector (Mansour et al., 1988), but other positive selectors, such as the hygromycin^R or puromycin^R genes (te Riele et al., 1990; Ramirez-Solis et al., 1995) have been used. Because of concerns about mutagenicity and toxicity of gancyclovir (Tybulewicz et al., 1991), other negative selectors have been developed such as the *Diphtheria* toxin fragment A (DTA) (Yagi et al., 1990) and *Hprt* (van der Lugt et al., 1991). The enrichment factors for the neo/tk PNS system range widely from 2 to 12,500 (reviewed by Doetschman, 1994), though most investigators suggest that the enrichment usually falls in a range of 4- to 10-fold. The DTA gene yielded an enrichment of 25-fold, and the *Hprt* mini gene yielded an enrichment of 6- to 7-fold. It should be noted that these enrichment scales are likely to vary from gene to gene. Nonetheless, it is clear that negative selection provides a significant enrichment for the number of targeted clones per positively selected clone.

Other enrichment schemes have used a promoterless positive selector gene (Doetschman et al., 1988). In this scheme, the selector gene will be expressed only if inserted in the correct orientation relative to the promoter of an endogenous gene, thereby reducing substantially the number of integrations capable of expressing the selector gene. The use of this targeting scheme has yielded enrichments estimated to range from 20 to 120. Similarly, another approach relies on the integration site providing the poly(A) addition signal for the positive selector gene (Thomas and Capecchi, 1987). The integration in intergenic regions or in the wrong orientation within a gene will limit expression. Such experiments have yielded an enrichment of from 2- to 5-fold. In a few studies both a promoterless and poly(A) addition signal-less positive selector gene were used in the same targeting construct (Charron et al., 1990; te Riele et al., 1990).

In another study, an AT-rich region in the arm of homology was hypothesized to improve the ratio of homologous to nonhomologous recombination by allowing for easier strand melting during the initiation of homologous recombination (Le et al., 1990). This approach has not become common procedure.

The orientation of the selectable marker gene usually does not affect the ability to target a gene. While making the targeting vector, it is important to think through all of the possible transcriptional and translational products that the targeted cell might be able to generate from the planned targeted allele. For example, if the plan is to inactivate a functionally critical exon, either by excision or interruption, then one must consider whether the products of a splice over the alteration (Moens et al., 1992) might have some other function or whether an alternative splice will

---

◀ target gene. The homologous ends of the vector are at the inside ends of the homologous ends of the targeted region. (C) PNS scheme. This scheme is similar to the replacement scheme but with the addition of the negative selector (NS) gene ligated to the end of one of homologous arms. If the vector undergoes random integration in the genome, the NS gene should remain intact. If the NS gene is a toxin gene, cell will be killed by the gene's stable integration and expression and subsequent expression. If the vector is inserted by homologous recombination, the NS gene will not become incorporated because it is outside of the region of homology and will be lost during homologous recombination.

render the transcript out of the reading frame. Potential fusion proteins between target and marker genes must also be considered, especially if the marker gene is promoterless or lacking a poly(A) addition signal.

It is extremely important to have the diagnostic tests for identifying targeted ESCs established before the gene targeting is done. Most drug-resistant ESC screening is done with the polymerase chain reaction (PCR). Ideally, this is done by having one primer located beyond the arm of homology in the flanking sequence and the other primer located in a unique sequence within the targeting vector such as the selection cassette. It is highly advisable to check both sides of the planned targeting event, because the resolution of recombination can be error prone. That is, it is not uncommon for us to identify clones that are targeted by PCR data from the 5′ side of the gene only to find that they have insertions or deletions when screened from the 3′ side of the recombination, and vice versa (Doetschman et al., 1988 and Sanford, personal observations).

B. Improving Gene-Targeting Vector Construction

Since the targeting vectors must contain sequences homologous to the target gene, these sequences ideally should be cloned from DNA that is "isogenic" to that of the ESCs being targeted. Early on, most ESC lines used for gene targeting were established from the strain 129 mouse. Now, C57Bl/6 ESCs lines are used widely. Some of the *Hprt* gene-targeting experiments in ESCs used isogenic DNA (Doetschman et al., 1987, 1988). When compared to similar experiments targeting the *Hprt* gene, but in which nonisogenic DNA was used (Thomas and Capecchi, 1987; Thompson et al., 1989), the targeting efficiencies were a minimum of five-fold higher when the targeting vector contained isogenic DNA. A later systematic study of the efficacy of isogenic DNA, carried out using the *Hprt* gene as the target, demonstrated that the use of isogenic DNA in the targeting vector increases targeting efficiency from four- to fivefold (Deng and Capecchi, 1992). Although the many substrains of 129 mice harbor a surprising degree of genetic diversity (Simpson et al., 1997), in general it does not appear to be necessary to match 129 substrains, though this would be beneficial for some loci. Some vectors produced from 129 strain DNA will not target certain loci in C57BL/6 strain no matter how many times the targeting is repeated, with no targeted clones out of >2500 screened drug-resistant ESC clones (Sanford, personal observations).

The size of homology in the targeting vector is another important consideration that has been systematically tested for its effect on targeting efficiency (Deng and Capecchi, 1992). It is generally advantageous to have from 4 to 10 kb of homology between targeting sequences and target gene, with no homologous arms <1 kb. Targeting experiments have been successful with less homology (Doetschman et al., 1988; Shull et al., 1992), but it is not advisable. Knowledge of the rate-limiting aspects of homologous recombination in mammalian cells would be useful for designing more efficient targeting constructs. Because targeting efficiency is not appreciably changed by the number of targeting molecules introduced into a cell (Thomas et al., 1986) or by the number of target loci per cell (Zheng and Wilson, 1990), it is likely that the cellular machinery for homologous recombination is rate limiting.

We have seen that a sequence in the targeting vector represented in multiple genomic loci such as repetitive sequence elements is less likely to recombine homologously with a specific target locus. This has been uncovered when attempting to target highly duplicated gene arrays (Sanford, personal observations). Therefore, choosing targeting fragments containing minimal repetitive sequences will improve efficiencies. Because intronic sequences often contain repetitive elements and usually constitute much of the homologous sequence in a targeting vector, large regions of homology may occasionally decrease targeting efficiency. An early review of homology versus targeting efficiency (Doetschman, 1994) suggested that >10 kb of homology can be detrimental presumably for the above reasons.

Some reports have argued that the use of very large gene-targeting vectors such as bacterial artificial chromosomes (BACs) not only improve gene-targeting efficiency but also eliminate DNA strain dependence (Valenzuela et al., 2003; Yang and Seed, 2003; Frendewey et al., 2010). We have always been concerned with the possibility of introducing additional mutations in microRNAs (miRs), the promoters, or the coding sequences of off-target genes in such a large vector. These mutations would be tightly linked to the intended mutation, virtually impossible to detect and could provide confounding phenotypes to the desired gene manipulation. Typically, BAC gene targeting is verified by Southern blots or real-time PCR; and neither of these methods will show the small mutations close to or large mutations distal to the gene of interest.

Furthermore, it has been our observation that well-designed short sequence gene-targeting vectors can routinely rival the targeting frequencies reported in the above citations and can be produced much more rapidly with today's technologies than in the recent past. There have been many advances contributing to rapid conventional vector construction that include the easy availability of genomic sequences for a wide variety of species (www.ensembl.org), vastly improved recombinant PCR polymerases, the commercial availability of inexpensive large synthetic DNA oligomers for immediate subcloning, and continued improvements in short-arm vector design. These advancements have enabled conventional vectors to occasionally exceed 50% targeting in the drug-resistant clones using vectors similar to the short-arm vectors described above (Sanford and Doetschman, personal observations; see also Chapter 17 for additional vector considerations).

C. Improving the Detection of Gene-Targeted ESC Clones

The screening of drug-resistant ESC candidate clones is done via PCR the vast majority of the time. We believe that failures to get gene-targeted clones from properly designed gene-targeting vectors are most often the result of poor screening methodology, resulting from a failure to make proper preparations for screening gene-targeted clones. Screening for the targeted locus should be inextricably intertwined with the vector design. Routinely we make two vectors for each gene-targeting project. The first is the actual gene-targeting vector and the second is the gene-targeting vector with 400–600 bp of additional sequence flanking both sides of the homologous sequences in the actual gene-targeting vector. These additional sequences are used to develop PCR screening strategies that use a primer outside of an arm of homology and a primer within the unique sequence of the targeted locus. These PCR screens are developed at 1, 0.5, and 0.1 copies of vector molecule per genome equivalent of ESC genomic DNA, so that the reaction is always done in the presence of excess genomic ESC DNA. PCR screening is done across both the 5′ and 3′ arms of homology. It is rare for ESC clones that provide expected bands from both 5′ and 3′ PCR screens to be untargeted by Southern blot analysis. We use only high-fidelity polymerases and frequently must use long-range PCR products to meet these requirements. When this type of screening and testing is worked out before electroporating ESCs, one's gene-targeting rates will be markedly improved.

While it is becoming less popular, Southern blots should always be done for at least two reasons: (i) to verify the gene-targeting event and (ii) to look for a second random insertion event. One of the diagnostic probes should be an "outside" probe (i.e., a probe that recognizes the target locus but binds outside of the region of homology between the targeting vector and target gene). This probe may contain intronic sequences where repetitive DNA may occur. We recommend avoiding these sequences by using software that identifies genomic repeats. The identification of a unique and useful outside probe is always valuable. Sometimes it is easier to alter the targeting construct to accommodate a good outside probe than it is to find a good probe that works with a given construct. A second probe should be directed toward unique sequence within the targeting vector. Blasting probes against the Ensembl mouse database will help determine the uniqueness of a Southern probe or a PCR primer.

With respect to verifying gene target events, there has been at least one report of small deletions occurring at one end of the targeting sequences (Doetschman et al., 1988), likely due to the short region of homology in that homologous arm (132 bp). One must be vigilant for such events because they are not rare (Reid et al., 1991; Sanford, personal observations). It would be unfortunate to ablate both the target gene and a neighboring gene or miR without knowing it and then report a phenotype, thinking it resulted solely from a single gene KO. This possibility is particularly pernicious when targeting gene clusters of very similar genes. Looking for a second random insertion event is essential because random integration can be mutagenic; the targeted allele is commonly detected via PCR directed against the selection cassette contained within the vector. Since the diagnostic PCR reactions used to identify the targeted locus are usually not used to screen the off-spring, failure to determine if a gene-targeted ESC clone contains a vector random integration could result in the loss of the targeted mouse line or the misidentification of targeted allele carriers from chimera breedings. In this regard, the use of electroporation is beneficial. Provided the amount of targeting DNA is not too high, usually only one integration site is present (Reid et al., 1991). However, one should always make this determination by using Southern blots.

D. Targeting Subtle Mutations into Genes

Gene targeting is being used more and more frequently to answer specific questions about gene function: questions concerning alternatively spliced gene products, alternative translational start products, the function of potentially significant polymorphic variants through the incorporation of subtle mutations in the targeted gene, and analysis of functional epitopes. One of the first targeting experiments introduced subtle changes into the target locus by simply incorporating those changes in one of the homologous arms (Doetschman et al., 1987). In the case of nonselectable target genes, this procedure would leave a positive selector gene in the targeted locus. Due to concerns over interference of the selector gene and its promoter with expression of a target gene (Jacks et al., 1994; Meyers et al., 1998), it has become common practice to remove the selection cassette using site-specific recombinases such as Cre and Flp site-specific recombinases. Additionally, since subtle gene modification requires sensitive phenotypic analysis, it is desirable that the final targeted allele contains only the subtle modifications without a marker gene.

The first approach of this type was termed "in—out" (Valancius and Smithies, 1991) or "hit and run" (Hasty et al., 1991a). These schemes use an insertional recombination vector containing positive and negative selection systems (either *Hprt* or *neo/tk*, respectively) and a subtle mutation in a homologous arm, resulting in a targeted allele with a duplication of target sequence. Spontaneous intrachromosomal recombination between the duplicated sequences removes the marker gene(s) and wild-type sequences, leaving the subtle mutation in the target locus. Negative selection allows the few cells with the loss of marker to survive.

Currently, the Cre-loxP or Flp-frt recombination systems are the method of choice for targeting subtle mutations into the genome (see Chapter 5 for detailed analysis of this methodology). Briefly, Cre recombinase mediates recombination between two 34-bp loxP sites. The loxP sites consist of 13-bp inverted repeats flanking an 8-bp sequence that determines the orientation of the recombination. If the 8-bp sequences are similarly oriented in two loxP sites, the sequences between those sites will be deleted; if the orientation is opposite, the sequences between the loxP sites will be inverted. Consequently, if loxP sites flank the marker gene and a subtle mutation is incorporated into one of the homologous arms, then the marker gene can be excised from the targeted locus by transient transfection with a Cre recombinase expression vector or by mating the targeted mouse with a transgenic mouse expressing Cre in the oocyte (Lewandoski et al., 1997) or zygote (Lasko et al., 1996). Many other types of modification, such as inversions and replacements (Torres and Kühn, 1997), translocations (van Deursen et al., 1995; Smith et al., 1995), large chromosomal deletions (Li et al., 1996; Lindsay et al., 1999), and the construction of mouse balancer chromosomes (Zheng et al., 1999) have been made using the Cre-loxP system.

For the introduction of SNPs or other small changes to a gene, a cleaner method is available. This approach uses a replacement-type targeting vector and is termed "tag and exchange" (Askew et al., 1993) or "double replacement" (Wu et al., 1994). It places both positive and negative selector gene(s) between the two homologous arms. Targeting with this vector replaces wild-type sequences, typically an entire

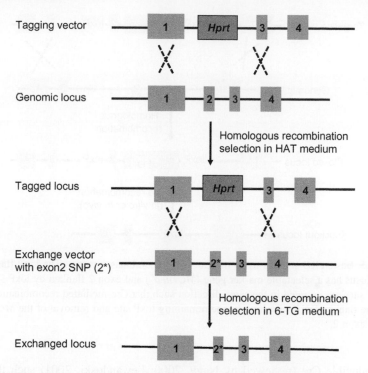

Figure 4.4 This scheme depicts the compound targeting that is required to introduce subtle mutations into a gene. This process begins with $Hprt^{-/-}$ ESCs, which can't grow in medium containing hypoxanthine, aminopterin, and thymidine (HAT). The targeting vector is designed to replace the exon of interest with the $Hprt$ mini gene; the presence of this gene in the targeted ESC clones permits growth in HAT medium. To exchange the $Hprt$ mini gene for the desired mutation, a second gene targeting is conducted to replace $Hprt$ with the altered exon. Cells expressing HPRT are killed in the presence of 6-TG (6-thioguanine). The second step is very efficient and shows a very high targeting rate due to the power of this negative selection. This tag and exchange targeting leaves no extraneous DNA sequences in the gene.

exon, with a selectable gene, thus knocking out the gene and tagging it with a nega-tively and positively selectable marker gene. In a second targeting step, the marker gene is exchanged with the original genomic sequences containing the subtle muta-tion. Negative selection in turn kills those cells in which the exchange reaction does not take place and has the advantage that it leaves no extraneous sequences (such as loxP sites) in the targeted locus (Figure 4.4). This method has been used to generate $Fgf2$ isoform-specific KO mice (Garmy-Susini et al., 2004; Azhar et al., 2009).

E. Conditional Gene Targeting

The most common type of gene targeting being done since the turn of the century is conditional gene targeting to produce cKOs. These involve using an inducible

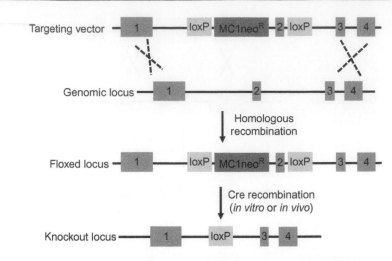

Figure 4.5 Basic cKO scheme using the Cre-loxP recombinase system. The loxP-flanked (floxed) locus has a selectable marker gene ($MC1neo^R$) and exon 2 flanked by loxP sites. The loxP sites are oriented in the same direction such that Cre-mediated recombination deletes the flanked DNA, resulting in one remaining loxP site and removal of the $MC1neo^R$ gene and exon 2.

or noninducible Cre (reviewed by Nagy, 2000; Lewandoski, 2001) such that Cre is expressed either spatially (tissue restricted) or temporally (activated by an exogenous agent) or both. Fundamentally, conditional gene targeting involves the flanking of an essential portion of a gene with loxP sites, resulting in a "floxed" allele (Figure 4.5). When a Cre gene is crossed into a mouse with a floxed allele, the resulting mice are called "deleter" mice. Deleter mice commonly express Cre in one of four ways: ubiquitous expression, tissue-specific expression, inducible expression, or a combination thereof. Inducible Cre proteins are fusion proteins between the estrogen receptor and the Cre recombinase that are activated by RU486 or tamoxifen (Feil et al., 1997; Wunderlich et al., 2001). The Jackson Labs has a sizable collection of these mice, including both ubiquitous and tissue-specific Cre expression.

Another Cre-based inducible system is the loxP-stop-loxP technology (Safran et al., 2003). Basically, one or more floxed poly(A) addition signals are placed in the 5′ untranslated sequence or in an early intron. If the loxP sites are in the same orientation, the expression of Cre will remove the stop signal usually in a tissue or inducible manner. An alternative to this is to orient asymmetric loxP sites pointing toward each other. The expression of Cre will cause a one-way inversion of the sequence between them. Depending on the orientation of the poly(A) signal(s), the expression of Cre can either turn a gene on or off. There are also asymmetric frt sites and an inducible Flp recombinase. A combination of the two systems could allow one to turn a gene turn off and back on again, or vice versa. These conditional methodologies will be covered in considerable detail in Chapter 5.

F. Targeted Transgenes

The targeted transgene idea is not a new concept (Bouhassira et al., 1997). It is basically the insertion of a single copy of a transgene either by recombinase cassette exchange or by gene targeting into a specific location in the genome. Two popular locations in the mouse are the *Hprt* and the *Thumpd3* (Rosa26) loci. These locations have been selected due to the near ubiquitous expression of the two genes, their nonessential function in the mouse, and the favorable gene expression environment of open chromatin structure associated with them. Targeted transgenesis has several sizable advantages over conventional transgenesis: gene expression is not variable between founders since all are identical; different isoforms of the same transgene can be compared because the single copy format provides equivalent gene expression; and it is readily reproducible. The production of a mouse line with an asymmetric loxP site in the 5′ untranslated regions of either of these genes can be used to make single copy transgenic mice via the pronuclear injection of a Cre recombinase-expressing vector and an asymmetric loxP-flanked transgene. The Cre recombinase mediates a single, nonrepeatable insertion of the transgene into the zygote's loxP site. Some transgene dosage studies could be done by breeding the mice to homozygosity if the *Thumpd3* gene contains the loxP site. The *Hprt* gene is X-linked and it can also serve as either a positive or negative selection system.

G. Gene Knockdowns

Since the discovery of short inhibitory RNA (siRNA) or miRs, the promise of gene knockdowns has been considered feasible. Early attempts at causing gene knockdowns by placing gene-complementary RNAs in the culture medium were frequently unsuccessful but became feasible if expressed as a transgene (Kunath et al., 2003). That paper reported that a wide range of gene expression mouse lines could be obtained as a byproduct of the uniqueness of transgenic insertion sites. These gene knockdowns could also be conditional (Chang et al., 2004; Yu and McMahon, 2006).

H. The Future of Gene Targeting (Homing Endonucleases)

A major advancement in gene targeting is emerging in the form of homing endonucleases. Three different methodologies have been successfully used to make genetic modifications in mice (reviewed by Gaj et al., 2013). The basic concept here is that the homing endonuclease produces a double-stranded cut in the DNA. Double-stranded cuts have been reported to stimulate homologous recombination by orders of magnitude (Smih et al., 1995). These approaches are zinc finger-FokI based, TALENS, and CRISPR/Cas-based. All of these approaches are designed to make a double-stranded break in or near the desired modification site. A coinjected gene-targeting DNA is able to undergo homologous recombination at the site in the double strand break at high frequency. These technologies can be used with pronuclear injection of zygotes to produce direct founders. The promise of these technologies is in the elimination of ESCs as an intermediate in gene-targeting experiments, the elimination of chimera breeding in gene-targeting projects, a rapid turnaround time for failed injections, and dramatically reduced costs for making genetically modified mice. CRISPRs have been reported to modify both alleles at frequencies up to 25% of the time in iPSCs (Deng et al., 2013). Compared to gene targeting in ESCs, these technologies are still in the very early days of development. However, the future application of these technologies is vast indeed.

III. Methods

The methods provided here are those used in our laboratories and are not to be considered the only techniques that have been successfully used. Other approaches are often used and can be found in the individual papers cited.

A. ESC Culture

The ability of ES cells to remain undifferentiated in cell culture is remarkable, as cells cultured continuously for as long as 5 months have been shown to colonize the germ line (Suda et al., 1987). However, to ensure germ line competency of ES cells, it is best to use the lowest passage ES cells possible. Like ES cells, mouse EG cells, when removed from their LIF-producing feeder layers, spontaneously differentiate into embryoid bodies and form tumors in nude mice (Matsui et al., 1992). Because PGCs can colonize the germ line, there is little reason to believe that their *in vitro* differentiation potential is any different than that of ES cells, unless the state of imprinting through differential allelic methylation affects *in vitro* differentiation.

1. Fetal Calf Serum Plus KOSR

ESC-grade fetal bovine serum (FBS) can be purchased from serum vendors. These lots of serum have been screened for their ability to support the undifferentiated state of ESCs in culture. It should be noted that ESC-grade sera are not necessarily optimal for differentiation culture. For differentiation culture, obtain aliquots from different lots of ESC-grade serum and have each vendor set aside the amount of serum you would want to purchase. Test each aliquot for its ability to support the type of differentiation you are interested in. When the test is done, purchase from the best lot and notify the vendors to release their hold on the other lots.

2. Culture Medium (ESC Medium)

Use 10% FBS (heat inactivated at 55°C for 30 min), 10% KOSR (Invitrogen) in high glucose containing DMEM with bicarbonate and 0.4 mM Glutamax (Life Technologies), nonessential amino acids (Life Technologies) 1 μM β-mercaptoethanol

(Sigma-Aldrich), 10 mg/mL insulin (Sigma-Aldrich), 1 μM PD0325901, and 3 μM CHIR99021 and 1 mM LIF (Sigma-Aldrich). For ESC expansion or single-cell subcloning, we use the above medium plus 0.8 μM PD184352. Cells are incubated in a rapid humidity- and CO_2-recovery incubator set to 90% humidity and 7.5% CO_2.

3. Passaging ESCs

Every two to three days, passage according to standard procedures using the same 0.25% trypsin/EDTA w/o Ca^{2+} and Mg^+ solution used for embryonic fibroblast preparation (see III. Section D). It is best to rinse the cultures three times in HBSS (Hank's balanced salt solution) before adding trypsin/EDTA. Cover the dish or well with the trypsin/EDTA and remove it. Place the culture in an incubator for 5 min, then suspend the cells in complete ESC culture medium and plate 1/10 to 1/20 of the cells onto a new feeder layer of cells. Try to keep about 10−20 ESC colonies per field (10× objective). Passage should be done before the ESC colonies express differentiated endoderm cells at their periphery.

4. Freezing and Thawing ESCs

Feed cells 4 h before freezing. Standard slow-freeze/rapid-thaw techniques work. Final concentration of DMSO (O.D. 275 < 0.3) in freezing medium is 10%. Freezing medium is made with the appropriate concentration of DMSO in FBS. If the ESCs are in a small volume of medium as they are added to the freezing medium, the DMSO concentration of the freezing medium must be increased to adjust for the dilution so that the final DMSO concentration is 10%. Freeze overnight in −80°C in Mr. Frosty freezing containers. Then transfer to vapor phase liquid nitrogen storage.

B. Differentiation Culture

This procedure is essentially the same as that described in (Doetschman et al., 1985).

1. Sedimentation to Remove Feeder Cells

Trypsinize cells in the late afternoon and allow to sediment 30 min in original dish (fibroblasts sediment preferentially). Remove unattached cells, put in a new gelatin-coated tissue culture dish, and let stand an additional hour. Remove nonattached cells (nearly all of these cells are now ESCs because most of the fibroblasts have attached). Count the ESCs that remained in suspension and plate in standard ES culture medium at about 500,000 cells per 100-mm bacterial culture dish for differentiation in suspension or on tissue culture dishes for substrate-attached differentiation.

2. Suspension Differentiation Culture

The method provided here should be seen as a standard procedure. To obtain enhancement of differentiation into specific lineages, these conditions should be changed according to the appropriate literature. Change medium (DMEM, high glucose plus beta-mercaptoethanol, Glutamax, nonessential amino acids, and 15% FBS) every 2 days of culture until embryonic ectoderm appears, sometime between 4 and 6 days of culture (Figure 4.2C and D). When ectoderm appears, switch to 20% FBS (other components as before). From this point on, change the medium every 2 days, but add one-half the volume of medium on the alternative days. This is necessary to prevent exhaustion of the medium, and it allows much better differentiation. Embryoid bodies with beating heart tissue should be observed at 8–12 days. Shortly thereafter, the embryoid bodies should become cystic (Figure 4.2E and F), with some containing blood islands. (*Note*: In some bacterial dishes, the embryoid bodies still attach to the substrate. When this happens, gently pipette them off the plates and passage them to fresh plates.)

C. Electroporation of ESCs

1. Buffer

This procedure is essentially the same as that previously described in detail (Doetschman et al., 1988). Electroporate 2 days after passage and 4 h after a medium change to ensure that the cells are growing as rapidly as possible before electroporation. This medium does not hinder the transfection or targeting efficiency and prevents unnecessary stress on the cells.

2. Electroporation Parameters

Electroporation at room temperature or on ice works well. Use from 10^7 to 10^8 cells per milliliter of culture medium. Use no more than 80 nM of targeting DNA fragment to minimize random integration of concatemers. Use 625 V/cm and 500 μF; this will yield about 50% survival. Leave the cells undisturbed for 5 min at room temperature and then distribute to 5–10 100-mm culture dishes with drug-resistant feeders. Begin drug selection after 24 h.

D. Preparation of Primary Mouse Fibroblasts

Under sterile conditions, remove embryos 13.5–14.5 days old (day of plug is 0.5 dpc). Discard liver and heart, rinse in HBSS to remove as much blood as possible, and tease embryos apart with forceps. Incubate in sufficient trypsin 0.25% plus EDTA, without calcium or magnesium to cover the tissue. Incubate for 10 min at 37°C. Gently run the softened tissue up and down in a sterile 25 mL serological pipette to release the cells. Transfer the cells to a sterile centrifuge tube and spin the cells down at $600 \times g$ for 5 min. Suspend the cells in DMEM plus 10% FBS and plate them in gelatin-treated 100 mm tissue culture plates at one embryo per two plates. This is passage zero. When the plates are confluent, freeze down each plate into two cryovials in FBS plus 9% DMSO. Thaw each vial onto two 100 mm plates. This is passage 2. Split at 1:2. More dilute splitting is possible, but may inhibit growth.

MEFs are suitable for use beginning at passage 3 but can be passaged to passage 6–12 before senescing. This is highly dependent on the split ratios.

When ready to use the cells as feeders, allow the plates to become heavily confluent, treat with mitomycin C (10 μg/mL) for 1 h (or irradiation at 3000 rads) in ESC medium. If mitomycin C is used to block feeder cell growth, the cells should receive two rinses to ensure that no trace amount of mitomycin C remains to affect the ESCs.

The cells can also be passaged once more after mitomycin C treatment or irradiation. However, because they will not grow, they must be plated at a confluent density: about 6×10^5 cells per 60 mm tissue culture dish or $1.5-2 \times 10^6$ cells per 100 mm dish. This method is preferred because the passaging selects against poorly metabolizing or dying cells, resulting in a healthier feeder layer.

E. Screening for Targeted Colonies

Wash plates containing selectable marker-resistant clones twice in DMEM and incubate in 10% KOSR medium without FBS in DMEM (this medium is free of trypsin inhibitor). Pick each $G418^R$ clone individually in 7 μL and place in 96-well plate containing 50 μL of KOSR. When done picking, add 10 μL of 0.25% Trypsin/EDTA solution. Incubate 10 min at 37°C, resuspend, and plate each resuspended clone in 0.5 mL ES cell medium (containing selection agent; any colonies that were not truly selection resistant will be eliminated by continued drug selection) in each well of a 24-well tray preseeded with inactivated drug-resistant feeder cells (one 100 mm plate of feeder cells to one 24-well tray). When ESCs have grown out (~4 days), rinse each well twice with Hank's balanced salt solution. Then incubate in 150 μL 0.25% Trypsin/EDTA at 37°C for 5 min. Add 0.5 mL freezing medium (13% DMSO in FBS), resuspend, freeze 0.5 mL of resuspension (final DMSO concentration will now be about 10%) and add remainder to fresh well of a 24-well tray preplated with feeders (ES cell medium without selective pressure). Identify each well with its respective frozen clone. Change medium the next day to wash out the residual DMSO. When these clones have expanded, remove medium and lyse in proteinase K buffer. The DNA from these samples is used to identify targeted clones by either PCR or Southern blots.

F. Estimated Size of Mouse Colony

This procedure provides an estimate of the mice that one would need to maintain a minimum-sized, self-contained mouse colony to support a blastocyst injection operation for a small lab. The first section contains procedures and the second section contains actual estimates for the number of animals needed. The numbers are given as a daily census. If your animals are counted in terms of cages rather than animals, our mouse colonies average 2.5 mice per cage in cages that hold a maximum of 4 adult mice.

1. Blastocyst Production for Blastocyst Injection Procedure

In order to minimize the costs of purchasing mice, a breeding colony capable of producing all of the experimental animals needed should be established and maintained. Assuming three blastocyst injection experiments a week (each experiment requiring 16 superovulated C57BL/6 females), 200 4- to 6-week-old females must be produced monthly, necessitating a breeding colony of about 40 paired males and females. We estimate a census of 600 mice to maintain this level of blastocyst production. Experienced personnel will be able to inject the blastocysts from 8 to 16 superovulated blastocyst donor mice. The number depends on a variety of factors, including shipping method, delay before using the mice, the quality of the superovulation hormones, and how mouse-friendly your mouse rooms are.

2. Pseudopregnant Females

On average, six pseudopregnant females per experiment will be used for blastocyst transfer. With high proficiency of the microinjectionist and excellent blastocyst production, this number can increase. Forty vasectomized males and 120 ICR females will be required at any given time to ensure that enough pseudopregnant females are available for three to four injection sessions per week. There will be an estimated turnover of about 21 females weekly. The pseudopregnant females that are not used will be recycled for pseudopregnancy after 2 weeks. We anticipate an average census of 150 females and 40 vasectomized males for the production of pseudopregnant females.

3. Embryos Required for Embryonic Fibroblast Production

We routinely prepare embryonic fibroblasts about once a year. Embryos from four litters are prepared each time. This requires an average yearly census of about four males and four females to ensure maintenance of the strain. We use 129/Sv mice for feeder cells because they usually do quite well as feeders, though we have used other strains with fairly good success. Because we also prepare embryonic fibroblasts from neo^R-expressing mice, we also maintain a small colony of transgenic neo^R mice (strain FVB/N). We maintain an average census of about 16 mice for feeder cell production.

4. Establish ESC Lines

Initially, a small colony of about 4 males and 16 females, with the females turning over at about one per week, should be sufficient for establishing ESC lines and for maintaining this small colony. To do so, and including the breeding colony necessary to replenish the females, will require about 30 animals.

5. Breeding Test for Germ Line Chimerism

As soon as the blastocyst injections start yielding chimeric animals, the animals will have to be mated with normal C57BL/6 females of a coat color that is recessive to that of the strain from which the ESCs were derived to test whether the targeted ESCs that were injected have colonized the germ line. The offspring of germ line chimeras will then be used to establish a breeding colony and produce experimental animals from that strain. Although it is difficult to estimate the number of animals that will be housed for these purposes, we estimate an additional 100 animals for each genetically modified strain.

6. Total Mouse Colony Census Requirement

About 1500 mice will be needed to establish a base operation as outlined above. From each germ line chimera, an additional 100 mice will be needed for breeding and experimental purposes.

References

Alt, F.W., Blackwell, T.K., DePinho, R.A., Reth, M.G., Yancopoulos, G.D., 1986. Regulation of genome rearrangement events during lymphocyte differentiation. Immunol. Rev. 89, 5−30.

Askew, G.R., Doetschman, T., Lingrel, J.B., 1993. Site-directed point mutations in embryonic stem cells: a gene-targeting tag-and-exchange strategy. Mol. Cell. Biol. 13, 4115−4124.

Aubert, J., Dessolin, S., Belmonte, N., Li, M., McKenzie, F.R., Staccini, L., et al., 1999. Leukemia inhibitory factor and its receptor promote adipocyte differentiation via the mitogen-activated protein kinase cascade. J. Biol. Chem. 274, 24965−24972.

Axelrod, H.R., 1984. Embryonic stem cell lines derived from blastocysts by a simplified technique. Dev. Biol. 101, 225−228.

Azhar, M., Yin, M., Zhou, M., Li, H., Mustafa, M., Nusayr, E., et al., 2009. Gene targeted ablation of high molecular weight fibroblast growth factor-2. Dev. Dyn. 238, 351−357.

Baribault, H., Kemler, R., 1989. Embryonic stem cell culture and gene targeting in transgenic mice. Mol. Biol. Med. 6, 481−492.

Beddington, R.S., Robertson, E.J., 1989. An assessment of the developmental potential of embryonic stem cells in the midgestation mouse embryo. Development 105, 733−737.

Benninger, Y., Marino, S., Hardegger, R., Weissmann, C., Aguzzi, A., Brandner, S., 2000. Differentiation and histological analysis of embryonic stem cell-derived neural transplants in mice. Brain Pathol. 10, 330−341.

Blackwell, T.K., Moore, M.W., Yancopoulos, G.D., Suh, H., Lutzker, S., Selsing, E., et al., 1986. Recombination between immunoglobulin variable region gene segments is enhanced by transcription. Nature 324, 585−589.

Borrell, V., Cardenas, A., Ciceri, G., Galceran, J., Flames, N., Pla, R., et al., 2012. Slit/Robo signaling modulates the proliferation of central nervous system progenitors. Neuron 76, 338−352.

Bouhassira, E.E., Westerman, K., Leboulch, P., 1997. Transcriptional behavior of LCR enhancer elements integrated at the same chromosomal locus by recombinase-mediated cassette exchange. Blood 90, 3332−3344.

Bradley, A., Robertson, E., 1986. Embryo-derived stem cells: a tool for elucidating the developmental genetics of the mouse. Curr. Top. Dev. Biol. 20, 357−371.

Bradley, A., Evans, M., Kaufman, M.H., Robertson, E., 1984. Formation of germ-line chimaeras from embryo-derived teratocarcinoma cell lines. Nature 309, 255−256.

Bradley, A., Hasty, P., Davis, A., Ramirez-Solis, R., 1992. Modifying the mouse: design and desire. Bio-Technology (New York) 10, 534−539.

Burdon, T., Stracey, C., Chambers, I., Nichols, J., Smith, A., 1999. Suppression of SHP-2 and ERK signalling promotes self-renewal of mouse embryonic stem cells. Dev. Biol. 210, 30−43.

Cannon, J.P., Colicos, S.M., Belmont, J.W., 1999. Gene trap screening using negative selection: identification of two tandem, differentially expressed loci with potential hematopoietic function. Dev. Genet. 25, 49−63.

Capecchi, M.R., 1989. The new mouse genetics: altering the genome by gene targeting. Trends Genet. 5, 70−76.

Chang, H.S., Lin, C.H., Chen, Y.C., Yu, W.C., 2004. Using siRNA technique to generate transgenic animals with spatiotemporal and conditional gene knockdown. Am. J. Pathol. 165, 1535−1541.

Charron, J., Malynn, B.A., Robertson, E.J., Goff, S.P., Alt, F.W., 1990. High-frequency disruption of the N-myc gene in embryonic stem and pre-B cell lines by homologous recombination. Mol. Cell. Biol. 10, 1799−1804.

Cheah, S.S., Behringer, R.R., 2000. Gene-targeting strategies. Methods Mol. Biol. 136, 455−463.

Chen, U., Mok, H., 1995. Development of mouse embryonic stem (ES) cells. IV. Differentiation to mature T and B lymphocytes after implantation of embryoid bodies into nude mice. Dev. Immunol. 4, 79−84.

Cheng, J., Dutra, A., Takesono, A., Garrett-Beal, L., Schwartzberg, P.L., 2004. Improved generation of C57BL/6J mouse embryonic stem cells in a defined serum-free media. Genesis 39, 100−104.

Colozza, G., Locker, M., Perron, M., 2012. Shaping the eye from embryonic stem cells: biological and medical implications. World J. Stem Cells 4, 80−86.

Conlon, F.L., Barth, K.S., Robertson, E.J., 1991. A novel retrovirally induced embryonic lethal mutation in the mouse: assessment of the developmental fate of embryonic stem cells homozygous for the 413.d proviral integration. Development 111, 969−981.

Deng, C., Capecchi, M.R., 1992. Reexamination of gene targeting frequency as a function of the extent of homology between the targeting vector and the target locus. Mol. Cell Biol. 12, 3365−3371.

Deng, L., Garrett, R.A., Shah, S.A., Peng, X., She, Q., 2013. A novel interference mechanism by a type IIIB CRISPR-Cmr module in Sulfolobus. Mol. Microbiol. 87, 1088−1099.

Dinkel, A., Aicher, W.K., Warnatz, K., Burki, K., Eibel, H., Ledermann, B., 1999. Efficient generation of transgenic BALB/c mice using BALB/c embryonic stem cells. J. Immunol. Methods 223, 255−260.

Doetschman, T., 1994. Gene transfer in embryonic stem cells. In: Pinkert, C.A. (Ed.), Transgenic Animal Technology: A Laboratory Handbook. Academic Press, New York, NY, pp. 115−146.

Doetschman, T., Gregg, R.G., Maeda, N., Hooper, M.L., Melton, D.W., Thompson, S., et al., 1987. Targetted correction of a mutant HPRT gene in mouse embryonic stem cells. Nature 330, 576−578.

Doetschman, T., Maeda, N., Smithies, O., 1988. Targeted mutation of the Hprt gene in mouse embryonic stem cells. Proc. Natl. Acad. Sci. USA 85, 8583−8587.

Doetschman, T.C., Eistetter, H., Katz, M., Schmidt, W., Kemler, R., 1985. The in vitro development of blastocyst-derived embryonic stem cell lines: formation of visceral yolk sac, blood islands and myocardium. J. Embryol. Exp. Morphol. 87, 27−45.

Drab, M., Haller, H., Bychkov, R., Erdmann, B., Lindschau, C., Haase, H., et al., 1997. From totipotent embryonic stem cells to spontaneously contracting smooth muscle cells: a retinoic acid and db-cAMP *in vitro* differentiation model. FASEB J. 11, 905−915.

Eggan, K., Akutsu, H., Loring, J., Jackson-Grusby, L., Klemm, M., Rideout III, W.M., et al., 2001. Hybrid vigor, fetal overgrowth, and viability of mice derived by nuclear cloning and tetraploid embryo complementation. Proc. Natl. Acad. Sci. USA 98, 6209−6214.

Evans, M.J., Kaufman, M.H., 1981. Establishment in culture of pluripotential cells from mouse embryos. Nature 292, 154−156.

Fedorov, L.M., Haegel-Kronenberger, H., Hirchenhain, J., 1997. A comparison of the germline potential of differently aged ES cell lines and their transfected descendants. Transgenic Res. 6, 223−231.

Feil, R., Wagner, J., Metzger, D., Chambon, P., 1997. Regulation of Cre recombinase activity by mutated estrogen receptor ligand-binding domains. Biochem. Biophys. Res. Commun. 237, 752−757.

Frendewey, D., Chernomorsky, R., Esau, L., Om, J., Xue, Y., Murphy, A.J., et al., 2010. The loss-of-allele assay for ES cell screening and mouse genotyping. Methods Enzymol. 476, 295−307.

Gaj, T., Gersbach, C.A., Barbas III, C.F., 2013. ZFN, TALEN, and CRISPR/Cas-based methods for genome engineering. Trends Biotechnol. 31, 397−405.

Garmy-Susini, B., Delmas, E., Gourdy, P., Zhou, M., Bossard, C., Bugler, B., et al., 2004. Role of fibroblast growth factor-2 isoforms in the effect of estradiol on endothelial cell migration and proliferation. Circ. Res. 94, 1301−1309.

George, S.H., Gertsenstein, M., Vintersten, K., Korets-Smith, E., Murphy, J., Stevens, M.E., et al., 2007. Developmental and adult phenotyping directly from mutant embryonic stem cells. Proc. Natl. Acad. Sci. USA 104, 4455−4460.

Gertsenstein, M., Nutter, L.M., Reid, T., Pereira, M., Stanford, W.L., Rossant, J., et al., 2010. Efficient generation of germ line transmitting chimeras from C57BL/6N ES cells by aggregation with outbred host embryos. PLoS One 5, e11260.

Ginsburg, M., Snow, M.H., McLaren, A., 1990. Primordial germ cells in the mouse embryo during gastrulation. Development 110, 521−528.

Gossler, A., Doetschman, T., Korn, R., Serfling, E., Kemler, R., 1986. Transgenesis by means of blastocyst-derived embryonic stem cell lines. Proc. Natl. Acad. Sci. USA 83, 9065−9069.

Gossler, A., Joyner, A.L., Rossant, J., Skarnes, W.C., 1989. Mouse embryonic stem cells and reporter constructs to detect developmentally regulated genes. Science 244, 463−465.

Handyside, A.H., O'Neill, G., Jones, M., Hooper, M.L., 1989. Use of BRL-conditioned medium in combination with feeder layers to isolate a diploid embryonal stem cell line. Roux's Arch. Dev. Biol. 198, 48−56.

Hanna, J., Wernig, M., Markoulaki, S., Sun, C.W., Meissner, A., Cassady, J.P., et al., 2007. Treatment of sickle cell anemia mouse model with iPS cells generated from autologous skin. Science 318, 1920−1923.

Hasty, P., Ramirez-Solis, R., Krumlauf, R., Bradley, A., 1991a. Introduction of a subtle mutation into the Hox-2.6 locus in embryonic stem cells (published erratum appears in Nature 1991 353(6339): 94). Nature 350, 243−246.

Hasty, P., Rivera-Perez, J., Chang, C., Bradley, A., 1991b. Target frequency and integration pattern for insertion and replacement vectors in embryonic stem cells. Mol. Cell. Biol. 11, 4509−4517.

Jacks, T., Shih, T.S., Schmitt, E.M., Bronson, R.T., Bernards, A., Weinberg, R.A., 1994. Tumour predisposition in mice heterozygous for a targeted mutation in Nf1. Nat. Genet. 7, 353−361.

Kanda, A., Sotomaru, Y., Shiozawa, S., Hiyama, E., 2012. Establishment of ES cells from inbred strain mice by dual inhibition (2i). J. Reprod. Dev. 58, 77−83.

Kaufman, M.H., Robertson, E.J., Handyside, A.H., Evans, M.J., 1983. Establishment of pluripotential cell lines from haploid mouse embryos. J. Embryol. Exp. Morphol. 73, 249−261.

Kawakami, M., Ishikawa, H., Tachibana, T., Tanaka, A., Mataga, I., 2013. Functional transplantation of salivary gland cells differentiated from mouse early ES cells in vitro. Hum. Cell 26, 80−90.

Kim, E.M., Manzar, G., Zavazava, N., 2013. Mouse ES cell-derived hematopoietic progenitor cells. Methods Mol. Biol. 1029, 109−117.

Kitani, H., Takagi, N., Atsunmi, T., Kawakura, K., Goto, S., Husakabe, M., et al., 1996. Isolation of a hermline-transmissable embryonic stem (ES) cell line from C3H/He mice. Zool. Sci. 13, 865−871.

Klinz, F., Bloch, W., Addicks, K., Hescheler, J., 1999. Inhibition of phosphatidylinositol-3-kinase blocks development of functional embryonic cardiomyocytes. Exp. Cell Res. 247, 79−83.

Klug, M.G., Soonpaa, M.H., Koh, G.Y., Field, L.J., 1996. Genetically selected cardiomyocytes from differentiating embronic stem cells form stable intracardiac grafts. J. Clin. Invest. 98, 216−224.

Koller, B.H., Smithies, O., 1992. Altering genes in animals by gene targeting. Annu. Rev. Immunol. 10, 705−730.

Kontgen, F., Stewart, C.L., 1993. Simple screening procedure to detect gene targeting events in embryonic stem cells. Methods Enzymol. 225, 878−890.

Kramer, J., Hegert, C., Guan, K., Wobus, A.M., Muller, P.K., Rohwedel, J., 2000. Embryonic stem cell-derived chondrogenic differentiation in vitro: activation by BMP-2 and BMP-4. Mech. Dev. 92, 193−205.

Kunath, T., Gish, G., Lickert, H., Jones, N., Pawson, T., Rossant, J., 2003. Transgenic RNA interference in ES cell-derived embryos recapitulates a genetic null phenotype. Nat. Biotechnol. 21, 559−561.

Labosky, P.A., Barlow, D.P., Hogan, B.L., 1994. Mouse embryonic germ (EG) cell lines: transmission through the germline and differences in the methylation imprint of insulin-like growth factor 2 receptor (Igf2r) gene compared with embryonic stem (ES) cell lines. Development 120, 3197−3204.

Lallemand, Y., Brulet, P., 1990. An in situ assessment of the routes and extents of colonisation of the mouse embryo by embryonic stem cells and their descendants. Development 110, 1241−1248.

Lasko, M., Pichel, J.G., Gorman, J.R., Sauer, B., Okamoto, Y., Lee, E., et al., 1996. Efficient in vivo manipulation of mouse genomic sequences at the zygote stage. Proc. Natl. Acad. Sci. USA 93, 5860−5865.

Lawson, K.A., Hage, W.J., 1994. Clonal analysis of the origin of primordial germ cells in the mouse. Ciba Found. Symp. 182, 68−84.

Le, M.H., Lallemand, Y., Brulet, P., 1990. Targeted replacement of the homeobox gene Hox-3.1 by the Escherichia coli lacZ in mouse chimeric embryos. Proc. Natl. Acad. Sci. USA 87, 4712−4716.

Lewandoski, M., 2001. Conditional control of gene expression in the mouse. Nat. Rev. Genet. 2, 743−755.

Lewandoski, M., Wassarman, K.M., Martin, G.R., 1997. Zp3-cre, a transgenic mouse line for the activation or inactivation of loxP-flanked target genes specifically in the female germ line. Curr. Biol. 7, 148–151.

Li, Z.W., Stark, G., Gotz, J., Rulicke, T., Gschwind, M., Huber, G., et al., 1996. Generation of mice with a 200-kb amyloid precursor protein gene deletion by Cre recombinase-mediated site-specific recombination in embryonic stem cells (published erratum appears in Proc. Natl. Acad. Sci. USA 1996 93(21):12052). Proc. Natl. Acad. Sci. USA 93, 6158–6162.

Lindsay, E.A., Botta, A., Jurecic, V., Carattini-Rivera, S., Cheah, Y.C., Rosenblatt, H.M., et al., 1999. Congenital heart disease in mice deficient for the DiGeorge syndrome region. Nature 401, 379–383.

Lowry, W.E., Quan, W.L., 2010. Roadblocks en route to the clinical application of induced pluripotent stem cells. J. Cell Sci. 123, 643–651.

Magnuson, T., Epstein, C.J., Silver, L.M., Martin, G.R., 1982. Pluripotent embryonic stem cell lines can be derived from tw5/tw5 blastocysts. Nature 298, 750–753.

Majumdar, A., Puri, N., Cuenoud, B., Natt, F., Martin, P., Khorlin, A., et al., 2003. Cell cycle modulation of gene targeting by a triple helix-forming oligonucleotide. J. Biol. Chem. 278, 11072–11077.

Mann, J.R., Gadi, I., Harbison, M.L., Abbondanzo, S.J., Stewart, C.L., 1990. Androgenetic mouse embryonic stem cells are pluripotent and cause skeletal defects in chimeras: implications for genetic imprinting. Cell 62, 251–260.

Mansour, S.L., 1990. Gene targeting in murine embryonic stem cells: introduction of specific alterations into the mammalian genome. Genet. Anal. Tech. Appl. 7, 219–227.

Mansour, S.L., Thomas, K.R., Capecchi, M.R., 1988. Disruption of the proto-oncogene int-2 in mouse embryo-derived stem cells: a general strategy for targeting mutations to non-selectable genes. Nature 336, 348–352.

Marais, A., Fabian, B., Kramer, B., 2011. Tooth development from mouse embryonic stem cells and jaw ectomesenchyme. SADJ. 66, 456, 458–461.

Martin, G.R., 1981. Isolation of a pluripotent cell line from early mouse embryos cultured in medium conditioned by teratocarcinoma stem cells. Proc. Natl. Acad. Sci. USA 78, 7634–7638.

Martin, G.R., Lock, L.F., 1983. Pluripotent cell lines derived from early mouse embryos cultured in medium conditioned by teratocarcinoma stem cells. In: Silver, L.M., Martin, G.R., Strickland., S. (Eds.), Teratocarcinoma Stem Cells. Cold Spring Harbor Press, Cold Spring Harbor, NY, pp. 635–663.

Martin, G.R., Silver, L.M., Fox, H.S., Joyner, A.L., 1987. Establishment of embryonic stem cell lines from preimplantation mouse embryos homozygous for lethal mutations in the t-complex. Dev. Biol. 121, 20–28.

Matsui, Y., Zsebo, K., Hogan, B.L., 1992. Derivation of pluripotential embryonic stem cells from murine primordial germ cells in culture. Cell 70, 841–847.

Matsunari, H., Nagashima, H., Watanabe, M., Umeyama, K., Nakano, K., Nagaya, M., et al., 2013. Blastocyst complementation generates exogenic pancreas in vivo in apancreatic cloned pigs. Proc. Natl. Acad. Sci. USA 110, 4557–4562.

Metzger, J.M., Lin, W.I., Samuelson, L.C., 1994. Transition in cardiac contractile sensitivity to calcium during the in vitro differentiation of mouse embryonic stem cells. J. Cell Biol. 126, 701–711.

Meyers, E.N., Lewandoski, M., Martin, G.R., 1998. An Fgf8 mutant allelic series generated by Cre- and Flp-mediated recombination. Nat. Genet. 18, 136–141.

Moens, C.B., Auerbach, A.B., Conlon, R.A., Joyner, A.L., Rossant, J., 1992. A targeted mutation reveals a role for N-myc in branching morphogenesis in the embryonic mouse lung. Genes Dev. 6, 691–704.

Morizane, R., Monkawa, T., Itoh, H., 2009. Differentiation of murine embryonic stem and induced pluripotent stem cells to renal lineage in vitro. Biochem. Biophys. Res. Commun. 390, 1334–1339.

Morrisey, E.E., Tang, Z., Sigrist, K., Lu, M.M., Jiang, F., Ip, H.S., et al., 1998. GATA6 regulates HNF4 and is required for differentiation of visceral endoderm in the mouse embryo. Genes Dev. 12, 3579–3590.

Mummery, C.L., van den Eijnden-van Raaij, A.J., Feijen, A., Freund, E., Hulskotte, E., Schoorlemmer, J., et al., 1990. Expression of growth factors during the differentiation of embryonic stem cells in monolayer (published erratum appears in Dev. Biol. 1991 145(1): 203). Dev. Biol. 142, 406–413.

Nagy, A., 2000. Cre recombinase: the universal reagent for genome tailoring. Genesis. 26, 99–109.

Nagy, A., Nichols, J., 2011. Derivation of murine ES cell lines. In: Pease, S., Saunders., T. (Eds.), Advanced Protocols for Animal Transgenesis. Springer-Verlag, Heidelberg, pp. 431–455.

Nagy, A., Vinterstein, K., 2006. Murine embryonic stem cells. Methods Enzymol. 418, 3–33.

Nagy, A., Gocza, E., Diaz, E.M., Prideaux, V.R., Ivanyi, E., Markkula, M., et al., 1990. Embryonic stem cells alone are able to support fetal development in the mouse. Development 110, 815–821.

Nagy, A., Rossant, J., Nagy, R., Abramow-Newerly, W., Roder, J.C., 1993. Derivation of completely cell culture-derived mice from early-passage embryonic stem cells. Proc. Natl. Acad. Sci. USA 90, 8424–8428.

Nasu, M., Takata, N., Danjo, T., Sakaguchi, H., Kadoshima, T., Futaki, S., et al., 2012. Robust formation and maintenance of continuous stratified cortical neuroepithelium by laminin-containing matrix in mouse ES cell culture. PLoS One 7, e53024.

Nichols, J., Evans, E.P., Smith, A.G., 1990. Establishment of germ-line-competent embryonic stem (ES) cells using differentiation inhibiting activity. Development. 110, 1341–1348.

Nichols, J., Jones, K., Phillips, J.M., Newland, S.A., Roode, M., Mansfield, W., et al., 2009. Validated germline-competent embryonic stem cell lines from nonobese diabetic mice. Nat. Med. 15, 814–818.

Nickoloff, J.A., Reynolds, R.J., 1990. Transcription stimulates homologous recombination in mammalian cells. Mol. Cell Biol. 10, 4837–4845.

Pan, F., Chi, L., Schlatt, S., 2013. Effects of nanostructures and mouse embryonic stem cells on in vitro morphogenesis of rat testicular cords. PLoS One 8, e60054.

Pease, S., Braghetta, P., Gearing, D., Grail, D., Williams, R.L., 1990. Isolation of embryonic stem (ES) cells in media supplemented with recombinant leukemia inhibitory factor (LIF). Dev. Biol. 141, 344–352.

Peng, X., Gao, H., Wang, Y., Yang, B., Liu, T., Sun, Y., et al., 2013. Conversion of rat embryonic stem cells into neural precursors in chemical-defined medium. Biochem. Biophys. Res. Commun. 431, 783–787.

Pettitt, S.J., Liang, Q., Rairdan, X.Y., Moran, J.L., Prosser, H.M., Beier, D.R., et al., 2009. Agouti C57BL/6N embryonic stem cells for mouse genetic resources. Nat. Methods 6, 493–495.

Poueymirou, W.T., Auerbach, W., Frendewey, D., Hickey, J.F., Escaravage, J.M., Esau, L., et al., 2007. F0 generation mice fully derived from gene-targeted embryonic stem cells allowing immediate phenotypic analyses. Nat. Biotechnol. 25, 91–99.

Prado, F., Piruat, J.I., Aguilera, A., 1997. Recombination between DNA repeats in yeast hpr1delta cells is linked to transcription elongation. EMBO J. 16, 2826–2835.

Ramirez-Solis, R., Liu, P., Bradley, A., 1995. Chromosome engineering in mice. Nature 378, 720–724.

Reid, L.H., Shesely, E.G., Kim, H.S., Smithies, O., 1991. Cotransformation and gene targeting in mouse embryonic stem cells. Mol. Cell. Biol. 11, 2769–2777.

Resnick, J.L., Bixler, L.S., Cheng, L., Donovan, P.J., 1992. Long-term proliferation of mouse primordial germ cells in culture. Nature 359, 550–551.

Reyrat, J.M., Pelicic, V., Gicquel, B., Rappuoli, R., 1998. Counterselectable markers: untapped tools for bacterial genetics and pathogenesis. Infect. Immun. 66, 4011–4017.

Robertson, E.J., 1991. Using embryonic stem cells to introduce mutations into the mouse germ line. Biol. Reprod. 44, 238–245.

Robertson, E.J., Evans, M.J., Kaufman, M.H., 1983a. X-chromosome instability in pluripotential stem cell lines derived from parthenogenetic embryos. J. Embryol. Exp. Morphol. 74, 297–309.

Robertson, E.J., Kaufman, M.H., Bradley, A., Evans, M.J., 1983b. Isolation, properties and karyotype analysis of pluripotent (EK) cell lines from normal and parthenogenetic embryos. In: Silver, L.M., Martin, G.R., Strickland, S. (Eds.), Teratocarcinoma Stem Cells, Cold Spring Harbor Conference on Cell Proliferation. Cold Spring Harbor Press, Cold Spring Harbor, NY, pp. 647–663.

Robertson, S.M., Kennedy, M., Shannon, J.M., Keller, G., 2000. A transitional stage in the commitment of mesoderm to hematopoiesis requiring the transcription factor SCL/tal-1. Development 127, 2447–2459.

Rohwedel, J., Guan, K., Zuschratter, W., Jin, S., Ahnert-Hilger, G., Furst, D., et al., 1998. Loss of beta1 integrin function results in a retardation of myogenic, but an acceleration of neuronal, differentiation of embryonic stem cells in vitro. Dev. Biol. 201, 167–184.

Safran, M., Kim, W.Y., Kung, A.L., Horner, J.W., DePinho, R.A., Kaelin Jr., W.G., 2003. Mouse reporter strain for noninvasive bioluminescent imaging of cells that have undergone Cre-mediated recombination. Mol. Imaging 2, 297–302.

Salminen, M., Meyer, B.I., Gruss, P., 1998. Efficient poly A trap approach allows the capture of genes specifically active in differentiated embryonic stem cells and in mouse embryos. Dev. Dyn. 212, 326–333.

Sasai, Y., 2013. Next-generation regenerative medicine: organogenesis from stem cells in 3D culture. Cell Stem Cell 12, 520–530.

Scherer, C.A., Chen, J., Nachabeh, A., Hopkins, N., Ruley, H.E., 1996. Transcriptional specificity of the pluripotent embryonic stem cell. Cell. Growth Differ. 7, 1393–1401.

Shamblott, M.J., Axelman, J., Wang, S., Bugg, E.M., Littlefield, J.W., Donovan, P.J., et al., 1998. Derivation of pluripotent stem cells from cultured human primordial germ cells. Proc. Natl. Acad. Sci. USA 95, 13726–13731.

Shull, M.M., Ormsby, I., Kier, A.B., Pawlowski, S., Diebold, R.J., Yin, M., et al., 1992. Targeted disruption of the mouse transforming growth factor-beta 1 gene results in multifocal inflammatory disease. Nature 359, 693–699.

Simpson, E.M., Linder, C.C., Sargent, E.E., Davisson, M.T., Mobraaten, L.E., Sharp, J.J., 1997. Genetic variation among 129 substrains and its importance for targeted mutagenesis in mice. Nat. Genet. 16, 19–27.

Smih, F., Rouet, P., Romanienko, P.J., Jasin, M., 1995. Double-strand breaks at the target locus stimulate gene targeting in embryonic stem cells. Nucleic Acids Res. 23, 5012–5019.

Smith, A.G., Hooper, M.L., 1987. Buffalo rat liver cells produce a diffusible activity which inhibits the differentiation of murine embryonal carcinoma and embryonic stem cells. Dev. Biol. 121, 1–9.

Smith, A.J., De Sousa, M.A., Kwabi-Addo, B., Heppell-Parton, A., Impey, H., Rabbitts, P., 1995. A site-directed chromosomal translocation induced in embryonic stem cells by Cre-loxP recombination. Nat. Genet. 9, 376–385.

Soriano, P., 1995. Gene targeting in ES cells. Annu. Rev. Neurosci. 18, 1–18.

Stewart, C.L., Gadi, I., Bhatt, H., 1994. Stem cells from primordial germ cells can reenter the germ line. Dev. Biol. 161, 626–628.

Suda, Y., Suzuki, M., Ikawa, Y., Aizawa, S., 1987. Mouse embryonic stem cells exhibit indefinite proliferative potential. J. Cell. Physiol. 133, 197–201.

Suemori, H., Kadodawa, Y., Goto, K., Araki, I., Kondoh, H., Nakatsuji, N., 1990. A mouse embryonic stem cell line showing pluripotency of differentiation in early embryos and ubiquitous beta-galactosidase expression. Cell. Differ. Dev. 29, 181–186.

Sugawara, A., Goto, K., Sotomaru, Y., Sofuni, T., Ito, T., 2006. Current status of chromosomal abnormalities in mouse embryonic stem cell lines used in Japan. Comp. Med. 56, 31–34.

Takahashi, K., Yamanaka, S., 2006. Induction of pluripotent stem cells from mouse embryonic and adult fibroblast cultures by defined factors. Cell 126, 663–676.

Tanimoto, Y., Iijima, S., Hasegawa, Y., Suzuki, Y., Daitoku, Y., Mizuno, S., et al., 2008. Embryonic stem cells derived from C57BL/6J and C57BL/6N mice. Comp. Med. 58, 347–352.

te Riele, H., Maandag, E.R., Clarke, A., Hooper, M., Berns, A., 1990. Consecutive inactivation of both alleles of the pim-1 proto-oncogene by homologous recombination in embryonic stem cells. Nature 348, 649–651.

Testa, G., Schaft, J., van der Hoeven, F., Glaser, S., Anastassiadis, K., Zhang, Y., et al., 2004. A reliable lacZ expression reporter cassette for multipurpose, knockout-first alleles. Genesis 38, 151–158.

Thomas, B.J., Rothstein, R., 1989. Elevated recombination rates in transcriptionally active DNA. Cell 56, 619–630.

Thomas, K.R., Capecchi, M.R., 1987. Site-directed mutagenesis by gene targeting in mouse embryo-derived stem cells. Cell 51, 503–512.

Thomas, K.R., Folger, K.R., Capecchi, M.R., 1986. High frequency targeting of genes to specific sites in the mammalian genome. Cell 44, 419–428.

Thompson, S., Clarke, A.R., Pow, A.M., Hooper, M.L., Melton, D.W., 1989. Germ line transmission and expression of a corrected HPRT gene produced by gene targeting in embryonic stem cells. Cell 56, 313–321.

Thomson, J.A., Itskovitz-Eldor, J., Shapiro, S.S., Waknitz, M.A., Swiergiel, J.J., Marshall, V.S., et al., 1998. Embryonic stem cell lines derived from human blastocysts. Science 282, 1145–1147.

Torres, R.M., Kühn, R., 1997. Laboratory Protocols for Conditional Gene Targeting. Oxford University Press, Oxford.

Tybulewicz, V.L., Crawford, C.E., Jackson, P.K., Bronson, R.T., Mulligan, R.C., 1991. Neonatal lethality and lymphopenia in mice with a homozygous disruption of the c-abl proto-oncogene. Cell 65, 1153–1163.

Valancius, V., Smithies, O., 1991. Testing an "in–out" targeting procedure for making subtle genomic modifications in mouse embryonic stem cells. Mol. Cell Biol. 11, 1402–1408.

Valenzuela, D.M., Murphy, A.J., Frendewey, D., Gale, N.W., Economides, A.N., Auerbach, W., et al., 2003. High-throughput engineering of the mouse genome coupled with high-resolution expression analysis. Nat. Biotechnol. 21, 652–659.

van der Lugt, N., Maandag, E.R., te Riele, H., Laird, P.W., Berns, A., 1991. A pgk::hprt fusion as a selectable marker for targeting of genes in mouse embryonic stem cells: disruption of the T-cell receptor delta-chain-encoding gene. Gene 105, 263–267.

van Deursen, J., Fornerod, M., Van Rees, B., Grosveld, G., 1995. Cre-mediated site-specific translocation between nonhomologous mouse chromosomes. Proc. Natl. Acad. Sci. USA 92, 7376–7380.

Wobus, A.M., Holzhausen, H., Jakel, P., Schoneich, J., 1984. Characterization of a pluripotent stem cell line derived from a mouse embryo. Exp. Cell Res. 152, 212–219.

Wobus, A.M., Kaomei, G., Shan, J., Wellner, M.C., Rohwedel, J., Ji, G., et al., 1997. Retinoic acid accelerates embryonic stem cell-derived cardiac differentiation and enhances development of ventricular cardiomyocytes. J. Mol. Cell Cardiol. 29, 1525–1539.

Wong, E.A., Capecchi, M.R., 1987. Homologous recombination between coinjected DNA sequences peaks in early to mid-S phase. Mol. Cell Biol. 7, 2294–2295.

Wu, H., Liu, X., Jaenisch, R., 1994. Double replacement: strategy for efficient introduction of subtle mutations into the murine Col1a-1 gene by homologous recombination in embryonic stem cells. Proc. Natl. Acad. Sci. USA 91, 2819–2823.

Wunderlich, F.T., Wildner, H., Rajewsky, K., Edenhofer, F., 2001. New variants of inducible Cre recombinase: a novel mutant of Cre-PR fusion protein exhibits enhanced sensitivity and an expanded range of inducibility. Nucleic Acids Res. 29, E47.

Yagi, T., Ikawa, Y., Yoshida, K., Shigetani, Y., Takeda, N., Mabuchi, I., et al., 1990. Homologous recombination at c-fyn locus of mouse embryonic stem cells with use of diphtheria toxin A-fragment gene in negative selection. Proc. Natl. Acad. Sci. USA 87, 9918–9922.

Yamanaka, S., 2012. Induced pluripotent stem cells: past, present, and future. Cell Stem Cell 10, 678–684.

Yamane, T., Kunisada, T., Yamazaki, H., Era, T., Nakano, T., Hayashi, S.I., 1997. Development of osteoclasts from embryonic stem cells through a pathway that is c-fms but not c-kit dependent. Blood 90, 3516–3523.

Yang, Y., Seed, B., 2003. Site-specific gene targeting in mouse embryonic stem cells with intact bacterial artificial chromosomes. Nat. Biotechnol. 21, 447–451.

Ying, Q.L., Smith, A.G., 2003. Defined conditions for neural commitment and differentiation. Methods Enzymol. 365, 327–341.

Ying, Q.L., Wray, J., Nichols, J., Batlle-Morera, L., Doble, B., Woodgett, J., et al., 2008. The ground state of embryonic stem cell self-renewal. Nature 453, 519–523.

Yu, H., Kessler, J., Shen, J., 2000. Heterogeneous populations of ES cells in the generation of a floxed Presenilin-1 allele. Genesis 26, 5–8.

Yu, J., McMahon, A.P., 2006. Reproducible and inducible knockdown of gene expression in mice. Genesis 44, 252–261.

Zheng, B., Sage, M., Cai, W.W., Thompson, D.M., Tavsanli, B.C., Cheah, Y.C., et al., 1999. Engineering a mouse balancer chromosome. Nat. Genet. 22, 375–378.

Zheng, H., Wilson, J.H., 1990. Gene targeting in normal and amplified cell lines. Nature 344, 170–173.

5 Gene Targeting in Embryonic Stem Cells, II: Conditional Technologies

Daniel J. Ledbetter[1], James G. Thomson[2], Jorge A. Piedrahita[3] and Edmund B. Rucker III[1]

[1]Department of Biology, University of Kentucky, Lexington, KY, [2]USDA–WRRC–ARS Crop Improvement and Utilization, Albany, CA, [3]Department of Molecular Biomedical Sciences, North Carolina State University, Raleigh, NC

I. Introduction

Genome modification via transgenesis has allowed researchers to link genotype and phenotype as an alternative approach to the characterization of random mutations through evolution. The synergy of technologies from the fields of embryonic stem (ES) cells, gene knockouts, and protein-mediated recombination has led to the development of exquisite models that have planned gene modifications fine-tuned down to the single base pair. In addition to gene targeting by homologous recombination, the use of zinc-finger nucleases (ZFNs), transcription activator-like effector nucleases (TALENs), or PiggyBac transposase has gained considerable traction in the genetic engineering field to edit the mouse genome (Menke, 2013). In gain-of-function studies, a variety of promoters have been used to drive transgene expression in a developmental-, tissue-, or cell-specific fashion. With the development of inducible systems (tetracycline and lactose based), it is also feasible to direct transgene expression in a time-dependent fashion. To circumvent the problems of secondary effects and embryonic lethality that arise from knockout mouse models (loss of function), the implementation of spatial and temporal deletion strategies has been both warranted and necessary. ES cells can be engineered with various recombination systems to generate conditional alleles for gene deletions, knock-ins, inversions, recombinase-mediated cassette exchange (RMCE), and chromosomal rearrangements (Garcia-Otin and Guillou, 2006; Otsuji et al., 2008; Schnutgen et al., 2006; Thomson and Ow 2006; Wirth et al., 2007; Zhang et al., 2012). As of October 2012, ES cell clones with conditional alleles for 8897 genes have been generated (Sung et al., 2012).

Transgenic Animal Technology. DOI: http://dx.doi.org/10.1016/B978-0-12-410490-7.00005-0

II. Conditional Modeling: A Brief History

The tyrosine recombinase family is comprised of more than 28 proteins from bacteria, phage, and yeast that have a common invariant His-Arg-Tyr triad (Abremski and Hoess, 1992). These proteins bind to a DNA recognition sequence and are involved in DNA recognition, synapsis, cleavage, strand exchange, and religation. Two of the most widely used site-specific recombination systems for eukaryotic applications include: (i) Cre-loxP from bacteriophage P1 (Sternberg et al., 1981) and (ii) FLP-FRT from the 2 μ plasmid of *Saccharomyces cerevisiae* (Andrews et al., 1985). The Cre-loxP and FLP-FRT systems require only the protein and its DNA recognition site. Although the wild-type systems can be used to promote DNA integration events, deletion events are kinetically favored and are used for this property in genetic engineering. In contrast, the serine integrase family is devoted to DNA integration. The most widely used member of this family is ϕC31, which recognizes the *attP/attB* sites for integration (Thorpe and Smith, 1998). Excision, insertion, and chromosomal rearrangements have been performed with these recombination systems in bacteria, plants, yeast, mammalian somatic and ES cells, as well as transgenic animals.

The Cre-lox system is a mechanism of site-specific recombination that was initially observed in the replication cycle of P1 bacteriophage and has since been adapted for use as a tool that enables a variety of controlled genomic manipulations in both cultured cells and transgenic plants and animals. The Cre recombinase facilitates recombination between pairs of specific 34-base pair sequences, known as *loxP* sites. Relative orientation of the *loxP* sites determines the outcome of the recombination event in that *loxP* sites oriented in the same direction on a chromosomal segment result in a cyclization and excision of one of the *loxP* sites and the intervening sequence. Alternatively, two *loxP* sites oriented in opposite directions on a chromosomal segment result in an inversion of the intervening segment upon Cre-mediated recombination (Figure 5.1). The former is commonly used to generate removable gene disruption cassettes as well as deletions of chromosomal segments for analysis of gene function in a variety of organisms by gene-targeting methods to flank a gene or promoter elements with *loxP* sites (Guldener et al., 1996; Orban et al., 1992; Sauer, 1987). Since the gene encoding Cre recombinase, *Cre*, is not found in eukaryotes, engineered regulation of transgenic *Cre* expression enables this system to delete sequences flanked by *loxP* sites, or "floxed," in an inducible as well as temporal- and tissue-specific manner. For example, by standard pronuclear injection methods, a *Cre* transgene placed under the control of the whey acidic protein (*WAP*) gene promoter in mice limits *Cre* expression to the mammary gland during lactation (Wagner et al., 1997). The *WAP-Cre* transgenic mouse has been bred with transgenic mice carrying numerous different floxed alleles to generate temporal- and tissue-specific deletions for analysis of gene function in mammary gland physiology and tumorigenesis.

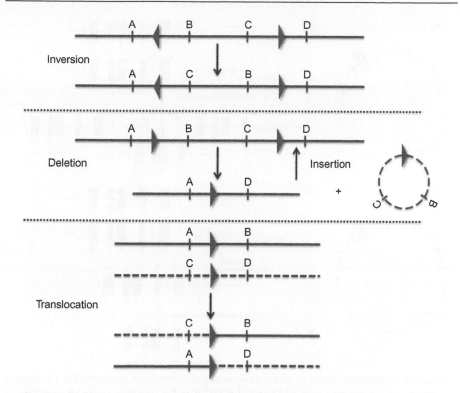

Figure 5.1 Overview of types of site-specific recombination events. Blue arrows indicate orientation and location of recombinase recognition sites in a genomic sequence containing loci A, B, C, and D. Recognition sites flanking loci in antiparallel orientation result in an inversion, whereas parallel orientation results in deletion. Recognition sites in trans result in a translocation.

III. Recombinase Systems in ES Cell Modification

The Cre-lox system in ES cells has been used for gene modification such as gene deletion, gene replacement (i.e., knock-in), subtle modification, inversions, translocations, and integration events (Figure 5.1) (reviewed by Misra and Duncan, 2002). Floxed ES cell resources can be generated by targeted homologous recombination strategies or through the use of retroviruses for random, nested chromosomal deletions (Su et al., 2000). Generation of floxed alleles within ES cells has been aided by the engineering of vectors that contain three *loxP* sites (two *loxP* sites flank a selectable marker for downstream removal) and loxP-FRT-based vectors that rely on the use of both Cre and FLP (Figure 5.2). With the three *loxP* strategy, floxed alleles can be generated by the removal of the marker in ES cells, but care has to

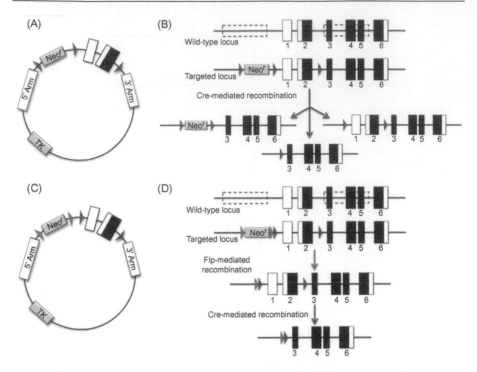

Figure 5.2 Conditional targeting vector designs and recombination products. The illustration depicts the 3 *loxP* targeting vector (A) and FLP-FRT/Cre-lox targeting vector (B) designs. The endogenous and targeted locus is also shown for each vector design (C–D). The 3 *loxP* targeted locus yields three possible products after Cre recombination: deletion of the Neo, deletion of exons 1 and 2, or deletion of Neo and exons 1 and 2 (C). The FLP-FRT/Cre-lox targeted locus yields only one product from each recombination event: deletion of Neo from FLP recombination, and deletion of exons 1 and 2 from Cre recombination. Red triangles indicate orientation and location of FRT sites. Blue triangles indicate orientation and location of *loxP* sites.

be taken to distinguish floxed cells from null cells from persistent Cre expression (Yu et al., 2000). Lower levels of Cre recombinase are not only permissive for effective gene deletions but also could reduce the risk of toxicity. High levels of Cre have been shown to inhibit G2-M progression and induce DNA damage (Loonstra et al., 2001).

A.   Recombinase Delivery Strategies

Most *in vitro* protocols utilize Cre- or FLP-encoding plasmids that are introduced into the cells by either electroporation or liposome-mediated transfer. However, viral-based approaches offer the advantage of high transduction efficiencies, which result in nearly 100% of the cells expressing Cre or FLP recombinase. Adenoviral

Cre can be used for *in vivo* deletions (Akagi et al., 1997; Wang et al., 1996), as well as deletion experiments in ES cells without compromising downstream applications such as germ line transmission (Shui and Tan, 2004). In addition, a sequential adenoviral approach can also be taken to achieve FLP-mediated recombination followed by Cre-mediated recombination within ES cells (Kondo et al., 2006). To a lesser extent than adenoviruses, retroviruses and lentiviruses have also been used to deliver Cre to ES cells (Psarras et al., 2004; Oka et al., 2006). In addition to viral-mediated delivery mechanisms, Cre RNA injections can be used to promote Cre-mediated effects (de Wit et al., 1998).

B.   Modified Recombinase Proteins

It was realized some time ago that FLP was less efficient than Cre, partly owing to an unsuitable optimum temperature for mammalian cells (Buchholz et al., 1996), which led to the application of molecular evolution to identify the thermostable variant called FLPe (Buchholz et al., 1998). Cycling mutagenesis was used to develop the thermostable FLPe, and was tested in ES cells (Buchholz et al., 1998). However, FLPe is still less efficient than Cre, although a recent codon-optimized FLPe, termed FLPo, has bridged some of the remaining gap (Raymond and Soriano, 2007). The FLPo and a codon-optimized ϕC31 (ϕC31o) have efficiencies similar to Cre (Raymond and Soriano, 2007). Mice have been generated from the FLPe protein, which were termed "flipper" or FLPeR (Farley et al., 2000; Rodriguez et al., 2000). In mammalian cells, FLPo appears to be about five times more active than FLPe, which in turn was about five times more active than wild-type FLP. The original FLP has a deletion efficiency of about 6% in ES cells (Schaft et al., 2001). A codon-improved Cre (iCre) has been generated which reduces the GC content found within the prokaryote sequence, originally designed to reduce DNA silencing of transgenes produced with this gene (Shimshek et al., 2002). A minimal increase in activity is found; however, studies have not detailed the resistance of the iCre transgene to DNA methylation.

While the Cre-lox system has made genomic modifications substantially easier and more accurate, there still exist certain technical issues that have motivated various modifications of the Cre protein itself. Generation of transgenic cells, plants, or animals allows for specific promoter-driven *Cre* expression, which has many benefits, yet can be a time-consuming and expensive process. The use of purified Cre protein has been shown to be effective when injected into one-cell-staged mouse embryos (Lauth et al., 2000). However, this would be a daunting task to inject single ES cells. Alternative Cre-delivery methods have been developed including lentiviral expression of *Cre* genes and transduction of cell-permeant Cre proteins. In order to increase the recombination efficiency of exogenous Cre, a nuclear localization sequence (NLS) peptide is often fused to the N-terminus of the Cre protein. Although the wild-type Cre recombinase has been shown to successfully permeate cells and enter the nucleus, recombination rates can be greatly increased (10% of cells with wild-type Cre versus >50% of cells with NLS-Cre) by the addition of the NLS (Will et al., 2002). As another example, using the 12-amino-acid membrane translocation sequence from the Kaposi fibroblast growth factor (FGF-4), the modified Cre was able to effectively enter cells and promote Cre-mediated recombination *in vitro* and *in vivo* (Jo et al., 2001). Other protein modifications have also been adapted for use with the Cre-lox system including the HIV-TAT protein transduction domain peptide, which has shown to drastically improve cellular uptake of labeled extracellular proteins (Peitz et al., 2002). When plasmid constructs carrying *Cre* transgenes are employed to remove floxed sequences in cells or tissues, the efficiency of recombination can be limited by the ability to transfect various cell types and ensure expression of plasmid DNA. Use of TAT and NLS peptides with Cre has allowed for simplified delivery of biologically active Cre recombinase to 95–100% of ES cells in culture and >60% of cells in specific tissues (Peitz et al., 2002; Joshi et al., 2002; Lin et al., 2004). The subsequent generation of a TAT-FLP shows greater than 75% recombination in fibroblasts, mouse, and human ES cells (Patsch et al., 2010).

C. Additional Recombinase Systems

The search for orthologs of Cre has led to the development of two additional systems: (i) vCre (from *Vibrio*), which recognizes a *vloxP* site and (ii) sCre (from *Shewanella*), which utilizes a *sloxP* site (Suzuki and Nakayama, 2011). The Cre-loxP, vCre-vloxP, and sCre-sloxP systems do not cross react with each other. The Cre and vCre systems were used concomitantly to specifically modify two separate alleles in ES cells. Dre was identified in a search through P1-like phages for a Cre-like enzyme that had diverged sufficiently to recognize an RT that is distinct from *loxP* (Sauer and McDermott, 2004). The Dre recognition target was termed *rox*. Thus, the addition of Cre homologs to the genetic engineering toolbox has greatly diversified the routes for genome modification. An example of this type of synergy was shown with the use of the ϕC31 to insert a transgene into a predetermined locus within ES cells, followed by FLP-mediated recombination for marker removal (Monetti et al., 2011). This approach would allow for downstream applications using the Cre-lox system.

D. Inducible Cre Systems

While the basic concept behind the Cre-lox system remains the same today, many other technologies have been incorporated to enhance the capabilities of this system as well as to increase throughput. In order to control the expression of *Cre*, the gene is often placed under the regulation of a specific promoter. However, depending on the organism, tissue, and experimental goals, it may not be feasible to identify a promoter that will drive *Cre* expression at all of the desired timepoints. To this end, the development and refinement of inducible Cre systems has allowed for much greater flexibility and tighter control of transgene expression *in vitro* and *in vivo*. Several different inducible Cre fusion proteins have been generated, including fusions with the glucocorticoid receptor (Cre-GC) (Brocard et al., 1998), estrogen receptor (Cre-ER^{Tam}) (Brocard et al., 1997). The most widely used inducible Cre system, Cre-ER^{T2}, developed by Feil et al. (1996), includes a mutant ligand-binding domain of the mouse estrogen receptor fused to Cre recombinase. Three mutations within the human ER ligand-binding domain (G400V/M543A/L544A) prevent the receptor from binding to its natural ligand, 17β-estradiol, but not the antiestrogen drug 4-hydroxy-tamoxifen. Using a LacZ reporter to detect Cre-mediated recombination of *loxP* sites, it was shown that relatively low levels of tamoxifen were sufficient to induce Cre activity in cultured ES cells (Feil et al., 1997). Cre-ER^{T2} transgenic mice have been used to investigate a variety of genes over several years. Most recently, Qu et al. (2013) described the role of *Myc* in hepatocyte proliferation using the tamoxifen-inducible Cre-ER^{T2}. In this investigation, an Alb-*Cre-ER^{T2}* mouse was bred with a *Myc$^{fl/fl}$* mouse to facilitate liver-specific Cre-ER^{T2} expression in the offspring. Tamoxifen-dependent Cre recombination enabled hepatocyte-specific *Myc* disruption at predetermined time-points during development as well as tumorigenesis (Qu et al., 2013). While the benefits of this system are numerous, a significant disadvantage occurs as a result of off-target effects of tamoxifen in transgenic animals. The increased ligand binding and activation of endogenous estrogen receptors can have varied neurological and physiological effects (Chen et al., 2002a,b). Use of the Cre-PR inducible system, which is responsive to RU486, in ES cells results in an 80% efficiency for deletion (Kellendonk et al., 1996). A modified and more sensitive Cre-progesterone receptor (Cre-PR) is based on a mutant Cre that lacks a potential cryptic splice donor site and an extended PR ligand-binding domain (Wunderlich et al., 2001).

An alternative inducible Cre system has also been developed that makes use of the tetracycline/doxycycline-inducible (or repressible) tet-O system for control of *Cre* expression. The tet-O system functions through a doxycycline-dependent transactivator (rtTA) or doxycycline-repressible transactivator (tTA), which bind to an operator sequence (tet-O) upstream of the gene of interest and activate transcription. The gene encoding the transactivator can be driven by strategic promoters in order to achieve tissue-specific, inducible expression of the tet-O-controlled gene of interest. The Tet-on gene expression system has been used to successfully regulate Cre activity *in vitro* (Guo et al., 2005) and tet-O-*Cre* transgenic lines have also been developed for inducible gene deletions (Lindeberg et al., 2002; Utomo et al., 1999). Alternatively, Cre-lox and tetracycline-inducible systems can be used separately to regulate gene function. Exposure of targeted ES cells with Cre can convert an inactive locus to an active one; subsequently, the transcription of the active gene can be turned off with the tet system (Mao et al., 2005).

This system was recently employed to generate a cardiac-specific deletion of the mechanistic target of rapamycin gene, *mTOR*, in an adult mouse. Using standard knockout or conditional knockout approaches, this model would not have been possible. However, mice carrying tet-O-*Cre* and α-MHC-*rtTA* alleles and floxed *mTOR* alleles were generated, which allowed normal cardiac development and a controlled loss of mTOR in mature mice by doxycycline injections at 8 weeks of age (Zhu et al., 2013). The tet-O system coupled with Cre-lox technology allows for highly specific control of expression of the gene of interest and has very few off-target effects, unlike the tamoxifen-inducible *Cre* system. As well, doxycycline can be administered to transgenic animals easily through food or drinking water (Cawthorne et al., 2007). However, it has been noted that the tet-O system suffers from relatively slow kinetics of induction/repression due to the requirement for gene expression in regulatory steps prior to expression of effector genes. In comparison, the tamoxifen-inducible system operates with much tighter control of *Cre* expression due to rapid induction upon ligand binding of the estrogen receptor.

Recently, another system has developed that results in inducible Cre activity by antibiotic-dependent stabilization of a mutant form of the recombinase (Sando et al., 2013). In this system, the Cre recombinase is fused to one of two tags: the human protein FKBP12 or the bacterial dihydrofolate reductase, ecDHFR. These protein tags will label the Cre recombinase for proteasomal degradation under normal conditions, thereby preventing Cre activity. Upon administration of the pharmacological agent shield-1 (for FKBP12 tag) or TMP (for ecDHFR tag) proteasomal degradation of tagged Cre is inhibited, enabling Cre-mediated recombination. The destabilized Cre (DD-Cre) system uses similar methodology to previous models in that the Cre transgene is driven by strategic promoters, ensuring tissue-specific expression. However, the DD-Cre system also enables extremely tight temporal control of Cre activity in cells or transgenic animals as the system hinges upon induced stabilization of proteins, rather than induced gene expression as seen in tet-O-*Cre* or Cre-ER^{T2} models. Another benefit to this system comes from the use of the antibiotic TMP, which has no endogenous targets in eukaryotes, limiting off-target effects, and quickly diffuses throughout peripheral and nervous tissues, ensuring rapid kinetics of induction (Sando et al., 2013; Iwamoto et al., 2010). As it is still an emerging method of inducible Cre activity, the DD-Cre system has only seen limited experimental use. However, it has been noted that constitutive background levels of Cre activity was detected *in vitro* and *in vivo*, likely due to reliance on proteasomal degradation of proteins and the rate-limiting kinetics therein (Sando et al., 2013). Another factor to consider when designing experiments using the inducible Cre systems is that induction and Cre-mediated recombination by synthetic hormone or other pharmacological agents appears to be dose dependent. Therefore, care should be taken to ensure that the administration method chosen, whether food, water, or intraperitoneal injection, will deliver the necessary dose to target tissues.

E. Recombinase-Mediated Deletion

Cre-mediated deletion can be performed in traditional tissue-specific deletion experiments, or to effect gene deletion events in ES cells. Proof-of-principle experiments showed the ability to delete large gene segments initially within ES cells, for example the 200 kb amyloid precursor protein gene (Li et al., 1996). The range of these deletions is limited by the lethality that is imposed upon ES cells. Analyzing deletions from *lox* sites positioned along chromosome 11 in ES cells, an 11% deletion efficiency was found at a distance of 4 Mb (Zheng et al., 2000). Although the efficiency decreased as the distance increased, deletions and rearrangements up to

75% were generated. The Cre-lox system can be used to recycle selectable markers. This strategy was used to target four alleles in ES cells (Abuin and Bradley, 1996). Mutant *loxP* sites can also be utilized in the marker removal design (Arakawa et al., 2001) or for target integration of DNA into a predetermined locus (Araki et al., 1997). This strategy would allow for a predictable expression pattern for the inserted marker or transgene (Day et al., 2000). Cre-mediated recombination in ES cells, in addition to its utility in the generation of null alleles prior to mouse production, can also be used for characterization of stem cell maintenance within ES cells. For example, Cre-mediated removal of the STAT3 gene, which encodes for a central signaling and transcription factor, reduces the self-renewal efficiency of ES cells and results in cell differentiation (Raz et al., 1999).

F. Detection of Recombinase Protein and Recombination

One common issue stems from experimental designs that do not include a reporter for visual confirmation of Cre expression/activity in cells. Often Cre-mediated recombination is employed solely for the removal of selectable markers, which cannot easily be confirmed for all cells in a culture or target tissue. One method that has shown to be highly effective and readily used with high-throughput systems is the fusion of *Cre* transgenes or proteins to enhanced green fluorescent protein (eGFP) (Gagneten et al., 1997). Attaching Cre to eGFP or other variants of GFP allows for rapid verification of expression in tissues as well as selection of Cre-positive cells in culture using fluorescence-activated cell sorting. The Cre-eGFP has been used in ES cells for efficient deletion of floxed elements, and remains a method for *in vivo* detection of Cre-modified cells without the necessity of using selectable markers (Gagneten et al., 1997). Further, this system may be coupled with other effector peptides (e.g., TAT, NLS) to ensure efficient protein transduction. This visual confirmation of *Cre* expression can also serve as a confirmation of specificity and the appropriate lack of *Cre* expression in transgenic animals that aim to have floxed alleles removed only in certain tissues. One drawback, however, is that compared to expression of reporter genes behind a flowed allele, the eGFP-Cre system only confirms the production or presence of eGFP-Cre in cells or tissues and not necessarily successful recombination of *loxP* sites. As a marker system, the Z/EG (lacZ/eGFP) ES cell line can be used for the detection of eGFP upon successful recombination (Novak et al., 2000). Subsequent generation of additional transgenic resources has expanded the Cre-activated reporters for detection of recombination. EYFP (yellow) and eCFP (cyan) markers have been targeted into the ROSA26 locus by homologous recombination, and could serve as potential reporter assays in cell culture (Srinivas et al., 2001). In addition, Z/RED (from the RFP variant DsRed.T3) ES cell lines and transgenic mice have been generated to cover the red fluorescent wavelength (Vintersten et al., 2004). While all of these technologies aim to enhance the capabilities of the Cre-lox system, they are not without unique drawbacks, which should be carefully weighed in each individual experimental design.

G. Recombinase-Mediated DNA Insertion

The use of wild-type *lox* sites to generate insertion events is not very efficient, so this approach requires marker-based selection strategies. For example, using a split marker design, where a functional marker is generated upon integration, the Cre system was shown to promote integration of a YAC into ES cells (Call et al., 2000). A more efficient methodology for site-specific integration using the Cre-lox system utilizes mutations within the *lox* inverted repeat regions. The strategy employs mutation of the left element (LE) inverted repeat in one *lox* site and the right element (RE) inverted repeat in the other *lox* site. Upon recombination, a double mutant *lox* site and native *lox* site are generated. The double mutant *lox* site binds Cre less efficiently, thus reducing the probability of another round of recombination. In ES cells, the efficacy of this model for Cre-mediated integration was tested with native and mutant *lox* sites. In ES cell lines containing a single target wild-type lox site, the insertional frequency compared to random integration approached 0.5% with wild-type or RE mutant *lox* (*lox66*)-containing plasmids. However, in ES cells harboring an LE mutant *lox* (*lox71*) target, this efficiency approached 16% when using the complementary RE mutant *lox66*-containing plasmid. Two additional mutant RE *lox* sites, *lox*JTZ17, and *lox*KR3, have been found to be as efficient as *lox*66 for Cre-mediated integration or inversion in ES cells. However, the double mutant *lox* sites generated from the use of these newer mutant RE *lox* sites are more stable and less likely to undergo excision events than the *lox66/lox71* double mutant (Araki et al., 2010).

Genetic engineering also has been done to develop a unidirectional Cre based on the properties of λ integrase. A chimeric Cre with a small N-terminal domain of λ integrase results in its dependence on facilitating accessory proteins (Warren et al., 2008). Future modifications could permit the generation of a Cre-based system that efficiently integrates transgenes into predetermined loci in ES cell and transgenic animals. The φC31 integrase is used for site-specific integration, and its use in ES cells does not preclude the ability of modified ES cells to contribute to germ line transmission (Belteki et al., 2003). This integrase from the *Streptomyces* phage utilizes sequences from the attachment sites in phage (*attP*) and bacteria (*attB*), and the system's efficiency is greater than 50% in human cells for extra chromosomal targets (Groth et al., 2000). Using randomly integrated genomic *attP* sites in mouse and human cells, the efficiency was approximately 15% (Thyagarajan et al., 2001).

H. Recombinase-Mediated Cassette Exchange

The ability to use Cre to catalyze a replacement event of one cassette for a preexisting cassette in the genome is based on the use of incompatible *lox* sites that flank the cassette. This strategy, called "RMCE," has been used in ES cells (Figure 5.3). Using thymidine kinase as a negative selectable marker, efficiency was found to vary between 10% and 50% (Feng et al., 1999; Soukharev et al., 1999).

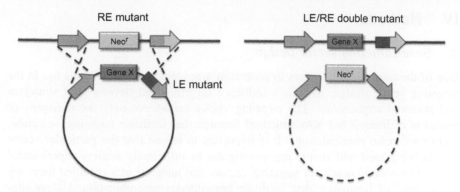

Figure 5.3 RMCE strategy. The inserted Neo cassette (silver) is flanked by a WT *loxP* site and a *lox71* site. The plasmid DNA carrying the gene to be inserted (blue) contains a WT *loxP* site and a *lox66* site. Cre-mediated recombination results in integration of the gene at the targeted locus, now flanked by a WT *loxP* site and an inactive, double mutant *lox* site. Arrows indicate orientation and location of WT *loxP* (gray), *lox71* (gray/green), *lox66* (purple/gray), and double mutant *lox* (purple/green) sites.

This suggests that there is a position effect such that the efficiency is dependent upon the locus. Cre mutants have been generated, for example, the N317A mutant, that recognize mutant *lox* sites with different affinities than the wild-type *loxP*; however, most practical applications involve the use of mutant, incompatible *lox* sites used in conjunction with a wild-type Cre recombinase (Hartung and Kisters-Woike, 1998). Gene trapping can be used in ES cells to insert a cassette with a mutant *lox* (e.g., *lox71*) and wild-type *loxP* site, followed by postinsertional Cre modification. Subsequent treatment of these modified ES cells with a cassette bearing a second mutant *lox* (e.g., *lox66*) and wild-type *loxP* site results in the generation of an inactive *lox* site and a *loxP* site. This gene trap-based technology is called flanked *lox* site insertion, or Floxin (Singla et al., 2010). The RMCE-ASAP (Adapted for targeting in Somatic cells to Accelerate Phenotyping) system was developed to reduce mouse sterility issues that accompany the use of hygromycin—thymidine kinase positive—negative selection. This protocol uses inverted *lox* sites and a puromycin—thymidine kinase selection cassette, and works in somatic cells and ES cells (Toledo et al., 2006). A comparison for FLPe and Cre for use in RMCE strategies has shown that FLPe is more efficient (13%) than Cre (5%), findings that may cause a shift in the biased utilization of Cre in many designs (Takata et al. 2011). FLP and Cre were both shown to work effectively at a modified ROSA26 locus designed to replace the endogenous promoter with ubiquitous or tissue-specific promoters (Sandhu et al., 2011; Tchorz et al., 2012). An RMCE approach in ES cells for such a modified locus would allow for the generation of a series of transgenic animals with high-level expression profiles.

IV. Methods

A. Gene-Targeting Vector Design

One of the most crucial aspects in generating transgenic cells or animals lies in the targeting vector design, as it will influence targeting and development strategies and potential applications. The targeting vector carries not only the transgene or mutation of interest but also structural features that facilitate targeting, selection, and downstream manipulations. It is important to ensure that the particular vector design being used will enable the investigator to sufficiently address experimental needs. Common to nearly all targeting vectors that integrate at a specified locus are the regions of homology that facilitate homologous recombination. Unless site-specific recombinases are used, it is a standard practice to include in the vector roughly 7 kb of sequence homologous to regions flanking the targeted genomic site. The homologous sequence is divided approximately equally into 5' and 3' regions of homology that flank all targeting vector features to be integrated into the ES cell genome. The most widely used marker for positive selection of targeted ES cells is the neomycin-resistance gene, Neo^r, which can be promoterless for use in promoter trap strategies or coupled with the PGK-1 promoter and poly(A) signal as a complete cassette. Also, the puromycin-resistance gene has successfully been used for positive selection of targeted ES cells. Even at high levels of expression, the Neo^r and $Puro$ markers show no effect on germ line transmission of transgenic ES cells. Beyond the flanking homology regions it is common to include a marker for negative selection, the herpes-simplex virus thymidine kinase (HSV-TK). When positioned beyond the homology regions, the TK cassette should not be integrated into ES cells through homologous recombination. When treated with the drug ganciclovir, cells carrying the TK cassette due to random integration are killed.

While targeting vectors for global deletions or specific mutations remains relatively straightforward in design, advances in site-specific recombination methods have resulted in variations in targeting vectors used for generation of conditional gene deletions. The complexity of the conditional targeting vectors can be partially attributed to the requirement for removal of selectable marker genes or cassettes. Often selectable markers integrated in a targeted locus can result in hypomorphic effects of adjacent genes (Jacks et al., 1994). Classically, conditional knockout targeting vectors employed the Cre-lox system for removal of the selectable marker as well as the gene to be deleted. By positioning three $loxP$ sites in the vector, Cre-mediated recombination results in three possible outcomes: deletion of the selectable marker, deletion of the coding region/transgene, or deletion of both the selectable marker and the coding region/transgene (Figure 5.2). While this is an effective means of marker removal and conditional gene disruption, it is not the most efficient method. One major disadvantage to this method is that these manipulations are performed by strategic mouse breeding, followed by screening for specified genotypes in the offspring. Since the recombination products should occur with equal frequency, only one-third of the offspring are expected to have the desired genotype for the targeted locus. Subsequent breeding may be required to remove the Cre transgene and breed mice to homozygosity. Mice may then be bred to obtain tissue-specific Cre expression for conditional deletion of the floxed allele. An alternative targeting vector design incorporates the FLP-FRT system in order to simplify the marker removal step. The *Neo*^r cassette is flanked by FRT sites and can be deleted by breeding with mice carrying the *Flp* gene, and the selectable marker can effectively be removed in all offspring. It is not necessary to breed out the *Flp* transgene. After breeding to homozygosity, tissue-specific Cre-mediated deletion can proceed (Figure 5.2). Another technological advancement that may be incorporated into the targeting vector is the use of heterologous *loxP* sites for use in RMCE (Figure 5.3). Both the combined FLP-FRT/Cre-lox system and heterologous *loxP*-based designs benefit from their ease of use *in vitro* as well as *in vivo*.

In addition to improved efficiency in generating conditional deletions, the FLP-FRT/Cre-lox-based targeting vector enables efficient generation of gene knock-in models. For example, Ledbetter and Rucker have engineered a model to investigate homologous functions of the apoptotic regulatory proteins BCL-2 and BCL-X$_L$ by inserting a floxed *GFP-Bcl-2* fusion gene immediately upstream of the translation initiation site at the endogenous *Bcl-x* locus (Figure 5.4). The inserted sequence also positions a cDNA encoding a red fluorescent protein upstream of the endogenous *Bcl-x* protein coding sequence. Following FLP-mediated removal of the *Neo*^r cassette, *GFP-Bcl-2* expression will be driven by the *Bcl-x* promoter. This model provides for additional flexibility in downstream applications as well upon Cre-mediated removal of the *GFP-Bcl-2* transgene, which positions the *RFP*-endogenous *Bcl-x* fusion under the *Bcl-x* promoter. Using the control enabled by FLP-FRT and Cre-lox-mediated recombination of transgenic elements, strategic breeding will allow for generation of mice concurrently expressing *GFP-Bcl-2* from one targeted *Bcl*-x allele and *RFP-Bcl-x* expression from the other targeted *Bcl-x* allele.

B. Validation of Lox and FRT-Containing Plasmids with AM-1, 294-Cre, and 294-FLP Bacterial Cells

If the targeting construct contains *lox* or FRT sites, it is important to validate them in plasmid DNA prior to transfection of stem cells, to save both time and money later. The AM-1 (Invitrogen) and 294-Cre (available from Gene Bridges) bacterial strains express Cre recombinase for confirmation of *lox* site function, while the 294-FLP (available from Gene Bridges) bacterial strain expresses the FLP recombinase for confirmation of FRT sites. These strains can be used to generate chemically competent bacterial cells using various kits available (Zymo Research Z-competent *E. coli* transformation kit) or by CaCl$_2$ treatment using the "Generation of chemically competent *E. coli*" protocol.

1. Generation of Chemically Competent E. coli (Beginning with Frozen Glycerol Stock)

1. Using a sterile toothpick, streak AM-1, 294-Cre, or 294-FLP bacteria on LB-agar plates (without antibiotic) using sterile technique. Incubate overnight at 37°C.

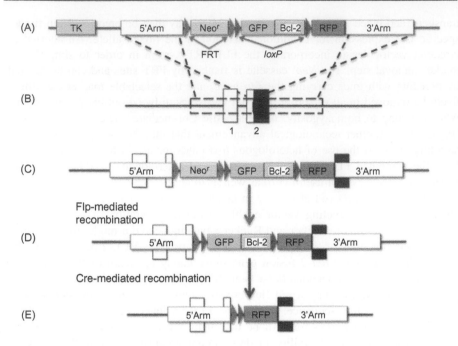

Figure 5.4 Bcl-2 knock-in targeting strategy and site-specific recombination products. The illustration depicts the floxed *Bcl-2* knock-in targeting vector (A) and the endogenous *Bcl-x* locus (B), with regions of homology and sites of homologous recombination indicated with dashed lines. *Bcl-x* exons 1 and 2 are shown with white- and black-filled regions indicating untranslated and translated regions, respectively. The targeted locus (C) expresses *GFP-Bcl-2* upon Flp-mediated excision of the selectable marker (D), which can be followed by Cre-mediated excision of the floxed *GFP-Bcl-2* transgene to express endogenous *Bcl-x* with an RFP tag (E). Red triangles indicate orientation and location of FRT sites. Blue triangles indicate orientation and location of *loxP* sites.

2. Using a sterile toothpick or inoculating loop, inoculate 2 mL fresh LB broth with single colonies using sterile technique. Incubate overnight at 37°C.
3. Inoculate 1 L fresh LB broth with 1 mL AM-1, 294-Cre, or 294-FLP culture. Incubate at 37°C until $OD_{600} = 0.5-0.7$.
4. Centrifuge cells at 4000 rpm for 10 min at 4°C.
5. Remove supernatant and gently resuspend pellet in 100 mL ice-cold 30 mM $CaCl_2$.
6. Centrifuge cells at 4000 rpm for 10 min at 4°C.
7. Remove supernatant and gently resuspend pellet in 10 mL ice-cold 30 mM $CaCl_2$.
8. Centrifuge cells at 4000 rpm for 10 min at 4°C.
9. Remove supernatant and gently resuspend pellet in 3 mL ice-cold 30 mM $CaCl_2$.
10. Aliquot 50–100 μL into 0.5-mL tubes and flash-freeze in EtOH-dry ice bath.
11. Store at −80°C.

2. Transfection of Chemically Competent AM-1, 294-Cre, or 294-FLP Bacterial Cells

1. Thaw competent cells on ice for 15 min.
2. Preheat heating block or water bath to 42°C.
3. Gently pipette 50–200 ng targeting vector plasmid DNA into competent cells and swirl gently with pipette tip. Incubate 30 min on ice.
4. Heat shock-competent cell/DNA mixture for 45 s at 42°C.
5. Incubate competent cell/DNA mixture on ice for 2 min.
6. Add 250 μL SOC (rich medium). Incubate in rolling wheel or shaking incubator for 1 h at 37°C.
7. Plate 100 μL on LB-agar plate containing appropriate antibiotic. Incubate plate overnight at 37°C.
8. Centrifuge remaining cells at 3 Krpm for 2 min.
9. Remove supernatant and resuspend pellet in 150 μL LB broth containing antibiotic.
10. Plate on LB-agar plate containing antibiotic. Incubate plate overnight at 37°C.

Colonies from overnight growth may be used to seed 2 mL cultures in LB broth for miniprep of DNA and RFLP analysis or screened for removal of floxed/ FRT-flanked sequence by colony PCR.

C. Targeting of mES Cells

Prior to transfection, the targeting vector should be linearized by restriction digest and restriction enzymes should be heat inactivated (when possible) and removed by extraction with phenol/chloroform/isoamyl alcohol (25:24:1), ethanol precipitated, and resuspended in TE, DPBS, or preferably electroporation buffer. In order to maximize the likelihood of germ line transmission in later steps, early-passage ES cells should be used for targeting. To increase targeting efficiency (due to increased frequency of homologous recombination in cells) ES cells should be growing exponentially when harvested for electroporation.

1. Electroporation of ES Cells for Transfection of Linear DNA

1. Twenty-four hours prior to harvesting cells for electroporation, plate mitotically inactive STO cells on a 100 mm plate (2×10^6 cells per plate in 8 mL STO medium). Incubate overnight at 37°C.
2. Three hours prior to harvesting, change ES cell medium.
3. Harvest ES cells by adding 3 mL 0.25% Trypsin-EDTA (prewarmed to room temperature). Incubate at 37°C for 5–7 min.
4. Stop trypsinization by adding 5 mL ES medium and pipette cells up and down 10–15 times, gently, to ensure all colonies are dissociated. Transfer cell/trypsin/medium mixture to a 14-mL tube.
5. Pellet cells by centrifugation for 5 min at 2000 rpm.
    During this time, replace medium on inactivated STO cells with 8 mL ES medium.
6. Remove supernatant without disturbing cell pellet and resuspend in 1 mL electroporation buffer (preferably) or DPBS.

7. Gently transfer cell suspension to 1.5 mL tube containing 25 µg linear targeting vector DNA; gently flick tube to mix.
8. Gently transfer 0.8 mL cell/DNA mixture to 4 mm electroporation cuvette.
   Ensure that there are no bubbles in cuvette and cell suspension does not fill the cuvette above the electrodes (narrow portion).
9. Apply single pulse from electroporator to cell/DNA mixture (standard conditions: 250 V, 500 µF capacitance).
10. Place cuvette on ice for 1−2 min.
11. Gently transfer cells to 100 mm plate containing inactivated STO cells using Pasteur pipette (rinse cuvette with 500 µL ES medium to collect any remaining cells). Incubate at 37°C for 24 h.

Selection of targeted ES cells can be carried out in a variety of ways. The selection strategy depends on the resistance genes used in the targeting vector as well as the targeting efficiency of different loci. The most commonly used strategies involve positive selection of an antibiotic resistance gene (e.g., neo^r selection using G418) often used in combination with negative selection of a marker (HSV-TK negative selection using ganciclovir) to eliminate cells carrying targeting vectors that have randomly integrated into the genome.

2.   Positive/Negative Selection of Targeted ES Cells

1. Twenty-four hours after electroporation, aspirate ES medium and wash cells with 3 mL DPBS to remove dead cells.
2. Replace with ES medium containing appropriate antibiotic (200 µg/mL G418, 1 µg/mL puromycin). Incubate at 37°C for 24 h.
3. Repeat Step 2.
4. Approximately 72 h after electroporation, replace medium with fresh ES medium containing antibiotic and ganciclovir (2 µM). Incubate at 37°C for 24 h.
   - Change medium daily; maintain selective pressure with antibiotic and ganciclovir for days 3−7 after electroporation, after which cells should be kept in medium containing antibiotic only. Ganciclovir is highly toxic to cells and prolonged exposure can reduce the germ line transmission potential of ES cells.
   - Selection with G418 should be maintained for 7 days to ensure that all remaining cells are drug resistant; puromycin eliminates nonresistant cells in 3−5 days.
   - ES cell colonies should be visible by 7 days and picked at 8−10 days after electroporation.

3.   Positive Selection of Targeted ES Cells (Promoter Trap Targeting Strategy)

1. Approximately 24−36 h after electroporation, aspirate ES medium and wash cells with 3 mL DPBS to remove dead cells.
2. Replace with ES medium containing appropriate antibiotic (100−150 µg/mL G418, 0.5−0.75 µg/mL puromycin). Incubate at 37°C for 24 h.
3. Replace with ES medium containing appropriate antibiotic (200 µg/mL G418, 1 µg/mL puromycin). Incubate at 37°C for 24 h.

- Change medium daily; maintain selective pressure with increased antibiotic concentrations for a minimum of 7 days.
- It is recommended that a kill curve be performed to determine appropriate antibiotic concentrations, since expression of the resistance gene will vary depending on the targeted locus.

At 8−10 days after electroporation, single ES cell colonies should be present and should be isolated when the appropriate size is reached (approximately the size of the end of a 200 μL pipette tip, when compared microscopically using 40−100× magnification). Undifferentiated ES cell colonies will have a characteristic "fried egg" appearance (slightly raised center and flat, clearly defined edges). Colonies will differentiate quickly if not isolated and dissociated, marked by dark regions and "cobblestone" appearance at edges.

4. Isolation/Expansion of Targeted ES Cells

1. Prepare two to three 96-well plates by plating inactivated STO feeder cells or coating wells with 0.1% gelatin[1] (add 100 μL 0.1% gelatin per well; incubate for 5 min at room temperature, then aspirate gelatin and immediately add 150−200 μL fresh ES medium per well).
2. Identify undifferentiated colonies of appropriate size using an inverted light microscope at 40−100× magnification.
3. With p200 pipettor (set to 50 μL) and a sterile p200 tip, gently dislodge the colony from the plate by pushing from the side at an angle of approximately 45° and immediately pull up the loose colony with the pipettor.
4. Transfer the colony to one well of the prepared 96-well plate.
5. Repeat Steps 2−4 for all remaining colonies.
6. Allow colonies to adhere in 96-well plate overnight at 37°C.
7. Aspirate medium from 96-well plate and add 50 μL 0.25% Trypsin-EDTA. Incubate at 37°C for 5 min.
8. Add 150 μL fresh ES medium per well to inactivate trypsin and dissociate colonies by pipetting up and down 10–15 times. Incubate overnight at 37°C. Replace with fresh medium[2] the following day.
9. After 2 days growth, aspirate medium and add 50 μL 0.25% Trypsin-EDTA. Incubate at 37°C for 5 min.
10. Add 150 μL fresh ES medium per well to inactivate trypsin and dissociate colonies by pipetting and transfer 100 μL from each well to the corresponding well on another 96-well plate.
    - Freeze one set of 96-well plates; expand one set of 96-well plates for DNA isolation for genotyping PCR.
    - Media should be changed daily and cells should be passaged every 2−3 days.

---

[1]0.1% gelatin (1.0% gelatin-Sigma) w/v solution in ddH₂O.
[2]Prior to Step 9, coat two to three 96-well plates with 0.1% gelatin and plate-inactivated STO feeder cells as previously described.

D. Removal of Floxed Selectable Markers in ES Cells

For targeting constructs containing a floxed selectable marker upstream of the transgene, it will be necessary to remove the marker, following selection, to observe/validate transgene expression *in vitro* prior to blastocyst injection. ES cells can be transiently transfected with plasmid DNA expressing *Cre* or *Flpe* genes under a eukaryotic promoter (pBS185/CMV-Cre (Sauer and Henderson, 1990), pCAGGS-FLPe (Schaft et al., 2001)) using standard electroporation methods (as previously described) or using polymer- or liposome-based transfection reagents (Clontech Xfect mESC Transfection Reagent). In either case, ES cell colonies should be dissociated by trypsinization prior to beginning to enable transfection of individual cells and prevent formation of chimeric colonies.

E. PCR-Based Strategies to Detect Targeting, Marker Removal, and Gene Deletion

It is important to be able to confirm accurate targeting and subsequent manipulations at the DNA level. The most efficient way to accomplish this is by PCR amplification of genomic DNA using primers specific to the targeted locus. It is best to use PCR primers that will anneal outside of the regions amplified for "homology regions" in the targeting vector. This will enable PCR amplification (and comparison by gel electrophoresis) of the endogenous locus and the targeted locus in a single reaction. Successful amplification of the entire targeted locus also enables easy comparison after manipulations such as marker removal and gene deletion. If the targeted allele is a particularly large region (>10 kb) it may be difficult to obtain such a large PCR product from genomic DNA, in which case nested PCR primers for both the 5' and 3' regions are sufficient to confirm targeting. For analysis of marker removal and gene deletions, it will be important to design primers that flank the region of the manipulation to compare the PCR products before and after manipulations. It is not recommended to use primers within the region of the manipulation, as a lack of PCR product is not sufficient to indicate deletion of genes or markers.

F. Cell-Based Detection of Transgene Expression

It may be worthwhile to assay for expression of the transgene in cultured cells prior to blastocyst injection, especially in the case of complex or experimental targeting constructs. While there are several molecular/biochemical techniques that may be used for this, analysis by fluorescence microscopy or immunocytochemistry, depending on the transgene, can provide rapid and accurate results. Often it is recommended to confirm results using both techniques (Figure 5.5). Live cells expressing fluorescent transgenes can often be visualized by inverted fluorescence microscopy in culture dishes. However, for analysis using ICC or use with epifluorescence microscopy, targeted ES cells must be grown on gelatin-coated coverslips, fixed, and mounted on slides.

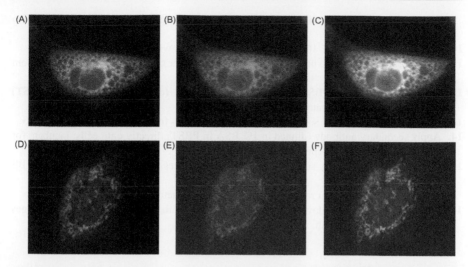

Figure 5.5 Confirmation of transgene expression by fluorescence microscopy and immunocytochemistry. Representative microscopy images of targeted ES cells following removal of selectable marker by Flp (A–C) and floxed *GFP-Bcl-2* by Cre expression (D–F). PFA-fixed cells were analyzed for expression of fluorescent transgenes *GFP-Bcl-2* (A) or *RFP-Bcl-X$_L$* (D) by fluorescence microscopy, and labeled with either α-Bcl-2 primary and Alexa594 secondary antibodies (B) or α-Bcl-X primary and Alexa488 secondary antibodies (E). Image overlays confirm the colocalization of GFP with *Bcl-2* (C) and RFP with *Bcl-X$_L$* (F), indicating successful expression of fusion transgenes. No primary antibody negative control cells lacked detectable fluorescence (not shown).

1. ICC of ES Cells Grown on Coverslips

Immunocytochemistry of targeted ES cells can be applied for several purposes. Antibody-based detection of proteins expressed from transgene is an important confirmation of targeting. It may also be necessary for detection of transgenes carrying fluorescent tags, because some genes may not be expressed highly enough in ES cells for visible fluorescence. Prior to blastocyst injection, ICC may also be used to confirm pluripotency by detecting markers such as Nanog, SSEA1, and Oct4.

Note: Antibodies should be diluted according to manufacturer recommendations.

1. Coat #1 coverslips with 0.1% gelatin in a 6-well plate.
2. Aspirate gelatin and replace with 1 mL complete ES cell growth medium.
3. Plate 5×10^5 ES cells (single cell suspension) onto gelatin-coated coverslips; incubate overnight at 37°C.
4. Aspirate medium, and wash cells twice with 1 mL DPBS.
5. Fix cells on coverslips in wells with 4% paraformaldehyde (500 μL/well); incubate at room temperature for 25 min.
6. Aspirate 4% paraformaldehyde, and wash cells twice with 1 mL DPBS.
7. Replace with 0.1% BSA in PBS; incubate for 1 h at 4°C.
8. Aspirate 0.1% BSA/PBS, wash cells with 1 mL DPBS.

9. Permeabilize membrane with 0.25% Triton X-100 in PBS; incubate for 10 min at room temperature.
10. Aspirate 0.25% Triton X-100 in PBS; wash three times with 1 mL DPBS.
11. Block cells in 1% BSA in PBST (with 0.3 M Glycine); incubate for 30 min at room temperature.
12. Aspirate 1% BSA in PBST, apply primary antibody (diluted in 1% BSA in PBST) directly onto cells; incubate overnight at 4°C.
13. Aspirate diluted antibody solution, wash cells three times for 5 min each with 1 mL PBS.
14. Apply secondary antibody (diluted in 1% BSA in PBST) directly onto cells; incubate for 1 h at room temperature (protected from light).
15. Aspirate diluted antibody solution, wash cells three times for 5 min each with 1 mL PBS (protected from light).
    Note: Apply detection reagents according to manufacturer recommendations.
16. Mount coverslips (face down) on microscope slides with mounting medium; seal edges with clear coat nail polish.

Abbreviations

BSA	bovine serum albumin
CMV	cytomegalovirus
DPBS	Dulbecco's phosphate-buffered saline
EDTA	Ethylenediaminetetraacetic acid
EYFP	enhanced yellow fluorescent protein
FLP	flippase
FRT	flippase recognition target
ICC	immunocytochemistry
MHC	major histocompatibility complex
PBS	phosphate buffered saline
PBST	phosphate buffered saline with Tween-20
PCR	polymerase chain reaction
RFLP	restriction fragment length polymorphism
RFP	red fluorescent protein
SIM	Sandoz inbred Swiss mouse
SOC	Super Optimal broth with Catabolite repression
STO	SIM immortalized cell line with thioguanine and ouabain-resistance
TAT	transactivator of transcription
TMP	trimethoprim
YAC	yeast artificial chromosome

References

Abremski, K.E., Hoess, R.H., 1992. Evidence for a second conserved arginine residue in the integrase family of recombination proteins. Protein Eng. 5, 87–91.

Abuin, A., Bradley, A., 1996. Recycling selectable markers in mouse embryonic stem cells. Mol. Cell. Biol. 16, 1851−1856.

Akagi, K., Sandig, V., Vooijs, M., Van der Valk, M., Giovannini, M., Strauss, M., et al., 1997. Cre-mediated somatic site-specific recombination in mice. Nucleic Acids Res. 25, 1766−1773.

Andrews, B.J., Proteau, G.A., Beatty, L.G., Sadowski, P.D., 1985. The FLP recombinase of the 2 micron circle DNA of yeast: interaction with its target sequences. Cell 40, 795−803.

Arakawa, H., Lodygin, D., Buerstedde, J.M., 2001. Mutant loxP vectors for selectable marker recycle and conditional knock-outs. BMC Biotechnol. 1, 7.

Araki, K., Araki, M., Yamamura, K., 1997. Targeted integration of DNA using mutant lox sites in embryonic stem cells. Nucleic Acids Res. 25, 868−872.

Araki, K., Okada, Y., Araki, M., Yamamura, K., 2010. Comparative analysis of right element mutant lox sites on recombination efficiency in embryonic stem cells. BMC Biotechnol. 10, 29.

Belteki, G., Gertsenstein, M., Ow, D.W., Nagy, A., 2003. Site-specific cassette exchange and germline transmission with mouse ES cells expressing phiC31 integrase. Nat. Biotechnol. 21, 321−324.

Brocard, J., Warot, X., Wendling, O., Messaddeq, N., Vonesch, J.L., Chambon, P., et al., 1997. Spatio-temporally controlled site-specific somatic mutagenesis in the mouse. Proc. Natl. Acad. Sci. USA 94, 14559−14563.

Brocard, J., Feil, R., Chambon, P., Metzger, D., 1998. A chimeric Cre recombinase inducible by synthetic, but not by natural ligands of the glucocorticoid receptor. Nucleic Acids Res. 26, 4086−4090.

Buchholz, F., Ringrose, L., Angrand, P.O., Rossi, F., Stewart, A.F., 1996. Different thermostabilities of FLP and Cre recombinases: implications for applied site-specific recombination. Nucleic Acids Res. 24, 4256−4262.

Buchholz, F., Angrand, P.O., Stewart, A.F., 1998. Improved properties of FLP recombinase evolved by cycling mutagenesis. Nat. Biotechnol. 16, 657−662.

Call, L.M., Moore, C.S., Stetten, G., Gearhart, J.D., 2000. A cre-lox recombination system for the targeted integration of circular yeast artificial chromosomes into embryonic stem cells. Hum. Mol. Genet. 9, 1745−1751.

Cawthorne, C., Swindell, R., Stratford, I.J., Dive, C., Welman, A., 2007. Comparison of doxycycline delivery methods for Tet-inducible gene expression in a subcutaneous xenograft model. J. Biomol. Tech. 18, 120−123.

Chen, D., Wu, C.F., Shi, B., Xu, Y.M., 2002a. Tamoxifen and toremifene impair retrieval, but not acquisition, of spatial information processing in mice. Pharmacol. Biochem. Behav. 72, 417−421.

Chen, D., Wu, C.F., Shi, B., Xu, Y.M., 2002b. Tamoxifen and toremifene cause impairment of learning and memory function in mice. Pharmacol. Biochem. Behav. 71, 269−276.

Day, C.D., Lee, E., Kobayashi, J., Holappa, L.D., Albert, H., Ow, D.W., 2000. Transgene integration into the same chromosome location can produce alleles that express at a predictable level, or alleles that are differentially silenced. Genes Dev. 14, 2869−2880.

de Wit, T., Drabek, D., Grosveld, F., 1998. Microinjection of Cre recombinase RNA induces site-specific recombination of a transgene in mouse oocytes. Nucleic Acids Res. 26, 676−678.

Farley, F.W., Soriano, P., Steffen, L.S., Dymecki, S.M., 2000. Widespread recombinase expression using FLPeR (flipper) mice. Genesis 28, 106−110.

Feil, R., Brocard, J., Mascrez, B., LeMeur, M., Metzger, D., Chambon, P., 1996. Ligand-activated site-specific recombination in mice. Proc. Natl. Acad. Sci. USA 93, 10887–10890.

Feil, R., Wagner, J., Metzger, D., Chambon, P., 1997. Regulation of Cre recombinase activity by mutated estrogen receptor ligand-binding domains. Biochem. Biophys. Res. Commun. 237, 752–757.

Feng, Y.Q., Seibler, J., Alami, R., Eisen, A., Westerman, K.A., Leboulch, P., et al., 1999. Site-specific chromosomal integration in mammalian cells: highly efficient CRE recombinase-mediated cassette exchange. J. Mol. Biol. 292, 779–785.

Gagneten, S., Le, Y., Miller, J., Sauer, B., 1997. Brief expression of a GFP cre fusion gene in embryonic stem cells allows rapid retrieval of site-specific genomic deletions. Nucleic Acids Res. 25, 3326–3331.

Garcia-Otin, A.L., Guillou, F., 2006. Mammalian genome targeting using site-specific recombinases. Front Biosci. 11, 1108–1136.

Groth, A.C., Olivares, E.C., Thyagarajan, B., Calos, M.P., 2000. A phage integrase directs efficient site-specific integration in human cells. Proc. Natl. Acad. Sci. USA 97, 5995–6000.

Guldener, U., Heck, S., Fielder, T., Beinhauer, J., Hegemann, J.H., 1996. A new efficient gene disruption cassette for repeated use in budding yeast. Nucleic Acids Res. 24, 2519–2524.

Guo, Z.M., Xu, K., Yue, Y., Huang, B., Deng, X.Y., Zhong, N.Q., et al., 2005. Temporal control of Cre recombinase-mediated *in vitro* DNA recombination by Tet-on gene expression system. Acta Biochim. Biophys. Sin. (Shanghai). 37, 133–138.

Hartung, M., Kisters-Woike, B., 1998. Cre mutants with altered DNA binding properties. J. Biol. Chem. 273, 22884–22891.

Iwamoto, M., Bjorklund, T., Lundberg, C., Kirik, D., Wandless, T.J., 2010. A general chemical method to regulate protein stability in the mammalian central nervous system. Chem. Biol. 17, 981–988.

Jacks, T., Shih, T.S., Schmitt, E.M., Bronson, R.T., Bernards, A., Weinberg, R.A., 1994. Tumour predisposition in mice heterozygous for a targeted mutation in Nf1. Nat. Genet. 7, 353–361.

Jo, D., Nashabi, A., Doxsee, C., Lin, Q., Unutmaz, D., Chen, J., et al., 2001. Epigenetic regulation of gene structure and function with a cell-permeable Cre recombinase. Nat. Biotechnol. 19, 929–933.

Joshi, S.K., Hashimoto, K., Koni, P.A., 2002. Induced DNA recombination by Cre recombinase protein transduction. Genesis 33, 48–54.

Kellendonk, C., Tronche, F., Monaghan, A.P., Angrand, P.O., Stewart, F., Schutz, G., 1996. Regulation of Cre recombinase activity by the synthetic steroid RU 486. Nucleic Acids Res. 24, 1404–1411.

Kondo, S., Takahashi, Y., Shiozawa, S., Ichise, H., Yoshida, N., Kanegae, Y., et al., 2006. Efficient sequential gene regulation via FLP-and Cre-recombinase using adenovirus vector in mammalian cells including mouse ES cells. Microbiol. Immunol. 50, 831–843.

Lauth, M., Moerl, K., Barski, J.J., Meyer, M., 2000. Characterization of Cre-mediated cassette exchange after plasmid microinjection in fertilized mouse oocytes. Genesis 27, 153–158.

Li, Z.W., Stark, G., Gotz, J., Rulicke, T., Gschwind, M., Huber, G., et al., 1996. Generation of mice with a 200-kb amyloid precursor protein gene deletion by Cre recombinase-mediated site-specific recombination in embryonic stem cells. Proc. Natl. Acad. Sci. USA 93, 6158–6162.

Lin, Q., Jo, D., Gebre-Amlak, K.D., Ruley, H.E., 2004. Enhanced cell-permeant Cre protein for site-specific recombination in cultured cells. BMC Biotechnol. 4, 25.

Lindeberg, J., Mattsson, R., Ebendal, T., 2002. Timing the doxycycline yields different patterns of genomic recombination in brain neurons with a new inducible Cre transgene. J. Neurosci. Res. 68, 248−253.

Loonstra, A., Vooijs, M., Beverloo, H.B., Allak, B.A., van Drunen, E., Kanaar, R., et al., 2001. Growth inhibition and DNA damage induced by Cre recombinase in mammalian cells. Proc. Natl. Acad. Sci. USA 98, 9209−9214.

Mao, J., Barrow, J., McMahon, J., Vaughan, J., McMahon, A.P., 2005. An ES cell system for rapid, spatial and temporal analysis of gene function *in vitro* and *in vivo*. Nucleic Acids Res. 33, e155.

Menke, D.B., 2013. Engineering subtle targeted mutations into the mouse genome. Genesis 51, 605−618.

Misra, R.P., Duncan, S.A., 2002. Gene targeting in the mouse: advances in introduction of transgenes into the genome by homologous recombination. Endocrine 19, 229−238.

Monetti, C., Nishino, K., Biechele, S., Zhang, P., Baba, T., Woltjen, K., et al., 2011. PhiC31 integrase facilitates genetic approaches combining multiple recombinases. Methods 53, 380−385.

Novak, A., Guo, C., Yang, W., Nagy, A., Lobe, C.G., 2000. Z/EG, a double reporter mouse line that expresses enhanced green fluorescent protein upon Cre-mediated excision. Genesis 28, 147−155.

Oka, M., Chang, L.J., Costantini, F., Terada, N., 2006. Lentiviral vector-mediated gene transfer in embryonic stem cells. Methods Mol. Biol. 329, 273−281.

Orban, P.C., Chui, D., Marth, J.D., 1992. Tissue- and site-specific DNA recombination in transgenic mice. Proc. Natl. Acad. Sci. USA 89, 6861−6865.

Otsuji, T., Matsumura, H., Suzuki, T., Nakatsuji, N., Tada, T., Tada, M., 2008. Rapid induction of large chromosomal deletions by a Cre/inverted loxP system in mouse ES cell hybrids. J. Mol. Biol. 378, 328−336.

Patsch, C., Peitz, M., Otte, D.M., Kesseler, D., Jungverdorben, J., Wunderlich, F.T., et al., 2010. Engineering cell-permeant FLP recombinase for tightly controlled inducible and reversible overexpression in embryonic stem cells. Stem Cells 28, 894−902.

Peitz, M., Pfannkuche, K., Rajewsky, K., Edenhofer, F., 2002. Ability of the hydrophobic FGF and basic TAT peptides to promote cellular uptake of recombinant Cre recombinase: a tool for efficient genetic engineering of mammalian genomes. Proc. Natl. Acad. Sci. USA 99, 4489−4494.

Psarras, S., Karagianni, N., Kellendonk, C., Tronche, F., Cosset, F.L., Stocking, C., et al., 2004. Gene transfer and genetic modification of embryonic stem cells by Cre- and Cre-PR-expressing MESV-based retroviral vectors. J. Gene. Med. 6, 32−42.

Qu, A., Jiang, C., Cai, Y., Kim, J.H., Tanaka, N., Ward, J.M., et al., 2014. Role of Myc in hepatocellular proliferation and hepatocarcinogenesis. J. Hepatol. 60, 331−338.

Raymond, C.S., Soriano, P., 2007. High-efficiency FLP and phiC31 site-specific recombination in mammalian cells. PLoS One 2, e162.

Raz, R., Lee, C.K., Cannizzaro, L.A., d'Eustachio, P., Levy, D.E., 1999. Essential role of STAT3 for embryonic stem cell pluripotency. Proc. Natl. Acad. Sci. USA 96, 2846−2851.

Rodriguez, C.I., Buchholz, F., Galloway, J., Sequerra, R., Kasper, J., Ayala, R., et al., 2000. High-efficiency deleter mice show that FLPe is an alternative to Cre-loxP. Nat. Genet. 25, 139−140.

Sandhu, U., Cebula, M., Behme, S., Riemer, P., Wodarczyk, C., Metzger, D., et al., 2011. Strict control of transgene expression in a mouse model for sensitive biological applications based on RMCE compatible ES cells. Nucleic Acids Res. 39, e1.

Sando III, R., Baumgaertel, K., Pieraut, S., Torabi-Rander, N., Wandless, T.J., Mayford, M., et al., 2013. Inducible control of gene expression with destabilized Cre. Nat. Methods 10, 1085–1088.

Sauer, B., 1987. Functional expression of the Cre-lox site-specific recombination system in the yeast *Saccharomyces cerevisiae*. Mol. Cell. Biol. 7, 2087–2096.

Sauer, B., Henderson, N., 1990. Targeted insertion of exogenous DNA into the eukaryotic genome by the Cre recombinase. New Biol. 2, 441–449.

Sauer, B., McDermott, J., 2004. DNA recombination with a heterospecific Cre homolog identified from comparison of the pac-c1 regions of P1-related phages. Nucleic Acids Res. 32, 6086–6095.

Schaft, J., Ashery-Padan, R., van der Hoeven, F., Gruss, P., Stewart, A.F., 2001. Efficient FLP recombination in mouse ES cells and oocytes. Genesis 31, 6–10.

Schnutgen, F., Stewart, A.F., von Melchner, H., Anastassiadis, K., 2006. Engineering embryonic stem cells with recombinase systems. Methods Enzymol. 420, 100–136.

Shimshek, D.R., Kim, J., Hubner, M.R., Spergel, D.J., Buchholz, F., Casanova, E., et al., 2002. Codon-improved Cre recombinase (iCre) expression in the mouse. Genesis 32, 19–26.

Shui, J.W., Tan, T.H., 2004. Germline transmission and efficient DNA recombination in mouse embryonic stem cells mediated by adenoviral-Cre transduction. Genesis 39, 217–223.

Singla, V., Hunkapiller, J., Santos, N., Seol, A.D., Norman, A.R., Wakenight, P., et al., 2010. Floxin, a resource for genetically engineering mouse ESCs. Nat. Methods 7, 50–52.

Soukharev, S., Miller, J.L., Sauer, B., 1999. Segmental genomic replacement in embryonic stem cells by double lox targeting. Nucleic Acids Res. 27, e21.

Srinivas, S., Watanabe, T., Lin, C.S., William, C.M., Tanabe, Y., Jessell, T.M., et al., 2001. Cre reporter strains produced by targeted insertion of EYFP and ECFP into the ROSA26 locus. BMC Dev. Biol. 1, 4.

Sternberg, N., Hamilton, D., Austin, S., Yarmolinsky, M., Hoess, R., 1981. Site-specific recombination and its role in the life cycle of bacteriophage P1. Cold Spring Harb. Symp. Quant. Biol. 45 (Pt. 1), 297–309.

Su, H., Wang, X., Bradley, A., 2000. Nested chromosomal deletions induced with retroviral vectors in mice. Nat. Genet. 24, 92–95.

Sung, Y.H., Baek, I.J., Seong, J.K., Kim, J.S., Lee, H.W., 2012. Mouse genetics: catalogue and scissors. BMB Rep. 45, 686–692.

Suzuki, E., Nakayama, M., 2011. VCre/VloxP and SCre/SloxP: new site-specific recombination systems for genome engineering. Nucleic Acids Res. 39, e49.

Takata, Y., Kondo, S., Goda, N., Kanegae, Y., Saito, I., 2011. Comparison of efficiency between FLPe and Cre for recombinase-mediated cassette exchange *in vitro* and in adenovirus vector production. Genes Cells 16, 765–777.

Tchorz, J.S., Suply, T., Ksiazek, I., Giachino, C., Cloetta, D., Danzer, C.P., et al., 2012. A modified RMCE-compatible Rosa26 locus for the expression of transgenes from exogenous promoters. PLoS One 7, e30011.

Thomson, J.G., Ow, D.W., 2006. Site-specific recombination systems for the genetic manipulation of eukaryotic genomes. Genesis 44, 465–476.

Thorpe, H.M., Smith, M.C., 1998. *In vitro* site-specific integration of bacteriophage DNA catalyzed by a recombinase of the resolvase/invertase family. Proc. Natl. Acad. Sci. USA 95, 5505–5510.

Thyagarajan, B., Olivares, E.C., Hollis, R.P., Ginsburg, D.S., Calos, M.P., 2001. Site-specific genomic integration in mammalian cells mediated by phage phiC31 integrase. Mol. Cell. Biol. 21, 3926−3934.

Toledo, F., Liu, C.W., Lee, C.J., Wahl, G.M., 2006. RMCE-ASAP: a gene targeting method for ES and somatic cells to accelerate phenotype analyses. Nucleic Acids Res. 34, e92.

Utomo, A.R., Nikitin, A.Y., Lee, W.H., 1999. Temporal, spatial, and cell type-specific control of Cre-mediated DNA recombination in transgenic mice. Nat. Biotechnol. 17, 1091−1096.

Vintersten, K., Monetti, C., Gertsenstein, M., Zhang, P., Laszlo, L., Biechele, S., et al., 2004. Mouse in red: red fluorescent protein expression in mouse ES cells, embryos, and adult animals. Genesis 40, 241−246.

Wagner, K.U., Wall, R.J., St-Onge, L., Gruss, P., Wynshaw-Boris, A., Garrett, L., et al., 1997. Cre-mediated gene deletion in the mammary gland. Nucleic Acids Res. 25, 4323−4330.

Wang, Y., Krushel, L.A., Edelman, G.M., 1996. Targeted DNA recombination *in vivo* using an adenovirus carrying the Cre recombinase gene. Proc. Natl. Acad. Sci. USA 93, 3932−3936.

Warren, D., Laxmikanthan, G., Landy, A., 2008. A chimeric Cre recombinase with regulated directionality. Proc. Natl. Acad. Sci. USA 105, 18278−18283.

Will, E., Klump, H., Heffner, N., Schwieger, M., Schiedlmeier, B., Ostertag, W., et al., 2002. Unmodified Cre recombinase crosses the membrane. Nucleic Acids Res. 30, e59.

Wirth, D., Gama-Norton, L., Riemer, P., Sandhu, U., Schucht, R., Hauser, H., 2007. Road to precision: recombinase-based targeting technologies for genome engineering. Curr. Opin. Biotechnol. 18, 411−419.

Wunderlich, F.T., Wildner, H., Rajewsky, K., Edenhofer, F., 2001. New variants of inducible Cre recombinase: a novel mutant of Cre-PR fusion protein exhibits enhanced sensitivity and an expanded range of inducibility. Nucleic Acids Res. 29, E47.

Yu, H., Kessler, J., Shen, J., 2000. Heterogeneous populations of ES cells in the generation of a floxed Presenilin-1 allele. Genesis 26, 5−8.

Zhang, J., Zhao, J., Jiang, W.J., Shan, X.W., Yang, X.M., Gao, J.G., 2012. Conditional gene manipulation: Cre-ating a new biological era. J. Zhejiang Univ. Sci. B. 13, 511−524.

Zheng, B., Sage, M., Sheppeard, E.A., Jurecic, V., Bradley, A., 2000. Engineering mouse chromosomes with Cre-loxP: range, efficiency, and somatic applications. Mol. Cell. Biol. 20, 648−655.

Zhu, Y., Soto, J., Anderson, B., Riehle, C., Zhang, Y.C., Wende, A.R., et al., 2013. Regulation of fatty acid metabolism by mTOR in adult murine hearts occurs independently of changes in PGC-1alpha. Am. J. Physiol. Heart Circ. Physiol. 305, H41−H51.

6 Retrovirus-Mediated Gene Transfer

Bon Chul Koo, Mo Sun Kwon and Teoan Kim

Department of Physiology, Catholic University of Daegu School of Medicine, Daegu, Republic of Korea

I. Retroviruses

Retroviruses are animal viruses with two identical strands of RNA in their virion (reviewed in Temin, 1987). As the prefix *retro-* indicates, when these viruses infect a host cell, the viral RNA is reverse transcribed in the cytoplasm to make linear double-stranded DNA, which is transported into the host cell nucleus and integrates into a chromosome directly without any change in its original linear form (Ellis and Bernstein, 1989). The viral genes are expressed from this integrated form of DNA, the provirus, and the progeny viruses are produced from the infected host cell as a result of proviral gene expression (Figure 6.1).

According to older taxonomic classifications centered on electromicroscopy-based morphology, genera of the family retroviridae were divided into four types (A, B, C, and D) (Goff, 2001). Among them, C-type retroviruses were extensively studied and therefore well characterized for use in vector construction. The most commonly used C-type retroviruses are murine leukemia virus (MLV), murine sarcoma virus (MSV), Gibbon ape leukemia virus (GaLV), avian leucosis virus (ALV), reticuloendotheliosis virus (REV), and Rous sarcoma virus (RSV). Like ALV, REV and RSV are avian retroviruses (Teich, 1985). As the number of retroviral genera increased significantly, the International Committee on Taxonomy of Viruses has more recently classified the retroviridae family into seven retrovirus genera: (i) alpharetrovirus, (ii) betaretrovirus, (iii) gammaretrovirus, (iv) deltaretrovirus, (v) epsilonretrovirus, (vi) lentivirus, and (vii) spumavirus. Based on the complexity of genomic structure, viruses of alpha-, beta-, and gammaretroviruses are categorized as "simple" retroviruses, whereas the viruses belonging to the remaining four retrovirus genera are called "complex" retroviruses (Goff, 2001). Among the seven genera of retroviruses, gammaretroviruses encompassing MLV and GaLV have been most widely used as virus vectors. Recently, lentiviruses, including human immunodeficiency virus type 1 (HIV-1, Pfeifer et al., 2002), feline immunodeficiency virus (FIV, Xu et al., 2013), and equine infectious anemia virus (EIAV, Whitelaw, et al., 2004; Lillico et al., 2007), are becoming more popular than gammaretroviruses in use as gene transfer vectors for their ability to insert foreign genes conveniently into the chromosomes of nondividing target cells (Naldini et al., 1996).

Transgenic Animal Technology. DOI: http://dx.doi.org/10.1016/B978-0-12-410490-7.00006-2

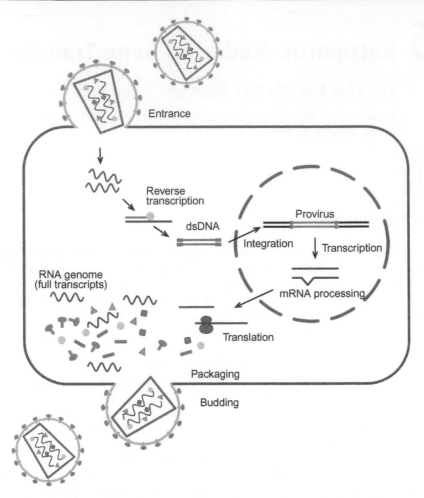

Figure 6.1 The life cycle of a retrovirus. The genome of the retrovirus consists of two identical copies of RNA. After infection, the reverse-transcribed DNA integrates into the host cell chromosome. The viral genes are expressed from this integrated form of DNA, provirus. Following transcription, both full-length and spliced RNAs are exported out of the nucleus, and then translation follows. Because spliced RNAs are devoid of encapsidation signal sequence, only nonspliced full-length RNAs are packaged by viral proteins, and the resulting virions are budding out of cells.

II. Strategies of Retrovirus Vector Systems

Simple retroviruses have three *trans*-acting protein-coding genes: *gag*, *pol*, and *env* (Figure 6.2). The proteins encoded by the *gag* gene are responsible for virus internal structure. The *pol* gene encodes enzymes, including the carboxyl portion of protease, for posttranslational cleavage of viral proteins; reverse transcriptase (RT), to

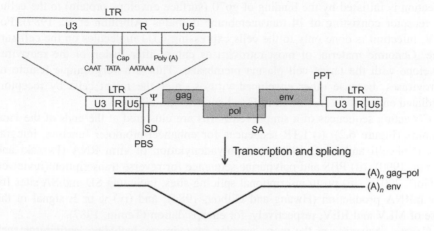

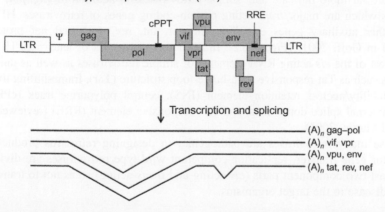

Figure 6.2 Structures of proviruses and their transcription patterns. Each of the two LTRs (long terminal repeats) consists of U3, R, and U5. Transcription of provirus begins at the first base of 5′ R and ends at the last base of 3′ R with Cap or polyadenylation at each end. Colored areas indicate *trans*-acting sequences. Each viral protein is shown at the right end of the corresponding transcript. *Abbreviations*: U3, the sequence unique to the 3′ end of mRNA; R, the sequence of direct repeats; U5, the sequence unique to the 5′ end of mRNA; PBS, primer binding site; SD, splicing donor site; ψ, encapsidation site; SA, splicing acceptor site; PPT, polypurine track; AATAAA, sequence of polyadenylation signal; poly (A), polyadenylation site; cPPT, central polypurine track; RRE, Rev-responsive element.

synthesize DNA from RNA template; and integrase, for the integration of reverse-transcribed viral DNA into the host cell chromosome (reviewed in Nishijima and Iijima, 2013). The *env* gene encodes viral envelope glycoproteins, which are part of the outside of the progeny virus and are the primary determinant of the retrovirus host range. For example, in the case of MoMLV (Moloney murine leukemia virus), which has been the most popular simple retrovirus applied in gene transfer, infection is initiated by the binding of gp70 (surface envelope protein) to the cellular receptor consisting of 14 transmembrane domains (Albritton et al., 1989). For HIV, infection is done only to the cells expressing CD4 molecules on the cell surface. Genomic material of most retroviruses enters after fusion of the retrovirus envelope with the target cell plasma membrane. This is a very unique feature of retroviruses, because other enveloped viruses enter the target cells by receptor-mediated endocytosis (reviewed in Freed and Martin, 2001).

Cis-acting sequences of a simple retrovirus are clustered at the ends of the viral genome (Figure 6.2): (i) LTR sequences for enhancer/promoter function, integration (Colicelli and Goff, 1988), and polyadenylation of viral RNA (Iwasaki and Temin, 1990); (ii) PBS and polypurine sequence for reverse transcription (reviewed in Goff, 2001); (iii) posttranscriptional splicing sites, including SD and SA sites for env mRNA production (Hwang and Gilboa, 1984); and (iv) ψ or E signal in the case of MLV and REV, respectively, for encapsidation (Temin, 1987).

Genetic sequences of the more complex retroviruses, including lentiviruses such as HIV, are far more intricate than simple retroviruses. In addition to gag, pol, and env genes which are major trans-acting protein-coding genes of retroviruses, HIV-1 carries other auxiliary genes encoding vif, vpr, tat, rev, vpu, and nef proteins (reviewed in Goff, 2001) (Figure 6.2). In cis-acting sequences of lentivirus, HIV-1 carries most of the cis-acting RNA elements of simple retroviruses as well as unique sequences such as Tat-responsive stem-bulge-loop structure (Tar), frameshifting motif (FS), instability/nuclear retention element (INS), central polypurine track (cPPT), multiple internal splice donor sites, and Rev-responsive element (RRE) (reviewed in Freed and Martin, 2001).

Because many retroviruses are pathogenic, in designing retrovirus production systems for gene transfer, replication-competent wild-type retroviruses are divided into at least two component parts (cis-acting and trans-acting) so as not to transmit possible disease to the target organisms.

A. Packaging Cells

Helper cells (or packaging cells) are usually constructed by the transfection of viral trans-acting sequences (gag, pol, and env) to an appropriate eukaryotic cell line. Because the viral RNAs produced from helper cells are devoid of retroviral cis-acting sequences, especially the encapsidation sites, the RNAs from helper cells cannot be encapsidated into retroviral virions. The purpose of a packaging cell is, therefore, to provide GAG, POL, ENV, and other essential proteins to the retrovirus vector having no trans-acting sequences. Unlike the construction of a replication-defective retrovirus vector, the promoters for gag, pol, and env gene expression need not be retroviral LTR promoters. Any promoter that is active in the helper cell can be used, because the helper cell does not require retrovirus cis-acting sequences, even though most helper cell lines available use the LTR promoter due to its strong activity. Most importantly, regardless of promoter, the ψ sequence (between the SD and the 5' end of gag) should be removed to prevent RNAs of *gag*, *pol*, and *env* from encapsidation. Summarized packaging cell lines capable of producing retrovirus vectors with high efficiency (around 10^6 infectious replication-defective retroviruses per milliliter of culture medium) are shown in Table 6.1.

Among retrovirus production systems for gene transfer in mammalian cells, NIH3T3 cells transformed with appropriate MLV genes were the most popular, due largely to the relatively simple structure and good characterization of MLV, as well as the permissiveness of NIH3T3 cells for MLV gene expression. The three kinds of MLV are ecotropic, amphotropic, and xenotropic (Weiss, 1982). Ecotropic MLV is able to infect only the cells of mice and rats, amphotropic MLV can infect not only murine cells but also other cells, and xenotropic MLV can infect only nonmurine cells. The host range difference is due primarily to differences in the envelope protein, which interacts with specific receptors on the host cell membrane. For most currently available NIH3T3 cell-based packaging cell lines, the *gag* and *pol* genes are from ecotropic MoMLV, while the *env* gene is from multitropic.

In earlier packaging cell line construction, NIH3T3 cells were transfected with the *gag*, *pol*, and *env* genes of MLV in a single transcriptional unit, causing production of the replication-competent helper virus, due primarily to homologous recombination between the vector sequence and helper virus sequences in a packaging cell (ψ-2, MXEN, and PA12 of Table 6.1) (Mann et al., 1983; Sleckman et al., 1987; Miller and Buttimore, 1986). Improvements in decreasing the possibility of homologous recombination were made in later works in several ingenious ways. For the *trans*-acting DNA sequence in a PA317 cell, deletion was made at the 5′

Table 6.1 Summary of Retrovirus Packaging Cell Lines

Packaging Cell Line	Line Derivation	Host Range	References
Ψ-2	NIH3T3	Ecotropic	Mann et al. (1983)
MXEN	NIH3T3	Xenotropic	Sleckman et al. (1987)
PA12	NIH3T3	Amphotropic	Miller and Buttimore (1986)
PA317	NIH3T3	amphotropic	Miller and Buttimore (1986)
PG13	NIH3T3	Multitropic[d]	Miller et al. (1991)
PT67[a]	NIH3T3	Multitropic[d]	Miller and Chen (1996)
293GP[b] + *VSV-G*	293	Pantropic[e]	Burns et al. (1993)
GP2-293[b] + *VSV-G*	293	Pantropic	Batchu et al. (2003)
293T[c] + *VSV-G*	293	Pantropic	Johnston et al. (2012)
293FT[c] + *VSV-G*	293	Pantropic	Javanbakht et al. (2003)

[a]PT67 is the commercial name of the 10A1 packaging cell line.
[b]Both 293GP and GP2-293 cells harbor only *gag/pol* genes of MLV. *VSV-G* is a substitute for the retrovirus envelope. Due to the cytotoxicity of *VSV-G*, transfection of this gene is done just before vector virus harvest. These cells are used for simple retroviral vector production.
[c]Neither 293T nor 293FT cells carry *trans*-acting sequences of retrovirus. To prevent possible biohazard risk and cytotoxicity, all *trans*-acting sequences including *gag/pol and VSV-G genes* are introduced to these cells by transient transfection. These cells are used for complex retroviral vector production.
[d]The viruses produced from PG13 packaging cells can infect most mammalian cells with exception of mouse cells. The *env* gene of PT67 cells is derived from 10A1 MLV, and the virions packaged with this Env can also infect mouse cells.
[e]*VSV-G* can infect all animal cells tested to date.

end of 5' LTR, and 3' LTR was replaced with an SV40 late polyadenylation sequence. Owing to continuous improvements, PG13 (ATCC CRL-10686), one of the most advanced NIH3T3 cell-based packaging cell lines, and features further progress in prevention of replication-competent virus production by separating the transcription unit of *env* from that of *gag/pol* and by substituting GaLV *env* gene for MLV *env* gene (Miller et al., 1991). Except for mouse cells, the viruses produced from PG13 cells can infect a variety of cells, even hamster and bovine cells, which have been known to be very resistant to amphotropic MLV infection (Kim et al., 1993). Later, further development in the PT67 cell line (ATCC CRL-12284) employed the *env* gene of 10A1 MLV, thus enabling the progeny viruses to infect mouse cells also (Miller and Chen, 1996).

The most popular packaging cell line at present is the 293-based packaging cell line, which takes advantage of far superior DNA transfection efficiency of adenovirus-transformed 293 human embryonic kidney cells (ATCC CRL-1573) to that of traditional mouse NIH3T3 cells. After transfection of retrovirus vector DNA to the 293-based packaging cells, therefore, there is no need to select the best virus-producing cells. In general, for the NIH3T3 cell-based retrovirus vector system, the subcloning of multiple numbers of individual virus-producing cells or at least pooling of drug-resistant (e.g., neomycin-resistant) colonies is necessary to prepare an efficient virus-producing cell line. Most of the 293-based packaging cell lines were designed to express *gag* and *pol* genes of MLV (or HIV) and the vesicular stomatitis virus glycoprotein G (*VSV-G*) gene (Burns et al., 1993; Johnston et al., 2012). Consequently, the viral vector is packaged with VSV-G instead of MLV (or HIV) envelope protein, and the resulting hybrid or so-called pseudotyped virus becomes significantly stable from external stress. For example, the stock of viral vector packaged with VSV-G can be centrifugally concentrated as much as 2000-fold, resulting in the titer of more than 10^9 transforming units per milliliter (Burns et al., 1993). Another characteristic of the MLV (or HIV) and VSV-G hybrid retrovirus vector is its pantropic nature, which enables the virus to infect almost every kind of animal cell including fish cells (Lu et al., 1997; Linney et al., 1999).

The *trans*-acting genes designed to express in the NIH3T3- or 293-based packaging cells that are the most popular at present are depicted in Figure 6.3. With complex retrovirus vector systems, further development is in progress. At present, the fourth generation of packaging cells is commercially available. One marked feature of the newest version is separate expression of *gag* and *pol* genes by placing them on two different plasmids.

B.   Retrovirus Vectors

To construct a virus that can be used as a transgene vector, only *cis*-acting DNA sequences should be retained with the gene sequence of interest. Consequently, the resultant virus is replication defective, but with the coordination of *trans*-acting functions provided from another source (i.e., a helper cell or packaging cell), the virus becomes infective to target cells. The target cell has no *trans*-acting elements

(A) NIH3T3-based packaging cell

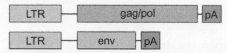

(B) 293-based pseudotype simple retrovirus packaging cell

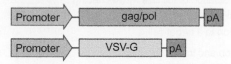

(C) 293-based pseudotype complex retrovirus packaging cell (third generation)

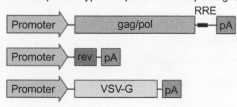

Figure 6.3 Three types of packaging cells. (A) PG13 and PT67 are the most popular NIH3T3-based packaging cell lines at present. In this system, the cells are designed to express *gag/pol* genes of MoMLV, and *env* gene of GaLV or 10A1 MLV. LTR, long terminal repeats of MoMLV; pA, SV40 early polyadenylation signal. (B) In the 293-based pseudotype packaging cell line (293GP and GP2-293) for simple retroviral vector system, the *gag/pol* genes are derived from simple retrovirus (mostly MLV). Full packaging ability of these cells is gained by transient transfection of the *VSV-G* gene sequence. (C) In the 293-based pseudotype packaging cell line for complex retroviral vector system (third generation), the *gag/pol* genes are derived from complex retroviruses such as HIV, EIAV, or FIV. The *rev* gene was included for efficient cytoplasmic export of the transcripts carrying RRE (Hammarskjold et al., 1989; Felber et al., 1989). Acquisition of full packaging ability of the complex retrovirus packaging cells is done by transient cotransfection of 293T or 293FT cells with all gene sequences identified.

of the retrovirus; therefore, the infected virus cannot produce its progeny, and no spread of pathogenic viruses is expected from the target cells.

The essential *cis*-acting sequences that should be included in the simple retrovirus vectors are ψ sequence for viral packaging; both 5′ and 3′ LTRs, PBS, and PPT for integration of reverse-transcribed vector DNA into the target cell chromosome (Figure 6.4). In addition, it is strongly recommended to include the 5′ end of the *gag* gene for efficient packaging of viral vector RNA (Adam and Miller, 1988). In complex retrovirus vectors, additional *cis*-acting sequences are included in the latest version, such as RRE and cPPT, which are not present in the simple retrovirus genome. RRE helps packaging and cytoplasmic export of full-length vector transcripts, while cPPT enhances nuclear import of the vector cDNA in the target cell (Pfeifer and Verma, 2001).

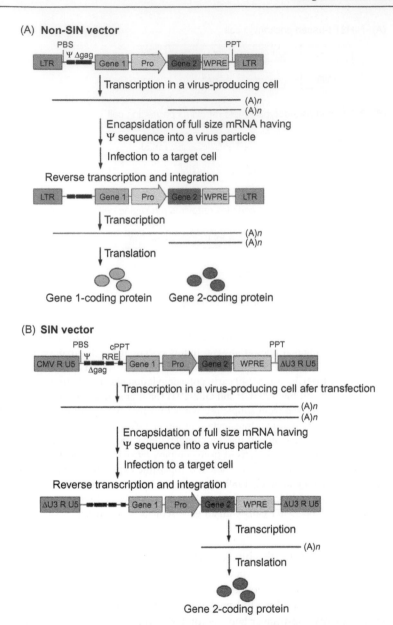

Figure 6.4 Schematic representations of the typical retroviral vectors and their expression pattern. In the virus-producing cell, regardless of non-SIN (A) or SIN vectors (B), the full-sized transcript driven from the 5′ end LTR or hybrid (indicated as "CMV-R-U5" in the figure) promoter is translated and encapsidated, while the small-sized transcript driven from the internal promoter (indicated as "pro") is translated but not encapsidated, due to the lack of an encapsidation sequence (designated "ψ"). In SIN vectors, U3 of 3′ LTR is partially deleted (indicated as "ΔU3"), mainly to prevent replication-competent virus production as a

To enhance transgene expression, nonretroviral *cis*-acting sequence is often added to the retrovirus vector. It was reported that insertion of woodchuck posttranscriptional regulatory element (WPRE) sequence downstream of the transgene increased expression of the transgene more than fivefold (Zufferey et al., 1999). In retrovirus vector construction, it should also be noted that unlike the WPRE sequences, a poly(A) signal of the transgene to be inserted inside a retrovirus vector should be deleted; otherwise, full-length transcription of the retrovirus vector can be hampered (Temin, 1987).

Among the many different types of the retrovirus vectors developed so far, non-SIN vectors having the structure LTR—cDNA of gene 1—internal promoter—cDNA of gene 2—LTR are most common (Figure 6.4A). From this structure, two exogenous genes are transferred simultaneously and expression of gene 2 is controllable by choosing an appropriate internal promoter. Usually a selection marker gene (e.g., neomycin resistance gene or hygromycin B resistance gene), facilitating selection of the best virus-producing cell colony during virus-producing cell construction is placed under the 5′ LTR promoter, while the gene of main interest to be transferred (gene 2, in this case) is located downstream of the internal promoter. The advantage of this type of vector is that transcriptional inactivation of the LTR promoter can be overcome in some target cells. One major drawback of this structure is, however, mutual suppression of two adjacent promoters—i.e., LTR and the internal promoter (Emerman and Temin, 1986; Soriano et al., 1991). Since the mutual suppression of two adjacent promoters is negligible in most cases, retrovirus vectors with this structure have been widely embraced. However, use of triple promoters (LTR and two internal promoters) in a single retrovirus vector is not recommended in light of likely transgene expression suppression.

Solutions for promoter interference were found from the self-inactivating (SIN; Figure 6.4B) and multicistronic vectors (Figure 6.5). As mentioned earlier, construction of a SIN vector is made by deleting the enhancer and part of the promoter from the U3 region of the 3′ LTR. In the target cell, however, the 5′ LTR of the provirus loses its promoter activity due to deletion of the U3 region in the processing of reverse transcription (Whitcomb and Hughes, 1992). For this reason, the virus-producing cells are to be constructed through transfection (not infection) of recombinant viral DNA to the packaging cells. Advantages other than overcoming promoter interference include decreased chance of the downstream endogenous

---

◀ result of homologous recombination. Use of a hybrid promoter (consisting of CMV promoter, R and U5) instead of intact 5′ LTR might further decrease homologous recombination events in the virus-producing cells. Δgag and WPRE indicate the 5′ portion of *gag* and woodchuck posttranscriptional regulatory element, respectively. The sequences of RRE and cPPT are employed in complex retrovirus vectors only. *In the target cell*, both full-sized and small-sized transcripts are made from the provirus of non-SIN vectors. With the SIN vector, however, only the small-sized transcript driven by the internal promoter is synthesized in the target cell, because the 5′ LTR loses promoter activity due to deletion of the U3 region in the processing of reverse transcription.

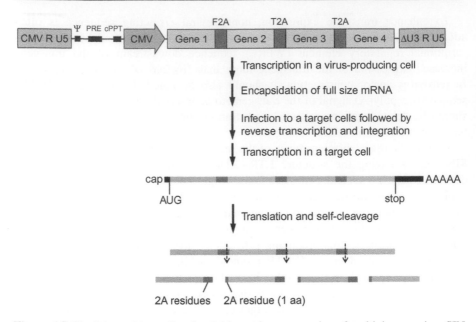

Figure 6.5 The 2A peptide-mediated stoichiometric coexpression of multiple genes in a SIN retrovirus vector. Through an enzyme-independent ribosomal skip mechanism, proteins linked by 2A peptide are cleaved between glycine and proline residues. As a result, a short 2A peptide (17–21 amino acids) is fused to the C-terminus of the upstream protein, while the remaining single proline residue is attached to the N-terminus of the downstream protein (Ryan and Drew, 1994; Provost et al., 2007). To prevent homologous recombination, which occurs frequently in retroviral vectors, it is recommended to use 2A peptide sequences derived from different picornavirus species, i.e., foot-and-mouth disease virus (F2A), equine rhinitis A virus (E2A), *Thosea asigna* virus (T2A), or porcine teschovirus-1 (P2A; Szymczak et al., 2004).

gene expression from the 3′ LTR (reviewed in Ponder, 2001) and prevention of some possible side effects accompanying lentivirus-based vectors (Miyoshi et al., 1998; Deglon et al., 2000).

Initial multicistronic vector was constructed simply by replacing the internal promoter of a non-SIN vector with an internal ribosomal entry site (IRES, Aran et al., 1998; Harries et al., 2000), enabling cap-independent translation of the downstream gene. However, a critical disadvantage of IRES is that the gene expression is nonstoichiometric. That is, expression levels of the genes in the same open reading frame (ORF) are usually disproportionate (Mizuguchi et al., 2000). In addition, in our experience, expression of the gene downstream of IRES has been frequently unsuccessful. Recently, 2A self-processing peptide derived from the viruses of the *Picornaviridae* family has suggested a promising alternative (Figure 6.5). Placement of a short DNA sequence encoding 2A peptide (~18−22 amino acids depending on the virus) as a linker between tandem cDNAs in a single ORF resulted in stoichiometric or equimolar coexpression of at least four different unfused protein products (Szymczak et al., 2004; Provost et al., 2007; Fang et al., 2005). Using the retrovirus or lentivirus vector designed for the 2A peptide-mediated stoichiometric coexpression of four transcription factors (*KLF4*, *OCT3/4*, *SOX2*, and *c-MYC*), successful generation of induced pluripotent stem (iPS) cells of murine and human was demonstrated (Shao et al., 2009; Carey et al., 2009).

C. Virus-Producing Cells

Virus-producing cells are constructed by either transfection of retrovirus vector plasmid to helper cells or infection of the helper cells with the viruses produced from other virus-producing cells. The advantage of the latter approach over the former is higher virus productivity in general, because of lower susceptibility to methylation of the provirus in the host cellular genome (Hwang and Gilboa, 1984). One thing to be noted in the latter approach is that env protein produced from the helper cell should not be identical to that of the virus intended to infect the helper cell. Retroviruses cannot infect cells producing envelope proteins of the same class, because the proteins block the interaction between cellular receptors and specific envelope proteins before infection. This phenomenon is called *superinfection interference* (Temin, 1988). However, this phenomenon does not need to be considered because most of the packaging cells being used at present are designed to express VSV-G as an envelope protein (Burns et al., 1993). When using SIN vector, as mentioned in the previous section, introduction of the retrovirus vector sequence to the packaging cell can be done only by the transfection approach; otherwise, the resultant cells cannot produce the progeny viruses due to deletion of the U3 region in the 5′ LTR during reverse transcription (Figure 6.4).

A general strategy for the construction of replication-defective, retrovirus-producing cells is summarized in Figure 6.6. For the production of complex retroviral vector (i.e., HIV-1 based vector), 293-based cells such as 293T (Gay et al., 2012; Benabdellah et al., 2011) or 293FT (Javanbakht et al., 2003; Kaleri et al., 2011) are transiently cotransfected with all the plasmids carrying *cis-* and *trans*-acting sequences for the cytotoxicity of lentiviral *gag/pol* gene product and VSV-G (Krausslich, 1992; Krausslich et al., 1993; Yee et al., 1994) (Figure 6.6A). For simple retrovirus production (Figure 6.6B), either 293GP (Burns et al., 1993) or GP2-293 cells (Batchu et al., 2003) are most widely used. Unlike 293T or 293FT cells, these cells already carry *gag/pol* genes of MLV, and introduction of only the sequences of vector and *VSV-G* is required for final construction of virus-producing cells. Frequently PG13 (Miller et al., 1991) or PT67 (Miller and Chen, 1996) cells harboring all necessary *trans*-acting gene sequences are also used. There are three choices applied in the construction simple retrovirus vector-producing cells: (i) transient cotransfection with vector sequence and the *VSV-G* gene (Figure 6.6B1); (ii) permanent transfection of vector sequence, followed by transient

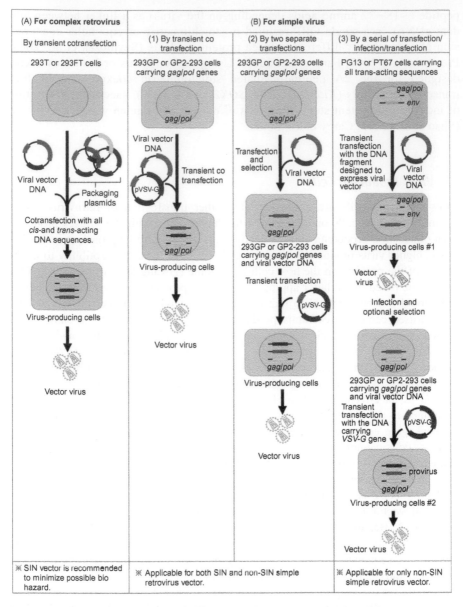

Figure 6.6 General strategy for the construction of replication-defective retrovirus-producing cells and introduction of genes in target cells.

transfection of *VSV-G* gene (Figure 6.6B2); and (iii) serial steps of transfection, infection, selection, and finishing with transient transfection of *VSV-G* gene (Figure 6.6B3). Both selection of transduced cells and infection-mediated introduction of viral vector sequences are expected to increase overall virus productivity.

III. Problems in the Use of Retrovirus Vectors in Gene Transfer

Although retrovirus vectors have several advantages, some drawbacks include size limitation, a high rate of recombination, and low titer. Because the most significant limitation of the retrovirus vector systems is low titer of the viruses, the following discussion focuses mainly on titer.

A. Size

The maximum permissible size for efficient encapsidation or reverse transcription of each retrovirus vector is approximately 10 kb (reviewed in Temin, 1987). Although this packaging limit might be extended further by using VSV-G as envelope protein, it is evident that the shorter the size is, the higher the titer obtained. It has been reported that increase in vector size results in semilogarithmical decrease in titer (Kumar et al., 2001). Because introns of foreign genes carried by retrovirus vectors are spliced out during replication, the insert does not have to contain introns. Considering that an insert of permissible size (around 8 kb) can code for a protein as big as 300 kDa, a maximum size limitation does not seem to be a significant obstacle in most cases. However, in the case of introns containing some regulatory sequences, removal of introns as the result of retrovirus-mediated gene transfer might affect the expression level in transgenic animals (Brinster et al., 1988). Modification of the RNA splicing sites and/or gathering dispersed transcriptional regulatory elements to shorten the enhancer/promoter size must be done for successful transgenic animal production (Lillico et al., 2007).

B. Recombination

Genetic stability of retroviruses is intrinsically very poor, and genetic variations, including deletion, base-pair substitution, insertion, and recombination, increase as the number of replication cycles increases (Temin, 1989). However, the provirus is as stable as the cellular genome because it is a part of the chromosome. Consequently, the genetic variation in retrovirus vector stock harvested from proviral transcripts in virus-producing cells is very low and tolerable, the exception to which is recombination.

The most serious effect of recombination is production of replication-competent retrovirus from virus-producing cells. Fortunately, most of the currently available retrovirus vector systems deal with this problem by reducing the homologous sequences between DNAs for packaging cells and vector and by using different plasmids to separate *gag*, *pol*, and *env* genes. In addition, use of SIN vector also significantly reduces homologous recombination.

C. Titer

Because, in most cases, the titer of viruses is measured by the expression level of progeny virions on the target cells, factors affecting the titer are reviewed in the following four sections.

1. Virus Productivity

When determining virus productivity from packaging cells, at least two factors are involved: (i) permissiveness of a specific cell line in regard to its use as a virus-producing cell line for the expression of *gag*, *pol*, and *env* genes as well as the genes of the retrovirus vector and (ii) nonviral, negative *cis*-acting sequences in the retrovirus vector.

An example of permissiveness is the poor expression of the avian spleen necrosis virus LTR in rat cells compared to that in dog or chicken cells (Emerman and Temin, 1984). In the case of *gag*, *pol*, and *env* genes, their expression can be optimized by using appropriate promoters that are active in the cells to be used as a packaging cell line. Likewise, transcription of the provirus can be optimized by substituting an appropriate promoter for the enhancer/promoter region of U3.

The best example of a nonviral, negative *cis*-acting sequence is the neomycin resistance gene of prokaryotic origin (Artelt et al., 1991), which is one of the most frequently used reporter genes in eukaryotic cells. Artelt et al. reported that the neomycin resistance gene acts as a silencer on proximal promoters, indicating suppression of the LTR promoter of the vector virus. Regardless of the unfavorable effect of the neomycin gene, many currently available retrovirus vectors use this selection marker gene, due primarily to its long history of use in retrovirus vector systems. To circumvent problems associated with the use of the neomycin resistance gene, Fernex et al. (1997) presented a Cre-loxP recombination system, by which the neomycin gene flanked by the *loxP* site is deleted. Substitution of the puromycin resistance gene for the neomycin resistance gene could be another, more simple solution (Hlavaty et al., 2004), although both genes are of prokaryotic origin.

2. Infectivity of Virus

The most critical problem in early transgenic animal production was low infectivity of the conventional amphotropic or xenotropic MLV-based retrovirus vector viruses, especially to the cells derived from ungulates. However, a remarkable breakthrough was made by using a pantropic VSV-G pseudotyped retrovirus vector system, from which the progeny viruses produced are encapsidated with VSV-G instead of retrovirus envelope protein (Burns et al., 1993). The biggest advantage of this hybrid vector system is that the virus titer can be increased more than 1000-fold by centrifugation of the viral stock. Chan et al. (1998) reported that injection of 10 pL of highly concentrated viral stock (10^9 CFU/mL) into the perivitelline space of the metaphase II bovine oocyte resulted in 100% efficiency of transgenic bovine production. The concentration of 10^9 CFU/mL viral stock, which permits the existence of a single infection unit of virus in a 10 pL volume, was accomplished by centrifugation of the viral stock harvested from the VSV-G pseudotyped retrovirus vector system. Using the same perivitelline space injection approach, successful production of transgenic monkeys was also reported (Chan et al., 2001). Preparation of highly concentrated viral stock is the most critical factor in transgenic avian studies as well. In transgenic poultry production, targeting blastodermal cells at stage X using retrovirus-mediated gene transfer is most commonly employed (Koo et al., 2006; Lillico et al., 2007). Stage X is the development stage of the embryo in the chicken egg just after laying (Eyal-Giladi and Kochav, 1976). Because the number of blastodermal cells at stage X is as high as 60,000, the more the viral stock is concentrated, the higher the expected transgenesis efficiency.

A major drawback in the use of VSV-G pseudotyped packaging cells is that it is impossible to make stable virus-producing cells because of cytotoxic *VSV-G* gene expression in mammalian cells. Therefore, virus-producing cells can be made only by transient transfection of the *VSV-G* gene to the 293-based cells just before virus harvest. Although solutions for establishing permanent packaging or virus-producing cells were made by controlling *VSV-G* expression employing a tetracycline-controllable system (Ory et al., 1996; Lo and Yee, 2007) or an ecdysone inducible system (Pacchia et al., 2001), use of transient transfection of the *VSV-G* gene to the 293-based cells is most prevalent at this time.

3. Integration into the Host Cell Genome

One of the significant advantages of the retrovirus vector system is integration of reverse-transcribed retrovirus vector DNA into the host cell genome. Without genomic integration, however, the transgene is destined to degrade in a short period of time. It has been reported that the intracellular half-life of MLV-derived retrovirus vector is in the range of 5.5−7.5 h (Andreadis et al., 1997) and that the integration occurs only in mitotically active cells (Stevenson et al., 1990) during the S phase (Savatier et al., 1989). Therefore, simple retrovirus-mediated gene transfer to nondividing cells was a problem at the early stage of the retrovirus vector system (Miller et al., 1990). This drawback has been overcome with the advent of the lentivirus vector system (Naldini et al., 1996).

4. Expression of the Transgene in the Target Cells

It has been generally believed that the MoMLV LTR becomes inactive after introduction into preimplantation mouse embryo (Savatier et al., 1990). This problem can be overcome by introduction of an internal or multicistronic promoter sequence (Tessanne et al., 2012; Ibrahimi et al., 2009) into the retrovirus vector. The next problem, which is more critical in determining the success of transgenic livestock studies, especially aiming for mass production of pharmaceuticals, is how to optimally control the expression of the transgene *in vivo*. For example, a low or constitutively high level of expression of the transgene results in no economical significance or serious physical disturbance of the transgenic animal, respectively. No gene transfer system employed in transgenic animal production has been free from these problems so far. The best solution is use of tissue-specific promoter (Lillico et al., 2007; Monzani et al., 2013) or tetracycline- or ecdysone-inducible promoter (Kim et al., 2011; Koo et al., 2012; Galimi et al., 2005) to achieve regulated expression. Of them, tissue-specific promoter is the most ideal, but the size is usually too big to use in a retrovirus vector. Therefore, tetracycline-inducible promoter is more widely used at present. An example of inducible expression of the foreign gene in the transgenic chicken is illustrated in Figure 6.7.

IV. Methods

A. Construction of Retrovirus Vector-Producing Cells

Schematic overall procedures are shown in Figure 6.6.

1. For Complex Retrovirus (HIV-1) Vector

The following protocol has been optimized for the third-generation lentivirus vector-producing system. Both SIN and non-SIN vectors can be used in this method.

Day 0: Preparation of cells Seed the 5×10^6 cells of 293T (ATCC CRL-11268) or 293FT (can be purchased from Invitrogen with catalog no. R700-07) in a 100-mm dish with 10 mL of high glucose DMEM (Dulbecco's Modified Eagle Medium), supplemented with 10% fetal bovine serum, penicillin (100 U/mL), streptomycin (100 μg/mL), 0.1 mM MEM nonessential amino acid, 6 mM L-glutamine, and 1 mM MEM sodium pyruvate. Allow 1 day of attachment.

Day 1: Calcium phosphate transfection
1. Feed cells 9 mL of fresh, complete medium 4 h before DNA addition.
2. DNA precipitation.
   a. Fill #1 polypropylene tube (15 mL) with total 30 μg of plasmid DNA in 450 μL ddH$_2$O (15 μg for vector plus 5 μg for each gene of *Rev*, *gag/pol*, and *VSV-G*) and 50 μL of 2.5 M CaCl$_2 \cdot$ 2H$_2$O (Sigma, St. Louis, MO, catalog no. C7902), then mix.
   b. Fill #2 polypropylene tube (15 mL) with 500 μL of 2X HBSS (Hank's balanced salt solution). 100 mL of 2X HBSS is made as follows: Dissolve NaCl (1.64 g), HEPES (1.19 g; Sigma, catalog no. H0891), and 0.1 mL of 1.5 M Na$_2$HPO$_4$ in H$_2$O, then adjust pH to 7.05 with 5N NaOH. For sterility, the solution is filtered through 0.45 μm pore size filter.
   c. Add the contents of the #1 tube slowly, dropwise, to the gently bubbling solution in the #2 tube, then vortex for 20 s before allowing the suspension to sit at room temperature for 20 min.
3. After several times of gentle pipetting, add 1 mL of the DNA mixture to the cells in a 100-mm dish containing 9 mL of complete medium. The addition must be done dropwise and evenly while gently swirling the medium in a dish.
4. Leave the dish in the CO$_2$ incubator for ~6–8 h.
5. Remove the medium and rinse the cells with PBS twice before the addition of 10 mL of fresh medium.

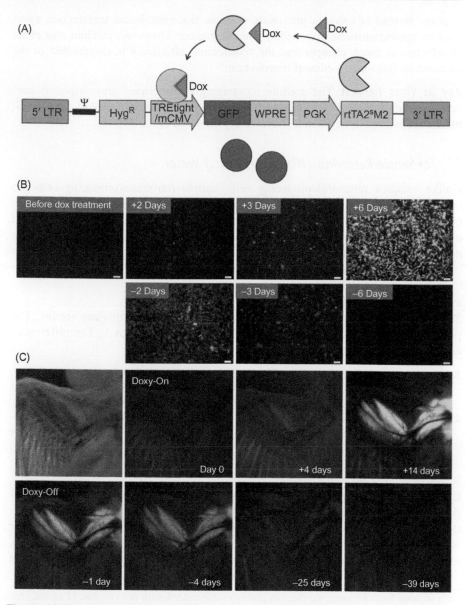

Figure 6.7 Tetracycline-mediated inducible expression of the *eGFP* gene in simple retrovirus vector. (A) Structure of the Tet2-GWPT provirus with schematic representation on the tetracycline-inducible expression mechanism. *Abbreviations*: Dox, doxycycline, a tetracycline derivative; LTR, long terminal repeat; *Hyg*^r, hygromycin B resistance gene; TREtight/mCMV, tetracycline response element fused to mini cytomegalovirus promoter; GFP, green fluorescent protein; PGK, phosphoglycerate kinase promoter; rtTA2^sM2, reverse tetracycline transactivator; WPRE, woodchuck hepatitis virus posttranscriptional regulatory element. (B) Induction of *GFP* expression *in vitro*. Chicken embryonic fibroblast (CEF)

Note: Instead of calcium phosphate solution, liposome-based transfection agent such as lipofectamine is also widely used at present. However, calcium phosphate transfection is much cheaper and the transfection efficiency is comparable to the commercial liposome-mediated transfection.

Day 3: Virus harvest The medium containing recombinant retroviruses is harvested approximately 48 h after transfection. Due to the cytotoxicity of the *VSV-G*, most of cells form syncytium and eventually die (Yee et al., 1994).

2. For Simple Retrovirus (MLV or GaLV, etc.) Vector

Unlike complex retrovirus-producing cells, simple retrovirus-producing cells are constructed by transducing 293GP (Burns et al., 1993) or GP2-293 cells (Batchu et al., 2003; can be purchased from Clontech, catalog no. 631458). As mentioned elsewhere in this chapter, both 293GP and GP2-293 carry *gag/pol* genes and have no need of *VSV-G* gene transfection, while 293T and 293FT cells used in complex retrovirus production carry no retrovirus sequences.

a. By Transient Cotransfection
This method is applicable to both SIN and non-SIN simple retrovirus vectors. The procedures are almost the same as described in Section 6.IV.A.1. The differences are as follows:

1. In cell preparation, we seed 2×10^6 cells of 293GP or its derivative GP2-293 cells in a 100-mm dish with 10 mL of high glucose (4.5 g/L) DMEM supplemented with 10% fetal bovine serum, penicillin (100 U/mL), and streptomycin (100 μg/mL).
2. In DNA preparation for transfection, 20 μg of total DNA (10 μg of retrovirus vector plasmid plus an equivalent concentration of *VSV-G* expression vector plasmid) is prepared in 450 μL of ddH$_2$O.

b. Use of Two Separate Transfections
Compared to the previous transient cotransfection method, this method consists of permanent transfection followed by selection and transient transfection. One advantage of this method is that only the cells harboring and expressing viral vector sequence are selected, resulting in higher virus productivity. Although high DNA transfection efficiency of 293-based cells was proven, it is unlikely that all cells

---

◄ cells isolated from a transgenic chicken embryo were treated with doxycycline (1 μg/mL) for 6 days and then cultured in the absence of doxycycline. A plus sign (+) indicates number of days in culture with doxycycline; A minus sign (−) indicates number of days in culture after removal of doxycycline from the culture medium. (C) Expression of *GFP* in a transgenic chicken fed doxycycline (50 mg/kg of formula feed). Doxycycline was supplied in the diet for 14 days, followed by a doxycycline-free diet for 39 days. A plus sign (+) indicates number of days post-doxycycline feeding; A minus sign (−) indicates number of days after removal of doxycycline from the diet.
Source: Courtesy of J. Reprod. Dev.

can be transduced by a single transient transfection. A disadvantage of this method is the extended incubation period; an additional 1 or 2 weeks is required for selection between the two sequential transfections. That said, this method is applicable to both SIN and non-SIN simple retrovirus vectors. However, if vectors are devoid of selection marker genes (e.g., neomycin resistance or hygromycin B resistance genes), this method is not suitable.

Day 0: Preparation of cells Plate 2×10^6 cells of 293GP or GP2-293 cells in a 100-mm dish with 10 mL of high glucose (4.5 g/L) DMEM supplemented with 10% fetal bovine serum, penicillin (100 U/mL), and streptomycin (100 μg/mL).

Day 1: Calcium phosphate transfection Except for DNA preparation, all other procedures are same as previously described. In this method, 20 μg of plasmid DNA carrying vector sequence in 450 μL of ddH$_2$O is prepared. With this infection procedure, the 293GP or GP2-293 cells carry retrovirus vector sequence. However, the cells cannot produce recombinant virus progeny because the cells are still devoid of *env* gene.

Day 2: Selection

1. Trypsinize and split the transfected cells with selection medium (in case of G418 selection, 600 μg/mL for 293-based cells). Selection usually takes 10–14 days. During this period, change with fresh medium is done every 2 or 3 days. Split the cells when cells are confluent.

Day 0 after selection: Preparation of cells For transfection of *VSV-G* gene, plate 2×10^6 selected cells in a 100-mm dish with 10 mL of high glucose (4.5 g/L) DMEM supplemented with 10% fetal bovine serum, penicillin (100 U/mL), and streptomycin (100 μg/mL).

Day 2–Day 4 postselection: Calcium phosphate transfection and virus harvest Except for DNA preparation for calcium phosphate transfection, all other procedures are the same as previously described. In this case, 20 μg of plasmid DNA carrying *VSV-G* sequence in 450 μL of ddH$_2$O is prepared.

c. Serial Transfection/Infection/Transfection

This method is a little time consuming, but very helpful for the preparation of high titer viral stock. This method needs an additional packaging cell line. For the first packaging cell line, we usually use NIH3T3 cell-based PG13 (Miller et al., 1991; ATCC CRL-10686) or PT67 (Miller and Chen, 1996; ATCC CRL-12284). The 293GP or GP2-293 cell line is used as a second packaging cell line. The first packaging cells carry all *trans*-acting sequences including *gag/pol* and *env* genes, while the second packaging cells are devoid of *env* gene. Because the vector sequence is introduced through infection, SIN vector is not used in this method. In addition, inclusion of an appropriate selection marker gene in the vector is recommended for efficient selection.

Day 0: Cell preparation Plate PG13 or PT67 cells (5×10^5 cells per 60 mm dish) with 5 mL of high glucose (4.5 g/L) DMEM supplemented with 10% fetal bovine serum, penicillin (100 U/mL), and streptomycin (100 μg/mL).

Day 1: DMSO/Polybrene transfection The gene transfer efficiency of this method is somewhat lower than that of its optimized calcium phosphate counterpart. However, due to the simplicity and consistent results of this method, we often use this method for transient transfection of the first packaging cells with retrovirus vector DNA. Viruses harvested after transient transfection are used to deliver vector sequence into the second packaging cells. The procedures are as follows:

1. Prepare 10 μg of plasmid in 1 mL of medium.
2. Aspirate the medium.
3. Add 1 mL of DNA medium mixture to one 60-mm dish.
4. Add 30 mL of polybrene stock (1 mg/mL) to make 30 μg/mL of final concentration.
5. Incubate at 37°C for 6 h.
6. Rock occasionally during the 6-h period.
7. DMSO shock after 6-h incubation. This is the most important step, and it must be done gently.
   a. Aspirate DNA/polybrene solution.
   b. Put 2 mL of 25% DMSO medium into a 60-mm dish using a 10-ml pipette and wait 1 min for NIH3T3-based cells.
   c. Aspirate DMSO medium.
   d. Rinse two to three times with medium. The first rinse should be done rapidly after the DMSO is removed. The resulting transduced cells correspond to "virus-producing cell #1" in Figure 6.6B3.
   e. Add 4−5 mL of fresh medium.

Day 2
1. Trypsinize and split the transfected cells in nonselection medium if necessary.
2. To use as second packaging cells, plate 293GP or GP2-293 cells (5×10^5 cells per 60 mm plate) with 4 mL of high glucose (4.5 g/L) DMEM supplemented with 10% fetal bovine serum, penicillin (100 U/mL), and streptomycin (100 μg/mL).

Day 3: Introduction of viral vector sequence into 293GP or GP2-293cells by infection
1. Harvest virus-containing medium from the transfected cells, and filter the medium through 0.45 μm pore size filter.
2. Feed the 293GP or GP2-293 cells prepared at *Day 2* with 4 mL of fresh medium mixed with small amount (10−100 μL) of virus-containing medium and 30 μL of polybrene stock (1 mg/mL). By this infection procedure, the 293GP or GP2-293 cells become to harbor retrovirus vector sequence. However, the cells cannot produce recombinant virus progeny because the cells are devoid of *env* gene yet.

Day 4: Selection
1. Trypsinize and split the infected 293GP or GP2-293 cells with 4 mL of selection medium (in case of G418 selection, 600 μg/mL for 293-based cells). Selection usually takes 10−14 days. During this period, change with fresh medium every 2 or 3 days. Split the cells when cells are confluent. It is recommended to freeze the selected cells for future similar experiments.

Day 0 after selection:
1. Plate the 2×10^6 cells that have undergone selection in a 100-mm dish with 10 mL of high glucose (4.5 g/L) DMEM supplemented with 10% fetal bovine serum, penicillin (100 U/mL), and streptomycin (100 μg/mL).

Day 2–Day 4 postselection: Cell prepatation, Calcium phosphate transfection, and virus harvest Using the same procedure described in the previous section, the cells are transfected with 20 μg of plasmid DNA carrying *VSV-G* sequence, then viruses are harvested.

B. Measurement of Virus

Day 0:

1. Plate target cells (5×10^5 cells per 60 mm plate).
2. Add 4 mL of nonselection fresh medium to the semiconfluent virus-producing cells in a 60-mm dish.

Day 1:

1. Harvest medium from semiconfluent virus-producing cells 48 h after medium change.
2. Filter medium through the 0.45 μm filters or centrifuge at $5000 \times g$ for 5 min to remove living cells.
3. Feed the target cells with 4 mL of fresh medium supplemented with a small amount of virus-containing medium and polybrene (8 μg/mL).

Day 2:

1. Trypsinize and split the target cells and add selection medium (in case of G418 selection, 800 μg/mL for NIH3T3-based cells). Selection usually takes 10–14 days. During this period, change with fresh medium every 2 or 3 days.

Colonies surviving after selection are stained with 1% crystal violet before counting using a dissecting microscope. When a nonselection marker gene such as *Lac Z* gene is used, counting LacZ$^+$ cells after X-gal staining is also done with the aid of a dissecting microscope. Virus titer is indicated by colony-forming units per milliliter (CFU/ml), that is, the number of colonies per milliliter of virus transformed from G418-sensitive phenotype to G418-resistant phenotype. In some cases, virus titer is indicated in a different way—for example, by *lacZ*+ transforming units per milliliter (TU/ml), a measure of the number of cells per milliliter of virus transformed from the *E. coli* β-galactosidase-negative phenotype to the *E. coli* β-galactosidase-positive phenotype. The virus titer is calculated by multiplying the counted number of phenotypes per dish by the dilution factor (1 mL divided by the volume of viral stock used to infect the target cells) and a factor of splitting done at day 2.

When other nonselection marker genes such as fluorescent protein genes are used, counting each cell expressing a nonselectable but visible marker gene under the UV microscope is very difficult and inaccurate. Therefore, virus titer is measured by real-time PCR using the specifically designed primer sets targeting provirus sequence in the target cell genome. Recently, titration kits for retrovirus and lentivirus vectors have been available from Clontech (catalog nos. 631453 and 631235). Regardless of selection marker, use of this real-time PCR kit is very convenient and produces accurate data.

For the establishment of optimized virus-producing cells, subcloning of cells surviving from the selection is recommended. The best clone is determined by comparing the titer of the viruses produced from each clone. Estimation of the titer also can be done by RNA dot blot analysis of the medium collected from each clone (Murdoch et al., 1997). After identification of the highest titer clone, expression of the main transgene of interest in the target cells should be verified.

C. Ultracentrifugal Concentration of Retrovirus Vectors

Retrovirus vectors coated with *VSV-G* can be pelleted by ultracentrifugation with almost no loss of infectivity.

1. Centrifuge the medium harvested from the virus-producing cells to pellet the viruses at $50,000 \times g$ for 2 h at $4°C$ using a vertical rotor (Beckman 70Ti).
2. Following complete removal of supernatant after centrifugation, the pellet was placed at $4°C$ overnight with 0.5–1% of the original volume of 0.1X HBSS (Yee et al., 1994) or culture medium for suspension.
3. If desired, run a second round of centrifugation following two steps above. Storage at $-70°C$ causes little decrease in titer.
4. Filter the concentrated viral stock through 0.45 μm pore size filter before applying to the target cells.

D. Detection of Replication-Competent Recombinant Retrovirus

Because most of retroviruses are pathogenic, it is very important to confirm no production of retrovirus from the target cells. Among the several ways of monitoring generation of replication-competent helper virus from the virus-producing cells, we usually use RT assay. Other methods such as mobilization assay are described in detail elsewhere (Markowitz et al., 1988; Danos and Mulligan, 1988).

1. RT (reverse transcriptase) Assay

In principle, if there is no helper virus (or replication-competent virus) in the vector viral stock, the *pol* gene encoding RT is not to be transferred to the target cells, and RT activity due to the *pol* gene expression is not to be detected. This RT assay was performed following the established protocols of the late Dr. Temin's laboratory.

1. 1×10^6 target cells carrying and expressing proviral sequences were seeded in a 60-mm dish and cultured for 2 days.
2. A volume of 10 μL of medium harvested from the culture was added to 50 μL of reaction solution to induce reverse transcription at $37°C$ for 60 min. Components of the reaction solution are: 50 mM Tris (pH 7.9), 0.6 mM $MnCl_2$, 60 mM NaCl, 0.5% NP-40, 1600 μg/mL RNasin (Promega), 50 μg/mL Poly $d(T)_{12-18}$, 100 μg/mL Poly A, 200–500 μCi ^{3}H-dTTT, 1 mM Dithithreitol.
3. A volume of 1 mL of ice-cold 10% TCA (trichloroacetic acid) was added, and the mixture was kept on ice for 30 min or more.

The TCA precipitated material was collected on a Whatman GF/B filter pre-washed with 10% TCA and washed several times with TCA solution. The dried filters were added to scintillation vials containing 5 mL of counting solution. Each filter was counted for 10 min, and the counts per minute were recorded.

Acknowledgments

We would like to express our sincere gratitude to Professor Neal L. First, who gave T.K. the opportunity to study retrovirus-mediated transgenic animal production nearly 25 years ago. This work was supported by the Bio-Industry Technology Development Program, Ministry for Food, Agriculture, Forestry and Fisheries, Republic of Korea, and by grants from the Next-Generation BioGreen 21 Program (No. PJ0096011), Rural Development Administration, Republic of Korea.

References

Adam, M.A., Miller, A.D., 1988. Identification of a signal in a murine retrovirus that is sufficient for packaging of nonretroviral RNA into virions. J. Virol. 62, 3802–3806.

Albritton, L.M., Tseng, L., Scadden, D., Cunningham, J.M., 1989. A putative murine ecotropic retrovirus receptor gene encodes a multiple membrane-spanning protein and confers susceptibility to virus infection. Cell 57, 659–666.

Andreadis, S.T., Brott, D., Fuller, A.O., Palsson, B.O., 1997. Moloney murine leukemia virus-derived retroviral vectors decay intracellularly with a half-life in the range of 5.5 to 7.5 hours. J. Virol. 71, 7541–7548.

Aran, J.M., Gottesman, M.M., Pastan, I., 1998. Construction and characterization of bicistronic retroviral vectors encoding the multidrug transporter and beta-galactosidase or green fluorescent protein. Cancer Gene Ther. 5, 195–206.

Artelt, P., Grannemann, R., Stocking, C., Friel, J., Bartsch, J., Hauser, H., 1991. The prokaryotic neomycin-resistance-encoding gene acts as a transcriptional silencer in eukaryotic cells. Gene 99, 249–254.

Batchu, R.B., Moreno, A.M., Szmania, S., Gupta, S.K., Zhan, F., Rosen, N., et al., 2003. High-level expression of cancer/testis antigen NY-ESO-1 and human granulocyte-macrophage colony-stimulating factor in dendritic cells with a bicistronic retroviral vector. Hum. Gene Ther. 14, 1333–1345.

Benabdellah, K., Cobo, M., Munoz, P., Toscano, M.G., Martin, F., 2011. Development of an all-in-one lentiviral vector system based on the original TetR for the easy generation of Tet-ON cell lines. PLoS One 6, e23734.

Brinster, R.L., Allen, J.M., Behringer, R.R., Gelinas, R.E., Palmiter, R.D., 1988. Introns increase transcriptional efficiency in transgenic mice. Proc. Natl. Acad. Sci. USA 85, 836–840.

Burns, J.C., Friedmann, T., Driever, W., Burrascano, M., Yee, J.K., 1993. Vesicular stomatitis virus G glycoprotein pseudotyped retroviral vectors: concentration to very high titer and efficient gene transfer into mammalian and nonmammalian cells. Proc. Natl. Acad. Sci. USA 90, 8033–8037.

Carey, B.W., Markoulaki, S., Hanna, J., Saha, K., Gao, Q., Mitalipova, M., et al., 2009. Reprogramming of murine and human somatic cells using a single polycistronic vector. Proc. Natl. Acad. Sci. USA 106, 157–162.

Chan, A.W., Homan, E.J., Ballou, L.U., Burns, J.C., Bremel, R.D., 1998. Transgenic cattle produced by reverse-transcribed gene transfer in oocytes. Proc. Natl. Acad. Sci. USA 95, 14028–14033.

Chan, A.W., Chong, K.Y., Martinovich, C., Simerly, C., Schatten, G., 2001. Transgenic monkeys produced by retroviral gene transfer into mature oocytes. Science 291, 309–312.

Colicelli, J., Goff, S.P., 1988. Sequence and spacing requirements of a retrovirus integration site. J. Mol. Biol. 199, 47–59.

Danos, O., Mulligan, R.C., 1988. Safe and efficient generation of recombinant retroviruses with amphotropic and ecotropic host ranges. Proc. Natl. Acad. Sci. USA 85, 6460–6464.

Deglon, N., Tseng, J.L., Bensadoun, J.C., Zurn, A.D., Arsenijevic, Y., Pereira de Almeida, L., et al., 2000. Self-inactivating lentiviral vectors with enhanced transgene expression as potential gene transfer system in Parkinson's disease. Hum. Gene Ther. 11, 179–190.

Ellis, J., Bernstein, A., 1989. Retrovirus vectors containing an internal attachment site: evidence that circles are not intermediates to murine retrovirus integration. J. Virol. 63, 2844–2846.

Emerman, M., Temin, H.M., 1984. Genes with promoters in retrovirus vectors can be independently suppressed by an epigenetic mechanism. Cell 39, 449–467.

Emerman, M., Temin, H.M., 1986. Quantitative analysis of gene suppression in integrated retrovirus vectors. Mol. Cell. Biol. 6, 792–800.

Eyal-Giladi, H., Kochav, S., 1976. From cleavage to primitive streak formation: a complementary normal table and a new look at the first stages of the development of the chick. I. General morphology. Dev. Biol. 49, 321–337.

Fang, J., Qian, J.J., Yi, S., Harding, T.C., Tu, G.H., VanRoey, M., et al., 2005. Stable antibody expression at therapeutic levels using the 2A peptide. Nat. Biotechnol. 23, 584–590.

Felber, B.K., Hadzopoulou-Cladaras, M., Cladaras, C., Copeland, T., Pavlakis, G.N., 1989. rev protein of human immunodeficiency virus type 1 affects the stability and transport of the viral mRNA. Proc. Natl. Acad. Sci. USA 86, 1495–1499.

Fernex, C., Dubreuil, P., Mannoni, P., Bagnis, C., 1997. Cre/loxP-mediated excision of a neomycin resistance expression unit from an integrated retroviral vector increases long terminal repeat-driven transcription in human hematopoietic cells. J. Virol. 71, 7533–7540.

Freed, E.O., Martin, M.A., 2001. HIVs and their Replication. In: Knipe, D.M., Howley, P.M. (Eds.), Fields Virology, fourth ed. Lippincott Williams & Wilkins, Philadelphia, PA, pp. 1971–2041.

Galimi, F., Saez, E., Gall, J., Hoong, N., Cho, G., Evans, R.M., et al., 2005. Development of ecdysone-regulated lentiviral vectors. Mol. Ther. 11, 142–148.

Gay, V., Moreau, K., Hong, S.S., Ronfort, C., 2012. Quantification of HIV-based lentiviral vectors: influence of several cell type parameters on vector infectivity. Arch. Virol. 157, 217–223.

Goff, S.P., 2001. Retroviridae: the retroviruses and their replication. In: Knipe, D.M., Howley, P.M. (Eds.), Fields Virology, fourth ed. Lippincott Williams & Wilkins, Philadelphia, PA, pp. 1871–1939.

Hammarskjold, M.L., Heimer, J., Hammarskjold, B., Sangwan, I., Albert, L., Rekosh, D., 1989. Regulation of human immunodeficiency virus env expression by the rev gene product. J. Virol. 63, 1959–1966.

Harries, M., Phillipps, N., Anderson, R., Prentice, G., Collins, M., 2000. Comparison of bicistronic retroviral vectors containing internal ribosome entry sites (IRES) using expression of human interleukin-12 (IL-12) as a readout. J. Gene Med. 2, 243–249.

Hlavaty, J., Portsmouth, D., Stracke, A., Salmons, B., Gunzburg, W.H., Renner, M., 2004. Effects of sequences of prokaryotic origin on titer and transgene expression in retroviral vectors. Virology 330, 351–360.

Hwang, L.H., Gilboa, E., 1984. Expression of genes introduced into cells by retroviral infection is more efficient than that of genes introduced into cells by DNA transfection. J. Virol. 50, 417–424.

Ibrahimi, A., Vande Velde, G., Reumers, V., Toelen, J., Thiry, I., Vandeputte, C., et al., 2009. Highly efficient multicistronic lentiviral vectors with peptide 2A sequences. Hum. Gene Ther. 20, 845–860.

Iwasaki, K., Temin, H.M., 1990. The efficiency of RNA 3′-end formation is determined by the distance between the cap site and the poly(A) site in spleen necrosis virus. Genes Dev. 4, 2299–2307.

Javanbakht, H., Halwani, R., Cen, S., Saadatmand, J., Musier-Forsyth, K., Gottlinger, H., et al., 2003. The interaction between HIV-1 Gag and human lysyl-tRNA synthetase during viral assembly. J. Biol. Chem. 278, 27644–27651.

Johnston, J.M., Denning, G., Doering, C.B., Spencer, H.T., 2012. Generation of an optimized lentiviral vector encoding a high-expression factor VIII transgene for gene therapy of hemophilia A. Gene Ther. doi:10.1038/gt.2012.76.

Kaleri, H.A., Xu, S.Y., Lin, H.L., 2011. Generation of transgenic chicks using an oviduct-specific expression system. Genet. Mol. Res. 10, 3046–3055.

Kim, M.J., Oh, H.J., Park, J.E., Kim, G.A., Hong, S.G., Jang, G., et al., 2011. Generation of transgenic dogs that conditionally express green fluorescent protein. Genesis 49, 472–478.

Kim, T., Leibfried-Rutledge, M.L., First, N.L., 1993. Gene transfer in bovine blastocysts using replication-defective retroviral vectors packaged with Gibbon ape leukemia virus envelopes. Mol. Reprod. Dev. 35, 105–113.

Koo, B.C., Kwon, M.S., Choi, B.R., Kim, J.H., Cho, S.K., Sohn, S.H., et al., 2006. Production of germline transgenic chickens expressing enhanced green fluorescent protein using a MoMLV-based retrovirus vector. FASEB J. 20, 2251–2260.

Koo, B.C., Kwon, M.S., Roh, J.Y., Kim, M., Kim, J.H., Kim, T., 2012. Quantitative analysis of tetracycline-inducible expression of the green fluorescent protein gene in transgenic chickens. J. Reprod. Dev. 58, 672–677.

Krausslich, H.G., 1992. Specific inhibitor of human immunodeficiency virus proteinase prevents the cytotoxic effects of a single-chain proteinase dimer and restores particle formation. J. Virol. 66, 567–572.

Krausslich, H.G., Ochsenbauer, C., Traenckner, A.M., Mergener, K., Facke, M., Gelderblom, H.R., et al., 1993. Analysis of protein expression and virus-like particle formation in mammalian cell lines stably expressing HIV-1 gag and env gene products with or without active HIV proteinase. Virology 192, 605–617.

Kumar, M., Keller, B., Makalou, N., Sutton, R.E., 2001. Systematic determination of the packaging limit of lentiviral vectors. Hum. Gene Ther. 12, 1893–1905.

Lillico, S.G., Sherman, A., McGrew, M.J., Robertson, C.D., Smith, J., Haslam, C., et al., 2007. Oviduct-specific expression of two therapeutic proteins in transgenic hens. Proc. Natl. Acad. Sci. USA 104, 1771–1776.

Linney, E., Hardison, N.L., Lonze, B.E., Lyons, S., DiNapoli, L., 1999. Transgene expression in zebrafish: a comparison of retroviral-vector and DNA-injection approaches. Dev. Biol. 213, 207–216.

Lo, H.L., Yee, J.K., 2007. Production of vesicular stomatitis virus G glycoprotein (VSV-G) pseudotyped retroviral vectors. Curr. Protoc. Hum. Genet. Chapter 12, Unit 12.7.

Lu, J.K., Burns, J.C., Chen, T.T., 1997. Pantropic retroviral vector integration, expression, and germline transmission in medaka (Oryzias latipes). Mol. Mar. Biol. Biotechnol. 6, 289–295.

Mann, R., Mulligan, R.C., Baltimore, D., 1983. Construction of a retrovirus packaging mutant and its use to produce helper-free defective retrovirus. Cell 33, 153–159.

Markowitz, D., Goff, S., Bank, A., 1988. A safe packaging line for gene transfer: separating viral genes on two different plasmids. J. Virol. 62, 1120–1124.

Miller, A.D., Buttimore, C., 1986. Redesign of retrovirus packaging cell lines to avoid recombination leading to helper virus production. Mol. Cell. Biol. 6, 2895–2902.

Miller, A.D., Chen, F., 1996. Retrovirus packaging cells based on 10A1 murine leukemia virus for production of vectors that use multiple receptors for cell entry. J. Virol. 70, 5564–5571.

Miller, A.D., Garcia, J.V., von Suhr, N., Lynch, C.M., Wilson, C., Eiden, M.V., 1991. Construction and properties of retrovirus packaging cells based on Gibbon ape leukemia virus. J. Virol. 65, 2220–2224.

Miller, D.G., Adam, M.A., Miller, A.D., 1990. Gene transfer by retrovirus vectors occurs only in cells that are actively replicating at the time of infection. Mol. Cell. Biol. 10, 4239–4242.

Miyoshi, H., Blomer, U., Takahashi, M., Gage, F.H., Verma, I.M., 1998. Development of a self-inactivating lentivirus vector. J. Virol. 72, 8150–8157.

Mizuguchi, H., Xu, Z., Ishii-Watabe, A., Uchida, E., Hayakawa, T., 2000. IRES-dependent second gene expression is significantly lower than cap-dependent first gene expression in a bicistronic vector. Mol. Ther. 1, 376–382.

Monzani, P.S., Sangalli, J.R., de Bem, T.H., Bressan, F.F., Fantinato-Neto, P., Pimentel, J.R., et al., 2013. Breeding of transgenic cattle for human coagulation factor IX by a combination of lentiviral system and cloning. Genet. Mol. Res. 12, doi:10.4238/2013. February.28.25.

Murdoch, B., Pereira, D.S., Wu, X., Dick, J.E., Ellis, J., 1997. A rapid screening procedure for the identification of high-titer retrovirus packaging clones. Gene Ther. 4, 744–749.

Naldini, L., Blomer, U., Gallay, P., Ory, D., Mulligan, R., Gage, F.H., et al., 1996. In vivo gene delivery and stable transduction of nondividing cells by a lentiviral vector. Science 272, 263–267.

Nishijima, K., Iijima, S., 2013. Transgenic chickens. Dev. Growth Differ. 55, 207–216.

Ory, D.S., Neugeboren, B.A., Mulligan, R.C., 1996. A stable human-derived packaging cell line for production of high titer retrovirus/vesicular stomatitis virus G pseudotypes. Proc. Natl. Acad. Sci. USA 93, 11400–11406.

Pacchia, A.L., Adelson, M.E., Kaul, M., Ron, Y., Dougherty, J.P., 2001. An inducible packaging cell system for safe, efficient lentiviral vector production in the absence of HIV-1 accessory proteins. Virology 282, 77–86.

Pfeifer, A., Verma, I.M., 2001. Virus vectors and their application. In: Knipe, D.M., Howley, P.M. (Eds.), Fields Virology, fourth ed. Lippincott Williams & Wilkins, Philadelphia, PA, pp. 469–491.

Pfeifer, A., Ikawa, M., Dayn, Y., Verma, I.M., 2002. Transgenesis by lentiviral vectors: lack of gene silencing in mammalian embryonic stem cells and preimplantation embryos. Proc. Natl. Acad. Sci. USA 99, 2140–2145.

Ponder, K.P., 2001. Vectors of gene therapy. In: Kresina, T.F. (Ed.), An Introduction to Molecular Medicine and Gene Therapy. Wiley-Liss, New York, NY, pp. 77–112.

Provost, E., Rhee, J., Leach, S.D., 2007. Viral 2A peptides allow expression of multiple proteins from a single ORF in transgenic zebrafish embryos. Genesis 45, 625–629.

Ryan, M.D., Drew, J., 1994. Foot-and-mouth disease virus 2A oligopeptide mediated cleavage of an artificial polyprotein. EMBO J. 13, 928–933.

Savatier, N., Rocancourt, D., Bonnerot, C., Nicolas, J.F., 1989. A novel system for screening antiretroviral agents. J. Virol. Methods 26, 229–235.

Savatier, P., Morgenstern, J., Beddington, R.S., 1990. Permissiveness to murine leukemia, virus expression during preimplantation and early postimplantation mouse development. Development 109, 655–665.

Shao, L., Feng, W., Sun, Y., Bai, H., Liu, J., Currie, C., et al., 2009. Generation of iPS cells using defined factors linked via the self-cleaving 2A sequences in a single open reading frame. Cell Res. 19, 296–306.

Sleckman, B.P., Peterson, A., Jones, W.K., Foran, J.A., Greenstein, J.L., Seed, B., et al., 1987. Expression and function of CD4 in a murine T-cell hybridoma. Nature 328, 351–353.

Soriano, P., Friedrich, G., Lawinger, P., 1991. Promoter interactions in retrovirus vectors introduced into fibroblasts and embryonic stem cells. J. Virol. 65, 2314–2319.

Stevenson, M., Stanwick, T.L., Dempsey, M.P., Lamonica, C.A., 1990. HIV-1 replication is controlled at the level of T cell activation and proviral integration. EMBO J. 9, 1551–1560.

Szymczak, A.L., Workman, C.J., Wang, Y., Vignali, K.M., Dilioglou, S., Vanin, E.F., et al., 2004. Correction of multi-gene deficiency in vivo using a single 'self-cleaving' 2A peptide-based retroviral vector. Nat. Biotechnol. 22, 589–594.

Teich, N., 1985. Taxonomy of retroviruses. In: Weiss, R., Teich, N., Varmus, H., Coffin, J. (Eds.), RNA Tumor Viruses. Cold Spring Harbor Laboratory, Cold Spring Harbor, NY, pp. 1–16.

Temin, H.M., 1987. Retrovirus vectors for gene transfer: efficient integration into and expression of exogenous DNA in vertebrate cell genomes. In: Kuchelapati, R. (Ed.), Gene Transfer. Plenum, New York, NY, pp. 149–187.

Temin, H.M., 1988. Mechanisms of cell killing/cytopathic effects by nonhuman retroviruses. Rev. Infect. Dis. 10, 399–405.

Temin, H.M., 1989. Retrovirus variation and evolution. Genome 31, 17–22.

Tessanne, K., Golding, M.C., Long, C.R., Peoples, M.D., Hannon, G., Westhusin, M.E., 2012. Production of transgenic calves expressing an shRNA targeting myostatin. Mol. Reprod. Dev. 79, 176–185.

Weiss, R., 1982. Experimental biology and assay of RNA tumor virus. In: Weiss, R., Teich, N., Varmus, H., Coffin, J. (Eds.), RNA Tumor Viruses. Cold Spring Harbor Laboratory, Cold Spring Harbor, NY, pp. 209–260.

Whitcomb, J.M., Hughes, S.H., 1992. Retroviral reverse transcription and integration: progress and problems. Annu. Rev. Cell Biol. 8, 275–306.

Whitelaw, C.B., Radcliffe, P.A., Ritchie, W.A., Carlisle, A., Ellard, F.M., Pena, R.N., et al., 2004. Efficient generation of transgenic pigs using equine infectious anaemia virus (EIAV) derived vector. FEBS Lett. 571, 233–236.

Xu, Y.N., Uhm, S.J., Koo, B.C., Kwon, M.S., Roh, J.Y., Yang, J.S., et al., 2013. Production of transgenic Korean native cattle expressing enhanced green fluorescent protein using a FIV-based lentiviral vector injected into MII oocytes. J. Genet. Genomics 40, 37–43.

Yee, J.K., Friedmann, T., Burns, J.C., 1994. Generation of high-titer pseudotyped retroviral vectors with very broad host range. Methods Cell Biol. 43 (Pt A), 99–112.

Zufferey, R., Donello, J.E., Trono, D., Hope, T.J., 1999. Woodchuck hepatitis virus posttranscriptional regulatory element enhances expression of transgenes delivered by retroviral vectors. J. Virol. 73, 2886–2892.

7 Nuclear Transfer Technologies

Yoko Kato and Yukio Tsunoda

Laboratory of Animal Reproduction, College of Agriculture, Kinki University, Nakamachi, Nara, Japan

I. Introduction and Discussion

Since the first successful nuclear transfer in animals by Briggs and King (1952), who reported the development of enucleated eggs receiving blastula-stage nuclei, nuclear transfer has been used extensively to examine the developmental potency of embryonic and somatic cell nuclei. A number of studies have demonstrated that at least some nuclei of embryos at different stages and from various somatic cells of tadpoles still have the potential to develop into fertile adults (DiBerardino, 1997). To date, however, no adult frogs have been produced following nuclear transfer of adult frog somatic cells. The development of nuclear transfer techniques in mammals has been delayed due to the lack of sufficient techniques and knowledge to manipulate mammalian oocytes, zygotes, and embryos. Because the earlier studies on nuclear transfer in mammals were performed without enucleation of recipient eggs, nuclear-transferred eggs did not develop well (Graham, 1969; Baranska and Koprowski, 1970; Lin et al., 1973; Bromhall, 1975; Modlinski, 1978). Illmensee and Hoppe (1981) reported that three live mice were produced after nuclear transfer of inner cell mass (ICM) cells of blastocysts into enucleated zygotes, but their results have not been replicated. Reliable success of nuclear transfer in mammals was reported by McGrath and Solter (1983), who exchanged female and male pronucleus karyoplasts (pronuclei with a small volume of oocyte cytoplasm) between zygotes using inactivated Sendai virus. Their finding was soon confirmed by several other laboratories (Surani et al., 1984; Mann and Lovell-Badge, 1984; Tsunoda et al., 1985). The pronuclear transplantation technique has been applied to clarify the reason for the developmental failure of mammalian parthenogenetic embryos, and it has demonstrated that both female and male genomes have different roles in normal development to term (Surani et al., 1984; McGrath and Solter, 1984a).

The developmental potency of nuclei from preimplantation embryos at advanced stages was examined by fusing them with enucleated zygotes. Although a few enucleated zygotes receiving nuclei from two-cell-stage embryos developed into blastocysts, none of the enucleated zygotes receiving nuclei from four- and eight-cell embryos or ICM cells of blastocysts developed into blastocysts (McGrath and Solter, 1984b). Fertile young were obtained after fusion of nuclear karyoplasts

Transgenic Animal Technology. DOI: http://dx.doi.org/10.1016/B978-0-12-410490-7.00007-4

from four- and eight-cell embryos but not from ICM cells into enucleated blastomeres of fertilized two-cell embryos (Tsunoda et al., 1987). The nuclei fused with enucleated two-cell embryos, however, were not fully reprogrammed, as compaction of nuclear-transferred two-cell embryos occurred at the four- to eight-cell stage but not at the eight- to 16-cell stage.

Willadsen (1986) reported that live lambs were produced after nuclear transfer of blastomeres from 8- to 16-cell-stage embryos to enucleated unfertilized oocytes but not to zygotes. The donor nuclei were considered to be reprogrammed, because nuclear-transferred oocytes developed into blastocysts in the same time frame as zygotes. Since then, a number of studies has attempted to increase the cloning efficiency in farm animals, and the nuclear transfer technique of preimplantation embryos has been established in cattle and sheep (see Chapter 14). On the other hand, successful nuclear transfer in mice has been limited. Live mice were produced after nuclear transfer of nuclei from two-cell-stage (Kono et al., 1991a; Cheong et al., 1993), four-cell-stage (Cheong et al., 1993; Kwon and Kono, 1996; Tsunoda and Kato, 1997), eight-cell-stage (Cheong et al., 1993), and compacted-morula-stage (Tsunoda and Kato, 1997) embryos into enucleated oocytes, or following serial nuclear transfer into enucleated zygotes or two-cell-stage embryos. The cell-cycle stage of the donor nuclei is important for the full-term development of nuclear-transferred oocytes. Nuclei at the G_1 (Cheong et al., 1993; Tsunoda and Kato, 1997), G_2 (Kono et al., 1991b, 1992), or M (Kwon and Kono, 1996) phase have been successfully used. When nuclei at the G_2 or M phase were used, emission of a second polar body was necessary after activation of nuclear-transferred oocytes. Tsunoda and Kato (1998) further reported the successful production of fertile mice after fusion of ICM cells and trophectoderm (TE) cells of blastocysts synchronized at the G_1 stage with enucleated unfertilized oocytes following serial nuclear transfer at the two-cell stage. The enucleated oocytes receiving fetal male germ cells on days 12.5–16.5 developed into blastocysts but did not develop into young (Tsunoda et al., 1989). The failure of development is probably due to the onset of gamete imprinting for the next generation in male fetal germ cells (Kato et al., 1999a). Table 7.1 shows the developmental ability of mouse enucleated oocytes, zygotes, and two-cell embryos receiving nuclei on the germ cells line.

Since the first successful report of the production of young from adult somatic cells by Wilmut et al. (1997), a number of lambs and calves have been produced (see Chapter 14); however, success has been limited in the mouse. For the nuclear transfer of somatic cells in farm animals, donor cells are fused with enucleated oocytes by electric stimulation. Somatic cell nuclear transfer in the mouse, however, is performed primarily by the direct injection method using a piezo-driven micromanipulator. The first success of somatic cell cloning in the mouse was achieved by direct injection of the cumulus cells surrounding ovulated oocytes into enucleated oocytes (Wakayama et al., 1998). So far, mice have been obtained following nuclear transfer of cumulus cells (Wakayama et al., 1998; Kishikawa et al., 1999), cultured follicular epithelial cells (Kato et al., 1999b), tail tip cells (Wakayama and Yanagimachi, 1999; Ogura et al., 2000a), and Sertoli cells (Ogura et al., 2000b). The yield to term was 1–4% of the number of transferred eggs (Table 7.2). Some of

Table 7.1 Developmental Ability of Mouse Enucleated Oocytes, Zygotes and Two-Cell Embryos Receiving Nuclei

Recipient	Donor Cell Stage	Number of Eggs Developing/Cultured to Two-Cell/Blastocyst (%)		Number of Young/Transferred Ova (%)	References
		Two-Cell	Blastocyst		
Zygote	Zygote	–	64/67 (96)	10/64 (16)	McGrath and Solter (1983)
Zygote	Zygote	43/43 (100)	–	22/32 (69)	Mann and Lovell-Badge (1984)
Zygote	Zygote	–	25/32 (78)	6/31 (19)	Tsunoda et al. (1985)
Zygote	2-cell	–	19/151 (13)	–	McGrath and Solter (1984b)
Zygote	2-cell	97/180 (54)	8/108 (7)	3/42 (7)	Tsunoda et al. (1987)
Zygote	ICM	96/142 (68)	23/142 (16)	3/16 (19)	Illmensee and Hoppe (1981)
	TE	34/68 (50)	1/68 (1)	–	
2-cell	4-cell	–	49/68 (72)	10/46 (22)	Tsunoda et al. (1987)
	8-cell		49/139 (35)	4/48 (8)	
2-cell	2-cell	–	79/83 (95)	53/76 (70)	Kono et al. (1991b)
	4-cell		84/118 (71)	18/61 (30)	
MII	2-cell	79/88 (90)	20/88 (23)	3/20 (15)	Kono et al. (1991a)
MII	2-cell	408/470 (87)	120/470 (26)	10/66 (15)	Kono et al. (1992)
MII	2-cell	45/46 (98)	36/46 (78)	10/34 (29)	Cheong et al. (1993)
	4-cell	42/42 (100)	30/42 (71)	6/27 (22)	
	8-cell	36/39 (92)	18/39 (46)	3/17 (18)	
MII	4-cell		58/70 (58)[a]	25/58 (57)	Kwon and Kono (1996)
MII	4-cell	71/88 (81)	20/88 (23)[b]	12/76 (16)	Tsunoda and Kato (1997)
	morula	–	15/128 (12)[b]	2/21 (10)	
MII	ICM		23/36 (64)[b,c]	2/18 (11)	Tsunoda and Kato (1998)
	TE		16/26 (62)[b,c]	2/25 (8)	

[a] Re-nuclear transfer to enucleated zygotes.
[b] Re-nuclear transfer to enucleated 2-cell embryos.
[c] Including morulae.

Table 7.2 Developmental Ability of Mouse Enucleated Oocytes Receiving Somatic Cells

Donor Cells	No. of Eggs Developing/Cultured to Two-Cell/Blastocyst (%)		No. of Young Transferred/ Ova (%)	Postnatal Death/Total Liveborn	References
	Two-cell	Blastocyst			
Cumulus [a]	—	433/778 (56)[b]	31/1385 (2)	9/31 (29)	Wakayama et al. (1998)
Follicular epithelial [c]	—	51/151 (34)[d]	1/30 (3)	—	Kato et al. (1999b)
Cumulus [a]	—	628/1568 (40)[b]	6/580 (1)	—	Kishikawa et al. (1999)
Tail tip [a]	—	—	3/274 (1)	2/3 (67)	Wakayama and Yanagimachi (1999)
Tail tip [a,e]	—	281/684 (41)[b]	8/301 (3)	0/8 (0)	Ogura et al. (2000a)
Sertoli [a]	1133/1846 (61)	436/1846 (34)[b]	16/446 (4)	1/16 (6)	Ogura et al. (2000b)
Fetal fibroblast [a]	—	278/938 (30)[b,f]	5/272 (2)	3/5 (60)	Ono et al. (2001)

[a]Injection.
[b]Including morulae.
[c]Fusion using HVJ.
[d]Re-nuclear transfer to enucleated 2-cell embryos.
[e]Fusion using electric stimulation.
[f]Re-nuclear transfer to enucleated zygotes.

the young also died soon after birth due to respiration problems, similar to bovine somatic cell clones (Kato et al., 1998, 2000). Morphologic abnormalities of young were not observed in all pups, except those obtained from Sertoli cells (Ogura et al., 2000b), but there is a significantly increased body weight in cloned mice (Tamashiro et al., 2000). In most cases, placentas from somatic cell cloned mice were two to three times heavier than those of the controls (Wakayama et al., 1998), and there was also hypertrophy of the placenta (Ono et al., 2001).

In our earlier study, enucleated oocytes receiving mouse embryonic stem (ES) cells, whose cell-cycle stage was not synchronized, developed into blastocysts but not into young (Tsunoda and Kato, 1993). Recently, live mice were produced by direct injection of ES cells, probably at the G_1 and G_2/M phase, into enucleated oocytes (Wakayama et al., 1999; Rideout et al., 2000) and also by fusing the ES cells at the M phase using HVJ (Amano et al., 2001a,b). The potential for nuclear-transferred oocytes to develop into young was low, however, except as reported by Rideout et al. (2000). Postnatal death of young was also observed due to morpho-logic abnormalities and respiration problems (Rideout et al., 2000; Amano et al., 2001a,b). Rideout et al. (2000) reported that the genetic background was an impor-tant factor in postnatal death of young. Amano et al. (2001b), however, demon-strated that the incidence was also different among ES cell lines with the same genetic background. Table 7.3 shows the developmental ability of mouse enucle-ated oocytes receiving ES cells.

The production of cloned mice following nuclear transfer of somatic cells and also ES cells is now possible. Cloned mice were also obtained from gene-targeted ES cells (Rideout et al., 2000). Production of gene-targeted mice by somatic cell nuclear transfer, however, might be difficult because live mice have only been obtained following nuclear transfer of primary cultured somatic cells or cells pas-saged several times. The success rate of clones is also generally low, and postnatal death or morphologic abnormalities of young are commonly observed. One possi-ble reason for such a deficiency is that the current nuclear transfer technique in the mouse is insufficient for remodeling of the donor nuclei. Considering that placentae of cloned mice are two to three times heavier than those of controls and implanta-tion rates are relatively high, the irreversible epigenetic modifications inhibiting normal fetal development, especially for imprinted genes, must be present in somatic cells and ES cells. Studies of gene expression and protein analysis in cloned embryos and conceptuses will reveal these deficiencies. Technical improve-ment of nuclear transfer is also required to develop more reliable cloning technol-ogy in the mouse.

II. Materials and Equipment

The materials and equipment necessary for nuclear transfer are presented here. Although the items used in our laboratory are listed, equivalents could also be used.

Table 7.3 Developmental Ability of Mouse Enucleated Oocytes Receiving ES Cells

Method	No. of Eggs Developed to Blastocysts/ Cultured (%)		No. of Live Young/ Transferred Ova (%)		Postnatal Death/Total Liveborn (%)		References
Injection	635/2358	(27)[a]	31/626	(5)	–		Wakayama et al. (1999)
Injection	34/227	(15)[a]	7/34	(21)	0/7	(0)	Rideout et al. (2000)
	76/418	(18)[a]	8/76	(11)	8/8	(100)	
Fusion	142/308	(46)	1/73	(1)	1/1	(100)	Amano et al. (2001a)
	63/78	(81)[b]	1/51	(2)	1/1	(100)	
	2007/3373	(60)	19/1903	(1)	14/19	(74)	Amano et al. (2001b)

[a]Including morulae.
[b]Re-nuclear transfer to enucleated 2-cell embryos.

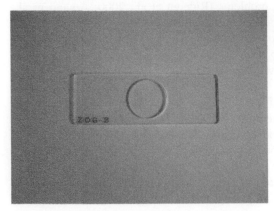

Figure 7.1 Manipulating chamber. Microdroplets are made on this chamber covered with mineral oil (see also Figure 7.11).

- Microforge: Narishige MF-9 or Technical Product International (TPI) with a 10× or 20× objective
- Platinum–iridium wire to attach the microforge: Narishige PT100 or PT150
- Pipette puller: Narishige PN-3 or Sutter Instrument Co. P-97/IVF
- Glass capillary: For holding pipette, Narishige G-1 (1 × 90 mm); for enucleation/injection pipette, Sutter Instrument B120-90-10 (1.2 mm outer diameter, 0.90 mm inner diameter); for needle, Drummond 10-mL Microdispenser (catalog no. 510G-310G-210G)
- Micromanipulator: Narishige
- Inverted microscope: Nikon TE300 or Olympus IX70
- Electrofusion system: BTX, Electro Cell Manipulator; BEX, Electro Cell Fusion Model LF101
- Warm/cool plate for inverted microscope: Tokai Hit Co., Thermoplate MATS 555RT
- CO_2 incubator
- Piezo-micropipette-driving unit: Prima Meat Packers, Piezo Micromanipulator Model PMM-01
- Manipulating chamber (Figure 7.1), for direct injection of ES and somatic cells

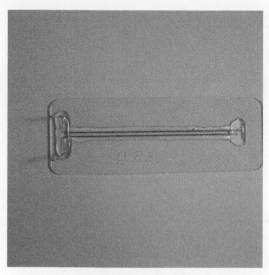

Figure 7.2 Fusion chamber. Fusion chamber has two stainless wires 1.0 mm apart attached with an electrofusion system.

Table 7.4 Composition of Zimmerman Cell Fusion Medium[a]

Sucrose	9.584 g/100 mL
Mg $(C_2H_3O_2)_2 \cdot 4H_2O$	0.011 g/100 mL
Ca $(C_2H_3O_2)$	0.002 g/100 mL
K_2HPO_4	0.017 g/100 mL
Glutathione	0.003 g/100 mL
BSA	1 mg

[a]pH 7.0.

- Fusion chamber (Figure 7.2)
- Transfer pipette for handling eggs
- 27-gauge needle
- 30-mm culture dish
- Female mice: F1 (C57BL/6 × CBA or C3H)
- Media and chemicals: medium for manipulation, M2 medium (Table 7.4; see also Chapter 2, Table 2.1); medium for *in vitro* culture, M16 medium (Table 7.4; see also Chapter 2, Table 2.1); and hyaluronidase, with a final concentration of 300 NF unit/mL M2 medium
- Cytoskeleton inhibitor: cytochalasin B, nocodazole; stock solution (1000×) is prepared with dimethylsulfoxide (DMSO) and stored at −20°C, or, once dissolved, it may be refrigerated at 4°C
- 0.05% trypsin with 0.05% EDTA/phosphate buffered saline (PBS), for disaggregation of blastomeres and cells
- Inactivated Sendai virus
- Light mineral oil
- Hoechst 33342 (catalog no. 382065, Calbiochem-Behring Corp., San Diego, CA); stock solution (1 mg/mL) is prepared using double-distilled water

Table 7.5 Strontium-Supplemented M16 Medium

		M16 (10 mL)
Stock Solution A (10×)	**g/100 mL**	
NaCl	5.534	1.0 mL
KCl	0.356	
$KH_2 \cdot PO_4$	0.162	
$MgSO_4 \cdot 7H_2O$	0.293	
Lactate	2.61	
Glucose	1.0	
Penicillin and Streptomycin		
Stock Solution B (10×)	**g/100 mL**	
$NaHCO_3$	2.101	1.0 mL
Phenol red	0.01	
Stock Solution C (100×)	**g/10 mL**	
Sodium pyruvate	0.036	0.1 mL
Stock Solution D (100×)	**g/10 mL**	
$SrCl_2 \cdot 6H_2O$	2.666	0.1 mL
H_2O (2× distilled)		7.8 mL
BSA		40 mg

- Zimmermann cell fusion medium (Table 7.5) or 0.3 M mannitol
- Aphidicolin: stock solution (1 mg/mL) is prepared with DMSO and stored at −20°C, or, once dissolved, it may be refrigerated at 4°C
- Polyvinylpyrrolidone (PVP): high molecular weight, such as 30,000, is preferred
- Polyvinylalcohol (PVA): when direct injection is performed, PVA replaces BSA to prevent cytoplasmic contents from sticking to donor cells
- Strontium-supplemented medium (Table 7.5)
- Polyclonal antibody against mouse tissues: for isolation of ICM
- Complement: low-tox guinea pig complement (Cedarlane Laboratories, Ltd.), for isolation of ICM
- Cell-proliferation kit: Amersham, code RPN 20, for detection of the cell cycle of synchronized embryos and cells at the G_1 phase
- Hormones for superovulation: pregnant mare serum gonadotropin (PMSG) and human chorionic gonadotropin (HCG)

III.   Preparation of Microtools

For nuclear transfer, three types of glass instruments are necessary: a holding pipette, an enucleation or injection pipette, and a fine glass needle. The holding pipette is used to hold the egg in place during microsurgery, the enucleation pipette is used to remove the chromosomes or nucleus from recipient oocytes or zygotes, the injection pipette is used for injection of the donor cell or nucleus into the perivitelline space or ooplasm, and the fine glass needle is used to cut the zona pellucidae of the eggs to facilitate nuclear transfer (Tsunoda et al., 1986).

Compared with the original nuclear transfer methods (MacGrath and Solter, 1983), zona cutting speeds up the micromanipulation. All glass pipettes for micromanipulation can be prepared from fine glass capillaries, but the diameter of each is slightly different. The microtools are fashioned using the microforge. In addition to those outlined in Chapter 2, the methods for preparation of microtools are also described in detail by Barton and Surani (1993), Hogan et al. (1994), and Latham and Solter (1993).

A. Holding Pipette

To make the holding pipettes, thick-walled glass tubes are used. As an alternative to the methods shown in Chapter 2, the taper or center of the capillary is heated over a flame and rapidly pulled by hand. The most appropriate outer diameter for holding eggs is the same as or slightly greater than the diameter of the egg (80−100 μm). The tip is examined under a microscope to determine if it is flat and whether the outer diameter is appropriate. It is useful to observe the tip under a microscope and fashion it on a microforge equipped with a platinum−iridium wire (100 μm) (Figure 7.3). The glass capillary is mounted vertically against the platinum−iridium filament wire on the microforge (Figure 7.4) and is briefly heated. The tip of the glass capillary is reduced in size by melting the glass (Figure 7.5). The resulting tip is slightly bent to a 30° angle.

B. Enucleation/Injection Pipette

For enucleation/injection pipettes, thin-walled glass capillaries are used. The glass capillary is first pulled using a pipette puller and then fashioned to the final shape using a microforge. The glass capillary is mounted horizontally against a

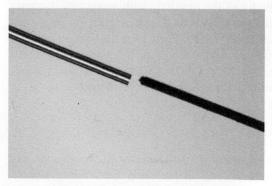

Figure 7.3 Glass capillary for holding pipette and platinum−iridium wire. When holding pipette (right) is prepared, it is useful to observe the tip under a microscope with a piece of platinum−iridium wire (left). Note that the tip of the holding pipette is flat, and the diameter is similar to the wire (100 μm).

Figure 7.4 Glass capillary for holding pipette attached to the microforge. Glass capillary is mounted vertically against a filament of platinum—iridium wire on the microforge. The arrowhead shows a glass capillary prepared for the holding pipette.

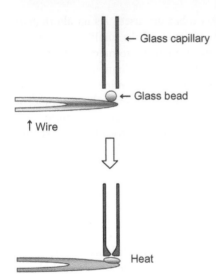

← Glass capillary

← Glass bead

↑ Wire

Heat

Figure 7.5 Method to prepare a holding pipette. By heating the wire of a microforge, the tip of the glass capillary is shrinked in size.

platinum—iridium wire filament. The tip of the capillary, with an outside diameter of $10-20 \, \mu m$, is touched lightly to the glass beads on the tip of the wire. By slowly heating the wire, the tip of pipette will begin to melt. Then, after the heat is removed, the wire will contract as it cools, bringing with it the tip of the pipette, which will result in a very fine glass tip (Figure 7.6). A suitable outer diameter for the injection pipette is similar to or slightly smaller than that of the donor cell diameter.

C. Needle

For the fine glass needle, the taper or center of the capillary is first heated over a flame and then rapidly pulled by hand, as for preparing the holding pipette.

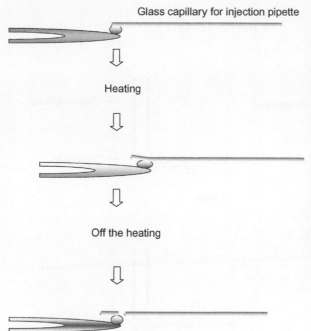

Glass capillary for injection pipette

Heating

Off the heating

Figure 7.6 Method to cutting the tip of enucleation/injection pipette. First, the tip of a capillary is contacted with the glass bead attached to the wire (upper). Then the wire is gradually heated. The appropriate temperature just fuses the pipette with the glass bead. When the tip bends slightly (middle), the heating must be stopped. As the wire cools and contracts, it breaks the capillary at the point of contact, leaving a fragment of the capillary on a glass bead of the wire (lower).

The diameter is not critical (120 μm). To sharpen the tip for the needle, the capillary is mounted vertically against the wire filament, and the tip of the glass capillary is pulled down during low heating of the wire filament (Figure 7.7). When this step is repeated two or three times, the tip of the glass capillary becomes very fine. All microtools are slightly bent (approximately 1 mm from the tip) by brief exposure of the tip to the heating wire, and, depending on the micromanipulator system used, one more bend must also be made by hand with a small flame.

IV. Nuclear Transfer of Preimplantation Embryos

For nuclear transfer of preimplantation embryos, enucleated zygotes and oocytes can be used for the recipient cytoplasm. Zygote cytoplasm can support development after nuclear transfer of karyoplasts from zygote or two-cell-stage embryos used as donor nuclei. After the four-cell stage, zygote cytoplasm does not support development following nuclear transfer. Oocyte cytoplasm at the second metaphase (MII phase), however, supports the development of reconstituted eggs receiving nuclei from four-cell-, eight-cell-, morula-, and blastocyst-stage embryos. Here, we present the method of nuclear transfer into enucleated zygotes and oocytes using blastomeres from preimplantation embryos as donor nuclei. The success rates of nuclear transfer of preimplantation embryos are described in Section 7.I. As shown in the flow chart (Figure 7.8), nuclear transfer consists of six principal steps: enucleation

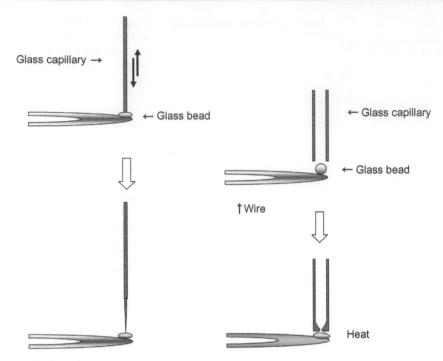

Figure 7.7 Method to produce a fine needle. The glass capillary is mounted vertically against a filament of platinum—iridium wire on the microforge. The tip is pulled up and down two or three times during low heating of the wire filament, resulting in a fine glass needle.

of recipient eggs, isolation and cell-cycle synchronization of donor cells, fusion with donor cells, activation, *in vitro* culture, and embryo transfer.

The activation process is necessary for nuclear transfer into enucleated oocytes, because recipient cytoplasm is not activated by sperm. Alternatively, artificial activation by electric pulses or culture with ethanol or strontium is effective for the further development of nuclear-transferred oocytes.

A. Nuclear Transfer into Enucleated Zygotes

1. Collection of Zygotes

Recipient zygotes for nuclear transfer are collected from hybrid mice (C57BL/ 6 × CBA or C3H), because the viability of eggs from hybrid mice is higher and they are released from the developmental block (two-cell block) during *in vitro* culture.

The materials and methods used to collect preimplantation embryos, including zygotes, are described in detail in Chapter 2. Briefly, mature female mice (6 weeks) are superovulated by injection of PMSG (5IU) and HCG (5IU) 48 h apart, and

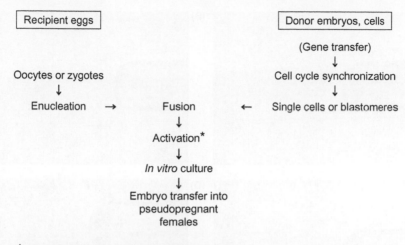

*Not necessary for zygote as recipient cytoplasm

Figure 7.8 Flow chart for nuclear transfer (see text).

oviducts are collected from females mated with same-strain males. Zygotes are collected from the dissected oviducts using a 27-gauge needle 20−22 h after HCG injection. Collected cumulus−oocyte complexes are treated with hyaluronidase (300 NF unit/mL) for 1−3 min at 37°C to dissociate the eggs from the cumulus cells. Denuded one-cell eggs are picked up using a transfer pipette and washed three times in fresh M2 medium. One-cell eggs with two pronuclei and the second polar body are selected and incubated in M16 medium in a 5% CO_2 incubator.

2. Enucleation of Zygotes

Collected zygotes are incubated in M2 medium supplemented with cytoskeleton inhibitors (cytochalasin B, 5 μg/mL; nocodazole, 1−3 μg/mL) prior to microsurgery. For microsurgery, zygotes are manipulated in a flat droplet of M2 medium in a micromanipulation chamber (Figure 7.1) under an inverted microscope using a micromanipulator. A cut (10−20% of the length of the zygote) is made in the zona pellucida with a fine glass needle (Figure 7.9; Tsunoda et al., 1986). As shown in Figure 7.9, the zygote is held with the holding pipette by applying negative pressure to the zona pellucida (Figure 7.9A). Then, the tip of the needle is carefully pushed into the perivitelline space (Figure 7.9B), so as not to damage the ooplasm membrane. The skewered zygotes are released from the holding pipette (Figure 7.9C), and the needle is pressed repeatedly against the wall of the holding pipette until the cut is produced (Figure 7.9D−F). Zona-cut zygotes are held with the holding pipette placed opposite to the slit in the zona pellucida (Figure 7.10A). Through the slit of the zona pellucida, the tip of the enucleation/injection pipette is inserted into the perivitelline space (Figure 7.10B). The tip is then advanced to a point adjacent to one of the two pronuclei. One pronucleus with a small volume of

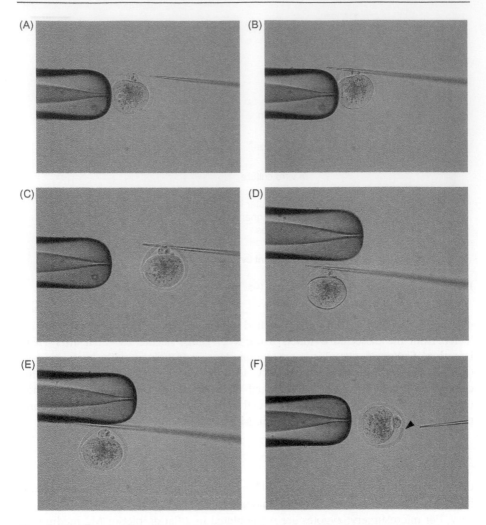

Figure 7.9 Zona cutting of a zygote. The arrowhead shows the slit of zona (see text).

cytoplasm is sucked into the injection pipette and this step is then repeated for the remaining pronucleus (Figure 7.10C). For this step, it is necessary to avoid penetrating the ooplasm membrane by using cytoskeleton inhibitors. After zygotes are enucleated (Figure 7.10D), they are rinsed with M2 medium and kept at room temperature or 37°C until use.

3. Preparation of Donor Embryos

Zygotes are collected as described above. Two-cell-stage embryos are collected from oviducts of superovulated females by flushing the oviducts with M2 medium 42–44 h after HCG injection. Collected embryos are incubated in M2 medium until use.

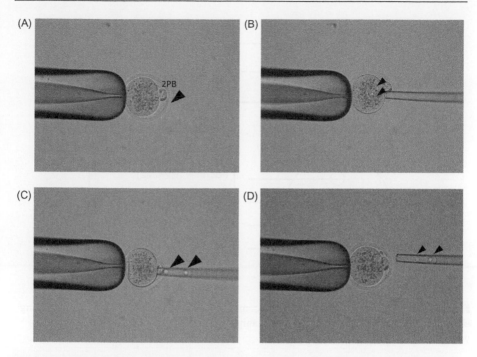

Figure 7.10 Enucleation of a zygote (see text). (A) The arrowhead shows the slit of zona. (B)−(D) The arrowheads show two pronuclei. 2pPb is the second polar body.

4. Fusion Procedure

Fusion of donor nuclei with recipient cytoplasm is induced by inactivated Sendai virus or electric pulses.

a. Inactivated Sendai Virus

To facilitate micromanipulation, three flat drops of M2 medium are placed on the manipulation chamber (Figure 7.1) and covered with light mineral oil. The top droplet is M2 medium with cytoskeleton inhibitors (cytochalasin B, 5 μg/mL; nocodazole, 1−3 μg/mL) for donor embryos; the second droplet is inactivated Sendai virus, and the bottom droplet is M2 medium without cytoskeleton inhibitors for recipient zygotes (Figure 7.11).

Refined Sendai virus is inactivated by ultraviolet (UV) exposure and diluted into 2500 units of hemagglutinating activity per milliliter with 200-mM, Ca-supplemented BSS medium (NaCl, 137 mM; KCl, 5.4 mM; $KH_2 \cdot PO_4$, 0.44 mM; $Na_2 \cdot HPO_4 \cdot 12H_2O$, 0.34 mM; Tris, 13 mM). There is no difference between chemical inactivation and UV irradiation on the inactivation rate and viability of the virus.

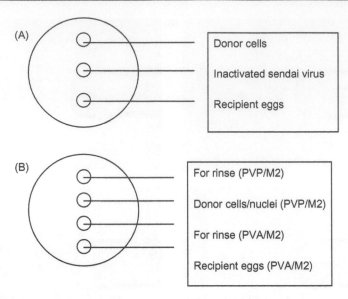

Figure 7.11 (A) Diagram of microdrop placement for micromanipulation on a manipulating chamber. (B) Diagram of microdrop on a manipulating chamber used for direct injection with piezo pulse.

When donor embryos are collected, the zonae pellucidae are cut as described above in the enucleation process. After cutting the zona, embryos are incubated for 10 min with M2 medium containing cytoskeleton inhibitors. Similar to the enucleation procedure, donor embryos are held in place with a holding pipette, and the pronuclei of the zygote or each nucleus of a two-cell-stage embryo with a small volume of cytoplasm (karyoplast) is carefully sucked into the injection pipette in the top droplet (Figure 7.12A). The pipette then moves to the next droplet, which contains inactivated Sendai virus. In this droplet, a tip of the karyoplast is exposed to the inactivated Sendai virus and a similar volume of virus is sucked into the injection pipette (Figure 7.12B). The pipette with karyoplast and inactivated Sendai virus is moved to the third droplet, and the karyoplast with Sendai virus is injected into the perivitelline space of the recipient zygote through the slit in the zona pellucida (Figure 7.12C). When the karyoplast is injected into the perivitelline space, the tip of the injection pipette is carefully pressed against the recipient ooplasm to promote fusion. This step is repeated with each embryo to produce reconstituted eggs. Reconstituted eggs are washed twice with M2 medium and then briefly incubated in M16 medium until fusion. Fused eggs (Figure 7.12D) are cultured *in vitro* for 4 days and transferred to recipient females, as described below.

If the fusion rate is low, drastic changes in temperature after injection are effective for fusion by Sendai virus. For example, donor cells can be injected into the

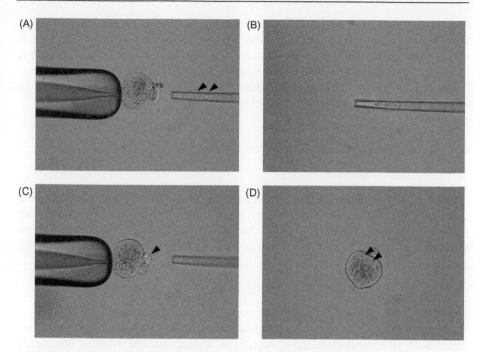

Figure 7.12 Nuclear transfer of zygote (pronuclei exchange, see text). (A) Karyoplast is obtained from a donor zygote (arrowheads show pronuclei). (B) Inactivated Sendai virus is injected into the pipette. (C) Donor karyoplast with inactivated Sendai virus is introduced into the perivitelline space of recipient zygote cytoplasm that was previously enucleated (arrowhead shows donor karyoplast). (D) Reconstituted zygote (arrowheads show donor karyoplast after fusion). 2pPb is the second polar body.

perivitelline space at 10°C and then, immediately after injection, the reconstituted zygotes are quickly transferred to warm M2 medium at 37°C for 15 min.

b. Electric Pulses

Reconstituted zygotes are transferred to 0.3 M mannitol or Zimmerman's cell fusion medium in a chamber with two wire electrodes mounted 1 mm apart on a glass slide (Figure 7.2). Reconstituted zygotes are given two direct current (DC) pulses of 50 V/mm for 50 ms three times at 20-min intervals. DC pulses are given after the donor cell is oriented parallel to the wires and recipient oocyte by hand. It is also useful to give alternating current (AC) pulses (100–500 kHz, 5 V/mm) before the DC pulses instead of orienting the cells by hand.

c. *In Vitro* Culture and Embryo Transfer

Reconstituted zygotes are cultured in M16 medium for 15–60 min and the success of fusion is verified. Fused eggs are further cultured for 4 days *in vitro* in a CO_2 incubator at 37°C, and embryos that develop into the blastocyst stage are

transferred to oviducts on day 0.5 or uteri on day 3.0–3.5 in recipient females previously mated with vasectomized males. Embryo transfer to recipient females is
not recommended until the embryos have reached the compacted morula stage,
because blastomeres of precompacted embryos often fall out of the slit in the zona,
and such embryos do not develop further. Recipient females are allowed to go to
term or are dissected on day 19.

B. Nuclear Transfer into Enucleated Oocytes

1. Collection of Oocytes

Female mice (3 weeks of age) are superovulated using PMSG and HCG, and
oocytes at the metaphase of the second meiosis (MII) are collected by dissecting
the oviducts with 27-gauge needles. Morphologically normal MII oocytes are
selected and incubated in M16 medium in a 5% CO_2 incubator until use.

2. Enucleation of Oocytes

a. Without Hoechst Staining

When denuded oocytes are observed under an inverted microscope with Nomarski
optics with $10\times$ ocular and $20\times$ objectives, a small area of swelling containing
MII chromosomes can be observed (Figure 7.13A). The zonae pellucidae around
the swollen areas of collected MII oocytes are cut as shown above in M2 medium
without cytoskeleton inhibitors at 37°C (Figure 7.13B–E). After cutting the zonae
pellucidae of the oocytes, the oocytes are incubated with cytochalasin B (5 µg/mL)
in M2 medium. The swelling areas become difficult to distinguish after cytochalasin B treatment. Oocytes are held opposite the slit in the zona pellucida, and the
enucleation pipette is inserted into the perivitelline space through the slit. An area
of translucent cytoplasm containing MII chromosomes (Figures 7.13F and 7.14)
moves when the enucleation pipette touches the ooplasm, and then the area is aspirated into the enucleation pipette while visually confirming the presence of croma-
tin. After enucleation of all the oocytes, the eggs are rinsed with M2 medium and
kept at room temperature or 37°C until use.

   If it is difficult to find the areas containing the MII chromosomes, they can first
be stained using Hoechst 33342 to visualize them with fluorescence microscopy
during the enucleation step (Tsunoda et al., 1988). The protocol of enucleation after
staining with Hoechst is shown below.

b. With Hoechst Staining

Collected oocytes are incubated with M2 medium containing Hoechst 33342
(0.5 µg/mL) for 3 min at 37°C. Stained oocytes are washed three times with M2
medium without Hoechst dye, and MII chromosomes are removed with an enucle-
ation/injection pipette under a fluorescence microscope (blue-violet, UV excitation
filter with 400–450 nm Nikon BV-2A). UV irradiation exposure time is critical to
the continued viability of the oocytes and should be less than 15 s.

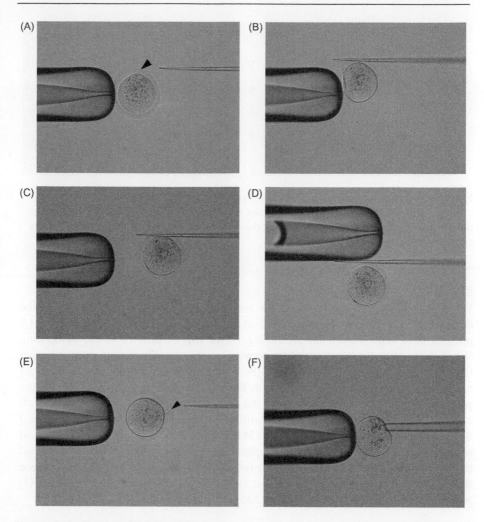

Figure 7.13 Zona cutting of a MII oocyte (see text). (A) Arrowhead shows a small area of swelling containing MII chromosomes. (B)(C)(D) Cutting the zona (E) Arrowhead shows a slit of the zona. (F) Oocyte enucleation.

3. Preparation of Donor Cells

a. Collection of Donor Embryos

Two-cell-stage embryos are collected from oviducts of superovulated females by flushing the oviducts with M2 medium 42–44 h after HCG injection and mating. Four- and eight-cell embryos are collected from *in vitro* culture of two-cell embryos, the latter being flushed from the oviducts and uterine horns 65 h after HCG injection. Compacted morula-stage embryos are obtained after 24-h culture of eight-cell-stage embryos. Collected embryos are incubated in M2 medium until use.

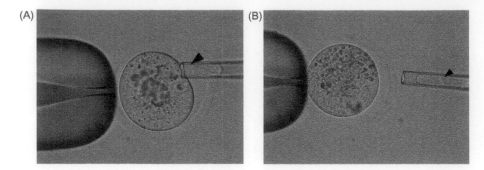

Figure 7.14 Enucleation from a MII oocyte (see text). Metaphase at the second meiosis can be observed. (A) Metaphase chromosomes are removed by an enucleation pipette from the slit of zona (arrowhead shows chromosomes). (B) Enucleated oocyte (arrowhead shows chromosomes).

b. Cell-Cycle Synchronization

When MII-phase oocytes are used for recipient cytoplasm, the cell cycle of the donor embryos must be synchronized at the G_1, G_2, or M phase because MII oocytes have high maturation promoting factor (MPF) activity in the cytoplasm. When donor cells are fused with MII oocytes, premature chromosome condensation (PCC) occurs due to MPF activity in the ooplasm. When the oocytes with donor nucleus in PCC are artificially activated, the nuclear envelope forms again and the pronuclear-like formation starts from the beginning of the G_1 phase of the cell cycle. If S-phase donor cells are fused with MII oocytes, the DNA will be abnormal due to PCC and subsequent activation. Thus, donor cells must be in G_1, G_2, or M phase. After synchronization at the G_1 phase as described below, the medium must be supplemented with aphidicolin until fusion with the oocyte cytoplasm, because the G_1 phase is very short in preimplantation embryos. If donor cells are at the G_2 phase, emission of the polar body after activation is necessary.

c. Four-Cell- to Morula-Stage Embryos

A method to synchronize the cell cycle of preimplantation mouse two-cell embryos to the M phase without adverse effects has been established. When cell-cycle synchronization of four-cell-stage embryos is required, two-cell embryos are incubated in M16 medium supplemented with nocodazole (3 μg/mL) for 12−14 h, then washed and cultured in M16 medium supplemented with aphidicolin (5 μg/mL) to suppress DNA synthesis. Division of both blastomeres synchronously occurs approximately 55 min after removing them from the nocodazole. When morula-stage embryos are required to synchronize at the G_1 phase, eight-cell embryos are treated with nocodazole-containing M16 medium for 6 h before incubation with aphidicolin-supplemented medium. When nuclei of four-cell-stage embryos are used as donors, enucleation pipettes are inserted into the perivitelline space to remove the karyoplast from each blastomere. When blastomeres are too small to remove the karyoplast, such as in the morula-stage embryos, the zona pellucida is removed, and the blastomeres are disaggregated into single blastomeres by pipetting in trypsin–EDTA solution.

After disaggregation, blastomeres are kept in aphidicolin-supplemented medium at room temperature until use.

d. ICM and Mural Trophectoderm Cells in Blastocysts

Blastocysts are recovered from females 4 days after naturally mating with males or are obtained after 24-h culture of morula-stage embryos. Blastocysts are cultured *in vitro* in M16 medium supplemented with nocodazole (3 μg/mL) for 12 h to synchronize the cell cycle at the M phase. After synchronization at the M phase, blastocysts are cultured in M16 medium without nocodazole but with aphidicolin (5 μg/mL) for 1 h to inhibit DNA synthesis. The zonae pellucidae of blastocysts, expected to be at the G_1 phase, are removed using acetic Tyrode's solution or pronase as described.

The ICMs of the blastocysts are isolated by immunosurgery. Blastocysts are treated with an appropriate concentration of polyclonal antibody against mouse liver and kidney for 30 min at 37°C. After the treatment, blastocysts are washed three times in M2 medium and then transferred to a complement (low-tox guinea pig complement) supplemented medium (1:7) for 30 min at 37°C. At this step, cell membranes of trophectoderms (TEs) are damaged and the ICM cells are isolated.

For preparation of murine TE cells, zona-free blastocysts at the G_1 phase are cultured further in M16 medium with aphidicolin (5 μg/mL) until the blastocoele cavity recovers. The murine TE cells are carefully separated from the blastocysts microsurgically.

Isolated ICM and murine TE cells are disaggregated into single cells by treatment with PBS containing trypsin (0.01%) and EDTA (0.02%) or collagenase (200 IU) and DNase (10 μg/mL) at 4°C for 30 min.

e. Determination of the Cell-Cycle Stage

To determine whether donor blastomeres are at the G_1 phase of the cell cycle, incorporation of the thymidine analog, 5-bromo-2′-deoxyuridine (BrdU) can be used. If BrdU is incorporated into the nuclei, the nuclei have entered into the S phase; thus, the presence of BrdU in nuclei indicates whether the cells have stopped at the G_1 phase. Embryos that might be in the M phase are cultured in M16 medium with BrdU for 4 h in the presence of aphidicolin (5 μg/mL). At the end of the culture, the embryos are fixed using acid ethanol and incubated with anti-BrdU monoclonal antibody for 1 h at room temperature. After washing, the embryos are treated with peroxidase anti-mouse IgG_{2a} for 30 min at room temperature, treated with substrate/intensifier, and observed in the presence of diaminobenzidine, following the manufacturer's protocol. As a positive control, embryos that might be in the M phase are cultured without aphidicolin in the presence of BrdU. For counterstain, 0.5% eosin solution is used for 2–3 min. BrdU-positive cells in S phase are stained blue-black.

4. Fusion Procedure

Using the injection pipette, karyoplasts or disaggregated single blastomeres at the G_1 phase are injected into the perivitelline space of enucleated oocytes with inactivated Sendai virus. The procedure for fusion of recipient ooplasm with donor karyoplast or blastomeres by Sendai virus is the same as described in the section on zygote nuclear transfer. When donor cells are fused with recipient oocytes using electric pulses, calcium ions must be removed from the Zimmerman's cell fusion medium to avoid oocyte activation.

5. Activation

Following fusion procedure, reconstituted oocytes are artificially activated by electric pulses or ethanol treatment. The electric pulses protocol has proved effective in our experience.

a. Electric Pulses

Reconstituted oocytes are transferred to Zimmerman's cell fusion medium in a chamber with two wire electrodes mounted 1 mm apart on a glass slide as described earlier. Reconstituted oocytes are given two DC pulses of 50 V/mm for 50 ms three times at 20-min intervals. If the donor cells are not fused, AC pulses (100−500 kHz, 5 V/mm) are administered before the DC pulses, or DC pulses are given after orienting the donor cell by hand rather than by AC pulses so that it is parallel to the wires and the recipient oocyte. After the electric pulses, reconstituted oocytes are cultured in M16 medium for 5 h and examined for pronuclear-like formation of donor nucleus derivation in ooplasm.

b. Ethanol Treatment

Reconstituted oocytes are cultured in M16 medium supplemented with 7% ethanol for 7 min to activate artificially.

6. In Vitro Culture and Embryo Transfer

After activation, reconstituted oocytes are cultured in M16 medium for 15−60 min and checked for fusion. Fused eggs are further cultured for 4 days in vitro, and embryos that develop into the blastocyst stage are transferred to day 0.5 oviducts or day 3.0–3.5 uterine horns of recipient females.

7. Serial Nuclear Transfer

After nuclear-transferred oocytes divide into the two-cell stage, during overnight culture in vitro, their nuclei are again transferred into enucleated blastomeres of two-cell embryos that were derived from in vivo or in vitro fertilization (Figure 7.15).

Serial nuclear transfer into fertilized embryo cytoplasm promotes the development of nuclear transplants (Tsunoda and Kato, 1998). The technique of serial nuclear transfer is basically the same as in the report of Tsunoda et al. (1987),

Nuclear transferred embryos   Fertilized embryos **Figure 7.15 Method for serial
nuclear transfer at the 2-cell
stage.** When nuclear -transferred
oocytes are cleaved (left), their
nuclei are again transferred into
enucleated blastomeres of 2-cell
embryos derived from fertilization
(right). Reconstituted 2-cell
embryos (lower figure) have a
cytoplasm from a fertilized 2-cell
embryo and nuclei from a nuclear-
transferred embryo.

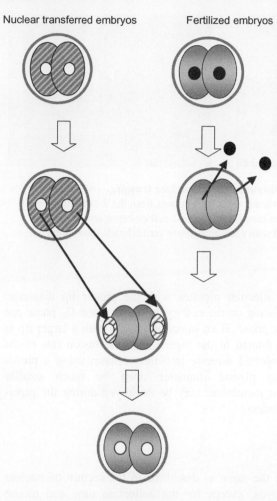

demonstrating the nuclear transfer of nucleus from four- to eight-cell embryos into
enucleated blastomeres of two-cell embryos. Figure 7.16 shows the reconstituted
two-cell embryos after serial nuclear transfer.

V. Nuclear Transfer of Embryonic Stem Cells and Somatic Cells

Due to the smaller size of embryonic stem (ES) cells and somatic cells, the nucleus
can be directly injected into ooplasm using a piezo-micropipette-driving unit
(Wakayama et al., 1999). Here, we present the protocol for direct injection of donor
nuclei into oocytes, in addition to the standard nuclear transfer method by cell
fusion. Successful nuclear transfer of ES cells and somatic cells requires

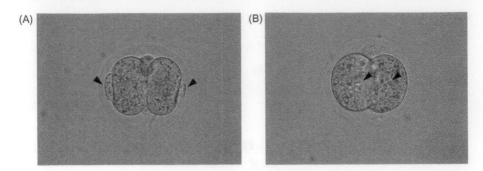

Figure 7.16 Reconstituted 2-cell embryos by serial nuclear transfer. (A) 2-cell embryos just after serial nuclear transfer (arrowheads show karyoplasts from the 2-cell embryos derived by nuclear transfer). (B) When reconstituted, the 2-cell embryos are cultured for 15–30 min, and karyoplasts are fused with each blastomere (arrowheads show fused karyoplasts).

appropriate microtools, such as injection pipettes with the proper tip diameter. The cell size varies widely, depending on the cell cycle. Cells at the G_0 phase are smaller than those at the G_2 M or phase. If an injection pipette with a larger tip is used, inactivated Sendai virus is diluted in the pipette, and the fusion rate might decrease. When the nucleus is injected directly into the ooplasm using a piezo-micropipette unit, the injection pipette diameter must be much smaller (10–20 μm); otherwise, the donor membrane may be destroyed during the pipetting, damaging the ooplasm membrane.

A. Collection of Oocytes

The materials and equipment are the same as described in the section on nuclear transfer of preimplantation embryos, except for the collection time and mouse strain of oocytes. When somatic cells are used as donor cells, oocytes are often collected earlier, such as 13–14 h after HCG injection. Recipient oocytes for nuclear transfer of cultured cells are also collected from B6C3F1 (C57BL/6 × C3H) and B6D2F1 (C57BL/6 × DBA) mice.

B. Enucleation of Oocytes

The oocyte enucleation procedure is the same as described in the section on nuclear transfer of preimplantation embryos.

When enucleation is performed with the piezo pulses, the zona-cutting process is not necessary. To insert the pipette through the zona pellucida, the pipette is touched to the surface of the zona and then piezo pulses are applied. When the injection pipette penetrates the zona, a small round fragment of the zona remains in the enucleation pipette. After this step, the enucleation method is the same as described above. After enucleation, oocytes must be incubated without CB cytochalasin B (CB) for up to 2 h at 37°C before direct injection by piezo pulses.

C. Preparation of Donor Cells

1. Embryonic Stem Cells

The basic protocol of maintenance of embryonic stem cells is described in other chapters. Briefly, ES cells are cocultured with a primary culture of mouse embryonic fibroblast (MEF) cells that were previously inactivated by mitomycin C (10 µg/mL for 2.5 h). At present, it is difficult to synchronize the cell cycle of ES cells at the G_0 and G_1 phases. In some reports, the cell cycles of ES cells are distinguished by the size of the ES cells. When ES cells in the M phase are used as donor cells, nuclear-transferred oocytes develop into the blastocyst stage at a high rate and some of them develop to term. For the synchronization of ES cells at the M phase, the cells are incubated in nocodazole-containing medium ($1-3$ µg/mL) for 3 h prior to use as nuclear transfer donors. The M phase ES cells are round and easy to distinguish.

2. Somatic Cells

At present, five kinds of somatic cells have successfully developed into full-term fetuses or adult mice by nuclear transfer into enucleated oocytes (Table 7.2), ovulated cumulus cells (Wakayama et al., 1998; Kishikawa et al., 1999), tail tip cells (Wakayama and Yanagimachi, 1999; Ogura et al., 2000a), fetal fibroblast cells (Ono et al., 2001), immature Sertoli cells (Ogura et al., 2000b), and follicle epithelial cells (Kato et al., 1999b). The developmental potential of cloned embryos into fetuses at full term or adults is extremely low. More basic research studies are necessary to characterize transgenic mouse development after nuclear transfer of somatic cells. Nuclear transfer of somatic cells produces conceptuses without any other cells, such as in the chimera experiments, suggesting that they are useful tools for analysis of gene expression mechanisms. Here we introduce an updated protocol of nuclear transfer of the cumulus cells surrounding the ovulated oocytes and follicle epithelial cells.

a. Cumulus Cells

Cumulus cells are obtained from the cells surrounding ovulated MII-phase oocytes. The collection method is the same as that for oocyte collection described above. Cumulus cells are stopped at the G_0 phase and can be used as donor cells without additional in vitro culture for cell-cycle synchronization.

b. Follicle Epithelial Cells

Follicle epithelial cells are obtained from the cells surrounding stage 5 oocytes, collected from ovaries of adult females, by pricking the follicles. Follicle epithelial cells with oocytes are cultured in Dulbecco's modified Eagle's medium (DMEM), modified for culturing mouse ES cells with 10% fetal bovine serum (FBS) excluding leukemia inhibitory factor (LIF) *in vitro* until the cells attach to the bottom of a culture dish. Attached and extended cells are subcultured for several passages and subjected to serum starvation (0.5% FBS) for 2−10 days before nuclear transfer. When the cells are used as nuclear donors, they are disaggregated with trypsin/EDTA solution, followed by rinsing with PBS. The BrdU incorporation test is useful to determine the cell-cycle stage of serum-starved cells, as described in Section 7.III.B.

c. Gene Transfer

Although gene-targeted transgenic sheep have been produced following nuclear transfer of cultured somatic cells, there has been limited success in the mouse. Transgenic mice, however, have been produced after nuclear transfer of gene-targeted ES cells. The methods for gene transfer into ES cells and cultured somatic cells are described in other chapters.

D. Procedure for Donor Nucleus Incorporation

1. Fusion Procedure

Basically the same methods as for nuclear transfer of preimplantation embryos are used. Because the cell size is smaller than blastomeres or karyoplasts of preimplantation embryos, an injection pipette must be constructed with the same diameter as the donor cells (10−20 μm) (Figure 7.17).

2. Piezo Micromanipulation

The method of direct injection of donor nuclei into oocytes by piezo pulses is basically the same as that for intracytoplasmic sperm injection (Kimura and Yanagimachi, 1995). Droplets of M2 medium supplemented with 10% PVP and M2 medium with PVA in place of BSA are placed in the micromanipulation chamber. Four droplets are created in the chamber: (1) rinse, (2) drop with donor cells/ nuclei, (3) rinse, and (4) drop with recipient eggs. Donor cells and the injection pipette are rinsed M2 medium containing 10% PVP. When the inside of the injection pipette becomes sticky during the microsurgery, it can be washed in this drop to clean the inside of the pipette.

Donor cells are released in 10% PVP-supplemented M2 medium, and the cells are moved in and out of an injection pipette repeatedly to destroy the cell membrane. Several nuclear donor cells are drawn into the injection pipette in the second drop, moved to rinse the outside of the pipette in the third drop to avoid PVP contamination of the fourth drop, and then moved to the fourth drop for injection. Holding the oocyte with the holding pipette, the injection pipette is inserted into the perivitelline space through the hole that is made in the enucleation process (Figures 7.18 and 7.21A). The injection pipette having donor cells (Figure 7.19) should be pushed into the ooplasm as deeply as possible (Figures 7.20A−C and 7.21B), and a piezo pulse is given to enter the ooplasm membrane. When the injection pipette breaks through the ooplasm and the shape of the oocyte becomes round again, the donor nucleus is injected inside the oocyte (Figures 7.20 D, E and 7.21C).

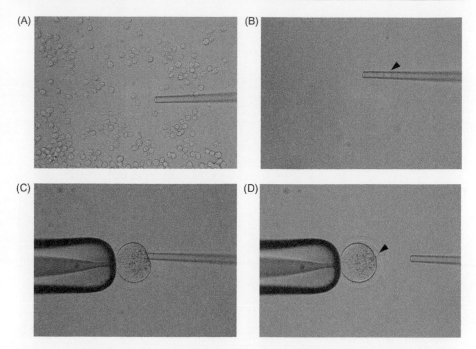

Figure 7.17 Nuclear transfer of cumulus cells by Sendai virus. (A) Ovulated cumulus cells are picked up by an injection pipette. (B) Inactivated Sendai virus is incorporated into the injection pipette (arrowhead shows a donor cell). (C) Donor cell with inactivated Sendai virus is injected into the perivitelline space of an enucleated oocyte. (D) Reconstituted oocyte before fusion with a donor cell (arrowhead shows donor cell).

E. Activation

1. Electric Pulses

Reconstituted oocytes are given two DC pulses of 50 V/mm for 50 ms three times at 20-min intervals. After the electric pulses, reconstituted oocytes are cultured in M16 medium supplemented with CB (5 μg/mL) for 5 h and examined for a pronuclear-like formation in the ooplasm. When M phase cells are used as donor cells, fused oocytes are activated in M16 without CB to release the polar body.

2. Strontium Treatment

Fused oocytes are cultured in M16 or CZB medium containing 10 mM $SrCl_2$ and cytochalasin B (5 μg/mL) for 5 h. After treatment with $SrCl_2$-containing medium, nuclear transfer oocytes are released from the drug by washing with M2 medium three times. They are then cultured further in standard M16 or CZB medium without $SrCl_2$. When M phase cells are used as donor cells, fused oocytes are activated with $SrCl_2$ but not with CB.

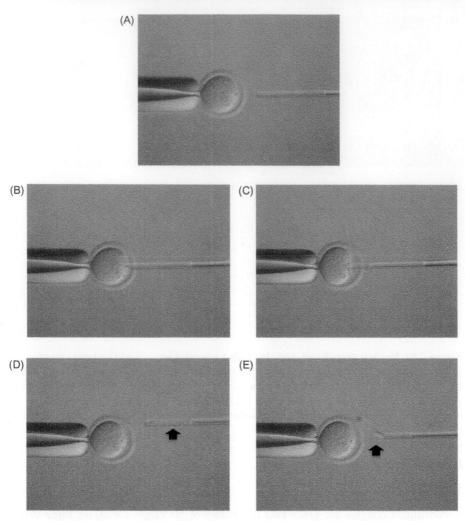

Figure 7.18 Method for enucleation by piezo pulses. (A) Before giving piezo pulses.
(B) After giving piezo pulses, the tip of the injection pipette can be seen in the perivitelline
space. (C) Pushing the surface of ooplasm and finding an area of MII plate, and then
enucleated. (D) and (E) Enucleated oocyte and karyoplast (arrow) within (D) and external to
(E) the pipette.

F. In Vitro *Culture and Embryo Transfer*

When nuclear transfer is performed using a piezo-micropipette-driving unit, nuclear-
transferred oocytes can be transferred to recipient females any time that is convenient, even in the precompaction stage, because they have only a very small hole in

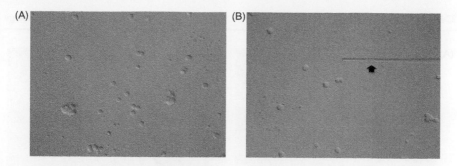

Figure 7.19 Donor cells for nuclear transfer. (A) Donor cells (cumulus cells). (B) Donor cells inside a pipette (arrow).

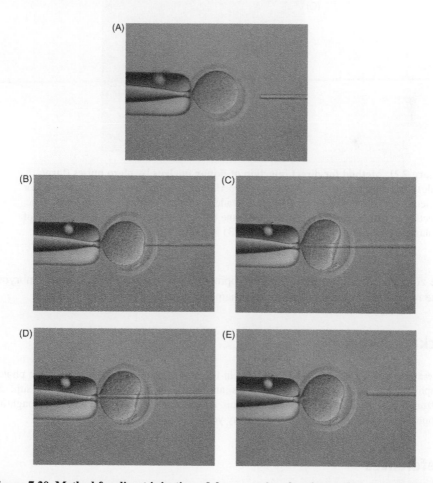

Figure 7.20 Method for direct injection of donor nucleus by piezo pulses. (A) Holding a recipient oocyte with holding pipette (left) and injection pipette (right). (B) Injection pipette is within the perivitelline space. (C) Injection pipette is not inserted into the ooplasm membrane. (D) The injection pipette enters the ooplasm membrane by piezo pulses. (E) Nuclear-transferred oocyte. In Figure 7.21, illustrates (B), (C), and (D) with greater magnification.

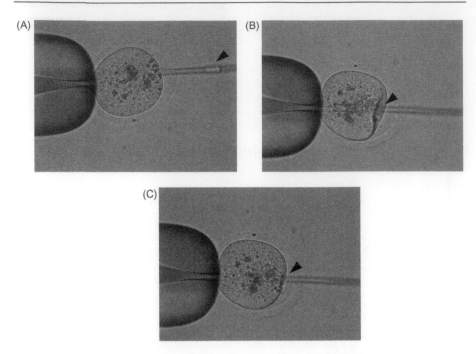

Figure 7.21 Method for direct injection of donor nucleus by piezo pulses (large scale). Injection pipette is inserted into the perivitelline space (arrowhead shows donor nucleus). (B) The injection pipette is not invaded into the ooplasm membrane. (C) The injection pipette enters the ooplasm membrane by piezo pulses. Compare the change of ooplasm membrane (arrowheads) between (B) and (C).

the zona that does not disturb the developmental potential of precompacted embryos. The rest of the protocol is basically the same as described in Section 7.IV.

Acknowledgments

The authors thank K. Ohata and A. Yabuuchi for their help in preparing many of the photographs of micromanipulation, and thanks are also extended to T. Tani, Y. Tsuji, S. Mizumoto, M. Nakano, T. Sugimoto, H. Tsuji, and all other lab members for their thoughtful cooperation and constant assistance for many years.

References

Amano, T., Tani, T., Kato, Y., Tsunoda, T., 2001a. Mouse cloned from embryonic stem (ES) cells synchronized in metaphase with nocodazole. J. Exp. Zool. 289, 139−145.
Amano, T., Kato, Y., Tsunoda, Y., 2001b. Full-term development of enucleated mouse oocytes fused with embryonic stem cells from different cell lines. Reproduction 121 (5), 729−733.

Baranska, W., Koprowski, H., 1970. Fusion of unfertilized mouse eggs with somatic cells. J. Exp. Zool. 174, 1–14.

Barton, S.C., Surani, M.A., 1993. Manipulations of genetic constitution by nuclear transplantation. In: Guide to Techniques in Mouse Development. Academic Press, pp. 732–746.

Briggs, R., King, T.J., 1952. Transplantation of living nuclei from blastula cells into enucleated frogs' eggs. Proc. Natl. Acad. Sci. USA 38, 455–463.

Bromhall, J.D., 1975. Nuclear transplantation in the rabbit egg. Nature 258, 719–721.

Cheong, H.T., Takahashi, Y., Kanagawa, H., 1993. Birth of mice after transplantation of early cell-cycle-stage embryonic nuclei into enucleated oocytes. Biol. Reprod. 48, 958–963.

DiBerardino, M.A., 1997. Nuclear potential of differentiated amphibian tissues and cells. In: Diberardino, M.A. (Ed.), Genomic Potential of Differentiated Cells. Columbia University Press, New York, NY, pp. 68–82.

Graham, C.E., 1969. The fusion of cells with one- and two-cell mouse embryo. In: Defendi, V. (Ed.), Heterospecific Genome Interaction. Wistar Institute Press, Philadelphia, PA, pp. 19–33.

Hogan, B., Beddington, R., Costantini, F., Lacy, E., 1994. Nuclear transplantation in the mouse embryo. In: Manipulating the Mouse Embryo. Cold Spring Harbor Laboratory, New York, pp. 209–216.

Illmensee, K., Hoppe, P.C., 1981. Nuclear transplantation in Mus musculus: developmental potential of nuclei from preimplantation embryos. Cell 23, 9–18.

Kato, Y., Tani, T., Stomomaru, Y., Kurokawa, K., Kato, J., Doguchi, H., et al., 1998. Eight calves cloned from somatic cells of a single adult. Science 282, 2095–2098.

Kato,Y., Rideout III, W.M., Hilton, K., Barton, S.C., Tsunoda,Y., and Surani, M.A., 1999a. Developmental potential of mouse primordial germ cells. Development 126, 1823–1832.

Kato, Y., Yabuuchi, A., Motosugi, N., Kato, J., Tsunoda, Y., 1999b. Developmental potential of mouse follicular epithelial cells and cumulus cells after nuclear transfer. Biol. Reprod. 61, 1110–1114.

Kato, Y., Tani, T., Tsunoda, Y., 2000. Female and male calves cloned from various somatic cell types of adults, newborns and fetuses. J. Reprod. Fertil. 120, 231–237.

Kimura, Y., Yanagimachi, R., 1995. Intracytoplasmic sperm injection in the mouse. Biol. Reprod. 52, 709–720.

Kishikawa, H., Wakayama, T., Yanagimachi, R., 1999. Comparison of oocyte activating agent for mouse cloning. Cloning 1, 153–159.

Kono, T., Kwon, O.Y., Nakahara, T., 1991a. Development of enucleated mouse oocytes reconstituted with embryonic nuclei. J. Reprod. Fertil. 93, 165–172.

Kono, T., Tsunoda, Y., Nakahara, T., 1991b. Production of identical twin and triplet mice by nuclear transplantation. J. Exp. Zool. 257, 214–219.

Kono, T., Kwon, O.Y., Watanabe, T., Nakahara, T., 1992. Development of mouse enucleated oocytes receiving a nucleus from different stages of the second cell cycle. J. Reprod. Fertil. 94, 481–487.

Kwon, O.Y., Kono, T., 1996. Production of identical sextuplet mice by transferring metaphase nuclei from four-cell embryos. Proc. Natl. Acad. Sci. USA 93, 13010–13013.

Latham, K.E., Solter, D., 1993. Transplantation of nuclei to oocytes and embryos. In: Wassarman, P.M., DePamphilis, M.L. (Eds.), Guide to Techniques in Mouse Development. Academic Press, San Diego, CA, pp. 719–731.

Lin, T.P., Florence, J., Oh, J.O., 1973. Cell fusion induced by a virus within the zona pellucida of mouse eggs. Nature 242, 47–49.

Mann, J.R., Lovell-Badge, R.H., 1984. Inviability of parthenogenones is determined by pronuclei, not egg cytoplasm. Nature 310, 66–67.

McGrath, J., Solter, D., 1983. Nuclear transplantation in the mouse embryo by microsurgery and cell fusion. Science 220, 1300–1302.

McGrath, J., Solter, D., 1984a. Completion of mouse embryogenesis requires both the maternal and paternal genomes. Cell 37, 179–183.

McGrath, J., Solter, D., 1984b. Inability of mouse blastomere nuclei transferred to enucleated zygotes to support development in vitro. Science 226, 1317–1319.

Modlinski, J.A., 1978. Transfer of embryonic nuclei to fertilized mouse eggs and development of tetraploid blastocysts. Nature 273, 466–467.

Ogura, A., Inoue, K., Takano, K., Wakayama, T., Yanagimachi, R., 2000a. Birth of mice after nuclear transfer by electrofusion using tail tip cells. Mol. Reprod. Dev. 57, 55–59.

Ogura, A., Inoue, K., Ogonuki, N., Noguchi, A., Takano, K., Nagano, R., et al., 2000b. Production of male cloned mice from fresh, cultured, and cryopreserved immature Sertoli cells. Biol. Reprod. 62, 1579–1584.

Ono, Y., Shimozawa, N., Ito, M., Kono, T., 2001. Cloned mice from fetal fibroblast cells arrested at metaphase by a serial nuclear transfer. Biol. Reprod. 64, 44–50.

Rideout III, W.M., Wakayama, T., Wutz, A., Egga, K., Jackson-Grusby, L., Dausman, J., et al., 2000. Generation of mice from wild-type and targeted ES cells by nuclear cloning. Nat. Genet. 24, 109–110.

Surani, M.A.H., Barton, S.C., Norris, M.L., 1984. Development of reconstituted mouse eggs suggests imprinting of the genome during embryogenesis. Nature 308, 548–550.

Tamashiro, K.L.K., Wakayama, T., Blanchard, R.J., Blanchard, D.C., Yanagimachi, R., 2000. Postnatal growth and behavioral development of mice cloned from adult cumulus cells. Biol. Reprod. 63, 328–334.

Tsunoda, Y., Kato, Y., 1993. Nuclear transplantation of embryonic stem cells in mice. J. Reprod. Fertil. 98, 537–540.

Tsunoda, Y., Kato, Y., 1997. Full-term development after transfer of nuclei from 4-cell and compacted morula stage embryos to enucleated oocytes in the mouse. J. Exp. Zool. 278, 250–254.

Tsunoda, Y., Kato, Y., 1998. Not only inner cell mass cell nuclei but also trophectoderm nuclei of mouse blastocysts have a developmental totipotency. J. Reprod. Fertil. 113, 181–184.

Tsunoda, Y., Yasui, T., Tokunaga, T., Uchida, T., Sugie, T., 1985. Pronuclear transplantation in the mouse. Jpn. J. Anim. Reprod. 31, 130–134.

Tsunoda, Y., Yasui, T., Nakamura, K., Uchida, T., Sugie, T., 1986. Effect of cutting the zona pellucida on the pronuclear transplantation in the mouse. J. Exp. Zool. 240, 119–125.

Tsunoda, Y., Yasui, T., Shioda, Y., Nakamura, K., Uchida, T., Sugie, T., 1987. Full-term development of mouse blastomere nuclei transplanted into enucleated two-cell embryos. J. Exp. Zool. 242, 147–151.

Tsunoda, Y., Shioda, Y., Onodera, M., Nakamura, K., Uchida, T., 1988. Differential sensitivity of mouse pronuclei and zygote cytoplasm to Hoechst staining and ultraviolet irradiation. J. Reprod. Fertil. 82, 173–178.

Tsunoda, Y., Tokunaga, T., Imai, H., Uchida, T., 1989. Nuclear transplantation of male primordial germ cells in the mouse. Development 107, 407–411.

Wakayama, T., Yanagimachi, R., 1999. Cloning of male mice from adult tail-tip cells. Nat. Genet. 22, 127–128.

Wakayama, T., Perry, A.C., Zuccotti, M., Johnson, K.R., Yanagimachi, R., 1998. Full-term development of mice from enucleated oocytes injected with cumulus cell nuclei. Nature 394, 369−374.

Wakayama, T., Rodriguez, I., Perry, A.C., Yanagimachi, R., Mombaerts, P., 1999. Mice cloned from embryonic stem cells. Proc. Natl. Acad. Sci. USA 96, 14984−14989.

Willadsen, S.M., 1986. Nuclear transplantation in sheep embryos. Nature 320, 63−65.

Wilmut, I., Schnieke, A.E., McWhir, J., Kind, A.J., Campbell, K.H., 1997. Viable offspring derived from fetal and adult mammalian cells. Nature 385, 810−813.

8 Gene Editing

David A. Dunn[1] and Carl A. Pinkert[2]

[1]Department of Biological Sciences, State University of New York at Oswego, Oswego, NY, [2]Department of Biological Sciences, College of Arts and Sciences, The University of Alabama, Tuscaloosa, AL

I. Introduction

While gene knockout/knock-in strategies involving embryonic stem (ES) cell approaches are standard in mouse laboratories, these techniques have until recently been restricted exclusively to mice (Dunn et al., 2005, 2012; Martin et al., 2010). Alternative methods for ablating specific gene function or knocking in alternative genetic sequences in other species, when available, are labor intensive and inefficient. Recently, the capacity to specifically target a given genetic locus in a much broader range of species has become a reality. This has been in large part driven by the development of chimeric enzymes consisting of sequence-specific DNA-binding domains tethered to endonuclease domains. Expression of DNA or RNA constructs coding for these engineered nucleases causes a double-strand break (DSB) in the target genomic locus. Endogenous cellular DNA repair mechanisms resolve these lesions, commonly leaving genetic mutations behind. Coupling these "designer" restriction enzymes with techniques already commonly used in transgenic animal production allows researchers the ability to create gene-targeted animal models that transcend the necessity for ES cell culture and associated methods. These technologies are widely and interchangeably labeled *genome engineering* or *genome editing*.

This chapter provides an introduction to strategies for performing genome editing *in vivo* via the use of synthetic engineered nucleases. These approaches include zinc finger nuclease (ZFN), transcription activator-like effector nuclease (TALEN), engineered homing endonuclease, and clustered regularly interspaced short palindromic repeats (CRISPR)/Cas nuclease. Methods for designing and producing these constructs are provided, as well as an overview of resources available for obtaining these technologies.

II. Sequence Modifications in Genome Editing

Engineered nucleases permit creation of DSBs at specific targetable genomic loci. In nature, there is a strong evolutionary drive to repair DSBs and maintain

Transgenic Animal Technology. DOI: http://dx.doi.org/10.1016/B978-0-12-410490-7.00008-6

chromosomal integrity (Pastink et al., 2001). Multiple DNA repair mechanisms have evolved to address DSBs, two of which, nonhomologous end joining (NHEJ) and homologous recombination (HR), are of special note to genome editing.

One mechanism by which genetic modification occurs is NHEJ. Here, two DNA ends are directly ligated without regard to homology (Pastwa and Błasiak, 2003). During this "error-prone" process, pieces of DNA are often added or removed resulting in the creation of small insertions or deletions (indels) of genetic material. When these indels occur in the coding region of a gene, they frequently result in frame-shift mutations, commonly inserting premature stop codons. In the context of genome engineering, DSB repair via NHEJ predominates when a sequence-specific nuclease creates a DSB in the absence of externally added homologous DNA. Taking advantage of this endogenous error-prone DNA repair mechanism following introduction of targeted DSBs via engineered endonucleases is an effective means for producing gene knockouts in transgenic animal experimentation.

Another mechanism by which repair of DSBs can occur is HR. In the presence of homologous donor DNA, either from the homologous chromosome in diploid organisms, or from an exogenous source in experimental situations, HR resolves DSBs by replacing the genetic material between two areas of homology (including the DSB) with sequence derived from the donor (Figure 8.1). HR using exogenous gene sequences has been used for years in gene targeting experiments with murine ES cells. The low efficiency of HR in traditional targeting procedures derives in part from the lack of DNA damage at the targeted locus. The introduction of DSBs at a targeted locus via engineered endonucleases can dramatically increase the rate

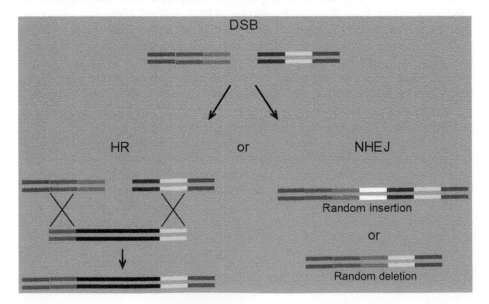

Figure 8.1 DSB resolution. DSBs generated by engineered nucleases can be resolved by endogenous DNA repair mechanisms to introduce exogenous sequences/alternate alleles via HR or create mutations via NHEJ.

at which HR occurs (Hockemeyer et al., 2011; Pruett-Miller et al., 2008). Another advantage of using genome engineering constructs in HR experiments is the ability to use a synthesized oligonucleotide as the exogenous donor molecule rather than plasmid DNA (Meyer et al., 2010). Together, NHEJ or HR resolution of DSBs created by engineered nucleases can result in inactivational or recombinational mutagenesis in their target cells.

As with other strategies for production of transgenic models, a number of methods are available for delivery of genome editing constructs (see Part 4). Methods reported include somatic cell nuclear transfer (SCNT) of modified somatic cells (Whyte et al., 2011), microinjection of ES cells (Wang et al., 2013a), and pronuclear microinjection of DNA (Tesson et al., 2011) and/or mRNA (Mashimo et al., 2010). For mammalian model creation, perhaps the most popular injection method is mRNA into the cytoplasm (Davies et al., 2013; Moreno et al., 2011; Wang et al., 2013b).

III. Approaches to Genome Editing

A. Zinc Finger Nuclease

1. Introduction

It has been recognized for several years that the ability to create unique genomic DSBs would afford the capability to replace one genomic segment for another both *in vitro* and *in vivo* (Chandrasegaran and Smith, 1999). The concept of using customized sequence-specific engineered endonucleases to perform genome editing was made possible on a large scale by the advent of ZFNs. In nature, zinc fingers (ZFs) are a well-known class of DNA-binding transcription factors (Evans and Hollenberg, 1988). In these proteins, one ZF domain binds to three nucleotides. Numerous attempts have been made to define the protein−DNA interactions (Choo and Klug, 1994) and to create libraries of recombinant ZFs that would bind many (if not all 64) possible DNA triplets (e.g., Jamieson et al., 1994 and Nardelli et al., 1992). It was eventually found that in order to create a ZF with optimal DNA binding strength, the first position of the DNA triplet usually harbors a guanine nucleotide (Jamieson et al., 1996).

2. ZFNs as Tools for Genome Editing

In order for ZFs (or any DNA-binding molecule) to be useful in genome editing, coupling DNA recognition with endonuclease activity is necessary. This is made possible through usage of type IIS restriction enzymes, usually FokI. The DNA binding and endonuclease functions of FokI are located on separate domains of the protein. This separation of the coding sequence of these two domains made possible the creation of fusion proteins in which the DNA cleavage fragment of FokI is coupled to DNA-binding molecules of various specificities (Kim et al., 1998; Podhajska and Szybalski, 1985).

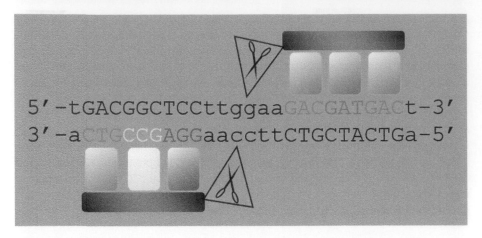

Figure 8.2 ZFN. Multiple ZFs, each recognizing a three-base DNA sequence, can be arrayed in tandem. When such a ZF array is coupled with the DNA cleavage domain of FokI endonuclease, it comprises one subunit of a ZFN. Two ZFN subunits can function as a dimer to recognize and cleave user-specified genomic loci.

Pioneering efforts in creation of hybrid ZF/FokI enzymes were carried out in the Chandrasegaran laboratory. An early report fused three different ZFs in tandem to produce a zinc finger array (ZFA) that recognized and specifically cleaved a nine-base-pair DNA sequence in the genome of lambda phage (Kim et al., 1996). Following this *in vitro* experiment, ZFN technology was further optimized (Bibikova et al., 2001) by creating paired ZFAs that recognized genomic elements on opposite DNA strands (Figure 8.2). This arrangement allowed FokI monomers to dimerize across the space between the two ZF recognition elements and provided increased genomic sequence specificity. An additional improvement to this strategy that resulted in decreased off-target binding of ZFNs was the introduction of mutant FokI domains to render them obligate heterodimers (Miller et al., 2007). This strategy increases sequence specificity, since the two arms of the ZFN must come together before endonuclease activity can occur, and self-dimerization is impossible.

Continued efforts in various research groups have extended the reach of ZFN-based gene targeting to various *in vivo* models via creation of transgenic animals. These include zebrafish (Doyon et al., 2008), xenopus (Nakajima et al., 2012), mouse (Carbery et al., 2010; Osiak et al., 2011), rat (Chu et al., 2012; Cui et al., 2011; Moreno et al., 2011), rabbit (Flisikowska et al., 2011), pig (Hauschild et al., 2011; Hauschild-Quintern et al., 2013; Lillico et al., 2013), and goat (Xiong et al., 2013).

3. Advantages/Drawbacks

Prior to the development of other genome editing technologies, ZFNs were perceived to be the only alternative to ES cells in performing gene knockout experiments in nonmurine species. The partial list above of ZFN-derived transgenic models is an indication of considerable success in this area. A major benefit for the use of ZFNs is their modular construction. Each ZF protein motif binds three nucleotides. Stringing several together leads to creation of a recombinant protein that recognizes a multiple of three genomic bases. While each ZF does indeed bind to three base pairs, a considerable amount of the binding specificity of any given ZF depends on the context of its neighboring ZFs (Cornu et al., 2008). This context dependence has resulted in a substantial amount of uncertainty with regard to the suitability of a given designed ZFN. Thus, for any given genomic locus multiple ZFNs must be constructed to ensure success in targeting a given genomic region. More importantly, decreases in ZFN specificity could lead to cellular toxicity via off-target cleavage. Much of the intellectual property revolving around ZFN technology is held by Sangamo Biosciences. Sangamo is directly pursuing human clinical applications deriving from ZFNs and ZF transcription factors, and has licensed Sigma Aldrich to provide reagents and customized solutions to the research community. Until the recent creation of open source reagents, this domination of ZFN capability appeared to stifle innovation among other laboratories in creating robust ZFNs (Sander et al., 2011a).

4. Methods for Construction

With the above limitations in mind, optimization efforts and published protocols by the Joung laboratory made ZFN technology and associated reagents available at reasonable cost to laboratories worldwide. Two methods for constructing ZFNs have been widely used. Oligomerized pool engineering (OPEN) involves creating a pool of ZFAs, then assessing them for use (Maeder et al., 2008). While this procedure improved on efficiency over then-available methods, efficiency was still rather low and the technique was very labor intensive. This was followed by development of a new ZFN construction technique termed context-dependent assembly (CoDA). In CoDA, individual ZFs were tested to assess their target specificity when placed in combination with other ZFs. This led to creation of an algorithm (ZiFiT, see Section 8.IV) into which a target DNA sequence is pasted and coding sequences for several ZFNs against that sequence are output (Sander et al., 2011a).

B. Transcription Activator-Like Effector Nuclease

1. Introduction

Plant pathogens of the genus *Xanthomonas* express a newly discovered class of DNA-binding proteins, the transcription activator-like effectors (TALEs) (Voytas and Joung, 2009). These proteins encompass a number of conserved repeats of 34 amino acids with two residues that vary in sequence at positions 12 and 13. The identity of the two varying amino acids, or repeat-variable diresidues (RVDs), dictate the DNA-binding specificity of the protein in a ratio of one repeat to one nucleotide. Formation of recombinant arrays of RVDs specific to any desired DNA sequence allows researchers the opportunity to build proteins that specifically bind to a desired genomic region. This enabled a new class of sequence-specific genome editing tools, similar in strategy and overall structure to ZFNs, but with a novel DNA-binding protein format.

2. TALENs as Tools for Genome Editing

The ability to utilize TALEs as customizable tools for genome engineering was opened up in 2009 by discovering the code by which TALE proteins recognize a given DNA sequence. In particular, determining which RVDs specifically bind to which nucleotides led to the ability to string together multiple TALE repeats for sequence-specific binding of genomic loci (Boch et al., 2009; Moscou and Bogdanove, 2009). Since many of the technical hurdles associated with creating fusion proteins with DNA-binding and -cleaving domains had already been cleared through ZFN research, development of tools for TALEN experimentation moved rapidly. This was first demonstrated using wild-type TALE sequences in a yeast assay system (Christian et al., 2010). Wild-type as well as recombinant TALEs were next used to demonstrate TALEN activity against endogenous cellular genes (Hockemeyer et al., 2011; Li et al., 2011; Miller et al., 2011; Mussolino et al., 2011; Sander et al., 2011b; Sun et al., 2012; Zhang et al., 2011). In addition to using TALEN capabilities to target nuclear-encoded genes, TALEN proteins harboring mitochondrial localization signals were used to specifically cleave mutant mitochondrial genomes, potentially paving the way for their use in a gene therapy strategy toward mitochondrial disease (Bacman et al., 2013).

TALEN technology has also been used to produce transgenic animals. The wide range of animals for which TALEN technology has been used to successfully achieve gene targeting include mouse (Davies et al., 2013; Panda et al., 2013; Qiu et al., 2013; Sung et al., 2013; Wang et al., 2013a), rat (Tesson et al., 2011), pig (Carlson et al., 2012; Lillico et al., 2013), rabbit (Song et al., 2013), bovine embryos (Carlson et al., 2012), xenopus (Ishibashi et al., 2012; Lei et al., 2012), tilapia (Li et al., 2013b), medaka (Ansai et al., 2013), and zebrafish (Cade et al., 2012; Dahlem et al., 2012; Moore et al., 2012).

3. Advantages/Drawbacks

The general design of TALENs is similar to that of ZFNs with two fusion proteins that bind DNA at adjacent sites and combine to create a DSB (Figure 8.3). One major difference is the comparative lack of dependence that adjacent nucleotide composition has on TALEN DNA-binding specificity. This lack of context dependence, the one-to-one nature of nucleotide binding to TALE repeats, together with the fact that ZFs do not exist for each DNA triplet result in TALENs displaying a much wider range of targetable genomic loci than ZFNs: estimated at one targetable site per 35 bp across a range of species for TALENs (Cermak et al., 2011) versus 1/400 for ZFNs (Sander et al., 2011a). Studies measuring the likelihood of newly designed constructs to cleave their targets found higher efficiency in TALENs versus ZFNs (Chen et al., 2013; Moore et al., 2012). TALENs are a

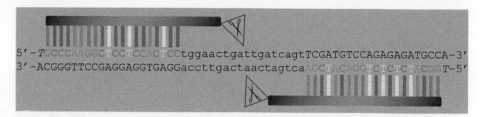

5′-T**GCCC**A**AGGC**T**CCTCCAC**T**CC**tggaactgattgatcagtTCGATGTCCAGAGAGATGCCA-3′
3′-ACGGGTTCCGAGGAGGTGAGGaccttgactaactagtca**AGC**T**AC**A**GGT**C**TCTCT**A**CGG**T-5′

Figure 8.3 TALEN. Engineered TALEs, containing multiple repeats, each of which specifically binds a single DNA base, are coupled with the DNA cleavage domain of FokI endonuclease to create TALENs. Like ZFNs, TALENS consist of two proteins, each possessing a DNA-binding domain and a DNA cleavage domain, that work together to target a specific genomic locus.

newer technology than ZFNs, and therefore have not yet been as fully characterized. Nevertheless, a consensus is growing that, of the two techniques, TALEN is the technique of choice for gene targeting experimentation, due to its superior flexibility, efficiency, and cytotoxicity.

4. Methods for Construction

Several TALEN construction platforms for which reagents are publicly available have been published. These represent a variety of cloning and construction strategies. Perhaps the simplest conceptually but most labor intensive involves traditional plasmid cloning via restriction digest and ligation of each individual TALE repeat (Sander et al., 2011b). In addition to cloning one TALE repeat at a time, TALENs can be constructed from plasmid libraries containing three or four TALE repeats (Ding et al., 2013; Reyon et al., 2012). However, iterative cloning of TALE repeats can become cumbersome and mistake prone. Therefore, different ways of assembling TALEs were described involving ligation-independent cloning (Schmid-Burgk et al., 2013) and the recently developed Golden Gate strategy (Engler et al., 2008), in which multiple fragments can be assembled at once in the same tube (Cermak et al., 2011; Sanjana et al., 2012; Zhang et al., 2011). These last two methods are also amenable to high-throughput approaches for large-scale TALEN undertakings.

C. Engineered Meganucleases/Homing Endonucleases

1. Introduction

Meganucleases or homing endonucleases (HEs) are endonuclease proteins that are found in a number of prokaryotes, archaea, and unicellular eukaryotes. These proteins can be encoded as free-standing genes, within an intron or as self-splicing inteins, and function in nature to support horizontal gene transfer of their coding sequences (Stoddard, 2005). As endonucleases, they display DNA binding and DNA cleavage domains that are fairly well characterized. Their DNA recognition sites of 20−30 bp (Stoddard, 2011) are considerably larger than those of the type II restriction endonucleases typically used in recombinant DNA technology. When a DSB is formed at the recognition site of an HE, the lesion is repaired by HR, during which the coding sequence for the meganuclease is inserted into the locus. This ability has led to the proposal of HEs as tools for targeted genome engineering (Hafez and Hausner, 2012).

2. Homing Endonucleases as Tools for Genome Editing

An understanding of the manner in which the DNA-binding motif of this class of enzymes functions and the ability to modify it was essential to establish the potential utility of HEs in genome editing. Though many different wild-type HEs exist, each with a different DNA sequence specificity, creation of enzymes with a much wider repertoire of binding specificities became necessary. To achieve this end, large libraries of clones encoding thousands of variants of the DNA-binding motif of I-CreI from *Chlamydomonas reinhardtii* were produced (Arnould et al., 2006; Smith et al., 2006). Using HEs with modified DNA-binding specificities, cell culture experiments specifically targeted the human Recombination activating gene 1 (RAG1) (Grizot et al., 2009; Munoz et al., 2011), Xeroderma pigmentosum, complementation group C (XPC) (Arnould et al., 2007; Redondo et al., 2008), and Monoamine oxidase B (MAO-B) (Takeuchi et al., 2011) genes.

3. Advantages/Drawbacks

Homing endonucleases have not achieved widespread adoption as tools for genome engineering. One reason for this is that while many modified homing endonucleases with different DNA-binding specificities have been created, they lack the modular DNA-binding domain architecture found in ZF or TALE proteins. Additionally, some sequence degeneracy is tolerated by homing endonucleases, increasing the likelihood of off-target binding (Argast et al., 1998; Gimble and Wang, 1996). One slight advantage to the use of meganucleases is that, despite their name, they are smaller proteins ($\sim$40 kD) than either ZFN or TALEN proteins and therefore might receive consideration in some viral vectors where shorter coding sequences are required. Nevertheless, the lack of readily available customizable reagents and the lack of clear rules for designing a construct that recognizes a specific DNA sequence make homing endonucleases less attractive for genome engineering than the other strategies discussed here.

4. Methods for Construction

As described above, meganuclease construction is unlike ZFN and TALEN architectures, where modular construction of repeating DNA-binding components are assembled, or CRISPR strategies, where a DNA sequence homologous to the targeted area can be simply cloned into a targeting vector. Development of a homing nuclease specific for a given genetic element is the result of screening large libraries of enzymes with diverse DNA-binding specificities. Thus, the other methods described here are preferred for design of the majority of genome editing endeavors.

D. CRISPR/Cas Nuclease

1. Introduction

The most recently described methods for genome engineering are based on the CRISPR and CRISPR associated (Cas) systems. CRISPR/Cas systems (often shortened simply to CRISPR) constitute a newly characterized mechanism for adaptive immunity in a diverse taxonomic array of bacteria and archaea. They function to protect their host from horizontal introduction of exogenous genetic material delivered by phages or plasmids. These systems share some similarities with RNAi in eukaryotes in that they involve a riboprotein complex where a protein with nuclease activity is bound to an RNA molecule that provides target specificity.

In nature, CRISPR/Cas systems function in three steps (Richter et al., 2012). First, foreign genetic elements are incorporated into a CRISPR array. The array is then transcribed and processed into CRISPR RNAs (crRNAs), and finally crRNAs are incorporated into crRNA-Cas protein ribonucleoprotein complexes where they can target and cleave exogenous DNA. A CRISPR array consists of alternating repeat and spacer elements. The repeats differ little from each other within a given bacterial genome; spacers are derived from exogenous DNA. As foreign DNA is detected in a cell, it is cleaved and short sequences of the foreign genetic material are incorporated into the host genome as a new spacer within the CRISPR array. CRISPR arrays are heritable and are derived from many introductions of exogenous DNA that occurred over the evolutionary lifetime of a bacterium and its ancestors. Second, spacer sequences are transcribed and processed into short RNA segments called CRISPR RNAs (crRNAs). In the final step, crRNAs combine with Cas proteins into a ribonucleoprotein complex that recognizes invading genetic material, leading to its degradation.

2. CRISPR/Cas Systems as Tools for Genome Editing

CRISPR/Cas systems are categorized into three types (I, II, and III), which are distinguished by different accessory RNAs and proteins. Of the three types, type II CRISPR/Cas systems have been identified for use in genome engineering. This choice stems from the fact that types I and III are composed of a number of Cas proteins, whereas the Cas9 protein is the only protein constituent of type II CRISPR/Cas systems (Makarova et al., 2011). Thus design and construction of novel CRISPR/Cas systems with researcher-defined targeting sequences is simplified.

Preliminary proof of concept *in vitro* experimentation established the potential of CRISPR/Cas systems for genome editing (Jinek et al., 2012; Gasiunas et al., 2012). Gasiunas and coworkers demonstrated the ability of a recombinant CRISPR/Cas system to cleave double-stranded DNA in which recognition is effected via RNA, and DSBs result from two domains in the Cas9 protein, each of which cleave a single DNA strand. Jinek and coworkers (2012) found that the Cas9 protein together with crRNA and an accessory RNA, the *trans*-activating crRNA (tracrRNA), could cleave DNA sequences that were homologous to the crRNA.

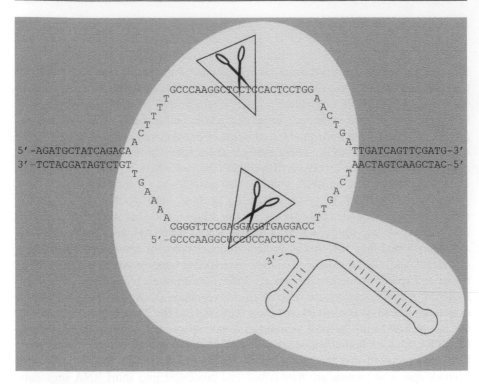

Figure 8.4 CRISPR. A crRNA, consisting of a sequence homologous to the targeted genomic region and accessory RNA sequence, complexes with a Cas protein to produce a construct capable of recognizing and cleaving a targetable genomic locus.

Furthermore, when they created a chimeric RNA containing both crRNA and tracrRNA sequences on a single RNA molecule, they found the engineered single-RNA system just as capable of targeting specific DNA sequences as the wild-type 2-RNA systems (Figure 8.4).

The first cellular uses of CRISPR/Cas for genome engineering were described early in 2013 (Cong et al., 2013; Mali et al., 2013). In these reports, mouse (Cong et al., 2013) and human (Cong et al., 2013; Mali et al., 2013) cells were transfected with DNAs coding for CRISPR RNAs and Cas9 protein and achieved mutational inactivation of multiple endogenous loci. Both groups developed CRISPR systems in which crRNA and tracrRNA were expressed as a single chimeric RNA (labeled gRNA by Mali et al., 2013), thus simplifying the construct assembly process. Additionally, in a subset of experiments a mutant version of Cas9 was used in both reports, in which only one strand of the target DNA was cleaved (as opposed to the wild-type activity of Cas9 which catalyzes the formation of DSBs). This creation of nicks in target loci restricted the resolution of genomic targeting events to HR only and not NHEJ.

Following these early accounts of the utility of CRISPR/Cas systems for genome editing, several groups have extended this technology to a number of animal species including zebrafish (Hwang et al., 2013a,b; Jao et al., 2013), mouse (Li et al., 2013a; Menke, 2013; Wang et al., 2013b; Yang et al., 2013), and rat (Li et al., 2013a,c).

3. Advantages/Drawbacks

The CRISPR/Cas strategy offers a number of advantages over other genome editing approaches. Unlike ZFN and TALEN, CRISPR methods rely on a single targeting molecule (gRNA) for DNA sequence recognition. This fact simplifies construction of vectors with multiple gRNAs for multiplexed gene targeting (Wang et al., 2013b). Another way in which construction of CRISPR plasmids is simplified in comparison to ZFNs or TALENs is that the DNA recognition sites are composed of nucleic acid rather than protein. Cloning of the DNA-recognition component of the gRNA sequence is comparatively simple. It involves ordering a pair of oligonucleotides containing the recognition sequence with sticky ends, annealing these oligonucleotides, phosphorylating them on the 5′ ends with T4 polynucleotide kinase, and ligating them into a CRISPR plasmid (e.g., pX330 from the Zhang laboratory; available through Addgene, see Section 8.IV).

One potential disadvantage to using CRISPR systems is a reported high incidence of nonspecific DNA cleavage (Cradick et al., 2013; Fu et al., 2013). While this has cooled some of the initial enthusiasm about this method, a potential remedy to the problem of off-site cleavage involves expression of two CRISPR modules with nickase activity against two genomic sites closely adjacent to one another (Ran et al., 2013).

4. Methods for Construction

Multiple resources are available to assist researchers in the design and construction of reagents for CRISPR-based experiments. As referenced below, a number of online design algorithms are available as well as plasmid backbones. As CRISPR is a comparatively newer technique than ZFN and TALEN models, design criteria are, as of this writing in late 2013, under considerable experimental scrutiny and thus subject to significant fine-tuning and modification. However, some sequence features are known to be crucial to success. Protospacer adjacent motifs (PAMs) are DNA sequences (5′-NGG-3′) at the 3′ end of genomic target sites that border but are not part of the recognition domain that is necessary for efficient targeting (Shah et al., 2013). Thus, any DNA sequence 5′-N(12-20)-NGG-3′ is a candidate for CRISPR targeting, although secondary structure of the target might present additional constraints (Cong et al., 2013).

IV. Online Resources

Addgene (http://www.addgene.org/genome_engineering): Addgene is a nonprofit organization whose mission is to serve as a depository for investigator-submitted plasmids, making them available to the research community. Addgene's collections are wide ranging, but include extensive special collections from numerous research groups focusing on genome editing with sections on CRISPR, TALEN, and ZFN technologies.

Cas9 Guide RNA Design (http://cas9.cbi.pku.edu.cn): The CRISPR/Cas9 design algorithm at this site allows users to input DNA sequence and obtain potential targets for use in targeting projects (Ma et al., 2013). A sequence may be scanned against human, mouse, rat, zebrafish, or drosophila genomes as part of the design process.

E-TALEN and E-CRISP (http://www.e-talen.org, http://www.e-crisp.org): These sister sites grew out of research performed in the Boutros laboratory at the German Cancer Research Center. They encompass design and evaluation tools for TALEN and CRISPR constructs, respectively. E-TALEN allows users to choose from the various TALEN assembly kits available from Addgene as part of the algorithm for TALEN design. This selection influences the number of repeats in the construct as well as which RVDs are picked. Both E-TALEN and E-CRISP allow users to mask results with predicted off-target recognition from the published genomes of several model organisms (Heigwer et al., 2013).

Genome-engineering.org (http://www.genome-engineering.org): This site contains design tools, protocols and discussion groups for TALEN and CRISPR projects using methods developed by the Zhang Laboratory at Massachusetts Institute of Technology.

Mojo Hand (http://www.talendesign.org): Mojo Hand contains web-based TALEN design software that outputs candidate TALENs for subsequent Golden Gate assembly (Neff et al., 2013).

RepeatMasker (http://www.repeatmasker.org): Accurate and precise binding specificity of any genome editing platform to the targeted genomic locus is essential for minimizing off-target events and for efficient gene targeting. RepeatMasker is an algorithm that screens DNA sequences for low complexity and repetitive elements (Tempel, 2012). The input to the program is a desired DNA sequence in FASTA format. Using alignment tools similar to Basic Local Alignment Search Tool (BLAST), the sequence is compared to transposons and DNA sequences of low complexity specific to one of several user-defined annotated genomes. Sequence is output with repetitive nucleotides masked as "N." Using RepeatMasker to screen DNA sequences prior to designing genome editing constructs will reduce the likelihood of targeting noncomplex sequences. Both web-based and standalone versions of the software are available.

TAL Effector Nucleotide Targeter (TALE-NT) 2.0 (https://tale-nt.cac.cornell.edu): Hosted and maintained by the Bogdanove Laboratory at Cornell University, TALE-NT is a downloadable or web-based software tool that assists in the design of TALENs or other TAL effectors (Cermak et al., 2011; Doyle et al., 2012).

TALengineering.org (http://www.talengineering.org): A site developed and maintained by the Joung Laboratory at Massachusetts General Hospital and Harvard Medical School, TALengineering.org provides an overview as well as links to publications and protocols covering a variety of TALEN platforms, including their own.

Zinc Finger Consortium (http://www.zincfingers.org): The consortium web site contains multiple resources that can guide researchers in the design and assembly of ZFN, TALEN, and CRISPR constructs. There are several members of the consortium, but the resources and web content are heavily influenced by the Joung Laboratory. One of the most valuable aspects of this web site is ZiFiT. ZiFiT is a software tool that allows users to design ZFN, TALEN, and CRISPR/Cas engineered nucleases from user-submitted DNA sequences (Sander et al., 2010). The output is linked to published protocols for construct assembly and cross-referenced to Addgene for purchase of required reagents.

V. Commercial Resources

Several commercial suppliers offer a variety of reagents and services aimed at facilitating genome engineering research. Available products range from design of genome engineering constructs, plasmid assembly, and functional *in vitro* verification to production of knockout animals. Since genome editing is a rapidly evolving area for producers of life sciences research reagents, the following list of companies that offer genome engineering products and/or services can be expected to rapidly evolve.

ZFN
Sigma Aldrich: http://www.sigmaaldrich.com/life-science/zinc-finger-nuclease-technology.html
TALEN
Allele Biotechnology: http://www.allelebiotech.com/genome-edit
BioCat GmbH: http://www.biocat.com/products/talenxtn-serv-trp
Cellectis: http://www.cellectis-bioresearch.com/talen-solutions
Excellgen: http://www.excellgen.com/genome-engineering-40
GeneCopoeia: http://www.genecopoeia.com/product/talen-tal-effector
ITSI Biosciences: http://www.itsibio.com/talens.html
Life Technologies: http://www.lifetechnologies.com/us/en/home/life-science/cloning/gene-synthesis/geneart-precision-tals.html
PNA Bio: http://www.pnabio.com/products/TALEN.htm
System Biosciences: http://www.systembio.com/ez-tal
Transposagen Biopharmaceuticals: http://transposagenbio.com/gene-modification-tools/xtn-talens/
CRISPR
BioCat GmbH: http://www.biocat.com/cgi-bin/page/sub2.pl?main_group=genomics&sub1=gene_editing&sub2=crispr_cas9_smartnuclease_genome_engineering_system
GeneCopoeia: http://www.genecopoeia.com/product/crispr-cas9/
Life Technologies: http://www.lifetechnologies.com/us/en/home/life-science/cloning/gene-synthesis/geneart-precision-tals/geneart-crispr.html
PNA Bio: http://www.pnabio.com/products/RGEN.htm
Sigma-Aldrich: http://www.sigmaaldrich.com/catalog/product/sigma/crispr
System Biosciences: http://www.systembio.com/genome-engineering-cas9-crispr-smartnuclease
Transposagen Biopharmaceuticals: http://transposagenbio.com/crisprcas/

Meganuclease
Cellectis: http://www.cellectis.com/technologies/protein-engineering
In Vivo **Model Creation**
Cyagen: http://www.cyagen.com/service/nuclease-mediated-knockout-rat.html   Custom
gene targeting via ZFN, TALEN, CRISPR in mouse and rat.
PolyGene   Transgenics: http://www.polygene.ch/talen-based-gene-targeting-services
Custom gene targeting via TALEN in mouse.
SAGE Labs: http://www.sageresearchlabs.com A variety of ZFN-derived off-the-shelf
gene knockout rats, as well as custom knockout model generation in mouse, rat, and
rabbit.
Transposagen Biopharmaceuticals: http://transposagenbio.com/custom-model-generation
A variety of TALEN-derived, off-the-shelf, gene knockout rats as well as custom knock-
out rat model generation.

VI. Summary

The advent of genome engineering technology has opened avenues not previously
available and generally made reverse genetics-based experimentation considerably
simpler. While the use of HR in murine ES cells has been the workhorse of gene
targeting experimentation for many years, it is a labor-intensive and inefficient process (Dunn et al., 2005, 2012; Martin et al., 2010). Efficiency of mouse transgenesis is increased compared to traditional ES methods with genome editing
constructs through knockout NHEJ-mediated DSB repair (Carbery et al., 2010;
Sung et al., 2013), as well as knock-in HR experiments where coinjection of a
knock-in construct bearing homology arms with the gene editing construct is carried out (Cui et al., 2011). Multiple factors contribute to this increased efficiency.
Use of ES cells is eliminated, since the construct (usually mRNA) is injected into
unicellular zygotes. Targeting events usually occur at the single-cell stage or soon
thereafter. Therefore, transgenic animals are more likely to carry targeted mutations
in their germ cells than chimeric mice derived from ES cells. Additionally, in
knock-in models, HR occurs at much higher rates in the presence of either a DSB
or a nick on one of the strands of DNA. Another appealing aspect of creating
knockout mice in this fashion is the direct cytoplasmic injection of mRNA, eliminating the ES-cell stage entirely.

Perhaps more excitingly, these techniques make gene targeting possible in a
wide range of species for the first time. For many years, the mouse was the only
species in which *in vivo* knockout models were possible. The development of
induced pluripotent stem cells in nonmurine species partially alleviated that situation (Hamanaka et al., 2011). Nevertheless, not until the creation of engineered
endonucleases did the ability to ablate gene expression in a diverse array of animals
come to reality.

Genome engineering has quickly progressed over the past few years, from a
future ideal for which more research was required, to a readily available technique
accessible   to   any   laboratory   with   basic   molecular   cloning   capabilities.

Nevertheless, as a very young field, refinements of existing techniques and discovery of new and enhanced approaches are certain. The fast pace of advances and the excitement generated just over the last few years show no signs of abating. Gene-targeted animal models for a range of biomedical and agricultural applications can now be established.

References

Ansai, S., Sakuma, T., Yamamoto, T., Ariga, H., Uemura, N., Takahashi, R., et al., 2013. Efficient targeted mutagenesis in medaka using custom-designed transcription activator-like effector nucleases. Genetics 193, 739–749.

Argast, G.M., Stephens, K.M., Emond, M.J., Monnat Jr., R.J., 1998. I-PpoI and I-CreI homing site sequence degeneracy determined by random mutagenesis and sequential *in vitro* enrichment. J. Mol. Biol. 280, 345–353.

Arnould, S., Chames, P., Perez, C., Lacroix, E., Duclert, A., Epinat, J.-C., et al., 2006. Engineering of large numbers of highly specific homing endonucleases that induce recombination on novel DNA targets. J. Mol. Biol. 355, 443–458.

Arnould, S., Perez, C., Cabaniols, J.-P., Smith, J., Gouble, A., Grizot, S., et al., 2007. Engineered I-CreI derivatives cleaving sequences from the human XPC gene can induce highly efficient gene correction in mammalian cells. J. Mol. Biol. 371, 49–65.

Bacman, S.R., Williams, S.L., Pinto, M., Peralta, S., Moraes, C.T., 2013. Specific elimination of mutant mitochondrial genomes in patient-derived cells by mitoTALENs. Nat. Med. 19, 1111–1113.

Bibikova, M., Carroll, D., Segal, D.J., Trautman, J.K., Smith, J., Kim, Y.-G., et al., 2001. Stimulation of homologous recombination through targeted cleavage by chimeric nucleases. Mol. Cell. Biol. 21, 289–297.

Boch, J., Scholze, H., Schornack, S., Landgraf, A., Hahn, S., Kay, S., et al., 2009. Breaking the code of DNA binding specificity of TAL-type III effectors. Science 326, 1509–1512.

Cade, L., Reyon, D., Hwang, W.Y., Tsai, S.Q., Patel, S., Khayter, C., et al., 2012. Highly efficient generation of heritable zebrafish gene mutations using homo- and heterodimeric TALENs. Nucleic Acids Res. 40, 8001–8010.

Carbery, I.D., Ji, D., Harrington, A., Brown, V., Weinstein, E.J., Liaw, L., et al., 2010. Targeted genome modification in mice using zinc-finger nucleases. Genetics 186, 451–459.

Carlson, D.F., Tan, W., Lillico, S.G., Stverakova, D., Proudfoot, C., Christian, M., et al., 2012. Efficient TALEN-mediated gene knockout in livestock. Proc. Natl. Acad. Sci. USA 109, 17382–17387.

Cermak, T., Doyle, E.L., Christian, M., Wang, L., Zhang, Y., Schmidt, C., et al., 2011. Efficient design and assembly of custom TALEN and other TAL effector-based constructs for DNA targeting. Nucleic Acids Res. 39, e82.

Chandrasegaran, S., Smith, J., 1999. Chimeric restriction enzymes: what is next? Biol. Chem. 380, 841–848.

Chen, S., Oikonomou, G., Chiu, C.N., Niles, B.J., Liu, J., Lee, D.A., et al., 2013. A large-scale *in vivo* analysis reveals that TALENs are significantly more mutagenic than ZFNs generated using context-dependent assembly. Nucleic Acids Res. 41, 2769–2778.

Choo, Y., Klug, A., 1994. Toward a code for the interactions of zinc fingers with DNA: selection of randomized fingers displayed on phage. Proc. Natl. Acad. Sci. USA 91, 11163–11167.

Christian, M., Cermak, T., Doyle, E.L., Schmidt, C., Zhang, F., Hummel, A., et al., 2010. Targeting DNA double-strand breaks with TAL effector nucleases. Genetics 186, 757–761.

Chu, X., Zhang, Z., Yabut, J., Horwitz, S., Levorse, J., Li, X., et al., 2012. Characterization of multidrug resistance 1a/P-glycoprotein knockout rats generated by zinc finger nucleases. Mol. Pharmacol. 81, 220–227.

Cong, L., Ran, F.A., Cox, D., Lin, S., Barretto, R., Habib, N., et al., 2013. Multiplex genome engineering using CRISPR/Cas systems. Science 339, 819–823.

Cornu, T.I., Thibodeau-Beganny, S., Guhl, E., Alwin, S., Eichtinger, M., Joung, J.K., et al., 2008. DNA-binding specificity is a major determinant of the activity and toxicity of zinc-finger nucleases. Mol. Ther. J. Am. Soc. Gene Ther. 16, 352–358.

Cradick, T.J., Fine, E.J., Antico, C.J., Bao, G., 2013. CRISPR/Cas9 systems targeting β-globin and CCR5 genes have substantial off-target activity. Nucleic Acids Res. 41, 9584–9592. Available from: http://dx.doi.org/10.1093/nar/gkt714. (Epub ahead of print).

Cui, X., Ji, D., Fisher, D.A., Wu, Y., Briner, D.M., Weinstein, E.J., 2011. Targeted integration in rat and mouse embryos with zinc-finger nucleases. Nat. Biotechnol. 29, 64–67.

Dahlem, T.J., Hoshijima, K., Jurynec, M.J., Gunther, D., Starker, C.G., Locke, A.S., et al., 2012. Simple methods for generating and detecting locus-specific mutations induced with TALENs in the zebrafish genome. PLoS Genet. 8, e1002861.

Davies, B., Davies, G., Preece, C., Puliyadi, R., Szumska, D., Bhattacharya, S., 2013. Site specific mutation of the Zic2 locus by microinjection of TALEN mRNA in mouse CD1, C3H and C57BL/6J oocytes. PLoS One 8, e60216.

Ding, Q., Lee, Y.-K., Schaefer, E.A.K., Peters, D.T., Veres, A., Kim, K., et al., 2013. A TALEN genome-editing system for generating human stem cell-based disease models. Cell Stem Cell 12, 238–251.

Doyle, E.L., Booher, N.J., Standage, D.S., Voytas, D.F., Brendel, V.P., VanDyk, J.K., et al., 2012. TAL Effector-Nucleotide Targeter (TALE-NT) 2.0: tools for TAL effector design and target prediction. Nucleic Acids Res. 40, W117–W122.

Doyon, Y., McCammon, J.M., Miller, J.C., Faraji, F., Ngo, C., Katibah, G.E., et al., 2008. Heritable targeted gene disruption in zebrafish using designed zinc-finger nucleases. Nat. Biotechnol. 26, 702–708.

Dunn, D.A., Kooyman, D.L., Pinkert, C.A., 2005. Transgenic animals and their impact on the drug discovery industry. Drug Discov. Today 10, 757–767.

Dunn, D.A., Cannon, M.V., Irwin, M.H., Pinkert, C.A., 2012. Animal models of human mitochondrial DNA mutations. Biochim. Biophys. Acta 1820, 601–607.

Engler, C., Kandzia, R., Marillonnet, S., 2008. A one pot, one step, precision cloning method with high throughput capability. PLoS One 3, e3647.

Evans, R.M., Hollenberg, S.M., 1988. Zinc fingers: gilt by association. Cell 52, 1–3.

Flisikowska, T., Thorey, I.S., Offner, S., Ros, F., Lifke, V., Zeitler, B., et al., 2011. Efficient immunoglobulin gene disruption and targeted replacement in rabbit using zinc finger nucleases. PLoS One 6, e21045.

Fu, Y., Foden, J.A., Khayter, C., Maeder, M.L., Reyon, D., Joung, J.K., et al., 2013. High-frequency off-target mutagenesis induced by CRISPR-Cas nucleases in human cells. Nat. Biotechnol. 31, 822–826.

Gasiunas, G., Barrangou, R., Horvath, P., Siksnys, V., 2012. Cas9-crRNA ribonucleoprotein complex mediates specific DNA cleavage for adaptive immunity in bacteria. Proc. Natl. Acad. Sci. USA 109, E2579–E2586.

Gimble, F.S., Wang, J., 1996. Substrate recognition and induced DNA distortion by the PI-SceI endonuclease, an enzyme generated by protein splicing. J. Mol. Biol. 263, 163–180.

Grizot, S., Smith, J., Daboussi, F., Prieto, J., Redondo, P., Merino, N., et al., 2009. Efficient targeting of a SCID gene by an engineered single-chain homing endonuclease. Nucleic Acids Res. 37, 5405–5419.

Hafez, M., Hausner, G., 2012. Homing endonucleases: DNA scissors on a mission. Genome 55, 553–569.

Hamanaka, S., Yamaguchi, T., Kobayashi, T., Kato-Itoh, M., Yamazaki, S., Sato, H., et al., 2011. Generation of germline-competent rat induced pluripotent stem cells. PLoS One 6, e22008.

Hauschild, J., Petersen, B., Santiago, Y., Queisser, A.-L., Carnwath, J.W., Lucas-Hahn, A., et al., 2011. Efficient generation of a biallelic knockout in pigs using zinc-finger nucleases. Proc. Natl. Acad. Sci. USA 108, 12013–12017.

Hauschild-Quintern, J., Petersen, B., Queisser, A.-L., Lucas-Hahn, A., Schwinzer, R., Niemann, H., 2013. Gender non-specific efficacy of ZFN mediated gene targeting in pigs. Transgenic Res. 22, 1–3.

Heigwer, F., Kerr, G., Walther, N., Glaeser, K., Pelz, O., Breinig, M., et al., 2013. E-TALEN: a web tool to design TALENs for genome engineering. Nucleic Acids Res. 41, e190.

Hockemeyer, D., Wang, H., Kiani, S., Lai, C.S., Gao, Q., Cassady, J.P., et al., 2011. Genetic engineering of human pluripotent cells using TALE nucleases. Nat. Biotechnol. 29, 731–734.

Hwang, W.Y., Fu, Y., Reyon, D., Maeder, M.L., Tsai, S.Q., Sander, J.D., et al., 2013a. Efficient genome editing in zebrafish using a CRISPR-Cas system. Nat. Biotechnol. 31, 227–229.

Hwang, W.Y., Fu, Y., Reyon, D., Maeder, M.L., Kaini, P., Sander, J.D., et al., 2013b. Heritable and precise zebrafish genome editing using a CRISPR-Cas system. PLoS One 8, e68708.

Ishibashi, S., Cliffe, R., Amaya, E., 2012. Highly efficient bi-allelic mutation rates using TALENs in *Xenopus tropicalis*. Biol. Open. 1, 1273–1276.

Jamieson, A.C., Kim, S.H., Wells, J.A., 1994. *In vitro* selection of zinc fingers with altered DNA-binding specificity. Biochemistry (Mosc.). 33, 5689–5695.

Jamieson, A.C., Wang, H., Kim, S.H., 1996. A zinc finger directory for high-affinity DNA recognition. Proc. Natl. Acad. Sci. USA 93, 12834–12839.

Jao, L.-E., Wente, S.R., Chen, W., 2013. Efficient multiplex biallelic zebrafish genome editing using a CRISPR nuclease system. Proc. Natl. Acad. Sci. USA 110, 13904–13909.

Jinek, M., Chylinski, K., Fonfara, I., Hauer, M., Doudna, J.A., Charpentier, E., 2012. A programmable dual-RNA-guided DNA endonuclease in adaptive bacterial immunity. Science 337, 816–821.

Kim, Y.G., Cha, J., Chandrasegaran, S., 1996. Hybrid restriction enzymes: zinc finger fusions to Fok I cleavage domain. Proc. Natl. Acad. Sci. USA 93, 1156–1160.

Kim, Y.G., Smith, J., Durgesha, M., Chandrasegaran, S., 1998. Chimeric restriction enzyme: Gal4 fusion to FokI cleavage domain. Biol. Chem. 379, 489–495.

Lei, Y., Guo, X., Liu, Y., Cao, Y., Deng, Y., Chen, X., et al., 2012. Efficient targeted gene disruption in Xenopus embryos using engineered transcription activator-like effector nucleases (TALENs). Proc. Natl. Acad. Sci. USA 109, 17484–17489.

Li, D., Qiu, Z., Shao, Y., Chen, Y., Guan, Y., Liu, M., et al., 2013a. Heritable gene targeting in the mouse and rat using a CRISPR-cas system. Nat. Biotechnol. 31, 681–683.

Li, M.-H., Yang, H.-H., Li, M.-R., Sun, Y.-L., Jiang, X.-L., Xie, Q.-P., et al., 2013b. Antagonistic roles of Dmrt1 and Foxl2 in sex differentiation via estrogen production in tilapia as demonstrated by TALENS. Endocrinology 154, 4814–4825. Available from: http://dx.doi.org/10.1210/en.2013-1451. (Epub ahead of print).

Li, W., Teng, F., Li, T., Zhou, Q., 2013c. Simultaneous generation and germline transmission of multiple gene mutations in rat using CRISPR-Cas systems. Nat. Biotechnol. 31, 684–686.

Li, T., Huang, S., Zhao, X., Wright, D.A., Carpenter, S., Spalding, M.H., et al., 2011. Modularly assembled designer TAL effector nucleases for targeted gene knockout and gene replacement in eukaryotes. Nucleic Acids Res. 39, 6315–6325.

Lillico, S.G., Proudfoot, C., Carlson, D.F., Stverakova, D., Neil, C., Blain, C., et al., 2013. Live pigs produced from genome edited zygotes. Sci. Rep. 3, 2847.

Ma, M., Ye, A.Y., Zheng, W., Kong, L., 2013. A guide RNA sequence design platform for the CRISPR/Cas9 system for model organism genomes. BioMed. Res. Int. 2013, 270805.

Maeder, M.L., Thibodeau-Beganny, S., Osiak, A., Wright, D.A., Anthony, R.M., Eichtinger, M., et al., 2008. Rapid "open-source" engineering of customized zinc-finger nucleases for highly efficient gene modification. Mol. Cell 31, 294–301.

Makarova, K.S., Haft, D.H., Barrangou, R., Brouns, S.J.J., Charpentier, E., Horvath, P., et al., 2011. Evolution and classification of the CRISPR-Cas systems. Nat. Rev. Microbiol. 9, 467–477.

Mali, P., Yang, L., Esvelt, K.M., Aach, J., Guell, M., DiCarlo, J.E., et al., 2013. RNA-guided human genome engineering via Cas9. Science 339, 823–826.

Martin, M.J., Dunn, D.A., Pinkert, C.A., 2010. Transgenic animals: secreted products. In: Ullrey, D.E., Baer, C.K., Pond, W.G. (Eds.), Encyclopedia of Animal Science, second ed. Dekker, Taylor & Francis, New York, NY, pp. 1047–1050.

Mashimo, T., Takizawa, A., Voigt, B., Yoshimi, K., Hiai, H., Kuramoto, T., et al., 2010. Generation of knockout rats with X-linked severe combined immunodeficiency (X-SCID) using zinc-finger nucleases. PLoS One 5, e8870.

Menke, D.B., 2013. Engineering subtle targeted mutations into the mouse genome. Genesis 51, 605–618.

Meyer, M., de Angelis, M.H., Wurst, W., Kühn, R., 2010. Gene targeting by homologous recombination in mouse zygotes mediated by zinc-finger nucleases. Proc. Natl. Acad. Sci. USA 107, 15022–15026.

Miller, J.C., Holmes, M.C., Wang, J., Guschin, D.Y., Lee, Y.-L., Rupniewski, I., et al., 2007. An improved zinc-finger nuclease architecture for highly specific genome editing. Nat. Biotechnol. 25, 778–785.

Miller, J.C., Tan, S., Qiao, G., Barlow, K.A., Wang, J., Xia, D.F., et al., 2011. A TALE nuclease architecture for efficient genome editing. Nat. Biotechnol. 29, 143–148.

Moore, F.E., Reyon, D., Sander, J.D., Martinez, S.A., Blackburn, J.S., Khayter, C., et al., 2012. Improved somatic mutagenesis in zebrafish using transcription activator-like effector nucleases (TALENs). PLoS One 7, e37877.

Moreno, C., Hoffman, M., Stodola, T.J., Didier, D.N., Lazar, J., Geurts, A.M., et al., 2011. Creation and characterization of a renin knockout rat. Hypertension 57, 614–619.

Moscou, M.J., Bogdanove, A.J., 2009. A simple cipher governs DNA recognition by TAL effectors. Science 326, 1501–1501.

Munoz, I.G., Prieto, J., Subramanian, S., Coloma, J., Redondo, P., Villate, M., et al., 2011. Molecular basis of engineered meganuclease targeting of the endogenous human RAG1 locus. Nucleic Acids Res. 39, 729–743.

Mussolino, C., Morbitzer, R., Lütge, F., Dannemann, N., Lahaye, T., Cathomen, T., 2011. A novel TALE nuclease scaffold enables high genome editing activity in combination with low toxicity. Nucleic Acids Res. 39, 9283–9293.

Nakajima, K., Nakajima, T., Takase, M., Yaoita, Y., 2012. Generation of albino *Xenopus tropicalis* using zinc-finger nucleases. Dev. Growth Differ. 54, 777−784.

Nardelli, J., Gibson, T., Charnay, P., 1992. Zinc finger-DNA recognition: analysis of base specificity by site-directed mutagenesis. Nucleic Acids Res. 20, 4137−4144.

Neff, K.L., Argue, D.P., Ma, A.C., Lee, H.B., Clark, K.J., Ekker, S.C., 2013. Mojo Hand, a TALEN design tool for genome editing applications. BMC Bioinformatics 14, 1.

Osiak, A., Radecke, F., Guhl, E., Radecke, S., Dannemann, N., Lütge, F., et al., 2011. Selection-independent generation of gene knockout mouse embryonic stem cells using zinc-finger nucleases. PLoS One 6, e28911.

Panda, S.K., Wefers, B., Ortiz, O., Floss, T., Schmid, B., Haass, C., et al., 2013. Highly efficient targeted mutagenesis in mice using TALENs. Genetics Available from: http://dx.doi.org/10.1534/genetics.113.156570. (Epub ahead of print).

Pastink, A., Eeken, J.C., Lohman, P.H., 2001. Genomic integrity and the repair of double-strand DNA breaks. Mutat. Res. 480−481, 37−50.

Pastwa, E., Błasiak, J., 2003. Non-homologous DNA end joining. Acta Biochim. Pol. 50, 891−908.

Podhajska, A.J., Szybalski, W., 1985. Conversion of the FokI endonuclease to a universal restriction enzyme: cleavage of phage M13mp7 DNA at predetermined sites. Gene 40, 175−182.

Pruett-Miller, S.M., Connelly, J.P., Maeder, M.L., Joung, J.K., Porteus, M.H., 2008. Comparison of zinc finger nucleases for use in gene targeting in mammalian cells. Mol. Ther. J. Am. Soc. Gene Ther. 16, 707−717.

Qiu, Z., Liu, M., Chen, Z., Shao, Y., Pan, H., Wei, G., et al., 2013. High-efficiency and heritable gene targeting in mouse by transcription activator-like effector nucleases. Nucleic Acids Res. 41, e120.

Ran, F.A., Hsu, P.D., Lin, C.-Y., Gootenberg, J.S., Konermann, S., Trevino, A.E., et al., 2013. Double nicking by RNA-guided CRISPR Cas9 for enhanced genome editing specificity. Cell 154, 1380−1389.

Redondo, P., Prieto, J., Muñoz, I.G., Alibés, A., Stricher, F., Serrano, L., et al., 2008. Molecular basis of xeroderma pigmentosum group C DNA recognition by engineered meganucleases. Nature 456, 107−111.

Reyon, D., Khayter, C., Regan, M.R., Joung, J.K., Sander, J.D., 2012. Engineering designer transcription activator-like effector nucleases (TALENs) by REAL or REAL-fast assembly. Curr. Protoc. Mol. Biol. 100, 12.15.1−12.15.14.

Richter, C., Chang, J.T., Fineran, P.C., 2012. Function and regulation of clustered regularly interspaced short palindromic repeats (CRISPR)/CRISPR associated (Cas) systems. Viruses 4, 2291−2311.

Sander, J.D., Maeder, M.L., Reyon, D., Voytas, D.F., Joung, J.K., Dobbs, D., 2010. ZiFiT (Zinc Finger Targeter): an updated zinc finger engineering tool. Nucleic Acids Res. 38, W462−W468.

Sander, J.D., Dahlborg, E.J., Goodwin, M.J., Cade, L., Zhang, F., Cifuentes, D., et al., 2011a. Selection-free zinc-finger nuclease engineering by context-dependent assembly (CoDA). Nat. Methods 8, 67−69.

Sander, J.D., Cade, L., Khayter, C., Reyon, D., Peterson, R.T., Joung, J.K., et al., 2011b. Targeted gene disruption in somatic zebrafish cells using engineered TALENs. Nat. Biotechnol. 29, 697−698.

Sanjana, N.E., Cong, L., Zhou, Y., Cunniff, M.M., Feng, G., Zhang, F., 2012. A transcription activator-like effector toolbox for genome engineering. Nat. Protoc. 7, 171−192.

Schmid-Burgk, J.L., Schmidt, T., Kaiser, V., Höning, K., Hornung, V., 2013. A ligation-independent cloning technique for high-throughput assembly of transcription activator-like effector genes. Nat. Biotechnol. 31, 76—81.

Shah, S.A., Erdmann, S., Mojica, F.J.M., Garrett, R.A., 2013. Protospacer recognition motifs: mixed identities and functional diversity. RNA Biol. 10, 891—899.

Smith, J., Grizot, S., Arnould, S., Duclert, A., Epinat, J.-C., Chames, P., et al., 2006. A combinatorial approach to create artificial homing endonucleases cleaving chosen sequences. Nucleic Acids Res. 34, e149.

Song, J., Zhong, J., Guo, X., Chen, Y., Zou, Q., Huang, J., et al., 2013. Generation of RAG 1- and 2-deficient rabbits by embryo microinjection of TALENs. Cell Res. 23, 1059—1062.

Stoddard, B.L., 2005. Homing endonuclease structure and function. Q Rev. Biophys. 38, 49—95.

Stoddard, B.L., 2011. Homing endonucleases: from microbial genetic invaders to reagents for targeted DNA modification. Struct. Lond. Engl. 19, 7—15.

Sun, N., Liang, J., Abil, Z., Zhao, H., 2012. Optimized TAL effector nucleases (TALENs) for use in treatment of sickle cell disease. Mol. Biosyst. 8, 1255—1263.

Sung, Y.H., Baek, I.-J., Kim, D.H., Jeon, J., Lee, J., Lee, K., et al., 2013. Knockout mice created by TALEN-mediated gene targeting. Nat. Biotechnol. 31, 23—24.

Takeuchi, R., Lambert, A.R., Mak, A.N.-S., Jacoby, K., Dickson, R.J., Gloor, G.B., et al., 2011. Tapping natural reservoirs of homing endonucleases for targeted gene modification. Proc. Natl. Acad. Sci. USA 108, 13077—13082.

Tempel, S., 2012. Using and understanding repeatmasker. Methods Mol. Biol. 859, 29—51.

Tesson, L., Usal, C., Ménoret, S., Leung, E., Niles, B.J., Remy, S., et al., 2011. Knockout rats generated by embryo microinjection of TALENs. Nat. Biotechnol. 29, 695—696.

Voytas, D.F., Joung, J.K., 2009. Plant science. DNA binding made easy. Science 326, 1491—1492.

Wang, H., Hu, Y.-C., Markoulaki, S., Welstead, G.G., Cheng, A.W., Shivalila, C.S., et al., 2013a. TALEN-mediated editing of the mouse Y chromosome. Nat. Biotechnol. 31, 530—532.

Wang, H., Yang, H., Shivalila, C.S., Dawlaty, M.M., Cheng, A.W., Zhang, F., et al., 2013b. One-step generation of mice carrying mutations in multiple genes by CRISPR/Cas-mediated genome engineering. Cell 153, 910—918.

Whyte, J.J., Zhao, J., Wells, K.D., Samuel, M.S., Whitworth, K.M., Walters, E.M., et al., 2011. Gene targeting with zinc finger nucleases to produce cloned eGFP knockout pigs. Mol. Reprod. Dev. 78, 2.

Xiong, K., Li, S., Zhang, H., Cui, Y., Yu, D., Li, Y., et al., 2013. Targeted editing of goat genome with modular-assembly zinc finger nucleases based on activity prediction by computational molecular modeling. Mol. Biol. Rep. 40, 4251—4256.

Yang, H., Wang, H., Shivalila, C.S., Cheng, A.W., Shi, L., Jaenisch, R., 2013. One-step generation of mice carrying reporter and conditional alleles by CRISPR/Cas-mediated genome engineering. Cell 154, 1370—1379.

Zhang, F., Cong, L., Lodato, S., Kosuri, S., Church, G., Arlotta, P., 2011. Programmable sequence-specific transcriptional regulation of mammalian genome using designer TAL effectors. Nat. Biotechnol. 29, 149—153.

Section Three

Production of Transgenic Laboratory and Domestic Animal Species

9 Production of Transgenic Rats

Philip Iannaccone and Vasiliy Galat

Department of Pediatrics, Northwestern University Feinberg School of Medicine and Developmental Biology Program, Ann & Robert H. Lurie Children's Hospital of Chicago Research Center, Chicago, IL

I. Introduction

It is now more than 30 years since the production of transgenic mice was achieved. It still seems amazing that the physical injection of linear sequence in buffer solution can find its way into the genome and that the random integration of such DNA results in functional genetic units. It is a humbling fact that these things occur. It is also enormously useful. In the past three decades transgenic animals have been used in an ever-expanding array of experimental paradigms which have greatly enriched our knowledge of gene function. The use of this technology has allowed the ethical study of human genes in intact animals, provided a new pharmaceutical approach to bioreactors, changed our food supply, and may even have provided an alternative end use for tobacco plants.

II. Importance of Rats

The procedures for production of transgenic mice are routine and widespread even if demanding. But transgenic rats, while widely used, are less available. This is a shame, for in many ways the rat is the model of choice for studies of human disease. More than a century of papers have been published with extensive phenotypic data from the rat. In the past 30 years alone more than 500,000 papers utilizing rat models are cited by PubMed. With more than 250 strains of rat, which include inbred strains, congenics, mutants, transgenics, and now targeted loss-of-function mutations, there are important animals available for the study of cardiopulmonary disease and physiology, renal disease and physiology, endocrinology/reproduction, immunology, toxicology, behavior, addiction, cancer, diabetes and other metabolic issues, arthritis, and neural development and disease. Indeed for many areas of biomedicine the rat is chosen for the extensive available data, the convenience of size, or because the physiology of the animal most closely resembles that of man. The rat

Transgenic Animal Technology. DOI: http://dx.doi.org/10.1016/B978-0-12-410490-7.00009-8

is the species of choice for gene therapy experiments, particularly when manipulation of the central nervous system is involved.

The use of the rat as a model is particularly important to studies of cardiac function and disease. In the area of stroke and hypertension, the rat has been used instructively for decades. For studies of various elements of hypertension, there are nine inbred strains with arterial pressure regulation phenotypes. Particularly important in this regard is the recent success with mutilplexed analysis of phenotypes, where hundreds of phenotypes and positional markers were analyzed together in cosegregation studies (Cowley et al., 2000; Stoll et al., 2000). This approach led to the identification of a large number of chromosomal locations that were significantly related to hypertension. Coupled with rapidly expanding knowledge based on the completion of the rat genome, predictions of human chromosomal regions and syntenic genes influencing the human disease were made.

The rat has been an important model for the study of arthritis and autoimmune diseases (Zhou et al., 1998). There are many inbred strains relevant to these studies, including congenic strains (particularly for major histocompatibility loci), relevant mutants such as athymic nude rats, and transgenic strains such as HLA-B27, TNF-alpha, and others. Disease penetrance is high in these genetic models and gender distribution (i.e., higher female susceptibility to arthritis) of disease in the rat is more like the human than in other animal models like mice.

The rat remains the most important model for the study of toxic exposure risk for both therapeutics and other newly developed chemicals. The growth of pharmacogenetics and the increased use of pharmaceuticals in an aging population make the use of the rat in these studies even more important. The need for multiple species in risk assessment further emphasizes the rat in toxicological experiments (Mayer et al., 1998; Schmezer and Eckert, 1999).

Classically important studies of carcinogenesis and cancer biology have been done in the rat. Hepatocarcinogenicity is a particularly important area in which the rat has been most informative. The rat liver responds with lesions that are much like those seen in the human. Chimeric rats (Weinberg et al., 1985) were used to establish the clonal origin of these lesions and of liver cancers (Weinberg et al., 1987; Weinberg and Iannaccone, 1988). Rat chimeras further lead to important models of organ development (Iannaccone, 1990; Khokha et al., 1994; Ng and Iannaccone, 1992). Breast cancer models in the rat are highly representative of the human disease, with highly similar histopathology and hormone response (Report of the NIH Rat Model Priority Meeting, May 3, 1999; Report of the NIH Rat Model Repository Workshop, August 19–20, 1998; http://www.nhlbi.nih.gov/meetings/model/index.htm).

The genomic resources for utilizing rat models of human disease conditions are robust and continue to grow (Jacob, 1999; Steen et al., 1999; Flicek et al., 2013). The complete rat genome is available with remarkable informatics tools behind it. For example, the Rat Genome Database (http://rgd.mcw.edu/) offers a wide range of functions related to comparative genomics and discovery (Hayman et al., 2013). Polymorphisms in genes relevant to human toxicological exposure have been studied in the rat for many decades. Several national initiatives to establish centers to maintain and distribute important rat strains of known genetic and microbiological quality including genome banking exist (Agca, 2012).

III. Contributions from Transgenic Rats

Transgenic rats have been used for a remarkably wide range of important experiments over the past 20 years or so. Here we will discuss only a few areas: hypertension research (reviewed in Bader et al., 1997), arthritis and autoimmunity research, toxicology (Mayer et al., 1998; Schmezer and Eckert, 1999), and cancer research (reviewed in Bader et al., 2000; Charreau et al., 1996; Mullins and Mullins, 1993, 1996.

A. Models of Hypertension

Hypertension and stroke are major killers worldwide. Animal models of this disease are critical to its elimination. Key in these studies is the renin angiotensin system, and transgenic rats have been particularly important (Bader et al., 1997; Sinn and Sigmund, 2000). There are lines that carry the mouse renin-2 gene and display fulminant hypertension. Transgenic rats with the human renin gene are being used to study gene expression profiling in hypertension and for evaluating drug therapy (Bader and Ganten, 1996; Springate et al., 1997). These animals were also used to study the role of human renin in hypertension during pregnancy (Bohlender et al., 2000). Animals doubly transgenic for human renin and human angiotensinogen were used to explore the causes of end organ damage in hypertension, determining that some of these effects are independent of blood pressure (Mervaala et al., 2000b) and establishing a protective effect of cyclosporin (Mervaala et al., 2000a). Further studies of the renin angiotensin system in rat have led to important new information about cardiovascular stress responses (Arnold et al., 2012).

B. Models of Arthritis

Humans with the HLA-B27 allele of the class I major histocompatibility system are very susceptible to developing arthritis as part of a multiorgan disease complex. These diseases, called spondyloarthropathies, have been studied using transgenic rats into which HLA-B27 has been integrated (Hammer et al., 1990; Taurog et al., 1999). The disease process that develops in the rat is like the human complex, particularly the axial peripheral arthropathy (Taurog and Hammer, 1996), and does not occur if a different allele type is used to make the transgenic or if HLA-B27 has a specific mutation. These animals also develop chronic inflammatory bowel disease (Lundin et al., 1997) and allow detailed studies of the specific infectious pathogens involved (Onderdonk et al., 1998). Transgenic rats bearing the *HLA-B27* allele have been used for mechanistic studies of arthritis, for example, by altering peptide transport in CTL recognition. Results of these studies suggest that HLA-B27-based disease is not related to binding peptides to the class I major histocompatibility complex (MHC) molecule (Simmons et al., 1996). Recent evidence from these rats has been used to elucidate the relationship between the GI tract and joint involvement in spondyloarthropathies (Milia et al., 2009).

C. Models of Carcinogenesis

Rats provide an important model of liver carcinogenesis (Weinberg and Iannaccone, 1988). A transgenic model of multistage liver carcinogenesis allowed comparative mutational analysis of tumor suppressor function and demonstrated that the liver cancers generated in these animals arose independently of mutations in the three tumor suppressor genes studied (Gomez-Angelats et al., 1999). However, when the animals were treated with carcinogens, mutations in p53 were identified (Haas and Pitot, 1998). The transgenic model was developed with a transgene consisting of the mouse albumin promoter and SV40 large T antigen. All of these transgenic animals develop hepatic lesions that give rise to malignant neoplasms displaying phenotypic similarities to chemically induced liver tumors; many of the transgenic rats develop tumors from pancreatic islet cells (Hully et al., 1994). Transgenic rats bearing the SV40 large T oncogene with the phosphoenolpyruvate carboxykinase promoter were developed; these animals develop pancreatic carcinomas from 5 to 8 months of age with metastases (Haas et al., 1999; Haas et al., 2000). Transgenic rats are being used currently to dissect the molecular aspects of mammary carcinogenesis (Matsuoka et al., 2007).

D. Other Models

The last 20 years have seen the development of an impressive array of transgenic models in the rat. A complete discussion of them is beyond the scope of this chapter, but a partial list may be of some use. Human apo A-I transgenics were used in the study of lipid metabolism and treatment of cholestasis (Chisholm et al., 2000). Polyamine metabolism has been analyzed in spermidine/spermine N(1)-acetyltransferase transgenics (Alhonen et al., 2000), and cardiac transplant models were developed where transgenic rat hearts were transplanted into primates, allowing the assessment of hyperacute rejection (Charreau et al., 1999).

Recently, the development of germ line competent ES cells from the rat has been combined with transgenics to make germ line competent GFP-positive ES cells, allowing access to important lineage tracers (Men et al., 2012).

IV.   Production

Production of transgenic rats is very much like that in the mouse. The methods for isolating eggs, determining pregnancy, and microinjecting DNA are the same (Charreau et al., 1996; Mullins and Mullins, 1996). The principal differences between production methods are in superovulation and culture. In general, the transformation rates of rat are lower than those of mice. The percentage of microinjected eggs that result in transgenic rats was reported to vary between 0.2% and 3.5% of injected eggs including the use of cyropreserved pronuclear stage zygotes (Charreau et al., 1996; Takahashi et al., 1999). This seems to be a function of the strain of rat used and the operator. There is a wide variation in the strain viability of eggs following microinjection of DNA.

A. Superovulation

Almost all facilities producing transgenic rats use hormone treatments to superovulate donor animals (Charreau et al., 1996; Mukumoto et al., 1995). The purpose is to increase the number of eggs available for manipulation. In the attempt to maximize the eggs available it is possible to reduce either the viability of the egg following microinjection or to reduce the developmental potential of the eggs obtained, or both. In general, superovulation involves administration of pregnant mare serum gonadotropin (PMSG) followed 48—56 h later with the administration of human chorionic gonadotropin (HCG). This is a well-established effective procedure in mice but works with more variable results in rats. The source and method of preparation of the hormones is critical to the result of the procedure, but often the precise details of source and production methods are not known, as the hormone may be obtained from third-party sources. An alternate procedure has been used successfully to increase the yield of eggs. This procedure uses various sources of follicle stimulating hormone (FSH) administered continuously by intraperitoneal (IP) osmotic minipump (Hamilton and Armstrong, 1991).

In our lab, we have used several approaches to superovulation (Table 9.1). In general, we have successfully employed pregnant mare serum (PMS; Sigma Chemical Co., St. Louis, G-4877) as a source of follicle stimulating activity, followed 47—51 h later by HCG (Sigma Chemical Co., CG-10) in young (4-week old) or adult (>8-week old) rats maintained on a 12-h light, 12-h dark light cycle (dark from 1800 to 0600). Following the second injection, the females were placed with fertile males and examined for copulatory plugs the following morning, when eggs were flushed from the oviduct. There are strain-, technique-, and age-dependent variations in the yield of one-cell eggs from mating with fertile or vasectomized males. Our usual protocol for young animals is: 15 IU PMS are injected IP between 1200 and 1500, followed by 15 IU of HCG injected IP 47—50 h later into female rats 28—30 days old. This results in an average of 25 eggs per inbred PVG female, with poorer results from outbred SD females mated with vasectomized SD males. For adult animals (>8 weeks old), we inject 20 IU of PMS at 1000—1130 subcutaneously. This is followed 4—6 h later with 30 IU of PMS subcutaneously and then 50 IU of HCG injected IP 47—49 h later. This yielded an average of 25 eggs per inbred PVG female mated with vasectomized males and an average of 18 eggs per PVG female mated with fertile males. Results with outbred SD rats are much poorer. We have also used subcutaneous osmotic minipumps (Alzet osmotic minipump, model 2001, Source Alza Corp., Palo Alto, CA) to deliver folltropin

Table 9.1 Response of Rats to Hormone Treatment

Strain	Male	Treatment	Age of Female	Number of Females	Eggs per Female, Mean (Range)
PVG	Fertile	PMS-HCG	2−6 months	11	18.5 (2−44)
	Vasectomized	PMS-HCG	2−6 months	12	24.8 (1−45)
	Fertile	PMS-HCG	29−32 days	14	24.7 (0−80)
	Fertile	Folltropin (minipump)-HCG	28−30 days	5	54.0 (35−72)
SD (Harland)	Fertile	PMS-HCG	2−6 months	9	19.0 (0−51)
	Vasectomized	PMS-HCG	2−6 months	4	17.2 (0−30)
SD (Charles River)	Fertile	PMS-HCG	2−6 months	12	10.8 (0−18)
	Vasectomized	PMS-HCG	2−6 months	16	9.6 (0−15)
	Vasectomized	PMS-HCG	29−32 days	6	8.8 (0−18)
SD (Charles River)	Vasectomized	PMS-HCG Intergonan/ Ovogest	23−30 days	15	55.8

Table 9.2 Light Cycle Relationship to Injection Times SD Rats

Light Cycle (Dark Period)	Midpoint of Dark Period	Endogenous LH Release	Time of Intergonan	Time of Ovogest
1800−0600	2400	1500−2000	1100−1300	1300−1500
0200−1400	0800	2300−0400	1900−2100	2100−2300

(FSH from Gonadotropin Kit#0372, Vetrepharm Inc., London, ON, Canada). These are inserted in 26−31-day-old female PVG rats between 1000 and 1200, followed by 11 IU of HCG injected IP 49−51 h later. This procedure yielded an average of 54 eggs per animal (Table 9.1).

Similar results were obtained in our lab with hormones from a different source and utilizing different light cycles (Table 9.2). An injection of Intergonan (PMS; Intervet GmbH), 20 IU injected IM, was followed 48−50 h later with Ovogest (HCG; Intervet GmbH) 20 IU injected IP into 23−31-day-old females. The second injection was made 2−3 h before the release of endogenous luteinizing hormone (LH). Ovulation occurred 12−13 h after the HCG injection and resulted in 40−70 eggs per female.

Another simpler procedure we have used successfully is to inject LHRH subcutaneously. This procedure can be used for both embryo isolation and recipients. The quality of eggs and embryos is comparable to natural mating, and the results are not as variable between different strains as superovulation. Female rats 5−10 weeks old are used for synchronization to generate rat embryos. Female SD rats 5–10 weeks old are used for synchronization to generate embryo recipients.

A single dose of LHRH (40 µg), when given to female rats subcutaneously 48 h before pairing with either a fertile or vasectomized male ensures ovulation and mating. The females are paired one to one with males. This is not superovulation and does not increase the number of embryos released. Female rats paired with vasectomized males are used successfully as embryo recipients. In our experience, a minimum of 50% of the female rats given the hormone will mate, as evidenced by the presence of a vaginal plug, allowing greater control in planning experiments.

1. Minipumps—HCG (and LHRH)

Osmotic minipumps (approximately 2.5 cm long, 1.0 cm in diameter) deliver a continuous flow of FSH (folltropin) after surgically inserted. Pumps are filled via a 27-gauge needle on a syringe with 0.2 mL of folltropin (supplied as lyophilized powder and reconstituted with 1.1 mL saline diluent from kit with folltropin). They are capped and left overnight in 0.9% saline (NaCl) at 37°C. The minipump is inserted subcutaneously (SC) through a 1.5 cm incision in the skin over the back of the neck. Under light anesthesia, the skin layer is separated from the muscle layer with scissor points to accommodate the minipump. After placing the minipump, the wound is closed with wound clips. Pumps are left in place until the mother is sacrificed for embryo obtainment. We do not reuse pumps.

B. Transfer of Manipulated Embryos

The surrogate mother is anesthetized with ketamine/xylazine injected IP (87 and 13 mg/kg, respectively) and a transverse incision is made in the dorsal lumbar region. A transverse lateral (approximately 4–5 mm) incision is made through the paraspinalis muscles and the abdominal musculature into the peritoneal cavity. The ovary, oviduct, and a portion of the uterus are withdrawn through these incisions by applying traction to the paraovarian fat pad. The infundibulum of the oviduct is identified to determine which direction within the distal loops of the oviduct is toward the uterus. The reconstructed ova are drawn into a sterile glass capillary pipette prepared with the aid of a De Fonbrune microforge in a very small volume. The pipette is inserted into the fimbriated end of the oviduct, and the ova is pushed into the oviduct toward the uterus. The skin incision is closed with stainless steel wound clips. All animals are observed following recovery for signs of discomfort, including abnormal movements and attempts to gain access to operative sites as by scratching and biting or inability to eat, drink, or groom. When necessary appropriate analgesics (Buprenex 2.0 mg/kg SC every 12 h for 48 h) are administered.

1. Vasectomy

The vasectomy is performed on anesthetized SD animals. The vas deferens is isolated through a mid-ventral lower abdominal incision. Sterile phosphate buffered saline (PBS) is injected distally into the vas to confirm location and to flush. One centimeter of the vas is removed bilaterally and both remaining ends tied off with suture. The skin incision is closed with stainless steel wound clips. The animals are observed for several hours following the operation for signs of discomfort. All animals are observed following recovery for signs of discomfort including abnormal movements and attempts to gain access to operative sites as by scratching and biting. When necessary appropriate analgesics (Buprenex 2.0 mg/kg SC every 12 h for 48 h) are administered.

C. Lentivirus

The most effective method for producing transgenic rats currently is by injecting lentivirus bearing the gene of interest (Dann, 2007; Lois et al., 2002). This method, provided the virus is of sufficient titer, will produce transgenics with every successful transfer. Indeed a 70% transformation rate at the egg injection stage is typical, with higher rates routinely obtainable.

1. Egg Obtainment for Lentivirus Injection

Mating and isolation of embryos: One male is paired with one female in the late afternoon to allow the animals to mate. Mating is established by the discovery of a copulatory plug. By using four female rats per experiment one should recover between 40 and 60 one-cell eggs each experiment. The female rats are sacrificed the day of plug discovery. The oviduct and attached segment of uterus is removed with fine scissors and placed into a small dish containing M2 medium. The cumulus mass containing the eggs is removed and transferred into M2 with 300-µg/mL hyaluronidase for 4−8 min to liberate single eggs. Single eggs are rinsed 2 × in fresh M2 and then transferred into KSOM medium and held in a CO_2 incubator at 5% CO_2 and 37°C until used for injection, usually within 3 h.

2. Lentivirus Injection of One-Cell Rat Eggs

Small 5-µm drops of the virus solution are put onto a plastic dish. The dish is then placed on the microinjection microscope. The virus is drawn into a fine electrode mounted in the micromanipulators of the microscope. The dish with the virus is disposed of into a container of 50% bleach solution. The rat embryos are transferred to droplets of M2 medium covered with mineral oil (embryo-tested) in a separate dish that is transferred to the microscope stage. During injection, the embryos are held in place with a minute amount of suction from a wider blunt-end electrode mounted in a micromanipulator. The sharp electrode containing the virus is forced through the zona pellucida and the virus is injected into the perivitelline space (Figure 9.1). This process is confirmed by visualizing swelling of the zona. The embryos are rinsed 3 × with the M2 medium and held to transfer. All dishes, gloves, pipettes, masks, the exposed media, etc., are disposed of into the 50% bleach solution. The platform and knobs on the microinjection scope are wiped

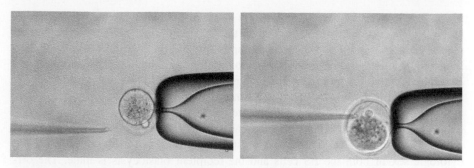

Figure 9.1 Lentivirus injection. Pipette containing the virus is inserted into the perivitelline space.

down with a 1% bleach solution. All of the viral work should be done at BL2 or BL2 + level in accordance with the Institutional Biosafety Committee guidelines.

D. Sources

At the present time, there are several commercial facilities that will produce transgenic rats on a commercial basis.

E. Rat Cloning

1. Gene Knockouts

Despite the extensive genomic tools that are available for the rat, a former major limitation was the inability to knockout genes by homologous recombination. Historically these procedures first came to fruition in the mouse. Embryo-derived stem cells (ES cells) are isolated from specific strains of mouse by culturing the preimplantation embryo. This is typically done from the blastocysts of 129-strain mice. Cells maintained appropriately retain a high degree of pluripotency; when surgically inserted into preimplantation embryos of some strains of mice the progeny of the ES cells contribute to many tissues in the developing fetus, if the combined embryo is transferred to a surrogate mother, creating a chimera. If progeny of the ES cells populate the germ line then it is possible to obtain functional gametes wholly derived from the cultured ES cells. Any genetic modification of those cells will be transmitted to the offspring of the chimera.

These cells are then targeted and selected *in vitro*. Cloned mutant ES cells are transplanted to the blastocyst cavity of normal recipient embryos. Homologous recombination, for example, will induce targeted mutations of desired genes by creating vectors with long regions of sequence homology to a given gene separated by various cassettes that disrupt the function of the gene. If the recombination events swap the normal endogenous gene for the targeting vector sequences, then one allele of the gene will be mutant and lose function. These are low-probability events in ES cells. The selection systems most widely used currently, developed by

Mario Cappechi and Oliver Smithies, rely on thymidine kinase-sensitivity to ganciclovir and neo cassette-based resistance to G418 antibiotic. The targeting vector places the neo cassette into the genome upon integration, and the TK cassettes are removed if the integration occurs by homologous recombination. This approach has generated many hundreds of mutant strains of mice. The loss of function is exploited for functional analysis of the gene.

Progeny of the mutant ES cells populate the fetal tissues of the recipient individual including the gonads, and in some cases functional gametes derived from the mutant ES cells develop. This works most efficiently, if not exclusively, when the cytologic sex of both the ES cells and the recipient blastocysts is male. The offspring of the chimera then include animals derived from the cultured ES cells with a normal allele, animals derived from cultured ES cells with a mutant allele, and animals derived from gametes derived from the recipient embryo strain.

By mating the resulting chimera to normal mice, offspring are derived from the ES cells. Since the targeted mutation in most cases is heterozygous, about half of the ES cell-derived offspring (identified most often with coat color markers) carry the mutation in the heterozygous state. These offspring can be identified by Southern analysis or other methods and then mated with each other to establish the mutation in the homozygous state.

Recently, the use of zinc finger endonucleases to create an allelic series of mutations in genes of interest has revolutionized the use of rats in loss-of-function studies (Zheng et al., 2012). Basically, a DNA recognition site is engineered in a protein containing the Fok I endonuclease. Two such proteins are designed such that Fok I can dimerize and cut the DNA. Literally hundreds of informative rat mutations have been made with these reagents. Transcription activator-like effector nuclease (TALEN) technology similarly targets DNA recognition sites to dimerize Fok I endonuclease and create DNA breaks. TALENs are arguably more flexible, easier to engineer, and have fewer off-target effects than zinc finger endonucleases. This technology has been successfully used to target genes in rat ES cells (Tong et al., 2012). Even newer powerful genome editing capabilities are coming on line. For example, clustered regularly interspaced short palindromic repeats (CRISPR) allow incredible flexibility in creating targeted mutations (Gaj et al., 2013). A complimentary sequence to the target of interest recognizing GNGG motifs ($N = 20$ mer) is cloned under the control of the U6 promoter followed by a sequence recognized by CAS9, a bacterial protein that protects the bacteria from phage. Expressed single-stranded guide RNA, in combination with CAS9, makes a double-strand break at the target site. Nonhomologous end joining results in small indels. These can be detected by sequencing, using restriction enzyme polymorphism or High Resolution Melt analysis. Further discussion of these methods can be found in Chapter 8.

2. Stem Cells

The two decades of efforts to establish embryonic stem cell lines from rats highlighted the developmental differences between mouse and rat species. Within the last 5 years essentially all barriers to knocking out genes in the rat have been removed by rapidly advancing technology (Iannaccone and Jacob, 2009). First of all, bona fide ES cells are available from the rat due to an inhibitor strategy worked out by Austin Smith and his colleagues (Blair et al., 2011). Since fibroblast growth factor 4 (FGF4) mediates differentiation through MAP kinase kinase extracellular regulated kinase (MEK-ERK) signaling, they used fibroblast growth factor receptor (FGFR) inhibition and inhibitors of MEK and glycogen synthase kinase-3 (GSK-3) to prevent differentiation. It is only with the help of these inhibitors blocking the activity of MEK, GSK-3, and the FGFR that it became possible to develop rat ES cells (Buehr et al., 2008; Li et al., 2008). This achievement finally allowed the production of knockout rats (Meek et al., 2010; Tong et al., 2010; Yamamoto et al., 2012). FGFR inhibition proved deleterious, and finally a two-inhibitor system proved robust for the isolation of pluripotent ES cells from rat blastocysts. These have proven capable of targeting by homologous recombination and are germ line competent. While rat ES cells are inherently interesting (see following paragraph), the new technologies for gene editing described above have largely made the use of rat ES cells for gene modification unnecessary.

Aside from requirements for maintenance and derivation, some other notable differences have been reported between mouse and rat ES cells. For example, rat ES cells express *Cdx2*, a gene involved in trophoblast fate determination (Hong et al., 2012). More broadly, the differences encompass other embryonic lineages. It has not been possible to establish a precursor of extraembryonic endoderm (XEN) directly from mouse blastocysts. Mouse XEN cells were developed with forced overexpression of *Oct4* in ES cells, which convert them into XEN cells (Niwa et al., 2000). On the contrary, from rat blastocysts we isolated the XEN-like cell lines expressing *Oct4*, SSEA1 at high levels and which proliferated in the presence of LIF, expressing the endodermal determinants *Gata6* and *Gata4* (Debeb et al., 2009). Interestingly, we found that the cell culture coexisted as mixed populations of two interconvertible phenotypes of flat and round cells, with preferential expression of stem cell markers Oct4 and SSEA1 in round cells. After injection into preimplantation embryos, the rat XEN-like cells colonize not only yolk sacs, but also randomly contribute to trophoblast lineages of postimplantation embryos following a transfer to surrogate mothers (Galat et al., 2009). Pluripotency of rat ES cells was also asserted by other groups that demonstrated ES cells can contribute to embryonic and extraembryonic lineages (Demers et al., 2011). These findings with rat ES cells are important for developing hypotheses of cell fate plasticity in the inner cell mass (ICM), metastable stage-specific heterogeneity of ES cell population, and defining pluripotency boundaries.

3. iPS Cells

Cloning requiring reactivation of somatic nuclei implied that isolating relevant factors from the oocyte might allow reactivation of somatic nuclei without the oocyte and suggested the possibility of developing pluripotent cells from adult somatic cells. Recent progress in derivation of induced pluripotent stem (iPS) cells, which are likely the functional equivalents of ES cells, from somatic cells by forced expression of exogenous transcription factors (e.g., Oct4, Sox2, c-Myc, and Klf4) will further facilitate transgenic technology. Rat iPS cell lines have been derived from a variety of somatic cells, including fibroblasts, neural precursor cells, bone marrow cells, and liver progenitor cells (Chang et al., 2010; Li et al., 2009; Liao et al., 2009; Liskovykh et al., 2011; Merkl et al., 2013). The status of pluripotency of extant rat iPS cell lines is still under investigation (Bui et al., 2013).

4. Cloning

Another approach to genetic modification is nuclear transfer (NT) or cloning. The technologies described previously mitigate its necessity, but cloning is inherently interesting. The biology of cloning is also important, leading for example to the process of developing pluripotent cells from adult cells. The process utilizes NT consisting of culturing a species's cells with or without genetic modification as a source of nuclei. These donor nuclei are surgically inserted into enucleated oocytes from that species. The new egg is activated and transferred to a surrogate for development to term. In general, this process will result in live offspring from about 2% of the manipulated eggs. The ability to clone animals from genetically modified nuclei has been established and provides both a way to establish targeted mutations and to produce transgenic animals. The process has been utilized to clone animals from many species (Latham, 2004).

There are important nuances to the procedure. The relationship of the cell cycle of the donor cells to that of the recipient egg is important but poorly understood. Most people in this field believe that the donor cell must be quiescent, in G0 or G1 of the cell cycle. The current opinion is that the nucleus of the donor cell must be inserted into the egg before the egg is activated. This is relatively straightforward in the case of cattle or sheep, but more difficult in the mouse. In the case of the rat, it is very difficult because the rat egg seems to be spontaneously activated upon isolation from the mother. Preventing the premature activation of the rat egg requires that eggs of extremely young ovular age be isolated. This is best achieved in animals that are light-reversed, using superovulation to define the ovular age of the egg. The method by which the cloned egg is developed is critical to the development of the cloned animal. In cattle and sheep cloned embryos are placed in intermediate surrogates and reisolated after several days of culture. In the mouse, it is possible to develop the egg *in vitro* because robust culture systems are available for this purpose. In the case of the rat, our lab has been working on modification of existing preimplantation culture systems (Miyoshi et al., 1997; Miyoshi and Niwa, 1997) that allow us to culture fertilized rat eggs from the one-cell stage to the blastocyst stage (Figure 9.2, Table 9.3). Up to 76% of one-cell embryos of some strains from fertile matings will develop to the blastocyst stage in these media. The ability to culture the embryos greatly facilitates the study of the many parameters that are important in NT by allowing longitudinal or developmental analysis of the embryo following manipulation (Iannaccone et al., 2001).

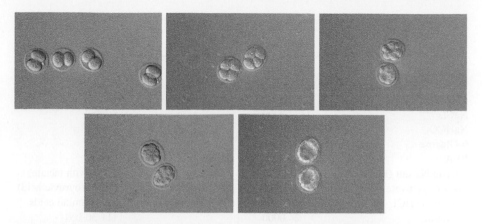

Figure 9.2 Photomicrographs of fertilized, cultured preimplantation rat embryos.
From left to right the panels show, top row: two-cell, four-cell, eight-cell; bottom row:
post-compacted eight-cell, and blastocyst stages of development. The embryos are all
approximately 100 μm in diameter.

5.   Cloning Steps

a.   Enucleation

In preliminary experiments, SD rat eggs were isolated from animals on a normal
light cycle (lights on at 0600 and off at 1800). The eggs were isolated at 1100 as
described above and then positioned into a droplet of M2 (Quinn et al., 1982;
Sturm and Tam, 1993) medium. The cytoplasmic bulge indicating the nascent second polar body was identified and a slit was made in the zona pellucida over it.
The second metaphase plate was removed by drawing the cytoplasmic bulge into a
suction pipette through the slit in the zona pellucida. This cytoplasmic fragment
containing the second metaphase plate pinches off spontaneously and is separated
from the enucleated egg (Fitchev et al., 1999).

b.   Reconstitution of Egg

Donor nuclei are injected using piezoactuated vibration in a droplet of M2. This
was done with glass pipettes with an internal diameter of 10 μm, which are acid
cleaned using 30% HF acid. The glass is washed extensively with distilled water
filtered through 0.2 μm millipore filters. The pipettes are backfilled with mercury
or Fluorinert FC-77 (Sigma F4758) and coated with 10% polyvinyl pyrrolidone
injection of the nuclei is done in the presence of cytochalasin B (5 μg/mL M2).
Following NT, the eggs are incubated for 30 min in M2 without cytochalasin B.
The eggs are then activated in CZB medium (Chatot et al., 1989) without Hepes or
$CaCl_2$ containing 5 mM $SrCl_2$. Activation periods ranged from 1.5 to 3 h. Eggs
cultured in mR1ECM (Miyoshi et al., 1997; Miyoshi and Niwa, 1997) for 5 days
following activation developed to the blastocyst stage.

Table 9.3 Formulation and Reagent Sources for Rat Embryo Culture Medium, mR1ECM

Reagent	mg/L	Source	Catalog
CaCl$_2$·2H$_2$O	294.0400	Sigma	C-7902
KCL	238.6000	Sigma	P-5405
MgCl$_2$ · 6H$_2$O	101.6500	Sigma	
NaCL	4482.0000	Sigma	S-9888
NaHCO$_3$	2100.2500	Gibco	11810
D-Glucose	1351.5000	Gibco	15023
PVA	1000.0000	Sigma	P-8136
Lactate Na salt (mL)	1.8683	Sigma	7900 (with lactate)
Sodium pyruvate	55.0000	Sigma	P4562 (pyruvic acid)
L-Arginine:HCl	126.4000	Gibco	MEM amino acids
L-Cystine	24.0000		111 30-051
L-Glutamine	14.6100		MEM now essentials
Glycine	7.500		AA
L-Histidine·HCl·H$_2$O	42.0000		1140-050
L-Isoleucine	52.5000		
L-Leucine	52.4000		
L-Lysine·HCl	72.5000		
L-Methionine	15.1000		
L-Phenylalanine	33.0000		
L-Serine	10.5000		
L-Theronine	46.6000		
L-Tryptophan	10.2000		
L-Tyrosine	36.0000		
L-Valine	46.8000		
L-Alanine	8.9000		
L-Asparagine·H$_2$O	15.0000		
L-Aspartic acid	13.3000		
L-Glutamic acid	14.7000		
L-Proline	11.5000		

246 m Osm; equilibrated pH = 7.4.

The efficiency of blastocyst development is not changed by genetic modification of the donor cells. We transfected two different green fluorescent protein coding transgenes into primary embryonic fibroblasts from SD rats and selected for integration with G418. The nuclei of these cells also support development to the blastocyst stage following NT (Iannaccone et al., 2001).

However, when eggs prepared in this way were cultured overnight in mR1ECM to the two-cell stage and then transferred to surrogate mothers, they failed to develop past gastrulation stage. The eggs, however, did implant. This result is suggestive of a genetic imprint failure and may indicate that the reprogramming of the transferred nucleus was not appropriate. In the mouse, nuclear reprogramming has been shown to be dependent on the time of NT relative to the time of activation. If the donor nucleus is inserted into the egg after activation, then reprogramming does not occur. It is reasonable to suppose that this is also true for the rat, but that is not known. Because the rat egg will activate spontaneously following removal from the female, the eggs used were activated prior to the insertion of the nucleus. It is possible to prevent the spontaneous activation of the rat egg by isolating eggs close to the time of ovulation; that may provide a method that supports reprogramming of the donor nucleus and postimplantation development.

For embryos that will be manipulated, the isolation may need to occur at a very early ovular age so that spontaneous activation is prevented. This allows the donor nucleus to be inserted prior to activation of the recipient egg. In order to do this, we have light-reversed rats with the lights going on at 1400 and out at 0200. The eggs are isolated at 1500, which is about 2 h following ovulation in SD rats. The eggs are activated and cultured. Parthenogenetic development to the blastocyst stage occurs in about 20% of eggs under these conditions. Oocytes may be collected even earlier or prior to ovulation, which may also provide an approach to delaying activation.

c. Lessons and Perspectives from Cloning

A breakthrough in rat cloning was reported with the application of proteosome inhibitor MG132, which prevents spontaneous activation of oocytes (Zhou et al., 2003a). It is believed that premature chromosome condensation (PCC) activity of the nonactivated cytoplasm contributes to nuclear reprogramming and improves development of NT embryos. However, other groups had difficulties reproducing the achievement, perhaps because of inherited developmental toxicity and strain dependency of this particular drug (Mizumoto et al., 2008, 2010; Nakajima et al., 2008; Tomioka et al., 2007; Webb et al., 2010). In search of alternatives, we discovered that the activation of rat oocytes could be reversibly suppressed with microtubule and microfilament disruptive agents such as cytochalasin B, cytochalasin D, nocodazole, and demecolcyne (Galat et al., 2007). Although the activation suppression was successful, these treatments decreased the parthenogenetic development as compared to controls.

The development to blastocyst stage of oocytes treated by nocodazole and demecolcine was 14.2−13.3%, respectively, and 28.4−29.8% for cytochalasin B and D. In this study, the eggs were retrieved at 14 h post-hCG, cultured in drug treatment for 4 h, then incubated for 2 h after washing. In a control group, we observed 11.4−59.6% blastocyst formation depending on ovular age of the oocytes. Increased efficiency of chemical activation in older oocytes was also reported by other groups (Krivokharchenko et al., 2003). Furthermore, nocodazole and demecolcyne did not result in improvement of development of reconstructed NT embryos (Mizumoto et al., 2010). This prompted us to take a more detailed look into specifics of spontaneous activation. We found that spontaneous activation as measured by polar body (PB) extrusion times vary dramatically with ovular age of oocytes (Galat et al., 2007). For example, it is complete by 40 min in oocytes isolated 22 h post-hCG, whereas it lasts 1 h 30 min in oocytes isolated 18 h post-hCG and up to 3 h in oocytes isolated 14 h post-hCG. Importantly, this observation means that the time of PB extrusion could be used as an "ovular chronometer" for a particular group of oocytes. Such a control could be helpful for studies where ovular age is important. This is particularly useful, since the time of ovulation is not well defined in rats (Kostyk et al., 1978).

Considering that PCC routinely takes place within 30–60 min, we determined whether PCC activity is retained in the cytoplasm of oocytes of different ovular ages in spite of spontaneous activation. Indeed, we found that PCC activity is preserved in young oocytes if another nucleus is introduced promptly after oocyte retrieval. We were further interested to find out whether enucleation can modify the course of PCC. The possible role of the spindle on activation was implied in earlier work of Winston (1995), who noted that the entry into interphase could be delayed if fertilization of mouse oocytes occurs in the presence of nocodazole. We assayed PCC after NT in intact and enucleated oocytes of different ovular ages and found that enucleation indeed decreased PCC capacity and enhanced nuclear activation. For example, PCC of the transferred nucleus was found in 100% of young (14 h post-hCG) intact oocytes and 66% of 18 h post-hCG, compared with 33% and 0% of enucleated oocytes at those respective ovular times. PCC was not observed in oocytes 22 h post-hCG, and all transferred nuclei were activated regardless of enucleation time. This result corroborates the conclusion that enucleation promotes inactivation of cytostatic factor activity in matured rat oocytes (Ito et al., 2007) and with a recent finding arguing that spindle removal can cause spontaneous exit from meiosis of human oocytes (Noggle et al., 2011; Tachibana et al., 2013). We further compared the developmental potential of reconstructed rat embryos enucleated prior to and after introduction of the donor nucleus, observing significantly improved *in vitro* development in the group when removal of the host nucleus was done after electrofusion with donor cells. Five percent of reconstructed embryos enucleated after NT developed to the blastocyst stage and 10% to morula stage, whereas only 3% developed to morulas with no blastocysts in the group enucleated before NT.

Additional considerations for NT optimization in rat could include an assessment of the methods of nuclear insertion. Electrofusion itself can provide a stimulus for activation, for example, three separate pulses were sufficient to efficiently drive parthenogenetic and NT development to the blastocyst in rabbit (Galat et al., 1999, 2002). Considering that PCC is important for NT of the rat, the electrofusion could promote unwarranted activation. This may explain a relatively high number of reconstructed embryos developed to blastocyst stage (12%) when NT was done with this injection technique (Iannaccone et al., 2001). The optimal protocol for activation, an essential component of NT, still remains debated. A strontium protocol was adapted from mouse studies and most frequently used. However, we and others (Hayes et al., 2001; Roh et al., 2003) observed embryolysis in Ca^{2+} free medium and in 10 mM Sr^{2+}. Subsequent optimizations included decreasing Sr^{2+} concentration to 5 mM Sr^{2+} (Iannaccone et al., 2001) or supplementing of activation medium with 1.7 mM calcium (Mizumoto et al., 2008). We developed an efficient protocol of parthenogenetic stimulation resulting in blastocyst formation in up to 60% of SD rat eggs based on electrical stimulation in combination with

1.5 mM dimethylaminopurine (DMAP) for 3 h (Galat et al., 2001, 2007), an improvement on the *in vitro* culture system (Zhou et al., 2003b).

An alternative way to avoid spontaneous activation could be developed using oocytes matured *in vitro*. Our experiments show that the maturation rate of oocytes recovered from ovarian follicles reaches over 90% and they remain arrested at metaphase II stage oocytes (MII) stage without entering spontaneous activation. Although parthenogenetic development of such oocytes is notably lower, an optimization of *in vitro* maturation conditions would allow establishing proper developmental competency.

V. Summary

The production of transgenic rats is now routine and commercially available. The methods employed are essentially the same as those used to produce mouse transgenics. The rate of transformation and the overall efficiency of the process is less than in the mouse, however. However, with lentivirus injection the efficiencies are very high, indeed much greater than for microinjection in the mouse. Moreover, many hundreds of peer-reviewed papers describing useful results from the genetic modification of rat models of human disease are published every year. New methods of transgenesis are being developed such as testicular tissue electroporation that if validated have the potential to greatly simplify the production of transgenic rats (Usmani, et al., 2013). Using embryonic stem cells to produce gene knockouts in the rat is now possible and the dream of knockouts in genes of choice has been realized with completely new technologies that allow genome editing and the generation of allelic series. Attempts to clone the rat by NT technology have not replicated the early reported results of success, but new information concerning activation of somatic nuclei following NT is available and more is coming. Practical cloning using adult cells as nuclear donors would also provide a mechanism for safe and easy transport of animals internationally and the potential of recovering valuable mutations from lost strains. Genetic and microbiological quality control, as well as the development of new technologies, are a priority for the US National Institutes of Health.

Acknowledgments

The authors are grateful to Greg Taborn for his expert technical skills with rat embryos, culture, and surgery. Jacek Topczewski provided valuable advice on new methods of genome editing. The work was supported in part by grants from the National Institutes of Health and by the George M. Eisenberg Foundation for Charities.

References

Agca, Y., 2012. Genome resource banking of biomedically important laboratory animals. Theriogenology 78, 1653–1665.

Alhonen, L., Parkkinen, J.J., Keinanen, T., Sinervirta, R., Herzig, K.H., Janne, J., 2000. Activation of polyamine catabolism in transgenic rats induces acute pancreatitis. Proc. Natl. Acad. Sci. USA 97, 8290−8295.

Arnold, A.C., Sakima, A., Kasper, S.O., Vinsant, S., Garcia-Espinosa, M.A., Diz, D.I., 2012. The brain renin-angiotensin system and cardiovascular responses to stress: insights from transgenic rats with low brain angiotensinogen. J. Appl. Physiol. 113, 1929−1936.

Bader, M., Ganten, D., 1996. Transgenic rats: tools to study the function of the renin-angiotensin system. Clin. Exp. Pharmacol. Physiol. 3 (Suppl. 3), S81−S87.

Bader, M., Paul, M., Ganten, D., 1997. Transgenic animal models in hypertension. In: Iannaccone, P.M., Scarpelli, D.G. (Eds.), Biological Aspects of Disease: Contributions from Animal Models. Harwood Academic Publishers, Amsterdam, pp. 165−200.

Bader, M., Bohnemeier, H., Zollmann, F.S., Lockley-Jones, O.E., Ganten, D., 2000. Transgenic animals in cardiovascular disease research. Exp. Physiol. 85, 713−731.

Blair, K., Wray, J., Smith, A., 2011. The liberation of embryonic stem cells. PLoS Genet. 7, e1002019.

Bohlender, J., Ganten, D., Luft, F.C., 2000. Rats transgenic for human renin and human angiotensinogen as a model for gestational hypertension. J. Am. Soc. Nephrol. 11, 2056−2061.

Buehr, M., Meek, S., Blair, K., Yang, J., Ure, J., Silva, J., et al., 2008. Capture of authentic embryonic stem cells from rat blastocysts. Cell 135, 1287−1298.

Bui, P., Rajanahalli, P., Hong, J., Weiss, M.L., 2013. Proof of pluripotency of rat iPSCs missing. Cell Prolif. 46, 119−120.

Chang, M.Y., Kim, D., Kim, C.H., Kang, H.C., Yang, E., Moon, J.I., et al., 2010. Direct reprogramming of rat neural precursor cells and fibroblasts into pluripotent stem cells. PLoS ONE 5, e9838.

Charreau, B., Tesson, L., Soulillou, J.P., Pourcel, C., Anegon, I., 1996. Transgenesis in rats: technical aspects and models. Transgenic Res. 5, 223−234.

Charreau, B., Menoret, S., Tesson, L., Azimzadeh, A., Audet, M., Wolf, P., et al., 1999. Protection against hyperacute xenograft rejection of transgenic rat hearts expressing human decay accelerating factor (DAF) transplanted into primates. Mol. Med. 5, 617−630.

Chatot, C.L., Ziomek, C.A., Bavister, B.D., Lewis, J.L., Torres, I., 1989. An improved culture medium supports development of random-bred 1-cell mouse embryos in vitro. J. Reprod. Fertil. 86, 679−688.

Chisholm, J.W., Paterniti, J.R., Dolphin, P.J., 2000. Accumulation of cholestatic lipoproteins in ANIT-treated human apolipoprotein A-I transgenic rats is diminished through dose-dependent apolipoprotein A-I activation of LCAT. Biochim. Biophys. Acta 1487, 145−154.

Cowley Jr., A.W., Stoll, M., Greene, A.S., Kaldunski, M.L., Roman, R.J., Tonellato, P.J., et al., 2000. Genetically defined risk of salt sensitivity in an intercross of Brown Norway and Dahl S rats. Physiol. Genomics 2, 107−115.

Dann, C.T., 2007. New technology for an old favorite: lentiviral transgenesis and RNAi in rats. Transgenic Res. 16, 571−580.

Debeb, B.G., Galat, V., Epple-Farmer, J., Iannaccone, S., Woodward, W.A., Bader, M., et al., 2009. Isolation of Oct4-expressing extraembryonic endoderm precursor cell lines. PLoS ONE 4, e7216.

Demers, S.P., Desmarais, J.A., Vincent, P., Smith, L.C., 2011. Rat blastocyst-derived stem cells are precursors of embryonic and extraembryonic lineages. Biol. Reprod. 84, 1128–1138.

Fitchev, P., Taborn, G., Garton, R., Iannaccone, P., 1999. Nuclear transfer in the rat: potential access to the germline. Transplant. Proc. 31, 1525–1530.

Flicek, P., Ahmed, I., Amode, M.R., Barrell, D., Beal, K., Brent, S., et al., 2013. Ensembl 2013. Nucleic Acids Res. 41, D48–D55.

Gaj, T., Gersbach, C.A., Barbas III, C.F., 2013. ZFN, TALEN, and CRISPR/Cas-based methods for genome engineering. Trends Biotechnol. 31 (7), 397–405. (Epub May 2013).

Galat, V., Lagutina, I., Mesina, M., Chernich, V., Prokofiev, M.I., 1999. Developmental potential of rabbit nuclear transfer embryos derived from donor fetal fibroblast. Theriogenology 51, 1.

Galat, V., Taborn, G., Bader, M., Iannaccone, P., 2001. Parthenogenic development of rat embryos. J. Mol. Med. 79, B5.

Galat, V., Lagutina, I., Mezina, M., Prokofiev, M.I., Zakhartchenko, V., 2002. Effect of donor cell age on the efficiency of nuclear transfer in rabbits. Reprod. Biomed. Online 4, 32–37.

Galat, V., Zhou, Y., Taborn, G., Garton, R., Iannaccone, P., 2007. Overcoming MIII arrest from spontaneous activation in cultured rat oocytes. Cloning Stem Cells 9, 303–314.

Galat, V., Binas, B., Iannaccone, S., Postovit, L.M., Debeb, B.G., Iannaccone, P., 2009. Developmental potential of rat extraembryonic stem cells. Stem Cells Dev. 18, 1309–1318.

Gomez-Angelats, M., Teeguarden, J.G., Dragan, Y.P., Pitot, H.C., 1999. Mutational analysis of three tumor suppressor genes in two models of rat hepatocarcinogenesis. Mol. Carcinog. 25, 157–163.

Haas, M.J., Pitot, H.C., 1998. Characterization of rare p53 mutants from carcinogen-treated albumin- simian virus 40 T-antigen transgenic rats. Mol. Carcinog. 21, 128–134.

Haas, M.J., Dragan, Y.P., Hikita, H., Shimel, R., Takimoto, K., Heath, S., et al., 1999. Transgene expression and repression in transgenic rats bearing the phosphoenolpyruvate carboxykinase-simian virus 40 T antigen or the phosphoenolpyruvate carboxykinase-transforming growth factor-alpha constructs. Am. J. Pathol. 155, 183–192.

Haas, M.J., Sattler, C.A., Dragan, Y.P., Gast, W.L., Pitot, H.C., 2000. Multiple polypeptide hormone expression in pancreatic islet cell carcinomas derived from phosphoenolpyruvatecarboxykinase-SV40 T antigen transgenic rats. Pancreas 20, 206–214.

Hamilton, G.S., Armstrong, D.T., 1991. The superovulation of synchronous adult rats using follicle-stimulating hormone delivered by continuous infusion. Biol. Reprod. 44, 851–856.

Hammer, R.E., Maika, S.D., Richardson, J.A., Tang, J.P., Taurog, J.D., 1990. Spontaneous inflammatory disease in transgenic rats expressing HLA-B27 and human beta 2 m: an animal model of HLA-B27-associated human disorders. Cell 63, 1099–1112.

Hayes, E., Galea, S., Verkuylen, A., Pera, M., Morrison, J., Lacham-Kaplan, O., et al., 2001. Nuclear transfer of adult and genetically modified fetal cells of the rat. Physiol. Genomics 5, 193–204.

Hayman, G.T., Jayaraman, P., Petri, V., Tutaj, M., Liu, W., De Pons, J., et al., 2013. The updated RGD Pathway Portal utilizes increased curation efficiency and provides expanded pathway information. Hum. Genomics 7, 4.

Hong, J., He, H., Weiss, M.L., 2012. Derivation and characterization of embryonic stem cells lines derived from transgenic Fischer 344 and Dark Agouti rats. Stem Cells Dev. 21, 1571–1586.

Hully, J.R., Su, Y., Lohse, J.K., Griep, A.E., Sattler, C.A., Haas, M.J., et al., 1994. Transgenic hepatocarcinogenesis in the rat. Am. J. Pathol. 145, 386–397.

Iannaccone, P., Taborn, G., Garton, R., 2001. Preimplantation and postimplantation development of rat embryos cloned with cumulus cells and fibroblasts. Zygote 9, 135–143.

Iannaccone, P.M., 1990. Fractal geometry in mosaic organs: a new interpretation of mosaic pattern. FASEB J. 4, 1508–1512.

Iannaccone, P.M., Jacob, H.J., 2009. Rats! Dis. Model. Mech. 2, 206–210.

Ito, J., Kato, M., Hochi, S., Hirabayashi, M., 2007. Effect of enucleation on inactivation of cytostatic factor activity in matured rat oocytes. Cloning Stem Cells 9, 257–266.

Jacob, H.J., 1999. Functional genomics and rat models. Genome Res. 9, 1013–1016.

Khokha, M.K., Landini, G., Iannaccone, P.M., 1994. Fractal geometry in rat chimeras demonstrates that a repetitive cell division program may generate liver parenchyma. Dev. Biol. 165, 545–555.

Kostyk, S.K., Dropcho, E.J., Moltz, H., Swartwout, J.R., 1978. Ovulation in immature rats in relation to the time and dose of injected human chorionic gonadotropin or pregnant mare serum gonadotropin. Biol. Reprod. 19, 1102–1107.

Krivokharchenko, A., Popova, E., Zaitseva, I., Vil'ianovich, L., Ganten, D., Bader, M., 2003. Development of parthenogenetic rat embryos. Biol. Reprod. 68, 829–836.

Latham, K.E., 2004. Cloning: questions answered and unsolved. Differentiation 72, 11–22.

Li, P., Tong, C., Mehrian-Shai, R., Jia, L., Wu, N., Yan, Y., et al., 2008. Germline competent embryonic stem cells derived from rat blastocysts. Cell 135, 1299–1310.

Li, W., Wei, W., Zhu, S., Zhu, J., Shi, Y., Lin, T., et al., 2009. Generation of rat and human induced pluripotent stem cells by combining genetic reprogramming and chemical inhibitors. Cell Stem Cell 4, 16–19.

Liao, J., Cui, C., Chen, S., Ren, J., Chen, J., Gao, Y., et al., 2009. Generation of induced pluripotent stem cell lines from adult rat cells. Cell Stem Cell 4, 11–15.

Liskovykh, M., Chuykin, I., Ranjan, A., Safina, D., Popova, E., Tolkunova, E., et al., 2011. Derivation, characterization, and stable transfection of induced pluripotent stem cells from Fischer344 rats. PLoS ONE 6, e27345.

Lois, C., Hong, E.J., Pease, S., Brown, E.J., Baltimore, D., 2002. Germline transmission and tissue-specific expression of transgenes delivered by lentiviral vectors. Science 295, 868–872.

Lundin, P.D., Ekstrom, G., Erlansson, M., Lundin, S., Westrom, B.R., 1997. Intestinal inflammation and barrier function in HLA-B27/beta 2- microglobulin transgenic rats. Scand. J. Gastroenterol. 32, 700–705.

Matsuoka, Y., Hamaguchi, T., Fukamachi, K., Yoshida, M., Watanabe, G., Taya, K., et al., 2007. Molecular analysis of rat mammary carcinogenesis: an approach from carcinogenesis research to cancer prevention. Med. Mol. Morphol. 40, 185–190.

Mayer, C., Klein, R.G., Wesch, H., Schmezer, P., 1998. Nickel subsulfide is genotoxic *in vitro* but shows no mutagenic potential in respiratory tract tissues of BigBlue rats and Muta Mouse mice *in vivo* after inhalation. Mutant Res. 420, 85–98.

Meek, S., Buehr, M., Sutherland, L., Thomson, A., Mullins, J.J., Smith, A.J., et al., 2010. Efficient gene targeting by homologous recombination in rat embryonic stem cells. PLoS ONE 5, e14225.

Men, H., Bauer, B.A., Bryda, E.C., 2012. Germline transmission of a novel rat embryonic stem cell line derived from transgenic rats. Stem Cells Dev. 21, 2606–2612.

Merkl, C., Saalfrank, A., Riesen, N., Kuhn, R., Pertek, A., Eser, S., et al., 2013. Efficient generation of rat induced pluripotent stem cells using a non-viral inducible vector. PLoS ONE 8, e55170.

Mervaala, E., Muller, D.N., Park, J.K., Dechend, R., Schmidt, F., Fiebeler, A., et al., 2000a. Cyclosporin A protects against angiotensin II-induced end-organ damage in double transgenic rats harboring human renin and angiotensinogen genes. Hypertension 35, 360–366.

Mervaala, E., Muller, D.N., Schmidt, F., Park, J.K., Gross, V., Bader, M., et al., 2000b. Blood pressure-independent effects in rats with human renin and angiotensinogen genes. Hypertension 35, 587–594.

Milia, A.F., Ibba-Manneschi, L., Manetti, M., Benelli, G., Messerini, L., Matucci-Cerinic, M., 2009. HLA-B27 transgenic rat: an animal model mimicking gut and joint involvement in human spondyloarthritides. Ann. N Y Acad. Sci. 1173, 570–574.

Miyoshi, K., Niwa, K., 1997. Stage-specific requirement of phosphate for development of rat 1-cell embryos in a chemically defined medium. Zygote 5, 67–73.

Miyoshi, K., Kono, T., Niwa, K., 1997. Stage-dependent development of rat 1-cell embryos in a chemically defined medium after fertilization *in vivo* and *in vitro*. Biol. Reprod. 56, 180–185.

Mizumoto, S., Kato, Y., Tsunoda, Y., 2008. The developmental potential of parthenogenetic and somatic cell nuclear-transferred rat oocytes *in vitro*. Cloning Stem Cells 10, 453–459.

Mizumoto, S., Kato, Y., Tsunoda, Y., 2010. The effect of the time interval between injection and parthenogenetic activation on the spindle formation and the *in vitro* developmental potential of somatic cell nuclear-transferred rat oocytes. Zygote 18, 9–15.

Mukumoto, S., Mori, K., Ishikawa, H., 1995. Efficient induction of superovulation in adult rats by PMSG and hCG. Exp. Anim. 44, 111–118.

Mullins, J.J., Mullins, L.J., 1993. Transgenesis in nonmurine species. Hypertension 22, 630–633.

Mullins, L.J., Mullins, J.J., 1996. Transgenesis in the rat and larger mammals. J. Clin. Invest. 97, 1557–1560.

Nakajima, N., Inomata, T., Ito, J., Kashiwazaki, N., 2008. Treatment with proteasome inhibitor MG132 during cloning improves survival and pronuclear number of reconstructed rat embryos. Cloning Stem Cells 10, 461–468.

Ng, Y.K., Iannaccone, P.M., 1992. Fractal geometry of mosaic pattern demonstrates liver regeneration is a self-similar process. Dev. Biol. 151, 419–430.

Niwa, H., Miyazaki, J., Smith, A.G., 2000. Quantitative expression of Oct-3/4 defines differentiation, dedifferentiation or self-renewal of ES cells. Nat. Genet. 24, 372–376.

Noggle, S., Fung, H.L., Gore, A., Martinez, H., Satriani, K.C., Prosser, R., et al., 2011. Human oocytes reprogram somatic cells to a pluripotent state. Nature 478, 70–75.

Onderdonk, A.B., Richardson, J.A., Hammer, R.E., Taurog, J.D., 1998. Correlation of cecal microflora of HLA-B27 transgenic rats with inflammatory bowel disease. Infect. Immun. 66, 6022–6023.

Quinn, P., Barros, C., Whittingham, D.G., 1982. Preservation of hamster oocytes to assay the fertilizing capacity of human spermatozoa. J. Reprod. Fertil. 66, 161–168.

Roh, S., Malakooti, N., Morrison, J.R., Trounson, A.O., Du, Z.T., 2003. Parthenogenetic activation of rat oocytes and their development (*in vitro*). Reprod. Fertil. Dev. 15, 135–140.

Schmezer, P., Eckert, C., 1999. Induction of mutations in transgenic animal models: BigBlue and Muta Mouse. IARC Sci. Publ. 146, 367–394.

Simmons, W.A., Leong, L.Y., Satumtira, N., Butcher, G.W., Howard, J.C., Richardson, J.A., et al., 1996. Rat MHC-linked peptide transporter alleles strongly influence peptide binding by HLA-B27 but not B27-associated inflammatory disease. J. Immunol. 156, 1661–1667.

Sinn, P.L., Sigmund, C.D., 2000. Transgenic models as tools for studying the regulation of human renin expression. Regul. Pept. 86, 77–82.

Springate, J.E., Feld, L.G., Ganten, D., 1997. Enalapril and renal function in hypertensive rats transgenic for mouse renin gene. Hypertension 30, 868–872.

Steen, R.G., Kwitek-Black, A.E., Glenn, C., Gullings-Handley, J., Van Etten, W., Atkinson, O.S., et al., 1999. A high-density integrated genetic linkage and radiation hybrid map of the laboratory rat. Genome Res. 9, AP1–AP8, insert.

Stoll, M., Kwitek-Black, A.E., Cowley, A.W., Harris, E.L., Harrap, S.B., Krieger, J.E., et al., 2000. New target regions for human hypertension via comparative genomics. Genome Res. 10, 473–482.

Sturm, K., Tam, P.L., 1993. Isolation and culture of whole postimplantation embryos and germ layer derivitives. In: Wassarman, P.M., DePamphilis, M.L. (Eds.), Methods in Enzymology: Guide to Techniques in Mouse Development, vol. 225. Academic Press, Inc., San Diego.

Tachibana, M., Amato, P., Sparman, M., Gutierrez, N.M., Tippner-Hedges, R., Ma, H., et al., 2013. Human embryonic stem cells derived by somatic cell nuclear transfer. Cell 153 (6), 1228–1238.

Takahashi, R., Hirabayashi, M., Ueda, M., 1999. Production of transgenic rats using cryopreserved pronuclear-stage zygotes. Transgenic Res. 8, 397–400.

Taurog, J.D., Hammer, R.E., 1996. Experimental spondyloarthropathy in HLA-B27 transgenic rats. Clin. Rheumatol. 15 (Suppl. 1), 22–27.

Taurog, J.D., Maika, S.D., Satumtira, N., Dorris, M.L., McLean, I.L., Yanagisawa, H., et al., 1999. Inflammatory disease in HLA-B27 transgenic rats. Immunol. Rev. 169, 209–223.

Tomioka, I., Mizutani, E., Yoshida, T., Sugawara, A., Inai, K., Sasada, H., et al., 2007. Spindle formation and microtubule organization during first division in reconstructed rat embryos produced by somatic cell nuclear transfer. J. Reprod. Dev. 53, 835–842.

Tong, C., Li, P., Wu, N.L., Yan, Y., Ying, Q.L., 2010. Production of p53 gene knockout rats by homologous recombination in embryonic stem cells. Nature 467, 211–213.

Tong, C., Huang, G., Ashton, C., Wu, H., Yan, H., Ying, Q.L., 2012. Rapid and cost-effective gene targeting in rat embryonic stem cells by TALENs. J. Genet. Genomics 39, 275–280.

Usmani, A., Ganguli, N., Sarkar, H., Dhup, S., Batta, S.R., Vimal, M., et al., 2013. A non-surgical approach for male germ cell mediated gene transmission through transgenesis. Sci. Rep. 3, 3430.

Webb, R.L., Findlay, K.A., Green, M.A., Beckett, T.L., Murphy, M.P., 2010. Efficient activation of reconstructed rat embryos by cyclin-dependent kinase inhibitors. PLoS ONE 5, e9799.

Weinberg, W.C., Iannaccone, P.M., 1988. Clonality of preneoplastic liver lesions: histological analysis in chimeric rats. J. Cell Sci. 89, 423–431.

Weinberg, W.C., Deamant, F.D., Iannaccone, P.M., 1985. Patterns of expression of class I antigens in the tissues of congenic strains of rat. Hybridoma 4, 27–36.

Weinberg, W.C., Berkwits, L., Iannaccone, P.M., 1987. The clonal nature of carcinogen-induced altered foci of gamma-glutamyl transpeptidase expression in rat liver. Carcinogenesis 8, 565−570.

Winston, N.J., McGuinness, O., Johnson, M.H., Maro, B., 1995. The exit of mouse oocytes from meiotic M-phase requires an intact spindle during intracellular calcium release. J. Cell Sci. 108, 143−151.

Yamamoto, S., Nakata, M., Sasada, R., Ooshima, Y., Yano, T., Shinozawa, T., et al., 2012. Derivation of rat embryonic stem cells and generation of protease-activated receptor-2 knockout rats. Transgenic Res. 21, 743−755.

Zheng, S., Geghman, K., Shenoy, S., Li, C., 2012. Retake the center stage−new development of rat genetics. J. Genet. Genomics 39, 261−268.

Zhou, M., Sayad, A., Simmons, W.A., Jones, R.C., Maika, S.D., Satumtira, N., et al., 1998. The specificity of peptides bound to human histocompatibility leukocyte antigen (HLA)-B27 influences the prevalence of arthritis in HLA-B27 transgenic rats. J. Exp. Med. 188, 877−886.

Zhou, Q., Renard, J.P., Le Friec, G., Brochard, V., Beaujean, N., Cherifi, Y., et al., 2003a. Generation of fertile cloned rats by regulating oocyte activation. Science 302, 1179.

Zhou, Y., Galat, V., Garton, R., Taborn, G., Niwa, K., Iannaccone, P., 2003b. Two-phase chemically defined culture system for preimplantation rat embryos. Genesis 36, 129−133.

10 Production of Transgenic Rabbits

*Tatiana Flisikowska, Alexander Kind and
Angelika Schnieke*

Livestock Biotechnology, Technische Universität München, Freising,
Germany

I. Introduction—Rabbits in Biomedicine

Genetically modified animals are important in many areas of biomedicine, for example, the study of the molecular basis of human diseases, modeling disease conditions, development and testing of new biopharmaceuticals, and the production of pharmaceutical proteins. The mouse has become the most intensively studied mammal because of the ease of producing precise gene-targeted modifications in embryonic stem (ES) cells (Capecchi, 2005). Mice are an invaluable tool, but their usefulness in representing humans in biomedicine, especially preclinical studies, is limited by their small body size, short lifespan, and differences in anatomy, physiology, and protein interactions. The limitations and shortcomings of mouse models have been demonstrated in a wide variety of areas including drug metabolism (Martignoni et al., 2006), cystic fibrosis (Guilbault et al., 2007), breast cancer (Vargo-Gogola and Rosen, 2007), and inflammatory conditions (Seok et al., 2013). Therefore, there is a strong need to establish techniques of precise genetic modification in other species.

Rabbits have long been an important laboratory animal and are currently the third most common after mice and rats (http://eur-lex.europa.eu/). Their key advantage is a body size intermediate between rodents and livestock species. Unlike the mouse model, it is relatively straightforward to monitor physiological changes in rabbits by standard methods such as cannulation of blood vessels (e.g., to measure hormone or glucose levels) or localization of specific nerves (e.g., to monitor injury following ischemia). The rabbit thus offers some of the advantages of larger species without the need for a specialized large animal facility. Rabbits rapidly attain sexual maturity, have a short gestation period, and yield large numbers of offspring. Housing conditions are relatively straightforward, and conditions to minimize the risk of infectious disease are well established, including designated pathogen-free facilities.

Transgenic Animal Technology. DOI: http://dx.doi.org/10.1016/B978-0-12-410490-7.00010-4

Rabbits are used for surgical and transplantation studies; for example, lung (Yoshida et al., 2005), heart (Furukawa et al., 2005), bone (Li and Li, 2005), and hepatocytes (Attaran et al., 2004). Rabbit early embryogenesis and gastrulation also resemble that in humans more closely than mouse. Rabbits have been used to study aspects of human pregnancy and fetal development (Fischer et al., 2012). Rabbits are also used as models in the study of cardiomyopathy (Lombardi et al., 2009), inflammation (Serhan et al., 2003), retinal degeneration (Kondo et al., 2009), cancer (Knight et al., 1988; Sethupathi et al., 1994), to evaluate human vaccines (Chentoufi et al., 2010), and in basic immunology such as the Basilea immunoglobulin mutant strain (Heidmann and Rougeon, 1983).

Rabbits are particularly useful in the study of cardiovascular disease. They are very sensitive to cholesterol feeding and like humans can develop severe hypercholesterolemia. This contrasts with mice, which are resistant to dietary cholesterol and atherosclerosis. Natural mutant strains such as the Watanabe heritable hyperlipidemic rabbit are used to model human hypercholesterolemia. These animals carry a frameshift mutation in the low-density lipoprotein receptor gene and develop atherosclerosis even when fed a normal diet (Watanabe, 1980). Several transgenic rabbit models for atherosclerosis have also been generated (Fan and Watanabe, 2003).

Transgenic farm animals, including rabbits, have been proposed as bioreactors for the production of pharmaceutical proteins in body fluids such as milk and blood, a process that has been termed "pharming" (reviewed by Kind and Schnieke, 2008). Production in milk has not been widely adopted, but rabbits could yet find a place as a niche producer. Two products produced in milk have so far gained regulatory approval, the first was the anticoagulant antithrombin III produced in goats, and the second is human recombinant C1 esterase inhibitor (RUCONEST or RHUCIN), produced in rabbits by Pharming BV for treatment of patients with hereditary angioedema. The small size of rabbits relative to goats and cattle allows complex proteins to be produced at low running cost, with the flexibility of raising or lowering the quantity produced by simple breeding. Rabbits also have the advantage that their milk protein content ($\sim$14%) is significantly higher than cow's milk ($\sim$5%) (Duby et al., 1993).

Rabbits are already widely used to produce antibodies in their blood, predominantly for diagnostic purposes. Extending this to therapeutic antibodies would be a major advance. The "humanization" of the mouse immune system was demonstrated two decades ago (Wagner et al., 1994); (Lonberg et al., 1994). Now human polyclonal antibodies generated in larger animals are set to become a reality. We recently demonstrated key steps toward the generation of rabbits that express a human immunoglobulin repertoire (Flisikowska et al., 2011).

Rabbits are thus a natural choice for genetic modification, and most of the requirements are in place. Rabbit genome data are rapidly improving; the most recent genome sequence assembly (coverage 7X) (http://www.ncbi.nlm.nih.gov/genome?term=oryctolagus%20cuniculus) is now comparable to that of the rat. But, as we describe later, the production of genetically modified rabbits, especially gene-targeted animals, has been hindered until very recently by technical difficulties.

II. Overview of Methods for Producing Genetically Modified Rabbits

Methods for producing genetically modified mammals can be broadly divided into three categories: direct transgenesis, cell-mediated genetic modification, and genome editing.

A. Direct Transgenesis

Direct transgenesis is the introduction of nucleic acids into the early embryo to add exogenous DNA sequences into the genome. Methods include DNA microinjection, transposon-mediated gene transfer, sperm-mediated DNA transfer, and viral transduction.

DNA microinjection into fertilized oocytes was first developed in mice (Gordon et al., 1980) and later extended to rabbits and other livestock species (Hammer et al., 1985) (see Chapters 1 and 2). Microinjection of DNA into the pronuclei of fertilized embryos has been by far the most common means of producing transgenic rabbits, with many transgenic lines generated. Microinjection has the advantage of being straightforward, but allows only transgene addition and no control over where the transgene integrates. It is also inefficient, with approximately 1% of injected rabbit eggs resulting in transgenic offspring (Besenfelder et al., 1998).

Transposon vectors are a newer development (Mates et al., 2009; Ivics et al., 2009; Carlson et al., 2011) that has only recently been applied in rabbits (Katter et al., 2013). DNA-based, or class II, transposons are mobile genetic elements that move around the host genome via a "cut-and-paste" mechanism. A variety of colorfully named transposon vectors are now available, including Sleeping Beauty (SB), PiggyBac, Frog Prince, Tol2, and Passport. Most DNA transposons have a simple structure: a transposase protein coding region flanked by inverted terminal repeats (ITRs) that include transposase binding sites. Separation of the transposase coding sequence from the ITRs has allowed the development of two-component vector systems. Exogenous DNA can be flanked by ITRs and will be recognized by the transposase and enzymatically integrated into the nuclear genome. While the capacity of most DNA transposons is limited to 10 kb, the PiggyBac transposon is able to mobilize DNA fragments over 200 kb (Li et al., 2013). The transposase can be expressed from a DNA construct or provided by coinjection of *in vitro* synthesized mRNA to avoid integration of transposase sequences. Transposon vectors provide a far higher rate of integration than DNA microinjection and a high proportion of single-copy insertion events, which is often desirable. Transposon-mediated integration also tends to favor euchromatic rather than heterochromatic regions, and is thus more supportive of transgene expression. The main drawback of transposon systems is the likelihood of multiple genomic integrations that must be resolved by segregation in subsequent generations.

Sperm-mediated DNA transfer began with observations that rabbit spermatozoa can associate *in vitro* with exogenous DNA and transfer it to an oocyte by fertilization (Brackett et al., 1971) (see also Chapter 23). This method raised considerable interest in the 1990s following a report of transgenic mice by Lavitrano et al. (1989) that offered a simple low-tech means of generating transgenic animals without the need for micromanipulation equipment. Subsequent investigations, however, revealed considerable problems with reproducibility and transgene rearrangement. This is consistent with findings that DNA that penetrates sperm nuclei becomes fragmented. There have been three reports of transgenic rabbits made using sperm-mediated DNA transfer (Wang et al., 2001; Li et al., 2006; Shen et al., 2006), but the method is not regarded as reliable and has not been widely adopted. Interested readers are directed to a book by Smith (2012).

The discovery that Moloney murine leukemia virus (MMLV) could infect early mouse embryos, integrate as a stable provirus into the genome, and transmit through the germ line launched the use of retroviral vectors to generate transgenic animals (Jaenisch, 1976) (see also Chapter 6). For many years cloned MMLV proviral sequences (Harbers et al., 1981) provided the basis for most retroviral vectors, but MMLV-based vectors transduce embryos inefficiently, and integrated transgenes tend to be silenced by epigenetic modification during development. Newer vectors based on a different class of retroviruses, lentiviruses, have been more successful. They are less subject to silencing and do not require breakdown of the nuclear envelope to gain access to the host genome, enabling integration in both replicating and nonreplicating cells at all stages from the unfertilized oocyte onward. Lentiviral vectors based on viruses such as the human immunodeficiency virus (HIV-1) and equine infectious anemia virus have been used to generate a variety of transgenic mammals. A brief review and useful protocols are provided by Pfeifer et al. (2010). Unfortunately, rabbit cells are difficult to transduce using common HIV-1 vectors, because of a block in HIV-1 protein trafficking (Cutino-Moguel and Fassati, 2006). There is one published report of a rabbit produced by lentiviral transduction, an animal expressing green fluorescent protein generated using a vector based on simian immunodeficiency virus. However, the founder animal failed to transmit the transgene through the germ line, probably due to mosaicism and a low germ line contribution (Hiripi et al., 2010).

Viral vectors have also been used to transduce rabbit somatic cells *in vivo*. Two papers describe adenoviral transduction into rabbit mammary gland *in vivo* (Han et al., 2008; Yang et al., 2012). Although not true transgenesis, in that the germ line of the treated animal is unaffected, this can be a useful means of gaining information about transgene expression and protein production in chosen tissues.

Viral vectors do have drawbacks including, as mentioned previously, mosaicism in founder animals due to delayed viral integration, restriction of the size of transgenes (typically <8 kb), the possibility of insertional disruption or activation of endogenous genes and, despite extensive precautions, the possibility of recombination with endogenous retroviral elements to form a replication-competent virus. The significance of these factors will of course depend on the particular transgenic experiment. Viruses tend to be avoided on a precautionary basis where any products are destined for human use.

Methods of direct transgenesis generally require that the presence of the transgene is detected in whole animals. In mice, the inefficiency of techniques such as microinjection is not a major problem because it is straightforward to inject large numbers of embryos and analyze the pups. However, in species where maintenance costs are higher, keeping large numbers of animals gestating nontransgenic fetuses is a significant waste of resources. Ethical considerations also require that the number of animals used in experiments should always be kept to a minimum. Efforts have therefore been made to identify transgenic animals at an early stage. Embryos can be screened for the presence of a transgene prior to transfer. A portion of the embryo, such as a single blastomere, can be extracted and the transgene detected by polymerase chain reaction (PCR) amplification. To our knowledge this has not been attempted in rabbits, but work in other species indicates that it is labor intensive, can reduce embryo viability, and is subject to false positives from residual nonintegrated DNA. Alternatively, a gene encoding a nontoxic fluorescent protein can be coinjected or incorporated into the transgene construct to visually identify transgenic embryos. Expression of a nonintegrated reporter construct may, however, produce a false positive signal, and the presence of additional DNA may be undesirable. Some efforts have also been made to detect and analyze fetal cells or DNA in the maternal circulation, but with limited success. The most common practice is therefore to screen animals shortly after birth by either PCR or Southern hybridization, using small samples taken from blood, tail, or ear tips.

The search for increased efficiency and a means of engineering precise genetic modifications in species other than mice lay behind the development of cell-mediated methods.

B. Cell-Mediated Genetic Modification

Genetic manipulation and analysis of cells in culture followed by conversion into whole animals is termed cell-mediated genetic modification. This includes incorporation of modified ES cells into a developing embryo to generate a chimeric animal and generation of cloned animals by nuclear transfer from modified cultured somatic cells.

Cell-mediated methods offer the advantage that genome analysis can be carried out before whole animals are generated. Cell clones with the desired genotype are identified and then converted to animals. This reduces the number of nontransgenic animals in gestation and also enables more sophisticated forms of genetic modification, notably gene targeting by homologous recombination. ES cells have been a mainstay of genetic modification in mice for almost three decades and were an obvious choice in the search for a method of cell-mediated transgenesis in rabbits and many other domestic animals. Several groups have described rabbit ES-like cells (Fang et al., 2006; Wang et al., 2007; Honda et al., 2008; Intawicha et al., 2009), but to date ours is the only group to demonstrate functionally pluripotent rabbit ES cells capable of forming a chimeric animal (Zakhartchenko et al., 2011). However, this is still not a practical means of generating genetically modified rabbits, because germ line transmission has not yet been achieved.

The difficulty of obtaining ES cells from domestic animals led to the development of nuclear transfer from cultured somatic cells as a functional alternative (Schnieke et al., 1997; McCreath et al., 2000). This was established in sheep, but rapidly extended to other domestic animals (reviewed in Niemann et al., 2011; see also Chapter 14). However, for reasons that remain unclear, rabbit nuclear transfer is still a challenge. Cloned rabbits have been produced from a number of cultured cell types, but the efficiency is consistently very low. Although a large proportion (up to 70−90%) of reconstructed embryos develop to blastocyst (Meng et al., 2009), few give rise to live offspring and fewer still survive to maturity. Collected published data show that of 31 live-born cloned rabbits, only 9 reached sexual maturity (Chesne et al., 2002; Challah-Jacques et al., 2003; Fang et al., 2006; Yang et al., 2007; Li et al., 2009; Meng et al., 2009).

There is evidence from mice that ES cells are more successful nuclear donors than other cell types in the proportion of viable offspring produced (Rideout et al., 2001). Together with our collaborator Eckhard Wolf (LMU, Munich), we carried out nuclear transfer with rabbit ES cells and another potentially good donor cell type, mesenchymal stem cells (MSCs). We performed almost 2000 embryo reconstructions using six independently derived rabbit ES cell lines, but obtained no ES-derived offspring (Zakhartchenko et al., 2011). We can thus state with reasonable confidence that ES cell usage is not advantageous for nuclear transfer in rabbits. MSCs were more successful, with a total of 18 live offspring born, including 8 from genetically modified cells (Zakhartchenko et al., 2011). However, our MSC-derived rabbits did not survive to adulthood (most lived about 1−2 weeks and 1−3 months), findings that accord with those of other researchers. In contrast, we have considerable success using MSCs in pig nuclear transfer (Flisikowska et al., 2012; Leuchs et al., 2012; Kurome et al., 2013), so further investigation into rabbit MSCs would seem worthwhile. Identifying the factors responsible for postpartum viability of nuclear transfer rabbits is probably key to further advances. It is noteworthy that the greatest number of cloned rabbits surviving to adulthood (eight) were derived from fresh, noncultured cells (Meng et al., 2009), indicating the damaging effects of prolonged culture.

So far this overview would appear to be a catalog of difficulties, but exciting new technical developments have now transformed the prospects for rabbit genetic modification. These are outlined in Section 10.II.C.

C. Genome Editing

Genome editing can be defined as the use of synthetic endonucleases to generate mutations or catalyze homologous recombination with exogenous DNA at a unique site in the genome (see Chapter 8). Four types of endonucleases are available: meganucleases (Arnould et al., 2011; Stoddard, 2011) and zinc finger nucleases (ZFNs; reviewed in Urnov et al., 2010) have been in use the longest. Transcriptional activator-like effector nucleases (TALENs) (see review by Mussolino and Cathomen, 2012) and RNA-guided endonucleases, particularly the

CRISPR/Cas system represent innovative improvements in this technology (see review by Mussolino and Cathomen, 2013).

Meganucleases, also called homing endonucleases, are rare-cutting specific endonucleases found in bacteriophages, bacteria, and various eukaryotes (Belfort and Roberts, 1997; Chevalier and Stoddard, 2001). They are characterized by a DNA recognition sequence substantially longer (12–40 bp) than typical restriction enzymes used in recombinant DNA methods. For example, the *I-SceI* meganuclease recognizes an 18 bp sequence. Meganucleases are the most specific of all naturally occurring endonucleases, but it is not easy to modify their DNA sequence specificity. Efforts have been made to reengineer the DNA recognition site (Seligman et al., 2002; Sussman et al., 2004; Ramos et al., 2006) or to generate chimeric meganucleases by fusing protein domains from different enzymes (Arnould et al., 2006; Smith et al., 2006), but the process is difficult.

ZFNs and TALENs have the great advantage that DNA binding domains can be customized to recognize a relatively long predetermined DNA sequence, 18–24 bp in ZFNs and 30–40 bp in TALENs. ZFNs and TALENs are both used as pairs. Cleavage at the target sequence requires binding of two monomers to adjacent half-sites on opposite DNA strands with correct orientation and spacing to form a functional *FokI* dimer (Figure 10.1A and B). The DNA binding domain of each ZFN monomer is composed of three to four zinc finger motifs, where each zinc finger recognizes a specific three-base DNA sequence (Figure 10.1A). The DNA binding domain of TALENs consists of repeated units of two amino acid residues (repeat variable diresidue, RVD), where one RVD recognizes one base in the DNA strand. Repeat units of 15–20 RVDs in each TALEN monomer are typical, allowing TALENs to recognize a 30–40 bp site, providing a high level of specificity (Figure 10.1B).

The most important difference between ZFNs and TALENs from the experimenter's point of view is the ease of production. Generating effective, highly specific ZFNs can be difficult for an average research group. The choice of DNA target site is restricted and the design, testing and optimization of a ZFN is not straightforward. Off-target DNA cleavage is a serious issue, because it can cause cytotoxicity and unintended alterations in the genome (DeFrancesco, 2011). TALENs are considerably easier for nonspecialists to make and can be engineered to recognize longer DNA sequences, reducing the likelihood of off-target cleavage.

The most recent tool offers yet further simplification. The type II CRISPR/Cas system (clustered regularly interspersed short palindromic repeats/CRISPR-associated) was originally discovered in *Streptococcus pyogenes* as a means of recognizing and cleaving foreign nucleic acids (Ishino et al., 1987; Barrangou et al., 2007). The endonuclease Cas9 forms a complex with two RNA transcripts, CRISPR RNA (crRNA) and *trans*-activating CRISPR RNA (tracrRNA) that is able to locate and cleave DNA at sites determined by base-pair complementary between the crRNA and the target DNA (protospacer) and a conserved sequence called the protospacer adjacent motif (Sapranauskas et al., 2011; Jinek et al., 2012). In the native type II CRISPR system, crRNA and transcrRNA are transcribed as

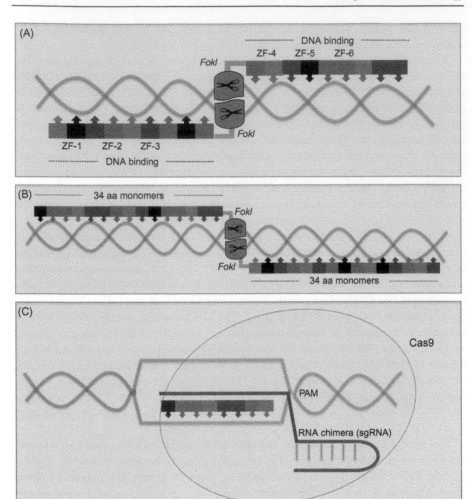

Figure 10.1 Basic structure of synthetic endonucleases: (A) ZFNs, (B) TALENs, and (C) RNA-guided Cas9.

two distinct RNA transcripts, but both can be fused to a single synthetic RNA (sgRNA) transcript that can very easily be modified to recognize different DNA sequences, hence the name RNA-guided endonucleases, or RNA-guided Cas9 (Figure 10.1C).

The CRISPR/Cas system has the advantage that construction involves only two steps: (i) identification of a target site and (ii) insertion of a target sequence (crRNA) into a guide RNA expression vector containing the tracrRNA, a suitable promoter, and all elements necessary for processing in eukaryotic cells. However, the target sequence is necessarily short (20 nucleotides) and off-target cleavage has already been observed in human cells (Fu et al., 2013). The same group described potential strategies for reducing off-target effects, including generation of more stable gRNA-genomic DNA hybrids by choosing target sites with high GC content or reducing the concentration of gRNA and Cas9 nuclease expressed in cells.

Genome editing has been rapidly adopted in many areas, and technical advances are being made very quickly. The most significant advantage for rabbits and many other species is that genome editing can be carried out directly in fertilized oocytes with very high efficiency. More interestingly, ZFN-mediated homologous recombination with an exogenous DNA vector has now been achieved in a number of species including rabbits (Flisikowska et al., 2011).

In short, genome editing is a major step forward for rabbits, as it allows a considerably greater range of modifications than direct transgenesis and circumvents the technical obstacles of cell-mediated methods.

III. Materials and Methods

This section provides practical information regarding methods of producing genetically modified rabbits, focusing on transposon-based transgene addition, nuclear transfer, and genome editing. All proposed animal experiments should be submitted for approval by the regulatory authorities for each country or region, and approved procedures performed in properly accredited institutions by trained personnel according to ethical guidelines.

A. Rabbit Breeds and Basic Reproduction

More than 80 different domestic rabbit breeds are recognized by the British Rabbit Council (http://www.thebrc.org/), all of which originate from the European rabbit (*Oryctolagus cuniculus*). The most common breeds used in research are New Zealand White and ZIKA hybrid rabbits, but local breeds are often used, for example, Japanese White rabbits in Japan and China.

The sexual development of rabbit varies with breed, sex, nutrition, and season of birth. Does of small- and medium-size breeds become sexually mature at $4-5$ months and larger breeds at $8-9$ months. Males reach sexual maturity one month later than females.

Rabbit reproduction has some unusual features compared to other mammals. Females do not have an estrus cycle, but are induced ovulators. Females show a $4-6$-day cycle of sexual receptivity followed by $1-2$ days nonreceptivity. Ovulation and formation of a functional *corpus luteum* are induced by mating, but also by contact with other females or cervical stimulation. Oocytes are released $9-13$ h after copulation (Hagen, 1974). Approximately 80% of matings result in pregnancy. To facilitate breeding activity the light cycle in rabbit housing is changed from a 12:12 light-dark cycle to a 16:8 light-dark cycle.

Rabbits are also distinct in that the preimplantation embryo has a thick mucin coat, termed the mucolemma, which is important for maintaining embryo viability *in vivo* (Figure 10.2). This is secreted by the epithelial cells of the tubal isthmus in response to estrogen. As the embryo passes down the reproductive tract and enters the uterus before implantation (day 6—7 post coitus, p.c.) the mucolemma thins and the zona pellucida disappears. Gestation lasts for 30—31 days.

B.   Pronuclear Microinjection and Transposon-Mediated Transgenesis

The basic protocol for pronuclear microinjection provided here applies to both classical DNA microinjection and to more efficient modern systems such as transposon-mediated transgenesis and genome editing (see Section 10.III.D).

Most transposon systems are binary systems, comprising a transposon donor plasmid containing the gene of interest flanked by ITRs, and an expression vector encoding the transposase helper protein. Both can be injected as DNA (circular plasmids) into the pronucleus of fertilized oocytes. However, to avoid integration of the expression plasmid into the genome, it is often preferable to inject the transposase component as mRNA.

Here we provide a protocol for a new hyperactive variant of the SB transposon, using SB100X transposase. A similar protocol can be used with other systems.

1.   Preparation of RNase-Free Genetic Material

RNases are very stable and active enzymes, so particular care should be taken to avoid introducing RNases into a sample.

To recover high purity, RNase-free material, we recommend performing phenol/chloroform extraction for transposon donor plasmid or template DNA for *in vitro* RNA synthesis and purification with Megaclear kit (Life Technologies) for mRNA transposase. Purified mRNA and DNA for microinjection should be then diluted in RNase-free injection buffer (0.1 mM EDTA, pH 8.0) to a concentration of 50 ng/μL, aliquoted in 5—10 μL samples, and stored at −80°C until use.

2.   In Vitro *Production of Transposase mRNA*

In vitro synthesis of mRNA requires the presence of a T7, T3, or SP6 bacteriophage polymerase promoter located upstream of the nuclease coding region. Modern cloning plasmids usually contain promoters for several different polymerases.

For *in vitro* transcription, RNase-free plasmid templates (see Section 10.III.B.1) should be linearized with a unique restriction enzyme, precisely downstream of the coding region's STOP codon. This allows the polymerase to run off the template generating an RNA transcript with a defined end. A poly(A) tail is then added to the 3′ end.

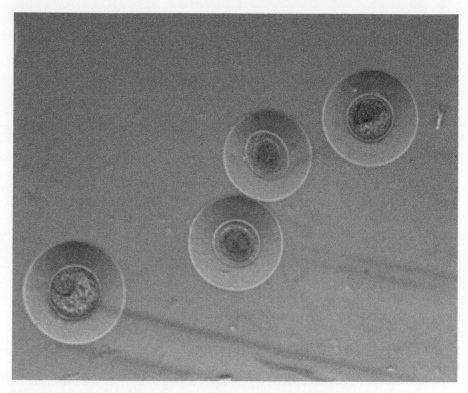

Figure 10.2 Rabbit blastocyst stage embryos 3.5 days postinsemination. The mucolemma is visible as the thick outer layer. The zona pelucida is visible as a thinner, more refractile layer directly surrounding the embryo.

Several kits are available for RNA production. Our laboratory uses the mMessage mMachine T7 for producing capped RNA, MEGAshortscript T7 for uncapped RNA, and poly(A) tailing kit, all from Life Technologies.

SB100X transposase mRNA can be synthesized from the expression plasmid pCMV(CAT)T7-SB100X, which contains the T7 promoter and is available from Addgene (http://www.addgene.org/34879).

3. Preparation of Microinjection Solution

Transposase mRNA and transposon vector DNA are thawed shortly before preparation of injection mixture. Various dilutions of the DNA and mRNA 50 ng/μL stock solutions described in Section 10.III.B.1 are used to identify the most suitable working concentration for microinjection, as described in Section 10.III.B.5. Once this has been established, 5–10 μL aliquots of ready to use microinjection mixture should be prepared and stored at −80°C until use.

4. Generation and Collection of Fertilized Oocytes

Timing of hormone priming, oocyte collection, and embryo transfer for microinjection is shown in the upper part of Figure 10.3.

Superovulation in sexually mature female rabbits (approximately 3.0 kg and 5−6 months old) can be induced by subcutaneous injection of gonadotropins, such as pregnant mare serum gonadotropin (PMSG) and human chorionic gonadotropin (HCG). On day 1, females used to provide fertilized oocyte are injected subcutaneously with 100 IU PMSG. On day 4, each female is fertilized by artificial insemination or by placing in a cage with male for an hour. Successful mating can be determined by observation of copulation or the presence of a gelatinous copulation plug in the vagina. Ovulation is induced directly p.c. by intravenous injection of 200 IU HCG.

On day 5 (19−20 h p.c.), donor females are humanely killed. Usually two donor females are sacrificed to obtain sufficient fertilized oocytes for transfer into one recipient. The abdomen is opened and the uterus, oviducts, and ovaries removed and placed onto a warming plate. The oviducts are separated and flushed using an 18-gauge needle connected to a 2-mL syringe containing 2 mL manipulation medium (Dulbecco's phosphate buffered saline (D-PBS), 20% fetal calf serum (FCS), 100 μg/mL penicillin/streptomycin (pen/strep), 10 mM HEPES). The needle is inserted into the fimbrial end of the oviduct, with the other end in a 60-mm round-bottom glass dish. Fertilized embryos are flushed into the dish by slow release of flushing medium from the syringe. Embryos are picked up and transferred to another 60-mm round-bottom dish containing fresh medium, and examined microscopically for the presence of two polar bodies as evidence of fertilization. At this stage, any embryos showing morphological abnormalities, such as cytoplasm defragmentation, are discarded. Immediately after collection, embryos are pooled into groups of approximately 30 embryos (or no more than can be injected within 30−45 min) and transferred to a drop of manipulation medium overlaid with embryo-quality mineral oil. If necessary, embryos can be stored in a humidified atmosphere at 37°C in 5% CO_2 until microinjection.

5. Microinjection of Transposon Vector and Transposase mRNA

The equipment and microinjection procedure are similar to that used to produce transgenic mice. Rabbit fertilized oocytes are slightly larger than mouse, have a thicker zona pellucida, and at 20 h post-hCG may have a very thin mucin coat. The microinjection and holding pipettes can be purchased readymade, or pulled from thin-walled borosilicate glass pipettes using a micropipette puller. The ends of holding pipettes are blunted with a microforge.

It might be advisable to perform initial microinjection experiments to optimize technical parameters that could affect embryo viability, for example, DNA and RNA purity, needle quality, and most importantly nucleic acid concentration in the microinjection solution. We recommend testing different concentrations, ranging between 3 and 9 ng/μL for mRNA and 5 and 15 ng/μL for DNA. A reasonable indication of viability can be gained by culturing microinjected embryos *in vitro* (see Section 10.III.B.6) and visually monitoring development to blastocyst stage.

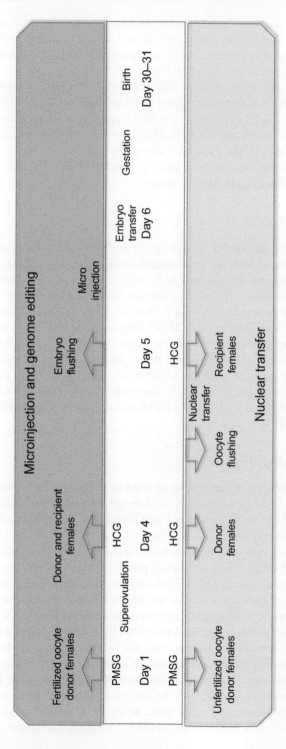

Figure 10.3 Timing of events for microinjection and genome editing (upper part) and nuclear transfer (lower part).

For microinjection, a drop of manipulation medium is added to a 100-mm Petri dish and overlaid with mineral oil. The microinjection pipette is loaded with 1 μL of DNA/RNA mixture diluted in injection buffer (see Section 10.III.B.1) to a desired concentration of transposon vector DNA and SB100X mRNA. Approximately 30 embryos are then transferred to the prepared drop. Microinjection is performed as two steps. One dose of the DNA/RNA mixture is injected into a pronucleus. Then, on withdrawing the needle, a second dose is injected into the cytoplasm to enable translation of transposase mRNA. Microinjection can be verified by observing slight swelling of the pronucleus, or disturbance in the cytoplasm caused by solution streaming from the capillary. After microinjection, embryos can be transferred directly into recipient females or cultured as described in the next section.

6. In Vitro *Embryo Culture*

Rabbit embryos are less susceptible to damage by *in vitro* culture than some other mammalian species, but if they are to complete gestation it is beneficial to minimize their time *in vitro* and usually best to transfer embryos to recipients directly after microinjection. However, if desired, embryos can be cultured up to the four- to eight-cell stage before transfer. A simple medium such as Krebs-Ringer bicarbonate buffer (Sigma) with 20% FCS, or M199 (Sigma) with 10% FCS can support rabbit embryo development up to blastocyst stage. However, for higher quality embryos and more successful establishment of pregnancy, embryos should be cultured in an upgraded medium such as Menezo B2 medium (INRA) containing 2.5% FCS. *In vitro* culture is carried out in four-well dishes, each well containing 250 μL medium (50 embryos/well) overlaid with mineral oil in a humidified atmosphere at 37°C in 5% CO_2.

7. *Laparoscopic Embryo Transfer*

Pseudopregnancy is induced in females to be used as recipients by hormonal synchronization. On day 1, recipient females are injected intravenously with 20 IU of PMSG and with 100 IU of HCG 72 h after PMSG injection. Embryo transfer is typically performed at two-cell stage, 21−22 h later.

Injected embryos are collected in a round-bottom glass dish containing D-PBS with 10% FCS, 10 mM HEPES, and 100 μg/mL pen/strep, then loaded into a capillary tube connected through a venous catheter (60−80 cm long) to a 2-mL syringe. To visualize the transfer of embryos, embryos are drawn up into the capillary in four groups in small amounts of medium each separated by 2 mm air bubbles. We usually transfer a total of 30−40 embryos into each recipient (15−20 embryos per oviduct).

Embryo transfer is carried out laparoscopically by the method of Besenfelder and Brem (1993). The area for surgery should be carefully disinfected and aseptic techniques used during embryo transfer. The pseudopregnant recipient female is anesthetised, the abdomen is shaved, and the animal placed in a dorsal recumbent position on a moveable operating table head down in a vertical position. This ensures that the stomach and intestines are shifted toward the diaphragm, exposing the fallopian tubes. A single 1 cm incision is made in the region of the navel and an endoscope trocar introduced through the abdominal wall. After the trocar is removed, the abdomen is inflated with air using an inflation bulb. The endoscope is then inserted, and the fallopian tubes should be visible as a downward pointing loop located between each ovary and the uterus. The number of visible *corpora lutea* is estimated as an indication of the success of hormone synchronization. Next, the status and position of the infundibulum and ampulla are determined. A second small incision in the abdomen is made 2−3 cm from the infundibulum as an entry point for the capillary carrying the embryos. The capillary is then inserted into the opening and guided by endoscopic observation. Premorula stage embryos are placed through the natural opening of the infundibulum into the ampulla (the upper part of the oviduct), taking care not to injure any tissue. Medium containing the embryos is injected by pressure using the syringe and progress monitored by movement of the air bubbles. Later-stage embryos are injected into the upper part of the uterus by piercing with the capillary. Figure 10.4 shows embryo transfer in process. Following transfer, the incision is closed with a Michel clamp.

Following surgery, recipient animals can be transferred directly to a breeding cage. The recipient female can be examined by palpation 14 days after embryo transfer to detect pregnancy.

8. Gestation and Parturition

Rabbit gestation lasts between 29 and 31 days. Approximately 3−5 days before parturition the doe begins to build a nest. Shortly before this stage a nest box and bedding materials should be placed in the cage. The female should also be monitored daily for nest building. When a nest becomes apparent, consisting of a large mound of hairs, parturition occurs a few hours later, usually at night. In normal breeding, a litter is usually 6−7 offspring (known as kits), but numbers are likely to be lower in transgenic experiments. Gestation that continues for longer than 32 days often results in stillborn offspring. So if pregnancy has been confirmed but spontaneous parturition fails to occur by day 32, kits should be delivered by cesarean section. Where there are few offspring (1−2 kits), survival is improved by combining two small litters to one mother. Fostered offspring are readily accepted by foster mothers if they are placed within a litter of the same age and provided with the mother's scent by rubbing with her nest material. It may thus be worthwhile setting up a normal mating in parallel with embryo transfer to provide a matched foster mother should the need arise.

9. Analysis of Offspring

Samples can be collected as soon as the litter is well established, approximately 1−2 weeks after birth. To verify the presence of the transgene, DNA for PCR can be isolated from ear biopsy. In animals identified as positive, transgene copy number can then be estimated by quantitative PCR or Southern blot analysis.

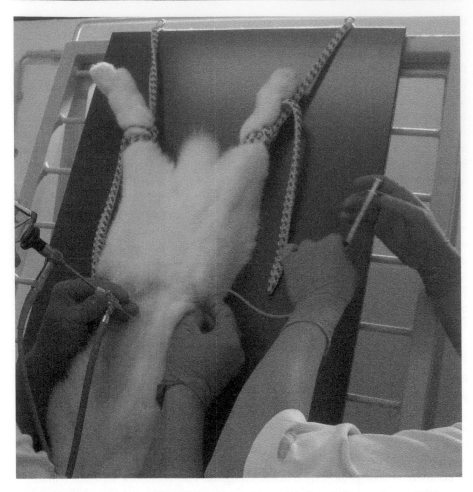

Figure 10.4 Laparoscopic embryo transfer, showing endoscope guidance of the capillary bearing the embryos, connected by a catheter to a syringe.

Where appropriate, the presence of transgenic proteins in blood can be tested by collection of samples from the marginal ear vein and used for Western blot analysis or enzyme-linked immunosorbent assay. Founder rabbits identified as transgenic are normally bred to verify transmission through the germ line and to provide F1 rabbits for more extensive analysis.

Classical DNA microinjection generally results in multiple copy transgene integration at a single locus, so approximately half of the offspring of a founder mated with a wild-type animal are expected to be transgenic. However, transposons (and also viral vectors) tend to integrate as single copies at multiple loci, so more than half of the offspring in the F1 generation can be transgenic as individual loci segregate. Transgene integration may also be delayed to the two-cell or later stage. This can result in germ line mosaicism, which reduces the rate of transmission and the proportion of transgenic F1 offspring.

C. Nuclear Transfer

Although somatic cell nuclear transfer in rabbits remains problematic, our group has obtained a high proportion of live births using MSCs as nuclear donors. We have also demonstrated that transfected MSCs support development *in vitro* and *in vivo* with similar efficiency as nonmanipulated MSCs, suggesting they are suitable for gene targeting experiments (Zakhartchenko et al., 2011). This approach is thus worthy of further investigation. Here we describe methods for rabbit MSC isolation, MSC transfection, and nuclear transfer. But other cell types such as fetal fibroblasts can also be used (Yang et al., 2007; Li et al., 2009).

At the outset we stress that stringent efforts should be made to minimize the total time cells destined to be nuclear donors spend in culture. Fresh preparations should be cryopreserved at an early stage, and genetic manipulations carried out using cells at the earliest possible cell passage. Replicate samples of genetically modified cell clones identified by primary screening should be cryopreserved at an early stage. Early passage samples should be retained as a frozen stock for nuclear transfer, while a replicate of each cell sample is expanded to provide material for analysis.

1. Isolation of Rabbit MSCs from Bone Marrow

Bone marrow MSCs can be isolated from the leg bones of juvenile or adult rabbits. Before isolation, the bones and all surgical instruments should be sterilized with 80% ethanol. All media should be prewarmed to 37°C. Media compositions are listed below.

MSC washing medium	Advanced DMEM, 10% FCS, 100 µg/mL pen/strep, 100 µg/mL amphotericin B.
MSC medium	Advanced DMEM, 10% FCS, 5 ng/mL recombinant human fibroblast growth factor-2, 2 mM glutamine, 0.1 mM NEAA, 0.1 mM β-ME. Optional: supplemented with antibiotics (100 µg/mL pen/strep), (100 µg/mL amphotericin B).
Freezing medium	40% MSC medium, 10% DMSO, 50% FCS.

Bones are opened with a saw, bone marrow flushed out with MSC washing medium into a 100-mm petri dish. The aspirate is filtered through a 100 µm cell strainer and centrifuged at $600 \times g$ for 10 min. The supernatant is aspirated off, the cell pellet washed with 30 mL MSC washing medium and again centrifuged at $600 \times g$ for 10 min. Centrifugation and washing is then repeated. Cells are then resuspended in 20 mL MSC medium supplemented with antibiotics, plated into three or four standard T-75 tissue culture flasks and cultured overnight. Flasks are not gelatin coated, because MSCs are selected on the basis of their adherence.

On the next day, all nonadherent cells are removed by washing with 5 mL D-PBS and fresh MSC medium with antibiotics is added. Washing and exchange of culture medium is repeated every day. After 3 days, antibiotics are removed from the MSC medium.

Cultures should be observed daily to detect potential microbial contamination. If contamination does occur, all affected flasks should be discarded. If not, the medium is replaced every day until cells reach 70–80% confluence. From that point, the medium should not be changed for 2 days to provide a medium sample for a mycoplasm test. Mycoplasm is a potential risk with any primary cell isolation and could severely affect the success of nuclear transfer. Several testing kits are commercially available, for example, LookOut Mycoplasma PCR Detection kit (Sigma). If mycoplasm contamination is detected, all cells should be discarded and new MSCs isolated. To continue culture, the 2-day-old medium is changed and cells allowed to reach confluence, then frozen stocks prepared for future use.

To prepare cryopreserved stocks, confluent layers of MSCs are detached by washing with D-PBS and incubation with Accutase (3 mL/T-75 flask) for 5 min at 37°C. Three milliliters MSC medium are then added to each flask, the cell suspension collected and pooled into one 50-mL tube. Cells are pelleted by centrifugation for 5 min at $340 \times g$, supernatant aspirated off, and cells resuspended in 15 mL freezing medium and aliquoted into cryovials. Cryovials are placed immediately in freezing containers, transferred to $-80°C$ for 24 h, then stored under liquid nitrogen.

2. Characterization of Rabbit MSCs by Differentiation

Before commencing work with a new MSC preparation it is worth performing some characterization to confirm their identity and exclude the possibility of large-scale contamination with other morphologically similar plastic-adherent cell types such as fibroblasts. Two types of criteria are commonly used to define MSCs. The most precise is the presence of cell surface markers. While human MSC-specific markers are reasonably well defined, there is however no definitive list available for rabbits (Boxall and Jones, 2012). Thus, it is more practical to check the identity of rabbit MSCs by testing their characteristic multipotent differentiation to cells of the osteogenic, adipogenic, and chondrogenic lineages.

a. Osteogenic Differentiation

MSCs should be seeded into six-well plates at a density of 3×10^4/well in MSC medium. When cells reach ~70% confluence, the MSC medium in half the wells is replaced with either osteogenic medium (MSC medium supplemented with 100 nM dexamethasone, 10 mM β-glycerophosphate, and 50 μg/mL L-ascorbate). The other wells serve as a control and are kept in standard MSC medium. MSCs are cultured for 14 days in osteogenic medium, with medium changed every other day. After 14 days the medium is aspirated off, cells washed three times with distilled water, fixed with 10% formalin (1 mL/well) at room temperature (RT) for 15 min, then washed three times with distilled water. Cells are stained by adding

5% silver nitrate solution (1 mL/well) and exposing to ultraviolet light for 30 min in the dark. After UV irradiation, the silver nitrate is removed and cells washed three times with distilled water. To visualize calcium salts, the cells are stained with 5% sodium thiosulfate solution for 5 min at RT. Calcium deposits should stain brown or black.

b. Adipogenic Differentiation

MSCs should be seeded into six-well plates at a density of 2×10^5/well in MSC medium. At 70% confluence, the medium in half the wells is replaced with adipogenic medium (MSC medium supplemented with 0.5 μg/mL isobutylmethyl-xanthine, 10 μg/mL ITS + 1 liquid medium supplement, 100 μM indomethacine, and 1 mM dexamethasone) and cultured for 21 days with medium changed every other day. After 21 days, cells are washed and fixed as above. Cells are stained by adding 1 mL/well Oil Red O solution and incubation for 30 min at RT. After incubation, cells are washed three times with dH$_2$O. Lipid droplets stain red.

c. Chondrogenic Differentiation

Samples of 5×10^5 MSCs are transferred to 15-mL falcon tubes and pelleted by centrifugation at $340 \times g$ for 5 min, the supernatant is aspirated off, cells washed with 500 μL D-PBS, and centrifuged at $3400 \times g$ for 3 min. The supernatant is aspirated off and 500 μL chondrogenic medium (MSC medium without FGF, supplemented with 10 ng/mL transforming growth factor β-1) added to half of the tubes. This medium is added very slowly, taking care not to disrupt the cell pellet. MSC medium is added to the other tubes as a control. Chondrogenic differentiation is induced by culture as a pellet for 21 days, with medium changed every other day. After 21 days, the pellet is carefully washed with D-PBS and centrifuged at $3400 \times g$ for 10 min. The supernatant is aspirated, and the cell pellet air-dried. If necessary, the dried cell pellet can be stored at $-20°C$ until further processing. The cell pellet is then digested overnight at $60°C$ with 200 μL 1 mg/mL papain. On the next day, the glycosaminoglycan (GAG) content is determined using a dimethylmethylene blue assay (Farndale et al., 1986), with chondroitin-6 sulfate from shark cartilage as a standard curve. The GAG quantity is normalized to the cell number in the pellet by measuring total DNA content using a PicoGreen assay according to manufacture's protocol (e.g., Quanti−iT PicoGreen dsDNA assay kit from Invitrogen).

3. Transfection and Drug Selection of Rabbit MSCs

Several methods can be used to transfect rabbit MSCs, but in our experience electroporation provides the highest efficiency, best cell viability, and does not seem to affect MSC multipotency or induce differentiation. Electroporation conditions should first be optimized for the cells used; precise details will vary with different electroporation systems. A straightforward means of identifying suitable conditions is to check transient transfection using an expression plasmid carrying a visual reporter gene (e.g., green fluorescence protein). Conditions are often a compromise between the proportion of cells transfected and overall cell viability. Results can vary with different DNA constructs and preparations, so when using a new DNA it is worth bracketing the optimized electroporation conditions.

Here we describe key parameters established for rabbit MSCs using the Eppendorf Multiporator system, based on protocols recommended by the manufacturer. About 48 h before transfection, MSCs are thawed into a T-75 or a T-150 tissue culture flask, depending on the number of cells. Once confluent, MSCs are harvested with Accutase, counted, pelleted by centrifugation at $340 \times g$ for 5 min, and resuspended in 400 µl electroporation buffer at 1×10^6 cells/mL. DNA (5−10 µg) is added, the mixture transferred to an electroporation cuvette (4 mm gap width), and electroporation performed with one pulse at 1200 V for 85 µs. After the electric pulse, cells are allowed to recover at RT for 10 min, then transferred into a T-75 flask. Where the intention is derive stable transfectants, cells are transferred 48 h after electroporation into four or five 15-cm plates in MSC medium with the appropriate selective drug and medium changed every other day. After 12−14 days, well-separated single cell clones can be isolated and expanded for cryopreservation and transgene analysis.

It is always necessary to determine the most appropriate concentration of drugs, such as G418 or blasticidin, needed to select for resistant cell clones. This usually varies between different MSC isolates, so a "killing curve" should be generated for each preparation. Typically, this is performed in a 12-well plate with each well containing medium with a different drug concentration. We usually seed cells directly in a selection medium at a density of 4×10^4/well. A usable drug concentration is that which causes substantial cell death within 3−4 days and kills all cells after 10−14 days. When applying this information to transfected cells it is good practice to use three concentrations of drug, one slightly higher and one slightly lower than that estimated by the killing curve.

4. Oocyte Collection and Enucleation, Nuclear Donor Cells, Embryo Reconstruction

Timing of hormone priming, oocyte collection, and embryo transfer for nuclear transfer is shown in the lower part of Figure 10.3.

Unfertilized oocytes to be used as recipient cytoplasts are obtained by superovulation (as in Section 10.III.B.4, but omitting the insemination step). Fifteen to sixteen hours after HCG injection, mature oocytes are flushed from the oviducts in warm D-PBS supplemented with 4 mg/mL bovine serum albumin. Cumulus cells are removed by treatment with 5 mg/mL hyaluronidase in M199 medium, 10% FCS for 15 min at 38.5°C followed by gentle pipetting with a small-bore pipette. Denuded oocytes are treated with 0.6 µg/mL demecolcine between 40 min and 2 h. The resulting metaphase II protrusion with little underlying cytoplasm is removed in M199 supplemented with 7.5 µg/mL cytochalasin B and 0.6 µg/mL demecolcine using an enucleation pipette. Enucleated oocytes are kept in M199 and later used as recipient cytoplasts.

Cells to be used as nuclear donors are prepared by culture in medium with reduced serum (0.5% FCS) for 3—5 days before nuclear transfer. Adherent cells are harvested by Accutase digestion, washed three times, and suspended in M199 medium supplemented with 10% FCS. An individual nuclear donor cell is introduced under the zona pellucida of an enucleated oocyte in M199. Karyoplast—cytoplast complexes (KCC) are manually aligned in a fusion chamber consisting of two wire electrodes 200 µm apart, overlaid with fusion medium (280 mM mannitol, 0.1 mM MgSO$_4$, 0.5 mM EDTA), and then fused with an Eppendorf Multiporator using double direct current of 1.95 kV/cm for 25 µs. After 30 min incubation in M199, fused KCC are activated by the same electric pulses as for fusion, then immediately incubated in M199 supplemented with 1.9 mM 6-dimethylaminopurine and 7.5 µg/mL cytochalasin B (activation medium). Thirty minutes later, fused embryos are treated again with the same electric pulses and cultured in activation medium for another 30 min. After activation, reconstructed embryos are washed and then cultured in Menezo B2 medium containing 2.5% FCS in a humidified atmosphere of 5% CO$_2$ in air at 38.5°C. Embryos destined for transfer to recipients are cultured overnight, while those used to assess *in vitro* development are cultured up to 5 days, with embryo stages scored by visual observation. Embryo culture conditions are as described in Section 10.III.B.6.

Details of embryo transfer, parturition, and analysis of offspring are as described previously in Sections 10.III.B.7—10.III.B.9, with the difference that the hormone treatment of donor and recipient females is asynchronous. Recipient females are injected with HCG 24 h later than the donor females.

D. Genome Editing

1. Design and Assembly of Synthetic Endonucleases

a. ZFNs and TALENs

ZFNs are available commercially from Sigma Life Science, who offer a custom ZFN service (CompoZr ZFN technology; http://www.sigmaaldrich.com/life-science/zinc-finger-nuclease-technology.html) of validated ZFN pairs custom-designed and assembled to edit a target gene. This does however come at a substantial cost. There is also the Zinc Finger Consortium (ZiFiT, http://zifit.partners.org), a group of academic experts who provide open source software and protocols for designing and synthesizing ZFNs.

TALENs are also available commercially, but are easily generated by experienced molecular biologists. TALENs can be designed for any chosen sequence using online tools, such as ZiFiT (http://zifit.partners.org/ZiFiT/), TAL Effector Nucleotide Targeter 2.0 (https://tale-nt.cac.cornell.edu/), The Hornung Laboratory TALEN Resources (http://www.hornunglab.de/TALEN.html), or the Zhang lab online resource (http://www.genome-engineering.org/). Several straightforward assembly platforms have been developed to generate DNA constructs that encode the customized TALE repeat arrays, for example, Golden Gate TALEN 2.0

(Cermak et al., 2011), Zhang Lab TALE Toolbox (Zhang et al., 2011), FLASH assembly kit (Reyon et al., 2012). The available platforms differ in the molecular techniques used. For example, Golden Gate cloning involves the use of type II restriction nucleases, which cut outside their recognition site to generate nonpalindromic 4-bp single-stranded DNA overhangs (Cermak et al., 2011). FLASH assembly is another approach to TALEN synthesis based on the capture and manipulation of nucleic acids attached to magnetic beads (Reyon et al., 2012). Each of these methods enables modular assembly typically within 1 week or less (e.g., FLASH assembly). In most cases, detailed protocols can be obtained from author-supported online websites.

Most importantly for practical application, current evidence suggests that TALENs show higher specificity and thus less cytotoxicity than ZFNs (Mussolino et al., 2011).

2. RNA-Guided Endonucleases

Suitable target sites for RNA-guide endonucleases can be identified using online tools. The ZiFiT website and Zhang lab online resources mentioned above have separate sections for CRISPR technology. There are also other open databases that enable scientists to choose CRISPR targeting sites in a wide range of species (e.g., CRISPRdb at http://crispr.u-psud.fr/crispr/). All vectors for CRISPR technology including a human codon-optimized version of Cas9 (humanized Cas9) enzyme expression plasmid can be purchased from Addgene (http://www.addgene.org).

There are several methods of building specific sgRNA expression vectors. All involve synthesis of a pair of oligonucleotides based on the target site sequence, which are then annealed and cloned into the CRISPR array.

3. Gene Disruption and Targeted Sequence Replacement in Fertilized Oocytes

As described previously, genes can be inactivated by deletions or insertions created during repair of a DSB by nonhomologous end joining. Where this is the intention, mRNA encoding a sequence-specific endonuclease can be injected into the cytoplasm of a zygote. This will be translated to the endonuclease protein that, if provided with a nuclear localization signal, will be transported into the nucleus of the developing embryo. Where more sophisticated modifications are required, such as targeted sequence replacement, this is best achieved by homologous recombination with a gene targeting vector. This requires two steps, injection of endonuclease mRNA into the cytoplasm, and injection of the gene targeting vector DNA into a pronucleus. Gene targeting vectors used in this system are often termed "donor vectors."

General guidelines and tips for the design of targeting vectors and DNA/mRNA preparation are described next.

a. *In Vitro* Production of Endonuclease mRNA
Kits for *in vitro* production of mRNA are described in Section 10.III.B.2.

For TALENs and ZFNs, capped mRNA for each pair is synthesized from two separate expression plasmids. mRNA for RNA-guided Cas9 is also produced from two plasmids, one encoding the Cas9 enzyme and one the sgRNA. Only the Cas9 mRNA needs to be capped and polyadenylated.

b. Targeting Donor Vector Design

Gene targeting constructs are designed to undergo homologous recombination at a chosen locus, to effect precise addition, deletion, or replacement of a particular DNA sequence. Classical gene targeting vectors, as used in mouse ES cells and also in livestock somatic cells, typically contain two regions of homology flanking a selection cassette. The length of homologous regions can vary between 3 and 5 kb for mouse ES cells and between 10 and 15 kb for porcine somatic cells. Donor vectors for endonuclease-induced homologous recombination directly in fertilized oocytes differ in that no selection marker is required and total homology as short as 0.5−2 kb is effective. Two important factors should be considered when designing a donor vector:

i) The position of homologous regions flanking the site of genetic modification should correspond to the genomic region directly upstream and downstream of the DSB.
ii) The homologous region of the targeting vector should not include the endonuclease recognition sequence, to avoid recleavage of the vector by the endonuclease.

Recent results show that synthetic single-stranded oligonucleotides, usually 100−300 bases, can be used instead of donor vectors for certain types of modification (Radecke et al., 2010; Chen et al., 2011; Wefers et al., 2013). Oligonucleotide synthesis is now simple and cheap, so this provides a further option to streamline the process. Because the oligonucleotide sequence extends across the endonuclease recognition sequence, it is necessary to incorporate base exchanges to avoid enzyme recleavage of the locus after homologous recombination.

c. Microinjection of Targeting Vector DNA and Endonuclease mRNA

As mentioned earlier, gene inactivation requires only cytoplasmic injection of mRNAs coding for the endonuclease. For this, purified RNase-free mRNA is diluted in injection buffer (see Section 10.III.B.1) to a total concentration of 50 ng/µL, aliquoted in 5−10 µL samples (see Section 10.III.B.5) and stored at −80°C until use. A mixture of mRNAs is prepared, using a 1:1 molar ratio for each ZFN or TALEN monomer, or 2:1 for Cas9/sgRNA, and stored in aliquots at −80°C. Recovery of RNase-free material and preparation of the microinjection mixture are described in Sections 10.III.B.1 and 10.III.B.3.

For homologous recombination in oocytes, the mRNAs and linearized donor vector DNA are diluted to a desired concentration (see Section 10.III.B.5) in injection buffer and microinjected into the cytoplasm and pronucleus following the same two-step microinjection procedure described for transposons in Section 10.III.B.5. All subsequent steps are as described in Sections 10.III.B.6−10.III.B.9. Timing of hormone priming, oocyte collection, and embryo transfer for genome editing is shown in the upper part of Figure 10.3.

IV. Conclusion

Rabbits are important laboratory animals that are widely used in many areas of bio-medical research, including the production of antibodies and recombinant proteins. Their usefulness would be significantly enhanced by the ability to generate the range of genetic modifications possible in mice. Unfortunately, rabbits have proven to be a difficult species that has not yielded easily to many current techniques. Here we have outlined the key features of the most relevant technologies. Most notably the power and simplicity of TALENs seems set to transform the prospects for genetic modification in rabbits. We are thus confident that rabbits will provide powerful new resources for research and medicine.

References

Arnould, S., Chames, P., Perez, C., Lacroix, E., Duclert, A., Epinat, J.C., et al., 2006. Engineering of large numbers of highly specific homing endonucleases that induce recombination on novel DNA targets. J. Mol. Biol. 355, 443−458.

Arnould, S., Delenda, C., Grizot, S., Desseaux, C., Paques, F., Silva, G.H., et al., 2011. The I-CreI meganuclease and its engineered derivatives: applications from cell modification to gene therapy. Protein Eng. Des. Sel. 24, 27−31.

Attaran, M., Schneider, A., Grote, C., Zwiens, C., Flemming, P., Gratz, K.F., et al., 2004. Regional and transient ischemia/reperfusion injury in the liver improves therapeutic efficacy of allogeneic intraportal hepatocyte transplantation in low-density lipoprotein receptor deficient Watanabe rabbits. J. Hepatol. 41, 837−844.

Barrangou, R., Fremaux, C., Deveau, H., Richards, M., Boyaval, P., Moineau, S., et al., 2007. CRISPR provides acquired resistance against viruses in prokaryotes. Science 315, 1709−1712.

Belfort, M., Roberts, R.J., 1997. Homing endonucleases: keeping the house in order. Nucleic Acids Res. 25, 3379−3388.

Besenfelder, U., Brem, G., 1993. Laparoscopic embryo transfer in rabbits. J. Reprod. Fertil. 99, 53−56.

Besenfelder, U., Aigner, B., Müller, M., Brem, G., 1998. Generation and application of transgenic rabbits. In: Cid-Arregui, A., Garcia-Carrancá, A. (Eds.), Microinjection and Transgenesis: Strategies and Protocol. Springer-Verlag, Berlin, pp. 561−586.

Boxall, S.A., Jones, E., 2012. Markers for characterization of bone marrow multipotential stromal cellsStem Cells Int. 2012:975871. Available from: http://dx.doi.org/10.1155/2012/975871. Epub 2012 May 14.

Brackett, B.G., Killen, D.E., Peace, M.D., 1971. Cleavage of rabbit ova inseminated *in vitro* after removal of follicular cells and zonae pellucidae. Fertil. Steril. 22, 816−828.

Capecchi, M.R., 2005. Gene targeting in mice: functional analysis of the mammalian genome for the twenty-first century. Nat. Rev. Genet. 6, 507−512.

Carbery, I.D., Ji, D., Harrington, A., Brown, V., Weinstein, E.J., Liaw, L., et al., 2010. Targeted genome modification in mice using zinc-finger nucleases. Genetics 186, 451−459.

Carlson, D.F., Geurts, A.M., Garbe, J.R., Park, C.W., Rangel-Filho, A., O'Grady, S.M., et al., 2011. Efficient mammalian germline transgenesis by cis-enhanced Sleeping Beauty transposition. Transgenic Res. 20, 29–45.

Cermak, T., Doyle, E.L., Christian, M., Wang, L., Zhang, Y., Schmidt, C., et al., 2011. Efficient design and assembly of custom TALEN and other TAL effector-based constructs for DNA targeting. Nucleic Acids Res. 39, c82.

Challah-Jacques, M., Chesne, P., Renard, J.P., 2003. Production of cloned rabbits by somatic nuclear transfer. Cloning Stem Cells 5, 295–299.

Chen, F., Pruett-Miller, S.M., Huang, Y., Gjoka, M., Duda, K., tauton, J., et al., 2011. High-frequency genome editing using ssDNA oligonucleotides with zinc-finger nucleases. Nat. Methods 8, 753–757.

Chentoufi, A.A., Dasgupta, G., Christensen, N.D., Hu, J., Choudhury, Z.S., Azeem, A., et al., 2010. A novel HLA (HLA-A*0201) transgenic rabbit model for preclinical evaluation of human CD8 + T cell epitope-based vaccines against ocular herpes. J. Immunol. 184, 2561–2571.

Chesne, P., Adenot, P.G., Viglietta, C., Baratte, M., Boulanger, L., Renard, J.P., 2002. Cloned rabbits produced by nuclear transfer from adult somatic cells. Nat. Biotechnol. 20, 366–369.

Chevalier, B.S., Stoddard, B.L., 2001. Homing endonucleases: structural and functional insight into the catalysts of intron/intein mobility. Nucleic Acids Res. 29, 3757–3774.

Cutino-Moguel, T., Fassati, A., 2006. A phenotypic recessive, post-entry block in rabbit cells that results in aberrant trafficking of HIV-1. Traffic 7, 978–992.

DeFrancesco, L., 2011. Move over ZFNs. Nat. Biotechnol. 29, 681–684.

Duby, R.T., Cunniff, M.B., Belak, J.M., Balis, J.J., Robol, J.M., 1993. Effect of milking frequency on collection of milk from nursing New Zealand white rabbits. Anim. Biotechnol. 4, 31–42.

Fan, J., Watanabe, T., 2003. Transgenic rabbits as therapeutic protein bioreactors and human disease models. Pharmacol. Ther. 99, 261–282.

Fang, Z.F., Gai, H., Huang, Y.Z., Li, S.G., Chen, X.J., Shi, J.J., et al., 2006. Rabbit embryonic stem cell lines derived from fertilized, parthenogenetic or somatic cell nuclear transfer embryos. Exp. Cell Res. 312, 3669–3682.

Farndale, R.W., Buttle, D.J., Barrett, A.J., 1986. Improved quantitation and discrimination of sulphated glycosaminoglycans by use of dimethylmethylene blue. Biochim. Biophys. Acta 883, 173–177.

Fischer, B., Chavatte-Palmer, P., Viebahn, C., Navarrete Santos, A., Duranthon, V., 2012. Rabbit as a reproductive model for human health. Reproduction 144, 1–10.

Flisikowska, T., Thorey, I.S., Offner, S., Ros, F., Lifke, V., Zeitler, B., et al., 2011. Efficient immunoglobulin gene disruption and targeted replacement in rabbit using zinc finger nucleases. PLoS One 6, e21045.

Flisikowska, T., Merkl, C., Landmann, M., Eser, S., Rezaei, N., Cui, X., et al., 2012. A porcine model of familial adenomatous polyposis. Gastroenterology 143, 1173–1175, e1–e7.

Fu, Y., Foden, J.A., Khayter, C., Maeder, M.L., Reyon, D., Joung, J.K., et al., 2013. High-frequency off-target mutagenesis induced by CRISPR-Cas nucleases in human cells. Nat. Biotechnol. 31, 822–826.

Furukawa, H., Oshima, K., Tung, T., Cui, G., Laks, H., Sen, L., 2005. Liposome-mediated combinatorial cytokine gene therapy induces localized synergistic immunosuppression and promotes long-term survival of cardiac allografts. J. Immunol. 174, 6983−6992.

Geurts, A.M., Cost, G.J., Freyvert, Y., Zeitler, B., Miller, J.C., Choi, V.M., et al., 2009. Knockout rats via embryo microinjection of zinc-finger nucleases. Science 325, 433.

Gordon, J.W., Scangos, G.A., Plotkin, D.J., Barbosa, J.A., Ruddle, F.H., 1980. Genetic transformation of mouse embryos by microinjection of purified DNA. Proc. Natl. Acad. Sci. USA 77, 7380−7384.

Guilbault, C., Saeed, Z., Downey, G.P., Radzioch, D., 2007. Cystic fibrosis mouse models. Am. J. Respir. Cell Mol. Biol. 36, 1−7.

Hagen, K.W., 1974. Colony husbandry. In: Weisbroth, S.H., Flatt, R.E., Kreus, A.L. (Eds.), The Biology of the Laboratory Rabbits. Academic Press, New York, NY, pp. 27−28.

Hammer, R.E., Pursel, V.G., Rexroad Jr., C.E., Wall, R.J., Bolt, D.J., Ebert, K.M., et al., 1985. Production of transgenic rabbits, sheep and pigs by microinjection. Nature 315, 680−683.

Han, Z.S., Li, Q.W., Zhang, Z.Y., Yu, Y.S., Xiao, B., Wu, S.Y., et al., 2008. Adenoviral vector mediates high expression levels of human lactoferrin in the milk of rabbits. J. Microbiol. Biotechnol. 18, 153−159.

Harbers, K., Schnieke, A., Stuhlmann, H., Jahner, D., Jaenisch, R., 1981. DNA methylation and gene expression: endogenous retroviral genome becomes infectious after molecular cloning. Proc. Natl. Acad. Sci. USA 78, 7609−7613.

Heidmann, O., Rougeon, F., 1983. Diversity in the rabbit immunoglobulin kappa chain variable regions is amplified by nucleotide deletions and insertions at the V-J junction. Cell 34, 767−777.

Hiripi, L., Negre, D., Cosset, F.L., Kvell, K., Czompoly, T., Baranyi, M., et al., 2010. Transgenic rabbit production with simian immunodeficiency virus-derived lentiviral vector. Transgenic Res. 19, 799−808.

Honda, A., Hirose, M., Inoue, K., Ogonuki, N., Miki, H., Shimozawa, N., et al., 2008. Stable embryonic stem cell lines in rabbits: potential small animal models for human research. Reprod. Biomed. Online 17, 706−715.

Intawicha, P., Ou, Y.W., Lo, N.W., Zhang, S.C., Chen, Y.Z., Lin, T.A., et al., 2009. Characterization of embryonic stem cell lines derived from New Zealand white rabbit embryos. Cloning Stem Cells 11, 27−38.

Ishino, Y., Shinagawa, H., Makino, K., Amemura, M., Nakata, A., 1987. Nucleotide sequence of the iap gene, responsible for alkaline phosphatase isozyme conversion in *Escherichia coli*, and identification of the gene product. J. Bacteriol. 169, 5429−5433.

Ivics, Z., Li, M.A., Mates, L., Boeke, J.D., Nagy, A., Bradley, A., et al., 2009. Transposon-mediated genome manipulation in vertebrates. Nat. Methods 6, 415−422.

Jaenisch, R., 1976. Germ line integration and Mendelian transmission of the exogenous Moloney leukemia virus. Proc. Natl. Acad. Sci. USA 73, 1260−1264.

Jinek, M., Chylinski, K., Fonfara, I., Hauer, M., Doudna, J.A., Charpentier, E., 2012. A programmable dual-RNA-guided DNA endonuclease in adaptive bacterial immunity. Science 337, 816−821.

Katter, K., Geurts, A.M., Hoffmann, O., Mates, L., Landa, V., Hiripi, L., et al., 2013. Transposon-mediated transgenesis, transgenic rescue, and tissue-specific gene expression in rodents and rabbits. FASEB J. 27, 930−941.

Kind, A., Schnieke, A., 2008. Animal pharming, two decades on. Transgenic Res. 17, 1025−1033.

Knight, K.L., Spieker-Polet, H., Kazdin, D.S., Oi, V.T., 1988. Transgenic rabbits with lymphocytic leukemia induced by the c-myc oncogene fused with the immunoglobulin heavy chain enhancer. Proc. Natl. Acad. Sci. USA 85, 3130–3134.

Kondo, M., Sakai, T., Komeima, K., Kurimoto, Y., Ueno, S., Nishizawa, Y., et al., 2009. Generation of a transgenic rabbit model of retinal degeneration. Invest. Ophthalmol. Vis. Sci. 50, 1371–1377.

Kurome, M., Geistlinger, L., Kessler, B., Zakhartchenko, V., Klymiuk, N., Wuensch, A., et al., 2013. Factors influencing the efficiency of generating genetically engineered pigs by nuclear transfer: multi-factorial analysis of a large data set. BMC Biotechnol. 13, 43.

Lavitrano, M., Camaioni, A., Fazio, V.M., Dolci, S., Farace, M.G., Spadafora, C., 1989. Sperm cells as vectors for introducing foreign DNA into eggs: genetic transformation of mice. Cell 57, 717–723.

Leuchs, S., Saalfrank, A., Merkl, C., Flisikowska, T., Edlinger, M., Durkovic, M., et al., 2012. Inactivation and inducible oncogenic mutation of p53 in gene targeted pigs. PLoS One 7, e43323.

Li, L., Shen, W., Min, L., Dong, H., Sun, Y., Pan, Q., 2006. Human lactoferrin transgenic rabbits produced efficiently using dimethylsulfoxide-sperm-mediated gene transfer. Reprod. Fertil. Dev. 18, 689–695.

Li, R., Zhuang, Y., Han, M., Xu, T., Wu, X., 2013. piggyBac as a high-capacity transgenesis and gene-therapy vector in human cells and mice. Dis. Model Mech. 6, 828–833.

Li, S., Guo, Y., Shi, J., Yin, C., Xing, F., Xu, L., et al., 2009. Transgene expression of enhanced green fluorescent protein in cloned rabbits generated from in vitro-transfected adult fibroblasts. Transgenic Res. 18, 227–235.

Li, Z., Li, Z.B., 2005. Repair of mandible defect with tissue engineering bone in rabbits. ANZ J. Surg. 75, 1017–1021.

Lombardi, R., Rodriguez, G., Chen, S.N., Ripplinger, C.M., Li, W., Chen, J., et al., 2009. Resolution of established cardiac hypertrophy and fibrosis and prevention of systolic dysfunction in a transgenic rabbit model of human cardiomyopathy through thiol-sensitive mechanisms. Circulation 17, 1398–1407.

Lombardo, A., Genovese, P., Beausejour, C.M., Colleoni, S., Lee, Y.L., Kim, K.A., et al., 2007. Gene editing in human stem cells using zinc finger nucleases and integrase-defective lentiviral vector delivery. Nat. Biotechnol. 25, 1298–1306.

Lonberg, N., Taylor, L.D., Harding, F.A., Trounstine, M., Higgins, K.M., Schramm, S.R., et al., 1994. Antigen-specific human antibodies from mice comprising four distinct genetic modifications. Nature 368, 856–859.

Martignoni, M., Groothuis, G.M., de Kanter, R., 2006. Species differences between mouse, rat, dog, monkey and human CYP-mediated drug metabolism, inhibition and induction. Expert Opin. Drug Metab. Toxicol. 2, 875–894.

Mashimo, T., Takizawa, A., Voigt, B., Yoshimi, K., Hiai, H., Kuramoto, T., et al., 2010. Generation of knockout rats with X-linked severe combined immunodeficiency (X-SCID) using zinc-finger nucleases. PLoS One 5, e8870.

Mates, L., Chuah, M.K., Belay, E., Jerchow, B., Manoj, N., Acosta-Sanchez, A., et al., 2009. Molecular evolution of a novel hyperactive Sleeping Beauty transposase enables robust stable gene transfer in vertebrates. Nat. Genet. 41, 753–761.

McCreath, K.J., Howcroft, J., Campbell, K.H., Colman, A., Schnieke, A.E., Kind, A.J., 2000. Production of gene-targeted sheep by nuclear transfer from cultured somatic cells. Nature 405, 1066–1069.

Meng, Q., Polgar, Z., Liu, J., Dinnyes, A., 2009. Live birth of somatic cell-cloned rabbits following trichostatin A treatment and cotransfer of parthenogenetic embryos. Cloning Stem Cells 11, 203–208.

Meyer, M., de Angelis, M.H., Wurst, W., Kuhn, R., 2010. Gene targeting by homologous recombination in mouse zygotes mediated by zinc-finger nucleases. Proc. Natl. Acad. Sci. USA 107, 15022–15026.

Moehle, E.A., Rock, J.M., Lee, Y.L., Jouvenot, Y., DeKelver, R.C., Gregory, P.D., et al., 2007. Targeted gene addition into a specified location in the human genome using designed zinc finger nucleases. Proc. Natl. Acad. Sci. USA 104, 3055–3060.

Mussolino, C., Cathomen, T., 2012. TALE nucleases: tailored genome engineering made easy. Curr. Opin. Biotechnol. 23, 644–650.

Mussolino, C., Cathomen, T., 2013. RNA guides genome engineering. Nat. Biotechnol. 31, 208–209.

Mussolino, C., Morbitzer, R., Lutge, F., Dannemann, N., Lahaye, T., Cathomen, T., 2011. A novel TALE nuclease scaffold enables high genome editing activity in combination with low toxicity. Nucleic Acids Res. 39, 9283–9293.

Niemann, H., Wilfried, A., Kues, W.A., Lucas-Hahn, A., Carnwath, J.W., 2011. Somatic cloning and epigenetic reprogramming in mammals. In: Atala, A. (Ed.), Principles in Regenerative Medicine, second ed. Academic Press, Burlington, MA, USA, pp. 129–158. ISBN-13: 978-01238142227.

Pfeifer, A., Lim, T., Zimmermann, K., 2010. Lentivirus transgenesis. Methods Enzymol. 477, 3–15.

Radecke, S., Cathomen, T., Schwarz, K., 2010. Zinc-finger nuclease-induced gene repair with oligonucleotides: wanted and unwanted target locus modifications. Mol. Ther. 18, 743–753.

Ramos, A.A., Yang, H., Rosen, L.E., Yao, X., 2006. Tandem parallel fragmentation of peptides for mass spectrometry. Anal. Chem. 78, 6391–6397.

Reyon, D., Tsai, S.Q., Khayter, C., Foden, J.A., Sander, J.D., Joung, J.K., 2012. FLASH assembly of TALENs for high-throughput genome editing. Nat. Biotechnol. 30, 460–465.

Rideout III, W.M., Eggan, K., Jaenisch, R., 2001. Nuclear cloning and epigenetic reprogramming of the genome. Science 293, 1093–1098.

Sapranauskas, R., Gasiunas, G., Fremaux, C., Barrangou, R., Horvath, P., Siksnys, V., 2011. The Streptococcus thermophilus CRISPR/Cas system provides immunity in *Escherichia coli*. Nucleic Acids Res. 39, 9275–9282.

Schnieke, A.E., Kind, A.J., Ritchie, W.A., Mycock, K., Scott, A.R., Ritchie, M., et al., 1997. Human factor IX transgenic sheep produced by transfer of nuclei from transfected fetal fibroblasts. Science 278, 2130–2133.

Seligman, L.M., Chisholm, K.M., Chevalier, B.S., Chadsey, M.S., Edwards, S.T., Savage, J.H., et al., 2002. Mutations altering the cleavage specificity of a homing endonuclease. Nucleic Acids Res. 30, 3870–3879.

Seok, J., Warren, H.S., Cuenca, A.G., Mindrinos, M.N., Baker, H.V., Xu, W., et al., 2013. Genomic responses in mouse models poorly mimic human inflammatory diseases. Proc. Natl. Acad. Sci. USA 110, 3507–3512.

Serhan, C.N., Jain, A., Marleau, S., Clish, C., Kantarci, A., Behbehani, B., et al., 2003. Reduced inflammation and tissue damage in transgenic rabbits overexpressing 15-lipoxygenase and endogenous anti-inflammatory lipid mediators. J. Immunol. 171, 6856–6865.

Sethupathi, P., Spieker-Polet, H., Polet, H., Yam, P.C., Tunyaplin, C., Knight, K.L., 1994. Lymphoid and non-lymphoid tumors in E kappa-myc transgenic rabbits. Leukemia 8, 2144−2155.

Shen, W., Li, L., Pan, Q., Min, L., Dong, H., Deng, J., 2006. Efficient and simple production of transgenic mice and rabbits using the new DMSO-sperm mediated exogenous DNA transfer method. Mol. Reprod. Dev. 73, 589−594.

Smith, J., Grizot, S., Arnould, S., Duclert, A., Epinat, J.C., Chames, P., et al., 2006. A combinatorial approach to create artificial homing endonucleases cleaving chosen sequences. Nucleic Acids Res. 34, e149.

Smith, K.R., 2012. Sperm-Mediated Gene Transfer: Concepts and Controversies. Bentham Science Publishers Ltd, Sharjah, UAE.

Song, J., Zhong, J., Guo, X., Chen, Y., Zou, Q., Huang, J., et al., 2013. Generation of RAG 1- and 2-deficient rabbits by embryo microinjection of TALENs. Cell Res. 23, 1059−1062.

Stoddard, B.L., 2011. Homing endonucleases: from microbial genetic invaders to reagents for targeted DNA modification. Structure 19, 7−15.

Sussman, D., Chadsey, M., Fauce, S., Engel, A., Bruett, A., Monnat Jr., R., et al., 2004. Isolation and characterization of new homing endonuclease specificities at individual target site positions. J. Mol. Biol. 342, 31−41.

Tesson, L., Usal, C., Menoret, S., Leung, E., Niles, B.J., Remy, S., et al., 2011. Knockout rats generated by embryo microinjection of TALENs. Nat. Biotechnol. 29, 695−696.

Urnov, F.D., Miller, J.C., Lee, Y.L., Beausejour, C.M., Rock, J.M., Augustus, S., et al., 2005. Highly efficient endogenous human gene correction using designed zinc-finger nucleases. Nature 435, 646−651.

Urnov, F.D., Rebar, E.J., Holmes, M.C., Zhang, H.S., Gregory, P.D., 2010. Genome editing with engineered zinc finger nucleases. Nat. Rev. Genet. 11, 636−646.

Vargo-Gogola, T., Rosen, J.M., 2007. Modelling breast cancer: one size does not fit all. Nat. Rev. Cancer 7, 659−672.

Wagner, S.D., Williams, G.T., Larson, T., Neuberger, M.S., Kitamura, D., Rajewsky, K., et al., 1994. Antibodies generated from human immunoglobulin miniloci in transgenic mice. Nucleic Acids Res. 22, 1389−1393.

Wang, H.J., Lin, A.X., Zhang, Z.C., Chen, Y.F., 2001. Expression of porcine growth hormone gene in transgenic rabbits as reported by green fluorescent protein. Anim. Biotechnol. 12, 101−110.

Wang, S., Tang, X., Niu, Y., Chen, H., Li, B., Li, T., et al., 2007. Generation and characterization of rabbit embryonic stem cells. Stem Cells 25, 481−489.

Watanabe, Y., 1980. Serial inbreeding of rabbits with hereditary hyperlipidemia (WHHL-rabbit). Atherosclerosis 36, 261−268.

Wefers, B., Meyer, M., Ortiz, O., Hrabe de Angelis, M., Hansen, J., Wurst, W. , et al., 2013. Direct production of mouse disease models by embryo microinjection of TALENs and oligodeoxynucleotides. Proc. Natl. Acad. Sci. USA 110, 3782−3787.

Yang, F., Hao, R., Kessler, B., Brem, G., Wolf, E., Zakhartchenko, V., 2007. Rabbit somatic cell cloning: effects of donor cell type, histone acetylation status and chimeric embryo complementation. Reproduction 133, 219−230.

Yang, H., Li, Q., Han, Z., Hu, J., 2012. High level expression of recombinant human anti-thrombin in the mammary gland of rabbits by adenoviral vectors infection. Anim. Biotechnol. 23, 89−100.

Yoshida, S., Sekine, Y., Saitoh, Y., Yasufuku, K., Iwata, T., Fujisawa, T., 2005. Surgical technique of experimental lung transplantation in rabbits. Ann. Thorac. Cardiovasc. Surg. 11, 7−11.

Zakhartchenko, V., Flisikowska, T., Li, S., Richter, T., Wieland, H., Durkovic, M., et al., 2011. Cell-mediated transgenesis in rabbits: chimeric and nuclear transfer animals. Biol. Reprod. 84, 229−237.

Zhang, F., Cong, L., Lodato, S., Kosuri, S., Church, G.M., Arlotta, P., 2011. Efficient construction of sequence-specific TAL effectors for modulating mammalian transcription. Nat. Biotechnol. 29, 149−153.

11 Production of Transgenic Fish

Rex A. Dunham[1] and Richard N. Winn[2]

[1]School of Fisheries, Aquaculture, and Aquatic Sciences, Auburn University, Auburn, AL, [2]Warnell School of Forest Resources, University of Georgia, Athens, GA

I. Introduction

Numerous transgenic fish have been developed using a variety of transgenes, methods, and species since the first transgenic fish were produced in China in 1985 (Zhu et al., 1985) (reviewed by Chen and Powers, 1990; Dunham, 2011; Fletcher and Davies, 1991; Hackett, 1993; Maclean, 1998). Since that time, transgenic gene knockout technology (Thresher et al., 2009; Wong and Van Eenennaam, 2008) and gene editing, which mutates the fish without inserting exogenous DNA using transgenic technology procedure (Tan et al., 2013), have been developed. Early research emphasized the refinement of transgenic methodology and development of novel strains of fish with commercially beneficial traits, with emphasis on growth enhancement. The production of transgenic fish has since expanded to research focused on five general applications: to enhance traits of commercially important species, to develop fish as bioreactors to produce biomedically important proteins, to enhance the utility of fish as indicators of adverse health effects associated with exposure to toxicants in aquatic environments, to develop new nonmammalian animal models for comparative biomedical research, and for functional genomics studies. Transgenic fish have been produced that could benefit the aquaculture industry, but regulatory agencies have yet to approve them for commercial application.

II. Discussion

Fish comprise the most diverse group among the vertebrates with an estimated 30,000 species. This diversity is reflected in wide differences in various features, including morphology, behavior, reproduction, development, generation times, and tolerance to environmental conditions, which present both opportunities and challenges to the production of a variety of transgenic fish and transgenic fish models. Numerous fish species offer beneficial attributes for transgenesis including cost-effective husbandry, short generation times, high fecundity, and large, easy-to-manipulate eggs. External fertilization for the majority of fish species precludes complex

Transgenic Animal Technology. DOI: http://dx.doi.org/10.1016/B978-0-12-410490-7.00011-6

manipulations such as *in vitro* culture and transfer of embryos in surrogate females as required for mammals. Several manipulations are possible or easier to accomplish, while difficult or impossible in mammals. Transparency of some fish embryos allows direct monitoring of development and, in some cases, evaluation of expression of reporter genes *in vivo*, especially with the advent of the green fluorescent protein (GFP) gene. Other genomic manipulations are also possible in some species using haploids, triploids, or homozygous gynogenetic/androgenetic lines.

Despite the large number of fish species, a much smaller number of species is used for research and commercial purposes. The major species cultured in the world include catfish, salmonids, tilapia, and carp, for meeting demands for high-quality protein as wild stocks continue to diminish worldwide. Numerous additional species are under evaluation for aquaculture or are cultured on a small scale, and culture of various marine species has increased during the last decade. The large size of some species, such as rainbow trout and carp, facilitates easy manipulation of organs or tissues, sample collection, and surgical procedures. The small size, rapid generation time, and cost-effective laboratory culture of species such as medaka (*Oryzias latipes*), swordtails, zebrafish (*Danio rerio*), and mummichog (*Fundulus heteroclitus*), contribute to these species being widely used as animal models for studies of human diseases, vertebrate development, toxicology, and genetics. Of these, zebrafish and medaka are the most commonly used model species.

The diversity of fish species also presents challenges to the routine production of transgenic fish. Among the challenges is the problem of readily adapting specific methods and procedures developed for one species to another. For example, eggs of some of the widely used fish species differ in size by nearly an order of magnitude, and generation times are as short as 6 weeks in some species and as long as several years in others. As a consequence, it is difficult to present procedures that can be applied universally to many species. However, many of the fundamental principles and general procedures that guide transgenic development for other vertebrate models apply generally to most fish species.

III. Methods of Gene Transfer

Transgenic fish research was established quickly in the mid-1980s by adapting a variety of procedures used successfully in developing transgenic mice. However, several important differences between mice and fish have dictated the extent to which protocols can be readily adapted to fish. Among the most important are the differences between the eggs of mice and fish. In contrast to mouse eggs, the eggs of fish are large, have pronuclei that are difficult to locate, and are surrounded by a hard chorion. To accomplish gene transfer in mice, the male pronucleus must be injected. However, for most fish species this is not possible, and random microinjection into the blastodisc will result in a significant percentage of transgenic fish, meaning either injection of fish pronuclei is not necessary to accomplish transformation or pronuclei are accidentally injected without being able to visualize them.

As a consequence of these differences in egg characteristics between mammals and fish, researchers have attempted a variety of approaches with varying degrees of success. Gene transfer methods include microinjection through the cytoplasm (Chourrout et al., 1986; Maclean et al., 1987; Winn et al., 1995, 2000, 2001; Zhu et al., 1985), microinjection through the germinal vesicle (Ozato et al., 1986), electroporation of embryos (Inoue et al., 1990; Powers et al., 1992) or sperm (Muller et al., 1992), retroviral infection (Gaiano et al., 1996; Lin et al., 1994; Lu et al., 1997; Sarmasik et al., 2001), particle-gun bombardment (Kolenikov et al., 1990; Nichols, 1998), sperm-mediated transfer (Lu et al., 2002), injection of gonads (mice, fish, and shellfish) (Chen et al., 2006; Lu et al., 2002; Yang et al., 2007), liposome-mediated transfer (Lu et al., 2002), nuclear transplantation using embryonic cells (Wakamatsu et al., 2001), and cell-mediated gene transfer (Ma et al., 2001).

A. Microinjection

Injection of DNA using drawn capillary needles via the cytoplasm of newly fertilized eggs has proven to be a reliable method of producing transgenic fish that have acceptable efficiencies in germ line transmission and expression of transgenes. This procedure remains the most commonly used method of introducing DNA into fish genomes but not necessarily the most effective. The general steps to producing transgenic fish by cytoplasmic microinjection of DNA are (1) prepare breeding stocks, (2) collect or produce unfertilized or fertilized eggs, (3) inject DNA into the cytoplasm of unfertilized eggs or one- to four-cell embryos (Hayat et al., 1991), (4) identify fish that carry the transgene, and (5) cross founders with each other or wild-type mates and verify germ line transmission (Figure 11.1).

The following methods were employed in the Aquatic Biotechnology and Environmental Laboratory at the University of Georgia for the production of transgenic medaka (Figure 11.2) and are intended to be representative of procedures that may guide the transgenic production of small model species. Medaka is a small aquaria fish used widely as a model species in a variety of fields including toxicology, genetics, medical technology, functional genomics, and developmental biology. This species has a number of beneficial characteristics, including relatively short generation time ($\sim$6 weeks), cost-effective husbandry, prolific capacity for reproduction, and transparent chorion, that make it well suited for transgenic studies.

1. Maintenance of Brood Stock and Collection of Eggs

A number of parameters related to culture of medaka, including photoperiod, temperature, density, sex ratios, and feeding, are optimized to ensure that the numbers and quality of eggs are sufficient to meet the needs of the study. Approximately, 2 weeks prior to the start of an injection series, three to four aquaria containing adult

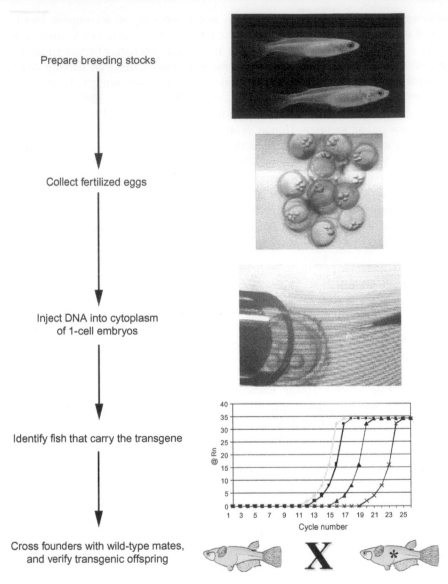

Figure 11.1 Generalized steps to developing transgenic medaka using microinjection into the cytoplasm of one-cell embryos.

fish (∼30 fish/40 L, 2−3 months of age) in a ratio of three females to one male are established. Fish are maintained on a photoperiod of 16 h light:8 h dark (e.g., timer set for daylight at 0900) and a temperature of 26−27°C. Fish are fed a minimum of twice a day with flake food and once a day with freshly hatched brine shrimp (∼24 h posthatch). Frequent feedings are preferable to large single meals.

Figure 11.2 Japanese medaka (*Oryzias latipes*). The medaka has been widely studied in the fields of toxicology, genetics, and developmental biology and has numerous desirable characteristics that make it suitable for transgenic development.

Supplementation with live brine shrimp improves egg production while decreasing potential water quality problems associated with overfeeding. Fish are monitored regularly, and tanks are cleaned as needed to ensure excess food or waste does not accumulate. Materials placed in the bottom of the aquaria, such as rocks and marbles, that would make cleaning more difficult are avoided.

Beginning about 1 h before the lights are turned on, eggs are collected directly from the abdomens of females and the bottom of the tanks, with the aid of a flashlight. A fish is captured and held with a fine-mesh net to reduce abrasion. Eggs are gently rubbed from the abdomen of the fish into the net and then are transferred using forceps into a watch glass filled with clean culture water. Eggs at the early one-cell stage are separated from the older eggs by using a dissection microscope. Chorionic fibers that entangle eggs are removed by cutting them with fine scissors or gently rolling eggs on a fine-mesh screen (1.5 mm mesh). Typically, more eggs are collected than can be used at one time. All eggs should be removed from the aquaria to reduce subsequent handling of older eggs not suited for injection. Individual females produce eggs once daily, and egg production from all females is not synchronous. Feeding the fish live brine shrimp within the first hour after the lights are turned on will stimulate further egg production. Eggs are obtained for up to 3 h, and a timer is set to go off every 15–20 min to facilitate the repeated collecting, sorting, and injecting of eggs.

2. Microinjection of Fertilized Eggs via the Cytoplasm

a Equipment

Various models of microscopes, micromanipulators, pipette pullers, microinjection systems, and other equipments are readily available and suitable for microinjection of fish eggs. Information presented in preceding chapters that guides the selection of equipment applies generally to fish. Exceptions to the generalizations are presented here.

A stereo dissection microscope equipped with an external light source, a base with transmitted light, and adjustable mirror and with a magnification of between 6 and 100× is suitable for microinjection of fish eggs (Figure 11.3). This configuration permits manipulation of the intensity of illumination and focus to aid in positioning the microinjection needle. Other microscopes, including upright and

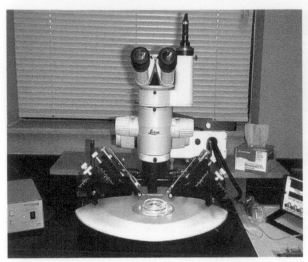

Figure 11.3 Microinjection equipment includes a stereozoom dissection microscope (6–100×) on a base with transmitted light and adjustable mirror, two micromanipulators attached to magnetic stands on a steel base, and an external light source. At the right is the gas-driven microinjector. This configuration allows for optimal manipulation of illumination intensity and focus to assist in positioning the microinjection needle.

inverted styles, are also acceptable if the requirements for low-power magnification and adequate working field are met.

To prepare microinjection needles, borosilicate glass capillaries with a 1-mm outside diameter and 0.5-mm inside diameter can be used. Capillaries are pulled with a pipette puller (Narishige model PB-7) to make needles with ~5-mm shaft and ~0.05-mm tips. Practice and patience are essential for optimizing the parameters for preparing suitable microinjection needles.

Two micromanipulators, one for the microinjection needle and the other for the egg-holding pipette, are attached to magnetic stands placed on a steel base. In our laboratory, we use microdispenser capillaries (Drummond) as holding pipettes. The 25-μL capillaries have an inside diameter of ~1 mm, which is slightly smaller than the diameter of medaka eggs. With the aid of a pipette holder attached to a micromanipulator, eggs are held in place on the bottom of a watch glass by applying gentle downward pressure (Figure 11.4). The holding pipette is adjusted to secure the egg while allowing it to be rotated into proper orientation with forceps for microinjection. After injection, the egg is removed and another is put in place using minimal hand movements and adjustments. Capillaries are available in a variety of sizes that are suitable for holding eggs of other fish species. For example, 50-μL capillaries are ideal for holding the eggs of *Fundulus*. Alternatively, microinjection plates can be made using 60-mm Petri dishes in which a microscope slide has been inserted at a 10–20° angle in molten agarose. After the agarose has solidified and the slide has been removed, embryos are positioned along the groove in the agarose for microinjection (Meng et al., 1999).

Systems designed for microinjecting DNA include oil-filled positive-displacement and gas-driven injectors. A gas-driven microinjector is preferred, as it has a feature for controlling the backpressure of the gas in the microinjection

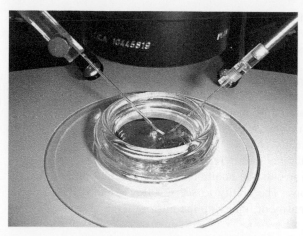

Figure 11.4 Microinjection setup. Eggs are placed in a watch glass filled with a salt solution (18 ppt). The holding pipette (capillary) is on the left and the microinjection needle is on the right.

needle to compensate for internal pressure in an egg, which may force cytoplasm into the needle after injection-killing the embryo.

b Microinjection Procedures

Guidelines for the preparation and handling of DNA for microinjection are essentially identical for mammals and fish. DNA constructs are prepared using standard methods to provide DNA of optimal purity to enhance survival of embryos and the overall efficiency of producing transgenic founders. DNA constructs are linearized and resuspended in microinjection buffers of low ionic strength, such as 5 mM Tris/0.1 mM EDTA, pH 7.4 (Winn et al., 2000) or phosphate buffered saline, 1.47 mM KH$_2$PO$_4$, 8.06 mM Na$_2$HPO$_4$, 137 mM NaCl, 2.68 mM KCl (Iwamatsu, 1994). However, transgenic fish can also be made by injecting circular DNA, although it is believed that integration rates are higher with linearized DNA. Alternatively, a DNA solution containing 0.1 M KCl and 0.125% of the dye tetramethylrhodamine dextran has been used to aid in monitoring the amount of DNA microinjected into the embryos (Meng et al., 1999). If dyes are used, precautions should be taken to ensure that the solution is free of particles that could clog the injection needles and to ensure that the dye does not have adverse effects on the developing embryo.

The efficiency of transgene integration is dependent upon the concentration of the injected DNA, whereas egg viability is inversely related to the DNA concentration. Consequently, a compromise is sought between embryo survival and integration efficiency. The amount of DNA injected into the cell is controlled by adjusting the concentration of the DNA rather than the volume of the injected DNA solution. To compensate for the dilution of the DNA microinjected into the large volume of cytoplasm and to achieve a reasonable probability of some of the injected DNA entering the nucleus while also maintaining an acceptable survival rate of injected embryos (>50%), DNA is prepared for injection at a concentration of 50–200 ng/mL ($\sim$10^4–10^6 transgene copies). Higher concentrations have been used, but once 10^8 transgene copies are in solution, the solution is too viscous for microinjection. Microinjection needles are loaded with DNA (~5 mL) by backfilling using microloader pipette tips (Eppendorf microloaders 5242956.003). Care is taken not to introduce bubbles into the needles, which can interfere with the flow of DNA solution.

Eggs (30), which have been batch sorted and cleaned, are transferred to a watch glass filled with 18 g/L (ppt) filtered salt solution (Instant Ocean). Immersion of the embryos in the salt solution during microinjection is not detrimental to embryo survival and likely promotes survival, adding to sterility as well as osmolarity benefits. The salt solution significantly enhances one's ability to visualize the flow of DNA from the tip of the needle due to differences in the densities of the two solutions, and use of the solution avoids the addition of dyes that may interfere with some gene transfer applications. The ability to see the flow of the DNA is also improved by adjusting the intensity of light with the mirror on the transmitted light base.

An egg is positioned under the holding pipette using gentle downward pressure sufficient to secure the egg while allowing manipulation of the egg with forceps. The egg is positioned in the center of field of view and oriented with the cell directed upward. In medaka, the location of the single cell for injection is observed opposite of the pole where the large oil droplets coalesce (Figure 11.5).

Prior to microinjection, the tip of the microinjection needle is opened by carefully positioning the needle and touching it gently to the side of the holding pipette. Touching the needle on the holding pipette while simultaneously engaging the injector or high-pressure mode facilitates this procedure. The DNA solution should flow freely from the tip of the needle when the backpressure (balance pressure) is set at 10 psi. These settings may vary depending upon the characteristics of the eggs of the species being used, and may also vary depending upon the season or

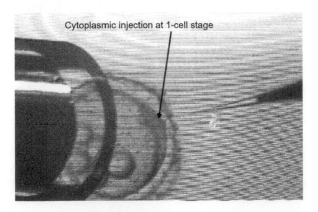

Figure 11.5 50× magnification of a medaka, *Oryzias latipes*, egg at the one-cell stage. The egg is positioned under the holding pipette so that the cell is directed upward. The single cell is clearly visible as a dark mass on the right, while the oil droplets are visible opposite the cell. The needle is positioned perpendicular to the egg surface to decrease the potential of the needle to break or not penetrate the chorion.

strain being utilized. The free flow of the DNA solution from the needle aids in reducing clogging. At the beginning of a series of injections, the gas injector is set for a new needle as follows: balance pressure, 5—10 psi (36—78 kPa); injection pressure, 50 psi (373 kPa); and injection time, 70 ms (Medical Systems Injector Model PL1-100). These settings are not fixed for all injections and may be adjusted as the flow from the needle changes with use.

The key factor to achieving a successful injection through the hard chorion is to position the needle perpendicular to the surface of the egg, thereby permitting smooth penetration by the needle while reducing the tendency of the needle to bend, break, or glance off the chorion. In the case of other egg types such as carp and catfish, the surface of the egg can be best penetrated by using a 45° angle for the needle's insertion. By using efficient hand movements, the focus on the needle is adjusted while the micromanipulators move the needle into position to pierce the chorion and enter the cell. The needle should not penetrate the yolk. An injector equipped with a foot pedal aids in this procedure by freeing the hands to focus and adjust the position of the needle. The cell is injected with ~100 pL of the DNA solution, which will cause discernible swelling of the cytoplasm and a darkening of the cytoplasm where the DNA solution is entering. Visualization of the DNA solution injected into the cytoplasm is aided by adjusting the intensity of the light transmitted through the bottom of the egg. DNA injected into the cytoplasm will disperse, whereas the DNA injected into the yolk will remain as a distinct bubble. The needle is retracted in a single continuous motion to reduce lysing of the cell. The backpressure of the gas on the needle is adjusted to a lower pressure (~4—5 psi) if the flow of DNA decreases due to accumulation of debris from the chorion. The flow of DNA solution should not be excessive but must be sufficient to prevent the cytoplasm from being forced back into the needle. A clogged needle is replaced if any material is not dislodged readily using the high-pressure ("clear") function.

After injection, the egg is transferred to a Petri dish (60 mm) filled with water, and another egg is moved into position. Eggs are injected within 20—30 min after collection. Typically, 40—100 eggs are injected in an hour. Eggs are incubated at 26°C and are inspected daily.

Lysed/nonviable embryos are promptly removed and the water is changed to reduce the incidence of bacterial infections. Between 40% and 90% of fry will survive to hatch within 12—15 days after injection. Addition of methylene blue (0.6 mg/L) to the egg culture water aids in distinguishing dead embryos (which stain blue) and further reduces bacterial growth (Kinoshita and Ozato, 1995). However, potential toxicity associated with this compound suggests that it may not be advisable for some transgenic applications.

3. Dechorionation and Other Techniques

A number of alternative procedures have been attempted to overcome some of the difficulties associated with microinjection through the hard chorion of fertilized eggs. Iwamatsu (1994) recommends incubation of medaka eggs prior to microinjection in an alkaline solution containing glutathione to prevent hardening of the chorion. To circumvent direct injection of a needle through the extremely hard chorion of some fish species such as salmonids, a small hole can be drilled in the chorion before insertion of the injection needle (Guyomard et al., 1989; Rokkones et al., 1985).

Various researchers have reported mixed success with mechanical or chemical removal of the chorion before proceeding with microinjection (Zhu et al., 1985; Hallerman et al., 1988; Yoon et al., 1990; Fletcher and Davies, 1991; Hackett, 1993). As a routine procedure, Meng et al. (1999) dechorionated newly fertilized zebrafish eggs by soaking them for several minutes in a diluted pronase solution (1 mg/mL in Holtfreter's solution: 7 g NaCl, 0.4 g sodium bicarbonate, 0.2 g anhydrous $CaCl_2$, pH 6.5−7.0). When selecting a method, the researcher must consider the relative effort, embryo survival, and transgene integration rates obtained from these procedures compared to microinjection through the chorion. Given sufficient time to master the techniques, most researchers using model species prefer microinjection of DNA through the chorion for gene transfer (Hackett, 1993), however, an increasing number of researchers using aquaculture species are using alternative methods (Chen et al., 2006; Dunham et al., 2002; Lu et al., 2002; Thresher et al., 2009).

B. Microinjection of Large Aquaculture Species

The primary disadvantage of gene transfer in large, aquaculture species is the generation interval. Of the major aquaculture species, tilapia may be the most like small, model species as they can have generation intervals of 6 months or less if maintained in warm water. Various species of catfish, salmonids, and carps have a 2−6-year generation interval, thus, multigeneration experiments, which is the norm for gene knock-in systems, will take longer in major aquaculture species compared to model species such as medaka or zebrafish.

Other disadvantages include the size of the major aquaculture species and the seasonality of their egg production. Larger, inherently more expensive holding facilities are required. Some species, such as channel catfish, have very short spawning seasons that can be as short as 2 months, which limits the time of year that transgenesis experiments can be initiated. However, other major aquaculture species, such as common carp and rainbow trout, can be spawned most of the year. Additionally, given sufficient facilities most species can be manipulated by temperature and/or photoperiod to allow collection of eggs and sperm year-round.

One advantage of these larger species is their high fecundity compared to model species, allowing production of large numbers of embryos in a short period of time for manipulation. The normal procedure for most species is to induce ovulation, hand-strip the eggs, and artificially fertilize them. Some species can be partially stripped of their eggs repeatedly for several hours, allowing the researcher to periodically fertilize the eggs and manipulate eggs for long periods of time. For some species, the eggs can be stripped at one time, and if they are of high quality, they can be held dry for 2−8 h, once again allowing manipulation of freshly fertilized eggs over extended periods of time.

Eggs of many large aquaculture species are amenable to microinjection and other gene transfer procedures. In most cases, techniques have been developed that do not require that the eggs be held in place with vacuum for manipulation, making the procedure faster and easier to perform. Various procedures have been used that allow the eggs to settle into wells so that they are held in place during microinjection, freeing one of the manipulator's hands (Figure 11.6). Adhesive eggs, such as those of channel catfish and common carp, adhere to a glass Petri dish, circumventing the necessity of holding them in place with vacuum and greatly speeding the microinjection process.

With these large species it is not necessary to clean the eggs and operate in a particularly sterile environment. However, if they are cultured in Hotzferter's solution with antibiotic treatment (Figure 11.7) after manipulation and until hatch, the survival rates can be greatly increased.

Sometimes age of sexual maturity of aquaculture-raised fish such as catfish can be manipulated, reducing the generation interval to make it more similar to that of model species. Some species such as catfish developmentally recognize years as change in temperature. By manipulating temperature, "years" can be less than 12 months and the fish age and mature based on the shortened years rather than the actual years (Davis, 2009).

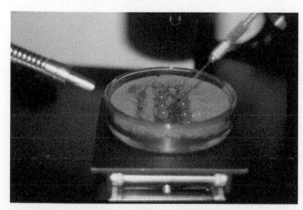

Figure 11.6 Microinjection of salmon eggs held in place in a matrix.
Source: Photo by Robert Devlin.

Figure 11.7 Static incubation of common carp, *Cyprinus carpio*, eggs in Hotzfreter's solution and doxycycline in tubs.
Source: Photo by Baofeng Su.

Not all research can be modeled, and sometimes transgenic studies need to be done in the aquaculture-raised species and not the model species. Some transgenes are going to react differently in different genetic backgrounds or genomes. Gene expression profile, biology, reproduction, and life histories can be quite different from one species to another, limiting how informative modeling experiments can be. Additionally, some culture-related or economic traits of aquaculture species cannot be measured in model species.

C. Alternative Methods

1. Electroporation

Electroporation has been used as an alternative to microinjection methods for gene transfer in fish (Inoue et al., 1990; Inoue, 1992; Buono and Linser, 1992; Muller et al., 1992; Powers et al., 1992; Dunham et al., 2002; Cheng, 2013). The method has particular appeal as a means of mass gene transfer for species such salmon, trout, and carp, which have limited seasonal availability of very large numbers of eggs. The wide range of success and failure rates reported by investigators using different species, protocols, and electroporators makes it difficult to identify the optimal conditions for protocols (reviewed by Hackett, 1993). From these reports, it appears that conditions for each application require systematic testing to obtain optimal results. Some researchers have used milligram quantities of DNA, which is a mistake. Electroporation of microgram quantities of DNA is quite effective (Powers et al., 1992). However, since this early research some researchers (Powers et al., 1992; Dunham et al., 2002; Chen et al., 2006; Thresher et al., 2009) have had good and consistent success utilizing both freshwater and marine fish such as zebrafish, channel catfish (*Ictalurus punctatus*), common carp (*Cyprinus carpio*), and silver sea bream (*Sparus sarba*), various constructs and more than one type of apparatus. Integration rates are typically 25−75%.

Electroporation is technically much simpler than microinjection and takes much less skill. Large numbers of P1 individuals can be generated quickly. Of course, this is not a great advantage if one does not have the resources to screen, evaluate, or breed this increased number of fish. In general, electroporated embryos are easier to keep alive than microinjected embryos during the first few days of incubation.

For the past 20 years, the Fish Genetic Enhancement Laboratory, School of Fisheries, Aquaculture, and Aquatic Sciences has routinely produced transgenic fish with electroporation. A Baekon 2000 macromolecule transfer system (Baekon, Inc., Saratoga, CA) is utilized; unfortunately, this company no longer exists. With this electroporation system, pulses of electricity are utilized. DNA can be transferred into either sperm or embryos (Powers et al., 1992). We utilize a double electroporation procedure for which electroporated sperm is used to fertilize eggs, and then the developing embryos are electroporated, increasing the probability of gene transfer.

The plasmid or DNA of interest is prepared separately in two tubes for the purpose of double electroporation. One tube is diluted in the 2.0 mL (0.9%) saline with a concentration of 50 μg/mL of the transgene into sperm. The other tube for the second electroporation of the embryos is prepared in 9.0 mL TE buffer (5 mM Tris−HCl, 0.5 M EDTA, pH 8.0) with a concentration of 50 μg/mL.

The electroporation parameters used are 6 kV, 2^7 pulses, 0.8 s burst, four cycles and 160 μs (Powers et al., 1992). One or two drops of sperm are placed in the DNA solution (1.0 mL) in saline and mixed (Figure 11.8). Theoretically, the sperm is partially dehydrated at 9 ppt for a freshwater fish. Sperm is kept in this solution a minimum of 5 min before use. Then DNA/sperm solution is poured into a 7-mL Petri dish and completely filled with freshwater. This hydrates the sperm, which also pulls DNA into the sperm and activates the sperm (Kang et al., 1999). Then the sperm is immediately electroporated and used for fertilization (Figure 11.9). All this must be done in a period of 2 min or the sperm and eggs deactivate and fertilization cannot occur. The second electroporation is conducted slightly before first cell division. The fertilized embryos are loaded into the Petri dishes and completely covered with the DNA solution in TE buffer (usually 2−3 mL). A total of 300−400 common carp eggs and 100−200 channel catfish eggs can easily fit into a plastic Petri dish of this size (Figure 11.10). The eggs remain in the DNA for 10 min. Then the fertilized eggs are electroporated, and the embryos transferred to 8.0 L tubs with 5.0 L Holtfreter's solution, and incubated statically. The embryos are gently agitated with compressed air delivered through airstones. Dead embryos are removed daily before changing Holtfreter's solution.

2. Other Gene Transfer Technologies

Several technologies have been introduced to address some of the limitations of current gene transfer methods. These technologies, described briefly here, exemplify the expansion and increase in sophistication of transgenic fish technology.

Retroviral vectors were among the methods being used to enhance the gene integration and transmission efficiency in several fish species during the 1990s (Amsterdam and Hopkins, 1999; Gaiano et al., 1996; Lin et al., 1994; Lu et al., 1997;

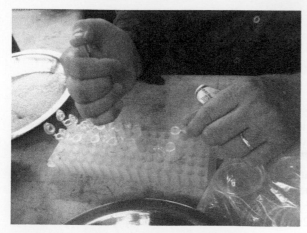

Figure 11.8 Mixing of sperm with a DNA/saline solution. Note the dry channel catfish, *Ictalurus punctatus*, eggs on the left waiting to be fertilized with electroporated sperm.
Source: Photo by Qi Cheng.

Figure 11.9 Electroporation of fish gametes or embryos.
Source: Photo by Qi Cheng.

Figure 11.10 Preparing to load DNA into a Petri dish sitting on the electroporation electrode and containing a few hundred fish eggs.
Source: Courtesy of Baofeng Su.

Sarmasik et al., 2001). The high frequencies (>83%) of injected zebrafish embryos that transmit proviral inserts to their offspring demonstrate the feasibility of this approach and its potential utility for use in large-scale insertional mutagenesis studies.

Stable transgenic medaka lineages were produced by electroporating broad-host-range (pantropic) replication-defective vectors into newly fertilized medaka embryos (Lu et al., 1997). These reports suggest that retroviral vectors might also serve as effective vehicles for transferring foreign DNA into the germ line of live-bearing fish. Sarmasik et al. (2001) demonstrated the feasibility of this approach by introducing a pantropic retroviral vector carrying a *neo*R into the immature gonads of the live-bearing fish *Poeciliopsis lucida*. They showed that 83% of the surviving males and 9% of the females carried the vector in their gonads.

Despite this promise the retroviral vector technology has not become widely used. During the last 20 years this procedure has only been sporadically utilized (Yee et al., 1994; Kurita et al., 2004; Sakai, 2006; Jao and Burgess, 2009). Most of these laboratories were coupling retroviral vector technology with sperm or cell culture to produce transgenic fish in model species.

One concern regarding the retroviral approach is the possibility of disease and health risks (FAO/WHO, 2003). The concern revolves around the theoretical possibility of recombination, leading to reconstitution of active and infectious viral particles. Some research has focused on the elimination of the expression of viral proteins that could lead to inflammatory responses. FAO/WHO (2003) concluded that there are some potential animal health hazards associated with the use of viral sequences, including the potential for recombination and subsequent expression, altered pathogenicity, and reverse transcription of RNA viral sequences. But they also came to three conclusions: developments on nonviral episomal vectors provide a means to overcome many of the concerns associated with viral-based vector systems, the principles of guidelines developed for human gene therapy should be utilized in transgenic fish research, and further research is needed.

Ivics et al. (1999) proposed an alternative means of inserting DNA in the genomes of fish by using a transposase gene to serve as a vector for gene delivery. Despite some remaining uncertainties related to the efficiency of transfer of different-sized transposons, this approach shows promise.

Ma et al. (2001) introduced a cell culture system for propagation of zebrafish embryo cells capable of producing germ cells *in vivo*, suggesting the potential utility of this approach for introducing targeted mutations in fish. Wakamatsu et al. (2001) established the basic techniques of nuclear transplantation in fish by transplanting embryonic cell nuclei from transgenic medaka carrying the GFP into unfertilized enucleated eggs. Intraspecific clones of individual fish have been produced (Lee et al., 2002). In the case of zebrafish, fertile transgenics were obtained by nuclear transfer using embryonic fibroblast cells from long-term cultures. The donor nuclei were modified with retroviral insertions expressing GFP and were transplanted into manually enucleated eggs. A shortcoming was the low survival rate, 2%, resulting in 11 adult transgenic zebrafish expressing GFP. These P1 nuclear-transplant transgenics produced functioning transgenic diploid F1/F2 progeny expressing GFP the same as the founder fish. This work represents an important first step toward generation of cloned fish from somatic or cultured cells. However, this work has not progressed much during the past decade.

Xenogenesis now is another avenue for gene transfer in fish. Diploid spermatogonial stem cells or primordial germ cells could be isolated, transformed, and transferred to sterile, triploid hosts. If these cells were first screened to ensure gene transfer, transgene research could be more rapid and mosacism could be reduced (Dunham, 2011).

IV. Establishing and Maintaining Transgenic Lineages

A. Detection of the Transgene

The procedures used to evaluate transgene integration in mammals are generally applicable to fish. Variable efficiencies in generating transgenic founders have been reported. This variability can be attributed to a number of factors, but may primarily result from the differences in microinjection procedures used among researchers and differences in the criteria applied to determine integration rates (Hackett, 1993). Size of the construct may also be important, with large transgenes being more difficult to transfer. Gene transfer efficiencies of ~10–20% can be used as a reasonable benchmark for comparison.

DNA isolated from fin clips can be used to detect the presence of a transgene in fish. Fins other than caudal fin are recommended as the caudal fin is essential for swimming, and a damaged caudal fin is more vulnerable to infection than other fins. If the fish has barbels, these can be sampled without compromising the fish and high-quality DNA is easily isolated from this tissue. For large aquaculture species, blood yields the highest quality DNA. To reduce the risk of mortality associated with the excision of tissue, fish of sufficient age and size to permit ease of handling (~1.5 cm in size) are held gently while sterile scissors are used to cut the fin (~2 mm in size). The tissue is placed in a microfuge tube for immediate DNA extraction or is flash-frozen in liquid nitrogen and stored at −80°C for subsequent analysis. In remote field locations, samples can be flash-frozen in a bath of dry ice and ethanol if liquid nitrogen in not available. DNA is recovered using standard methods of proteinase digestion, organic solvent extraction, and ethanol precipitation.

The polymerase chain reaction (PCR) remains the most commonly used method to detect the presence of a transgene in presumptive transgenic founders and offspring. Similar procedures and precautions described in preceding chapters for performing PCR can be applied to fish. Southern blot hybridization remains the definitive method to demonstrate gene integration, because it allows detection of junction fragments containing bases from the transgene and the adjacent, native genomic DNA.

The Southern blot and quantitative PCR are also important to elucidate the classification of integration event. Many types of integration are possible. Single or multiple copies of the transgene can be integrated. These integration events can be at a single or multiple insertion sites. After injection into an embryo, the transgene usually forms concatamers prior to integration. This can result in the integration of tandem arrays as short as two copies or as long as 100 copies. These tandem arrays can be different orientations such as head–head, tail–tail, and head–tail. In some cases, the transgene can be cleaved prior to insertion, resulting in integration of a partial copy.

B. Assays of Gene Expression

Many of the important considerations in the analysis of transgene expression described in preceding chapters apply equally to fish. However, in contrast to studies using rodent models, numerous researchers have reported difficulties with sustaining gene expression in lineages of transgenic fish. Expression of foreign DNA, commonly observed in the founder fish, can be rare or unpredictable in subsequent generations (Chen and Powers, 1990; Fletcher and Davies, 1991; Hackett, 1993; Iyengar et al., 1996). Suboptimal expression may result from various factors, including the use of heterologous transcription units derived from nonfish species (Foecking and Hofstetter, 1986; Gorman et al., 1982a,b), from methylation of CG dinucleotides (Boyes and Bird, 1991; Ehrlich et al., 1990), from incorporation of multiple transgenes into a single site (Dunham, 2011), or from integration of transgenes in loci susceptible to gene silencing (Eissenberg and Elgin, 1991).

Progress in stabilizing expression of transgenes in fish by using such methods as insulating border elements is reported (Stief et al., 1989; Udvardy et al., 1985; Gibbs and Schmale, 2000). Other researchers report success in monitoring stable gene expression by using GFP in medaka (Tanaka et al., 2001; Wakamatsu et al., 2001). By using GFP as a reporter gene, patterns of expression in embryos and adults can be easily monitored *in vivo* under a fluorescence microscope. Jessen et al. (1998) developed a technique for modifying bacterial artificial chromosomes (BACs) through homologous recombination to insert a GFP reporter gene in fish. Embryos microinjected with modified BAC clones were less mosaic and had improved GFP expression compared to those injected with smaller plasmid constructs. Another method shows promise for enhancing uniform expression of transgenes in the F0 generation, preventing gene silencing and unstable transmission in subsequent generations. Hsiao et al. (2001) found that by incorporating adeno-associated virus with two inverted terminal repeats (ITRs) into a DNA construct they were able to reduce mosaic transgene integration and increase stable germ line transmission in zebrafish.

Contrary to some of the disappointing results above, there are several examples of stable integration, expression, and phenotypic alterations lasting into the F1 and F2 generation and beyond for multiple species, medaka, common carp, channel catfish, salmon, and zebrafish and in multiple genes, growth hormone, cecropin, desaturase and short hair pin RNA interference (shRNAi) and complementary DNA (cDNA) against bone morphogenetic protein 2 (Thresher et al., 2009; reviewed in Dunham, 2011). AquaAdvantage growth hormone transgenic salmon propagated by AquaBounty have had stable integration and dramatic growth enhancement maintained for several generations during the past 20 years.

C. Establishment of Transgenic Lineages

The process of establishing lineages of transgenic fish is not technically challenging or fundamentally different from that used in other vertebrate groups. Transgenic founders are placed with a wild-type mate or another transgenic founder, the F1 offspring are collected and DNA extracted from embryos, fry, or excised caudal fins is screened for the presence of the transgene using PCR or other techniques.

Although the process of establishing transgenic lineages is not technically difficult, it is made more complex because even though an organism, mammal or fish, is injected at the one-cell stage, mosaic integration of transgenes is universal. Late integration of transgenes in rapidly dividing fish cells contributes to the generation of transgenic animals that carry the transgene in only a portion of their tissues and a portion of their cells in the transgenic tissues. Naturally, the germ cells are also highly mosaic, such that a fish that harbors transgene sequences in fin tissues or other tissues may not transmit the transgene, or may transmit it only at very low frequencies. Expected germ line transmission frequencies of less than 10% can serve as a guide to determining the effort that may be required to produce multiple lineages of transgenic fish (Hackett, 1993), although transmission rates as high as 75% are possible (Hayat, 1989).

To compensate for high rates of mosaicism in fish, relatively large numbers of eggs (hundreds) are injected in most transgenic fish projects, which in turn necessitates the screening of large numbers of founders and their offspring. Investigators are advised to consider the scale of the tasks required to establish and maintain transgenic lineages and whether existing resources and personnel are sufficient to meet the challenge.

D. Derivation of Homozygous Fish Lineages

The process of deriving a transgenic lineage of fish is similar to that of other animals described in preceding chapters. Positively identified transgenic F1 brothers and sisters produced from a single transgenic founder are mated. The F2 generation fish are grown to a size sufficient to safely perform biopsy or bleeding for DNA extraction and PCR screening. The group of F2 fish should include 25% nontransgenics, 25% homozygotes, and 50% hemizygotes. Quantitative PCR methods can be used instead of DNA hybridization techniques to distinguish homozygous from hemizygous transgenic fish (Winn et al., 2001). The method involves detecting the twofold difference in gene copy number between the hemizygotes and homozygotes. To assess zygosity, replicate samples of DNA from the F2 generation fish (minimum of three samples per fish), are analyzed. These methods, as with DNA hybridization procedures, are not foolproof. Prior to breeding the presumptive homozygote with a homozygous mate, the individual should be tested genetically by crossing with a wild-type fish. The offspring from this cross are then tested as previously described using PCR. Homozygosity of the fish is confirmed if 100% of the offspring are positive. Homozygous F2 fish are then mated to one another to produce a homozygous population.

Generation of homozygous lines can be more complicated and can require additional screening in some cases. As discussed in Section 11.IV.A, multiple insertion sites in the genome are possible, thus it may take additional generations of progeny testing to clear the additional loci.

Another complication is inbreeding. Brother–sister matings result in inbreeding, which will lead to inbreeding depression for traits such as growth, viability, reproduction, and various disorders. Thus, additional generations of outbreeding and screening may be required to generate homozygous lines that are not inbred for the other loci. An alternative to prevent inbreeding from compromising performance evaluation is to outbreed the homozygous line and only test noninbred hemizygous fish. This may not be appropriate for all transgenes, especially knockout constructs.

V. Gene Knockout

A. Gene Knockout

In mice, it is possible to establish embryonic stem (ES) cell lines, accomplish gene knockout via homologous recombination (Capecchi, 1989), and then using transplantation generate transgenic mice (Babinet and Cohen-Tannoudji, 2001) for study of gene function and genomics, which has led to tremendous advances in genomics and human medicine (Muller, 1999). The combination of ES cell culture, homologous recombination, and transplantation to produce knockout transgenic fish has not been accomplished to date (Liu, 2007; Wong and Van Eenennaam, 2008). Hong et al. (2000) were able to establish ES lines in fish, but they lost their totipotency and ability to contribute to the germ line during culture. Major stumbling blocks to homologous recombination in fish have been a lack of ES cell culture and a lack of technology to transfer such cells while keeping germ cell potential. Fan et al. (2006) accomplished targeted plasmid insertion by homologous recombination in ES cells of the zebrafish *Danio rerio*. Nontransgenic, temporary gene knockdown with techniques such as morpholinos is common in model species such as zebrafish (Weidinger et al., 2003).

Homologous recombination is not a routine procedure, but it has many advantages compared to other knockout strategies (Fish and Kruithof, 2004; Liu, 2007; Wong and Van Eenennaam, 2008). Homologous recombination is very clean compared to other methods. Other technologies have much higher efficiency but result in random integration, greater knockdown in some tissues than others, instability, usually result in knockdown of a fraction of expression and unstable knockdown in subsequent generations. With that said, homologous recombination has not been accomplished in fish. shRNAi and various forms of antisense strategies are alternatives to homologous recombination (Wong and Van Eenennaam, 2008).

Various knockout approaches exist for sterilization and are under evaluation, including antisense against reproductive genes(Uzbekova et al., 2000a,b; Hu et al., 2006). Repressible Tet-OFF-based systems have been developed to knockdown genes essential for embryonic development or reproduction coupled with RNAi and RNA overexpression approaches (Thresher et al., 2009). A disadvantage of these systems is their relative complexity.

Gonad-specific transgenic excision utilizes site-specific recombinases, such as Cre and flippase (FLP), to exact excision, insertion, or inversion events in DNA at specific recognition sequences (Rodriguez et al., 2000; Wong and Van Eenennaam, 2008). This technology has been successfully demonstrated in zebrafish (Dong and Stuart, 2004; Le et al., 2007). Unfortunately, this strategy has several significant disadvantages. It generates new forms of heritable transgenics such as recombinase transgenics (Wong and Van Eenennaam, 2008), requires fertile transgenic brood stock, and the escaped recombinase-expressing transgenic fish could cause chromosomal rearrangements and reduce the fertility in native fish due to pseudorecognition sites in the genome (Thyagarajan et al., 2000; Wong and Van Eenennaam, 2008).

B. Gene Editing

A type of gene knockout that results in a mutant or nontransgenic individual is targeted gene editing using transfer of mRNA from specialized nucleases. Since these fish are not technically transgenic, any valuable phenotypic change should be able to be commercialized without regulatory approval. Although the genetic change is not transgenic, the basic procedures of transgenesis are utilized.

Technology for targeted gene editing is advancing rapidly. Zinc finger nucleases (ZFNs) were a major breakthrough that allowed targeted gene editing. Transcription activator-like effectors (TALEs) combined with the nuclease domain of FokI restriction enzyme, TALE nucleases (TALENs) were an improvement on both complexity and transformation rates. Modified clustered regularly interspaced short palindromic repeats (CRISPRs) systems are much simpler, with the potential to cause high rates of biallelic mutagenesis, allowing phenotypic evaluation in the original transformed fish (Jao et al., 2013).

ZFNs are fusion proteins of a modular DNA-binding domain yoked to a FokI endonuclease monomer. When a pair of ZFNs binds to their target in the correct orientation, FokI monomers can dimerize and introduce a DNA double-strand DNA break (Kim et al., 1996), which can be repaired by nonhomologous end joining. The result is a small insertion and/or deletion (indel) (Porteus and Carroll, 2005), which causes frameshift-disabling encoded proteins in 67% of the indels. In the case of ZFNs and TALENs, genes are disabled without introduction of exogenous DNA. ZFN knockout can be limited by available target sites (Miller et al., 2011; Carlson et al., 2012a; Reyon et al., 2012).

TALENs are nucleases that join FokI endonuclease with the modular DNA-binding domain of TALEs. TALENs are more easily manufactured with straightforward design and assembly strategies (Cermak et al., 2011), are less expensive, and can result in high knockout rates (2–75%, often >50%) in primary cells and embryos of livestock (Carlson et al., 2012b). Knockout can be homozygous for the same indel in up to 67% of the embryos for both ZFNs and TALENs (Urnov et al., 2005; Santiago et al., 2008; Kim et al., 2009; Liu et al., 2010; Carlson et al., 2012b), which exceeds predictions if cleavage/repair were independent events, suggesting that gene conversion of mediated genetic changes from the sister chromatid is common. Multiple deletions (knockout alleles) can be generated in a founder individual (Carbery et al., 2010; Nakajima et al., 2012) with up to six alleles found in a single founder. Higher doses of TALENs can stimulate higher mutation rates compared to low doses, 43% up to 75%, but has been found to reduce survival of the blastocysts threefold (Carlson et al., 2012b). In the case of fish, Dong et al. (2011) were able to edit the myostatin gene of the yellow catfish, (*Pelteobagrus fulvidraco*) using ZFN. Both ZFN knockout (Doyon et al., 2008) and TALEN knockout (Huang et al., 2011) have been accomplished in zebrafish.

The CRISPRs/CRISPR-associated protein (CRISPRs/Cas) system in eubacteria and archaea targets foreign DNA. Endonuclease Cas9 with CRISPR RNA (crRNA) and *trans*-acting RNA (tracrRNA) can cleave of foreign DNA, and a single guide RNA chimera mimicking the crRNA:tracrRNA complex in conjunction with Cas9 can produce site-specific DNA double-strand breaks in fish (Chang et al., 2013; Hwang et al., 2013; Jao et al., 2013) A modified Cas9 system optimized with zebrafish codons appears to be another quantum improvement compared to TALENs, both in increased simplicity and in the apparent production of homozygous/biallelic founder individuals that possess the mutation in almost every cell (Jao et al., 2013).

Recently, TALEN technology was used to produce single nucleotide alterations or small indels in mammals that mimic naturally occurring performance-enhancing or disease-resistance alleles (Tan et al., 2013). This allows meiosis-free intraspecific introgression of specific alleles. These approaches would not introduce exogenous DNA or introduce one or minimal base pairs of naturally existing alleles to fish. Again, "transgenic-like" technology is used, but the alleles are intraspecific and the fish technically nontransgenic, allowing commercial application in today's regulatory environment. This approach overcomes one of the limitations of marker-assisted selection, if the actual trait locus is known, as the valuable genotype can be gene-edited rather than introgressed through long-term, more complicated breeding made less efficient by strain-specific markers.

VI. Performance of Transgenic Aquaculture-Generated Species of Fish

Transgenic fish have been developed that have improved growth, color, disease resistance (Dunham, 2008), tolerance of heavy metals, survival in cold, body composition, and ability to produce pharmaceutical proteins (reviewed by Dunham and Liu, 2006; Dunham, 2011). The great potential of transgenic technology was demonstrated in transgenic salmon, with an average 11-fold increase of growth compared to nontransgenic controls (Devlin et al., 1995) and a 30× improvement of growth of mud loach, *Misgurnus mizolepis* (Nam et al., 2001), growth that probably can never be achieved by traditional breeding. Transgenic alteration of the nutritional characteristics of fish could be beneficial for consumers, and it is now possible to directly alter body composition via transgenesis. Zebrafish transfected with B-actin-salmon desaturase genes had enhanced levels of omega-3 fatty acids, docosahexaenoic acid (DHA), and eicosapentaenoic acid (EPA) in their flesh (Alimuddin et al., 2005, 2007).

The most effective mechanism for increasing disease resistance in catfish is transgenesis via transfer of cecropin genes (Dunham et al., 2002). The two- to four-fold increase in bacterial disease resistance exhibited by cecropin transgenics is much better than what is obtained through selection for disease resistance in channel catfish, which is sporadically successful in some, but not all strains of channel catfish (Dunham et al., 1993; Waters, 2001). The cecropin gene also imparts bacterial disease resistance to transgenic medaka and rainbow trout (Chiou et al., 2002, 2013). There is evidence that cecropin might also have antiviral and antifungal properties (Chiou et al., 2002, 2013) in rainbow trout, *Oncorhyncus mykiss.*

A. Pleiotropic Effects and Changes in Gene Expression

When a transgene is transferred to fish, it may affect more than the targeted trait, the pleiotropic effect. This is especially true with growth hormone gene transfer, which can affect feed conversion efficiency, body shape, and tolerance of low oxygen, disease resistance, color, reproductive traits, behavior, gut length, carcass yield, body composition, and muscle ultrastructure.

Additionally, the integration of a transgene can also affect the expression of entire suites of other endogenous genes. This has been observed for the transfer of growth hormone, cecropin, and desaturase genes in fish.

B. Commercial Application of Transgenic Fish

The first commercialization of transgenic fish was the Glo-fish, a transgenic zebrafish containing fluorescent protein genes from jellyfish, in the aquarium fish trade. These were not a concern for most regulatory agencies as they were not considered a food fish. They were not considered a risk to the environment because they were less fit than normal zebrafish, which have never been able to establish in the natural environment in the United States.

Applications are under consideration for approval of transgenic fish as food in Canada, China, Cuba, and the United States. In November 2013, for the first time commercial production of a transgenic food fish was approved but with limitations. Canada has approved the production of triploid, growth hormone transgenic, Atlantic salmon (*Salmo salar*) embryos. However, the embryos must be exported and only to countries that have appropriate confinement and approve the consumption of transgenic fish.

Two major concerns have prevented widespread use of transgenic fish in aquaculture: food safety and environmental risk. Major scientific organizations such as the US National Academy of Sciences, the Royal Society of London, and FAO/WHO have concluded that under the vast majority of cases transgenic fish would pose no human health risk. In certain cases, if transgenes were derived from certain organisms, allergenicity is the most likely health issue with consumption of transgenic fish.

The primary concern regarding transgenic fish is the potential risk to native gene pools and natural ecosystems. Several studies have demonstrated that transgenic fish have lower fitness than nontransgenic conspecifics (reviewed by Dunham, 2011). Thus, if they were to escape or were intentionally stocked, the transgene would be selected against and likely eliminated, resulting in no long-lasting environmental effects. The theory of the Trojan gene effect (Muir and Howard, 1999) contradicts this conclusion. However, the Trojan gene effect is an overly simplistic model using a very unnatural habitat without cover, with a fish (medaka) possessing atypical reproductive behavior, and the model does not include foraging, swimming, predator avoidance, or genotype-environment interactions. Regardless of the outcome of environmental risk assessment, it is likely that transgenic fish will not be allowed in aquaculture unless they are absolutely confined. The best option in this regard is transgenic sterilization.

VII. Care and Containment of Transgenic Fish

Maintenance of wild-type and transgenic fish under conditions that optimize their health is vitally important to the ultimate success of the transgenic fish project. Variations in numerous conditions of fish culture such as density, behavior, diet, and water quality can have profound and unpredictable influences on the reproducibility of measurements of gene function or other traits. Loss of an individual fish at any point during the process due to disease or poor care may, at a minimum, hamper efficient production, and availability of transgenic fish lineages or, worse, can be catastrophic to an entire study.

Novices may initially give issues of animal care a low priority; however, investigators should be keenly aware of the importance of proper animal care and should make concerted efforts to improve training of personnel; to increase the body of research on diet, disease diagnosis, and prevention; and to seek improved institutional support for fish culture facilities (Nairn et al., 2001). With the current Animal Welfare guidelines for publicly supported research institutions, proper animal care is a requirement.

Procedures and facilities used to maintain transgenic fish should be reviewed to ensure that sufficient precautions are being taken to safeguard against releases of transgenic fish into the environment. Guidance in this review is provided by the guidelines proposed for the development, care, and use of genetically modified fish and shellfish (USDA, 1995). In addition, state regulations should be examined related to the introduction, long-term culture, and transportation of exotic or nonnative species. Methods of containment should be in place to ensure that there is no direct access of eggs, juveniles, or adult fish from a facility to the environment. Typically, acceptable measures include the placement of screens or other physical barriers over drains and restricting passage of screened culture water only into permitted septic or water-treatment facilities.

If transgenic fish are ever to be approved for commercial food production, absolute, infallible confinement will be required. The ultimate method would be to have complete reproductive control of the fish. This might be accomplished through transgenic sterilization, including gene editing approaches. A transgenic sterilization system would need to be repressible to allow production of the transgenic genotype. The most logical options include antisense approaches, knockout with shRNAi or cDNA overexpression, or gene editing, all addressing either survival genes or essential reproductive genes. Progress has been made with these approaches but has not yet been perfected (Uzebekova et al., 2000a, 2000b; Thresher et al., 2009; Su et al., 2014). In these systems, the fish are not able to breed without the intervention of man.

VIII.   Future of Transgenic Fish

The field of transgenic fish research is maturing, as evidenced by the increasing numbers of transgenic fish being produced. Examination of the most recent research addressing transgenic fish (via PubMed search, among others) shows that the majority, >80%, of transgenic fish research during 2013 was conducted using model species for medical studies, developmental studies, and functional genomics research. The remaining 20% of research in 2013 examined aquaculturally relevant fish species. As genetic research moves from descriptive to functional genomics over the coming years, fish will offer appealing opportunities. In particular, the tremendous diversity of biology exhibited in fish should provide valuable resources to aid in discerning the relative importance of various environmental factors on gene function.

It is likely that transgenic fish and gene-edited fish will be widely used in the future. The best genotypes will be developed by using combinations of traditional, biotechnological, and transgenic approaches.

References

Alimuddin, Y.G., Kiron, V., Satoh, S., Takeuchi, T., 2005. Enhancement of EPA and DHA biosynthesis by over-expression of masu salmon delta 6-*desaturase*-like gene in zebrafish. Transgenic Res. 14, 159–165.

Alimuddin, Y.G., Kiron, V., Satoh, S., Takeuchi, T., 2007. Expression of masu salmon delta5-desaturase-like gene elevated EPA and DHA biosynthesis in zebrafish. Mar. Biotechnol. (NY) 9, 92–100.

Amsterdam, A., Hopkins, N., 1999. Retrovirus-mediated insertional mutagenesis in zebrafish. In: Detrich, H., Westerfield, M., Zon, L. (Eds.), The Zebrafish: Genetics and Genomics. Academic Press, San Diego, CA, pp. 87–98.

Babinet, C., Cohen-Tannoudji, M., 2001. Genome engineering via homologous recombination in mouse embryonic stem (ES) cells: an amazingly versatile tool for the study of mammalian biology. An. Acad. Bras. Cienc. 73, 365–383.

Boyes, J., Bird, A., 1991. DNA methylation inhibits transcription indirectly via a methyl-CpG binding protein. Cell 64, 1123–1134.

Buono, R.J., Linser, P.J., 1992. Transient expression of RSVCAT in transgenic zebrafish made by electroporation. Mol. Mar. Biol. Biotechnol. 1 (4/5), 271–275.

Capecchi, M.R., 1989. Altering the genome by homologous recombination. Science 244, 1288–1292.

Carbery, I.D., Ji, D., Harrington, M., Brown, V., Weinstein, E.J., Liaw, L., et al., 2010. Targeted genome modification in mice using zinc-finger nucleases. Genetics 186, 451–459.

Carlson, D.F., Fahrenkrug, S.C., Hackett, P.B., 2012a. Targeting DNA with fingers and TALENs. Mol. Ther. Nucleic Acids 1, e3.

Carlson, D.F., Tana, W., Lillico, S.G., Stverakova, D., Proudfoot, C., Christiana, M., et al., 2012b. Efficient TALEN-mediated gene knockout in livestock. Proc. Natl. Acad. Sci. USA 109, 17382–17387.

Cermak, T., Doyle, E.L., Christian, M., Wang, L., Zhang, Y., Schmidt, C., et al., 2011. Efficient design and assembly of custom TALEN and other TAL effector-based constructs for DNA targeting. Nucleic Acids Res. 39, e82.

Chang, N., Sun, C., Gao, L., Zhu, D., Xu, X., Zhu, X., et al., 2013. Genome editing with RNA-guided Cas9 nuclease in zebrafish embryos. Cell Res. 23, 465–472.

Chen, H.L., Yang, H.S., Huang, R., Tsai, H.J., 2006. Transfer of a foreign gene to Japanese abalone (*Haliotis diversicolor supertexta*) by direct testis-injection. Aquaculture 253, 249–258.

Chen, T.T., Powers, D.A., 1990. Transgenic fish. Trends Biotechnol. 8, 209–215.

Cheng, Q., 2013. The effect of diet and the masou salmon delta5-desaturase transgene on delta6-desaturase and stearoyl-coa desaturase gene expression and n-3 fatty acid level in common carp (*Cyprinus carpio*). M.S. thesis. Auburn University, Auburn, AL.

Chiou, P.P., Chen, M.J., Lin, C.M., Khoo, J., Larson, J., Holt, R., et al., 2013. Production of homozygous transgenic rainbow trout with enhanced disease resistance. Mar. Biotechnol. (NY). Available from: http://dx.doi.org/10.1007/s10126-013-9550-z. (Epub ahead of print).

Chiou, P.P., Lin, C.-M., Perez, L., Chen, T.T., 2002. Effect of cecropin B and a synthetic analogue on propagation of fish viruses *in vitro*. Mar. Biotechnol. 4, 294–302.

Chourrout, D., Guyomard, R., Houdebine, L.M., 1986. High efficiency gene transfer in rainbow trout (*Salmo gairdneri* Rich.) by microinjection into egg cytoplasm. Aquaculture 51, 143–150.

Davis, K.B., 2009. Age at puberty of channel catfish, *Ictalurus punctatus*, controlled by thermoperiod. Aquaculture 299, 244–250.

Devlin, R.H., Yesaki, T.Y., Donaldson, E.M., Du, S., Hew, C., Du, S.J., et al., 1995. Production of germline transgenic Pacific salmonids with dramatically increased growth performance. Can. J. Fish. Aquat. Sci. 52, 1376–1384.

Dong, J., Stuart, G.W., 2004. Transgene manipulation in zebrafish by using recombinases. Methods Cell. Biol. 77, 363–379.

Dong, Z., Ge, J., Li, K., Xu, Z., Liang, D., Li, J., et al., 2011. Heritable targeted inactivation of myostatin gene in yellow catfish (*Pelteobagrus fulvidraco*) using engineered zinc finger nucleases. PLoS ONE 6, e28897. Available from: http://dx.doi.org/10.1371/journal.pone.0028897.

Doyon, Y., McCammon, J.M., Miller, J.C., Faraji, F., Ngo, C., Katibah, G.E., et al., 2008. Heritable targeted gene disruption in zebrafish using designed zinc-finger nucleases. Nat. Biotechnol. 26, 702–708.

Dunham, R.A., 2008. Transgenic fish resistant to infectious diseases, their risk and prevention of escape into the environment and future candidate genes for disease transgene manipulation. Comp. Immunol. Microbiol. Infect. Dis. 32, 139–161. Available from: http://dx.doi.org/10.1016/j.cimid.2007.11.006.

Dunham, R.A., 2011. Aquaculture and Fisheries Biotechnology: Genetic Approaches. second ed. CABI Publishing, Wallingford, UK.

Dunham, R.A., Liu, Z., 2006. Transgenic fish: where we are and where do we go? Israeli J. Aquacult. 58, 297–319.

Dunham, R.A., Brady, Y., Vinitnantharat, S., 1993. Response to challenge with *Edwardsiella ictaluri* by channel catfish, *Ictalurus punctatus*, selected for resistance to *E. ictaluri*. J. Appl. Aquacult. 3, 211–222.

Dunham, R.A., Warr, G., Nichols, A., Duncan, P.L., Argue, B., Middleton, D., et al., 2002. Enhanced bacterial disease resistance of transgenic channel catfish, *Ictlaurus punctatus*, possessing cecropin genes. Mar. Biotechnol. (NY) 4, 338–344.

Ehrlich, M., Zhang, X., Asiedu, C.K., Khan, R., Supakar, P.C., 1990. Methylated DNA-Binding Protein from Mammalian Cells. Alan R. Liss, New York, NY.

Eissenberg, J.C., Elgin, S.C.R., 1991. Boundary functions in the control of gene expression. Trends Genet. 7, 335–340.

Fan, L., Moon, J., Crodian, J., Collodi, P., 2006. Homologous recombination in zebrafish ES cells. Transgenic Res. 15, 21–30.

FAO/WHO, 2003. FAO/WHO expert consultation on safety assessment of foods derived from genetically modified animals including fish. <http://www.fao.org/es/ESN/food/risk_biotech_animal_en.stm>. (accessed 21.11.13).

Fish, R.J., Kruithof, E.K., 2004. Short-term cytotoxic effects and long-term instability of RNAi delivered using lentiviral vectors. BMC Mol. Biol. 5, 9.

Fletcher, G.L., Davies, P.L., 1991. Transgenic fish for aquaculture. Genet. Eng. 13, 331–370.

Foecking, M.K., Hofstetter, H., 1986. Powerful and versatile enhancer-promoter unit for mammalian expression vectors. Gene 45, 101–105.

Gaiano, N., Amsterdam, A., Kawakami, K., Allende, M., Becker, T., Hopkins, N., 1996. Insertional mutagenesis and rapid cloning of essential genes in zebrafish. Nature 383, 829–832.

Gibbs, P.D., Schmale, M.C., 2000. GFP as a genetic marker scorable throughout the life cycle of transgenic zebra fish. Mar. Biotechnol. (NY) 2, 107–125.

Gorman, C.M., Merlino, G.T., Willingham, M.C., Pastan, I., Howard, B.H., 1982a. The Rous sarcoma virus long terminal repeat is a strong promoter when introduced into a variety of eukaryotic cells by DNA-mediated transfection. Proc. Natl. Acad. Sci. USA 79, 6777–6781.

Gorman, C.M., Moffat, L.F., Howard, B.H., 1982b. Recombinant genomes which express chloramphenicol acetyltransferase in mammalian cells. Mol. Cell. Biol. 2, 1044–1051.

Guyomard, R., Chourrout, D., Leroux, C., Houdebine, L.M., Pourrain, F., 1989. Integration and germ line transmission of foreign genes microinjected into fertilized trout eggs. Biochimie 71, 857–863.

Hackett, P.B., 1993. The molecular biology of transgenic fish. In: Hochachka, Mommsen (Eds.), Biochemistry and Molecular Biology of Fishes. Elsevier Science, Amsterdam, pp. 207–240.

Hallerman, E.M., Schneider, J.F., Gross, M., Faras, A.J., Hackett, P.B., Guise, K.S., et al., 1988. Enzymatic dechorionation of goldfish, walleye, and northern pike eggs. Trans. Am. Fish. Soc. 117, 456–460.

Hayat, M., 1989. Transfer, expression and inheritance of growth hormone genes in channel catfish (Ictalurus punctatus) and common carp (Cyprinus carpio). Ph.D. dissertation. Auburn University, Auburn, AL.

Hayat, M., Joyce, C.P., Townes, T.M., Chen, T.T., Powers, D.A., Dunham, R.A., 1991. Survival and integration rate of channel catfish and common carp embryos microinjected with DNA at various developmental stages. Aquaculture 99, 249–255.

Hong, Y., Chen, S., Schartl, M., 2000. Embryonic stem cells in fish: current status and perspectives. Fish Physiol. Biochem. 22, 165–170.

Hsiao, C., Hsieh, F., Tsai, H., 2001. Enhanced expression and stable transmission of transgenes flanked by inverted terminal repeats from adeno-associated virus in zebrafish. Dev. Dyn. 220, 323–336.

Hu, W., Wang, Y., Zhu, Z., 2006. A perspective on fish gonad manipulation for biotechnical applications. Chin. Sci. Bull. 51, 1–6.

Huang, P., Xiao, A., Zhou, M., Zhu, Z., Lin, S., Zhang, B., 2011. Heritable gene targeting in zebrafish using customized TALENs. Nat. Biotechnol. 29, 699–700.

Hwang, W.Y., Hwang, W.Y., Fu, Y., Reyon, D., Maeder, M.L., Tsai, S.Q., et al., 2013. Efficient genome editing in zebrafish using a CRISPR-Cas system. Nat. Biotechnol. 31, 227–229.

Inoue, K., 1992. Expression of reporter genes introduced by microinjection and electroporation in fish embryos and fry. Mol. Mar. Biol. Biotechnol 1, 266–270.

Inoue, K., Yamashita, S., Hata, J.I., Kabeno, S., Asada, S., Nagahisa, E., et al., 1990. Electroporation as a new technique for producing transgenic fish. Cell Differ. Dev. 29 (2), 123–128.

Ivics, Z., Izsvak, Z., Hackett, P., 1999. Genetic applications of transposons and other repetitive elements in zebrafish. In: Detrich, H., Westerfield, M., Zon, L. (Eds.), The Zebrafish: Genetics and Genomics. Academic Press, San Diego, CA, pp. 99–131.

Iwamatsu, T., 1994. Stages of normal development in the medaka *Oryzias latipes*. Zool. Sci. 11, 825–839.

Iyengar, A., Muller, F., Maclean, N., 1996. Regulation and expression of transgenes in fish—a review. Transgenic Res. 5, 147–166.

Jao, L.E., Burgess, S.M., 2009. Production of pseudotyped retrovirus and the generation of proviral transgenic zebrafish. Methods Mol. Biol. 546, 13–30. Available from: http://dx.doi.org/10.1007/978-1-60327-977-2_2.

Jao, L.E., Wente, S.R., Chen, W., 2013. Efficient multiplex biallelic zebrafish genome editing using a CRISPR nuclease system. Proc. Natl. Acad. Sci. USA 110, 13904–13909. 10.1073/pnas.1308335110.

Jessen, J.R., Meng, A., McFarlane, R.J., Paw, B.H., Zon, L.I., Smith, G.R., et al., 1998. Modification of bacterial artificial chromosomes through chi-stimulated homologous recombination and its application in zebrafish transgenesis. Proc. Natl. Acad. Sci. USA 95, 5121–5126.

Kang, J.H., Yoshizaki, G., Homma, O., Strussmann, C.A., Takashima, F., 1999. Effect of an osmotic differential on the efficiency of gene transfer by electroporation of fish spermatozoa. Aquaculture 173, 297–307.

Kim, H.J., Lee, H.J., Kim, H., Cho, S.W., Kim, J.S., 2009. Targeted genome editing in human cells with zinc finger nucleases constructed via modular assembly. Genome Res. 19, 1279–1288.

Kim, Y.G., Cha, J., Chandrasegaran, S., 1996. Hybrid restriction enzymes: zinc finger fusions to Fok I cleavage domain. Proc. Natl. Acad. Sci. USA 93, 1156–1160.

Kinoshita, M., Ozato, K., 1995. Cytoplasmic microinjection of DNA into fertilized medaka (*Oryzias latipes*) eggs. Medaka 7, 59–64.

Kolenikov, V.A., Alimov, A.A., Barmintsev, V.A., Benyumov, A.O., Zelenina, I.A., Krasnov, A.M., et al., 1990. High velocity mechanical injection of foreign DNA into fish eggs. Genetika 26, 2122–2126.

Kurita, K., Burgess, S.M., Sakai, N., 2004. Transgenic zebrafish produced by retroviral infection of *in vitro*-cultured sperm. Proc. Natl. Acad. Sci. USA 101, 1263–1267.

Le, X., Langenau, D.M., Keefe, M.D., Kutok, J.L., Neuberg, D.S., Zon, L.I., 2007. Heat shock-inducible Cre/Lox approaches to induce diverse types of tumors and hyperplasia in transgenic zebrafish. Proc. Natl. Acad. Sci. USA 104, 9410–9415.

Lee, K.-Y., Huang, H., Ju, B., Yang, Z., Shuo, 2002. Cloned zebrafish by nuclear transfer from long-term-cultured cells. Nat. Biotechnol. 20, 795–799.

Lin, S., Gaiano, N., Culp, P., Burns, J.C., Friedman, T., Yee, J.K., et al., 1994. Integration and germ-line transmission of a pseudotyped retroviral vector in zebrafish. Science 265, 666–668.

Liu, P.Q., Chan, E.M., Cost, G.J., Zhang, L., Wang, J., Miller, J.C., et al., 2010. Generation of a triple-gene knockout mammalian cell line using engineered zinc-finger nucleases. Biotechnol. Bioeng. 106, 97−105.

Liu, Z., 2007. Aquaculture Genome Technologies. Blackwell Publishing, Ames, IA.

Lu, J.K., Burns, J.C., Chen, T., 1997. Pantropic retroviral vector integration, expression, and germline transmission in medaka (*Oryzias latipes*). Mol. Mar. Biol. Biotechnol. 6, 289−295.

Lu, J.-K., Fu, B.-H., Wu, J.-L., Chen, T.T., 2002. Production of transgenic silver sea bream (*Sparus sarba*) by different gene transfer methods. Mar. Biotechnol. (NY) 4, 328−337.

Ma, C., Fan, L., Ganassin, R., Bols, N., Collodi, P., 2001. Production of zebrafish germ-line chimeras from embryo cell cultures. Proc. Natl. Acad. Sci. USA 98, 2461−2466.

Maclean, N., 1998. Regulation and exploitation of transgenes in fish. Mutat. Res. 399, 255−266.

Maclean, N., Penman, D., Zhu, Z., 1987. Introduction of novel genes into fish. Biotechnology 5, 257−261.

Meng, A., Jessen, J.R., Lin, S., 1999. Transgenesis. In: Detrich, H., Westerfield, M., Zon, L. (Eds.), The Zebrafish: Genetics and Genomics. Academic Press, San Diego, CA, pp. 133−148.

Miller, J.C., Tan, S., Qiao, G., Barlow, K.A., Wang, J., Xia, D.F., et al., 2011. A TALE nuclease architecture for efficient genome editing. Nat. Biotechnol. 29, 143−148.

Muir, W.M., Howard, R.D., 1999. Possible ecological risks of transgenic organism release when trans-genes affect mating success: sexual selection and the Trojan gene hypothesis. Proc. Natl. Acad. Sci. USA 96, 13853−13856.

Muller, F., Ivics, Z., Erdelyi, F., Papp, T., Varadi, L., Horvath, L., et al., 1992. Introducing foreign genes into fish eggs with electroporated sperm as a carrier. Mol. Mar. Biol. Biotechnol. 1 (4/5), 276−281.

Müller, U., 1999. Ten years of gene targeting: targeted mouse mutants, from vector design to phenotype analysis. Mech. Dev. 82, 3−21.

Nairn, R.S., Schmale, M.C., Stegeman, J., Winn, R.N., Water, R.B., 2001. Aquaria fish models of human disease: reports and recommendations from the working groups. Mar. Biotechnol. (NY) 3, S248−S258.

Nakajima, K., Nakajima, T., Takase, M., Yaoita, Y., 2012. Generation of albino *Xenopus tropicalis* using zinc-finger nucleases. Dev. Growth Differ. 54, 777−784. Available from: http://dx.doi.org/10.1111/dgd.1200.

Nam, Y.K., Noh, J.K., Cho, Y.S., Cho, H.J., Cho, K.-N., Kim, C.G., et al., 2001. Dramatically accelerated growth and extraordinary gigantism of transgenic mud loach *Misgurnus mizolepis*. Transgenic Res. 10, 353−362.

Nichols, A., 1998. Comparison of electroporation and particle bombardment gene transfer techniques in channel catfish, *Ictalurus punctatus* and common carp, *Cyprinus carpio*. M.S. thesis. Auburn University.

Ozato, K., Kondoh, H., Inohara, H., Iwamatsu, T., Wakamatsu, Y., Okada, T.S., 1986. Production of transgenic fish: introduction and expression of chicken d-crystallin gene in medaka embryos. Cell Differ. 19, 237−244.

Porteus, M.H., Carroll, D., 2005. Gene targeting using zinc finger nucleases. Nat. Biotechnol. 23, 967−973.

Powers, D.A., Hereford, L., Cole, T., Chen, T.T., Lin, C.M., Kight, K., et al., 1992. Electroporation: a method for transferring genes into the gametes of zebrafish (*Brachydanio rerio*), channel catfish (*Ictalurus punctatus*), and common carp (*Cyprinus carpio*). Mol. Mar. Biol. Biotechnol. 1, 301−308.

Reyon, D., Tsai, S.Q., Khayter, C., Foden, J.A., Sander, J.D., Joung, J.K., 2012. FLASH assembly of TALENs for high-throughput genome editing. Nat. Biotechnol. 30, 460–465.

Rodriguez, C.I., Buchholz, F., Galloway, J., Sequerra, R., Kasper, J., Ayala, R., et al., 2000. High-efficiency deleter mice show that FLPe is an alternative to Cre-*loxP*. Nat. Genet. 25, 139–140.

Rokkones, E., Alestrom, P., Skjervold, H., Gautvik, K.M., 1985. Development of a technique for microinjection of DNA into salmonid eggs. Acta Physiol. Scand. 124 (Suppl. 542), 417.

Sakai, N., 2006. *In vitro* male germ cell cultures of zebrafish. Methods 39, 239–245.

Santiago, Y., Chan, E., Liu, P.Q., Orlando, S., Zhang, L., Urnov, F.D., et al., 2008. Targeted gene knockout in mammalian cells by using engineered zinc-finger nucleases. Proc. Natl. Acad. Sci. USA 105, 5809–5814.

Sarmasik, A., Chun, C.Z., Jang, I.K., Lu, J.K., Chen, T.T., 2001. Production of transgenic live-bearing fish and crustaceans with replication-defective pantropic retroviral vectors. Mar. Biotechnol. (NY) 3, S177–S184.

Stief, A., Winter, D.M., Stratling, W.H., Sippel, A.E., 1989. A nuclear DNA attachment element mediates elevated and position-independent gene activity. Nature 341, 343–345.

Su, B., Peatman, E., Shang, M., Thresher, R., Grewe, P., Patil, J., et al., 2014. Expression and knockdown of primordial germ cell genes, vasa, nanos and dead end in common carp (*Cyprinus carpio*) embryos for transgenic sterilization and reduced sexual maturity. Aquaculture 420–421 (S1), S72–S84. http://dx.doi.org/10.1016/j.aquaculture.2013.07.008.

Tan, W., Carlson, D.F., Lancto, C.A., Garbe, J.R., Webster, D.A., Hackett, P.B., et al., 2013. Efficient nonmeiotic allele introgression in livestock using custom endonucleases. Proc. Natl. Acad. Sci. USA 110, 16526–16531. <http://www.ncbi.nlm.nih.gov/pubmed/24014591>.

Tanaka, M., Kinoshita, M., Kobayashi, D., Nagahama, Y., 2001. Establishment of medaka (*Oryzias latipes*) transgenic lines with the expression of green fluorescent protein fluorescence exclusively in germ cells: a useful model to monitor germ cells in a live vertebrate. Proc. Natl. Acad. Sci. USA 98, 2544–2549.

Thresher, R., Dunham, R., Grewe, P., Whyard, S., Patil, J., Templeton, C.M., et al., 2009. Development of repressible sterility to prevent the establishment of feral populations of exotic and genetically modified animals. Aquaculture 290, 104–109.

Thyagarajan, B., Guimaraes, M.J., Groth, A.C., Calos, M.P., 2000. Mammalian genomes contain active recombinase recognition sites. Gene 244, 47–54.

Udvardy, A., Maine, E., Schedl, P., 1985. The 87A7 chromomere: identification of novel chromatin structures flanking the heat shock locus that may define the boundaries of higher order domains. J. Mol. Biol. 185, 341–358.

Urnov, F.D., Miller, J.C., Lee, Y.L., Beausejour, C.M., Rock, J.M., Augustus, S., et al., 2005. Highly efficient endogenous human gene correction using designed zinc-finger nucleases. Nature 435, 646–651.

USDA, 1995. Performance standards for safely conducting research with genetically modified fish and shellfish (final draft 4/15/95). Prepared by the working group on aquatic biotechnology and environmental safety. Agricultural Biotechnology Research Advisory Committee, US Department of Agriculture, Washington, DC.

Uzbekova, S., Chyb, J., Ferriere, F., Bailhache, T., Prunet, P., Alestrom, P., et al., 2000a. Transgenic rainbow trout expressed sGnRH-antisense RNA under the control of sGnRH promoter of Atlantic salmon. J. Mol. Endocrinol. 25, 337–350.

Uzbekova, S., Alestrom, P., Ferriere, F., Bailhache, T., Prunet, P., Breton, B., 2000b. Expression of recombinant sGnRH-antisense RNA under the control of either specific or constitutive promoters in transgenic rainbow trout. ISFE Abstracts (W-296).

Wakamatsu, Y., Pristyaznhyuk, I., Niwa, K., Ladygine, T., Kinoshita, M., Araki, K., et al., 2001. Fertile and diploid nuclear transplants derived from embryonic cells of a small laboratory fish, medaka (*Oryzias latipes*). Proc. Natl. Acad. Sci. USA 98 (3), 1071−1076.

Waters, P.J., 2001. Response and correlated response to selection of channel catfish, *Ictalurus punctatus* for resistance to *Edwardsiella ictaluri, Flexibacter columnaris*, low levels of dissolved oxygen and toxic levels of nitrite. M.S. thesis. Auburn University, AL.

Weidinger, G., Stebler, J., Slanchev, K., Dumstrei, K., Wise, C., Lovell-Badge, R., et al., 2003. *dead end,* a novel vertebrate germ plasm component, is required for zebrafish primordial germ cell migration and survival. Curr. Biol. 13, 1429−1434.

Winn, R.N., Van Beneden, R.J., Burkhart, J.G., 1995. Transfer, methylation and spontaneous mutation frequency of fX174*am3cs70* sequences in medaka (*Oryzias latipes*) and mummichog (*Fundulus heteroclitus*): implications for gene transfer and environmental mutagenesis in aquatic species. Mar. Environ. Res. 40 (3), 247−265.

Winn, R.N., Norris, M.B., Brayer, K.J., Torres, C., Muller, S.L., 2000. Detection of mutations in transgenic fish carrying a bacteriophage lambda *cII* transgene target. Proc. Natl. Acad. Sci. USA 97 (23), 12655−12660.

Winn, R.N., Norris, M., Muller, S., Torres, C., Brayer, K., 2001. Bacteriophage lambda and plasmid pUR288 transgenic fish models for detecting *in vivo* mutations. Mar. Biotechnol. (NY) 3, S185−S195.

Wong, A.C., Van Eenennaam, A.L., 2008. Transgenic approaches for the reproductive containment of genetically engineered fish. Aquaculture 275, 1−12. Available from http://dx.doi.org/10.1016/j.aquaculture.2007.12.026.

Yang, S.Y., Wang, J.G., Cui, H.X., Sun, S.G., Li, Q., Gu, L., et al., 2007. Efficient generation of transgenic mice by direct intraovarian injection of plasmid DNA. Biochem. Biophys. Res. Commun. 358, 266−271.

Yee, J.K., Friedmann, T., Burns, J.C., 1994. Generation of high-titer pseudotyped retroviral vectors with very broad host range. Methods Cell Biol. 43 (Pt A), 99−112.

Yoon, S.J., Liu, Z., Kapuscinski, A.R., Hackett, P.B., Faras, A., Guise, K.S., 1990. Successful gene transfer in fish. In: Svrjcek, R.S. (Ed.), Genetics in Aquaculture. Proceedings of the 16th US−Japan Meeting on Aquaculture, Charleston, South Carolina, USA pp. 39−44.

Zhu, Z., Li, G., He, L., Chen, S., 1985. Novel gene transfer into the fertilized eggs of gold fish (*Carassius auratus* L. 1758). Z. Angew. Ichthyol. 1, 31−34.

12 Production of Transgenic Poultry

James N. Petitte and Paul E. Mozdziak

Prestage Department of Poultry Science, College of Agriculture and Life Sciences, North Carolina State University, Raleigh, NC

I. Introduction and Discussion

A. Introduction

The use of transgenic animals of various species has increased significantly during the last 30 years and is now an established tool to expand our understanding of almost every biological process including vertebrate development, immunology, and cancer. At the same time, the application of transgenic technology to economically important species of livestock, poultry, and fish is expanding in animal agriculture. In 2012, the value of broilers, eggs, and turkeys was $38 billion. Given the huge economic worth of the commercial poultry industry in the United States and globally, interest in genetic engineering of the chicken remains strong. The goal for gene transfer for commercial stocks is simply to develop genetically better birds for the production of meat and eggs, which has been the same purpose of conventional selection programs for the past 50 years. The most often-cited application of transgenics in poultry is the introduction of disease resistance and, to a lesser extent, an increase in the efficiency of production. In addition to the obvious applications in agriculture, the domestic laying hen is being viewed as a dual purpose research animal with other applications in developmental biology, biomedicine, and biomanufacturing. Hence, transgenic technology in poultry can influence entirely new industries outside of agriculture.

B. Opportunities for Intervention

Because of the considerable differences between mammals and birds in reproductive biology, many procedures for gene transfer of poultry have required adaptation of the standard techniques used for mammals; and in some cases, the development of specific approaches for birds. The most obvious reproductive difference is the oviposition of an egg, consisting of a large yolk-filled ovum surrounded by layers of albumen encased in a calcified shell. Construction of the avian egg begins with ovulation of the mature ovum, which is engulfed by the infundibulum of the oviduct. It is here that fertilization takes place. The fertilized egg enters the

Transgenic Animal Technology. DOI: http://dx.doi.org/10.1016/B978-0-12-410490-7.00012-8

magnum, which secretes the albumen. Subsequently, the ovum surrounded by a firm albumen capsule enters the isthmus, where the outer and inner shell membranes are deposited in preparation for deposition of the eggshell. At this point, the fertilized ovum has spent about 3.5–4 h navigating the oviduct. The first cleavage divisions occur upon entry of the ovum into the shell gland. It takes about 20–22 h for complete formation of the eggshell. During that period, a considerable amount of cell division takes place and the embryo acquires its polarity (e.g., anterior/posterior), yet it is visually radially symmetric. By the time the egg is laid, the embryo or blastoderm comprises about 50,000–60,000 cells on the surface of the yolk mass. At this point, the blastoderm can be divided into two regions, a peripheral ring of cells attached to the yolk called the area opaca and a central more translucent region, the area pellucida, which is suspended above a non-yolky fluid deposited by the embryo (Figure 12.1). The area pellucida can be divided further into the marginal zone at the periphery and the central disk. The area opaca will contribute only to extraembryonic structures, and the bulk of the embryo will develop from the area pellucida. Upon incubation, the area pellucida differentiates into two layers, an upper epiblast and a lower hypoblast. Only the epiblast will give rise to the embryo proper, while the hypoblast contributes to some extraembryonic tissues. This period of development (i.e., fertilization through hypoblast formation) has been classified into a series of 14 stages by Eyal-Giladi and Kochav (1976) for

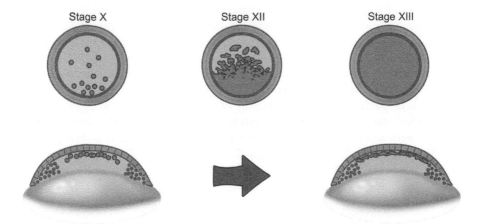

Figure 12.1 Diagrammatic representation of the structure of the early avian embryo at the time of oviposition and during formation of the hypoblast (stages X-XIII; Eyal-Giladi and Kochav, 1976). Upper panels are ventral views, lower panels are cross sections. Generally, a freshly laid egg contains an embryo at stage X, characterized by complete formation of the area pellucida. Only the area opaca is in contact with the yolk, creating a subgerminal cavity beneath the area pellucida. Upon incubation, growth of the hypoblast begins from the posterior marginal zone and by stage XII covers about half of the central disk of the area pellucida. At stage XIII, the area pellucida becomes a bilayered structure with a distinctive epiblast and hypoblast. To produce transgenic chickens using vectors, it is best to inject viral particles beneath the epiblast of a stage X embryo.

the domestic hen and 11 stages by Bakst et al. (1997) for the turkey, indicated using Roman numerals. Subsequent stages for both species are classified using the staging system of Hamburger and Hamilton (1951), using Arabic numerals. All references to stages of development will refer to those of the domestic hen.

The next period of embryo development that is significant for the production of transgenic poultry is during the establishment of the germ line. The development of the avian germ line has been examined for a century. Figure 12.2 summarizes the developmental pathway of avian germ cells from incubation onwards. Swift (1914) was the first to show the presence of primordial germ cells (PGCs), in an

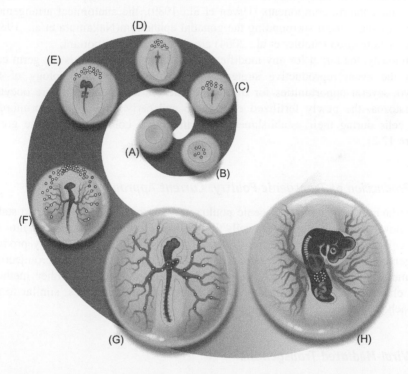

Figure 12.2 A schematic representation of the developmental history of primordial germ cells (PGCs) in the chick embryo. Committed germ cells have not been demonstrated in the stage X embryo (A), although some cells express a chicken VASA homolog and begin to express SSEA-1. At stage XIII, SSEA-1 marks a population of cells on the hypoblast that can give rise to germ cells (B). These SSEA-1-positive hypoblast cells move anteriorly during gastrulation and headfold stages (stages 3−8) to form the germinal crescent (C-E) described by Swift (1914). During the formation of blood islands and the vasculature (F), the germ cells enter the embryonic circulation at stages 14−16 (G) until they colonize the gonadal ridge at stages 20−22 (H). PGCs can be readily identified after gastrulation using periodic acid-Schiff's staining or immunohistochemistry for SSEA-1 or chicken VASA homolog. The circuitous journey of avian germ cells is thought to involve periods of passive and active migration guided by morphogenetic movements, chemotaxis, extracellular matrix components, and the vascular configuration.

extraembryonic region called the germinal crescent, well before development of the gonad. Swift's observations, which were based on the morphological character-istics of PGCs, were later confirmed by several investigators (Goldsmith, 1926; Willier, 1937). The migration of PGCs from the germinal crescent to the gonadal ridge occurs in two phases. First, PGCs are carried to the vicinity of the germinal ridge passively through the intra- and extraembryonic circulation (Swift, 1914; Meyer, 1964; Fujimoto et al., 1976a; Fujimoto et al., 1976b). Second, the blood-borne PGCs leave the vessels and migrate actively into the germinal epithelium via the dorsal mesentery. In this second and active phase of migration, chemotactic sig-nals released from the gonad (Dubois and Croisille, 1970; Kuwana et al., 1986), extracellular matrix components (Urven et al., 1989), the anatomical arrangement of the vascular system surrounding the gonadal epithelium(Nakamura et al., 1988), and chemoattraction (Stebler et al., 2004) are thought to be important.

Obviously, the target for any modification of the avian genome is the germ cell. Given the avian reproductive strategy and the developmental biology of the embryo, several opportunities for transgenic intervention include mature oocytes/spermatozoa, the newly fertilized egg/zygote and early embryo, and primordial germ cells during their establishment, migration, and colonization of the gonad (Figure 12.2).

C. Production of Transgenic Poultry: Current Approaches

All methods of producing transgenic poultry rely on techniques designed to stably insert novel genetic material into cells that will give rise to germ cells or into the germ cells *per se*. Currently, three methods are available to successfully produce transgenic poultry through targeting the embryo directly: DNA microinjection, viral-mediated transfection, and transposon-mediated integration. Other methods use a chimeric intermediate through the transfer of cultured PGCs, similar to the approaches in mouse using embryonic stem cells.

D. Viral-Mediated Transgenesis

The use of retroviruses for gene transfer is a common procedure and forms one approach for gene therapy strategies in humans and transgenesis in laboratory and domestic animals. For poultry, retroviral gene transfer is the most successful meth-odology to date. This is due mainly to the features of the retroviral life cycle.

Retroviruses have an RNA genome encased in a protein core containing inte-grase, reverse transcriptase, and a protease which is coated by a protein envelope. For infection, viral envelope proteins bind to specific proteins on the host cell membrane and are internalized by receptor-mediated endocytosis. The envelope is removed by cellular enzymes, and viral reverse transcriptase copies the RNA into DNA. The DNA moves to the nucleus and is integrated into a chromosome of the host cell through the activity of an integrase on the long terminal repeats at each end. With integration, the provirus is replicated with the chromosome and is inherited in a Mendelian manner. It is this aspect of the retroviral life cycle that permits successful transgenesis.

In addition to replication, the proviral DNA can be transcribed into viral RNA for the synthesis of new viral proteins. These RNAs encode three classes of proteins: Pol for polymerases, Gag for group-associated antigens, and Env for envelope proteins. Once translated, Pol and Gag proteins associate with the specific packaging sequences on the RNA and assemble into new viral cores. The Env proteins are transported to the host cell membrane, and the viral core buds from the cell and produces a new infectious particle.

In the case of replication-competent vectors, the viral structural genes and packaging sequences are intact, allowing for the continuous production of infectious particles. With replication-defective vectors, deletions are made in the structural genes *(pol, gag, env)*. This allows the virus to infect a host cell, but the provirus will not generate new infectious virions. To produce infectious particles for transgenesis, helper cell lines were developed to package the defective vector. The helper cell lines were generated using a proviral plasmid that is missing the encapsidation site but contains the *gag, pol*, and *env* regions of the virus. When this cell line is transfected with a replication-defective vector containing the exogenous gene of interest, the helper cells can package the recombinant viral RNA into infectious particles that do not in turn produce other infectious particles. The need for helper packaging cell lines makes replication-defective retroviral vectors more difficult to work with than replication-competent systems. In addition, because viral titers are reduced, considerable concentration is required to yield sufficient material for infection. However, replication-defective vectors allow for larger exogenous genes (about 10 kb versus 2.5 kb) and can encode multiple transgenes. Over the last 20 or more years, retroviral and lentiviral vectors have been used to develop lines of transgenic chicken expressing beta-glactosidase (Mozdziak et al., 2003), beta-lactamase (Harvey et al., 2002), GFP (McGrew et al., 2004; Chapman et al., 2005), human interferon alfa-2b (Rapp et al., 2003), single chain Fv-Fc fusion protein (Kamihira et al., 2005), human granulocyte colony-stimulating factor (Kwon et al., 2008), human erythropotein (Koo et al., 2010), human extracellular oxide dismustase (Byun et al., 2013), and inducible GFP (Kwon et al., 2011). While the list is not comprehensive, it does indicate the range of transgenic lines generated through viral infection.

E. Microinjection of DNA

In addition to retroviral vectors, microinjection of DNA has been demonstrated to produce transgenic poultry. Injection of DNA into the pronucleus of the newly fertilized egg is a common procedure for the production of transgenic laboratory animals and livestock (see Chapter 2). Unfortunately, these techniques were not readily adapted to the chicken because of the large yolky ovum and the formation of the calcified egg. Before DNA microinjection could be attempted in birds, a complete *ex ovo* culture system from fertilization to hatch needed to be developed that would yield sufficient numbers of hatchlings to screen for gene integration. The basic method currently in use is a three-step system developed by Perry (1988) using a combination of methods employed by both Ono and Wakasugi (1984) and Rowlett and Simkiss (1987) for postovipositional stages of development. In the first step, newly fertilized eggs surrounded with a capsule of albumen are removed from the magnum and cultured for about 18−24 h in synthetic oviductal fluid without a shell. Step 2 requires transfer of the egg to an eggshell, completely sealed with no simulated air space. After 2−4 days, the final step places the embryo into a larger shell with an upper air space for the remaining period of incubation. Such procedures have also been adapted to quail embryos using chicken eggshells (Ono et al., 1994). For the production of transgenic poultry, DNA expression vectors are injected into the cytoplasm of the germinal disk of the ovum upon recovery from the magnum prior to culture. Similar to DNA injection of mammalian pronuclei, the injected DNA forms concatemers and remains mainly episomal (Sang and Perry, 1989). However, one mosaic rooster was produced that transmitted a bacterial beta-galactosidase gene to about 3.4% of its offspring (Love et al., 1994). Transgene copy number averaged about 6, apparently in a single chromosomal location. Test mating of one transgenic rooster showed predictable Mendelian inheritance of the reporter gene. This demonstrated that it is possible to produce transgenic poultry using DNA injection. Unfortunately, no expression data were reported. The efficiency of integration, which is often a limiting variable for the production of transgenic animals (mammalian or otherwise), appears to be low but is not lower than that for other agricultural livestock. To help improve integration, a plasmid containing the *Drosophila* transposable element mariner was injected into early zygotes. Analysis of embryos and adults indicated several transposition events, suggesting that active transposable elements could help to improve gene integration (Sherman et al., 1998). Unfortunately, microinjection is now rarely used owing to the specialized equipment and skill requirements that make routine use of this approach impractical for general laboratory use.

F. Primordial Germ Cells

Although retroviral/lentiviral vectors and direct DNA injection can be used to produce transgenic poultry, neither method can take advantage of current technology using DNA constructs for gene targeting, which is the ability to make locus-specific modifications to the genome. For several years, lines of transgenic mice have been produced with relatively precise changes to particular loci to examine the genetic basis of disease, to be used as therapeutic models, and to study various aspects of development. The use of gene targeting in the production of transgenic animals of commercial importance is particularly attractive because the chances of obtaining a predictable phenotype appear greater than having to evaluate several lines of animals produced through random integration of the transgene. The basic procedure for the production of chickens with targeted modifications to the genome is illustrated in Figure 12.3. The first step is to obtain primordial germ cells, which are essentially germ line stem cells, and to culture them to allow for gene transfer *in vitro*. These cells are transfected; e.g., using electroporation or cationic polyamines, with constructs designed to allow for homologous recombination, and cells in which

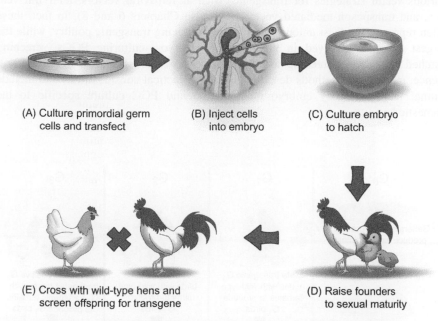

(A) Culture primordial germ cells and transfect

(B) Inject cells into embryo

(C) Culture embryo to hatch

(E) Cross with wild-type hens and screen offspring for transgene

(D) Raise founders to sexual maturity

Figure 12.3 The technical scheme for the production of transgenic poultry using chimeric intermediates. PGCs are cultured using conditions that allow proliferation without de-differentiation into embryonic stem cells (A). Such cultures are transfected with DNA constructs that can undergo homologous recombination. Cells with the correct integration event are expanded using the appropriate selectable markers and transferred to recipient embryos (B and C). The resulting chimeras are bred and the offspring screened for the presence of the transgene (D and E).

the correct (and rare) recombination event has occurred are selected and expanded *in vitro*. Finally, the cells are returned to an embryo, where they integrate and become part of the cellular makeup of the recipient germ line. Such germ line chimeras should have germ cells at sexual maturity that derive from the transgenic donor cells. When the chimeras are bred, a portion of the offspring will be transgenic (Figure 12.3). This technical scheme represents the coordinated effort of three technologies: (1) the ability to produce germ line chimeras, (2) the development of suitable DNA constructs for transfection, and (3) the means of culturing germ cells. Recently, cultured PGCs have been used to generate transgenic poultry expressing reporter genes and have been used for gene targeting (van de Lavoir et al., 2006; Leighton et al., 2008; Macdonald et al., 2010; van de Lavoir et al., 2012).

II. Methods

The methodologies described here were chosen because of their specificity to avian transgenesis. It is beyond the scope of this chapter to go into vector constructs and various vector strategies for transgenics such as retroviral vectors, lentiviral vectors, and transposon-mediated vectors (See also Chapters 6 and 8). In fact, these often represent the least amount of effort in producing transgenic poultry, while the largest effort is involved in manipulating embryos, culturing PGCs, screening hatched chicks, and breeding to homozygosity to establish a flock (Figure 12.4). Hence, most of the methods described here are practical considerations and include animal facilities, early embryo manipulation and PGC culture specific to the domestic hen.

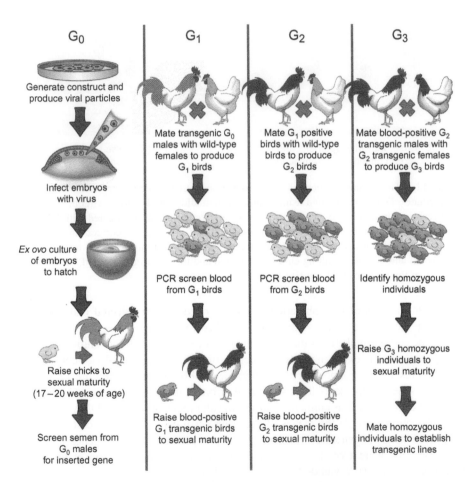

Figure 12.4 General steps in the production and establishment of a line of transgenic chickens using viral vectors and the subsequent breeding to homozygosity. Generation 0 (G_0) involves the construction and production of virus and the infection of embryos followed by *ex ovo* culture, hatching, rearing, and screening of putative mosaics for breeding. Generations 1–3 (G_1–G_3) require the majority of time, cost, and facilities to establish and characterize a usable line of birds.

A. Animal Facilities and Flock Management

A full program for the production of transgenic poultry requires a significant investment in animal and laboratory facilities. The main reason for this is the breeding required once G_0 founder birds are generated. Figure 12.4 shows the generations of breeding required from G_0 to the breeding of a homozygous population of transgenic birds. Hence, the animal facilities required for the generation of transgenic birds and the subsequent breeding and maintenance of the transgenic lines needs to be considered before embarking on any effort to produce transgenic poultry.

1. The Hatchery

One aspect that is often overlooked in a transgenic program is adequate incubation and hatching space. Equipment needs for incubation and hatching are independent of the specific methods used to produce transgenic poultry. Common experience suggests that there is never enough space. Incubation needs should be divided into two categories: (1) facilities for the incubation and hatching of manipulated embryos for the production of chicks that potentially are transgenic, and (2) facilities for the incubation and hatching of eggs from test mating efforts and for the generation of replacement stocks. Small incubators and hatchers that hold up to 360 eggs serve well for manipulated eggs. Whatever the model used, they must have good digital temperature and humidity controls and must be easily altered to provide different angles of rotation that may be needed for various manipulation procedures. Other advantages include their small size, which allows placement on a low table and transfer to another room for cleaning and disinfection. It is very important to locate the incubators and the hatchers in separate rooms. Hatching is a dirty process. Chick down and dander transfer bacteria and other microorganisms to the incubator which can decrease the viability of the manipulated embryos. As the microbial load of the incubator increases, the survival and hatchability of the manipulated embryos will decrease. All avian transgenic projects require some period of test mating. The eggs generated from this effort can be unwieldy without sufficient incubator/hatching space. In addition, unless the eggs used for producing the transgenic birds are purchased, flocks will have to be maintained to provide a continuous supply of fertile eggs for the project. Hence, larger capacity incubators and hatchers dedicated for this purpose are needed. In addition, eggs from progeny testing of founder transgenics must be pedigree hatched, usually in hatching trays designed for this purpose.

2. Brooding and Rearing

Brooding and rearing should be carried out in separate facilities. Once chicks have hatched they must be properly brooded. In a research setting, tiered battery brooders are convenient for brooding birds for about 4 weeks, depending upon the size of the birds. Soon after hatching, all chicks must be individually identified either through wing bands or neck tags. Neck tags, which are quick and easy to apply, are most useful for identification of chicks prior to screening for insertion of the transgene. Once a transgenic chick is identified, sequentially numbered wing bands should be used on each wing.

After the birds outgrow the brooder, they must be reared in a light-tight facility with 8 hours of light. Chickens, like all commercial poultry, are long-day breeders and are sensitive to changes in day length. For good reproductive performance, the lighting of birds prior to sexual maturity must be managed properly. A simple and basic minimal lighting schedule is to transfer the birds from the rearing facility with 8 h of light to the production facility with 14−18 h of light. While this may not be the most efficient lighting program from a commercial standpoint, the rearing facility will often contain birds between 5 and 18 weeks of age, which precludes custom light schedules over the rearing period. Once the chicks are placed in the rearing cages, it is also important to train the birds to drink from the waterers to avoid losses from dehydration. Normally, the birds will be housed individually in cages during the test mating phase; therefore, they should be reared in cages with similar style drinkers, preferably the nipple type.

3. Pedigree Mating Facilities

When the birds reach 18−20 weeks of age, depending upon the size and strain used, they should be transferred to the production facility. Test mating is best performed using artificial insemination rather than a system of natural mating. This allows monitoring of fertility of individual males and females, and the regular handling is a good means for ongoing supervision of bird health. Commercial cages designed specifically for poultry artificial insemination are available, or custom caging can be used. Cage sizes are specific for roosters and laying hens. During the first week, it is important to monitor the birds to be sure that they are eating and drinking properly. If they were reared using the same type of watering system, then this aspect usually is not a problem. In producing transgenic chickens, male G_0 birds are the most valuable, as they can be mated to several females and quickly generate large numbers of offspring that can be screened for the transgene. In general, semen can be collected three times a week from an individual male and can be diluted with a commercial extender, depending upon the number of hens to be inseminated. Up to 10 females can be inseminated with extended semen.

B. Embryo Manipulation and Culture

Regardless of the approach to the production of transgenic birds, access to the avian embryo is essential for approaches such as injection of viral vectors, production of germ-line chimeras, or the injection of transposon-mediated vectors. Even with access to the early embryo, a considerable amount of work can be fruitless if the effort does not result in live chicks. Hence, various techniques have been developed for the culture of avian embryos for different periods of incubation, with and without shells, *in vitro* and *ex ovo*. For the development of transgenic chickens using DNA microinjection, it is necessary to begin *ex ovo* culture of the ovum soon after fertilization. The three-stage system derived by Perry (1988) is the best method available to date. Briefly, three culture systems are used sequentially. The first spans the oviductal period of development. The ovum is removed from the magnum and cultured in a container with an artificial oviductal fluid and thin albumen. After 24 h, the egg and albumen are transferred to the shell of a surrogate egg, sealed with plastic cling film, and incubated for 3 days with a full 90-degree rotation. Subsequently, the eggs are then transferred to larger surrogate eggshells and again sealed with plastic film but with an artificial air space provided. The eggs are incubated with only 60-degree turning until 18 days, when they are transferred to the hatcher. The hatching process is highly stereotyped, and the embryos are examined for position and penetration of the chorioallantoic membrane. At that time, the cling film is removed, and the shell is loosely covered with a 35-mm Petri dish, which allows the viable chicks to emerge on their own.

When using retroviral/lentiviral vectors, the embryo is normally infected with virus at the time the egg is laid. During early work using viral vectors, a window was simply made in the egg and sealed with some type of adhesive or a piece of the shell membrane, but this simplicity often resulted in very poor production of live chicks. Optimal hatchability of windowed eggs is important for the efficient production of transgenic poultry. Speksnijder and Ivarie (2000) described a windowing technique in which the shell is ground off and the opening is covered in phosphate-buffered saline (PBS) before the shell membrane is cut. This allows no air to enter the egg but permits access for the injection of cells or virus into the embryo. After injection, the egg is sealed with a piece of wet shell membrane, allowed to dry, and then covered with adhesive cement (e.g., nitrocellulose based Duco® cement; ITW Devcon, Danvers, MA). This procedure improved hatch from about 8% for the usual windowing protocol to about 34%. Bednarczyk et al. (2000) reported improvements in hatchability after eggs were stored for 5−7 days with the window made in the blunt end of the shell over the air cell. An alternative to the windowing techniques described above is to use *ex ovo* culture of freshly laid eggs. Borwornpinyo et al. (2005) optimized the hatchability of cultured chicken embryos from freshly laid eggs using surrogate eggshells. This was based upon the procedures of Perry (1988), where two surrogate eggshells are used to culture the newly laid egg to hatch, a method which improved hatchability from the usual 20−50% to over 75%. Hence, to access the *unincubated* embryo for microinjection of virus, cells, or DNA constructs, a complex culture system is required.

1. Ex Ovo *Culture of Freshly Laid Eggs to Hatch*

Day 0 to Day 3

1. Prepare recipient eggshells from freshly laid eggs that are 3−4 g heavier than donor eggs.
2. Swab with 70% ETOH.
3. Draw a 32-mm circle on the small end of the egg using a template.
4. With a small Dremel-style drill, cut and remove the shell along the drawn circle and discard the yolk and albumen.

5. In a Biological Safety Cabinet, rinse the insides with distilled water, and place the shells open-end down on bench paper to drain.
6. For short-term storage, spray the eggshells with 70% ETOH and cover with moistened bench paper followed by plastic cling film to prevent the shells from drying.
7. Match donor eggs with recipient eggs and swab with 70% ETOH.
8. Open the donor eggs into a 200-ml plastic drinking cup angled at about 45 degrees.
9. Transfer the contents to a recipient eggshell and fill with additional thin albumen; at this point, virus, cells or DNA constructs can be microinjected beneath the blastoderm.
10. Seal with a 5 × 5-cm piece of plastic wrap and secure with a pair of plastic rings and rubber bands (see Figure 12.5).
11. Incubate the eggs window side down at 37°C, 60% relative humidity, with rocking through 90 degrees every hour for 3 days.

Day 3 to Hatch

1. Choose recipient turkey eggshells (alternatively eggshells from double- yolked eggs) that are 40 g heavier than the original donor egg weight. Swab with 70% ETOH and remove the blunt end as above to create a 42- to 45-mm window.
2. Discard the yolk and albumen, and rinse with distilled water and store as indicated above.

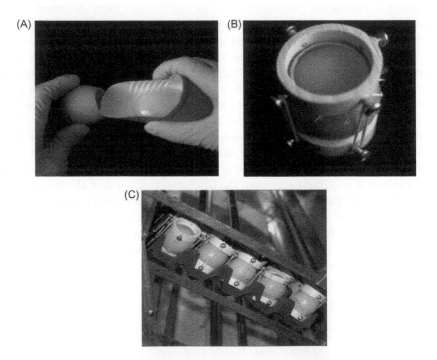

Figure 12.5 *Ex ovo* culture from day 0 to day 3 of incubation. The recipient eggshell is prepared and the donor egg is transferred to the new shell (A). After injection of cells or virus, the egg is sealed with plastic film and secured with rings and rubber bands (B). The egg is incubated window side down for 3 days with continuous rocking (C). After 3 days, the embryo is ready for transfer to the next eggshell (see Figure 12.6).

3. To transfer the day 3 embryo to the turkey eggshell, remove the plastic film from a small region of the shell and transfer about 5 ml of albumen to the turkey shell. Roll the turkey shell to wet the shell membrane with albumen. Remove the remaining plastic film from the chicken eggshell and hold the turkey and chicken eggshells side by side.

5. Cut the edge of the chicken eggshell if needed to remove any jagged edges.

6. Gently pour the contents into the turkey eggshell (this requires some practice). It is essential that the embryo appears uppermost after transfer.

7. Seal the shell with a 6.5 × 6.5-cm piece of plastic cling film. Albumen is a good glue for this purpose.

8. Secure the film with plastic rings (50 mm in diameter for the window and 35 mm in diameter for the small end) (Figure 12.6).

9. Place the eggs with the window uppermost in an incubator at 37.5°C, 60% relative humidity, with rocking through an angle of 30 degrees every hour.

10. After 15 days of incubation in the turkey eggshell, move the eggs to a hatcher at 37°C, 60% relative humidity.

11. Remove the plastic film at the first sign of penetration of the chorioallantoic membrane, usually at 19.5 days, and then cover with a 60-mm Petri dish lid.

12. Allow the chicks to hatch and crawl out of the shells independently.

C. Retroviral/Lentiviral Infection

The procedures used for the production of transgenic poultry using viral vectors generally encompass vector construction, transfection of the vector into a suitable packaging cell line, and harvesting the virus (see Chapters 6 and 8).

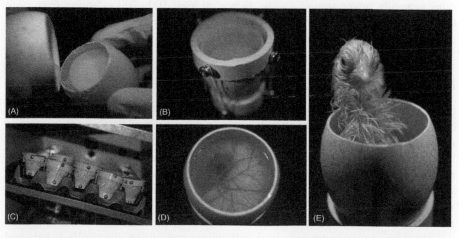

Figure 12.6 *Ex ovo* culture from day 4 to hatch. Recipient turkey eggshells are prepared, and the contents of the egg incubated for 3 days (see Figure 12.5) are transferred to a new shell (A). The embryo should lie uppermost after transfer (B) and then is sealed with plastic film, leaving an air space. The eggs are incubated for 18−19 days, with hourly rocking through 30°C before transfer to a hatcher (C). A 9-day embryo (D). A newly emerging chick (E). It is recommended that the incubation facility be located separately from the hatching facility.

In preparation for producing transgenic poultry, the virus must be concentrated. Methods to concentrate supernatant from virus-producing cells include ultracentrifugation, a stirred ultrafiltration cell (Millipore, Model 8050), or products based upon filter absorption and elution of virus. Yields of $10^7 - 10^8$ virus particles/ml or more are recommended to infect cells of the unincubated embryo.

1. Injection of Viral Vectors

1. At this point, unincubated eggs should be prepared ahead of time, either using the procedures for *ex ovo* culture described in the previous section (highly recommended), or if using a simple windowing procedure, the eggs need to be positioned vertically (blunt end up) for at least 3 h to allow the yolk to rotate up so that the blastoderm lies beneath the shell. In the latter case, remove the shell using a small drill with a grinding bit, but manually remove the shell membrane with a scalpel.
2. Position the windowed egg on a dissection scope with lateral illumination and locate the white blastoderm at the surface of the yolk. Visualization of the white blastoderm on the yellow background can be improved with the use of blue-filtered light for illumination.
3. Microinject 10 μl virus into the subgerminal cavity of the blastoderm. If phenol red is used in the medium, the center of the blastoderm will become pink.
4. Seal the egg and place in an incubator blunt end up, or proceed to the next steps of *ex ovo* culture as described previously.

With good viral titers, the proportion of male chicks with PCR-detectable transgenes in the semen can reach 50%. Generally, no correlation between transgenic blood and transgenic semen has been observed. In general, germ line transmission rates generally hover around $1-2\%$. Hence, at least $100-200$ offspring from each G_0 bird are required to obtain a few G_1 individuals for further analysis. This requires extensive PCR screening of chicks.

D. Primordial Germ Cell Culture

Since primordial germ cells are the target cell type for the production of transgenic birds, considerable progress has been made in the culture of PGCs so that genomic manipulations can be done *in vitro*. This would be followed by the production of "putative" germ-line chimeras and test mating (Figure 12.3). PGC culture has allowed the production of transgenic poultry with modifications to the avian genome including gene targeting. Currently, the domestic hen is the only species where primordial germ cells can be cultured and used for this purpose. The culture system to do this is complicated; most of the procedures are similar to those described for mouse embryonic stem cell culture and utilize mitotically inactivated feeder layers, growth factors, and conditioned medium.

1. Preparation of BRL Conditioned Medium

1. Initiate a culture of BRL (Buffalo rat liver cells; ATCC #CRL-1442) in DMEM (Dulbecco's modified Eagle medium) with 10% fetal bovine serum (FBS).

2. Pass the cells until there are about 10 confluent T75 confluent culture flasks. Use more or less as needed.
3. Remove medium and discard. Add 30 ml DMEM/10% FBS to each flask and incubate 72 h.
4. Collect medium in 50 ml polypropylene centrifuge tubes, add 30 ml fresh DMEM/10FBS to each flask and return to the incubator. Repeat twice more for a total of three collections.
5. After each collection, pellet cell debris by centrifuging the medium in the 50 ml tubes at *3000-5000xg* for 5 min. Remove the medium, being careful not to disturb the pellet.
6. Pool all of the collected medium and adjust pH to 7.4.
7. Aliquot BRL conditioned medium in 50-ml tubes, and store at $-70°C$. Thaw as needed.

2. Irradiation, Storage, and Preparation of STO Feeder Layers

1. Culture STO cells (ATCC #CRL-1S03) in DMEM with 10% FBS until there are six confluent 100 mm culture dishes (about $4-6 \times 10^6$ cells/dish).
2. Wash twice with PBS and add 1 ml trypsin/EDTA (O.05% trypsin/0.25 mM EDTA).
3. Incubate until cells dissociate, add 3 ml culture medium/dish, and pool cells.
4. Place cells on ice and irradiate with 4000—8000 rads from a gamma source.
5. Pellet cells by centrifuging at about $210 \times g$ for 5 min.
6. Remove medium, and add DMEM/20% FBS to resuspend the cells at $8-12 \times 10^6$ cells/ml. Place the cells on ice for 5 min.
7. Slowly add an equal volume of cold DMEM with 20% DMSO (dimethylsulfoxide)/20% FBS to the cells while mixing.
8. Aliquot the desired amount into screw-cap cryo vials; place vials in a freezing container such as a Nalgene Cryo freezer box.
9. Store the container at $-80°C$ overnight to generate $1°C$ per min freezing rate, and the next day, transfer the vials to liquid nitrogen storage.
10. To prepare feeder cells, gelatinize a tissue culture plate by adding 0.1% gelatin to each well. Incubate at room temperature for at least 30 min, then remove gelatin.
11. Thaw frozen feeder cells in the $37°C$ water bath. As soon as thawed, add dropwise to 10 ml DMEM/10% FBS and centrifuge at $210 \times g$ for 5 min.
12. Remove and discard medium. Add 2—5 ml fresh culture medium, resuspend pellet, and count the cells on a hemocytometer.
13. Add approx. 20,000 cells/cm^2 in DMEM/10% FBS (each frozen batch will have to be tested to determine the correct amount to seed to produce the desired monolayer the next day).
14. Incubate the cells at $37°C$ at 5% CO$_2$ for 24 h before use.

3. Embryonic Blood Collection for PGC Culture

Materials and Supplies:

- Fertile eggs incubated 53—60 h to about Stage 15
- 100 mm Petri dish
- phosphate-buffered saline (PBS)
- Filter paper disk (Whatman #1)
- Standard and fine forceps
- Scissors

- 0.5 ml microtubes with 200 μl of Medium 199/10% FBS
- 0.5 ml tubes with 45 μl PBS
- Large weigh boat
- Micropipette needles, 50−60 u I.D. for blood collection
- Dissection microscope with lateral and ventral illumination

a.   Blood Collection

1. Gently crack the egg in the weigh boat, making sure the embryo is upper-most.
2. Lay a filter paper ring over embryo with embryo in center of cutout area (Figure 12.7A).
3. Cut yolk membrane around outside of the ring (Figure 12.7B).
4. Flip ring over and transfer to a dry Petri dish with the ventral side of the embryo upper-most (Figure 12.7C).
5. Place under dissection microscope.
6. Focus microscope so that heart and blood vessels can be visualized.
7. Rinse the embryo with PBS to remove any excess yolk material and remove the PBS from the dish so that the embryo is still wet but sticks to the dish and does not float. Tilt the dish if necessary to pool the PBS.
8. Gently insert glass needle into the dorsal aorta (Figure 12.7D).

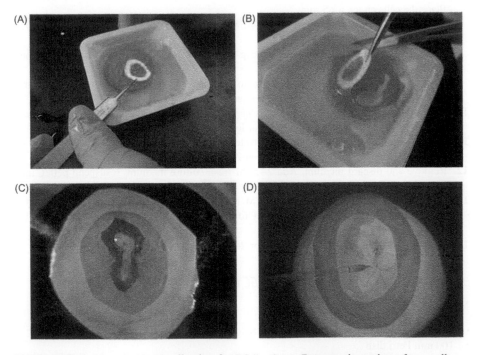

Figure 12.7 Embryonic blood collection for PGC culture. Remove the embryo from yolk with filter paper ring (A, B). Place embryo ventral side up, clean off yolk with PBS, and remove PBS (C). Aspirate blood with micropipette at the junction of the vitelline blood vessels (D). Note: Embryo is older than stage 15 for illustration purposes.

9. Capillary action will draw the blood up the needle once in place. Gentle suction can be applied to speed collection.

10. Using gentle pressure, transfer the blood into a 0.5 ml tube containing 200 µl Medium 199/10% FBS.

b. Initiation of PGC Culture

1. 24 h prior to initiation, prepare 48-well plates with feeder layers of irradiated STO cells.
2. Prepare 10 ml of PGC culture medium as follows or expand volumes as needed:
   a. 2 ml BRL-conditioned medium
   b. 0.75 ml FBS
   c. 0.25 ml chicken serum (Sigma)
   d. 0.1 ml antibiotic/antimycotic
   e. 0.1 ml 100X GlutaMAX
   f. 0.2 ml 50X GS nucleoside supplement
   g. 0.024 ml Mercaptoethanol
   h. 6 µl mSCF (from 10 ng/µl stock)
   i. 4 µl hFGFb (from 10 ng/µl stock)
   j. 6.58 ml KO-DMEM (Invitrogen)
3. Pellet the embryonic blood sample @ 400 g for 4 min.
4. Remove culture medium from feeder layer and discard.
5. Remove the supernatant from the blood sample and discard.
6. Resuspend the blood sample in 300 µl of the PGC culture medium and transfer to one well of the 48-well plate, giving one well/embryo, and incubate at 5%CO_2 and 37°C.
7. Feed every day for 2 days by removing 130 µl of medium and replace with fresh medium.
8. From day 3 onwards, feed the culture every 48 hours.
9. Between 5−7 days of culture, subculture the cells onto a 48-well plate with fresh STO feeder cells.
10. After 14 days, score the wells for the presence of PGCs.

Initially, it will be difficult to identify PGCs among the red blood cells (Figure 12.8A). Eventually, the blood cells die and the PGCs appear as smooth, round, translucent cells, larger than the blood cells, and remain unattached to the plate. Typically, the PGCs will "appear" after about 14 days of culture as small groups of round cells (Figure 12.8B). Successful cultures of PGCs are sequentially expanded from 48-well to 24-, 12-, and finally 6-well plates (Figure 12.8C). In a 6-well plate about 500,000 to 1,000,000 PGCs per well is of a sufficient density for cryopreservation of PGC lines or chimera production (Figure 12.8D). For successful cultures, it is imperative that the FBS and the chicken serum be tested and screened for quality. In addition, fresh growth factors should always be used and replaced regularly, even before official expiration dates. It is recommended that large batches of BRL-conditioned medium be made to limit variability in the culture system. The key to PGC culture is observation and regular care. At a minimum, the cells must be fed every 48 h. Expansion and passaging of cultures must be timed so that the culture does not become so overgrown that the medium becomes exhausted before 48 h. Transfection of PGCs can be accomplished using standard electroporation or cationic-liposome/polyamine reagents. In these cases, some type of selectable marker is needed to identify and grow PGCs that stably express the gene of interest.

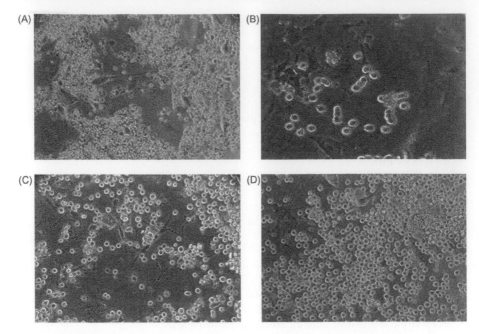

Figure 12.8 Phase contrast of PGC culture initiation. Initial embryonic blood sample on feeder layer (A). Cluster of PGCs after 14 days of culture, note significant reduction in blood cells. Subsequent culture undergoing expansion (C). Fully establish line of PGCs derived from a single embryo (D).

c. Injection of PGCs into Stage 17 Embryos
Materials:

- Fertile eggs incubated for 52–63 h (stage 17 H & H)
- Curved forceps
- Superfine forceps
- 70% alcohol
- Pasteur pipettes
- Microinjection needles (50–60 u diameter with bevel and spike) for embryo injection
- Sterile PBS
- Squares (1–2 cm^2) of Parafilm
- Alcohol burner
- Scalpel with round #10 blade

Prepare PGCs for Injection

1. Resuspend PGCs from a 6-well dish and count a 10 μl sample.
2. Pellet the PGCs @ 400 g for 4 min and discard supernatant.
3. Resuspend cell pellet in 2% trypan blue in complete PGC culture medium at about 20,000 cells/10 μl.

4. Aliquot 10,000–20,000 cells in 10–15 μl drops of the cell suspension on a 100 mm culture dish.
5. Gently cover the drops with 11 ml of light white mineral oil.

Prepare Eggs for Injection

1. Do not allow eggs to turn for 1 h prior to injection, keeping the blunt end up.
2. Candle the egg to locate air cell and outline with pencil (Figure 12.9A).
3. Make a ~1–2 cm^2 window in the middle of the air cell by carefully removing the shell and the outer shell membrane. Be careful not to damage the inner shell membrane (Figure 12.9B). Clean away the shell dust with alcohol. Eggs can be stored in the incubator for 2–3 h in this manner.
4. Place an egg under dissection microscope and using lateral illumination locate the beating heart of the embryo lying under the inner shell membrane.
5. Cover the shell membrane just above the heart with 0.5–1 ml PBS. This will make the inner shell membrane more transparent (Figure 12.9C).
6. Using the superfine forceps, remove a small portion of the inner shell membrane to visualize the heart.
7. Using the microinjection needle, aspirate the PGCs from the micro-drop.
8. Insert the microinjection needle into the dorsal aorta or directly into the heart (Figure 12.9D).
9. Deliver the cells by gently blowing into the desired region of the heart using the glass microneedle. The delivered volume should be between 3 and 5 μl. You should be able to see the cells clear the heart when the heart beats.

Sealing the window

1. Flame the scalpel over the alcohol burner so that the blade is at a medium heat.
2. Quickly and carefully pass one piece of Parafilm wax square through the flame 3–4 times until a corner has slightly melted and curled.
3. Beginning with the curled corner, place the wax film across window. Using the warmed scalpel blade, carefully scrape the edges of the paraffin downward and away from the window so the wax gradually begins to form a seal. Work around the paraffin until the entire window has been sealed (Figure 12.9E).
4. Repeat the process using a second piece of paraffin to create a double layer, but overlay the second piece at a 90-degree angle from the first (Figure 12.9F).
5. Incubate the eggs in a vertical position with window side up.

Following injection, the eggs should not be allowed to rotate for approximately 1 h. Incubate eggs until day 18. At this time the eggs should be moved to a clean and properly disinfected hatcher still in an upright position.

At day 20 eggs can be removed from the tray and carefully placed in the horizontal position and allowed to hatch.

Another consideration when producing chimeras is the genetics of the donor and recipient lines. Currently, Barred Plymouth Rock fowl are most frequently used as a donor line coupled to White Leghorn birds as the recipient line. It is important to be sure that the While Leghorn line is homozygous dominant (I,I) at the I locus, which inhibits dermal melanin. Barred Plymouth Rocks are homozygous recessive (i,i) at the I locus. The advantage of this pairing is that germ line chimerism is

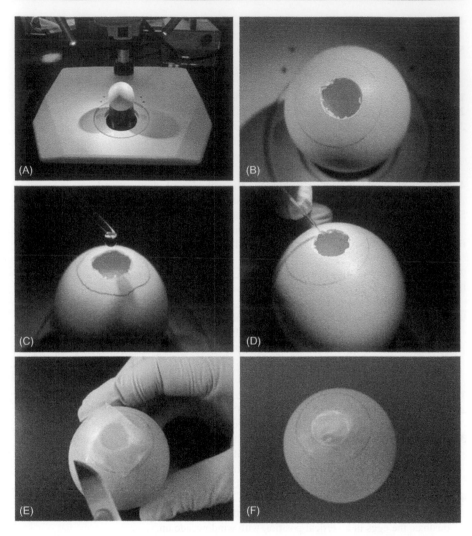

Figure 12.9 Injection of PGCs into stage 17 embryos. (A) Outline air cell and place egg on dissection microscope with lateral illumination. (B) Remove shell over air cell and locate embryo. (C) Add PBS to air cell and open inner shell membrane over heart of embryo. (D) Inject cells into heart of embryo. Seal with warmed scalpel blade and Parafilm (E and F).

easily assessed as Barred Plymouth Rock chicks have predominantly black down, and White Leghorn chicks are yellow. Chimeric males that have the genotype (I,I/i,i) are then mated to Barred Rock hens (i,i). A normal cross will yield heterozygotes (I,i), which are yellow chicks with black flecks. However, any Barred Plymouth Rock offspring (i,i) from the mating will be derived from the donor cells. An added advantage is if transgenic embryonic stem cells are made from Barred

Plymouth Rock embryos the screening for transgenic offspring from the chimeric G_0 birds is significantly reduced, because only the Barred Plymouth Rock chicks must be screened. Hence, if the frequency of germ-line chimerism is low, hundreds of offspring per male can be obtained without the need to screen every chick.

III. Summary

Strong interest still remains in the development of transgenic poultry despite the seemingly slow progress compared to other agriculturally important species. Viral vectors continue to be the most successful means of developing transgenic chickens. However, with the advent of the ability to culture avian primordial germ cells and the use of transposon-mediated integration, the future of transgenic poultry looks very promising. Given the current state of the art and the significant interest in commercial applications for transgenic poultry, the future decade holds considerable promise for the refinement of techniques to generate transgenic birds of value to developmental biology, agriculture, and medicine.

Acknowledgments

The authors wish to acknowledge Julie Angerman and Rebecca Barnes Wysocky for their technical assistance and Jennifer Petitte for providing the scientific illustrations.

References

Bakst, M.R., Gupta, S.K., Akuffo, V., 1997. Comparative development of the turkey and chicken embryo from cleavage through hypoblast formation. Poult. Sci. 76, 83–90.

Bednarczyk, M., Lakota, P., Siwek, M., 2000. Improvement of hatchability of chicken eggs injected by blastoderm cells. Poult. Sci. 79, 1823–1828.

Borwornpinyo, S., Brake, J., Mozdziak, P.E., Petitte, J.N., 2005. Culture of chicken embryos in surrogate eggshells. Poult. Sci. 84, 1477–1482.

Byun, S.J., et al., 2013. Human extracellular superoxide dismutase (EC-SOD) expression in transgenic chicken. BMB. Rep. 46, 404–409.

Chapman, S.C., Lawson, A., MacArthur, W.C., Wiese, R.J., 2005. Ubiquitous GFP expression in transgenic chickens using a lentiviral vector. Development 132, 935–940.

Dubois, R., Croisille, Y., 1970. Germ-cell line and sexual differentiation in birds. Philos. Trans. R. Soc. Lond. B. Biol. Sci. 259, 73–89.

Eyal-Giladi, H., Kochav, S., 1976. From cleavage to primitive streak formation: a complementary normal table and a new look at the first stages of the development of the chick. I. General morphology. Dev. Biol. 49, 321–337.

Fujimoto, T., Ninomiyya, T., Ukeshima, A., 1976a. Observations of the primoridal germ cells in blood samples from the chick embryo. Dev. Biol. 49, 278–282.

Fujimoto, T., Ukeshima, A., Kiyofuji, R., 1976b. The origin, migration and morphology of the primordial germ cells in the chick embryo. Anat. Rec. 185, 139–145.

Goldsmith, J.B., 1926. The history of the germ cells in the domestic fowl. J. Morphol. Physiol. 41, 275–313.

Hamburger, V., Hamilton, H.L., 1951. A series of normal stages in the development of the chick embryo. J. Morphol. 88, 49–92.

Harvey, A.J., Speksnijder, G., Baugh, L.R., Morris, J.A., Ivarie, R., 2002. Expression of exogenous protein in the egg white of transgenic chickens. Nat. Biotechnol. 20, 396–399.

Kamihira, M., et al., 2005. High-level expression of single-chain Fv-Fc fusion protein in serum and egg white of genetically manipulated chickens by using a retroviral vector. J. Virol. 79, 10864–10874.

Koo, B.C., et al., 2010. Tetracycline-dependent expression of the human erythropoietin gene in transgenic chickens. Transgenic Res. 19, 437–447.

Kuwana, T., Maeda-Suga, H., Fujimoto, T., 1986. Attraction of chick primordial germ cells by gonadal anlage in vitro. Anat. Rec. 215, 403–406.

Kwon, M.S., et al., 2008. Generation of transgenic chickens that produce bioactive human granulocyte-colony stimulating factor. Mol. Reprod. Dev. 75, 1120–1126.

Kwon, M.S., et al., 2011. Production of transgenic chickens expressing a tetracycline-inducible GFP gene. Biochem. Biophys. Res. Commun. 410, 890–894.

Leighton, P.A., Van De Lavoir, M.C., Diamond, J.H., Xia, C.Y., Etches, R.J., 2008. Genetic modification of primordial germ cells by gene trapping, gene targeting, and phi C31 integrase. Mol. Reprod. Dev. 75, 1163–1175.

Love, J., Gribbin, C., Mather, C., Sang, H., 1994. Transgenic birds by DNA microinjection. Biotechnol. (N. Y). 12, 60–63.

Macdonald, J., Glover, J.D., Taylor, L., Sang, H.M., McGrew, M.J., 2010. Characterisation and Germline Transmission of Cultured Avian Primordial Germ Cells. PLoS ONE 5.

McGrew, M.J., et al., 2004. Efficient production of germline transgenic chickens using lenti-viral vectors. EMBO Rep. 5, 728–733.

Meyer, D.B., 1964. The migration of primordial germ cells in the chick embryo. Dev. Biol. 10, 154–190.

Mozdziak, P.E., Borwornpinyo, S., McCoy, D.W., Petitte, J.N., 2003. Development of trans-genic chickens expressing bacterial beta-galactosidase. Dev. Dyn. 226, 439–445.

Nakamura, M., Kuwana, T., Miyayama, Y., Fujimoto, T., 1988. Extragonadal distribution of primordial germ cells in the early chick embryo. Anat. Rec. 222, 90–94.

Ono, T., et al., 1994. A complete culture system for avian transgenesis, supporting quail embryos from the single-cell stage to hatching. Dev. Biol. 161, 126–130.

Ono, T., Wakasugi, N., 1984. Mineral content of quail embryos cultured in mineral-rich and mineral-free conditions. Poult. Sci. 63, 159–166.

Perry, M.M., 1988. A complete culture system for the chick embryo. Nature 331, 70–72.

Rapp, J.C., Harvey, A.J., Speksnijder, G.L., Hu, W., Ivarie, R., 2003. Biologically active human interferon alpha-2b produced in the egg white of transgenic hens. Transgenic Res. 12, 569–575.

Rowlett, K., Simkiss, K., 1987. Explanted embryo culture: in vitro and in ovo teachniques for the domestic fowl. Br. Poult. Sci. 28.

Sang, H., Perry, M.M., 1989. Episomal replication of cloned DNA injected into the fertilised ovum of the hen, Gallus domesticus. Mol. Reprod. Dev. 1, 98–106.

Sherman, A., et al., 1998. Transposition of the Drosophila element mariner into the chicken germ line. Nat. Biotechnol. 16, 1050–1053.

Speksnijder, G., Ivarie, R., 2000. A modified method of shell windowing for producing somatic or germline chimeras in fertilized chicken eggs. Poult. Sci. 79, 1430—1433.

Stebler, J., et al., 2004. Primordial germ cell migration in the chick and mouse embryo: the role of the chemokine SDF-1/CXCL12. Dev. Biol. 272, 351—361.

Swift, C.H., 1914. Origin and early history of primordial germ cells in the chick. Am. J. Physiol. 15, 483—516.

Urven, L.E., Abbott, U.K., Erickson, C.A., 1989. Distribution of extracellular matrix in the migratory pathway of avian primordial germ cells. Anat. Rec. 224, 14—21.

van de Lavoir, M.C., et al., 2012. Interspecific Germline Transmission of Cultured Primordial Germ Cells. PLoS ONE 7.

van de Lavoir, M.C., et al., 2006. Germline transmission of genetically modified primordial germ cells. Nature 441, 766—769.

Willier, B.H., 1937. Experimentally produced sterile gonads and the problem of the origin of germ cells in the chick embryo. Anat. Rec. 70, 89—112.

13 Production of Transgenic Nonhuman Primates

Anthony W.S. Chan

Yerkes National Primate Research Center, Department of Human
Genetics, Emory University School of Medicine, Emory University,
Atlanta, GA

I. Introduction

A. Importance of the Transgenic Nonhuman Primate Model of Human Diseases

Transgenic animal modeling of human diseases has led to biomedical breakthroughs that have significant impact on the diagnoses, treatments, and interventions in human diseases (Chan, 2004; Chan et al., 2001; Yang and Chan, 2011; Yang et al., 2008; Brosh et al., 2000; Cayzac et al., 2011; Hitz et al., 2009; Lassnig et al., 2005; Ramaswamy et al., 2009; Shavlakadze et al., 2005; Tsunematsu et al., 2011). Since the development of the first transgenic rodents in the early 1980s, thousands of transgenic rodents that captured conditions of human diseases have been developed. The advancement of genetic engineering technologies and the completed sequencing of the human genome, as well as in other species such as rodent and rhesus macaque, have opened new venues in transgenic animal modeling. Technologies such as high-throughput sequencing and gene expression microarray have accelerated the process in identifying candidate genes related to specific disease conditions. Together with novel technologies such as gene silencing by small interference RNA (siRNA) (Hitz et al., 2009; Raymond et al., 2010; Seibler et al., 2005, 2007; Van Pham et al., 2012; Xia et al., 2006), gene targeting by zinc finger nucleases (ZFNs) (Carbery et al., 2010; Ellis et al., 2013; Kobayashi et al., 2012; Passananti et al., 2010; Strange and Petolino, 2012), and transcription activator-like effector nucleases (TALENs) (Cermak et al., 2011; Li et al., 2012; Liu et al., 2012; Mahfouz et al., 2011; Sung et al., 2012), efficient reduction of gene transcripts (knockdown) or complete knockout of the gene functions can be achieved by interfering in the transcription process by siRNA or by targeted

Transgenic Animal Technology. DOI: http://dx.doi.org/10.1016/B978-0-12-410490-7.00014-1

disruption of the gene of interest using ZFN or TALEN. These novel genetic engineering tools open new opportunities in animal modeling of human genetic diseases, not only in dominant genetic disorders but also in recessive and dominant negative genetic diseases. Although significant advancement in the transgenic rodent model of human diseases has been made using these novel technologies, there are fundamental limitations in rodents that suggest the need for a model system that could better capture clinical features, or an animal model with higher anatomical/physiological similarity to humans. Due to physiologic differences between rodents and higher primates such as cellular metabolism (Finger et al., 1988; MacDonald et al., 2011), life span (Gilley et al., 2011), brain size and complexity (Yang et al., 2008; Chen et al., 2000; Hsiao et al., 1996; Miller et al., 2010; Polymeropoulos et al., 1997), and motor repertoire (Rice, 2012; Courtine et al., 2007), as well as the availability of cognitive behavioral testing (Bachevalier et al., 2001, 2011; Bachevalier and Nemanic, 2008; Ewing-Cobbs et al., 2012), nonhuman primates (NHPs) are considered one of the best animal models for complex and neurological disorders such as those correlated with aging, mental, and psychiatric dysfunctions. It is also important to note that the prolonged life span of NHPs (e.g., rhesus macaque) (Makris et al., 2010) allows for longitudinal study of disease trajectories. Additionally, in contrast to many mouse models, NHPs are not inbred and will therefore resemble the heterogeneity of human populations (Table 13.1).

A study in autism spectrum disorder (ASD) children suggests that the principle of neurological windows of developmental opportunity is linked to stages of brain development (Minshew and McFadden, 2011). Indeed, this study suggests the importance of the correlation between early brain development, when brain structures and functions develop higher order abilities with the emergence of neuropsychiatric symptoms. Studies in rodents have suggested that modulation of ASD susceptibility genes can directly lead to phenotypic changes that parallel those seen in human patients (Bangash et al., 2011; Bozdagi et al., 2010; Kumar et al., 2011; Patterson, 2011; Peca et al., 2011; Wang et al., 2011). Although transgenic rodent models have been widely used for evaluating the genetic abnormalities observed in human patients, there are limitations in nonprimate models for studying psychiatric disorders. Indeed, many human-associated behaviors cannot be recapitulated in rodents (Robertson and Feng, 2011), simply due to the physiological differences between rodents and humans (Yang et al., 2008; Finger et al., 1988; MacDonald et al., 2011; Gilley et al., 2011; Chen et al., 2000; Hsiao et al., 1996; Miller et al., 2010; Polymeropoulos et al., 1997; Rice, 2012; Courtine et al., 2007). An ASD transgenic NHP model will not only mimic patient conditions but most importantly it will aid in determining the longitudinal trajectory of ASD, thus advancing our understanding of ASD development and the critical need for new diagnostics and therapeutic options. Due to the progressive development and deleterious behavioral and psychiatric alterations associated with ASD, an ASD transgenic NHP model will provide accessibility to a wide range of testing methods designed for infant and adult NHPs, including the possibility for high-resolution brain imaging such as functional magnetic resonance imaging. In addition to longitudinal

Table 13.1 Breakthroughs in Nonhuman Primate Research

Event	Species	Year
Frozen/thawed semen/AI	Gorilla	Douglass (1981)
IVF/cleavage	Rhesus macaque	Bavister et al. (1983)
IVF/fresh/ET/live birth	Rhesus macaque	Bavister et al. (1984)
In vivo/frozen embryo/ET/live birth	Baboon	Pope et al. (1984)
IVF/frozen embryo/ET/live birth	Cynomolgus macaque	Balmaceda et al. (1986)
IVF/fresh/ET/live birth	Marmoset	Lopata et al. (1988)
Frozen/thawed semen/AI	Chimpanzee	Gould and Styperek (1989)
Frozen/thawed semen/AI	Cynomolgus macaque	Tollner et al. (1990)
IVF/frozen/thawed embryo/ET/twin	Rhesus macaque	Lanzendorf et al. (1990)
Embryonic stem cell	Rhesus macaque	Thomson et al. (1995)
Embryonic stem cell	Marmoset	Thomson et al. (1996)
Cloning/blastomere/NT/live birth	Rhesus macaque	Meng et al. (1997)
ICSI/ET/live birth	Rhesus macaque	Hewitson et al. (1999)
Cloning/embryo splitting/live birth	Rhesus macaque	Chan et al. (2000)
Frozen/thawed semen/AI/live birth	Rhesus macaque	Sanchez-Partida et al. (2000)
Transgenic/live birth	Rhesus macaque	Chan et al. (2001)
SCNT-derived embryonic stem cells	Rhesus macaque	Byrne et al. (2007)
Transgenic/human disease model	Rhesus macaque	Yang et al. (2008)
Transgenic/germ line transmission	Marmoset	Sasaki et al. (2009)
Mitochondrial transfer/live birth	Rhesus macaque	Tachibana et al. (2009)
Chimeric monkey/live birth	Rhesus macaque	Tachibana et al. (2012)

AI: Artificial insemination; IVF: *In vitro* fertilization; ET: Embryo transfer; NT: Nuclear transplantation; SCNT: Somatic cell nuclear transplantation.

cognitive evaluation and noninvasive imaging, molecular characterization of peripheral blood and other biomaterials from NHP models will also be critical for identifying biomarkers for early diagnosis and the development of novel treatments.

Besides complex disorders such as ASD, retinal disorders are medical conditions that often involve structures (e.g., macular and fovea) that are anatomically unique in the retina of humans and NHPs. Rodent eyes only remotely resemble human ocular anatomy, but the availability of genetic engineering tools such as transgenic and gene targeting in rodents has made it one of the most popular animal models in ocular research (Grossniklaus et al., 2010; Ou et al., 2010; Ramaesh et al., 2003; Zeiss, 2010). Transgenic rodents play a key role in dissecting the genetic cause of ocular diseases, even when they do not fully recapitulate human conditions, especially macular and fovea functions (Grossniklaus et al., 2010; Lassota, 2008; Sommer et al., 2011). On the other hand, canines with natural retinal degenerative diseases have been successfully used in the development of gene therapy, while transgenic pigs have also been developed because of the similarity in eye size to humans as well as their availability for preclinical trials. Since there is no perfect model for ocular research, most animal models recapitulate some of the human conditions and advance the development of novel therapeutic strategies such as gene therapy and surgical intervention. However, the lack of macular and fovea functions in most species limits the therapeutic efficacy on retinal disorders related to these functions in humans. Although NHPs are known to be the best model of human ocular disorders because of high similarity in anatomical structures, genetics, and immunity (Stieger et al., 2009), ocular diseases are uncommon in NHPs, and the applications of the NHP model have been limited by the high costs and the availability of genetic engineering tools. An NHP model with mutations that leads to macular degeneration such as age-related macular degeneration (AMD), the most common cause of the loss of vision, is expected to capture key pathological features throughout the course of disease development that may advance our knowledge of this disease and the development of novel therapeutics.

Our group created the first transgenic NHP in 2001 (Chan et al., 2001) and the first transgenic NHP model of inherited neurodegenerative Huntington's disease (HD) in 2008 (Yang et al., 2008) (Figure 13.1). HD-NHPs capture clinical features, including pathology and motor impairment, that no other animal model can recapitulate (Yang et al., 2008), strongly suggesting the importance and potential of transgenic monkey models for human inherited genetic diseases. We will use HD-NHPs as an example, because it is the only reported transgenic NHP model of human disease (Yang and Chan, 2011; Yang et al., 2008), and HD has well-defined genetics with breakthrough development in therapeutics (Matsui et al., 2013; Yu et al., 2012; Boudreau et al., 2009; Davidson, 2012; McBride et al., 2008, 2011).

One of the key rationales in developing a transgenic animal model is the inheritance of the transgene through the germ cells, which mimics the inheritable pattern

Figure 13.1 The first transgenic NHP model of HD expressed green fluorescent protein at 1 month of age.

of human genetic disorders (Chan, 2004; Chan et al., 2001; Yang et al., 2008; Bates et al., 1998; Schilling et al., 1999). Unlike most laboratory animal species, NHPs such as rhesus macaque reach pubertal age at around 4 years old. This is a relatively long physiological event compared to most laboratory species and the small NHPs such as marmosets (Sasaki et al., 2009). While there is no perfect animal model for humans, it is important to identify appropriate model systems to address specific underlying pathogenic mechanisms and clinical manifestations of inherited neurodegenerative disorders, such as HD, Alzheimer's disease (AD), and Parkinson's disease (PD), that are also influenced by aging events (Rice, 2012; Courtine et al., 2007), and their systemic impact on the body, which includes motor functions, psychiatric disturbance, and metabolic disturbance (Yang et al., 2008; Paulsen, 2011; Tabrizi et al., 2011,2012). To capture the systemic impact of complex diseases such as HD, NHPs can effectively model the progression of disease through longitudinal studies applying similar clinical measurements performed on human patients. Moreover, macaques also share a similar motor repertoire with humans that allows the evaluation of fine movement control, something that cannot be done with most of the currently available model systems including smaller primates such as marmoset (Courtine et al., 2007). Most importantly, neurobehavioral impact can be evaluated with a sophisticated battery of cognitive behavioral tests developed for macaques that are not as well developed in other species (Bachevalier et al., 2001, 2011; Ewing-Cobbs et al., 2012).

Discrepancies due to overestimation of the sensitivity of measurements in cross-sectional studies suggest the importance of unbiased longitudinal studies for precise interpretation of the results and for determining possible clinical applications (Tabrizi et al., 2011,2012; Hobbs et al., 2010a,b; Solomon et al., 2008). For example, recent longitudinal studies on premanifest and early-stage HD patients demonstrate significant progressive atrophy in the caudate, putamen, thalamus, and nucleus accumbens (Tabrizi et al., 2011; Hobbs et al., 2010a,b; Aylward et al., 2011a,b; Aylward et al., 2012; Paulsen et al., 2010; Wolf et al., 2011). The progression in brain atrophy parallels the decline in cognition and motor functions in early HD (Tabrizi et al., 2011; Scahill et al., 2011). These studies suggest that quantitative longitudinal evaluation is a powerful tool for monitoring disease status and progression rate (Tabrizi et al., 2011; Hobbs et al., 2010a,b; Aylward et al., 2011a,b; Aylward et al., 2012; Paulsen et al., 2010; Wolf et al., 2011). Compared to rodents, NHPs are considered the best animal model for longitudinal study, especially for neurodegenerative diseases and disorders that are linked to stages of brain development (Minshew and McFadden, 2011). This is because of their long life span and high similarity to human brain development (which can be assessed with a battery of tests specifically designed for evaluating brain development) and clinical manifestation of neurological dysfunctions from infancy to adulthood. In addition to the possibility and advantages of long-term longitudinal study, it is also important that clinical measurements similar to those offered to humans are available, to enable direct comparison with human data.

While there is no perfect model for humans with HD, it is important to be able to correlate critical pathological and clinical events that occur in animal models with human clinical data, enabling a reliable projection of clinical development between animal models and human HD patients. The application of human clinical measurements to an animal model is particularly important for direct comparisons. An NHP model such as the HD-NHP can potentially be used for preclinical and nonclinical research with a set of clinical measurements for determining therapeutic efficacy. Additionally, potential biomarkers and therapeutic targets can be identified through a longitudinal molecular profiling study. Thus the development of a transgenic NHP model will not only provide important information for developing clinically relevant measurements to determine the efficacy of new therapies but is also important for developing a comparable platform for studying human disease leading to the development of novel treatments.

B. Emerging Technologies for Genetic Manipulation

Most transgenic NHPs reported today are primarily generated by the overexpression of the gene of interest using viral vector as a vehicle. Due to the significant resources required for the creation of a germ line of transgenic NHPs (Chan et al., 2001; Yang et al., 2008; Sasaki et al., 2009; Niu et al., 2010; Sun et al., 2008), focal transgenesis (Kordower et al., 2000; Mittoux et al., 2000; Palfi et al., 2007), widely used in rodents, is also commonly used in NHPs as a short-term and relatively low-cost method for evaluating gene functions in higher primates. Although a wide range of genetic manipulation tools have been successfully used in rodents and other livestock species such as cattle and swine (Hauschild et al., 2011; Hauschild-Quintern et al., 2013; Tessanne et al., 2012; Golding et al., 2006), the production of transgenic NHPs has been limited to the overexpression approach (Chan et al., 2001; Yang et al., 2008; Sasaki et al., 2009; Niu et al., 2010; Sun et al., 2008) and primarily focused on dominant genetic diseases such as HD (Chan, 2004; Chan et al., 2001; Yang and Chan, 2011; Yang et al., 2008). While functional knockdown and functional knockout rodent models of human diseases have been developed by (i) homologous recombination in embryonic stem cells followed by blastocyst injection to create chimeric rodents (Bertelli et al., 2009; Lin et al., 2001; Woodman et al., 2007), (ii) nuclear transplantation of genetically modified donor cells to create cloned rodents (Rideout et al., 2002), (iii) silencing of the target gene expression using shRNA (Hitz et al., 2009; Raymond et al., 2010; Seibler et al., 2005; Tessanne et al., 2012), or (iv) gene-targeted disruption using ZFN and TALEN (Carbery et al., 2010; Sung et al., 2012, 2013; Hauschild et al., 2011; Hauschild-Quintern et al., 2013), a similar technique has not been reported in the creation of transgenic NHPs.

A transgenic NHP was first reported in 2001 (Chan et al., 2001), two decades after the first transgenic mouse was created. The development of the first transgenic NHP model of human inherited disease was not achieved until 2008 (Yang et al., 2008). The major bottleneck in the development of gene-targeted NHPs is gene targeting efficiency and permanent disruption of the target gene. Although the creation of a chimeric monkey was reported using an embryonic blastomere, the generation of genetically modified chimeric monkeys has yet to be achieved.

Alternatively, several revolutionary genetic tools have shed light on transgenic NHP modeling of diseases that are caused by the loss of gene function or haplo-insufficiency. shRNA has been successfully used to target reduction of gene transcripts (Hitz et al., 2009; Raymond et al., 2010; Seibler et al., 2005, 2007; Van Pham et al., 2012; Xia et al., 2006; Hauschild et al., 2011; Hauschild-Quintern et al., 2013; Tessanne et al., 2012; Golding et al., 2006; Whyte and Prather, 2012; Zschemisch et al., 2012), while ZFN and TALEN have been successfully used to permanently disrupt the target gene and have resulted in the loss of function (Carbery et al., 2010; Ellis et al., 2013; Kobayashi et al., 2012; Passananti et al., 2010; Strange and Petolino, 2012; Li et al., 2012; Liu et al., 2012; Mahfouz et al., 2011). While gene-targeted transgenic rodents and livestock have been successfully created by these methods (Carbery et al., 2010; Hauschild et al., 2011; Hauschild-Quintern et al., 2013; Tessanne et al., 2012; Golding et al., 2006; Whyte and Prather, 2012; Zschemisch et al., 2012), their translation to NHPs has yet to be determined. These technologies have opened the door for manipulating the NHP genome by silencing or targeted disruption. To optimize the gene targeting efficiency of these approaches prior to the successful production of a gene-targeted NHP model of human diseases, a great deal of work is still necessary. While the application of these emergent technologies in NHPs remains in its infancy stage, this chapter will be focused on the lentiviral gene transfer method in creating transgenic NHP models of human diseases.

II. Methods

A. Production of High-Titer VSV-G Pseudotyped Lentiviral Vector

1. Vector Construction

A fragment of the gene of interest (e.g., exon 1 of the huntingtin gene with expanded polyglutamine (CAG) repeats) was inserted into a lentiviral vector such as pFUW [F: human immunodeficiency virus-1 flap element (HIV-flap), U: ubiquitin promoter, W: woodchuck hepatitis virus post-transcriptional regulatory element (WRPE)] (Yang et al., 2008; Lois et al., 2002). HIV-flap and WRPE enhance transcription level and RNA stability, minimize position effect, and increase lentiviral titer. The enhancement of gene expression and attenuation of position effect by composing an F and W sequence in the lentiviral vector has been demonstrated in tissue culture, in mice, rat, and NHPs (Yang et al., 2007, 2008; Kordower et al., 2000; Lois et al., 2002). Ubiquitin promoter has been successfully used to create transgenic rat, mice, and NHPs (Yang et al., 2007, 2008; Lois et al., 2002; Agca et al., 2008).

2. Production of High-Titer Lentiviral Vector

In order to produce high-titer VSV-G pseudotyped lentiviral vector, four components are required: (i) a packaging cell, e.g., 293FT (Life Technologies, Gaithersburg, MO), (ii) lentiviral vector, e.g., pFUW, (iii) a lentiviral packaging construct, e.g., pΔ8.9 (composed of the structural genes for virion assembly), and (iv) expression plasmid of vesicular stomatitis virus glycoprotein-G (VSV-G) envelope protein (pVSVG; Life Technologies). To produce high-titer lentiviral vectors, pFUW composed of the gene of interest is to be cotransfected with pΔ8.9 (composed of the structural genes for virion assembly) and pVSVG (plasmid DNA expressing VSV-G envelope protein; Invitrogen) into a 293FT packaging cell (Invitrogen) at 60−70% confluency. 293FT packaging cells were freshly passaged the day before transfection. The transfection ratio of plasmid DNA in a 15 cm petri dish is lentiviral vector (25 μg): pΔ8.9 (18.75 μg): pVSVG (12.5 μg) (a ratio of 2:1.5:1). Culture medium is collected at 48 h post-transfection for two consecutive days. Each daily collection is filtered through a 0.45 μm filter (Millopore Inc.) to remove cell debris followed by ultracentrifugation. Supernatant is centrifuged at $25,000 \times g$ for 90 min and supernatant was carefully removed without disturbing the pellet. The viral pellet should be resuspended in 50−100 μL of phosphate-buffered saline (PBS). Lentiviruses are then aliquoted in 3−5 μL, titered, and kept frozen at −80°C. All plasticware and instruments that made contact with the lenti-viruses must be decontaminated with 10% bleach solution or autoclaved before disposal.

3.   Determination of Lentiviral Vector Titer

Titer represents the number of infectious vector particles and is determined by the formation of colonies expressing the protein of interest or resistance to an antibiotic selection marker such as neomycin. 293FT or the target cell of interest are plated in triplicates in a 96-well plate to achieve 80−90% confluency on the next day. The following day, the cells are inoculated with serial dilutions (ranging from 10^{-5} to 10^{-9}) of pseudotyped lentiviral vector in the presence of 8 μg/mL polybrene (hexadimethrine bromide; Sigma, St. Louis, MO). Polybrene is a polycation that neutralizes the surface charge of the target cell's surface and the vector particles in order to enhance efficient viral adsorption (Andreadis and Palsson, 1997; Cornetta and Anderson, 1989; Toyoshima and Vogt, 1969). A stock solution of polybrene can be prepared by dissolving 1 mg of polybrene in 1 mL sterile water with 10% DMSO; the stock solution is aliquoted and is stable at −20°C. Fresh medium is replaced the day after inoculation; titer can be determined by the expression of a fluorescent tag such as green fluorescent protein or immunostaining using the anti-body specifically recognized as the protein of interest. For titering using antibiotic selection method, fresh medium with antibiotic such as neomycin (e.g., 400 μg/mL of G418) should be replaced the day after inoculation and repeated every 3 days up to 14 days, when G418 resistant colonies appear. The titer is calculated by multi-plying the number of positive cells or antibiotic-resistant colonies by the dilution factor. Vector titer is expressed as colony forming units (cfu/mL), and the average titer of pseudotyped vector collected from the culture medium before concentration is about 1×10^5 to 1×10^6 cfu/mL. Vector titer can be concentrated 1000-fold and reaches $10^8 - 10^9$ cfu/mL after ultracentrifugation.

B. Superovulation and Recovery of Rhesus Macaque Oocytes

1. Follicle Stimulation

Female rhesus monkeys exhibiting regular menstrual cycles are induced with exogenous gonadotropins (Vandevoort et al., 1989; Zelinski-Wooten et al., 1995). Monkeys are to be down-regulated at the beginning of menses by daily subcutaneous injections of GnRH antagonist (Acyline 75 mg/kg, NICHD) for 6 days and by twice daily injection of recombinant human FSH (r-FSH; Organon Inc., West Orange, NJ; 30 IU, i.m.) concomitantly. This is followed by 1, 2, or 3 days of r-FSH + r-hLH (r-hLH Ares Serono; 30 IU each, i.m., twice daily). Ultrasonography is performed on day 7 of the stimulation to confirm follicular response. An i.m. injection of 1000 IU r-HCG (Serono, Randolph, MA) is administered for ovulation at approximately 37 h prior to oocyte retrieval, when there are follicles at 3−4 mm in diameter.

2. Oocyte Aspiration

Follicular aspiration is performed 37 h post-HCG. Oocytes are aspirated from follicles using a needle suction device lined with Teflon tubing modified by Bavister et al. (1983) and Chan and Yang (2009). Briefly, a 10 mm trocar is placed through the abdominal wall and a telescope is introduced. Ovaries are visualized by a monitor attached to the telescope. Two small skin incisions are made to facilitate the insertion of 5 mm trocars bilaterally. Grasping forceps are introduced through each trocar to fixate the ovary at two points. Once stabilized, a 20-gauge stainless steel hypodermic needle connected with teflon tubing is attached to a vacuum regulator for oocyte aspiration. The tubing is first flushed with sterile TALP-Hepes (Bavister et al., 1983; Chan and Yang, 2009), supplemented with 5 IU/mL of heparin in order to prevent blood clots in the tubing during aspiration. Follicles are aspirated with continuous vacuum pressure adjusted at approximately 70 mm Hg, into a 15 mL conical tube containing 1 mL of TALP-Hepes supplemented with 5 IU/mL of heparin and maintained at 37°C.

3. Collection and Evaluation of Rhesus Macaque Oocytes

Collection tubes with follicular fluid should be diluted in TALP-Hepes supplemented with 2 mg/mL hyaluronidase. Follicular fluid with buffer is transferred into a 60 mm dish containing 5 mL of TALP-Hepes with hyaluronidase. Oocytes were picked up under a dissecting microscope located inside a biosafety cabinet and transferred to a dish containing fresh TALP-Hepes with no hyaluronidase. Oocytes are rinsed and then transferred to pre-equilibrated maturation medium [Connaught Medical Research Laboratories medium 1066 (CMRL-1066; Invitrogen) supplemented with 10% heat-inactivated fetal bovine serum (HyClone), 40 µg/mL sodium pyruvate, 150 µg/mL glutamine, 550 µg/mL calcium lactate, 100 ng/mL estradiol, and 3 µg/mL of progesterone under mineral oil (Sigma)]. Metaphase II-arrested oocytes with expanded cumulus cells, distinct perivitelline space (PVS), and the first polar body are immediately used for PVS injection of lentiviral vector, followed by culture in maturation medium for up to 8 h before fertilization. Immature oocytes are matured in oocyte maturation medium for up to 24 h.

C.   PVS Injection of Oocytes, Fertilization of Oocytes, and Culture of Embryos

1.   Preparation of Oocytes for PVS Injection

Oocytes at MII stage with first polar body were selected and cumulus cells were removed by gentle pipetting using a fire-polished pipette with inner diameter (i.d.) about the same size of an oocyte. Oocytes were then washed in TALP-Hepes buffer before being transferred into oocyte maturation medium and returned to 37°C with 5% CO_2 and 90% N_2 until injection.

2.   Preparation of PVS Injection Apparatus

A holding pipette (outer diameter (o.d.) 100 µm; i.d. 20 µm) was filled with mineral oil and connected to a CellTram microinjector with tubing filled with mineral oil, and mounted on a Narishigi manipulator. An injection needle (tip of 100–200 µm in diameter) was made by a Sutter micropipette puller. A sharp, pointed pipette was then transferred into a Biosafety level II hood for vector loading. A vial of pseudotyped vector (4 µL) was thawed on ice after removal from −80°C. One microliter of polybrene solution was added to the viral vector solution with a final concentration of 5 µg/mL. The vector/polybrene mixture was then loaded into the injection pipette from the back using a microloader (Eppendorf, Westbury, NY). The mixture was loaded at the tip of the pipettes, avoiding bubble formation.

3.   PVS Injection of Mature Oocyte

Injection is to be carried out in a 30 µL drop of TALP-Hepes buffer covered with mineral oil on the cover of a 3.5 cm petri dish. An injection pipette loaded with vector mixture is mounted on the other side of the Narishigi manipulator and connected to a 50 mL glass syringe with Tygon tubing. The tip of the injection pipette is placed and hit slightly against the holding pipette to break the tip. Oocytes are then transferred into the injection drop after a wash in a TALP-Hepes buffer. Oocytes are held by holding pipette with the polar body at the 12 o'clock position and followed by the insertion of injection needle with vector mixture into the zona pellucida at the 6 o'clock position. Damage to the oolemma membrane should be avoided. The vector mixture is then injected into the space slowly until excess solution forces the vector solution out, then the pipette is slowly pulled out (Figure 13.2A). Successful injection can be confirmed by a brief exposure to fluorescent excitation light source if a fluorescent tag was used (see inserts of Figure 13.2A). However, one should be aware that the fluorescence is emitted from fluorescent protein expressed in packaging cells encapsulated into the viral particle

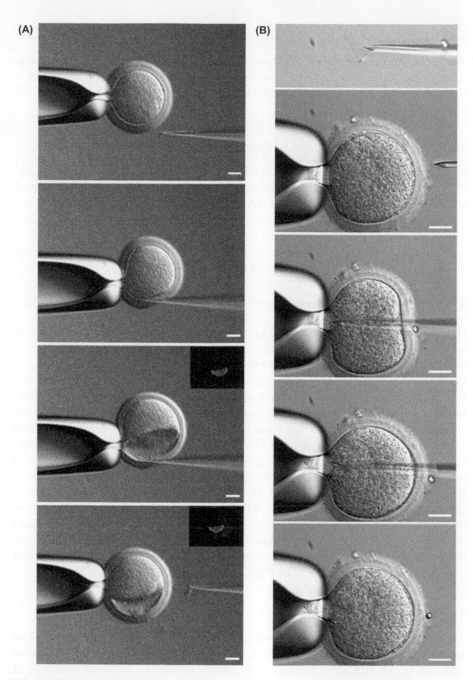

Figure 13.2 Micromanipulation of rhesus macaque oocyte. (A) PVS injection of mature rhesus oocyte. Vector solution is injected into the PVS with a fine needle and the deformation of the ooplasm is caused by the injected vector. The insert pictures show the green fluorescent derived from the green fluorescent protein enclosed in the lentiviral vector particles injected into the PVS. (B) Intracytoplasmic sperm injection (ICSI) of rhesus oocyte. Sperm is picked up by injection needle followed by injection into the cytoplasm of a mature rhesus oocyte. Scale bar = 25 μm.

during the viral packaging process instead of the expression of the fluorescent tag. After injection, oocytes are transferred back to Hamster Embryo Culture Medium 9 (HECM-9) (Zheng et al., 2001a,b) after two washes and cultured at 37°C with 5% CO_2 and 90% N_2.

4. Collection and Preparation of Ejaculated Rhesus Sperm

Semen samples were collected by penile electroejaculation (Lanzendorf et al., 1990). After liquefaction of the coagulated ejaculate, the liquid semen was removed and washed three times in 10 mL of TALP-Hepes by centrifugation at $400 \times g$ for 5 min. The pellet was resuspended in 1 mL of TALP-Hepes, and a small sample was removed for analysis of motility and morphology. The remaining suspension was counted, diluted to a concentration of 20×10^6 sperm/mL in equilibrated TALP, and incubated at 37°C with 5% CO_2 prior to ICSI.

5. ICSI of Mature Rhesus Oocytes

Holding pipette (o.d. 100 μm; i.d. 20 μm) and microinjection needle (o.d. 7 μm; i.d. 4–5 μm) with a 50-degree beveled tip (Humagen, Charlottesville, VA) are mounted on an Olympus IX71 inverted microscope equipped with Modulation Contrast optics. The holding and injection pipettes are filled with mineral oil and held on a Narishigi manipulator. They are then connected to CellTram microinjector (Eppendorf). Injection is carried out in a 30 μL drop of TALP-Hepes buffer covered with mineral oil on the cover of a 3.5 cm Petri dish. Sperm are diluted 1:10 in 10% polyvinylpyrrolidone (PVP) to reduce motility. A single sperm is aspirated, tail first, from the sperm-PVP drop into the injection needle and transferred into the oocyte-containing drop (Figure 13.2B). Oocytes are held by holding pipette with the polar body at 6 o'clock or 12 o'clock position, followed by insertion of the injection needle with a sperm through the zona into the cytoplasm. The oolemma is then penetrated by gentle cytoplasmic aspiration, and the sperm is expelled into the oocyte. It is important to immobilize the sperm by gently squashing the midpiece of the sperm against the bottom of the dish using the injection pipette and by injecting a minimum amount of PVP-containing buffer into the oocyte.

6. Culture of Rhesus Oocytes After ICSI

After ICSI, oocytes are washed twice in TALP-Hepes buffer followed by two washes in HECM-9 before being transferred into a pre-equilibrated four-well plate with 500 μL of HECM-9, covered with 300 μL of mineral oil, and incubated at 37°C with 5% CO_2 and 90% N_2 until the next morning (Figure 13.3). In brief, for the first 48 h post-ICSI, embryos are to be cultured in HECM-9 supplemented with amino acids/pantethonate (AAP) stock (Zheng et al., 2001a,b) without FBS (HyClone). Fresh HECM-9/AAP supplemented with 10% FBS should be replaced at 48 h post-ICSI or at the four- to eight-cell stage, then cultured until the blastocyst stage.

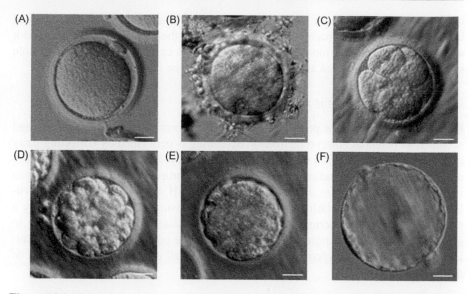

Figure 13.3 The development of a rhesus embryo: (A) mature oocyte, (B) fertilized oocyte, (C) four to six cells, (D) early morula, (E) compact morula, and (F) blastocyst. Scale bar = 25 μm.

7. Assisted Hatching

At the four- to eight-cell stage (48 h post-ICSI), embryos with clear nuclei and equal blastomere size are identified and submitted for assisted hatching to improve hatching, implantation, and pregnancy rate. Selected embryos based on morphology are transferred into manipulation conditions similar to those described for PVS injection and ICSI. A small opening at the zona pellucida is made by XYClone laser (Hamilton Throne, Inc.). These embryos are then used for oviductal transfer.

D. Selection of Recipient and Embryo Transfer

1. Conscious Bleeding of Females for Hormonal Assay

Bleeding of female macaques is performed using the "Bleed Tower Station" (Carter2 Systems). Animals are transferred from a transfer box to the bleed tower. After the necessary restraint, the door located at the posterior of the animal is opened just enough to allow access to the animal leg. When the animal is secured, a technician can reach into the tower, wearing appropriate personal protection equipment (PPE), through the provided doorway at the posterior of the animal. With the animal's foot and leg in hand, the leg is pulled out of the tower in preparation for blood collection. The leg is restrained in the areas of the ankle and knee to allow for maximum visualization of the saphenous vein for blood collection. The leg is then wiped with an alcohol swab to sanitize the area. About 3 mL of blood was taken by using a sterile 3 mL syringe with a 23-gauge needle followed by transferring into a heparinized blood tube. After the monkey has formed a clot at the puncture site, the monkey was returned to its cage. A total of 3 mL whole blood is used for hormonal assay to determine estrogen and progesterone level.

2. Selection of Recipient for Embryo Transfer

Rhesus females with prior pregnancy and normal menstrual cycles are selected and screened as potential embryo recipients. Screening is performed by collecting daily blood samples beginning on day 8 of the menstrual cycle (day 1 is the day of menses) and analyzed for serum progesterone and estrogen. When serum estrogen increases two to four times that of basal levels, the LH surge has occurred and ovulation usually follows within 12−24 h. Timing of ovulation can be detected by a significant decrease in serum estrogen and an increase in serum progesterone to greater than 1 ng/mL. Surgical embryo transfers are performed on days 2−3 following ovulation by transferring two four- to eight-cell embryos into the oviduct of the recipient.

3. Embryo Transfer by Laparotomy

Surgical embryo transfers are performed by midventral laparotomy. The oviduct is annulated using a tomcat catheter containing two four- to eight-cell stage embryos in TALP-Hepes buffer supplemented with 3 mg/mL BSA. Embryos are expelled from the catheter into the oviduct with about 50 μL of medium while the catheter is withdrawn slowly. The catheter is flushed with medium following removal to ensure that the embryos are successfully transferred.

4. Confirmation and Monitoring of Pregnancy

Implantation is confirmed by daily blood collection for serum estrogen and progesterone analysis after the next expected menstrual cycle of the recipient female (Lanzendorf et al., 1990). Pregnancy is further confirmed by transabdominal ultrasonography on at least 40 days post-transfer, when hormone level indicates possible pregnancy. During ultrasonography, fetal cardiac activity will also be confirmed.

5. Births and Infants

All babies in our case were delivered naturally except for multiple pregnancies when caesarean section at approximately 155 days of pregnancy was scheduled to avoid potential complications at birth. Females at 155 gestation days were monitored at night by infrared camera streaming online for remote monitoring. Babies were weighed; physical measurement such as head circumference was recorded and the newborns were housed under an infant warmer at the nursery. Toys and towels were provided to infants for enrichment purposes. The infants were raised by a principal human caregiver who fed and handled them several times a day from the day they arrived in the primate nursery. At 3 months of age until approximately 9 months of age, all infants received daily social interactions (3−4 h, 5 days/week) with other age- and sex-matched peers of the same cohort and in the presence of one to three of the familiar human caregivers. From 3−4 months, socialization with peers took place in a playpen/cage located in the primate nursery and containing toys and towels. From 5−9 months, socialization with peers took place in a large enclosure containing perches and toys and located in the nursery. From 10 months to 2 years, socialization took place in a large social enclosure located in a room in the housing facility. Nonetheless, the housing and social arrangements as the animal ages will be dependent on research goals; appropriate cognitive behavioral tasks can also be arranged as soon as the first week after birth or earlier.

E. Confirmation of Transgenic Status

Given that a newborn monkey is about 500 g, the amount of tissue that can be retrieved at the time of birth was very limited. Samples like cord blood and placenta were collected at birth if possible, while buccal swab, peripheral blood, and ear punch samples were collected a day after birth for genomic and gene expression analyses.

1. Preparation of Genomic DNA

Whole blood: 50−100 µL of whole blood collected in a heparinized tube is used for genomic DNA extraction with the Maxwell 16 Blood DNA Purification Kit (Promega).

   Buccal swab: Buccal cells are collected by a buccal swab kit followed by washes with PBS and centrifugation. Total genomic DNA is then isolated by using Maxwell 16 Buccal Swab LEV DNA Purification Kit.

   Ear punch: Ear punch sample is collected under sterile conditions and genomic DNA is extracted by using Maxwell 16 Tissue DNA Purification Kit.

2. Preparation of RNA

Tissue samples (e.g., ear punch) are homogenized in TRIZOL reagent (Life Technologies) (1 mL per 100 mg tissues) using a glass TEFFLON. Homogenized samples are incubated for 5 min at room temperature (RT) to allow complete dissociation of nucleoprotein complexes, and 200 µL of chloroform is added to 1 mL TRIZOL reagent. The samples are then mixed vigorously and centrifuged at $12,000 \times g$ for 15 min at 4°C. The RNA is precipitated from the aqueous phase by mixing with 500 µL of isopropanol. The samples are incubated at RT for 10 min and centrifuged at $12,000 \times g$ for 10 min at 4°C. Finally, the RNA pellet is washed twice with 75% ethanol, the pellet air-dried, and then dissolved in diethyl pyrocarbonate treated water.

3. PCR Analysis

Genomic DNA was extracted as described and subjected for PCR analysis using two sets of specific primers. In the case of HD-NHPs, to detect the mutant hunting-tin (*mHTT*) transgene *HTT* gene, ubiquitin-F forward primer (5′-GAGGCGTC AGTTTCTTTGGTC-3′) and *htt*-R reverse primer (5′-GCTGGGTCACTCTG TCTCTG-3′) were used to yield 818 bp products after amplification of genomic DNA from the HD tissue (Yang et al., 2008). However, variation in the size of PCR products resulted from the variable numbers of polyglutamine (CAG) repeats. Genomic DNA (100 ng) was subjected to PCR with 1.65 M Betaine (Sigma) at 96°C for 5 min first; 96°C for 45 s, 62°C for 45 s, 72°C for 150 s for 35 cycles, and then 72°C for 7 min. To determine the number of CAG repeats in HD monkeys, the PCR products were sequenced using HD exon 1 forward primer (5′-GGCGACCCTGGAAAAGCTGA-3′).

4. Southern Blotting Analysis

Ten micrograms of genomic DNA were digested with the restriction enzyme that digests once within the lentiviral vector. The DNA fragments are separated by electrophoresis on a 0.8% agarose gel; this is followed by acid depurination (by soaking the gel in 0.25 N HCl and shaking for 15 min) and denaturation (by soaking the gel twice in fresh 1.5 M NaCl, 0.5 M NaOH for 20 min) at RT. The DNA fragments were then downward transferred to Hybond-N + nylon membranes (Amersham Pharmacia Biotech, Piscataway, NJ) at RT for 2−3 h using 10X SSC. After transfer, the membrane was soaked in neutralizing solution with 1 M Tris Cl pH 8.0 and 1.5 M NaCl for 15 min. The membrane was air-dried, followed by cross-linking the DNA fragments with the membrane using the Crosslinker (Stratagene, La Jolla, CA). The membrane was transferred into a hybridization tube and 6 mL of preheated Rapid Hybridization Buffer (Amersham Pharmacia Biotech) was added. The membrane was incubated in a hybridization oven at 65°C, rolling for 1 h. A 1×10^6 cpm/mL of ^{32}P-labeled fragment of the mHTT lentiviral vector (Random Primed DNA Labeling Kit; Roche Molecular Biochemicals, Indianapolis, IN) was used as a probe, mixed with the hybridization solution, and hybridized at 65°C for another 40−60 min. After hybridization, the membrane was washed at 65°C with high stringency buffer (2X SSC and 0.1% SDS twice; 1X SSC and 0.1% SDS twice; 0.1X SSC and 0.1% SDS once). The membrane was then wrapped in plastic cling film and exposed to X-ray film at −80°C for 2−3 weeks or by exposing (^{32}P) hybridized member to the phosphor screen and then scanned by phosphorimager.

5. Sequencing of Polyglutamine (polyQ) Repeats

For polyQ sequencing from the transgenic *HTT* transcripts of HD-NHP samples, 500 ng of total RNA was extracted and reverse transcribed to cDNA using the High Capacity Reverse Transcription Kit (Applied Biosystems). The *HTT* transcript is amplified from the cDNA by PCR at an annealing temperature of 67°C for 40

cycles with the following primers; HD32-Forward 5′-CTACGAGTCCCTCAAG
TCCTTCCAGC-3′ and MD177-Reverse 5′-GACGCAGCAGCGGCTGTGCCTG-3′.
All PCR products were electrophoresed on a 1.5% agarose gel and target bands were
gel purified, cloned into the pGEM-T easy vector (Promega), and subsequently
sequenced with T7 and SP6 primers.

6. Quantitative RT-PCR (qPCR) Analysis

For total RNA isolation from the samples collected from the transgenic NHPs, a
Trizol-based protocol was used. Briefly, each sample is homogenized in 500 μL of
Trizol (Invitrogen). A phenol-chloroform extraction of the RNA is done by addition
of 100 μL of chloroform to the Trizol homogenates followed by centrifugation at
12,000 × g for 10 min at 4°C. The aqueous layer is removed for RNA precipitation
overnight with isopropanol and 40 μg of glycogen (Invitrogen) at −20°C. The precipitated RNA is pelleted at 12,000 × g for 30 min at 4°C. All RNA pellets are
washed twice with 75% ethanol and then subsequently dissolved in water (RNase/
DNase free). A volume of 200 ng of total RNA for each sample was reverse transcribed to cDNA using the High Capacity RNA to cDNA kit (Life Technologies),
following the manufacturer's protocol. For qPCR, 1 μL of cDNA is primed with
custom-designed Taqman assays from Applied Biosystems for the target mRNA in
a final reaction volume of 1X Taqman Universal PCR Master Mix (Life
Technologies). qPCR was performed using Gene Expression Master Mix (Applied
Biosystems) and TaqMan® gene expression primers on CFX96 Real-Time
Detection System (Bio-Rad, Hercules, CA).

7. Detection of Protein by Western Blot Analysis

Total protein was extracted from ear punch or accessible tissues such as placenta
by using the appropriate volume of 1X RadioImmuno Precipitation Assay buffer
(RIPA buffer; 50 mM Tris−HCl pH 8.0, 150 mM NaCl, 1 mM EDTA pH 8.0,
1 mM EGTA pH 8.0, 0.1% SDS, 0.5% deoxycholate, 1% Triton) with 1X protease
inhibitor cocktail (Roche). Peripheral blood cells can be used when a sufficient
amount of blood can be collected. Total protein was quantified by Bio-Rad DC
Protein Assay (Pierce, Rockford, IL). Equal amounts of protein extract are loaded
and separated by electrophoresis in 9% SDS-PAGE gel. Proteins are transferred to
a polyvinylidene fluoride membrane (Bio-Rad Inc.) and probed with primary antibodies (e.g., mEM48 detection of mHTT with expanded polyQ and γ-tubulin at
1:50 and 1:2000, respectively) overnight. After three washes, the membrane is
incubated with peroxidase secondary antibody followed by detection using Western
Lightning Chemiluminescence Reagent Plus (Perkin-Elmer, Norwalk, CT).

8. Detection of Protein Expression by Immunohistochemistry

Placental and other biopsied tissues are fixed by using 4% paraformaldehyde, equilibrated in 30% sucrose, embedded in O.C.T. compound and snap frozen in liquid
nitrogen followed by 10 μm sections. Tissue sections are first blocked in 10% goat

serum in PBS for 20 min at RT on a shaking platform. Primary monoclonal antibody (e.g., mEM48 1:50, Chemicon) is diluted in PBS with 1.5% goat serum and applied for 60 min on a shaking platform at RT. After extensive PBS rinse, the primary antibody is detected using rhodamine-conjugated anti-mouse (IgG) secondary antibody diluted in PBS with 1.5% goat serum (45 min; RT) on a shaking platform in the dark. After repeated rinses in PBS, the DNA is counterstained with Hoechst 33342 (5 μg/mL) for 2 min prior to mounting in Vectashield antifade (Vector Labs, Burlingame, CA). Slides were examined with an Olympus IX71 epifluorescent microscope equipped with appropriate filters, and images were captured by ORCA CCD camera (Hamamatsu Inc.) using CellSens software (Olympus).

9. Determination of Integration Sites

Genomic DNA from transgenic NHPs was digested with a restriction enzyme cut once within the transgene (Figure 13.4). The partially purified digestion mixture is diluted followed by ligation with T4 DNA ligase in order to prepare a circular form DNA fragment. A nested inverse PCR primer set is designed to amplify the junction between the transgene and the genome (Figure 13.4). The nested PCR product is then subcloned into a cloning vector pGEM-T Easy (Promega), followed by transformation and plating onto an agar plate. Individual colonies were chosen and plasmid DNA was submitted for sequence analysis. Integration sites are to be determined by sequence alignment and comparison with the genomic database.

III.   Summary

It has been an exciting decade of genetic engineering with major advancement in gene targeting technologies such as ZFN and TALEN, and with great success in creating gene-targeted animal models including rodents and livestock. While its application in NHPs has yet to be demonstrated, it is foreseeable that a gene-targeted NHP model will soon be developed, and significant impact in modeling of human diseases is expected. The increased development of a better animal model for human diseases, such as a transgenic NHP model, has been clearly demonstrated by the increased numbers of primate facilities around the globe, especially in Southeast Asia. The establishment of the consortium for developing transgenic marmoset models for neuroscience research in Japan and the increase in the newly established primate facilities in China are two good examples of the growing interest in NHP research.

With the advancement in genetic tools and increased interest in developing genetically modified NHPs to model human diseases, it is expected that transgenic NHPs with different genetic mutations will be reported in the near future. The major roadblocks in developing gene-targeted or functional knockdown/knockout NHPs are (i) targeting efficiency, (ii) high cost, and (iii) availability of primates. While technical challenges such as targeting efficiency can be overcome and the generation of gene-targeted NHP models of human diseases is a matter of time, it

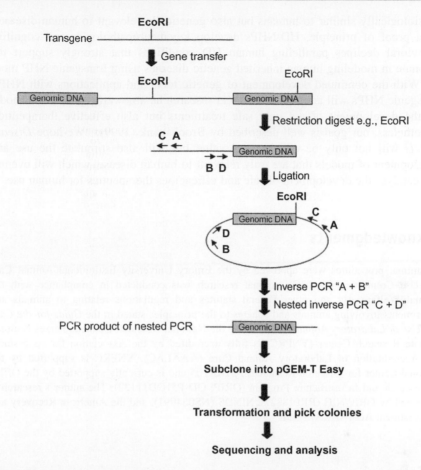

Figure 13.4 Cloning and determination of lentiviral integration sites.

is important to determine their appropriateness and whether newly developed transgenic NHPs could significantly advance our knowledge of disease conditions over other model systems. Additionally, the potential unique applications of the NHP model include the possibility of longitudinal study and its use as a preclinical model for determining therapeutic efficacy. Therefore, a specific research plan for each disease model, such as cognitive behavioral evaluation schemes with sampling schedules and methods, should be in place to maximize the usage of the animal model in developing biomarkers, early diagnostics, and novel treatments. Finally, to maximize the applications of the transgenic NHP model, a strategic breeding plan should also be developed for quick expansion of the animal model for investigators.

We have presented the protocol that was successfully used to produce the first transgenic NHP model of human inherited neurodegenerative diseases. This is a step forward in modeling human diseases using the NHP model that is not only physiologically similar to humans but also genetically relevant to human diseases. As a proof of principle, HD-NHPs develop key neuropathologies and cognitive behavioral declines paralleling human HD conditions that strongly support the rationale in modeling human inherited genetic diseases using transgenic NHP models. With the continued development of genetic tools and applications with NHPs, transgenic NHPs will advance biomedical research as a powerful preclinical model for the development of not only safe treatments but also effective therapeutics. Nonetheless, our goal is well described by Brooksbank (1999): "We hope *Disease models* will not only be of education value but will also stimulate the use and development of models that are truly relevant to human disease, which will eventually catalyse the development of safe and efficacious therapeutics for human use."

Acknowledgments

All animal procedures were approved by the Emory University Institutional Animal Care and Use Committee (IACUC). Animal research was conducted in compliance with the Animal Welfare Act and other Federal statutes and regulations relating to animals and experiments involving animals and adheres to the principles stated in the *Guide for the Care and Use of Laboratory Animals* prepared by the National Research Council. Yerkes National Primate Research Center (YNPRC) is fully accredited by the Association for Assessment and Accreditation of Laboratory Animal Care (AAALAC). YNPRC is supported by the National Center for Research Resources P51RR165 and is currently supported by the Office of Research and Infrastructure Program (ORIP; OD P51OD11132). The author's research is supported by ORIP/NIH (RR018827), NINDS (NS064991), and the American Recovery and Reinvestment Act (ARRA) Fund.

References

Agca, C., Fritz, J.J., Walker, L.C., Levey, A.I., Chan, A.W., Lah, J.J., et al., 2008. Development of transgenic rats producing human beta-amyloid precursor protein as a model for Alzheimer's disease: transgene and endogenous APP genes are regulated tissue specifically. BMC Neurosci. 9, 28.

Andreadis, S., Palsson, B.O., 1997. Coupled effects of polybrene and calf serum on the efficiency of retroviral transduction and the stability of retroviral vectors. Hum. Gene Ther. 8, 285–291.

Aylward, E., Mills, J., Liu, D., Nopoulos, P., Ross, C.A., Pierson, R., et al., 2011a. Association between age and striatal volume stratified by CAG repeat length in prodromal Huntington disease. PLoS Curr. 3, RRN1235.

Aylward, E.H., Nopoulos, P.C., Ross, C.A., Langbehn, D.R., Pierson, R.K., Mills, J.A., et al., 2011b. Longitudinal change in regional brain volumes in prodromal Huntington disease. J. Neurol. Neurosurg. Psychiatry 82, 405–410.

Aylward, E.H., Liu, D., Nopoulos, P.C., Ross, C.A., Pierson, R.K., Mills, J.A., et al., 2012. Striatal volume contributes to the prediction of onset of Huntington disease in incident cases. Biol. Psychiatry 71, 822–828.

Bachevalier, J., Nemanic, S., 2008. Memory for spatial location and object-place associations are differently processed by the hippocampal formation, parahippocampal areas TH/TF and perirhinal cortex. Hippocampus 18, 64—80.

Bachevalier, J., Malkova, L., Mishkin, M., 2001. Effects of selective neonatal temporal lobe lesions on socioemotional behavior in infant rhesus monkeys (Macaca mulatta). Behav. Neurosci. 115, 545—559.

Bachevalier, J., Machado, C.J., Kazama, A., 2011. Behavioral outcomes of late-onset or early-onset orbital frontal cortex (areas 11/13) lesions in rhesus monkeys. Ann. N. Y. Acad. Sci. 1239, 71—86.

Balmaceda, J.P., Heitman, T.O., Garcia, M.R., Pauerstein, C.J., Pool, T.B., 1986. Embryo cryopreservation in cynomolgus monkeys. Fertil Steril 45 (3), 403—406.

Bangash, M.A., Park, J.M., Melnikova, T., Wang, D., Jeon, S.K., Lee, D., et al., 2011. Enhanced polyubiquitination of Shank3 and NMDA receptor in a mouse model of autism. Cell 145, 758—772.

Bates, G.P., Mangiarini, L., Davies, S.W., 1998. Transgenic mice in the study of polyglutamine repeat expansion diseases. Brain Pathol. 8, 699—714.

Bavister, B.D., Boatman, D.E., Leibfried, L., Loose, M., Vernon, M.W., 1983. Fertilization and cleavage of rhesus monkey oocytes in vitro. Biol. Reprod. 28 (4), 983—999.

Bavister, B.D., Boatman, D.E., Collins, K., Dierschke, D.J., Eisele, S.G., 1984. Birth of rhesus monkey infant after in vitro fertilization and nonsurgical embryo transfer. Proc. Natl. Acad. Sci. U S A. 81 (7), 2218—2222.

Bertelli, M., Alushi, B., Veicsteinas, A., Jinnah, H.A., Micheli, V., 2009. Gene expression and mRNA editing of serotonin receptor 2C in brains of HPRT gene knock-out mice, an animal model of Lesch-Nyhan disease. J. Clin. Neurosci. 16, 1061—1063.

Boudreau, R.L., McBride, J.L., Martins, I., Shen, S., Xing, Y., Carter, B.J., et al., 2009. Nonallele-specific silencing of mutant and wild-type huntingtin demonstrates therapeutic efficacy in Huntington's disease mice. Mol. Ther. 17, 1053—1063.

Bozdagi, O., Sakurai, T., Papapetrou, D., Wang, X., Dickstein, D.L., Takahashi, N., et al., 2010. Haploinsufficiency of the autism-associated Shank3 gene leads to deficits in synaptic function, social interaction, and social communication. Mol. Autism 1, 15.

Brooksbank, C., 1999. Disease models: relevance is everything. Mol. Med. Today 5, 274.

Brosh, S., Boer, P., Sperling, O., Zoref-Shani, E., 2000. Elevated UTP and CTP content in cultured neurons from HPRT-deficient transgenic mice. J. Mol. Neurosci. 14, 87—91.

Byrne, J.A., Pedersen, D.A., Clepper, L.L., Nelson, M., Sanger, W.G., Gokhale, S., et al., 2007. Producing primate embryonic stem cells by somatic cell nuclear transfer. Nature 450 (7169), 497—502.

Carbery, I.D., Ji, D., Harrington, A., Brown, V., Weinstein, E.J., Liaw, L., et al., 2010. Targeted genome modification in mice using zinc-finger nucleases. Genetics 186, 451—459.

Cayzac, S., Delcasso, S., Paz, V., Jeantet, Y., Cho, Y.H., 2011. Changes in striatal procedural memory coding correlate with learning deficits in a mouse model of Huntington disease. Proc. Natl. Acad. Sci. USA 108, 9280—9285.

Cermak, T., Doyle, E.L., Christian, M., Wang, L., Zhang, Y., Schmidt, C., et al., 2011. Efficient design and assembly of custom TALEN and other TAL effector-based constructs for DNA targeting. Nucleic Acids Res. 39, e82.

Chan, A.W., Dominko, T., Luetjens, C.M., Neuber, E., Martinovich, C., Hewitson, L., et al., 2000. Clonal propagation of primate offspring by embryo splitting. Science 287 (5451), 317—319.

Chan, A.W., Chong, K.Y., Martinovich, C., Simerly, C., Schatten, G., 2001. Transgenic monkeys produced by retroviral gene transfer into mature oocytes. Science 291 (5502), 309−312.

Chan, A.W., 2004. Transgenic nonhuman primates for neurodegenerative diseases. Reprod. Biol. Endocrinol. 2, 39.

Chan, A.W., Yang, S.H., 2009. Generation of transgenic monkeys with human inherited genetic disease. Methods 49, 78−84.

Chen, G., Chen, K.S., Knox, J., Inglis, J., Bernard, A., Martin, S.J., et al., 2000. A learning deficit related to age and beta-amyloid plaques in a mouse model of Alzheimer's disease. Nature 408, 975−979.

Cornetta, K., Anderson, W.F., 1989. Protamine sulfate as an effective alternative to polybrene in retroviral-mediated gene-transfer: implications for human gene therapy. J. Virol. Methods 23, 187−194.

Courtine, G., Bunge, M.B., Fawcett, J.W., Grossman, R.G., Kaas, J.H., Lemon, R., et al., 2007. Can experiments in nonhuman primates expedite the translation of treatments for spinal cord injury in humans? Nat. Med. 13, 561−566.

Davidson, B.L., 2012. Taking a break from huntingtin. Mol. Ther. Nucleic Acids 1.

Douglass, E.M., 1981. First Gorilla born using artificial insemination. In Zoo News 28, 9−15.

Ellis, B.L., Hirsch, M.L., Porter, S.N., Samulski, R.J., Porteus, M.H., 2013. Zinc-finger nuclease-mediated gene correction using single AAV vector transduction and enhancement by Food and Drug Administration-approved drugs. Gene Ther. 20 (1):35−42.

Ewing-Cobbs, L., Prasad, M.R., Swank, P., Kramer, L., Mendez, D., Treble, A., et al., 2012. Social communication in young children with traumatic brain injury: relations with corpus callosum morphometry. Int. J. Dev. Neurosci. 30, 247−254.

Finger, S., Heavens, R.P., Sirinathsinghji, D.J., Kuehn, M.R., Dunnett, S.B., 1988. Behavioral and neurochemical evaluation of a transgenic mouse model of Lesch-Nyhan syndrome. J. Neurol. Sci. 86, 203−213.

Gabriel Sanchez-Partida, L., Maginnis, G., Dominko, T., Martinovich, C., McVay, B., Fanton, J., et al., 2000. Live rhesus offspring by artificial insemination using fresh sperm and cryopreserved sperm. Biol. Reprod. 63 (4), 1092−1097.

Gilley, J., Adalbert, R., Coleman, M.P., 2011. Modelling early responses to neurodegenerative mutations in mice. Biochem. Soc. Trans. 39, 933−938.

Golding, M.C., Long, C.R., Carmell, M.A., Hannon, G.J., Westhusin, M.E., 2006. Suppression of prion protein in livestock by RNA interference. Proc. Natl. Acad Sci. USA 103, 5285−5290.

Gould, K.G., Styperek, R.P., 1989. Improved methods for freeze preservation of chimpanzee sperm. Am. J. Primotol. 18, 275−284.

Grossniklaus, H.E., Kang, S.J., Berglin, L., 2010. Animal models of choroidal and retinal neovascularization. Prog. Retin. Eye Res. 29, 500−519.

Hauschild, J., Petersen, B., Santiago, Y., Queisser, A.L., Carnwath, J.W., Lucas-Hahn, A., et al., 2011. Efficient generation of a biallelic knockout in pigs using zinc-finger nucleases. Proc. Natl. Acad Sci. USA 108, 12013−12017.

Hauschild-Quintern, J., Petersen, B., Queisser, A.L., Lucas-Hahn, A., Schwinzer, R., Niemann, H., 2013. Gender non-specific efficacy of ZFN mediated gene targeting in pigs. Transgenic Res. 22, 1−3.

Hewitson, L., Dominko, T., Takahashi, D., Martinovich, C., Ramalho-Santos, J., Sutovsky, P., et al., 1999. Unique checkpoints during the first cell cycle of fertilization after intracytoplasmic sperm injection in rhesus monkeys. Nat. Med. 5 (4), 431−433.

Hitz, C., Steuber-Buchberger, P., Delic, S., Wurst, W., Kuhn, R., 2009. Generation of shRNA transgenic mice. Methods Mol. Biol. 530, 101−129.

Hobbs, N.Z., Barnes, J., Frost, C., Henley, S.M., Wild, E.J., Macdonald, K., et al., 2010a. Onset and progression of pathologic atrophy in Huntington disease: a longitudinal MR imaging study. AJNR Am. J. Neuroradiol. 31, 1036−1041.

Hobbs, N.Z., Henley, S.M., Ridgway, G.R., Wild, E.J., Barker, R.A., Scahill, R.I., et al., 2010b. The progression of regional atrophy in premanifest and early Huntington's disease: a longitudinal voxel-based morphometry study. J. Neurol. Neurosurg. Psychiatry 81, 756−763.

Hsiao, K., Chapman, P., Nilsen, S., Eckman, C., Harigaya, Y., Younkin, S., et al., 1996. Correlative memory deficits, Abeta elevation, and amyloid plaques in transgenic mice. Science 274, 99−102.

Kobayashi, M., Horiuchi, H., Fujita, K., Takuhara, Y., Suzuki, S., 2012. Characterization of grape C-repeat-binding factor 2 and B-box-type zinc finger protein in transgenic Arabidopsis plants under stress conditions. Mol. Biol. Rep. 39 (8), 7933−7939.

Kordower, J.H., Emborg, M.E., Bloch, J., Ma, S.Y., Chu, Y., Leventhal, L., et al., 2000. Neurodegeneration prevented by lentiviral vector delivery of GDNF in primate models of Parkinson's disease. Science 290, 767−773.

Kumar, A., Wadhawan, R., Swanwick, C.C., Kollu, R., Basu, S.N., Banerjee-Basu, S., 2011. Animal model integration to AutDB, a genetic database for autism. BMC Med. Genomics 4, 15.

Lanzendorf, S.E., Zelinski-Wooten, M.B., Stouffer, R.L., Wolf, D.P., 1990. Maturity at collection and the developmental potential of rhesus monkey oocytes. Biol. Reprod. 42 (4), 703−711.

Lanzendorf, S.E., Gliessman, P.M., Archibong, A.E., Alexander, M., Wolf, D.P., 1990. Collection and quality of rhesus monkey semen. Mol. Reprod. Dev. 25, 61−66.

Lassnig, C., Sanchez, C.M., Egerbacher, M., Walter, I., Majer, S., Kolbe, T., et al., 2005. Development of a transgenic mouse model susceptible to human coronavirus 229E. Proc. Natl. Acad. Sci. USA 102, 8275−8280.

Lassota, N., 2008. Clinical and histological aspects of CNV formation: studies in an animal model. Acta Ophthalmol. 86 Thesis 2, 1−24.

Li, T., Liu, B., Spalding, M.H., Weeks, D.P., Yang, B., 2012. High-efficiency TALEN-based gene editing produces disease-resistant rice. Nat. Biotechnol. 30, 390−392.

Lin, C.H., Tallaksen-Greene, S., Chien, W.M., Cearley, J.A., Jackson, W.S., Crouse, A.B., et al., 2001. Neurological abnormalities in a knock-in mouse model of Huntington's disease. Hum. Mol. Genet. 10, 137−144.

Liu, J., Li, C., Yu, Z., Huang, P., Wu, H., Wei, C., et al., 2012. Efficient and specific modifications of the Drosophila genome by means of an easy TALEN strategy. J. Genet. Genomics 39, 209−215.

Lois, C., Hong, E.J., Pease, S., Brown, E.J., Baltimore, D., 2002. Germline transmission and tissue-specific expression of transgenes delivered by lentiviral vectors. Science 295, 868−872.

Lopata, A., Summers, P.M., Hearn, J.P., 1988. Births following the transfer of cultured embryos obtained by *in vitro* and *in vivo* fertilization in the marmoset monkey (Callithrix jacchus). Fertil Steril 50 (3), 503−509.

MacDonald, M.J., Longacre, M.J., Stoker, S.W., Kendrick, M., Thonpho, A., Brown, L.J., et al., 2011. Differences between human and rodent pancreatic islets: low pyruvate carboxylase, ATP citrate lyase, and pyruvate carboxylation and high glucose-stimulated acetoacetate in human pancreatic islets. J. Biol. Chem. 286, 18383−18396.

Mahfouz, M.M., Li, L., Shamimuzzaman, M., Wibowo, A., Fang, X., Zhu, J.K., 2011. *De novo*-engineered transcription activator-like effector (TALE) hybrid nuclease with novel DNA binding specificity creates double-strand breaks. Proc. Natl. Acad. Sci. USA 108, 2623–2628.

Makris, N., Kennedy, D.N., Boriel, D.L., Rosene, D.L., 2010. Methods of MRI-based structural imaging in the aging monkey. Methods 50, 166–177.

Matsui, M., Prakash, T.P., Corey, D.R., 2013. Transcriptional silencing by single-stranded RNAs targeting a noncoding RNA that overlaps a gene promoter. ACS Chem. Biol. 8, 122–126.

McBride, J.L., Boudreau, R.L., Harper, S.Q., Staber, P.D., Monteys, A.M., Martins, I., et al., 2008. Artificial miRNAs mitigate shRNA-mediated toxicity in the brain: implications for the therapeutic development of RNAi. Proc. Natl. Acad. Sci. USA 105, 5868–5873.

McBride, J.L., Pitzer, M.R., Boudreau, R.L., Dufour, B., Hobbs, T., Ojeda, S.R., et al., 2011. Preclinical safety of RNAi-mediated HTT suppression in the rhesus macaque as a potential therapy for Huntington's disease. Mol. Ther. 19, 2152–2162.

Meng, L., Ely, J.J., Stouffer, R.L., Wolf, D.P., 1997. Rhesus monkeys produced by nuclear transfer. Biol. Reprod. 57 (2), 454–459.

Miller, J.A., Horvath, S., Geschwind, D.H., 2010. Divergence of human and mouse brain transcriptome highlights Alzheimer disease pathways. Proc. Natl. Acad. Sci. USA 107, 12698–12703.

Minshew, N., McFadden, K., 2011. Commentary for special issue of Autism research on mouse models in ASD: a clinical perspective. Autism Res. 4, 1–4.

Mittoux, V., Joseph, J.M., Conde, F., Palfi, S., Dautry, C., Poyot, T., et al., 2000. Restoration of cognitive and motor functions by ciliary neurotrophic factor in a primate model of Huntington's disease. Hum. Gene Ther. 11, 1177–1187.

Niu, Y., Yu, Y., Bernat, A., Yang, S., He, X., Guo, X., et al., 2010. Transgenic rhesus monkeys produced by gene transfer into early-cleavage-stage embryos using a simian immunodeficiency virus-based vector. Proc. Natl. Acad. Sci. USA 107, 17663–17667.

Ou, J., Lowes, C., Collinson, J.M., 2010. Cytoskeletal and cell adhesion defects in wounded and Pax6 +/− corneal epithelia. Invest. Ophthalmol. Vis. Sci. 51, 1415–1423.

Palfi, S., Brouillet, E., Jarraya, B., Bloch, J., Jan, C., Shin, M., et al., 2007. Expression of mutated huntingtin fragment in the putamen is sufficient to produce abnormal movement in non-human primates. Mol. Ther. 15, 1444–1451.

Passananti, C., Corbi, N., Onori, A., Di Certo, M.G., Mattei, E., 2010. Transgenic mice expressing an artificial zinc finger regulator targeting an endogenous gene. Methods Mol. Biol. 649, 183–206.

Patterson, P.H., 2011. Modeling autistic features in animals. Pediatr. Res. 69, 34R–40R.

Paulsen, J.S., 2011. Cognitive impairment in Huntington disease: diagnosis and treatment. Curr. Neurol. Neurosci. Rep. 11, 474–483.

Paulsen, J.S., Nopoulos, P.C., Aylward, E., Ross, C.A., Johnson, H., Magnotta, V.A., et al., 2010. Striatal and white matter predictors of estimated diagnosis for Huntington disease. Brain. Res. Bull. 82, 201–207.

Peca, J., Feliciano, C., Ting, J.T., Wang, W., Wells, M.F., Venkatraman, T.N., et al., 2011. Shank3 mutant mice display autistic-like behaviours and striatal dysfunction. Nature 472, 437–442.

Polymeropoulos, M.H., Lavedan, C., Leroy, E., Ide, S.E., Dehejia, A., Dutra, A., et al., 1997. Mutation in the alpha-synuclein gene identified in families with Parkinson's disease. Science 276, 2045–2047.

Pope, C.E., Pope, V.Z., Beck, L.R., 1984. Live birth following cryopreservation and transfer of a baboon embryo. Fertil Steril 42 (1), 143–145.

Ramaesh, T., Collinson, J.M., Ramaesh, K., Kaufman, M.H., West, J.D., Dhillon, B., 2003. Corneal abnormalities in Pax6 +/− small eye mice mimic human aniridia-related keratopathy. Invest. Ophthalmol. Vis. Sci. 44, 1871−1878.

Ramaswamy, S., McBride, J.L., Han, I., Berry-Kravis, E.M., Zhou, L., Herzog, C.D., et al., 2009. Intrastriatal CERE-120 (AAV-Neurturin) protects striatal and cortical neurons and delays motor deficits in a transgenic mouse model of Huntington's disease. Neurobiol. Dis. 34, 40−50.

Raymond, C.S., Zhu, L., Vogt, T.F., Shin, M.K., 2010. *In vivo* analysis of gene knockdown in tetracycline-inducible shRNA mice. Methods Enzymol. 477, 415−427.

Rice, J., 2012. Animal models: not close enough. Nature 484, S9.

Rideout III, W.M., Hochedlinger, K., Kyba, M., Daley, G.Q., Jaenisch, R., 2002. Correction of a genetic defect by nuclear transplantation and combined cell and gene therapy. Cell 109, 17−27.

Robertson, H.R., Feng, G., 2011. Annual Research Review: transgenic mouse models of childhood-onset psychiatric disorders. J. Child. Psychol. Psychiatry 52, 442−475.

Sasaki, E., Suemizu, H., Shimada, A., Hanazawa, K., Oiwa, R., Kamioka, M., et al., 2009. Generation of transgenic non-human primates with germline transmission. Nature 459 (7246), 523−527.

Scahill, R.I., Hobbs, N.Z., Say, M.J., Bechtel, N., Henley, S.M., Hyare, H., et al., 2011. Clinical impairment in premanifest and early Huntington's disease is associated with regionally specific atrophy. Hum. Brain Mapp.

Schilling, G., Becher, M.W., Sharp, A.H., Jinnah, H.A., Duan, K., Kotzuk, J.A., et al., 1999. Intranuclear inclusions and neuritic aggregates in transgenic mice expressing a mutant N-terminal fragment of huntingtin. Hum. Mol. Genet. 8, 397−407.

Seibler, J., Kuter-Luks, B., Kern, H., Streu, S., Plum, L., Mauer, J., et al., 2005. Single copy shRNA configuration for ubiquitous gene knockdown in mice. Nucleic Acids Res. 33, e67.

Seibler, J., Kleinridders, A., Kuter-Luks, B., Niehaves, S., Bruning, J.C., Schwenk, F., 2007. Reversible gene knockdown in mice using a tight, inducible shRNA expression system. Nucleic Acids Res. 35, e54.

Shavlakadze, T., Winn, N., Rosenthal, N., Grounds, M.D., 2005. Reconciling data from transgenic mice that overexpress IGF-I specifically in skeletal muscle. Growth Horm. IGF Res. 15, 4−18.

Solomon, A.C., Stout, J.C., Weaver, M., Queller, S., Tomusk, A., Whitlock, K.B., et al., 2008. Ten-year rate of longitudinal change in neurocognitive and motor function in pre-diagnosis Huntington disease. Mov. Disord. 23, 1830−1836.

Sommer, J.R., Estrada, J.L., Collins, E.B., Bedell, M., Alexander, C.A., Yang, Z., et al., 2011. Production of ELOVL4 transgenic pigs: a large animal model for Stargardt-like macular degeneration. Br. J. Ophthalmol. 95, 1749−1754.

Stieger, K., Lheriteau, E., Moullier, P., Rolling, F., 2009. AAV-mediated gene therapy for retinal disorders in large animal models. ILAR J. 50, 206−224.

Strange, T.L., Petolino, J.F., 2012. Targeting DNA to a previously integrated transgenic locus using zinc finger nucleases. Methods Mol. Biol. 847, 391−397.

Sun, Q., Dong, J., Yang, W., Jin, Y., Yang, M., Wang, Y., et al., 2008. Efficient reproduction of cynomolgus monkey using pronuclear embryo transfer technique. Proc. Natl. Acad. Sci. USA 105, 12956−12960.

Sung, Y.H., Baek, I.J., Seong, J.K., Kim, J.S., Lee, H.W., 2012. Mouse genetics: catalogue and scissors. BMB Rep. 45, 686−692.

Sung, Y.H., Baek, I.J., Kim, D.H., Jeon, J., Lee, J., Lee, K., et al., 2013. Knockout mice created by TALEN-mediated gene targeting. Nat. Biotechnol. 31, 23−24.

Tabrizi, S.J., Scahill, R.I., Durr, A., Roos, R.A., Leavitt, B.R., Jones, R., et al., 2011. Biological and clinical changes in premanifest and early stage Huntington's disease in the TRACK-HD study: the 12-month longitudinal analysis. Lancet Neurol. 10, 31–42.

Tabrizi, S.J., Reilmann, R., Roos, R.A., Durr, A., Leavitt, B., Owen, G., et al., 2012. Potential endpoints for clinical trials in premanifest and early Huntington's disease in the TRACK-HD study: analysis of 24 month observational data. Lancet Neurol. 11, 42–53.

Tachibana, M., Sparman, M., Sritanaudomchai, H., Ma, H., Clepper, L., Woodward, J., et al., 2009. Mitochondrial gene replacement in primate offspring and embryonic stem cells. Nature 461 (7262), 367–372.

Tachibana, M., Sparman, M., Ramsey, C., Ma, H., Lee, H.S., Penedo, M.C., et al., 2012. Generation of chimeric rhesus monkeys. Cell 148 (1–2), 285–295.

Tessanne, K., Golding, M.C., Long, C.R., Peoples, M.D., Hannon, G., Westhusin, M.E., 2012. Production of transgenic calves expressing an shRNA targeting myostatin. Mol. Reprod. Dev. 79, 176–185.

Thomson, J.A., Kalishman, J., Golos, T.G., Durning, M., Harris, C.P., Becker, R.A., et al., 1995. Isolation of a primate embryonic stem cell line. Proc. Natl. Acad. Sci. U S A. 92 (17), 7844–7848.

Thomson, J.A., Kalishman, J., Golos, T.G., Durning, M., Harris, C.P., Hearn, J.P., 1996. Pluripotent cell lines derived from common marmoset (Callithrix jacchus) blastocysts. Biol. Reprod. 55 (2), 254–259.

Tollner, T.L., VandeVoort, C.A., Overstreet, J.W., Drobnis, E.Z., 1990. Cryopreservation of spermatozoa from cynomolgus monkeys (Macaca fascicularis). J Reprod Fertil 90 (2), 347–352.

Toyoshima, K., Vogt, P.K., 1969. Enhancement and inhibition of avian sarcoma viruses by polycations and polyanions. Virology 38, 414–426.

Tsunematsu, T., Kilduff, T.S., Boyden, E.S., Takahashi, S., Tominaga, M., Yamanaka, A., 2011. Acute optogenetic silencing of orexin/hypocretin neurons induces slow-wave sleep in mice. J. Neurosci. 31, 10529–10539.

Van Pham, P., Vu, N.B., Duong, T.T., Nguyen, T.T., Truong, N.H., Phan, N.L., et al., 2012. Suppression of human breast tumors in NOD/SCID mice by CD44 shRNA gene therapy combined with doxorubicin treatment. Onco Targets Ther. 5, 77–84.

Vandevoort, C.A., Baughman, W.L., Stouffer, R.L., 1989. Comparison of different regimens of human gonadotropins for superovulation of rhesus monkeys: ovulatory response and subsequent luteal function. J. In Vitro Fertil. Embryo Transf. 6, 85–91.

Wang, X., McCoy, P.A., Rodriguiz, R.M., Pan, Y., Je, H.S., Roberts, A.C., et al., 2011. Synaptic dysfunction and abnormal behaviors in mice lacking major isoforms of Shank3. Hum. Mol. Genet. 20, 3093–3108.

Whyte, J.J., Prather, R.S., 2012. Cell Biology Symposium: zinc finger nucleases to create custom-designed modifications in the swine (Sus scrofa) genome. J. Anim. Sci. 90, 1111–1117.

Wolf, R.C., Sambataro, F., Vasic, N., Wolf, N.D., Thomann, P.A., Landwehrmeyer, G.B., et al., 2011. Longitudinal functional magnetic resonance imaging of cognition in pre-clinical Huntington's disease. Exp. Neurol. 231, 214–222.

Woodman, B., Butler, R., Landles, C., Lupton, M.K., Tse, J., Hockly, E., et al., 2007. The Hdh(Q150/Q150) knock-in mouse model of HD and the R6/2 exon 1 model develop comparable and widespread molecular phenotypes. Brain Res. Bull. 72, 83–97.

Xia, X.G., Zhou, H., Samper, E., Melov, S., Xu, Z., 2006. Pol II-expressed shRNA knocks down Sod2 gene expression and causes phenotypes of the gene knockout in mice. PLoS Genet. 2, e10.

Yang, S.H., Chan, A.W., 2011. Transgenic animal models of Huntington's disease. Curr. Top. Behav. Neurosci. 7, 61–85.

Yang, S.H., Agca, Y., Cheng, P.H., Yang, J.J., Agca, C., Chan, A.W., 2007. Enhanced transgenesis by intracytoplasmic injection of envelope-free lentivirus. Genesis 45, 177–183.

Yang, S.H., Cheng, P.H., Banta, H., Piotrowska-Nitsche, K., Yang, J.J., Cheng, E.C., et al., 2008. Towards a transgenic model of Huntington's disease in a non-human primate. Nature 453 (7197), 921–924.

Yu, D., Pendergraff, H., Liu, J., Kordasiewicz, H.B., Cleveland, D.W., Swayze, E.E., et al., 2012. Single-stranded RNAs use RNAi to potently and allele-selectively inhibit mutant huntingtin expression. Cell 150, 895–908.

Zeiss, C.J., 2010. Animals as models of age-related macular degeneration: an imperfect measure of the truth. Vet. Pathol. 47, 396–413.

Zelinski-Wooten, M.B., Hutchison, J.S., Hess, D.L., Wolf, D.P., Stouffer, R.L., 1995. Follicle stimulating hormone alone supports follicle growth and oocyte development in gonadotrophin-releasing hormone antagonist-treated monkeys. Hum. Reprod. 10, 1658–1666.

Zheng, P., Bavister, B.D., Ji, W., 2001a. Energy substrate requirement for in vitro maturation of oocytes from unstimulated adult rhesus monkeys. Mol. Reprod. Dev. 58, 348–355.

Zheng, P., Wang, H., Bavister, B.D., Ji, W., 2001b. Maturation of rhesus monkey oocytes in chemically defined culture media and their functional assessment by IVF and embryo development. Hum. Reprod. 16, 300–305.

Zschemisch, N.H., Glage, S., Wedekind, D., Weinstein, E.J., Cui, X., Dorsch, M., et al., 2012. Zinc-finger nuclease mediated disruption of Rag1 in the LEW/Ztm rat. BMC Immunol. 13, 60.

14 Nuclear Transfer Technology in Cattle, Sheep, and Swine

Satoshi Akagi[1], Kazutsugu Matsukawa[2] and Akira Onishi[3]

[1]Animal Breeding and Reproduction Research Division, Institute of Livestock and Grassland Science, National Agriculture and Food Research Organization, Tsukuba, Ibaraki, Japan, [2]Multidisciplinary Science Cluster, Life and Environmental Medicine Science Unit, Kochi University, Nankoku, Kochi, Japan, [3]Transgenic Pig Research Unit, Genetically Modified Organism Research Center, National Institute of Agrobiological Sciences, Tsukuba, Ibaraki, Japan

I. Introduction

Nuclear transfer (NT), as noted in Chapter 7, is a technique used for producing cloned animals. The first mammalian clones were produced by transferring the nuclei of 8- or 16-cell sheep embryos into enucleated ovulated oocytes (Willadsen, 1986). Soon after the embryonic cell cloning of sheep, similar results were reported in cattle (Prather et al., 1987), rabbits (Stice and Robl, 1988), and pigs (Prather et al., 1989). At that time, NT technology was expected to be useful for the production of large numbers of animals with genetically superior traits for agricultural purposes (Marx, 1988). Wilmut et al. (1997) reported the birth of "Dolly," the first cloned animal derived from a cultured somatic cell from an adult ewe. Subsequently, NT technology has also been used for the genetic modification of livestock. Previously, transgenic livestock have mainly been produced by injection of the gene of interest into the pronuclei of zygotes (Eyestone and Campbell, 1999). NT using cultured cells offers several advantages for the production of transgenic livestock compared with the pronuclear microinjection method (Wilmut et al., 2000; Campbell, 2002; Hodges and Stice, 2003). For example, the use of preselected transgenic cells as a source of nuclei for NT increases the efficiency of transgenic animal production to 100% and overcomes the problem of mosaicism observed after pronuclear microinjection (Baguisi et al., 1999). The first report of transgenic sheep obtained by NT demonstrated that the efficiency of transgenic animal production was improved more than twofold compared with pronuclear microinjection (Schnieke et al., 1997). NT technology was also used to produce the first gene-targeted mammal other than the mouse (McCreath et al., 2000). Furthermore, a porcine model of severe combined immunodeficiency (SCID) was

Transgenic Animal Technology. DOI: http://dx.doi.org/10.1016/B978-0-12-410490-7.00015-3

recently generated by NT using fibroblasts containing a targeted disruption of the X-linked interleukin-2 receptor gamma chain gene (Suzuki et al., 2012).

Despite the production of several species of cloned animals by NT, the efficiency of cloning is still extremely low. This can probably be attributed to incomplete reprogramming of the donor nucleus. Moreover, most of the developmental problems observed in the clones are believed to be due to epigenetic defects (Morgan et al., 2005). Recently, in mice, the enhancement of histone acetylation levels in cloned embryos (Kishigami et al., 2006; Rybouchkin et al., 2006) and correction of X-linked gene expression (Inoue et al., 2010; Matoba et al., 2011) has led to a significant improvement in cloning efficiency.

This chapter summarizes the protocols used for cloning in livestock (Figure 14.1). Electric fusion (Wilmut et al., 1997) or microinjection techniques (Wakayama et al., 1998) are commonly used for NT. We describe a protocol based on cell fusion in cattle and a protocol based on microinjection in swine and sheep.

II. Protocol for Bovine NT

A. *Preparation of Donor Cells*

The cell cycle status of donor cells is an important factor affecting the success of cloning, because cell cycle coordination between the donor nucleus and recipient

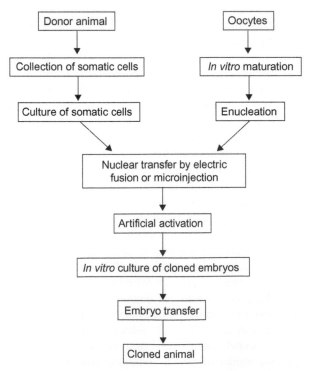

Figure 14.1 Cloning procedure in livestock.

cytoplast is essential to maintain ploidy and prevent DNA damage (Campbell and Alberio, 2003). Donor cells in the G_0/G_1 phase with metaphase II recipient oocytes have been used in almost all successful reports (Kato and Tsunoda, 2010). In bovine NT, although fresh noncultured cells (Akagi et al., 2003) or cycling cells (Cibelli et al., 1998) can be used for the production of cloned calves, the donor cell cycle stage is usually synchronized by serum starvation (Kato et al., 1998), contact inhibition (Kato et al., 2000), and drug treatment (Gibbons et al., 2002; Urakawa et al., 2004) to increase the population of cells in the G_0/G_1 phase. We describe a protocol for the preparation of ear skin fibroblast cells synchronized in the G_0/G_1 phase by serum starvation.

1. Isolation of Bovine Ear Skin Fibroblast Cells

Ear skin biopsy is performed on a donor animal. The tissue biopsy sample is dissected into small pieces. Tissue explants are transferred into a culture dish containing Dulbecco's modified Eagle's medium (DMEM, Sigma-Aldrich, St. Louis, MO) with 10% fetal bovine serum (FBS), 100 IU/mL penicillin G potassium (Meiji Seika, Tokyo, Japan), and 100 μg/mL streptomycin (Meiji Seika). The explants are incubated at 38.5°C under 5% CO_2 in air. The cell culture medium is changed every 2 days. Fibroblast cells begin to grow from the tissue explants within 2–4 days. The tissue explants are removed after approximately 7 days. Incubation is continued until cells reach 80% confluence.

2. Subculture of Fibroblast Cells

The cells are washed with phosphate-buffered saline (PBS) and treated with 0.1% trypsin-EDTA for 5–10 min. The trypsin is inactivated by the addition of cell culture medium. The cell suspension is collected, transferred into a conical tube, and centrifuged for 5 min at 1200 rpm. The supernatant is discarded and the pellet is resuspended in the cell culture medium. The cell suspension is transferred into a new culture dish.

3. Cryopreservation of Fibroblast Cells

The fibroblast cell pellet is obtained by trypsinization and centrifugation as described in Section 14.II.A.2. After discarding the supernatant, the pellet is resuspended in cell cryopreservation medium (Cell Banker 1, Nippon Zenyaku Kogyo Co., Fukushima, Japan) at a concentration of 5×10^5–5×10^6 cells/mL. A total of 0.5–1 mL of cell suspension is transferred into each cryovial. The vials are stored in a freezer at $-80°C$.

4. Preparation of G_0/G_1 Phase Cells

After thawing or subculture, fibroblast cells are seeded onto a culture dish in the cell culture medium and incubated until cells reach 70–80% confluence at 38.5°C under 5% CO_2 in air. The cell culture medium is replaced with a serum-starved medium (DMEM with 0.5% FBS) and cells are incubated for 5 days. The cells are pelleted by following the trypsinization and centrifugation steps described in Section 14.II.A.2. After discarding the supernatant, the cells are suspended in PBS with 0.5% FBS.

B. Preparation of Recipient Oocytes

Recipient oocytes are collected from slaughterhouse-derived ovaries or live cows by ovum pick-up (OPU). The oocytes are generally used for bovine NT after *in vitro* maturation (IVM); however *in vivo*-matured oocytes obtained by OPU can be also used (Akagi et al., 2008). We describe the preparation of oocytes obtained from slaughterhouse-derived ovaries only.

1. Collection of Cumulus-Oocyte Complexes

Ovaries obtained from a slaughterhouse are transported to the laboratory in saline containing 100 IU/mL penicillin G potassium and 100 µg/mL streptomycin. After washing the ovaries with saline, cumulus-oocyte complexes (COCs) are aspirated from antral follicles, 2−8 mm in diameter, using an 18-gauge needle attached to a 5-mL syringe.

2. In Vitro *Maturation*

Maturation medium consists of TCM-199 (Gibco, Invitrogen, Grand Island, NY) supplemented with 10% FBS, 100 IU/mL penicillin G potassium and 100 µg/mL streptomycin. COCs are washed three times in the maturation medium and subsequently transferred to each well of a four-well multidish (Nunc, Roskilde, Denmark) containing 700 µL maturation medium (50−70 COCs per well). The COCs are incubated for 18−20 h at 38.5°C under 5% CO_2 in air.

3. Removal of Cumulus Cells

After IVM, COCs are transferred into a 15-mL conical tube (BD Falcon, Becton, Dickinson & Co., Franklin Lakes, NJ) containing 700 µL of M2 medium with 0.1% hyaluronidase (Sigma-Aldrich) and vortex-agitated for 3 min. Subsequently, cumulus cells are completely removed from oocytes by gentle pipetting.

C. Production of Cloned Embryos

Cloned embryos are produced from donor cells and recipient oocytes prepared as described above. The enucleation, insertion of donor cells, and oocyte-cell fusion procedures are performed using micromanipulators mounted on an inverted microscope. The cloned embryos require artificial activation after fusion to trigger further development.

1. Enucleation

After the removal of cumulus cells, oocytes with the first polar body are placed into a mineral oil-covered drop of HEPES-buffered TCM-199 with 20% FBS in a culture dish on a microscope stage. An oocyte is rotated with an enucleation needle until the polar body is in a position at 12 o'clock and held with a holding pipette (Figure 14.2A). The enucleation needle is inserted into the perivitelline space above the first polar body (Figure 14.2B) and the oocyte is released from the holding pipette. The enucleation needle is withdrawn from the oocyte by rubbing the holding pipette and enucleation needle together to cut the zona pellucida. The cytoplasm underlying the first polar body is extruded through this slit in the zona pellucida with the enucleation needle (Figure 14.2C). Successful enucleation is confirmed by staining the extruded cytoplasm with 5 μg/mL Hoechst 33342 (Sigma-Aldrich) under a fluorescence microscope.

2. Insertion of Donor Cells

A drop of HEPES-buffered TCM-199 with 20% FBS and 50 μg/mL phytohemagglutinin-P (PHA-P, Wako Pure Chemical Industries, Ltd., Osaka, Japan) and a drop of PBS with 0.5% FBS are placed in a culture dish and covered

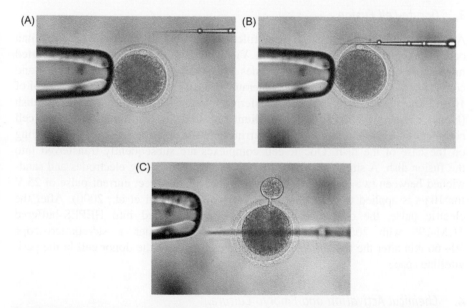

Figure 14.2 Enucleation of bovine oocytes. (A) An oocyte is held with a holding pipette with the polar body at the 12 o'clock position. (B) The enucleation needle is inserted into the perivitelline space. The oocyte is released from the holding pipette and rubbed against the holding pipette until a slit is produced. (C) The cytoplasm underlying the first polar body is extruded through the slit with the enucleation needle. Successful enucleation is confirmed by staining the extruded cytoplasm with Hoechst 33342.

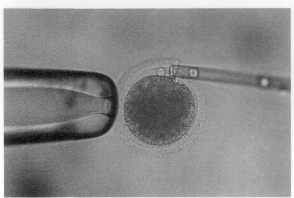

Figure 14.3 Insertion of donor cells into the perivitelline space of an enucleated oocyte. Donor cells are picked up with an injection pipette. An enucleated oocyte is held with a holding pipette and a donor cell is inserted into the perivitelline space through the slit in the zona pellucida.

with mineral oil. Donor cells are transferred into a drop of PBS with 0.5% FBS and enucleated oocytes are transferred into a drop of HEPES-buffered TCM-199 with 20% FBS and 50 μg/mL PHA-P in a culture dish. Donor cells are picked up with an injection pipette. An enucleated oocyte is held with a holding pipette and a donor cell is inserted into the perivitelline space through the slit in the zona pellucida (Figure 14.3).

3. Oocyte-Cell Fusion

A 100-mm culture dish (fusion dish) filled with 25 mL Zimmermann mammalian cell fusion medium (Zimmerman and Vienken, 1982) is placed on the inverted microscope. Two needle-type electrodes are attached to micromanipulators and connected to an electro cell fusion generator (LF101, Nepa Gene, Japan). A total of 10−15 oocyte-cell complexes are transferred into one well of a four-well multidish (Nunc) containing 500 μL fusion medium prepared for equilibration. Oocyte-cell complexes are washed twice by transferring into the neighboring well after settling on the base of the dish. Oocyte-cell complexes are subsequently transferred into the fusion dish. A single oocyte-cell complex is aligned using electrodes and sandwiched between two electrodes (Figure 14.4). A single direct current pulse of 25 V for 10 μs is applied to an oocyte-cell complex (Takahashi et al., 2000). After the electric pulse, the oocyte-cell complexes are transferred into HEPES-buffered TCM-199 with 20% FBS. Fusion is confirmed under a stereomicroscope 30−60 min after the transfer by observing the absence of the donor cell in the perivitelline space.

4. Chemical Activation and Embryo Culture

At 1−2 h after fusion, cloned embryos are placed into 10 μM calcium ionophore A23187 (Calbiochem, Merck KGaA, Darmstadt, Germany) in PBS for 5 min. The embryos are transferred into 2.5 μg/mL cytochalasin D (Sigma-Aldrich) and 10 μg/mL cycloheximide (Sigma-Aldrich) in TCM-199 with 10% FBS and incubated at 38.5°C under 5% CO_2 in air (Takahashi et al., 2000). After 1 h, the embryos are

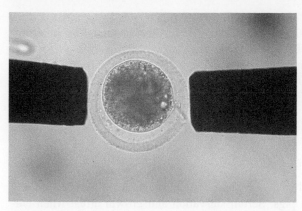

Figure 14.4 Oocyte-cell fusion by needle-type electrodes. An oocyte-cell complex is aligned using electrodes and sandwiched between two electrodes. An electric pulse is applied to induce fusion of an oocyte-cell complex.

transferred into 10 μg/mL cycloheximide in TCM-199 with 10% FBS and incubated for 4 h at 38.5°C under 5% CO_2 in air (Takahashi et al., 2000). After chemical activation, the embryos are transferred into IVD-101 medium (Research Institute for the Functional Peptides, Yamagata, Japan) and incubated for 7−8 days under 5% O_2, 5% CO_2, and 90% N_2 at 38.5°C.

D. Embryo Transfer

Cloned embryos at the blastocyst stage are nonsurgically transferred to recipient cows. Recipient cows are synchronized by an intramuscular injection of 0.15 mg D-cloprostenol (2 mL, Dalmazin, Kyouritsu Seiyaku, Tokyo). Each synchronized recipient cow receives a single blastocyst into the uterine horn ipsilateral to the corpus luteum. Pregnancy status is monitored at regular intervals by ultrasonography from 30−90 days of gestation and subsequently by rectal palpation until calving.

III. Protocol for Porcine NT

A. Preparation of Donor Cells

1. Isolation of Fetal Fibroblast

Fetal fibroblasts are generally used as donor cells in pig cloning. Cumulus cells, adult cells derived from heart, kidney, oviduct, skin, and ear have also been successful for production of cloned pigs. However, it is still unclear which cell type is the most successful for cloning; fetal cells have been preferred because the reverse of chromatin status is believed to be easier in fetal cells than adult cells.

Fetal fibroblasts were isolated from a fetus at 40−70 days of gestation. After removing the head and visceral tissue, this was washed in PBS (-) (Takara bio, Otsu, Japan), minced with scissors and suspended in 20 mL PBS(-). Tissue was collected by low-speed centrifugation and resuspended in 50-mL tubes with 20 mL PBS(-) containing 0.1% (w/v) of trypsin (Gibco). Tubes were incubated at 4°C overnight for the dispersion of cells. After removing trypsin by centrifugation and incubation at 37°C for 30 min, the cells were further dispersed by pipetting in 20 mL DMEM (Nacalai Tesque, Kyoto, Japan) containing 10% (v/v) fetal calf serum (FCS, Gibco). Single cells were filtered through a cell strainer (BD Falcon), collected by centrifugation, resuspended in 25 mL DMEM containing 10% FCS and seeded into a 75 cm^2 culture flask (Corning, Corning, NY). Cells were cultured to confluence at 37°C in a humidified atmosphere of 5% (v/v) CO_2 in air.

2. Preparation of G_0/G_1 Phase Cells

Quiescent donor cells arrested in G_0/G_1 phases of cell cycle have been commonly used for production of cloned pigs. Serum starvation and growth arrest when cultured cells reach confluence are the usual methods for synchronization at the G_0/G_1 cell cycle stage. Donor cells that reach confluence are left in the same medium without replacement for about 2 weeks to synchronize in G_0/G_1 phase.

B. Preparation of Recipient Oocytes

1. Collection of COCs

Porcine ovaries were obtained from prepubertal crossbred gilts at a local slaughterhouse and transported to the laboratory at 30−35°C within 1 h of slaughter. COCs were collected from follicles (3−5 mm in diameter) in the outer layer of the ovaries by aspiration with an 18-gauge needle into a 10-mL disposable syringe. COCs surrounded by not less than three layers of cumulus cells were selected and washed three times with TALP-HEPES (Research Institute for the Functional Peptides), and then washed three times in mNCSU-37.

2. IVM and Removal of Cumulus Cells

The COCs were cultured in 100 μL of mNCSU-37 supplemented with 10 IU hCG (Sankyo, Tokyo), 10 IU pregnant mare serum gonadotropin (Sankyo), and 1 mM dbcAMP, covered with liquid paraffin (light mineral oil, Nacalai Tesque) for 20 h, then cultured in mNCSU-37 omitting dbcAMP and gonadotrophin for 24 h at 39°C in an atmosphere of 5% CO_2 and 5% O_2 in N_2 gas. After maturation, COCs were transferred into TALP-HEPES supplemented with 150 IU/mL hyaluronidase, and oocytes then mechanically freed from cumulus cells by repeated pipetting using a fine glass pipette.

C. Production of Cloned Embryos and Embryo Transfer

The metaphase II oocytes were enucleated by gentle aspiration of the first polar body and adjacent cytoplasm using a beveled pipette (25−30 μm) in porcine zygote medium 3 (PZM3) medium containing 5.0 μg/mL cytochalasin B (CB). Enucleated oocytes were washed in PZM3 medium lacking CB. A donor cell is aspirated into a glass pipette with a diameter smaller than the cell to rupture the cell membrane, and the nucleus is directly injected into the cytoplasm of enucleated oocyte using a piezo-actuated micromanipulator (Prime Tech., Tsuchiura, Japan). The piezo-actuated micromanipulator greatly improves the efficiency and success rate for both enucleation and injection of nucleus. Oocytes injected with donor cell nuclei were transferred to an activation solution consisting of 0.28 M D-mannitol, 0.05 mM $CaCl_2$, 0.1 mM $MgSO_4$, and 0.01% (w/v) bovine serum albumin (BSA; Sigma-Aldrich), and washed once. Oocytes were then stimulated with a direct current pulse of 1.5 kV/cm for 100 μS using a somatic hybridizer and transferred for 2 h to PZM3 supplemented with CB to prevent extrusion of a pseudo-second polar body. The reconstructed (nuclear transferred) oocytes were then cultured in PZM3 medium in an atmosphere of 5% CO_2, 5% O_2, and 90% air at 38.5°C for 2 days. The embryos developed at two- to four-cell stage are transferred into the oviduct of anesthetized pseudopregnant surrogates. Since the NT embryos are generally of a low quality in development and at least four fetuses are required for successful pregnancy in pigs, it is recommended that a large number (>100) of embryos are transferred per surrogate.

IV. Protocol for Ovine NT

The basic protocol for cloning in sheep is similar to that described above for cattle and swine. Here we briefly describe the method used to produce ovine cloned embryos derived from cumulus cells.

A. Donor Cell Collection

COCs collected from the ovaries of slaughtered ewes are matured for 24 h in TCM-199 with 10% FBS, 5 μg/mL FSH (Ovagen, ICP, Auckland, New Zealand), 5 μg/mL LH (Sigma-Aldrich), and 1 μg/mL estradiol-17β (Sigma-Aldrich) in an incubator at 38.5°C with 5% CO_2. Expanded COCs are briefly incubated in hyaluronidase (300 USP units/mL; Sigma-Aldrich) and mechanically dissociated into a single cell population (Loi et al., 2008).

B. Oocyte Maturation

Ovine ovaries are collected from a local slaughterhouse and transported to the laboratory in physiological saline at approximately 25°C. COCs are aspirated from antral follicles; only COCs with evenly granulated cytoplasm and at least three layers of cumulus cells are used. COCs are matured in vitro in bicarbonate-buffered TCM-199 (275 mOsm) containing 2 mM glutamine (Sigma-Aldrich), 100 μM cysteamine (Sigma-Aldrich), 0.3 mM sodium pyruvate (Sigma-Aldrich), 10% FBS, 5 μg/mL FSH (Ovagen), 5 μg/mL LH (Sigma-Aldrich), and 1 μg/mL estradiol-17β

(Sigma-Aldrich) in a humidified atmosphere of 5% CO_2 in air at 39°C for 24 h (Ptak et al., 2002).

C. Oocyte Enucleation and NT

After IVM, cumulus cells are completely removed from oocytes by pipetting. Oocytes are incubated for 15 min at 39°C in HEPES-buffered TCM-199 containing 4 mg/mL BSA (Sigma-Aldrich), 7.5 µg/mL CB (Sigma-Aldrich), and 5 µg/mL of Hoechst 33342 (Sigma-Aldrich). Enucleation is performed using a micromanipulator in HEPES-buffered TCM-199 with 4 mg/mL BSA and 7.5 µg/mL CB. The metaphase II plate with a small amount of surrounding cytoplasm is aspirated into the enucleation pipette under UV light exposure. Enucleated oocytes are allowed to recover from CB treatment and then directly injected with cumulus cells. Reconstructed oocytes are activated in HEPES-buffered TCM-199 containing 5 µg/mL ionomycin (Sigma-Aldrich) for 5 min and subsequently incubated for 3−5 h in synthetic oviductal fluid (SOF) medium containing 4 mg/mL BSA, 10 mM 6-dimethylaminopurine (Sigma-Aldrich), and 7.5 µg/mL CB (Loi et al., 2008).

D. Embryo Culture

Reconstructed embryos are transferred into 20-µL drops of SOF enriched with 1% (v/v) basal medium Eagle essential amino acids (Sigma-Aldrich), 1% (v/v) minimum essential medium nonessential amino acids (Gibco), 1 mM glutamine (Sigma-Aldrich), and 8 mg/mL fatty acid-free BSA (Sigma-Aldrich). The embryos are incubated in a humidified atmosphere of 5% CO_2, 7% O_2, and 88% N_2 at 39°C (Ptak et al., 2002). On days 3 and 5 (day 0 = day of NT), the medium is changed to SOF with 5% charcoal-stripped FBS (Ptak et al., 2002). Embryo cultures are maintained for 7 days, and embryos that develop into the blastocyst stage are used for embryo transfer.

V. Summary

NT consists of the removal of the nucleus from an oocyte and replacing it with the nucleus of a donor cell. Electric fusion or microinjection techniques are commonly used to transfer a donor nucleus into an enucleated oocyte. After NT, reconstructed embryos are artificially activated to initiate development and cultured *in vitro* until the optimal stage for embryo transfer. Animal cloning by NT has been successful in several species, although the cloning efficiency is extremely low. NT technology has been useful for genetic modification of livestock because of several advantages conferred by the use of cultured cells compared with the pronuclear microinjection method.

References

Akagi, S., Takahashi, S., Adachi, N., Hasegawa, K., Sugawara, T., Tozika, T., et al., 2003. *In vitro* and *in vivo* developmental potential of nuclear transfer embryos using bovine cumulus cells prepared in four different conditions. Cloning Stem Cells 5, 101−108.

Akagi, S., Kaneyama, K., Adachi, N., Tsuneishi, B., Matsukawa, K., Watanabe, S., et al., 2008. Bovine nuclear transfer using fresh cumulus cell nuclei and *in vivo-* or *in vitro-* matured cytoplasts. Cloning Stem Cells 10, 173−180.

Baguisi, A., Behdoodi, E., Melican, D.T., Pollock, J.S., Destrempes, M.M., Cammuso, C., et al., 1999. Production of goats by somatic cell nuclear transfer. Nat. Biotechnol. 17, 456−461.

Campbell, K.H., 2002. A background to nuclear transfer and its applications in agriculture and human therapeutic medicine. J. Anat. 200, 267−275.

Campbell, K.H., Alberio, R., 2003. Reprogramming the genome: role of the cell cycle. Reprod. Suppl. 61, 477−494.

Cibelli, J.B., Stice, S.L., Golueke, P.J., Kane, J.J., Jerry, J., Blackwell, C., et al., 1998. Cloned transgenic calves produced from non-quiescent fetal fibroblasts. Science 280, 1256−1258.

Eyestone, W.H., Campbell, K.H., 1999. Nuclear transfer from somatic cells: applications in farm animal species. J. Reprod. Fertil. Suppl. 54, 489−497.

Gibbons, J., Arat, S., Rzucidlo, J., Miyoshi, K., Waltenburg, R., Respess, D., et al., 2002. Enhanced survivability of cloned calves derived from roscovitine-treated adults somatic cells. Biol. Reprod. 66, 895−900.

Hodges, C.A., Stice, S.L., 2003. Generation of bovine transgenics using somatic cell nuclear transfer. Reprod. Biol. Endocrinol. 1, 81.

Inoue, K., Kohda, T., Sugimoto, M., Sado, T., Ogonuki, N., Matoba, S., et al., 2010. Impeding Xist expression from the active X chromosome improves mouse somatic cell nuclear transfer. Science 330, 496−499.

Kato, Y., Tsunoda, Y., 2010. Role of the donor nuclei in cloning efficiency: can the ooplasm reprogram any nucleus? Int. J. Dev. Biol. 54, 1623−1629.

Kato, Y., Tani, T., Sotomaru, Y., Kurokawa, K., Kato, J., Doguchi, H., et al., 1998. Eight calves cloned from somatic cells of a single adult. Science 282, 2095−2098.

Kato, Y., Tani, T., Tsunoda, Y., 2000. Cloning of calves from various somatic cell types of male and female adult, newborn and fetal cows. J. Reprod. Fertil. 120, 231−237.

Kishigami, S., Mizutani, E., Ohta, H., Hikichi, T., Thuan, N.V., Wakayama, S., et al., 2006. Significant improvement of mouse cloning technique by treatment with trichostatin A after somatic nuclear transfer. Biochem. Biophys. Res. Commun. 340, 183−189.

Loi, P., Matsukawa, K., Ptak, G., Clinton, M., Fulka Jr., J., Nathan, Y., et al., 2008. Freeze-dried somatic cells direct embryonic development after nuclear transfer. PLoS ONE 3, 2978.

Marx, J.L., 1988. Cloning sheep and cattle embryos. Science 239, 463−464.

Matoba, S., Inoue, K., Kohda, T., Sugimoto, M., Mizutani, E., Ogonuki, N., et al., 2011. RNAi-mediated knockdown of Xist can rescue the impaired postimplantation development of cloned mouse embryos. Proc. Natl. Acad. Sci. USA 108, 20621−20626.

McCreath, K.J., Howcroft, J., Campbell, K.H., Colman, A., Schnieke, A.E., Kind, A.J., 2000. Production of gene-targeted sheep by nuclear transfer from cultured somatic cells. Nature 405, 1066−1069.

Morgan, H.D., Santos, F., Green, K., Dean, W., Reik, W., 2005. Epigenetic reprogramming in mammals. Hum. Mol. Genet. 15, R47–R58.

Prather, R.S., Barnes, F.L., Sims, M.M., Robl, J.M., Eyestone, W.H., First, N.L., 1987. Nuclear transplantation in the bovine embryo: assessment of donor nuclei and recipient oocyte. Biol. Reprod. 37, 859–866.

Prather, R.S., Sims, M.M., First, N.L., 1989. Nuclear transplantation in early pig embryos. Biol. Reprod. 41, 414–418.

Ptak, G., Clinton, M., Tischner, M., Barboni, B., Mattioli, M., Loi, P., 2002. Improving delivery and offspring viability of *in vitro*-produced and cloned sheep embryos. Biol. Reprod. 67, 1719–1725.

Rybouchkin, A., Kato, Y., Tsunoda, Y., 2006. Role of histone acetylation in reprogramming of somatic nuclei following nuclear transfer. Biol. Reprod. 74, 1083–1089.

Schnieke, A.E., Kind, A.J., Ritchie, W.A., Mycock, K., Scott, A.R., Ritchie, M., et al., 1997. Human factor IX transgenic sheep produced by transfer of nuclei from transfected fetal fibroblasts. Science 278, 2130–2133.

Stice, S.L., Robl, J.M., 1988. Nuclear reprogramming in nuclear transplant rabbit embryos. Biol. Reprod. 39, 657–664.

Suzuki, S., Iwamoto, M., Saito, Y., Fuchimoto, D., Sembon, S., Suzuki, M., et al., 2012. Il2rg gene-targeted severe combined immunodeficiency pigs. Cell Stem Cell 10, 753–758.

Takahashi, S., Kubota, C., Shimizu, M., Tabara, N., Izaike, Y., Imai, H., 2000. Production of cloned calves by somatic cell-nuclear transplantation. In: Roberts, R.M., Yanagimachi, R., Kariya, T., Hashizume, K. (Eds.), Cloned Animal and Placentation. Yokendo Publishers, Tokyo, pp. 30–35.

Urakawa, M., Ideta, A., Sawada, T., Aoyagi, Y., 2004. Examination of a modified cell cycle synchronization method and bovine nuclear transfer using synchronized early G1 phase fibroblast cells. Theriogenology 62, 714–728.

Wakayama, T., Perry, A.C.F., Zuccotti, M., Johnson, K.R., Yanagimachi, R., 1998. Full-term development of mice from enucleated oocytes injected with cumulus nuclei. Nature 394, 369–374.

Willadsen, S.M., 1986. Nuclear transplantation in sheep embryos. Nature 320, 63–65.

Wilmut, I., Schniek, A.E., McWhir, J., Kind, A.J., Campbell, K.H., 1997. Viable offspring derived from fetal and adult mammalian cells. Nature 385, 810–813.

Wilmut, I., Young, L., DeSousa, P., King, T., 2000. New opportunities in animal breeding and production: an introductory remark. Anim. Reprod. Sci. 60–61, 5–14.

Zimmerman, U., Vienken, J., 1982. Electric field-induced cell-to-cell fusion. J. Membr. Biol. 67, 165–182.

15 Alternative Methods for Transgenesis in Domestic Animal Species

Michael J. Martin[1] and Carl A. Pinkert[2]

[1]Spring Point Project, Minneapolis, MN, [2]Department of Biological Sciences, College of Arts and Sciences, The University of Alabama, Tuscaloosa, AL

I. Introduction and Discussion

A. Introduction

In the first edition of this chapter in 1994, we stated that DNA microinjection was the only method successfully used by a number of laboratories to create transgenic swine and other farm animal species. Within less than a decade, methodologies improved, and a number of procedures were employed that finally allowed the development of models for both gain-of-gene-function and loss-of-function developmental endpoints. Within the last decade, nuclear transfer technologies (and with some species-specific constraints, retroviral-based protocols) have clearly become the methodologies of choice, as described elsewhere in this edition (see Chapters 6, 7, and 14). That said, this chapter will highlight the contributions, utility, and protocols involving DNA microinjection and some lesser used procedures in use and in validation today.

The production of transgenic mice that grew twice as large as their nontransgenic littermates was the first evidence that genetic engineering could be used to greatly modify an animal's phenotype (Palmiter et al., 1982). With swine production representing a multibillion dollar per year industry (Pursel et al., 1989), the litter-bearing pig with a moderate-length gestation period (~114 days) is an obvious candidate for the application of genetic engineering. Genetic engineering is currently being used to enhance livestock performance in two major areas: growth/development (Pursel et al., 1989), as well as food safety and disease resistance (Pinkert et al., 1989a,b; Lo et al., 1991; Weidle et al., 1991). Pilot studies have also pointed to methodology to solve environmental pollution problems posed by large-scale livestock production practices (Golovan et al., 2001). However, in addition to livestock production considerations, applications of genetic engineering have grown to include the modification of livestock to produce human proteins related to

Transgenic Animal Technology. DOI: http://dx.doi.org/10.1016/B978-0-12-410490-7.00013-X

various biomedical applications. These applications include the use of livestock as bioreactors to produce human pharmaceuticals in various body fluids or tissues (e.g., protein harvest from blood or milk; Clark et al., 1987; Van Brunt, 1988; Swanson et al., 1992) and disease models such as macular degeneration (Sommer et al., 2011). Additionally, within the past decade, genetically modified pigs have been used to generate tissues and organs for xenotransplantation; the advent of such technology would have dramatic impact on human health initiatives (Pinkert, 1994; Sharma et al., 1996; Tucker and White, 1998).

Genes that have been introduced into swine are listed in Table 15.1. The majority of genes utilized in swine experiments have been growth hormone (GH) or growth hormone releasing factor (GRF) fusion constructs. In contrast to mice, which express foreign GH genes, neither sheep (Rexroad et al., 1989, 1990; Ward et al., 1989) nor swine (Ebert et al., 1988; Miller et al., 1989; Pursel et al., 1990a,b; Wieghart et al., 1990) grew at an accelerated rate when compared to nontransgenic control animals. Vize et al. (1988), however, reported the production of several transgenic pigs containing a porcine growth hormone (pGH) construct, one of which grew substantially faster than littermate controls. Serum concentrations of GH in this animal were twice those observed in nontransgenic littermates. These researchers have speculated that the enhanced growth exhibited by this pig was due to the incorporation of a gene construct that expressed a homologous (pGH) as opposed to heterologous (bovine: bGH or human: hGH), GH gene. In contrast, Ebert et al. (1990) found no difference in phenotype between transgenic swine that expressed a homologous versus heterologous GH construct.

A comparison of endocrine profiles between transgenic swine and sheep producing bGH or hGH and their respective nontransgenic littermate controls revealed the presence of elevated plasma concentrations of insulin and insulin-like growth factor-1 (IGF-1) in the transgenic individuals (Rexroad et al., 1989; Ward et al., 1989; Miller et al., 1989; Pinkert, 1991). The physiological consequences of circulating hGH and bGH in transgenic swine and sheep are similar to those observed following the exogenous administration of GH, that is, body fat is reduced (Hammer et al., 1986; Pursel et al., 1987; Ward et al., 1989) while feed efficiency is enhanced (Evock et al., 1988; Campbell et al., 1988; Pursel et al., 1989).

Unfortunately, transgenic livestock that expressed a heterologous GH construct demonstrated a host of health-related problems. Transgenic swine expressing GH genes exhibited lameness, peptic ulcers, lethargy, and impaired reproductive performance (Pursel et al., 1989, 1990a,b; Wieghart et al., 1990). The occurrence of deleterious phenotype in hGH, bGH, pGH, and rat GH (rGH) transgenic swine may have resulted from the animal's continuous exposure to elevated concentrations of GH (Pursel et al., 1990a; Ebert et al., 1990). The identification of regulatory sequences (i.e., promoters and enhancers) that restricted expression of the GH transgene exclusively during the rapid growth phase or in an episodic fashion was proposed as a means to reduce the health-related concerns associated with GH fusion constructs (Pursel et al., 1990a). Indeed, Pursel et al. (2000) found that a functional α-actin driven IGF-1 cDNA in swine resulted in greater average daily gain and lean tissue accretion and a reduction in fat tissue accretion.

Table 15.1 Genes that have been Introduced into Swine[a]

Gene	Eggs Injected	Transgenic Offspring	Functional Transgenics	Citations
mMT/hGH	268	1	[b]	Brem et al. (1985)
	2035	20	11/18	Hammer et al. (1985)
hMT/bGH	423	6	1/6	Vize et al. (1988)
hMT/pGH	3162	88	20/88	Nottle et al. (1999)
mMT/pGH	2198	11	8/11	Pursel et al. (1987)
bPRL/bGH	289	4	2/4	Polge et al. (1989)
rPEPCK/bGH	1057	7	5/7	Wieghart et al. (1990)
CMV/pGH	372	15	[b]	Ebert et al. (1990)
MLV/rGH	59	1	1/1	Ebert et al. (1988)
MLV/pGH	410	6	[b]	Ebert et al. (1990)
mMT/hGRF	2627	8	2/8	Pinkert et al. (1987); Pursel et al. (1989)
	1041	6	[b]	Brem et al. (1988)
hALB/hGRF	968	5	3/3	Pursel et al. (1989)
mMT/hIGF-1	387	4	¼	Pursel et al. (1989)
mMT/MX	1083	6	[b]	Brem et al. (1988)
mMX-SV	376	6	1/6	Pinkert et al. (2001)
hLCR/ααβ	709	3	3/3	Swanson et al. (1992)
MSV/c-ski	1091	29	10/29	Pursel et al. (1992)
mμIg	119	0	–	Pinkert (1990)
mκγIg	[a]	3	1/3	Weidle et al. (1991)
mαIg	542	2	2/2	Lo et al. (1991)
mMT/hβIFN	848	2	2[c]/2	Pinkert (1990)
mWAP	850	5	3/3	Shamay et al. (1991)
mWAP/hProtein C	775	12	6/8	Williams et al. (1992); Van Cott et al. (2001)
mWAP/hFVIII	227	1	1/1	Paleyanda et al. (1997)
mWAP/hFIX	557	4	2/3	Van Cott et al. (1999); Velander et al. (2001)
mWAP/hFibrinogen[d]	730	6	3/4	Butler et al. (1997)
H2/CD59	250	3	1/3	Fodor et al. (1994)
H2/CD55 + HTF	1047	20	4/20	Nottle et al. (2001)
H2/CD55 + CD59 + HTF[d]	1460	16	11/16	Nottle et al. (2001)
ICAM/HTF	1137	185	8/185	Nottle et al. (2001)
ICAM/ CD46 + CD55 + CD59[d]	2822	94	2/94	Nottle et al. (2001)
GT	1610	39	5	Martin et al. (2000)
CTLA4-Ig	2663	15	15/15	Phelps et al. (2009)
Anti-*gag*/anti-*pol* shRNAs	967	25	18/25	Ramsoondar et al. (2009)

[a]For chimeric constructs, a slash separates a promoter or enhancer sequence from the structural gene. The species derivation is indicated by a lowercase letter before the abbreviation of the gene: b, bovine; c, chicken; h, human; m, mouse; o, ovine; p, porcine; r, rat; ALB, albumin; ααβ (α, α, and β) globin chains; β-IFN, β-interferon; CD46/55/59, cell surface complement regulatory proteins antigens; CTLA4-Ig, cytotoxic T lymphocyte-associated antigen fused to the hinge and CH2/CH3 regions of human IgG1; FVIII/IX, blood factor VIII/IX; GT, galactosyl transferase; GH, growth hormone; GRF, growth hormone releasing factor; H2, mouse MHC class II; HTF, H-substance glycosyltransferase; Ig, immunoglobulin (κ, α, and/or μ chain); ICAM, intercellular adhesion molecule; LCR, human β-globin gene control locus; MLV, Moloney murine leukemia virus; MSV, mouse sarcoma virus; MT, metallothionein; MX, myxovirus resistant, PEPCK, phosphoenolpyruvate carboxykinase; PRL, prolactin; snRNAs, small interfering RNAs (siRNAs) were expressed as short hairpin RNAs (shRNA) against *gag* and *pol* PERV (porcine endogenous retrovirus) genes; SV, SV40 T antigen; WAP, whey acidic protein. A functional transgene indicates the number of founder transgenic animals (or their offspring) expressing a transgene encoded mRNA or protein product, divided by the total number of founder animals (or offspring within a line) evaluated.
[b]Data incomplete.
[c]Pigs died at birth.
[d]Coinjection of two or three constructs. Where one promoter was identified, each construct included the same regulatory sequence.

These differences were most evident after swine had reached maturity (e.g., 90 kg body weight; Pursel et al., 2000).

The hypothesis that farm animals may be utilized as bioreactors received major support with the production of transgenic mice that expressed the sheep β-lactoglobulin gene in the mammary gland (Simons et al., 1987). β-Lactoglobulin concentrations in the milk of these mice were fivefold higher than those estimated in sheep milk. Gordon et al. (1990) have further shown that two human proteins, tissue plasminogen activator and protein C, can be produced in the mammary gland of lactating transgenic mice.

The first livestock species to be utilized as a bioreactor was sheep (Simons et al., 1988). Using the β-lactoglobulin gene as a promoter, Simons et al. (1988) produced transgenic sheep that secreted human clotting factor IX or α1-antitrypsin in the milk. Researchers have now been able to create transgenic swine (Wall et al., 1991), sheep (Wright et al., 1991), and goats (Denman et al., 1991; Ebert et al., 1991) that secrete heterologous milk proteins as well. While the mammary gland has been the major target for the so-called "biopharming" of heterologous proteins, the production of transgenic pigs that synthesize and secrete human hemo-globin indicated that erythroid tissues could be utilized in a similar manner (Swanson et al., 1992).

B. The Production of Transgenic Swine by DNA Microinjection: Current Efficiencies

In contrast to the mouse, the efficiency associated with the production of transgenic livestock, including swine, is generally low (Hammer et al., 1985; Pursel et al., 1989, 1990a,b; Wieghart et al., 1990; Wall, 1997). The development of modified estrous synchronization and superovulation protocols in swine, however, have improved the transgenic rate in this species (Pinkert et al., 2000). Since the first transgenic swine were reported in 1985, DNA microinjection has been the only successful method identified to produce transgenic pigs. While other technologies are in development, the focus of this chapter will describe the use and refinement of DNA microinjection technology.

An initial problem encountered during the creation of transgenic swine and other farm animal species concerned the visualization of the pronuclei or nuclei within the ova. Wall et al. (1985) found that centrifugation of pig ova at $15,000 \times g$ for 3–5 min results in stratification of the cytoplasm, which renders the pronuclei visible using differential interference contrast microscopy (Figure 15.1). Unfortunately, centrifugation failed to reveal the pronuclei in 15% (Wall et al., 1985) to 33% (Brem et al., 1989) of the fertilized ova. After experiencing a similar difficulty in observing pronuclei following centrifugation, Hammer et al. (1986) attempted to produce transgenic swine by injecting DNA into the cytoplasm of one- and two-cell ova. None of the fetuses derived from these ova were transgenic. Introduction of the same gene construct into the pronuclei or nuclei of centrifuged one- and two-cell ova, however, yielded an integration frequency in offspring of 10%.

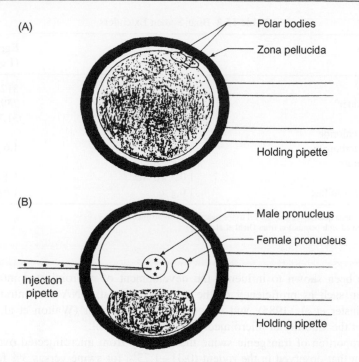

Figure 15.1 Microinjection of a pig zygote. While the cytoplasm in murine eggs does not inhibit visualization of the pronuclei, lipid-rich pig eggs are relatively opaque (A). To facilitate visualization of nuclear structures, one- and two-cell eggs are centrifuged to stratify lipids in the cytoplasm (B). After centrifugation, the eggs are placed in a microdrop of medium overlaid with silicone oil and held in place, sequentially, using the large bore holding pipette. A small bore injection pipette containing DNA in a buffer solution is inserted through the zona pellucida and plasma membrane into a pronucleus (or nuclei of two-cell eggs) that readily expands (~50% greater in volume) as the DNA solution is delivered. The diameter of the egg is ~125 μm. If two-cell eggs are used, generally both blastomeres are filled with DNA.
Source: The figure is reproduced with permission from Pinkert (1987) and Pinkert et al. (1990).

These findings indicate that genomic incorporation of the transgene takes place only when the gene is introduced directly into the pronucleus or nucleus of the ovum.

The proportion of transferred microinjected ova that develop into viable off-spring in swine varies from 6% to 11.7% (Pursel et al., 1989, 1990b). The survival to term of microinjected porcine ova is related to several factors, which include the developmental stage of ova injected (Pursel et al., 1987), the duration of *in vitro* culture (Davis, 1985; Brem et al., 1989), synchrony of donors and recipients (Polge, 1982; Brem et al., 1989), the number of ova transferred (Brem et al., 1989), and the age of the donor (Pinkert et al., 1989a,b; Brem et al., 1989). Other factors

Table 15.2 Boar Semen Extenders

Ingredient	BL-1 (1 quart)	Egg Yolk (1 quart)
Egg yolk		317.0 mL
Distilled water[a]		739.0 mL
Glucose	27.4 g	31.7 g
Potassium chloride	0.3 g	
Sodium bicarbonate	1.9 g	1.6 g
Sodium citrate	9.5 g	
Penicillin	1.0×10^6 IU	1.1×10^6 IU
Streptomycin sulfate	1.0 g	1.6 g

[a]Put salts in a clean quart container and fill to the quart line with distilled water.
Source: Reprinted with permission from Diehl et al. (1979).

that have been shown to influence the development of microinjected mouse and sheep ova, such as proficiency of the microinjectionist, DNA concentration and form (Brinster et al., 1985), and injection pipette diameter (Walton et al., 1987), may affect the viability of microinjected swine ova as well.

The proportion of transgenic swine that develop from microinjected ova is also lower than that observed in the rodent (0.31−1.73% for swine versus 3% for mice: Pursel et al., 1990b; Table 15.2). The low frequency at which integration of the transgene occurs in the pig genome may be associated with the age of the host ovum or the composition of its chromosomes (Pursel et al., 1989). The frequency of transgene integration and the efficiency of gene transfer (proportion of microinjected ova that develop into transgenic pigs) may also be influenced by the size of the DNA molecule. A recent study by Martin et al. (2000) reported that both the efficiency of gene transfer and the transgene integration rate were lower for ova injected with 60−90 kb as opposed to 2.5−18 kb DNA molecules. When integration does occur, the number and orientation (head-to-head versus head-to-tail) of transgene copies inserted into the genome varies greatly as well (Hammer et al., 1985; Vize et al., 1988; Miller et al., 1989; Polge et al., 1989; Wieghart et al., 1990).

In swine, approximately 70% of the transgenic pigs containing an exogenous GH construct expressed the integrated transgene's protein product (Pursel et al., 1990b). Failure of the remaining 30% of transgenic swine to express the foreign gene has been attributed to integration of the transgene into an inactive chromosomal locus or alteration of the transgene sequence during its integration (Pursel et al., 1990a). Once the gene has been integrated, the level of expression appears to vary greatly among individuals regardless of the source of the GH construct (bGH, pGH, hGH; Hammer et al., 1985; Miller et al., 1989; Pursel et al., 1989). According to Pursel et al. (1990b), transgene transcription rates are probably regulated by the level of activity present at the locus or site of integration, as well as the properties of the enhancer sequences located in genes flanking the fusion gene.

The existence of an interaction between gene copy number and its level of expression is most likely dependent upon the particular construct or transgene. Hammer et al. (1985) saw no relationship between the number of gene (hGH) copies present per cell and plasma concentrations of hGH. Miller et al. (1989), however, found a positive correlation between plasma concentrations of bGH and the number of bGH copies per cell in transgenic swine.

II. Methods

A. Donor and Recipient Management

1. Estrous Synchronization

The successful transfer of ova in swine depends greatly upon synchronization of estrus between the donor and recipient. Webel et al. (1970) found no difference in the pregnancy rate between recipients whose estrus periods were 24 h ahead or behind those of the donors. Polge (1982), however, noted that the pregnancy rate was highest (70–86%) among recipients that had expressed estrus 24–40 h later than the donors and lowest (10–56%) among recipients that had expressed estrus earlier than the donors. Blum–Reckow and Holtz (1991) observed a similar increase in both the pregnancy and embryo survival rates when ova were transferred to recipients whose estrous cycles were 24 h behind that of the donors. The current recommendation for embryo transfer in swine is to utilize recipient females whose onset of estrus is at least synchronous or 12–24 h later than the occurrence of estrus in the donors.

The majority of methods used for synchronizing estrus in sexually mature gilts or sows involve manipulation of the luteal phase of the estrous cycle. These procedures include (a) pseudopregnancy induction followed by exogenously induced luteal regression, (b) induction of accessory or secondary corpora lutea (CL) followed by exogenously induced CL regression, and (c) maintenance of an artificial luteal phase through the administration of an orally active progestin such as altrenogest.

2. Pseudopregnancy

Pseudopregnancy can be induced by the administration of estradiol benzoate (EB) or estradiol valerate (EV) during the period of pregnancy recognition. Injections of 5 or 10 mg of EB (Guthrie, 1975) or EV (Zavy et al., 1984) on days 11–15 of the estrous cycle have proven to be effective in significantly prolonging luteal function. The administration of EB for 20 days also prolonged luteal function and suppressed estrus (Kraeling and Rampacek, 1977). The return of pseudopregnant gilts to estrus is achieved through prostaglandin $F_{2\alpha}$ ($PGF_{2\alpha}$)-induced luteal regression. The time of $PGF_{2\alpha}$ treatment is apparently unimportant as its administration on days 1, 5, 10, or 20 following cessation of EB treatment appears to be equally effective in causing gilts to exhibit synchronized estrus 4–6 days later (Guthrie, 1975).

3. Accessory CL Induction

Ovulation induction and subsequent formation of accessory CL during the luteal phase of the estrous cycle has been accomplished in the pig through the use of pregnant mare serum gonadotropin (PMSG) and human chorionic gonadotropin (HCG; Neill and Day, 1964; Caldwell et al., 1969). Because $PGF_{2\alpha}$ is not luteolytic in the pig until at least day 11 or 12 of the estrous cycle (Diehl and Day, 1974; Guthrie and Polge, 1976), estrus induction in animals possessing accessory CL is most effective when $PGF_{2\alpha}$ is given on days 12 and 13 (Guthrie and Polge, 1967) or 13 and 14 (Guthrie, 1979) after HCG administration (d 0). The incidence of estrus expression can be increased by administering a second injection of PMSG alone on the second day of $PGF_{2\alpha}$ treatment (Guthrie and Polge, 1967) or by administering PMSG followed 96 h later by HCG (Guthrie, 1979). Accessory CL induction does not appear to be detrimental to fertilization as 80% conception rates have been achieved following a single insemination (Guthrie and Polge, 1967).

4. Orally Active Progestagen

Perhaps the most effective estrous synchronization method involves the feeding of an orally active synthetic progestin (allyl trenbolone or altrenogest: AT; 15 mg/h/d; both Regumate and Matrix, manufactured by Intervet/Schering-Plough, Millsboro, DE, are 0.22% solutions of altrenogest in oil; Regumate is used to "suppress estrus in mares," while Matrix is used to "synchronize estrus in sexually mature gilts.") for 14—21 days (Webel, 1976; Knight et al., 1976; O'Reilly et al., 1979; Davis et al., 1979; Redmer and Day, 1981; Pursel et al., 1981; Flowers, 1999) to sexually mature female swine. Estrous synchronization rates, the proportion of females that exhibit estrus between 4 and 7 days following AT withdrawal, of 89% or better were reported (Knight et al., 1976; Redmer and Day, 1981; Pursel et al., 1981). Altrenogest (allyl trenbolone) also appears to have no detrimental effect on ovulation, fertilization or gestation. In fact, AT consumption may enhance the ovulation rate (Davis et al., 1979).

B. Superovulation

Superovulation of donors in a transgenic swine program provides two major benefits: it increases the number of ova that can be obtained from each donor and it establishes the onset of ovulation. The latter benefit enables one to determine the optimum time at which to recover pronuclear stage ova.

Because the likelihood of producing mosaic individuals increases when the nuclei of two-cell ova are microinjected with DNA, it is advantageous to recover a high percentage of pronuclear stage ova. One-cell ova can be obtained from swine that have ovulated naturally or donors that have been superovulated using PMSG and HCG. In the former case, the start of ovulation and the optimum time at which to collect pronuclear stage ova are based solely upon the onset of estrus.

Natural ovulation in swine begins between 30 and 57 h after the onset of estrus (Pope et al., 1989; Town et al., 1999) and takes at least 3—6.5 h to complete

(Burger, 1952; Betteridge and Raeside, 1962). Unfortunately, the relationship between the onset of estrus and the start of ovulation in swine is quite variable. Pope et al. (1989) found that the majority of gilts begin to ovulate somewhere between 30 and 38 h after the onset of estrus. Furthermore, when ovaries from gilts at 34 h after the onset of estrus were examined, ovulation appeared to have been completed in some cases, while in others ovulation had not yet begun. Additional data indicate that neither ovulation nor early embryonic development are synchronous processes in swine (Pope et al., 1989; Didion et al., 1990; Martin et al., 1990; Xie et al., 1990a,b). It is clear that a great amount of variation exists throughout the chronology of events associated with estrus, ovulation, and early embryonic development in swine.

In contrast to natural ovulation, a superovulatory regimen including PMSG and HCG enables one to program the onset and reduce the asynchronicity of ovulation (Dziuk and Baker, 1972; Pope et al., 1989). These benefits should increase the likelihood of selecting the proper time at which to recover ova at the pronuclear stage of development.

Superovulation of sexually mature gilts can be achieved through either the administration of PMSG alone (Hunter, 1964, 1966) or in combination with HCG (Day et al., 1965, 1967). Gilts-given PMSG alone following luteal regression (day 15−16 of the estrous cycle) show estrus 4−7 days later following treatment and ovulate without receiving HCG (Day et al., 1965). Ovulation can be more precisely controlled when HCG is administered between 72 and 80 h following PMSG (Wall et al., 1985; Hammer et al., 1985; Pope et al., 1989). Gilts should exhibit estrus within 24 h following HCG administration.

The dose of PMSG to administer depends upon both the target ovulation rate and hormone source one chooses. As the ovulatory response to PMSG/HCG approaches or exceeds 40 ova, the probability of obtaining immature, degenerate, or otherwise abnormal ova may increase as well (Holtz and Schlieper, 1991). Pregnant mare serum gonadotropin should be administered at a dosage that results in an average ovulation rate of between 25 and 35 ova. This dosage must be determined for each new source of PMSG (Martin et al., 1989).

Exogenous gonadotropin treatment can stimulate ovulation in prepuberal gilts as well (Casida, 1935; Dziuk and Gehlbach, 1966; Baker and Coggins, 1968; Amet et al., 1991). However, Pinkert et al. (1989a,b) found that fertilized ova recovered from superovulated prepuberal gilts exhibited reduced development *in vitro* as compared to zygotes recovered from sexually mature gilts. French et al. (1991) further demonstrated that both the pregnancy rate and litter size were lower following the transfer of microinjected ova obtained from prepuberal as opposed to puberal gilts. In contrast, no reduction in the *in vitro* development of one-cell ova following microinjection was noted in German studies (Brem et al., 1989) that utilized prepuberal gilts exclusively as ova donors. Differences in the developmental capacity of one-cell ova recovered from prepuberal gilts, illustrated in these studies, may be due to variation in the genetic composition of the donor animals, that is, European versus US versus Australian breeds of swine. In addition to domestic swine, methodologies have been explored for superovulation and transgenic pig production in a number of miniature pig strains. Interestingly, the same protocols used in domestic animals have been applied to miniature swine, although the relative efficiency of embryo production has been lower (Pinkert and Murray, 1993).

C. Recipient Management

Prospective recipients must be structurally and reproductively sound. Females that fall into this category include sows and puberal gilts. Prepuberal swine should not be utilized as recipients of microinjected ova. Both Rampacek et al. (1979) and Segal and Baker (1973) found that the pregnancy rate was low among prepuberal gilts that had been induced to ovulate, unless HCG or EB and progesterone were administered after breeding. Luteinizing hormone receptor studies of CL from prepuberal versus puberal swine (Estienne et al., 1988) further suggest that CL formation and function following induced ovulation may be abnormal in the prepuberal pig.

An alternate approach to the two-pool (one pool of donor animals and one pool of recipients) system for creating transgenic swine is the use of a single animal pool where ovum donors also serve as recipients of microinjected ova. This dual-purpose approach reduces the number of animals required for the production of transgenic swine and appears to have no detrimental effect on pregnancy rate or litter size (Pursel and Wall, 1991).

D. Breeding Management

A factor critical to the production of transgenic swine is the recovery of fertilized ova from donor animals. Because fertilization of ova is dependent upon the presence of sperm in the reproductive tract of the female shortly before the onset of ovulation, the timing and frequency of insemination performed during the estrus period is important. Inseminations performed too early or late during estrus have been shown to result in low fertility (Boender, 1966). If females are exposed to a mature boar once daily, gilts/sows should be bred each day they exhibit standing estrus. If estrus detection is conducted twice daily, females should be bred at 12 and 24 h after the onset of estrus (Diehl et al., 1990).

When artificial insemination is utilized, females should be inseminated with at least 3 billion live sperm in a total volume of 100 mL (semen + extender). Since a mature boar will produce an average of 100–300 million sperm/mL (Foote, 1980), a typical ejaculate should contain a sufficient number of sperm to breed six to eight females when properly extended (Diehl et al., 1979).

Semen collected from a boar of proven fertility can usually be extended at a ratio of 1 part semen to 4–5 parts extender and still yield acceptable conception rates. If a 1:10 ratio of semen to extender is required, one should determine the actual sperm concentration before the semen is extended, in order to make sure each female receives at least 3 billion live sperm. Several commercial extenders, some of which maintain sperm viability for up to 1 week, are available. The formulas for two commonly used semen extenders are presented in Table 15.2 (Diehl et al., 1979).

E. Ova Recovery

In order to obtain ova at the pronuclear or two-cell stage for microinjection, surgical recovery is performed between 60 and 66 h after the administration of HCG (Hammer et al., 1985; Wall et al., 1985; Pinkert et al., 1989a,b). General anesthesia may be induced in swine by administering one of the following drug combinations through a peripheral ear vein:

a. 1 g sodium thiopental/100 kg body weight (bw)
b. 1.3 mg xylazine/kg bw + 1.3 mg ketamine/kg bw
c. 0.4 mg acepromazine/kg bw + 7.5 mg ketamine/kg bw +0.1 mL 0.9% saline/kg bw

The use of a or b requires that anesthesia be maintained by a closed-circuit system of O_2 (600–1000 mL/min) and halothane (4–5%; Webel et al., 1970).

Once the animal has been anesthetized, the reproductive tract is exteriorized via a midventral laparotomy. A drawn glass cannula (outer diameter 5 mm, length 8 cm) is then inserted into the ostium of the oviduct a distance of 3 cm and anchored in place using a single 2–O silk tie through the mesosalpinx (Vincent et al., 1964; Day, 1979). The oviduct is flushed toward the infundibulum by inserting a 20-g needle into the lumen of the oviduct at a point just superior to the uterotubal junction and infusing 10 mL sterile, warm (38°C) Beltsville Embryo Culture Medium-3 supplemented with 10 mM HEPES and 2 mM $NaHCO_3$ (Dobrinsky et al., 1996). The medium is collected in 15 mL sterile plastic tubes. Flushings are transferred to 15×100 mm petri dishes and searched at low (50X) power using a stereomicroscope. After washing the ova twice in G medium (Pinkert et al., 1989a,b), the ova are transferred to microdrops of BECM-3 medium, which have been overlaid with silicone oil. The ova are maintained in this manner at 38°C under a 5% CO_2, 5% O_2 environment until just prior to microinjection.

F. Ova Culture

In order to create transgenic swine, one- and/or two-cell ova must be recovered, injected with DNA and subsequently transferred back to suitable recipients. The developmental capacity of porcine one- and two-cell ova in various media has been reviewed by Wright and Bondioli (1981). Until recently, however, the ability of one- and two-cell ova to undergo more than one or two cleavages *in vitro* has been difficult to achieve. Culture of one- and two-cell ova to the two- to four-cell stage (short-term culture: 12–24 h) can be accomplished using Modified Kreb's Ringer Bicarbonate medium (mKRB) or modified BMOC-3 supplemented with bovine serum albumin (BSA) and EDTA (G medium: Pinkert et al., 1989a,b; Wieghart et al., 1990), oviductal fluid (Archibong et al., 1989) or hypotaurine and taurine (NCSU-23 medium: Reed et al., 1992). Long-term culture (>72 h) of one- and two-cell ova to the blastocyst stage has also been demonstrated using Whitten's medium supplemented with 1.5% BSA (Beckmann et al., 1990; Beckmann and Day, 1991), NCSU-23 medium (Long et al., 1999) and BECM-3 (Dobrinsky et al., 1996). The birth of live pigs from one- or two-cell ova cultured in G, Whitten's and NCSU-23 media indicates that any of these media may be used to culture porcine ova prior to and immediately following microinjection.

The use of a bicarbonate buffer to maintain the pH of a culture media such as mKRB, Whitten's and NCSU-23 between 7.2 and 7.4 requires that the medium be stored under an atmosphere which includes at least a 5% CO_2 gas phase. The atmospheres commonly used are 5% CO_2 in air or a mixture of 5% CO_2, 5% O_2, and 90% N_2. Wright (1977) found that a reduced O_2 atmosphere was superior to a 5% CO_2 in air atmosphere in supporting embryonic development, while Niemann et al. (1983) observed just the opposite experimental outcome. A review of several studies in which transgenic swine were successfully produced indicates that the most viable atmosphere for maintaining the pH of bicarbonate buffered media is 5% CO_2 and 5% O_2.

Coculture (Allen and Wright (1984); Krisher et al., 1989) or xenogeneic (Papaioannou and Ebert, 1988) culture systems have also been shown to support development of one-cell porcine ova. The use of such elaborate systems, however, is not necessary for the production of transgenic swine, since the recovery, microinjection, and transfer processes should require only a short ova culture period.

In the past, the duration of the culture period greatly affected porcine ovum viability. However, development of improved media such as NCSU-23 has reduced the adverse effect(s) of long-term culture on an embryo's ability to develop to term *in vivo*. Machaty et al. (1998) transferred embryos that had been cultured in NCSU-23 medium or *in vivo* for 96 h to recipients and found no significant difference in the proportion of embryos that formed conceptuses by day 40 of pregnancy. This result was noted in spite of the fact that embryos cultured for 96 h *in vivo* had significantly higher cell numbers and ratio of inner cell mass (ICM) to trophoectoderm (TE) cells (ICM:TE cell ratio) when compared to embryos that were cultured *in vitro* in NCSU-23.

G. Microinjection

1. Equipment

The equipment used to create transgenic swine is similar to that used to create transgenic mice (see Chapter 2 for a detailed breakdown). A brief list of instruments and supplies currently used in our laboratories to create transgenic swine is identified in Table 15.3.

2. DNA Preparation

Because of variation associated with the technician's injection ability and the dimensions of the injection pipette tip, the volume of DNA solution introduced into the pronucleus or nucleus is difficult to regulate initially. Most investigators

Table 15.3 Examples of Specialty Equipment and Supplies Used for Gene
Transfer in Swine

- Differential interference contrast microscope (Smith or Nomarski DIC; e.g., Leica
  Laborlux or Nikon Diaphot Ti series)
- Micromanipulators (2 units; e.g., Leica or Narishige, Nikon NT-88-V3, Eppendorf
  TransferMan mechanical manipulators)
- Micromanipulator baseplate and antivibration table (e.g., Micro-G or ServaBench)
- Microcentrifuge and microfuge tube carriers (e.g., Eppendorf, Brinkman)
- Microforge (De Fonbrune-type; TPI, Inc. or Valiant Instruments)
- Pipette puller (e.g., Sutter, P-87)
- Stereomicroscope (e.g., Wild M3 or M8 with transillumination base)
- Microsyringe assemblies (2 units; e.g., P-3 system, see Chapter 2)
- Microinjectors (manual: Eppendorf CellTram vario, Nikon IM-9B; electronic: Eppendorf
  FemtoJet and FemtoJet Express)
- Glass capillary tubing (e.g., inner diameter 0.027, outer diameter 0.037; Garner Glass, La
  Jolla, CA, or World Precision Instruments, Sarasota, FL)
- Silicone or paraffin/mineral oil
- Incubator with 5% CO_2, 5% CO_2, and 90% N_2 or 5% CO_2 in air environment
- Surgical suite (with surgical supplies for multiple laparotomies)

estimate that $1-2$ pL is injected per nucleus for mouse ova (Hogan et al., 1994),
we estimate that 4 pL is injected per pig ovum according to micrometer-based estimates. Two factors that appear to affect embryo viability are DNA concentration
and injection buffer composition. In mouse ova, optimal integration was obtained
when DNA was introduced in a linear form at a concentration of 2 ng/μL in
Tris—HCl buffer that contained between 0.1 and 0.3 mM EDTA (Brinster et al.,
1985). DNA integration and embryo survival were significantly reduced when the
DNA and EDTA concentrations were increased beyond 10 ng/mL and 5 mM,
respectively.

In swine, the relationship of DNA concentration, buffer composition, and DNA
form (linear versus circular) to gene integration and embryo survival has not been
well characterized. A recent study found that the transgene integration rate and efficiency of gene transfer were greater when $2.5-18$ versus $60-90$ kb genes were
microinjected into porcine ova. The viability of ova also appeared to be affected by
the molecular concentration of DNA that was microinjected (Martin et al., 2000).
Williams et al. (1992), however, found that the detrimental effect of pronuclear
microinjection (PM) on subsequent development of porcine one-cell ova was
caused by the injection of DNA and/or contaminants, not by the mechanical process associated with microinjection *per se*.

3.   Ova Preparation

One- and/or two-cell ova are initially recovered using BECM-3 supplemented with
10 mM HEPES and 2 mM $NaHCO_3$ (Dobrinsky et al., 1996). Ova are then transferred to 10×15 mm Petri dishes that contain microdrops of BECM-3 overlaid

with mineral oil. The ova are stored under a humidified atmosphere of 5% CO_2, 5% O_2. While mineral oil can be equilibrated with culture medium prior to its use as an overlay, we have observed no detrimental effect of unequilibrated oil on ova viability during short-term culture under a humidified atmosphere.

Prior to microinjection, 10–50 ova are transferred to a 2-mL Eppendorf tube which contains 1 mL of BECM-3 supplemented with 10 mM HEPES and 2 mM $NaHCO_3$. The ova are centrifuged at $15,000 \times g$ for 6 min (Wall et al., 1985 see Figure 15.3 and 15.4). As previously mentioned, nuclear structures will remain hidden in one-third of the ova following initial centrifugation. This percentage can be reduced by culturing ova for an additional 1–4 h before recentrifugation (personal observation).

4. Microinjection Procedure

Initial steps include attachment of the holding and injection pipettes to their respective lines, followed by the uptake of silicone or paraffin/mineral oil. Care must be taken to ensure that no air bubbles are left in any portion of the microinjection apparatus. The holding pipette tip should have an outer and inner diameter of 25–50 and 10–15 μm, respectively. The injection pipette tip should be 1–2 μm in outer diameter and initially closed. The tip is then opened by breaking it on the edge of the holding pipette. Bevelled tip injection pipettes can also be made by grinding the ends with a diamond dust-coated wheel. However, in our experience pipettes prepared in this manner do not appear to be as sharp as those with broken tips. Pipettes fabricated for DNA microinjection may also be obtained from commercial suppliers, for example, Humagen Fertility Diagnostics. Charlottesville, VA.

Ova to be microinjected are held in a microdrop of BECM-3 medium supplemented with 10 mM HEPES and 2 mM $NaHCO_3$ (10–15 ova per drop) that has been placed on the middle of a depression slide. Another drop (2–5 μL) containing the DNA buffer solution is placed above or below the ova-containing drop. Both drops are covered with silicone or paraffin oil. The injection pipette is loaded by drawing up HEPES medium first, followed by the DNA buffer solution.

Once the DNA solution has stopped flowing into the injection pipette, the syringe plunger is screwed down until the DNA begins to flow out of the tip slowly. The injection pipette is then transferred to the microdrop containing the ova. The pipette tip is inserted through the zona pellucida and vitelline membrane and into the pronucleus of a one-cell ovum or nucleus of a two-cell ovum using one continuous motion. When the pipette tip has penetrated the pronuclear or nuclear membrane, the pronucleus or nucleus should immediately begin to expand (Figure 15.2). The instant expansion ceases, the injection pipette is withdrawn, again using one continuous motion. DNA should be injected into both nuclei of two-cell ova in order to enhance the rate of transgenic animal production (Rexroad et al., 1988).

In addition to standard hollow capillary tubes, filament fiber capillary tubes (World Precision Instruments) can also be used for DNA microinjection (see Chapter 2). While filament fiber capillary tubes negate the need for a DNA drop on

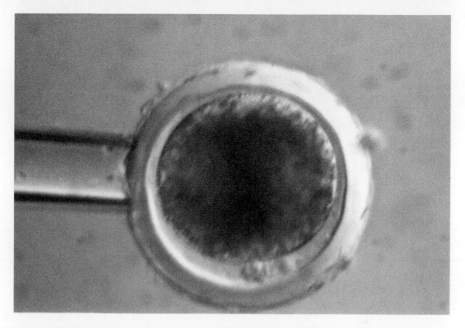

Figure 15.2 One-cell pig egg before centrifugation. Note opaque cytoplasm.

the microinjection slide, a greater volume of DNA for microinjection is required over the course of an experiment.

H. Ova Transfer

Ova collected at the one- or two-cell stage for microinjection are transferred to the oviducts of sexually mature recipients (Dziuk et al., 1964; Pope and Day, 1977). Following anesthetization, the reproductive tract is exposed via a midventral laparotomy. Oviductal transfer entails aspiration of the ova along with 1–2 mL BECM-3 medium into the tubing obtained from a 21 g × 3/4 in. butterfly infusion set. The tube is then fed through the ostium of the oviduct until it reaches the lower third or isthmus of the oviduct. The ova are expelled as the tubing is slowly withdrawn.

The pregnancy rate and the number of ova transferred appear to be directly related. Pope et al. (1972), using noninjected ova, observed that the pregnancy rate at 26–29 days of gestation improved when the number of ova transferred was increased from 12 to 24 (pregnancy rates were 71.4% and 100%, respectively). A similar finding was noted by Brem et al. (1989) and Wei et al. (1993) following the transfer of various numbers of microinjected ova to recipients (Table 15.4). Wei et al. (1993) found that the efficiency of gene transfer (number of

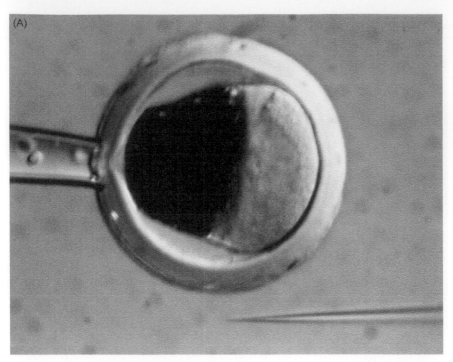

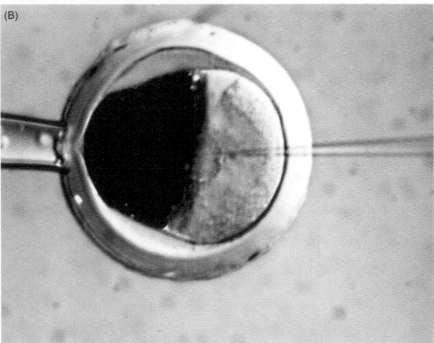

Figure 15.3 One-cell pig egg after centrifugation. (A) After 6 min, the cytoplasmic lipids are stratified and the pronuclei are visible. A nucleolus and pronuclear membrane are evident at the center of the "lipid-cytoplamic" interface. The injection pipette is focused in the same plane as the visible pronucleus, then brought in line to pierce the egg. (B) After impaling the pronucleus, expansion is seen as the pronucleus fills with the DNA solution.

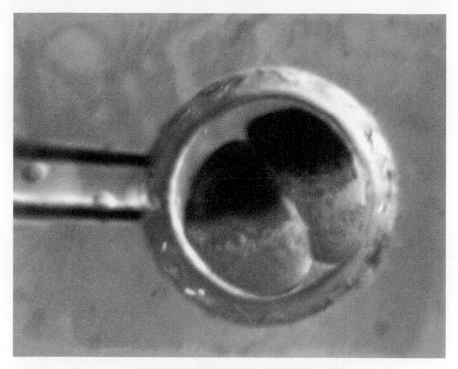

Figure 15.4 Two-cell pig egg after centrifugation. Even after maneuvering the egg, the nuclei are not always in focus in the same plane. However, both nuclei are injected with DNA and show similar enlargement ($\sim 50\%$) as the DNA is delivered.

liveborn/number of embryos transferred) was greatest when 20–29 microinjected ova were transferred per recipient.

The transplantation of noninjected or control ova has proven useful in ensuring pregnancy maintenance (Pinkert, unpublished). In addition, the transfer of control ova, from a strain of swine exhibiting a unique coat color pattern, allows one to differentiate easily between offspring that have developed from control versus microinjected ova. However, this procedure does increase the labor and overall cost associated with the present production system.

III. Alternative Methods for the Production of Transgenic Swine

In addition to DNA microinjection (outlined in Figure 15.5), nuclear transfer (described in Chapters 7 and 14) and retroviral methods (described in Chapter 6) a variety of techniques have been employed to genetically modify swine. Discussion of three alternative methodologies follows.

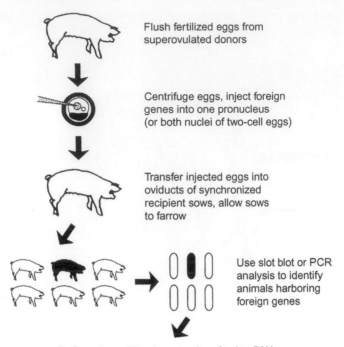

Flush fertilized eggs from superovulated donors

Centrifuge eggs, inject foreign genes into one pronucleus (or both nuclei of two-cell eggs)

Transfer injected eggs into oviducts of synchronized recipient sows, allow sows to farrow

Use slot blot or PCR analysis to identify animals harboring foreign genes

• Perform tissue biopsies—analyze foreign DNA integration, mRNA transcription, and protein production

• Establish transgenic lines to study gene regulation in progeny

Figure 15.5 Scheme for production of transgenic swine. The methodology employed in the production and evaluation of transgenic pigs is illustrated.
Source: Adapted with permission from Pinkert (1987) and Pinkert et al. (1990).

A. Cell-Mediated DNA Transfer

According to Chan (1999), two major barriers stand in the way of progress in the transgenic livestock industry. These include the inefficiency of PM as a means of gene delivery and a limited knowledge of those mechanisms that regulate transgene expression. Not surprisingly, our poor understanding of what regulates transgene expression is due primarily to the low efficiency of PM and the resulting paucity of transgenic livestock made to date. In order to remove these barriers, new and/or improved method(s) of DNA delivery must be developed.

Several groups have attempted to develop carrier cell lines, for example, embryonic stem (ES) cell and embryonic germ (EG) cell lines, which could be transfected with DNA and used to generate chimeras (Stewart et al., 1994; Gerfen and Wheeler, 1995; Piedrahita et al., 1999). Unfortunately, while porcine-derived cell lines with ES-like morphology have been reported, the ability of these cells to

Table 15.4 Effect of Number of Injected Embryos Transferred on the Pregnancy Rate and Proportion of Pigs Born

Number of Embryos Transferred Per Recipient	Total Number of Embryos	Pregnancy Rate (%)	Number of Liveborn Pigs	Mean Litter Size	Mean overall Efficiency (%)
10–19	109	57.1 (4/7)	9	2.25[a]	8.3[a]
20–29	28	66.7 (8/12)	63	7.88[a]	21.9[c]
30–39	407	75 (9/12)	64	7.11[b]	15.7[b]

[a-c]Different superscripts within each column reflect differences between means.
Source: Wei et al. (1993).

contribute to the germ line of chimeric individuals has not been demonstrated (Piedrahita, 2000). EG cells share many similarities with ES cells, for example, morphology, pluripotency, and the ability to contribute to a chimera after blastocoele injection (Stewart et al., 1994; Piedrahita, 2000). Shim et al., (1997) demonstrated that porcine EG cells could differentiate and participate in the development of a chimeric pig. Piedrahita (2000) further showed that genetically transformed EG cells could contribute to term development of a chimeric pig as well. In neither case, however, was germ line transmission of the EG cell genotype confirmed. Drawbacks associated with current EG cell technology include a low efficiency of isolation and difficulty establishing long-term culture of EG cell lines due to suboptimal *in vitro* systems. These limitations greatly restrict one's ability to genetically transform EG cells (Piedrahita, 2000).

B. Sperm-Mediated DNA Transfer

An alternative approach to the use of ES and EG cells for the delivery of DNA is the incorporation of the sperm cell itself as a DNA vector. This method has been used to produce transgenic fish, mice (Maioni et al., 1998; Lavitrano et al., 1989), and swine (Lavitrano et al., 1997, 1999). This method involves culturing of viable sperm with exogenous DNA. DNA binds to the postacrosomal region of sperm and is rapidly internalized. Delivery of bound exogenous DNA into the oocyte, and ultimately the oocyte genome, occurs as a result of fertilization. According to a model presented by Spadofora (1998) binding of exogenous DNA to sperm cells utilizes a 30–35 kD DNA binding protein (DBP) receptor present on the sperm cell's surface and may be dependent upon the expression of class II major histocompatibility complex (MHC) molecules. Following DNA binding, nuclear internalization of DNA mediated by CD4 molecules takes place almost immediately. The resulting DNA-DBP-CD4 complex leaves the sperm head surface and penetrates into the sperm nucleus through its pores. Upon reaching the nuclear matrix DNA dissociates from the DNA-DBP-CD4 complex.

While sperm-DNA association has been observed and confirmed in a variety of species (see Gandolfi, 2000, for a review), the likelihood that exogenous DNA delivered by a sperm cell is incorporated into the oocyte's genome in an unperturbed, intact state is low and very unpredictable. Results from studies in a variety of species (cattle: Schellander et al., 1995; mice: Bachiller et al., 1991; swine: Gandolfi et al., 1989), suggest that the majority of exogenous DNA that is transferred into an ovum by a sperm cell is completely or partially degraded and/or rearranged. The small fraction of DNA that survives these alterations exists outside the chromosome as episomic DNA. Sperm-mediated DNA transfer can occasionally generate individuals that express an intact transgene (Gandolfi, 2000). However, the events associated with these successes are "unpredictable" and have neither been identified nor characterized. This lack of knowledge is a major drawback to the use of live sperm as a DNA vector.

Surprisingly, "killed" sperm cells may be a superior DNA vector when compared to live sperm. When Perry et al. (1999) freeze-thawed, freeze-dried, or exposed mouse sperm to Triton X-100, coincubated these sperm with plasmid DNA, and subsequently injected a single sperm cell into the oocyte cytoplasm, the proportion of resulting mouse embryos that expressed plasmid DNA 84 h later ranged from 64% to 94%. Seventeen to 21% of the pups born were transgenic and 8/11 (73%) of the founders transmitted the gene to their progeny. In this scenario, treatment of the sperm head using one of the previously listed methods effectively disrupts the sperm plasma membrane and damages the nuclear DNA. These events appear to facilitate the association of exogenous DNA with nuclear structures in the sperm and stabilize or protect the transgene after it is carried into the oocyte (Gandolfi, 2000).

C. Intracytoplasmic Sperm Injection

Because "killed" sperm are immotile, intracytoplasmic injection must be used to introduce the sperm into the oocyte cytoplasm. Early studies in *Xenopus* Brun (1974) and mice and rats (Thadani, 1980; Markert, 1983) demonstrated that direct injection of a spermatozoan into the cytoplasm of an oocyte could result in fertilization. Since these pioneering efforts intracytoplasmic sperm injection (ICSI) has yielded live human (Tesarik, 1996), rabbit (Hosoi et al., 1988), murine (Kimura and Yanagimachi, 1995), ovine (Catt et al., 1996), bovine (Hamano et al., 1999), equine (Grondahl et al., 2000) and, most recently, porcine (Martin, 2000) offspring.

In swine, the events associated with ICSI-induced fertilization have been well characterized (Kim et al., 1999; Kim and Shim, 2000). The proportion of ICSI porcine ova that develop into blastocysts *in vitro*, however, is low (<40%) and may be due to a low incidence of oocyte activation and/or failure of the injected sperm to generate a pronucleus (PN). In the bovine, these deficiencies were addressed by pretreating sperm with dithiothreitol (DTT) and activating oocytes with ionomycin and 6-dimethylaminopurine (Rho et al., 1998). These treatments significantly increased the rates of PN and blastocyst formation but failed to yield live calves. Exposure of porcine sperm to DTT prior to ICSI did not improve the cleavage rate or the proportion of embryos that developed into blastocysts (Cho et al., 2001).

Considerable interest in the use of ICSI as a potential means of modifying the genome of domestic species was generated following the birth of transgenic mice that expressed a green fluorescent protein (GFP) reporter gene (Perry et al., 1999). Since this report, ICSI has also been used to produce transgenic porcine embryos (Kim and Shim, 2000) and fetuses (Umeyama et al., 2012) that express a GFP reporter gene. In this study, porcine sperm were treated with Triton X-100 as described by Perry et al. (1999) and coincubated with pEGFP-N1 for 1 min. Seventeen of 65 (26%) ova injected with coincubated sperm developed beyond the four-cell stage and ten (58.8%) of these embryos expressed GFP. The expression of GFP in porcine embryos produced by ICSI suggests that sperm-mediated DNA transfer may also be possible in swine.

IV. Summary

The procedures outlined in this chapter represent state-of-the-art technology during the first two editions of this text but are now relegated to alternative technologies in light of a host of nuclear transfer/homologous recombination studies used in the production of transgenic swine and ruminant species (see Niemann et al., 2005). It is envisioned that these and other procedures will be further modified and enhanced as data from future transgenic farm animal research become available. Central to short-term enhancement will be (a) optimization of alternate DNA delivery systems, (b) refinement of existing embryo and cell culture systems, and (c) enhancing the efficiency of successful germ line modification of embryos; thereby reducing the number of ova required for transfer per recipient female and per experiment.

References

Allen, R.L., Wright Jr., R.W., 1984. *In vitro* development of porcine embryos in coculture with endometrial cell monolayers or culture supernatants. J. Anim. Sci. 59, 1657–1661.

Amet, T.M., Li, J., Markert, C.L., 1991. Estrus, ovulation, and fertilization in immature gilts treated with PG 600. J. Anim. Sci. 69 (Suppl. 1), 444. (abstract).

Archibong, A.E., Petters, R.M., Johnson, B.H., 1989. Development of porcine embryos from one- and two-cell stages to blastocysts in culture medium supplemented with porcine oviductal fluid. Biol. Reprod. 41, 1076–1083.

Bachiller, D., Schellander, K., Peli, J., Ruther, U., 1991. Liposome-mediated DNA uptake by sperm cells. Mol. Reprod. Dev. 30, 194–200.

Baker, R.D., Coggins, E.G., 1968. Control of ovulation rate and fertilization in prepuberal gilts. J. Anim. Sci. 27, 1607–1610.

Beckmann, L.S., Day, B.N., 1991. Culture of the one- and two-cell porcine embryo: effects of varied osmolarity in Whitten's and Kreb's Ringer bicarbonate media. Theriogenology 35, 184. (abstract).

Beckmann, L.S., Cantley, T.C., Rieke, A.R., Day, B.N., 1990. Development and viability of one- and two-cell porcine embryos cultured through the "four-cell block". Theriogenology 33, 193. (abstract).

Betteridge, K.J., Raeside, J.I., 1962. Observation of the ovary by peritoneal cannulation in pigs. Res. Vet. Sci. 3, 390–398.

Blum–Reckow, B., Holtz, W., 1991. Transfer of porcine embryos after 3 days of *in vitro* culture. J. Anim. Sci. 69, 3335–3342.

Boender, J., 1966. The development of AI in pigs in the Netherlands and the storage of boar semen. World Rev. Anim. Prod. 2, 29–44.

Brem, G., Brenig, B., Goodman, H.M., Selden, R.C., Graf, F., Kruff, B., et al., 1985. Production of transgenic mice, rabbits and pigs by microinjection into pronuclei. Zuchthygiene 20, 251–252.

Brem, G., Springmann, K., Meier, E., Kraußlich, H., Brenig, B., Muller, M., et al., 1989. Factors in the success of transgenic pig programs. In: Church, R.B. (Ed.), Transgenic Models in Medicine and Agriculture. Wiley-Liss, New York, NY, pp. 61–72.

Brinster, R.L., Chen, H.Y., Trumbauer, M.E., Yagle, M.K., Palmiter, R.D., 1985. Factors affecting the efficiency of introducing foreign DNA into mice by microinjecting DNA. Proc. Natl. Acad. Sci. USA 82, 4438–4442.

Brun, R.B., 1974. Studies on fertilization in *Xenopus laevis*. Biol. Reprod. 11, 513–518.

Burger, J.F., 1952. Sex physiology of pigs. Onderspoort J. Vet. Res. 25 (Suppl. 1), 22–131.

Butler, S.P., van Cott, K., Subramanian, A., Gwazdauskas, F.C., Velander, W.H., 1997. Current progress in the production of recombinant human fibrinogen in the milk of transgenic animals. Thromb. Haemost. 78, 537–542.

Caldwell, B.V., Moor, R.M., Wilmut, I., Polge, C., Rowson, L.E.A., 1969. The relationship between day of formation and functional lifespan of induced corpora lutea in the pig. J. Reprod. Fertil. 18, 107–113.

Campbell, R.G., Steele, N.C., Caperna, T.J., McMurtry, J.P., Solomon, M.B., Mitchell, A.D., 1988. Interrelationships between energy intake and endogenous porcine growth hormone administration on the performance, body composition and protein and energy metabolism of growing pigs weighing 25 to 55 kilograms live weight. J. Anim. Sci. 66, 1643–1655.

Casida, L.E., 1935. Prepuberal development of the pig ovary and its relation to stimulation with gonadotropic hormones. Anat. Rec. 61, 389–396.

Catt, S.L., Catt, J.W., Gomez, M.C., Maxwell, W.M.C., Evans, G., 1996. The birth of a male lamb derived from an *in vivo* matured oocyte fertilized by intra-cytoplasmic injection of a single presumptive male sperm. Vet. Rec. 139, 494–495.

Chan, A.W.S., 1999. Transgenic animals: current and alternative strategies. Cloning 1, 25–46.

Cho, S.K., Cho, S.Y., Park, M.R., Kim, J.H., 2001. Development and expression of porcine embryos injected with sperm treated with exogenous DNA: effect of DTT. Theriogenology 55, 520.

Clark, A.J., Simons, P., Wilmut, I., Lathe, R., 1987. Pharmaceuticals from transgenic livestock. Trends Biotechnol. 5, 20–24.

Davis, D.L., 1985. Culture and storage of pig embryos. J. Reprod. Fertil. (Suppl.) 33, 115–124.

Davis, D.L., Knight, J.W., Killian, D.B., Day, B.N., 1979. Control of estrus in gilts with a progestogen. J. Anim. Sci. 49, 1506–1509.

Day, B.N., 1979. Embryo transfer in swine. Theriogenology 11, 27–31.

Day, B.N., Oxenreider, S.L., Waite, A.B., Lasley, J.F., 1965. Use of gonadotropins to synchronize estrus in swine. J. Anim. Sci. 24, 1075–1079.

Day, B.N., Longnecker, D.E., Jaffey, S.C., Gibson, E.W., Lasley, J.F., 1967. Fertility in swine following superovulation. J. Anim. Sci. 21, 697–699.

Denman, J., Hayes, M., O'Day, C., Edmunds, T., Bartlett, C., Hirani, S., et al., 1991. Transgenic expression of a variant of human tissue-type plasminogen activator in goat milk: purification and characterization of the recombinant enzyme. Biotechnology (NY) 6, 839–843.

Didion, B.A., Martin, M.J., Markert, C.L., 1990. Characterization of fertilization and early embryonic development of naturally ovulated pig ova. Theriogenology 33, 284. (abstract).

Diehl, J.R., Day, B.N., 1974. Effect of prostaglandin $F_{2\alpha}$ on luteal function in swine. J. Anim. Sci. 39, 392–396.

Diehl, J.R., Day, B.N., Stevermer, E.J., Pursel, V., Holden, K. (1979). Artificial insemination in swine. Iowa State University Pork Industry Handbook, Reproduction Fact Sheet #64.

Diehl, J.R., Danion, J.R., Thompson, L.H. (1990). Managing sows and gilts for efficient reproduction. Iowa State University Pork Industry Handbook, Reproduction Fact Sheet #8.

Dobrinsky, J.R., Johnson, L.A., Rath, D., 1996. Development of a culture medium (BECM-3) for porcine embryos: effects of bovine serum albumin and fetal bovine serum on embryo development. Biol. Reprod. 55, 1069–1074.

Dziuk, P.J., Baker, R.D., 1972. Induction and control of ovulation in swine. J. Anim. Sci. 21, 697–699.

Dziuk, P.J., Gehlbach, G.D., 1966. Induction of ovulation and fertilization in the immature gilt. J. Anim. Sci. 25, 410–413.

Dziuk, P.J., Polge, C., Rowson, L.E.A., 1964. Intra-uterine migration and mixing of embryos in swine following egg transfer. J. Anim. Sci. 23, 37–42.

Ebert, K.M., Low, M.J., Overstrom, E.W., Buonomo, F.C., Baile, C.A., Roberts, T.M., et al., 1988. A Moloney MLV-rat somatotropin fusion gene produces biologically active somatotropin in a transgenic pig. Mol. Endocrinol. 2, 227–283.

Ebert, K.M., Smith, T.E., Buonomo, F.C., Overstrom, E.W., Low, M.J., 1990. Porcine growth hormone gene expression from viral promoters in transgenic swine. Anim. Biotechnol. 1, 145–159.

Ebert, K.M., Selgrath, J.P., DiTullio, P., Denman, J., Smith, T.E., Memon, M.A., et al., 1991. Transgenic production of a variant of human tissue–type plasminogen activator in goat milk: generation of transgenic goats and analysis of expression. Biotechnology (NY) 6, 835–838.

Estienne, C.E., Rampacek, G.B., Kraeling, R.R., Estienne, M.J., Barb, C.R., 1988. Luteinizing hormone receptor number and affinity in corpora lutea from prepuberal gilts induced to ovulate and spontaneous corpora lutea of mature gilts. J. Anim. Sci. 66, 917–922.

Evock, C.M., Etherton, T.D., Chang, C.S., Ivy, R.E., 1988. Pituitary porcine growth hormone (pGH) and a recombinant pGH analog stimulate pig growth performance in a similar manner. J. Anim. Sci. 66, 1928–1941.

Flowers, W., 1999. Dose confirmation study in sexually mature gilts orally administered altrenogest (Regu-Mate solution 0.22%) to suppress and synchronize estrus. Summary and Statistical Analyses Report. Sponsor Study Number: HRV 97-0014. June 29, 1999. Intervet/Schering-Plough Animal Health, Millsboro, DE.

Fodor, W.L., Williams, B.L., Matis, L.A., Madri, J.A., Rollins, S.A., Knight, J.W., et al., 1994. Expression of a functional human complement inhibitor in a transgenic pig as a model for the prevention of xenogeneic hyperacute organ rejection. Proc. Natl. Acad. Sci. USA 91, 11153–11157.

Foote, R.H., 1980. Artificial insemination. In: Hafez, E.S.E. (Ed.), Reproduction in Farm Animals. Lea and Febiger, Philadelphia, PA, p. 525.

French, A.J., Zviedrans, P., Ashman, R.J., Heap, P.A., Seamark, R.F., 1991. Comparison of prepubertal and postpubertal young sows as a source of one-cell embryos for microinjection. Theriogenology 35, 202. (abstract).

Gandolfi, F., 2000. Sperm-mediated transgensis. Theriogenology 53, 127–137.

Gandolfi, F., Lavitrano, M., Camaioni, A., Spadafora, C., Siracusa, G., Lauria, A., 1989. The use of sperm-mediated gene transfer for the generation of transgenic pigs. J. Reprod. Fertil. 4, 10. (abstract).

Gerfen, R.W., Wheeler, M.B., 1995. Isolation of embryonic cell lines from porcine blastocysts. Anim. Biotechnol. 6, 1–14.

Golovan, S.P., Hayes, M.A., Phillips, J.P., Forsberg, C.W., 2001. Transgenic mouse models demonstrating reduced pollution potential in monogastric animals. Nat. Biotechnol. 19, 429–433.

Gordon, K., Vitale, J., Roberts, B., Monastersky, G., DiTullio, P., Moore, G., 1990. Expression of foreign genes in the lactating mammary gland of transgenic animals. In: Church, R.B. (Ed.), Transgenic Models in Medicine and Agriculture. Wiley-Liss, New York, NY, pp. 55–59.

Grondahl, G., Hansen, T.H., Hossaini, A., Heinze, I., Greve, T., Hyttel, P., 2000. Intracytoplasmic sperm injection of in vitro matured equine oocytes. Biol. Reprod. 57, 1495–1501.

Guthrie, H.D., 1975. Estrus synchronization and fertility in gilts treated with estradiol benzoate and prostaglandin $F_{2\alpha}$. Theriogenology 4, 69–78.

Guthrie, H.D., 1979. Fertility and estrous cycle control using gonadotropin and prostaglandin $F_{2\alpha}$ treatment of sows. J. Anim. Sci. 49, 158–162.

Guthrie, H.D., Polge, C., 1967. Control of oestrus and fertility in gilts treated with accessory corpora lutea by prostaglandin analogues, ICI 29,939 and ICI 80,996. J. Reprod. Fertil. 48, 427–430.

Guthrie, H.D., Polge, C., 1976. Luteal function and oestrus in gilts treated with a synthetic analogue of prostaglandin $F_{2\alpha}$ (ICI–79,939) at various times during the oestrus cycle. J. Reprod. Fertil. 48, 423–425.

Hamano, K., Li, X., Qian, X.-Q., Funauchi, K., Furudate, M., Minato, Y., 1999. Gender reselection in cattle with intracytoplasmically injected, flow cytometrically sorted heads. Biol. Reprod. 60, 1194–1197.

Hammer, R.E., Pursel, V.G., Rexroad Jr., C.E., Wall, R.J., Bolt, D.J., Ebert, K.M., et al., 1985. Production of transgenic rabbits, sheep and pigs by microinjection. Nature 315, 680–683.

Hammer, R.E., Pursel, V.G., Rexroad Jr., C.E., Wall, R.J., Bolt, D.J., Palmiter, R.D., et al., 1986. Genetic engineering of mammalian embryos. J. Anim. Sci. 63, 269–278.

Hogan, B., Beddington, R., Costantini, F., Lacy, E., 1994. Manipulating the Mouse Embryo. Cold Springs Harbor Laboratory, New York, NY.

Holtz, W., Schlieper, B., 1991. Unsatisfactory results with the transfer of embryos from gilts superovulated with PMSG and hCG. Theriogenology 35, 1237–1249.

Hosoi, Y., Miyake, M., Utsumi, K., and Iritani, A. (1988). Development of rabbit oocytes after microinjection of a spermatozoon. In: Proceedings of the 11th International Congress on Animal Reproduction and Artificial Insemination, Dublin, Ireland. 3, pp. 3331–3333.

Hunter, R.H.F., 1964. Superovulation and fertility in the pig. Anim. Prod. 6, 189–194.

Hunter, R.H.F., 1966. The effect of superovulation on fertilization and embryonic survival in the pig. Anim. Prod. 8, 457–465.

Kim, N.-H., Shim, H., 2000. Intracytoplasmic sperm injection in pigs. ETS Newsl. 18, 10–13.

Kim, N.-H., Jun, S., Do, J., Uhm, S., Lee, H., Chung, K., 1999. Intracytoplasmic injection of porcine, bovine, mouse, or human spermatozoa into porcine oocytes. Mol. Reprod. Dev. 53, 84–91.

Kimura, Y., Yanagimachi, R., 1995. Intracytoplasmic sperm injection in the mouse. Biol. Reprod. 52, 709–720.

Knight, J.W., Davis, D.L., Day, B.N., 1976. Estrus synchronization in gilts with a progestogen. J. Anim. Sci. 42, 1358. (abstract).

Kraeling, R.R., Rampacek, G.B., 1977. Synchronization of estrus and ovulation in gilts with estradiol and prostaglandin $F_{2\alpha}$. Theriogenology 8, 103.

Krisher, R.L., Petters, R.M., Johnson, B.H., Bavister, B.D., Archibong, T.E., 1989. Development of porcine embryos from the one-cell stage to blastocyst in mouse oviducts maintained in organ culture. J. Exp. Zool. 249, 235–239.

Lavitrano, M., Camaioni, A., Fazio, V.M., Dolci, A., Farace, M.G., Spadafora, C., 1989. Sperm cells as vectors for introducing foreign DNA into eggs: genetic transformation of mice. Cell 57, 717–723.

Lavitrano, M., French, D., Varzi, V., Pucci, L., Bacci, M.L., Di, S.C., et al., 1997. Sperm-mediated gene transfer: production of pigs transgenic for a human regulator of complement activation. Transpl. Proc. 29, 3508–3509.

Lavitrano, M., Stoppacciaro, A., Bacci, M.L., Forni, M., Fioretti, D., Pucci, L., et al., 1999. Human decay accelerating factor transgenic pigs for xenotransplantation obtained by sperm-mediated gene transfer. Transpl. Proc. 31, 972–974.

Lo, D., Pursel, V., Linton, P.J., Sandgren, E., Behringer, R., Rexroad, E., et al., 1991. Expression of mouse IgA by transgenic mice, pigs and sheep. Eur. J. Immunol. 21, 1001–1006.

Long, C.R., Dobrinsky, J.R., Johnson, L.A., 1999. In vitro production of pig embryos: comparisons of culture media and boars. Theriogenology 51, 1375–1390.

Machaty, Z., Day, B.N., Prather, R.L., 1998. Development of early porcine embryos in vitro and in vivo. Biol. Reprod. 59, 451–455.

Maioni, B., Lavitrano, M., Spadafora, C., Kiessling, A.A., 1998. Sperm mediated gene transfer in mice. Mol. Reprod. Dev. 50, 406–409.

Markert, C.L., 1983. Fertilization of mammalian eggs by sperm injection. J. Exp. Zool. 228, 195–201.

Martin, M.J., 2000. Development of in vivo-matured porcine oocytes following intracytoplasmic sperm injection. Biol. Reprod. 63, 109–112.

Martin, M.J., Didion, B.A., Markert, C.L., 1989. Effect of gonadotropin administration on estrus synchronization and ovulation rate following induced abortion in swine. Theriogenology 32, 929–937.

Martin, M.J., Didion, B.A., Markert, C.L., 1990. Characterization of fertilization and early embryonic development of naturally ovulated pig ova. Theriogenology 33, 284. (abstract).

Martin, M.J., Adams, C., Cottrill, F., Houtz, J., Keirns, J., Thomas, D., et al. (2000). The effect of size and molecular concentration of DNA on the production of transgenic swine by pronuclear microinjection. In: Third International Conference on Transgenic Animals, Beijing, China, p. 26, (abstract).

Miller, K.F., Pursel, V.G., Hammer, R.E., Pinkert, C.A., Palmiter, R.D., Brinster, R.L., 1989. Expression of human or bovine growth hormone gene with a mouse metallothionein-1 promoter in transgenic swine alters the secretion of porcine growth hormone and insulin-like growth factor-I. J. Endocrinol. 120, 481—488.

Neill, J.D., Day, B.N., 1964. Relationship of developmental stage to regression of the corpus luteum in swine. Endocrinology 74, 355—360.

Niemann, H., Illera, M.J., Dziuk, P.J., 1983. Developmental capacity size and number of nuclei in pig embryos cultured in vitro. Anim. Reprod. Sci. 5, 311—321.

Niemann, H., Kues, W.A., Carnwath, J.W., 2005. Transgenic farm animals: present and future. Rev. Sci. Tech. 24, 285—298.

Nottle, M.B., Verma, P.J., Du, Z.T., Grupen, C.G., McIlfatrick, S.M., Ashman, R.J., et al., 1999. Production of pigs expressing multiple transgenes for use in xenotransplantation studies. Theriogenology 51, 422. (abstract).

Nottle, M.B., Haskard, K.A., Verma, P.J., Du, Z.T., Grupen, C.G., McIlfatrick, S.M., et al., 2001. Effect of DNA concentration on transgenesis rates in mice and pigs. Transgenic Res. 10, 523—531.

O'Reilly, P.J., McCormack, R., O'Mahoney, K., Murphy, C., 1979. Estrus synchronization and fertility in gilts using a synthetic progestagen (allyl trenbolone) and inseminated with fresh stored or frozen semen. Theriogenology 12, 131—137.

Paleyanda, R.K., Velander, W.H., Lee, T.K., Scandella, D.H., Gwazdauskas, F.C., Knight, J.W., et al., 1997. Transgenic pigs produce functional human factor VIII in milk. Nat. Biotechnol. 15, 971—975.

Palmiter, R.D., Brinster, R.L., Hammer, R.E., Trumbauer, M.E., Rosenfeld, M.G., Birnberg, N.C., et al., 1982. Dramatic growth of mice that develop from eggs microinjected with metallothionein-growth hormone fusion genes. Nature 300, 611—615.

Papaioannou, V.E., Ebert, K.M., 1988. The preimplantation pig embryo: cell number and allocation to trophectoderm and inner cell mass of the blastocyst in vivo and in vitro. Development 102, 793—803.

Perry, A.C.F., Wakayama, T., Kishikawa, H., Kasai, T., Okabe, M., Toyoda, Y., et al., 1999. Mammalian transgenesis by intracytoplasmic sperm injection. Science 284, 1180—1183.

Phelps, C.J., Ball, S.F., Vaught, T.D., Vance, A.M., Mendicino, M., Monahan, J.A., et al., 2009. Production and characterization of transgenic pigs expressing porcine CTLA4-Ig. Xenotransplantation 16, 477—485.

Piedrahita, J.A., 2000. Targeted modification of the domestic animal genome. Theriogenology 53, 105—116.

Piedrahita, J.A., Dunne, P., Lee, C.-K., Rucker, E., Vazquez, J.C., 1999. Use of embryonic and somatic cells for production of transgenic domestic animals. Cloning 1, 73—87.

Pinkert, C.A., 1987. Gene transfer and the production of transgenic livestock. Proc. U.S. Anim. Health Assn. 91 (129), 41.

Pinkert, C.A. 1990. Transgenic livestock development in the year 2000. In: Proceedings of the Seventh FAVA Congress, pp. 20—30.

Pinkert, C.A., 1991. Transgenic animals as models for metabolic and growth research. J. Anim. Sci. 69 (Suppl. 3), 49—55.

Pinkert, C.A., 1994. Transgenic swine models for xenotransplantation. Xeno 2, 10—15.

Pinkert, C.A., 2000. Genetic engineering of farm mammals. In: Hafez, E.S.E., Hafez, B. (Eds.), Reproduction in Farm Animals, 7th ed. Williams & Wilkins, Baltimore, pp. 318—330. Chapter 21.

Pinkert, C.A., Murray, K.A., 1993. Superovulation and egg transfer in Yucatan miniature swine. Anim. Reprod. Sci. 31, 155—163.

Pinkert, C.A., Kooyman, D.L., Baumgartner, A., Keisler, D.H., 1989a. *In vitro* development of zygotes from superovulated prepuberal and mature gilts. J. Reprod. Fertil. 87, 63–66.

Pinkert, C.A., Manz, J., Linton, P.J., Klinman, N.R., Storb, U., 1989b. Elevated PC responsive B cells and anti-PC antibody production in transgenic mice harboring anti-PC immunoglobulin genes. Vet. Immunol. Immunopath. 23, 321–332.

Pinkert, C.A., Johnson, L.W., Irwin, M.H., Wong, S.W., Baetge, E.E., Wolfe, D.F., et al., 2001. Optimization of superovulation and fertilization protocols in the production of transgenic swine. Adv. Reprod. 5, 45–53.

Pinkert, C.A., Dyer, T.J., Kooyman, D.L., Kiehm, D.J., 1990. Characterization of transgenic livestock production. Dom. Anim. Endocrinol. 7, 1–18.

Polge, C., 1982. Embryo transplantation and preservation. In: Cole, D.J.A., Foxcroft, G.R. (Eds.), Control of Pig Reproduction. Butterworth, London, pp. 283–285.

Polge, E.J.C., Barton, S.C., Surani, M.H.A., Miller, J.R., Wagner, T., Elsome, K., et al., 1989. Induced expression of a bovine growth hormone construct in transgenic pigs. In: Heap, R.B., Prosser, C.G., Lamming, G.E. (Eds.), Biotechnology of Growth Regulation. Butterworth, London, pp. 189–199.

Pope, C.E., Day, B.N., 1977. Transfer of preimplantation pig embryos following *in vitro* culture for 24 or 48 h. J. Anim. Sci. 44, 1036–1040.

Pope, C.E., Christenson, R.K., Zimmerman–Pope, V.A., Day, B.N., 1972. Effect of number of embryos on embryonic survival in recipient gilts. J. Anim. Sci. 35, 805–808.

Pope, W.F., Wilde, M.H., Xie, S., 1989. Effect of electrocautery of nonovulated day 1 follicles on subsequent morphological variation among d 11 porcine embryos. Biol. Reprod. 39, 882–887.

Pursel, V.G., Wall, R.J., 1991. Use of donor gilts as recipients of microinjected ova. J. Anim. Sci. 67 (Suppl. 1), 375. (abstract).

Pursel, V.G., Elliot, D.O., Newman, C.W., Staigmiller, R.B., 1981. Synchronization of estrus in gilts with allyl trenbolone: fecundity after natural service and insemination with frozen semen. J. Anim. Sci. 52, 130–133.

Pursel, V.G., Miller, K.F., Pinkert, C.A., Palmiter, R.D., Brinster, R.L., 1987. Development of 1-cell and 2-cell pig ova after microinjection of genes. J. Anim. Sci. 65 (Suppl. 1), 402. (abstract).

Pursel, V.G., Pinkert, C.A., Miller, K.F., Bolt, D.J., Campbell, R.G., Palmiter, R.D., et al., 1989. Genetic engineering of livestock. Science 244, 1281–1288.

Pursel, V.G., Hammer, R.G., Bolt, D.J., Palmiter, R.D., Brinster, R.L., 1990a. Integration, expression and germ-line transmission of growth-related genes in pigs. J. Reprod. Fertil. (Suppl.) 41, 77–87.

Pursel, V.G., Bolt, D.J., Miller, K.F., Pinkert, C.A., Hammer, R.E., Palmiter, R.D., et al., 1990b. Expression and performance in transgenic pigs. J. Reprod. Fertil. (Suppl.) 40, 235–245.

Pursel, V.G., Michell, A.D., Wall, R.J., Solomon, M.B., Coleman, M.E., Schwartz, R.J. (2000). Transgenic research to enhance growth and lean carcass composition in swine. In: Toutant, J.P., Balazs, E. (Eds.), Molecular Farming. Proceedings of OECD Workshop, Montpellier, France.

Rampacek, G.B., Kraeling, R.R., Kiser, T.E., Barb, C.R., Benyshek, L.L., 1979. Prostaglandin F concentrations in uterovarian vein plasma of prepuberal and mature gilts. Prostaglandins 18, 247–255.

Ramsoondar, J., Vaught, T., Ball, S., Mendicino, M., Monahan, J., Jobst, P., et al., 2009. Production of transgenic pigs that express porcine endogenous retroviral small interfering RNAs. Xenotransplantation 16, 164–180.

Redmer, D.A., Day, B.N., 1981. Ovarian activity and hormonal patterns in gilts fed allyl trenbolone. J. Anim. Sci. 53, 1089–1094.

Reed, M.L., Illera, M.J., Petters, R.M., 1992. In vitro culture of pig embryos. Theriogenology 37, 95–109.

Rexroad Jr., C.E., Pursel, V.G., Hammer, R.E., Bolt, D.J., Miller, K.F., Mayo, K.E., et al., 1988. Gene insertion: role and limitations of technique in farm animals as a key to growth. In: Steffens, G.L., Rumsey, T.S. (Eds.), Biomechanisms Regulating Growth and Development, vol. 12. Kluwer, Dordrecht, The Netherlands, pp. 87–97.

Rexroad Jr., C.E., Hammer, R.E., Bolt, D.J., Mayo, K.E., Frohman, L.A., Palmiter, R.D., et al., 1989. Production of transgenic sheep with growth regulating genes. Mol. Reprod. Dev. 1, 164–169.

Rexroad Jr., C.E., Hammer, R.E., Behringer, R.R., Palmiter, R.D., Brinster, R.L., 1990. Insertion, expression and physiology of growth-regulating genes in ruminants. J. Reprod. Fertil. (Suppl.) 41, 119–124.

Rho, G.J., Kawarsky, S., Johnson, W., Kochhar, K., Betteridge, K., 1998. Sperm and oocyte treatments to improve fertilization of male and female pronuclei and subsequent development following intracytoplasmic sperm injection into bovine oocytes. Biol. Reprod. 59, 918–924.

Schellander, K., Peli, J., Schmoll, F., Brem, G., 1995. Artificial insemination in cattle with DNA-treated sperm. Anim. Biotechnol. 6, 41–50.

Segal, D.H., Baker, R.D., 1973. Maintenance of corpora lutea in prepuberal gilts. J. Anim. Sci. 37, 762–767.

Shamay, A., Sabina, S., Pursel, V.G., McKnight, R.A., Alexander, L., Beattie, C., et al., 1991. Production of the mouse whey acidic protein in transgenic pigs during lactation. J. Anim. Sci. 69, 4552–4562.

Sharma, A., Okabe, J., Birch, P., McClellan, S., Martin, M., Platt, J., et al., 1996. Reduction in the level of Gal(α1,3) Gal in transgenic mice and pigs by the expression of an α(1,2) fucosyltransferase. Proc. Natl. Acad. Sci. USA 93, 7190–7195.

Shim, H., Gutierrez-Adan, A., Chen, L.-R., BonDurant, R.H., Behboodi, E., Anderson, G.B., 1997. Isolation of pluripotent stem cells from cultured porcine primordial germ cells. Biol. Reprod. 57, 1089–1095.

Simons, J.P., McClenaghan, M., Clark, J.A., 1987. Alteration of the quality of milk by expression of sheep β-lactoglobulin in transgenic mice. Nature 328, 530–532.

Simons, J.P., Wilmut, I., Clark, A.J., Archibald, A.L., Bishop, J.O., Lathe, R., 1988. Gene transfer into sheep. Biotechnology (NY) 6, 179–183.

Sommer, J.R., Estrada, J.L., Collins, E.B., Bedell, M., Alexander, C.A., Yang, Z., et al., 2011. Production of ELOV4 transgenic pigs: a large animal model for Stargardt-like macular degeneration. Br. J. Ophthalmol. 95, 11749–11754.

Spadofora, C., 1998. Sperm cells and foreign DNA: a controversial relation. Bioessays 20, 955–964.

Stewart, C.L., Gadi, I., Blatt, H., 1994. Stem cells from primordial germ cells can reenter the germ line. Dev. Biol. 161, 626–628.

Swanson, M.E., Martin, M.J., O'Donnell, J.K., Hoover, K., Lago, W., Huntress, V., et al., 1992. Production of functional human hemoglobin in transgenic swine. Biotechnology (NY) 10, 557–559.

Tesarik, J., 1996. Fertilization of oocytes by injecting spermatozoa, spermatids and spermatocytes. Rev. Reprod. 1, 149–152.

Thadani, V.M., 1980. A study of hetero-specific sperm-egg interactions in the rat, mouse, and deer mouse using in vitro fertilization. J. Exp. Zool. 212, 435–453.

Town, S.C., Almeida, F.R.C.L., Novak, S., Foxcroft, G.R., 1999. Optimizing timing of artificial insemination in gilts. J. Reprod. Fertil. 23, 40. (abstract).

Tucker, A.W., White, D.J.G. (1998). Transgenic pigs as organ donors for man. In: Proceedings of the 15th IPVS Congress, Birmingham, England, July 5−9, pp. 175−180.

Umeyama, K., Saito, H., Kurome, M., Matsunari, II., Watanabe, M., Nakauchi, H., et al., 2012. Characterization of the ICSI-mediated gene transfer method in the production of transgenic pigs. Mol. Reprod. Dev. 79, 218−228.

Van Brunt, J., 1988. Molecular farming: transgenic animals as bioreactors. Biotechnology (NY) 6, 1149−1154.

Van Cott, K., Butler, S.P., Russell, C.G., Subramanian, A., Lubon, H., Gwazdauskas, F.C., et al., 1999. Transgenic pigs as bioreactors: a comparison of gamma-carboxylation of glutamic acid in recombinant human protein C and factor IX by the mammary gland. Genet. Anal. 15, 155−160.

Van Cott, K.E., Lubon, H., Gwazdauskas, F.C., Knight, J., Drohan, W.N., Velander, W.H., 2001. Recombinant human protein C expression in the milk of transgenic pigs and the effect on endogenous milk immunoglobulin and transferrin levels. Transgenic Res. 10, 43−51.

Velander,W.H., Sickey, T.K., Butler, S.P., Cadiz, A., Joo, L., Limonta, M., et al. (2001). A potential treatment for type B Hemophilia in developing countries using the milk of transgenic pigs. Hematologia, Havana, Cuba, 14−18.

Vincent, C.K., Robison, O.W., Ulberg, L.C., 1964. A technique for reciprocal embryo transfer in swine. J. Anim. Sci. 23, 1084−1088.

Vize, P.D., Michalska, A.E., Ashman, R., Lloyd, B., Stone, B.A., Quinn, P., et al., 1988. Introduction of a porcine growth hormone fusion gene into transgenic pigs promotes growth. J. Cell Sci. 90, 295−300.

Wall, R.J., Pursel, V.G., Hammer, R.E., Brinster, R.L., 1985. Development of porcine ova that were centrifuged to permit visualization of pronuclei and nuclei. Biol. Reprod. 32, 645−651.

Wall, R.J., Pursel, V.G., Shamay, A., McKnight, R.A., Pettius, C.W., Hennighausen, L., 1991. High-level synthesis of a heterologous protein in the mammary glands of transgenic swine. Proc. Natl. Acad. Sci. USA 88, 1696−1700.

Wall, 1997. A new lease on life for transgenic livestock. Nat. Biotechnol. 15, 416−417.

Walton, J.R., Murray, J.D., Marshall, T.T., Nancarrow, C.D., 1987. Zygote viability in gene transfer experiments. Biol. Reprod. 37, 957−967.

Ward, K.A., Nancarrow, C.D., Murray, J.D., Wynn, P.C., Speck, P., Hales, J.R.S., 1989. The physiological consequences of growth hormone fusion gene expression in transgenic sheep. J. Cell. Biochem. 13B, 164. (abstract).

Webel, S.K., 1976. Estrus control in swine with a progestogen. J. Anim. Sci. 42, 1358.

Webel, S.K., Peters, J.B., Anderson, L.L., 1970. Synchronous and asynchronous transfer of embryos in the pig. J. Anim. Sci. 30, 565−568.

Wei, Q., Fan, J., Chen, D., 1993. Effect of number of transferred microinjected embryos on pregnancy rate and litter size of pigs. Theriogenology 39, 338. (abstract).

Weidle, U.H., Lenz, H., Brem, G., 1991. Genes encoding a mouse monoclonal antibody expressed in transgenic mice, rabbits and pigs. Gene 98, 185−191.

Wieghart, M., Hoover, J.L., McGrane, M.M., Hanson, R.W., Rottman, F.M., Holtzman, S.H., et al., 1990. Production of transgenic swine harboring a rat phosphoenolpyruvate carboxykinase-bovine growth hormone fusion gene. J. Reprod. Fertil. (Suppl.) 41, 89−96.

Williams, B.L., Sparks, A.E.T., Canseco, R.S., Knight, J.W., Johnson, J.L., Velander, W.H., et al., 1992. *In vitro* development of zygotes from prepubertal gilts after microinjection of DNA. J. Anim. Sci. 70, 2207–2211.

Wright, C., Carver, A., Cottom, D., Reeves, D., Scott, A., Simons, P., et al., 1991. High level expression of active human alpha-1-antitrypsin in the milk of transgenic sheep. Biotechnology (NY) 9, 830–834.

Wright Jr., R.W., 1977. Successful culture *in vitro* of swine embryos to the blastocyst stage. J. Anim. Sci. 44, 854–858.

Wright, R.W., Bondioli, K.R., 1981. Aspects of *in vitro* fertilization and embryo culture in domestic animals. J. Anim. Sci. 53, 702–729.

Xie, S., Broermann, D.M., Nephew, K.P., Bishop, M.D., Pope, W.F., 1990a. Relationships between oocyte maturation and fertilization on zygote diversity in swine. J. Anim. Sci. 68, 2027–2033.

Xie, S., Broermann, D.M., Nephew, K.P., Ottobre, J.S., Pope, W.F., 1990b. Changes in follicular endocrinology during maturation of porcine oocytes. Anim. Reprod. Sci. 7, 75–82.

Zavy, M.T., Buchanan, D., Geisert, R.D., 1984. The combination of estrogen-induced prolonged luteal function (PLF) and $PGF_{2\alpha}$ administration as a means of estrus synchronization in swine. J. Anim. Sci. 59 (Suppl. 1), 327. (abstract).

Section Four

Molecular Biology, Analyses and Enabling Technologies

16 Analysis of Phenotype

Cory F. Brayton[1], Colin McKerlie[2] and Steve Brown[3]

[1]Johns Hopkins University, Baltimore, MD, [2]The Hospital for Sick Children and University of Toronto, Toronto, ON, Canada, [3]MRC Mammalian Genetics Unit, Harwell, Oxford, UK

I.   Introduction

The term *phenotyping* is used frequently in functional genomics research and literature to refer to assessing phenotypes (features, traits, abnormalities) in genetically engineered animals. In this context it is done to provide data and insight about the functions of genes, or gene products in a living system (Brown and Moore, 2012a,b). Tests used for phenotyping are used to diagnose disease, or to assess responses to therapeutic interventions in clinical settings, or are used to assess responses to experimental interventions in other preclinical research settings. Their use is not restricted to phenotyping of genetically engineered mice (GEM). High-throughput-testing pipelines, developed by the International Mouse Phenotyping Consortium (IMPC), aim to be efficient and broadly informative to diverse research areas. This chapter discusses phenotyping *sensu lato*, but highlights tests and resources included in the IMPC's 10-year plan to phenotype every protein-coding gene, as well as developing practical pathology to support phenotyping and other translational research. Depending on the aims and research settings, phenotyping tests and strategies vary widely, but the approaches can be characterized as primarily: (1) hypothesis-driven, (2) purpose-driven, or (3) hypothesis-generating, or some combination of these.

A.   Hypothesis-Driven (or Hypothesis-Testing) Phenotyping

Hypothesis-driven (or hypothesis-testing) phenotyping aims to test specific hypotheses. Simple and specific, robust tests are favored by editors, reviewers, and granting agencies. This approach continues to dominate academic research, where GEM phenotyping correlates with the aims, hypotheses, and resources of the investigator. Thus phenotyping strategies, methodology, terminology, and the "nature and nurture" of the animals tested (their genetic backgrounds, diet, microbial status, and other housing and test conditions), vary substantially by laboratory. The advantage of this approach is that it should answer an important question without wasting resources.

Transgenic Animal Technology. DOI: http://dx.doi.org/10.1016/B978-0-12-410490-7.00016-5

Disadvantages include difficulties in comparing or replicating studies from different laboratories (Crabbe et al., 2006; Wahlsten et al., 2006). Studies that cannot be replicated are justifiably criticized and contribute to concerns about failures of animal models to translate to human conditions or therapies (Kilkenny et al., 2009, 2010; ILAR-NRC, 2011). Also, the limited breadth of the phenotyping, while sufficient to test the hypothesis, may sacrifice opportunities to identify, characterize, and associate additional phenotype traits resulting from unrecognized gene pleiotropy.

B. Purpose-Driven Phenotyping

Purpose-driven phenotyping generates data sets relevant to a specific area of investigation. In pharmaceutical settings such data sets are collected for the purpose of determining or defining the safety, toxicity, and carcinogenicity of a compound or device. Final testing is conducted according to US Food and Drug Administration (FDA) Good Laboratory Practices (GLP) or other international standards, and is subject to extensive quality assurance (QA) and quality control (QC) measures. Protocols, procedures, terminology, and reporting tend to be comprehensive and standardized. Examples of these types of protocols, study design, and results are available on line from the National Toxicology Program (NTP) (http://ntp.niehs.nih.gov/). Best practices, diagnostic criteria, and terminology are reviewed and updated periodically (Crissman et al., 2004). Influences of diverse microbial, dietary, and other factors on study results have been discussed (Haseman et al., 1989; Ward et al., 1994; Hailey et al., 1998; Rao and Crockett, 2003; Martin et al., 2010), as has the utility of historical control data (Elmore and Peddada, 2009; Keenan et al., 2009a,b). Commercial producers of research animals also may collect data sets on growth, longevity, fertility, and production (phenotypes), sometimes clinical chemistry and pathology, as well as on QA and QC procedures such as microbial surveillance or genetic monitoring, for the purpose of characterizing (and selling) their products (mice), or other research animals. The information can provide useful benchmarks and husbandry suggestions for maintenance and breeding, or baseline data to guide experimental design.

C. Systematic (Hypothesis-Generating) Phenotyping

Systematic (hypothesis-generating) phenotyping is exemplified by the IMPC, from which broad based, unbiased, and standardized phenotyping is expected to provide insights across many areas of biology to expose new functions of well-characterized genes and offer insight to genes with little or no known function (Gailus-Durner et al., 2009; Abbott, 2010; Guan et al., 2010; Moore and IMPC and SteeringCommittee, 2010; Nature, 2010; Wurst and de Angelis, 2010; Brown and Moore, 2012a,b).

Beginning in 2002, the EUMORPHIA program developed standard operating procedures for comprehensive, high throughput primary phenotyping pipelines termed EMPReSS (European Mouse Phenotyping Resource of Standardized Screens), and a database for EMPReSS data called EuroPhenome (http://www.europhenome.org/). From these efforts, the EUropean MOuse DIsease Clinic (EUMODIC) consortium (http://www.eumodic.org/) was developed to conduct primary phenotyping of 500 mutant mouse lines produced from the C57BL/6N ES cell resource generated by the International Knockout Mouse Consortium (IKMC; http://knockoutmouse.org) (Brown et al., 2005; Hrabe de Angelis et al., 2006). The EUMODIC consortium of four sites with expertise in mouse genetics, functional genomics, and analysis conducted the pilot studies from which the IMPC phenotyping pipeline was developed and is now implemented at 13 phenotyping centers within the consortium (http://www.mousephenotype.org/).

IMPC pipeline and protocol information is available through IMPReSS (International Mouse Phenotyping Resource of Standardized Screens, formerly EMPReSS) https://www.mousephenotype.org/impress/. Data are generated and quality-reviewed by each center, uploaded to the IMPC Data Coordinating Centre as soon as it is available, and subjected to IMPC QC prior to presentation in the IMPC web portal http://www.mousephenotype.org/, as well as to the Mouse Genome Informatics (MGI) site http://www.informatics.jax.org/. MGI already includes large-scale data from the ethyl nitrosourea (ENU) mutagenesis phenotyping initiatives, and also accepts data derived from individual laboratories and the biomedical literature (Smith and Eppig, 2012).

The IKMC is nearing completion of its goal of generating (lac-z) reporter-tagged, knockout first, conditional-ready, targeted alleles in C57BL/6N embryonic stem (ES) cells for every protein-coding gene. As of April 2013, approximately 15,000 targeted ES cell lines are available, and complete genome coverage for approximately 20,000 protein-coding genes is expected within a few years. Production and characterization of mouse lines produced from this ES cell resource will be done by IMPC members. The IMPC aims to produce and phenotype all 20,000 mouse lines by 2021. Phase 1 is funded and in progress, expecting to produce and phenotype 5,000 mutant mouse lines by 2016. Phase 2 is expected to run from 2016 to 2021, to generate and phenotype the remaining 15,000 mutant mouse lines. Importantly, all of the resources (ES cells, mice, frozen germplasm) and data (production and phenotyping) are publically available to the research community.

D. Combined Approaches

Hypothesis-testing and hypothesis-generating approaches can be combined. In the United States, The Jackson Laboratory developed and continues to grow the Mouse Phenome Database (MPD) http://phenome.jax.org/. MPD includes data from hypothesis-driven testing, plus broader based hypothesis-generating screening of multiple inbred, hybrid, and outbred strains. Detailed characterization of inbred and GEM strains, including negative or normal findings, enhance their value as genetic research tools. MPD prioritizes phenotype data from 36 common and important mouse strains. Strong emphasis on standardized, well-described protocols provides a resource for experimental and statistical design, and facilitates replication and comparison with other studies. Many of the more than 100 contributing studies in MPD are published and peer reviewed (Bogue, 2003; Bult, 2012; Maddatu et al., 2012).

Tang et al. (2010) offers another example of a combined approach, where systematic (hypothesis-generating) phenotyping was used as a platform from which to select GEM lines for hypothesis-driven testing. The broad-based phenotyping screen identified potential defects in general metabolism, bone metabolism, cardiovascular, immune, or neural systems in the majority of 472 GEM lines selected for their engineered defects in secreted and transmembrane proteins. More than 50% of the mouse lines from this selected set of genes were annotated with two or more phenotypes, confirming expectations of substantial gene pleiotropy. The positive findings identified lines that warrant repeat tests (confirmation), follow-up validation using larger cohorts of mice, and further testing.

II. Methods for Robust, Relevant, and Reproducible Phenotyping

A. *Primary, Secondary, Tertiary Phenotyping*

Primary, secondary, and tertiary phenotyping tests are defined variably by different sites or sources. In general, primary tests are less specific, less invasive, and higher throughput; secondary and tertiary tests are more specialized, invasive, expensive, and lower throughput.

The IMPC is doing *primary* phenotyping on all mouse lines that are viable as adults. Some sites will work on heterozygous cohorts from lines that are embryo lethal. The primary pipeline aims to be sensitive, high throughput, and minimally invasive (until the terminal assessment). The objective is to attain the most *hits* (deviant results compared to controls), relevant to broad areas of interest, for reasonable cost and effort, be relatively rapid and humane, and generate compelling and biologically relevant data sets to the end-user community for secondary and tertiary follow-up testing. The tests selected for the primary pipelines (Table 16.1) reflect survey results from the United States and Europe, and prioritize the following broad research areas (Gailus-Durner et al., 2009; Moore and IMPC and SteeringCommittee, 2010; Wurst and de Angelis, 2010):

1. Neuroscience including sensory
2. Cardiovascular function and disease
3. Immunology
4. Metabolic disorders including obesity and diabetes
5. Musculoskeletal
6. Development and congenital disease.

Technology development and pilot studies to assess test modalities for important disease areas, such as cardiovascular disease or lung function, are underway, and additional tests relevant to these areas are expected to be incorporated in the IMPC pipeline soon.

Secondary phenotyping tends to be more complex or specialized than primary testing. Secondary phenotyping conducted at some of the IMPC sites is also

Table 16.1 IMPC Pipeline Tests

Mandatory *in vivo* Tests (9–15 Weeks of Age)	In Life Tests Under Development	Mandatory Terminal Tests (16 Weeks of Age)	Terminal Tests Under Consideration	Optional Tests
1. Body weight (weekly)	1. Echocardiography	1. Hematology	1. Gene expression (using *lacZ* reporter genes)	In life:
2. Dysmorphology and modified SHIRPA	2. Electrocardiography	2. Clinical blood chemistry	2. Blood insulin	Open Field/ activity monitoring
3. Grip strength	3. Plethysmography	3. Flow cytometry (spleen)		Terminal:
4. Acoustic startle, and pre pulse inhibition (PPI)	4. Acute pain (nociception)	4. Heart weight		Tissue embedding, banking, histopathology
5. Calorimetry		5. Gross pathology and tissue collection		
6. Intraperitoneal Glucose tolerance tests (IPGTT)				
7. Radiography (Faxitron®)				
8. Body composition (lean/fat mass by DXA[a])				
9. Auditory brain stem response (ABR)				
10. Eye morphology				

[a]Dual energy X-ray absorptiometry.
Source: Summarized from http://www.mousephenotype.org/impress, accessed May 1, 2013.

standardized, with protocols in IMPRESS, and sufficient throughput to test multiple mouse lines simultaneously or in sustained sequence. Secondary phenotyping can include challenge tests such as dietary challenges (responses to high fat and carbo-hydrate diets), microbial, allergen, or exercise challenges, or other stressors.

Tertiary phenotyping tends to be even more highly specialized, invasive, or expensive, and usually is reserved for testing hypotheses about the gene, muta-tion or mutant, based on positive results (hits) in primary or secondary tests. Examples include brain electrophysiology, or metabolic or cardiovascular testing that involves surgical manipulations, and aging studies (Gailus-Durner et al., 2009; Moore and IMPC and SteeringCommittee, 2010; Wurst and de Angelis, 2010). In the German Mouse Clinic (GMC), "GMC III" testing refers to systemic analysis of compounds and drugs (preclinical drug screening), aiming to improve and accel-erate understanding of molecular mechanisms on a general and more personalized (genotype) level, and to make these data available, as with all IMPC generated data (Fuchs et al., 2012).

B.   Phenotyping Pipelines and Protocols (Brown and Moore, 2012a,b)

The IMPC currently includes 16 large mouse production and phenotyping and/or bioinformatics sites (in 12 countries):

1. (AU) Australian Phenomics Network
2. (CA) Toronto Centre for Phenogenomics (TCP)
3. (CN) Model Animal Research Center (MARC), Nanjing University
4. (CZ) Czech Centre for Phenogemomics (CCP)
5. (DE) Helmholtz Zentrum Munich (GMC)
6. (FR) Institut Clinique de la Souris (ICS)
7. (IT) CNR Monterotondo (IMC)
8. (JA) RIKEN BioResource Center (JMC)
9. (KR) Korea Mouse Phenotype Consortium
10. (TW) National Laboratory Animal Center, National Applied Research Laboratories (NLAC NARLabs), Taiwan
11. (UK) European Bioinformatics Institute (EBI)
12. (UK) Medical Research Council (MRC) Harwell
13. (UK) Wellcome Trust Sanger Institute (WTSI)
14. (US +) BaSH (Baylor College of Medicine, Sanger, Harwell)
15. (US +) The Davis, Toronto (TCP), Charles River and CHORI Consortium (DTCC)
16. (US) The Jackson Laboratory

Each production and phenotyping site has capacity and expertise for large-scale production (or reanimation) of mice from ES cell lines, and for large-scale primary phenotyping of adult mice, and collects breeding (fertility and viability) data. The current IMPC pipeline tests seven homozygous mutant mice of each sex, starting from birth with viability analysis, body weight at 4 weeks, and a battery of *in vivo* tests in a defined test sequence starting at 9 weeks of age, concluding with a terminal panel of *ex vivo* tests at 16 weeks of age. The protocols, test schedule, and cohort size were validated with data generated in the EUMORPHIA and the

EUMODIC programs, and were further refined prior to IMPC implementation in September 2011. It is also expected that individual sites will continue to incorporate additional tests relevant to local research interests as long as those tests do not interfere with or invalidate the standardized IMPC pipeline.

Age-related disease conditions such as cancer, cardiovascular disease, dementia, and degenerative conditions are important targets of biomedical research and are not expected to be represented by phenotyping pipelines that terminate at 16 weeks of age. It is recognized that aging or longevity studies involving older GEM will be necessary to assess gene influences on cancers or other age-related phenotypes, but they exceed current IMPC scope and budgets. Such studies require substantial investment and expertise to optimize study design, conduct analysis, and reporting (Ladiges et al., 2009; Conover et al., 2010; Sundberg et al., 2011; Yuan et al., 2011; Brayton et al., 2012).

C. Bioinformatics: Data Management, Terminology

Underlying bioinformatics infrastructure is widely recognized to be critical to the utility of large-scale omics projects such as the IMPC (see also Chapter 24). Europhenome, MGI, and MPD offer insights for optimizing bioinformatics infrastructure for phenotype and genotype data, as well as into the use of these resources by the scientific community. IMPC data are generated, recorded, and locally curated in each production and phenotyping site using local laboratory information management systems, and then exported to the IMPC Data Coordination Centre (DCC). The DCC is responsible for validation, QC, deposition into the IMPC central data archive, and display and access to the research community. DCC developed the IMPReSS database to hold and update information on the standard phenotyping protocols and to interface with the mammalian phenotype (MP) ontology (Hancock and Mallon, 2008; Field et al., 2009; Grubb et al., 2009; Hancock et al., 2009; Abbott, 2010; Blake et al., 2010; Moore and IMPC and SteeringCommittee, 2010; Morgan et al., 2010; Nature, 2010; Brown and Moore, 2012b).

Terminology, or a common vocabulary, is a critical element of large-scale phenotype initiatives such as the IMPC, with data generators and end users distributed across the globe (Hancock and Mallon, 2008; Field et al., 2009; Hancock et al., 2009; Bult et al., 2010; Wilkinson et al., 2010). Inadequate or unclear reporting of biomedical research has attracted criticism (Kilkenny et al., 2009, 2010; ILAR-NRC, 2011), as has inadequate comparative mapping and translation of animal research data and phenotypes to human conditions (Hartung and Leist, 2008; Hartung, 2009). Large-scale data and project management, as well as accurate and useful reporting of robust, relevant, and reproducible research, require attention to terminology. Concerns about mouse and gene nomenclature have been discussed (Linder, 2003; Bult et al., 2010). International guidelines for mouse and gene nomenclature are accessible online at http://www.informatics.jax.org/mgihome/nomen/, which links conveniently to the Mouse Genome Informatics (MGI) database.

The MP ontology was developed largely to support GEM (also rat) phenotyping. MP is a structured vocabulary, used by IMPC and MGI for describing MPs, and for consistent annotation and effective mining of phenotype data. A structured hierarchy of broad and specific phenotype terms facilitates annotation of data from initial observations to complex test results. The MP ontology is subject to continuous assessment and revision by users and expert reviewers. The interface of a species-neutral, MP vocabulary with genotype annotations aims to facilitate phenotype and genotype comparisons, and to align mouse mutant phenotypes with human disease phenotypes, to improve and expedite identification of relevant mouse genetic models (Gkoutos et al., 2012; Mallon et al., 2012; Smedley et al., 2013). Each MP term (phenotype) has a unique name, unique accession identification number (MP id), synonyms, and a definition. As of July 1, 2013, MP has 27 top nodes or "parental" categories (e.g., cardiovascular phenotype, behavior/neurologic phenotype), with about 10,000 more specific "child" terms (e.g., ventricular septal defect, seizures), with about 25,000 annotations to more than 5,000 genes. Terms can be searched by entity such as anatomic site (using terms from the Mouse Anatomy (MA) ontology), with or without quality terms (e.g., normal/abnormal, presence, absence), or with MPATH (see below) process or diagnosis terms (e.g., inflammation, necrosis, carcinoma). The MA ontology also annotates the tissue/entity with gene expression results via the mouse anatomical dictionary browser at http://www.informatics.jax.org/searches/anatdict_form.shtml.

The MPATH (mouse pathology) ontology was developed, beginning in 1999, to annotate mouse pathology images maintained in the Pathbase mouse pathology database at http://www.pathbase.net/. MPATH process or diagnosis terms (e.g., inflammation, necrosis, carcinoma) are an additional annotation layer to the MP ontology. The MPATH ontology (intentionally) shares many terms with International Harmonization of Nomenclature and Diagnostic Criteria for Lesions in Rats and Mice (INHAND) (http://www.toxpath.org/nomen/), and terminology developed from the NCI Mouse Models of Human Cancer Consortium (MMHCC) http://www.nih.gov/science/models/mouse/resources/hcc.html consensus papers (Schofield et al., 2010, 2011).

INHAND terminology is widely used in toxicologic pathology, and is especially relevant to spontaneous and chemically induced lesions in contemporary toxicity, safety, and carcinogenicity studies (Renne et al., 2009; Thoolen et al., 2010; Creasy et al., 2012; Kaufmann et al., 2012; Frazier et al., 2012; Rudmann et al., 2012; Keenan and Goodman, 2013). MMHCC has developed GEM cancer resources, including consensus papers on diagnostic criteria and terminology for neoplasia of the hematopoietic system (Kogan et al., 2002; Morse et al., 2002), nervous system (Gutmann et al., 2003), breast (mammary gland) (Cardiff et al., 2000; Jeffrey et al., 2002), prostate (Shappell et al., 2004; Ittmann et al., 2013), lung (Nikitin et al., 2004), intestine (Boivin et al., 2003), and on precancers (Cardiff et al., 2006). In addition, MMHCC has developed a Cancer Models Database at http://cancermodels.nci.nih.gov/, and a Cancer Images Database: http://emice.nci. nih.gov/caimage (also http://spectrum.ucdavis.edu/Welcome.php) (Wehling, 2009; Wendler and Wehling, 2010). MMHCC terminology diverges from INHAND terminology when mechanisms, morphology, and behavior of genetically engineered lesions differ from spontaneous and chemically induced lesions (Cardiff et al., 2004). INHAND, MMHCC, and MPATH have some traction in the GEM pathology literature. These three systems (MPATH, INHAND, and MMHCC) were developed in different research contexts and different contexts of scientific communication, but they serve similar aims and share many terms. As long as the terminology used in reporting pathology phenotypes is referenced clearly and the findings are characterized adequately, translation between the systems should be feasible.

D. Pathology in Phenotyping

In medical and veterinary clinical practice, pathology is a primary test to characterize disease phenotypes, to diagnose or identify diverse causes and contributors, and to supplement the clinical tests with morphologically and biologically relevant data. The role of histopathology in the study of gene function in disease is well established. In phenotyping, pathology should play important roles in confirming and characterizing phenotypes, and validating GEM models, as well as in troubleshooting and crisis-driven ("my mice are dying!") diagnostic assessments (Brayton et al., 2001, 2012; Barthold, 2004; Cardiff et al., 2004; Barthold et al., 2007; Cardiff et al., 2008; Ince et al., 2008; Bronson, 2009; Schofield et al., 2011; Brayton and Treuting, 2012).

Pathology was identified as a critical need for phenotyping and translational research by the PRIME initiative (Schofield et al., 2009), and pathology and imaging were identified as important for phenotyping by surveys in the EU and United States. A complete necropsy with collection of a standardized panel of tissues is a mandatory test in the IMPC pipeline. However, further processing of tissue for histopathology testing is currently optional, and there are not resources for this level of evaluation (Fuchs et al., 2011; Schofield et al., 2011; Ayadi et al., 2012; Brown and Moore, 2012b; Fuchs et al., 2012; Adissu et al., 2014).

Primary phenotyping tests are intended to be sensitive, likely to identify deviations from normal (hits). Tests such as retinal examination or echocardiography are more clearly relevant to specific systems or areas of interest; that is, ophthalmology or cardiology. Many tests are less specific. Increased auditory brain stem response (ABR), abnormal prepulse inhibition (PPI), or acoustic startle responses could indicate a genetically determined deafness related to the genetic manipulation or the genetic background (Willott and Erway, 1998; Willott et al., 2003; Keithley et al., 2004; Zhou et al., 2006). But excessive noise, or ear infections associated with immune deficiencies, or with high environmental bioburdens in suboptimal environments, also may be significant contributors to or accelerators of deafness (Mitchell et al., 1997; Ohlemiller et al., 2000; Yoshida et al., 2000; Davis et al., 2001; Turner et al., 2005; MacArthur et al., 2006; White et al., 2009). Primary neurologic, primary musculoskeletal problems, possibly vestibular syndromes could affect SHIRPA or Rotarod results, open field activity or tests like grip strength (Rogers et al., 1997; Crawley, 2000; Rafael et al., 2000; Bailey et al., 2006; Zhou et al., 2006; Brown and Wong, 2007). A heart-specific test such as echocardiogram could be altered by atrial thrombi, infections, drugs, or possibly by a primary muscular dystrophy with myocardial involvement (Barthold, 2004; Wenzel et al., 2007; Spurney et al., 2008; Fayssoil et al., 2013). Appendix Table A.16.1 summarizes some contributors to phenotypes in different research areas, and lists tests, including pathology, to evaluate or diagnose causes and contributors to phenotypes. Clearly, influences of different genetic backgrounds are more relevant to hypothesis-driven phenotyping involving different genetic backgrounds than to IMPC phenotyping on the C57BL/6N background. Differences among C57BL/6 substrains or sources may be relevant also (Zurita et al., 2011; Chang et al., 2012; Simon et al., 2013).

The IMPC continues to assess their testing strategies to increase "hit rate," and improve the value of hits. Preliminary data from primary phenotyping programs that have included histopathology demonstrate that it adds value to the pipeline by identifying and characterizing tissue-level changes associated with *in vivo* phenotype annotations, providing a morphological context to clinical phenotyping assays. Histopathology has provided additional morphological annotations that are not directly associated with the *in vivo* annotations (pleiotropy), and has added pathophysiologically meaningful data that can lead to hypothesis-driven research (Adissu et al., 2014). Thus it is likely that histopathology will uncover unique and additional phenotypes not detected by the battery of tests currently in place, and will enrich the phenotype discovery and hit rate in phenotyping pipelines.

In addition to planned data-generating pathology, diagnostic pathology complements microbial surveillance in QA/QC for large-scale, and even for smaller scale, phenotyping and translational research initiatives. Prompt assessment of clinical conditions or other concerning disease phenotypes can help to identify problems and causes early, to minimize loss of animals, money and time, and to prevent serious compromise of the colony, project, or program. Systematic implementation of pathology in experimental design includes:

1. Diagnostic pathology for unexpected and concerning disease phenotypes in control and experimental mice, as part of health surveillance and QC.
2. Evaluation of suspected phenotypes identified by other tests, to include assessment for likely confounders and contributors, validation of models, and confirmation of absence of phenotypes, when important to the model.
3. Preservation/archiving of specimens (tissues, fluids, blocks, slides, digital slides) not evaluated immediately, so that they will be suitable and available for pathology evaluation or other testing in the future.
4. Terminology, data management, analysis, and reporting.

Additional considerations include protocols for clinical pathology, image capture, radiography (if these were not included in previous phenotyping), and specific procedures for examination, dissection, weighing, fixation or perfusion, processing, and reporting, as well as for preservation and archiving of fluids, tissues, blocks, slides, digital images, and data (Barthold, 2002, 2004; Bolon et al., 2008; Fiette and Slaoui, 2011; Schofield et al., 2011; Slaoui and Fiette, 2011; Brayton and Treuting, 2012).

A strategic and systematic plan for pathology to assess phenotypes identified by other tests and to validate models can identify causes and contributors to phenotypes, and characterize the similarities and differences between a potential model and the condition it is hypothesized to model. Virtual (digital) slide technologies offer possibilities for central slide repositories, and remote analysis by expert pathologists or virtual pathology working groups for the IMPC, or for collaborating groups or institutions (Brayton et al., 2001; Barthold, 2002; Cardiff et al., 2004; Sundberg, 2005; Hrabe de Angelis et al., 2006; Valli et al., 2007; Bolon et al., 2008; Bronson, 2009; Cardiff, 2009; Schofield et al., 2009).

Table 16.2 includes examples of two 10-slide protocols that examine more than 40 tissues including the 29 tissues required by IMPC, which are listed with their MA ontology terms. Figure 16.1 demonstrates the glass slides obtained from the Toronto Centre for Phenogenomics (TCP) protocol. Figure 16.2 demonstrates the glass slides from another protocol (JH) (per Brayton and Treuting, 2012). Once "blocked" or processed to paraffin, tissues also can be assessed by immunohistochemistry, PCR, or other techniques. Additional detail and options are reviewed elsewhere (Brayton et al., 2001; Fiette and Slaoui, 2011; Slaoui and Fiette, 2011; Brayton and Treuting, 2012).

E. Embryo Phenotyping (Wong et al., 2012; Brown and Moore, 2012a; Adams et al., 2013; Mohun et al., 2013)

Data from EUMODIC and other production and phenotyping projects using mouse lines produced from IKMC ES cells indicate that more than 30% of null or knockout gene mutations are embryo or perinatal lethal. The IMPC centers have defined lethality as the lack of homozygous animals at genotyping (usually near weaning). At least 28 animals must be genotyped prior to calling a line embryo lethal at 95% confidence. Mutants with fewer than 50% of the expected homozygotes (or 12.5% total) are called *subviable* (Adams et al., 2013). Phenotyping options are (1) to phenotype heterozygous adults (when viable) or (2) to apply an embryo phenotyping protocol. The standard adult phenotyping pipelines are applied to heterozygous adults from nonviable or subviable lines at some centers. Embryo phenotyping by viability and morphology assessments at specified time points is also being done at some sites. Determining the approximate age of embryo death is an essential first step in embryo phenotyping. The IMPC recommends a mid-gestation stage of embryonic day 12.5 (E12.5) as the initial reference time point. Staining and annotation of expression of *lacZ* can also be assessed at this time point. If viable at E12.5, E14.5/15.5 embryos are examined; if not viable at E12.5, E9.5 embryos are examined. Embryo morphology is assessed grossly, and by an imaging modality. Optical projection tomography (OPT) is suggested at E9.5 and E12.5, and microcomputed tomography (μCT) is favored for use at E14.5/15.5. Histology (including the placenta), magnetic resonance imaging, and high-resolution episcopic microscopy are other assessment options (Figure 16.3).

Table 16.2 Two Examples (JH and TCP) of 10 Slide Histopathology Protocols That Examine More Than 40 Tissues Including 29 Tissues Required by IMPC, Listed Here with Their MA Ontology Terms

JH	TCP	IMPC 2013 Required Tissue Collection and MA Term
Cassette/Slide 1	**Cassette/Slide 1** (Decalcified head)	1. Brain MA:0000168
1. **Heart** (hemisected, both halves, all chambers)	1. Nasal sinuses	2. Eye with optic nerve MA:0000261
2. **Thymus** (both lobes)	2. **Forebrain**	3. Spinal cord MA:0000216
3. Tongue (hemisected, one half)	3. **Cerebrum**	4. Thymus MA:0000142
Skeletal muscle	4. **Cerebellum with brainstem**	5. Thyroid MA:0000129
4. Sternum (entire)		6. Heart MA:0000072
+/– Mediastinal **lymph nodes** (attached)		7. Trachea MA:0000441
		8. Esophagus MA:0000352
Cassette/Slide 2	**Cassette/Slide 2**	9. Lung MA:0000415
5. **Lungs** (entire)	5. **Thymus**	10. Liver MA:0000358
6. **Trachea** (entire)	6. **Lymph Node**	11. Gall bladder MA:0000356
7. Larynx (attached)	7. with **Mammary Gland**	12. Stomach MA:0000353
8. **Esophagus** (attached)	8. **Pancreas**	13. Small intestine MA:0000337
9. **Thyroid** (attached) with	9. **Spleen**	14. Large intestine MA:0000333
10. Parathyroid	10. **Liver** Median Lobe	15. Pancreas MA:0000120
+/– Mediastinal lymph nodes	11. w/ **Gall Bladder**	16. Spleen MA:0000141
+/– Aorta (attached)	12. Liver Left Lobe	17. Kidney MA:0000368
		18. Adrenal gland MA:0000116
Cassette/Slide 3	**Cassette/Slide 3**	19. Mammary gland MA:0000145
11. **Kidney** Right (2 transverse sections)	13. Left **Kidney**	20. Lymph node MA:0000139
12. Kidney Left (2 sagittal sections)	14. Right Kidney	21. Skin MA:0000151
13. **Adrenal** glands (entire)	15. with **Adrenal**	22. Skeletal muscle MA:0000165
+/– Lymph nodes, renal (attached)	16. **Heart**	23. Urinary bladder MA:0000380
+/– Ganglion (attached)	17. Tongue	

Cassette/Slide 4

14. Submandibular Salivary glands
15. Sublingual Salivary glands
16. Parotid glands
17. Lymph nodes (Superficial and deep cervical)
18. Fat (adiposa) brown
19. Fat, white
    +/− Exorbital lacrimal glands
    +/− **Mammary Glands**

Cassette/Slide 5

20. **Pancreas** exocrine
21. Pancreas endocrine
22. Mesentery/vasculature (attached)
    Lymph nodes, pancreatic, mesenteric

Cassette/Slide 6

23. Forestomach
24. Glandular stomach
25. **Small Intestine**, Duodenum
26. **Small Intestine**, Jejunum
27. **Small Intestine**, Ileum
28. Cecum
29. **Large Intestine**, proximal
30. **Large Intestine**, distal (Rectum–Anus)
    +/− Lymph nodes

Cassette/Slide 4

18. **Trachea**
19. with **Thyroid**
20. Salivary Glands
21. **Left Lung**
22. Right Lung Lobes

Cassette/Slide 5

23. **Stomach**
24. Duodenum
25. **Jejunum**
26. Ileum
27. Cecum
28. **Colon/Rectum**

Cassette/Slide 6
Female

29. **Urinary bladder**
30. **Ovary**
31. **Uterus**

Cassette/Slide 6
Male

Urinary bladder
Prostate
Seminal vesicles
Testes
Epididymides

24. Testis   MA:0000411
25. Epididymis   MA:0000397
26. Prostate gland   MA:0000404
27. Seminal vesicle   MA:0000410
28. Ovary   MA:0000384
29. Uterus   MA:0000389

(Continued)

Table 16.2 (Continued)

JH	TCP	IMPC 2013 Required Tissue Collection and MA Term
Cassette/Slide 7	**Cassette/Slide 7**	
31. **Liver** (2 sections: left lateral and median lobes)	32. Sciatic nerve	
32. **Gall bladder** (with median lobe)	33. w/ **skeletal muscle**	
33. **Spleen** (entire or hemisected)	34. **Spinal cord:**	
+/− Lymph nodes (attached)	35. Cervical	
	36. Thoracic	
	37. Lumbar	
Cassette/Slide 8	**Cassette/Slide 8**	
Female	38. Sternum	
34. **Ovaries** (entire)	39. Femur (decalcified)	
35. Oviduct		
36. **Uterus**		
37. Cervix		
38. Vagina		
39. **Urinary bladder**		
40. Ureter		
41. Urethra		
Cassette/Slide 9	**Cassette/Slide 9**	
42. **Skin** (head, face)	40. Eyes	
43. Skin (inguinal with Preputial or Clitoral glands)	with optic nerves	
+/− **Mammary glands**		
+/− Decalcified leg, arm or spine		

Cassette/Slide 8
Male
Testes (entire)
Epididymis
Seminal vesicle with
   Coagulating gland
Prostate
Urinary Bladder
Ureter
Penis
Urethra

Cassette/Slide 10 Head (decalcified)

44. **Brain** (*in situ*)
45. Pituitary gland
46. Temporomandibular joint
47. Bone, marrow
48. Nose, sinuses
49. Oral cavity
50. Molar teeth
51. Incisor teeth
52. **Eyes (w optic nerve)**
53. Harderian gland
54. Lacrimal glands
55. Middle ear
56. Inner ear

Additional slides

Spine—decalcified
Arm or leg—decalcified
Lesions

Cassette/Slide 10

41. **Skin**—Tail base
42. Skin—Snout
43. Skin—Eyelid
44. Skin—Dorsal skin
45. Skin—Ear/Pinna

Bolded tissues in the protocols are also required by IMPC.

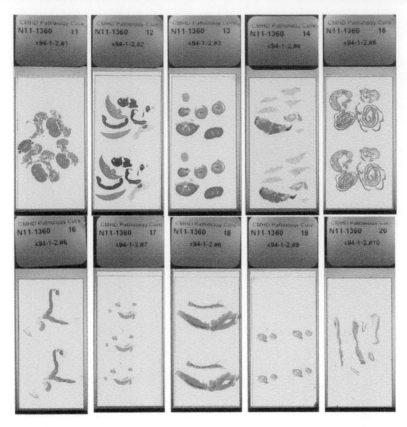

Figure 16.1 Glass slides from TCP protocol per Table 16.2. Images were obtained from whole slide scans of representative cases.

III.   Conclusions

Phenotyping tests are widely applied in translational research, often in hypothesis-driven, academic research settings. The IMPC's program of large-scale, broad spectrum, standardized phenotyping of conditional-ready knockout mouse lines is progressing in its aim to provide the biomedical research community with a genome-wide, publicly accessible resource of mammalian gene function information. This effort will provide GEM models, GEM phenotype data and protocol detail, to facilitate and accelerate robust, relevant, and reproducible translational research. Pathology continues to be critical to diagnosing and understanding disease, to confirm and characterize phenotypes, and to validate translational models. It seems certain that genome-wide phenotyping efforts, including pathology, will offer tools and strategies to improve hypothesis-driven experimental design, and lead to many important breakthroughs in translational science.

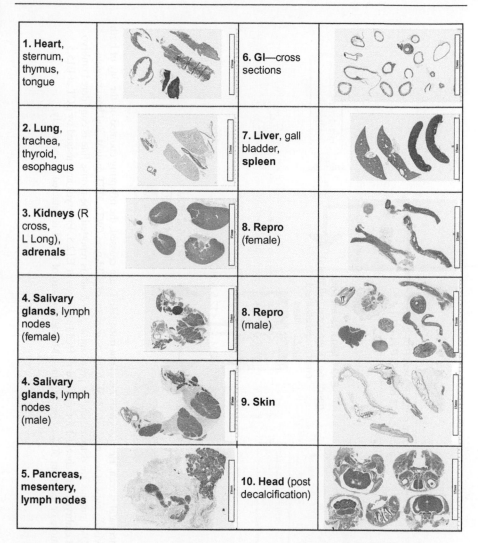

1. Heart, sternum, thymus, tongue		**6. GI**—cross sections	
2. Lung, trachea, thyroid, esophagus		**7. Liver**, gall bladder, **spleen**	
3. Kidneys (R cross, L Long), **adrenals**		**8. Repro** (female)	
4. Salivary glands, lymph nodes (female)		**8. Repro** (male)	
4. Salivary glands, lymph nodes (male)		**9. Skin**	
5. Pancreas, mesentery, lymph nodes		**10. Head** (post decalcification)	

Figure 16.2 Glass slides from slides from JH protocol per Table 16.2 (Brayton and Treuting, 2012). Images were obtained from whole slide scans of representative cases. *Source*: Dissections by N. Forbes; histology by B. Gambicher, scanning courtesy of Flagship Biosciences LLC, Boulder, CO.

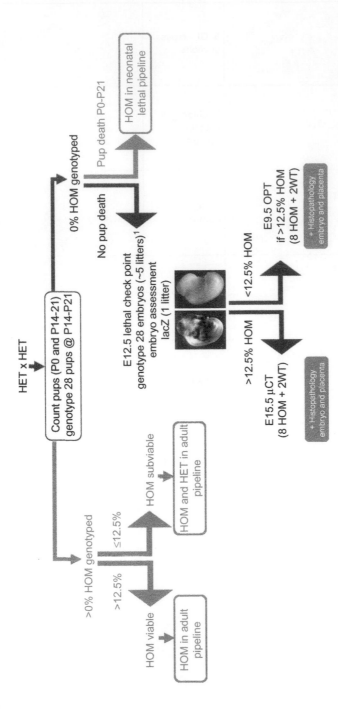

Figure 16.3 The TCP embryonic and neonatal screen. Heterozygous knockout mice (HET) are crossed, the offspring counted on the day of birth (P0), followed by genotyping of 28 surviving pups at weaning to establish the proportion of wild-type (WT), HET mutant, and homozygous (HOM) mutant pups produced. If no HOM mutants are detected and no preweaning pup death is reported, the mouse line is considered embryo lethal and approximately five litters (28 embryos) are checked for survival of HOM embryos at E12.5. Embryos are imaged by μCT at E15.5 or OPT at E9.5 depending on whether more or less than 50% of the expected Mendelian ratio of HOM is detected, respectively.

References

Abbott, A., 2010. Mouse project to find each gene's role: International Mouse Phenotyping Consortium launches with a massive funding commitment. Nature. 465, 410.

Adams, D., Baldock, R., Bhattacharya, S., Copp, A.J., Dickinson, M., Greene, N.D., et al., 2013. Bloomsbury report on mouse embryo phenotyping: recommendations from the IMPC workshop on embryonic lethal screening. Dis. Model Mech. 6 (3), 571–579.

Adissu, H.A., Estabel, J., Sunter, D., Tuck, E., Hooks, Y., Carragher, D.M., et al., S. M. Wellcome Trust Sanger Institute 2014. Histopathology reveals correlative and unique phenotypes in a high throughput mouse phenotyping screen. Dis. Model Mech.

Ayadi, A., Birling, M.C., Bottomley, J., Bussell, J., Fuchs, H., Fray, M., et al., 2012. Mouse large-scale phenotyping initiatives: overview of the European Mouse Disease Clinic (EUMODIC) and of the Wellcome Trust Sanger Institute Mouse Genetics Project. Mamm. Genome. 23 (9–10), 600–610.

Bailey, K.R., Rustay, N.R., Crawley, J.N., 2006. Behavioral phenotyping of transgenic and knockout mice: practical concerns and potential pitfalls. ILAR J. 47 (2), 124–131.

Barthold, S.W., 2002. "Muromics": genomics from the perspective of the laboratory mouse. Comp. Med. 52 (3), 206–223.

Barthold, S.W., 2004. Genetically altered mice: phenotypes, no phenotypes, and faux phenotypes. Genetica. 122 (1), 75–88.

Barthold, S.W., Borowsky, A.D., Brayton, C., Bronson, R., Cardiff, R.D., Griffey, S.M., et al., 2007. From whence will they come?—A perspective on the acute shortage of pathologists in biomedical research. J. Vet. Diagn. Invest. 19 (4), 455–456.

Blake, J.A., Bult, C.J., Kadin, J.A., Richardson, J.E., Eppig, J.T., 2010. The mouse genome database (MGD): premier model organism resource for mammalian genomics and genetics. Nucleic Acids Res.

Bogue, M., 2003. Mouse Phenome Project: understanding human biology through mouse genetics and genomics. J. Appl. Physiol. 95 (4), 1335–1337.

Boivin, G.P., Washington, K., Yang, K., Ward, J.M., Pretlow, T.P., Russell, R., et al., 2003. Pathology of mouse models of intestinal cancer: consensus report and recommendations. Gastroenterology. 124 (3), 762.

Bolon, B., Brayton, C., Cantor, G.H., Kusewitt, D.F., Loy, J.K., Sartin, E.A., et al., 2008. Editorial: best pathology practices in research using genetically engineered mice. Vet. Pathol. 45 (6), 939–940.

Brayton, C., Treuting, P., 2012. Phenotyping. In: Treuting, P., Dintzis, S. (Eds.), Comparative Anatomy and Histology: A Mouse and Human Atlas. Elsevier (Academic Press), London, pp. 361–381. (Chapter 2).

Brayton, C., Justice, M., Montgomery, C.A., 2001. Evaluating mutant mice: anatomic pathology. Vet. Pathol. 38 (1), 1–19.

Brayton, C.F., Treuting, P.M., Ward, J.M., 2012. Pathobiology of aging mice and GEM: background strains and experimental design. Vet. Pathol. 49 (1), 85–105.

Bronson, R.T., 2009. Pathologic phenotyping of mutant mice. Methods Mol. Biol. 530, 435–461.

Brown, R.E., Wong, A.A., 2007. The influence of visual ability on learning and memory performance in 13 strains of mice. Learn. Mem. 14 (3), 134–144.

Brown, S.D., Moore, M.W., 2012a. The International Mouse Phenotyping Consortium: past and future perspectives on mouse phenotyping. Mamm. Genome. 23 (9-10), 632–640.

Brown, S.D., Moore, M.W., 2012b. Towards an encyclopaedia of mammalian gene function: the International Mouse Phenotyping Consortium. Dis. Model Mech. 5 (3), 289−292.

Brown, S.D., Chambon, P., de Angelis, M.H., 2005. EMPReSS: standardized phenotype screens for functional annotation of the mouse genome. Nat. Genet. 37 (11), 1155.

Bult, C.J., 2012. Bioinformatics resources for behavior studies in the laboratory mouse. Int. Rev. Neurobiol. 104, 71−90.

Bult, C.J., Kadin, J.A., Richardson, J.E., Blake, J.A., Eppig, J.T., 2010. The Mouse Genome Database: enhancements and updates. Nucleic Acids Res. 38 (Database issue), D586−D592.

Cardiff, R.D., 2009. How to phenotype a mouse. Dis. Model Mech. 2 (7−8), 317−321.

Cardiff, R.D., Anver, M.R., Gusterson, B.A., Hennighausen, L., Jensen, R.A., Merino, M.J., et al., 2000. The mammary pathology of genetically engineered mice: the consensus report and recommendations from the Annapolis meeting. Oncogene. 19 (8), 968−988.

Cardiff, R.D., Rosner, A., Hogarth, M.A., Galvez, J.J., Borowsky, A.D., Gregg, J.P., 2004. Validation: the new challenge for pathology. Toxicol. Pathol. 32 (Suppl. 1), 31−39.

Cardiff, R.D., Anver, M.R., Boivin, G.P., Bosenberg, M.W., Maronpot, R.R., Molinolo, A.A., et al., 2006. Precancer in mice: animal models used to understand, prevent, and treat human precancers. Toxicol. Pathol. 34 (6), 699−707.

Cardiff, R.D., Ward, J.M., Barthold, S.W., 2008. "One medicine—one pathology": are veterinary and human pathology prepared? Lab. Invest. 88 (1), 18−26.

Chang, H.Y., Mitzner, W., Watson, J., 2012. Variation in airway responsiveness of male C57BL/6 mice from 5 vendors. J. Am. Assoc. Lab. Anim. Sci. 51 (4), 401−406.

Conover, C.A., Bale, L.K., Mader, J.R., Mason, M.A., Keenan, K.P., Marler, R.J., 2010. Longevity and age-related pathology of mice deficient in pregnancy-associated plasma protein-A. J. Gerontol. A Biol. Sci. Med. Sci. 65 (6), 590−599.

Crabbe, J.C., Metten, P., Ponomarev, I., Prescott, C.A., Wahlsten, D., 2006. Effects of genetic and procedural variation on measurement of alcohol sensitivity in mouse inbred strains. Behav. Genet. 36 (4), 536−552.

Crawley, J.N., 2000. What's Wrong with My Mouse? Behavioral Phenotyping of Transgenic and Knockout Mice. Wiley Liss, New York, NY.

Creasy, D., Bube, A., de Rijk, E., Kandori, H., Kuwahara, M., Masson, R., et al., 2012. Proliferative and nonproliferative lesions of the rat and mouse male reproductive system. Toxicol. Pathol. 40 (6 Suppl.), 40S−121S.

Crissman, J.W., Goodman, D.G., Hildebrandt, P.K., Maronpot, R.R., Prater, D.A., Riley, J.H., et al., 2004. Best practices guideline: toxicologic histopathology. Toxicol. Pathol. 32 (1), 126−131.

Davis, R.R., Newlander, J.K., Ling, X.-B., Cortopassi, G.A., Krieg, E.F., Erway, L.C., 2001. Genetic basis for susceptibility to noise-induced hearing loss in mice. Hear. Res. 155 (1-2), 82.

Elmore, S.A., Peddada, S.D., 2009. Points to consider on the statistical analysis of rodent cancer bioassay data when incorporating historical control data. Toxicol. Pathol. 37 (5), 672−676.

Fayssoil, A., Renault, G., Guerchet, N., Marchiol-Fournigault, C., Fougerousse, F., Richard, I., 2013. Cardiac characterization of mdx mice using high-resolution Doppler echocardiography. J. Ultrasound Med. 32 (5), 757−761.

Field, D., Sansone, S.A., Collis, A., Booth, T., Dukes, P., Gregurick, S.K., et al., 2009. Megascience. Omics data sharing. Science. 326 (5950), 234−236.

Fiette, L., Slaoui, M., 2011. Necropsy and sampling procedures in rodents. Methods Mol. Biol. 691, 39−67.

Frazier, K.S., Seely, J.C., Hard, G.C., Betton, G., Burnett, R., Nakatsuji, S., et al., 2012. Proliferative and nonproliferative lesions of the rat and mouse urinary system. Toxicol. Pathol. 40 (4 Suppl), 14S−86S.

Fuchs, H., Gailus-Durner, V., Adler, T., Aguilar-Pimentel, J.A., Becker, L., Calzada-Wack, J., et al., 2011. Mouse phenotyping. Methods. 53 (2), 120–135.

Fuchs, H., Gailus-Durner, V., Neschen, S., Adler, T., Afonso, L.C., Aguilar-Pimentel, J.A., et al., 2012. Innovations in phenotyping of mouse models in the German Mouse Clinic. Mamm. Genome. 23 (9-10), 611–622.

Gailus-Durner, V., Fuchs, H., Adler, T., Aguilar Pimentel, A., Becker, L., Bolle, I., et al., 2009. Systemic first-line phenotyping. Methods Mol. Biol. 530, 463–509.

Gkoutos, G.V., Schofield, P.N., Hoehndorf, R., 2012. Computational tools for comparative phenomics: the role and promise of ontologies. Mamm. Genome. 23 (9-10), 669–679.

Grubb, S.C., Maddatu, T.P., Bult, C.J., Bogue, M.A., 2009. Mouse phenome database. Nucleic Acids Res. 37 (Database issue), D720–D730 (and <http://phenome.jax.org/>).

Guan, C., Ye, C., Yang, X., Gao, J., 2010. A review of current large-scale mouse knockout efforts. Genesis. 48 (2), 73–85.

Gutmann, D.H., Baker, S.J., Giovannini, M., Garbow, J., Weiss, W., 2003. Mouse models of human cancer consortium symposium on nervous system tumors. Cancer Res. 63 (11), 3001–3004.

Hailey, J.R., Haseman, J.K., Bucher, J.R., Radovsky, A.E., Malarkey, D.E., Miller, R.T., et al., 1998. Impact of Helicobacter hepaticus infection in B6C3F1 mice from twelve National Toxicology Program two-year carcinogenesis studies. Toxicol. Pathol. 26 (5), 602–611.

Hancock, J.M., Mallon, A.-M., 2008. Phenobabelomics mouse phenotype data resources. Brief. Funct. Genomic. Proteomic. 6 (4), 292–301.

Hancock, J.M., Mallon, A.M., Beck, T., Gkoutos, G.V., Mungall, C., Schofield, P.N., 2009. Mouse, man, and meaning: bridging the semantics of mouse phenotype and human disease. Mamm. Genome. 20 (8), 457–461.

Hartung, T., 2009. Toxicology for the twenty-first century. Nature. 460 (7252), 208–212.

Hartung, T., Leist, M., 2008. Food for thought... on the evolution of toxicology and the phasing out of animal testing. ALTEX. 25 (2), 91–102.

Haseman, J.K., Huff, J.E., Rao, G.N., Eustis, S.L., 1989. Sources of variability in rodent carcinogenicity studies. Fundam. Appl. Toxicol. 12 (4), 793–804.

Hrabe de Angelis, M., Chambon, P., Brown, S.D.M., 2006. Standards of Mouse Model Phenotyping. Wiley-VCH, Weinheim.

ILAR-NRC, 2011. Guidance for the Description of Animal Research in Scientific Publications. The National Academies Press. National Academy of Sciences, Washington, DC (URL: <http://www.nap.edu/catalog.php?record_id = 13241>).

Ince, T.A., Ward, J.M., Valli, V.E., Sgroi, D., Nikitin, A.Y., Loda, M., et al., 2008. Do-it-yourself (DIY) pathology. Nat. Biotechnol. 26 (9), 978–979 (discussion 979).

Ittmann, M., Huang, J., Radaelli, E., Martin, P., Signoretti, S., Sullivan, R., et al., 2013. Animal models of human prostate cancer: the consensus report of the New York meeting of the Mouse Models of Human Cancers Consortium Prostate Pathology Committee. Cancer Res. 73 (9), 2718–2736.

Jeffrey, E.G., Robert, C., Lothar, H., Lalage, W., Ulrike, W., Eva, L., et al., 2002. Validation of transgenic mammary cancer models: goals of the NCI Mouse Models of Human Cancer Consortium and the Mammary Cancer CD-ROM. Transgenic. Res. V11 (6), 635.

Kaufmann, W., Bolon, B., Bradley, A., Butt, M., Czasch, S., Garman, R.H., et al., 2012. Proliferative and nonproliferative lesions of the rat and mouse central and peripheral nervous systems. Toxicol. Pathol. 40 (4 Suppl.), 87S–157S.

Keenan, C., Elmore, S., Francke-Carroll, S., Kemp, R., Kerlin, R., Peddada, S., et al., 2009a. Best practices for use of historical control data of proliferative rodent lesions. Toxicol. Pathol. 37 (5), 679–693.

Keenan, C., Elmore, S., Francke-Carroll, S., Kerlin, R., Peddada, S., Pletcher, J., et al., 2009b. Potential for a global historical control database for proliferative rodent lesions. Toxicol. Pathol. 37 (5), 677−678.

Keenan, C.M., Goodman, D.G., 2013. Regulatory Forum Commentary*: Through the Looking Glass—SENDing the Pathology Data We Have INHAND. Toxicol. Pathol.

Keithley, E.M., Canto, C., Zheng, Q.Y., Fischel-Ghodsian, N., Johnson, K.R., 2004. Age-related hearing loss and the ahl locus in mice. Hear. Res. 188 (1−2), 21−28.

Kilkenny, C., Parsons, N., Kadyszewski, E., Festing, M.F., Cuthill, I.C., Fry, D., et al., 2009. Survey of the quality of experimental design, statistical analysis and reporting of research using animals. PLoS One. 4 (11), e7824.

Kilkenny, C., Browne, W.J., Cuthill, I.C., Emerson, M., Altman, D.G., 2010. Improving bioscience research reporting: the ARRIVE guidelines for reporting animal research. PLoS Biol. 8 (6), e1000412.

Kogan, S.C., Ward, J.M., Anver, M.R., Berman, J.J., Brayton, C., Cardiff, R.D., et al., 2002. Bethesda proposals for classification of nonlymphoid hematopoietic neoplasms in mice. Blood. 100 (1), 238−245.

Ladiges, W., Van Remmen, H., Strong, R., Ikeno, Y., Treuting, P., Rabinovitch, P., et al., 2009. Lifespan extension in genetically modified mice. Aging Cell. 8 (4), 346−352.

Linder, C.C., 2003. Mouse nomenclature and maintenance of genetically engineered mice. Comp. Med. 53 (2), 119−125.

MacArthur, C.J., Hefeneider, S.H., Kempton, J.B., Trune, D.R., 2006. C3H/HeJ mouse model for spontaneous chronic otitis media. Laryngoscope. 116 (7), 1071−1079.

Maddatu, T.P., Grubb, S.C., Bult, C.J., Bogue, M.A., 2012. Mouse Phenome Database (MPD). Nucleic Acids Res. 40 (Database issue), D887−D894.

Mallon, A.M., Iyer, V., Melvin, D., Morgan, H., Parkinson, H., Brown, S.D., et al., 2012. Accessing data from the International Mouse Phenotyping Consortium: state of the art and future plans. Mamm. Genome. 23 (9-10), 641−652.

Martin, B., Ji, S., Maudsley, S., Mattson, M.P., 2010. "Control" laboratory rodents are metabolically morbid: why it matters. Proc. Natl. Acad. Sci. USA. 107 (14), 6127−6133.

Mitchell, C.R., Kempton, J.B., Scott-Tyler, B., Trune, D.R., 1997. Otitis media incidence and impact on the auditory brain stem response in lipopolysaccharide-nonresponsive C3H/HeJ mice. Otolaryngol. Head. Neck. Surg. 117 (5), 459−464.

Mohun, T., Adams, D.J., Baldock, R., Bhattacharya, S., Copp, A.J., Hemberger, M., et al., 2013. Deciphering the Mechanisms of Developmental Disorders (DMDD): a new programme for phenotyping embryonic lethal mice. Dis. Model Mech. 6 (3), 562−566.

Moore, M., IMPC and SteeringCommittee (2010) The International Mouse Phenotyping Consortium: Initial Business Plan Published online at <http://nihroadmap.nih.gov/KOMP2/IMPC-Business-Plan.pdf. http://commonfund.nih.gov/sites/default/files/IMPC_Business_Plan.pdf>.

Morgan, H., Beck, T., Blake, A., Gates, H., Adams, N., Debouzy, G., et al., 2010. EuroPhenome: a repository for high-throughput mouse phenotyping data. Nucleic Acids Res. 38 (Database issue), D577−D585.

Morse III, H.C., Anver, M.R., Fredrickson, T.N., Haines, D.C., Harris, A.W., Harris, N.L., et al., 2002. Bethesda proposals for classification of lymphoid neoplasms in mice. Blood. 100 (1), 246−258.

Nature, 2010. Editorial: mouse megascience. Nature. 465 (7298), 526.

Nikitin, A.Y., Alcaraz, A., Anver, M.R., Bronson, R.T., Cardiff, R.D., Dixon, D., et al., 2004. Classification of proliferative pulmonary lesions of the mouse: recommendations of the mouse models of human cancers consortium. Cancer Res. 64 (7), 2307−2316.

Ohlemiller, K.K., Wright, J.S., Heidbreder, A.F., 2000. Vulnerability to noise-induced hearing loss in 'middle-aged' and young adult mice: a dose–response approach in CBA, C57BL, and BALB inbred strains. Hear. Res. 149 (1–2), 239.

Rafael, J.A., Nitta, Y., Peters, J., Davies, K.E., 2000. Testing of SHIRPA, a mouse phenotypic assessment protocol, on Dmd(mdx) and Dmd(mdx3cv) dystrophin-deficient mice. Mamm. Genome. 11 (9), 725–728.

Rao, G.N., Crockett, P.W., 2003. Effect of diet and housing on growth, body weight, survival and tumor incidences of B6C3F1 mice in chronic studies. Toxicol. Pathol. 31 (2), 243–250.

Renne, R., Brix, A., Harkema, J., Herbert, R., Kittel, B., Lewis, D., et al., 2009. Proliferative and nonproliferative lesions of the rat and mouse respiratory tract. Toxicol. Pathol. 37 (7 Suppl.), 5S–73S.

Rogers, D.C., Fisher, E.M., Brown, S.D., Peters, J., Hunter, A.J., Martin, J.E., 1997. Behavioral and functional analysis of mouse phenotype: SHIRPA, a proposed protocol for comprehensive phenotype assessment. Mamm. Genome. 8 (10), 711–713.

Rudmann, D., Cardiff, R., Chouinard, L., Goodman, D., Kuttler, K., Marxfeld, H., et al., 2012. Proliferative and nonproliferative lesions of the rat and mouse mammary, Zymbal's, preputial, and clitoral glands. Toxicol. Pathol. 40 (6 Suppl.), 7S–39S.

Schofield, P.N., Brown, S.D., Sundberg, J.P., Arends, M., Warren, M.V., Dubus, P., et al., 2009. PRIME importance of pathology expertise. Nat. Biotechnol. 27 (1), 24–25.

Schofield, P.N., Gruenberger, M., Sundberg, J.P., 2010. Pathbase and the MPATH ontology. Community resources for mouse histopathology. Vet. Pathol. 47 (6), 1016–1020.

Schofield, P.N., Dubus, P., Klein, L., Moore, M., McKerlie, C., Ward, J.M., et al., 2011. Pathology of the laboratory mouse: an international workshop on challenges for high throughput phenotyping. Toxicol Pathol.

Shappell, S.B., Thomas, G.V., Roberts, R.L., Herbert, R., Ittmann, M.M., Rubin, M.A., et al., 2004. Prostate pathology of genetically engineered mice: definitions and classification. The consensus report from the Bar Harbor meeting of the Mouse Models of Human Cancer Consortium Prostate Pathology Committee. Cancer Res. 64 (6), 2270–2305.

Simon, M.M., Greenaway, S., White, J.K., Fuchs, H., Gailus-Durner, V., Sorg, T., et al., 2013. A comparative phenotypic and genomic analysis of C57BL/6J and C57BL/6N mouse strains. Genome. Biol. 14 (7), R82.

Slaoui, M., Fiette, L., 2011. Histopathology procedures: from tissue sampling to histopathological evaluation. Methods Mol. Biol. 691, 69–82.

Smedley, D., Oellrich, A., Kohler, S., Ruef, B., Westerfield, M., Robinson, P., et al., 2013. PhenoDigm: analyzing curated annotations to associate animal models with human diseases. Database (Oxford). 2013. 10.1093/database/bat025.

Smith, C.L., Eppig, J.T., 2012. The Mammalian Phenotype Ontology as a unifying standard for experimental and high-throughput phenotyping data. Mamm. Genome. 23 (9-10), 653–668.

Spurney, C.F., Knoblach, S., Pistilli, E.E., Nagaraju, K., Martin, G.R., Hoffman, E.P., 2008. Dystrophin-deficient cardiomyopathy in mouse: expression of Nox4 and Lox are associated with fibrosis and altered functional parameters in the heart. Neuromuscul. Disord. 18 (5), 371–381.

Sundberg, J., 2005. Handbook on Genetically Engineered Mice. Taylor & Francis (CRC Press), Boca Raton, FL.

Sundberg, J.P., Berndt, A., Sundberg, B.A., Silva, K.A., Kennedy, V., Bronson, R., et al., 2011. The mouse as a model for understanding chronic diseases of aging: the histopathologic basis of aging in inbred mice. Pathobiol. Aging Age Relat. Dis. 1, 10.

Tang, T., Li, L., Tang, J., Li, Y., Lin, W.Y., Martin, F., et al., 2010. A mouse knockout library for secreted and transmembrane proteins. Nat. Biotechnol. 28 (7), 749−755.

Thoolen, B., Maronpot, R.R., Harada, T., Nyska, A., Rousseaux, C., Nolte, T., et al., 2010. Proliferative and nonproliferative lesions of the rat and mouse hepatobiliary system. Toxicol. Pathol. 38 (7 Suppl.), 5S−81S.

Turner, J.G., Parrish, J.L., Hughes, L.F., Toth, L.A., Caspary, D.M., 2005. Hearing in laboratory animals: strain differences and nonauditory effects of noise. Comp. Med. 55 (1), 12−23.

Valli, T., Barthold, S.W., Ward, J.E., Brayton, C., Nikitin, A., Borowsky, A.D., et al., 2007. Over 60% of NIH extramural funding involves animal-related research. Vet. Pathol. 44 (6), 962−963 (author reply 963−964).

Wahlsten, D., Bachmanov, A., Finn, D.A., Crabbe, J.C., 2006. Stability of inbred mouse strain differences in behavior and brain size between laboratories and across decades. Proc. Natl. Acad. Sci. USA. 103 (44), 16364−16369.

Ward, J.M., Fox, J.G., Anver, M.R., Haines, D.C., George, C.V., Collins Jr., M.J., et al., 1994. Chronic active hepatitis and associated liver tumors in mice caused by a persistent bacterial infection with a novel Helicobacter species. J. Natl. Cancer Inst. 86 (16), 1222−1227.

Wehling, M., 2009. Assessing the translatability of drug projects: what needs to be scored to predict success? Nat. Rev. Drug. Discov. 8 (7), 541−546.

Wendler, A., Wehling, M., 2010. The translatability of animal models for clinical development: biomarkers and disease models. Curr. Opin. Pharmacol. 10 (5), 601−606.

Wenzel, K., Geier, C., Qadri, F., Hubner, N., Schulz, H., Erdmann, B., et al., 2007. Dysfunction of dysferlin-deficient hearts. J. Mol. Med. 85 (11), 1203.

White, C.H., Ohmen, J.D., Sheth, S., Zebboudj, A.F., McHugh, R.K., Hoffman, L.F., et al., 2009. Genome-wide screening for genetic loci associated with noise-induced hearing loss. Mamm. Genome. 20 (4), 207−213.

Wilkinson, P., Sengerova, J., Matteoni, R., Chen, C.K., Soulat, G., Ureta-Vidal, A., et al., 2010. EMMA—mouse mutant resources for the international scientific community. Nucleic Acids Res. 38 (Database issue), D570−D576.

Willott, J.F., Erway, L.C., 1998. Genetics of age-related hearing loss in mice. IV. Cochlear pathology and hearing loss in 25 BXD recombinant inbred mouse strains. Hear. Res. 119 (1-2), 27−36.

Willott, J.F., Tanner, L., O'Steen, J., Johnson, K.R., Bogue, M.A., Gagnon, L., 2003. Acoustic startle and prepulse inhibition in 40 inbred strains of mice. Behav. Neurosci. 117 (4), 716−727.

Wong, M.D., et al., 2012. A novel 3D mouse embryo atlas based on micro-CT. Development. 139, 3248−3256.

Wurst, W., de Angelis, M.H., 2010. Systematic phenotyping of mouse mutants. Nat. Biotechnol. 28 (7), 684−685 (News and Views).

Yoshida, N., Hequembourg, S.J., Atencio, C.A., Rosowski, J.J., Liberman, M.C., 2000. Acoustic injury in mice: 129/SvEv is exceptionally resistant to noise-induced hearing loss. Hear. Res. 141 (1-2), 97−106.

Yuan, R., Peters, L.L., Paigen, B., 2011. Mice as a mammalian model for research on the genetics of aging. ILAR J. 52 (1), 4−15.

Zhou, X., Jen, P.H., Seburn, K.L., Frankel, W.N., Zheng, Q.Y., 2006. Auditory brainstem responses in 10 inbred strains of mice. Brain. Res. 1091 (1), 16−26.

Zurita, E., Chagoyen, M., Cantero, M., Alonso, R., González-Neira, A., López-Jiménez, A., et al., 2011. Genetic polymorphisms among C57BL/6 mouse inbred strains. Transgenic. Res. 20 (3), 481−489.

Appendix

Table A.16.1 Examples of Contributors to "Hits" in Primary Phenotyping, and Tests for Further Assessment or Characterization

Primary Research Area[a] (Some Tests That May Be Affected)	Potential Genetic Background (Nature) Contributors (In Common Strains)	Potential Environmental Especially Microbial (Nurture) Contributors	Tests to Confirm, Characterize, Validate a Phenotype, or to Diagnose Contributors or Confounders
Neuroscience, sensory—Visual (dysmorphology SHIRPA, tests involving visual cues, eye morphology, pathology)	Strain, sex, age variation reviewed in [1–6] Blindness (*rd1*) retinal degeneration in C3H, CBA, FVB/N, SJL/J, some Swiss [7–12] Microphthalmia in C57BL, etc. [13,14] C57BL/6J vs C57BL/6N [15,16] Albinism [4,17,18] Glaucoma in DBA/2J [19–23] Cataracts [24–26] Corneal opacities, especially BALB/c, C3H, DBA/2 [27–30]	Traumatic or infectious eye damage Excessive light [31] Cold, anesthesia, etc. induced cataracts or corneal opacities [32–36] Mutagens causing mutations that result in cataracts [37,38].	Slit lamp microscopy for anterior chamber Fundoscopy (ophthalmoscopy) Molecular, genetic assessments (for relevant genotypes) Imaging—MRI, interferometry, etc. Pathology, macroscopic and histopathology
Neuroscience, sensory—Auditory (SHIRPA, ABR, PPI and tests involving auditory cues; hematology; pathology)	Strain, sex, age variation in hearing, ABR, PPI [3,5,39–42] (MPD[b]) BALB/c [43] C57BL/6J & C57BL/6N [15,40] DBA/2 Hearing and vestibular syndromes [44,45] Immune deficiency or other genetically determined susceptibility to infection/ otitis [46], e.g., homozygosity for *Tlr4*[lps] in C3H/HeJ [47,48]	Infectious agents in otitis [49] Gram positive, e.g., *Enterococcus* spp. [50]; *Streptococcus* spp. [51–54] Gram negative, e.g., *Klebsiella oxytoca* [55]; *Pasteurella pneumotropica* [56,57]; *Pseudomonas* [50] Viral infections [51,58,59] Noise [60–64] Diet [65,66]	Molecular, genetic assessments (for relevant genotypes) Microbial assessments, culture, PCR, serology Pathology, macroscopic and histopathology

(Continued)

Table A.16.1 (Continued)

Primary Research Area[a] (Some Tests That May Be Affected)	Potential Genetic Background (Nature) Contributors (In Common Strains)	Potential Environmental Especially Microbial (Nurture) Contributors	Tests to Confirm, Characterize, Validate a Phenotype, or to Diagnose Contributors or Confounders
Neuroscience, sensory—Vestibular (SHIRPA, tests of balance, coordination, activity; hematology; pathology)	Hearing and vestibular syndromes in DBA/2 [39,69,70] Immune deficiency or other genetically determined susceptibility to infection (otitis interna) [46], e.g., homozygosity for *Tlr4lps* in C3H/HeJ [55] Arteritis, polyarteritis [71–73] Infarcts [74]	Ototoxic drugs, compounds, e.g., Gentamycin, Kanamycin, Cisplatin [67,68] Otitis interna with infectious causes similar to those listed for otitis also [75,76] Viral infections [58,59] Vestibulotoxic drugs, compounds [68,77–79]	Molecular, genetic assessments (for relevant genotypes) Microbial assessments, culture, PCR, serology Pathology, macroscopic and histopathology
Neuroscience—Seizures (possibly SHIRPA or other behavioral tests; pathology)	Strain sex age variation [80–86] DBA/2 [39,80,87] FVB/N [81,88]	Noise [61,89,90] Infections [91] Epileptogenic drugs or stimuli [80–86]	Seizure challenge Molecular, genetic assessments (for relevant genotypes) Imaging, MRI [92] PET [93], etc. Histopathology for seizure associated lesions [80,88,94]
Neuroscience—Brain morphology (Possibly SHIRPA or other behavioral tests; pathology)	Strain variation in brain and ventricle size [95,96] (MPD) Strain variation in cerebellum anatomy [97–99] Hypocallosity, acallosity 129, BALB/c [100–103] [104] Hydrocephalus C57BL [96,105]; MRL/lpr [106]	Environmental contributors to hypocallosity [110] Experimental viral infections cause hydrocephalus [111–117] Parvovirus (MMV) cerebellum development [118] Perinatal irradiation or drugs [119]	Imaging, MRI, etc. Pathology, macroscopic and histopathology Molecular, genetic assessments (for relevant genotypes) Imaging (MRI, CT) Serology, PCR to assess for viral contributors

Neuroscience and/or Musculoskeletal—Weakness, paralysis, tremor (SHIRPA, tests of balance, coordination, activity; hematology; pathology)	Incomplete or incompetent posterior communicating artery C57BL/6, CD-1 [107–109] A/J, SJL/J dysferlin (*Dysf*) mutations, [120–122] Dystrophin mutations [120] [123] Laminin alpha2 (*Lama2*) mutations, [120,124–126] Spine/CNS neoplasia, usually in older mice Hematopoietic [127] Spine tumors [128] CNS tumors [129] Infected or chronically ill [130] Mouse Hepatitis virus [131–139] (MHV) encephalitis, demyelination [131–134] Theiler's mouse encephalitisvirus (TMEV), encephalitis, demyelination [135–139] Parvovirus (MMV) cerebellum development [118] Diet, drugs may modulate some myopathy or muscular dystrophy phenotypes [124,140]	Molecular, genetic assessments (for relevant genotypes) Microbial assessments, culture, PCR, serology Pathology, macroscopic and histopathology
Cardiovascular (ECG; BP; heart weight; hematology; pathology)	Strain, sex variation [141–145] BALB/c thrombi, "cardiomyopathy" [146,147] BALB/C, C3H, DBA cardiac calcinosis [148–150] B6.129 [151] C57BL/6J vs C57BL/6N [15] C57BL/6 Hypertension Spontaneous mutants obese, diabetes, fat, tubby, and lethal yellow [152] C57BL/6, A/J, FVB/N hemostasis thrombosis atherosclerosis [153,154] Arteritis, polyarteritis [73] possibly associated with autoimmunity, [71,72,155] or with hypertension [156] Muscular dystrophy mice [120,121,125] Infections in susceptible mice [157–164] Diet [165–170]	Imaging, MRI, echo/ultrasound, isproterenol challenge, aortic banding; blood pressure, diet challenge (fat, salt) [171–174] Microbial assessments, culture, PCR, serology Pathology, macroscopic and histopathology

(Continued)

Table A.16.1 (Continued)

Primary Research Area[a] (Some Tests That May Be Affected)	Potential Genetic Background (Nature) Contributors (In Common Strains)	Potential Environmental Especially Microbial (Nurture) Contributors	Tests to Confirm, Characterize, Validate a Phenotype, or to Diagnose Contributors or Confounders
Immunology, immunity (hematology; FACS; pathology)	Strain/sex variation in hematology [175–178] (MPD) Strain variation in susceptibility to infection, disease [46], e.g., Sendai virus, [179,180] MHV (SJL/J lack Ceacam1) [181], Mycoplasma pulmonis, [182,183] Helicobacters [184,185], Salmonella [186] C3H/HeJ [55,187]	Infections, viral, bacterial, parasites [49] Diet, environment [15,188–192] Irradiation Immunotoxic agents [193]	Infection challenges; cytokine, chemokine measurements Molecular, genetic assessments (for relevant genotypes) Microbial assessments, culture, PCR, serology, parasitology Pathology, macroscopic and histopathology
Metabolic—General (growth curves; IPGTT; DXA; clinical chemistry; hematology, pathology)	Strain/sex variations [194–197] (MPD)	Infection, inflammation [198–202] Diet, environment [166,167,203–206]	Molecular, genetic assessments (for relevant genotypes) Microbial assessments, culture, PCR, serology, parasitology Macroscopic and microscopic pathology, histopathology
Metabolic—Diabetes, obesity, metabolic syndrome (growth curves; IPGTT; DXA; clinical chemistry; hematology; pathology)	Age, strain, sex variation (MPD) [207–211] MPD C57BL/6J vs C57BL/6N [15] C57BL/6 vs C57BLKS [195,212,213] C57BL/6 vs other [212,214–217]	Microbiome [218,219] Infection [220–234] Diet [203–205,235–242] Environment, other [206,230,243–245]	Molecular, genetic assessments (for relevant genotypes) Microbial assessments, culture, PCR, serology, parasitology Pathology, macroscopic and histopathology
Metabolic—Wasting phenotypes "Premature aging" (growth curves; DXA; clinical chemistry; hematology; pathology)	Age, strain, sex associated conditions [1,246,247] Amyloidosis in amyloid susceptible strains ApoA alleles, [248,249] Reactive amyloid [250] Acidophilic macrophage pneumonia [251,252] Renal disease [73,151,253]	Infections, dermatitis, environmental factors may contribute to Amyloidosis [254,255] Glomerulonephritis [256,257] Virus infections in susceptible strains [258–261]	Molecular, genetic assessments (for relevant genotypes) Microbial assessments, culture, PCR, serology, parasitology Pathology, macroscopic and histopathology

Phenotype (screen)	Confounding factors	Infection/environmental factors	Assessments
Metabolic bowel disease or neoplasia (growth curves; DXA; clinical chemistry; hematology; pathology)	Strain/sex variation in susceptibility to microbial disease, [46,262] or IBD phenotypes [263–265]	Infections [49]; Helicobacters [184,266–268]; MHV [260,269,270]; Cl piliforme [271–273], C rodentium [274–277]; Microbiome [218,262]	Molecular, genetic assessments (for relevant genotypes); Microbial assessments, culture, PCR, serology; Pathology, macroscopic and histopathology
Metabolic hepatobiliary disease (growth curves; DXA; clinical chemistry; hematology; pathology)	Strain/sex variation in susceptibility to microbial disease [185,278–280] or hepatotoxicity [281] or steatohepatitis [282,283]	Infections [49]; Helicobacters [284–287], Cl piliforme [271–273], MHV [270,288,289]; Hepatotoxins [290–292]	Molecular, genetic assessments (for relevant genotypes); Microbial assessments, culture, PCR, serology; Pathology, macroscopic and histopathology
Development (viability, fertility)	Strain variation [293–295]; Inbreeding [295–297]; Vaginal septa / imperforate vagina [298,299]	Infections [300] [301,302] [303–307]; Environment, diet (endocrine disruptors) [308–316]	Embryo imaging (OPT, uCT, MRI, HREM, Histology)²; Molecular, genetic assessments (for relevant genotypes); Microbial assessments, culture, PCR, serology; Pathology, macroscopic and histopathology
Development—Small size developmental delay (growth curves; dysmorphology; hematology; pathology)	Strain, sex variation (MPD); Malocclusion (incisor overgrowth) [317,318]	Diet, environment [319]; Infection [320–322]	Molecular, genetic assessments (for relevant genotypes); Microbial assessments, culture, PCR, serology; Pathology, macroscopic and histopathology
Respiratory (plethysmography; hematology; pathology)	Strain/sex variation [323,324] MPD; Age [325,326]	Respiratory infections in susceptible strains [49]; Diet, environment [327,328]	Molecular, genetic assessments (for relevant genotypes); Microbial assessments, culture, PCR, serology; Pathology, macroscopic and histopathology

(Continued)

Table A.16.1 (Continued)

Primary Research Area[a] (Some Tests That May Be Affected)	Potential Genetic Background (Nature) Contributors (In Common Strains)	Potential Environmental Especially Microbial (Nurture) Contributors	Tests to Confirm, Characterize, Validate a Phenotype, or to Diagnose Contributors or Confounders
Aging—Life span	Strain/sex variation in longevity and conditions that contribute to morbidity or mortality [1,246,247] (MPD) Amyloid [73,248,329,330] Cancer [331—336] (MTB)[d] Cardiovascular disease [73,151,337] Kidney disease [73,151,253,337]	Viruses in susceptible strains [49,338,339] Opportunist infections in susceptible mice [340—344] Contaminated biomaterials [345—348] Diet, environment [254,349—353]	Molecular, genetic assessments (for relevant genotypes) Microbial assessments, culture, PCR, serology Pathology, macroscopic and histopathology
Aging—Neoplasia (growth curves; life span; DXA; hematology; pathology)	Strain, sex variation [1,246,247,331—335] (MTB) Oncogenes, tumor suppressors [354,355]	Infections: Helicobacters [287,356—359] Polyoma virus [360—365] Retroviruses [366—371] Parvoviruses as oncolytic agents [372—374] Diet or diet restriction [351,375—381] Environment, other (inc carcinogens) [351,354,382—384] Irradiation [385—388]	Molecular, genetic assessments (for relevant genotypes) Serology or PCR for viral contributors Pathology, macroscopic and histopathology, and Immunohistochemistry
Aging—Neoplasia Hematopoietic (growth curves; life span; DXA; hematology; pathology)	AKR [368,389,390] BALB/c [391—394] C58 [395] NODscid [396,397] SJL/J [398—400] Swiss [127,253,370,401] Endogenous retroviruses [366—371]	Environment, diet, carcinogens [351,354,382,402—404] Irradiation [386—388]	Molecular, genetic assessments (for relevant genotypes) Pathology, macroscopic and histopathology, and Immunohistochemistry [390,405—408]

Aging—Neoplasia Lung (growth curves; life span; DXA; hematology; pathology)	Reviewed in [409,410] A/J [411–413] BALB/c [413] C57BL/6 resistant [411] FVB/N [414,415] Kras, Tp53 etc mutations [354,410,416,417]	Environment, diet, carcinogens [351,354,382,403]	Molecular, genetic assessments (for relevant genotypes) Histopathology, immunohistochemistry [418,419]
Aging—Neoplasia Mammary (growth curves; life span; DXA; hematology; pathology)	C3H [420–423] C57BL/6 [424] CBA resistant [425] FVB/N [424,426–428] Endogenous retroviruses (Mtv's) [429–431] Wnt Fgf Tp53 myc, Kras [354,432] Callahan, 1996 #2518}	MMTV [422,429,433,434] Infections [435] Environment, diet, carcinogens [424,436–441]	Molecular, genetic assessments (for relevant genotypes) Microbial assessments, PCR, serology Pathology, macroscopic and histopathology [442–447]
Aging—Neoplasia Liver (growth curves; life span; DXA; clinical chemistry; hematology; pathology)	C3H CBA [448–450] Hcs, Ras etc mutations [354,451,452]	Helicobacters [287,359] Environment, diet, carcinogens [351,382,441]	Molecular, genetic assessments (for relevant genotypes) Microbial assessments, culture, PCR, serology Histopathology, immunohistochemistry [453–455]
General—Integument (dysmorphology, SHIRPA or behavioral tests with severe pruritus; hematology, FACS, pathology)	C57BL/ etc. Mouse ulcerative dermatitis (MUD) [456,457] C3H alopecia areata [458–460]	Barbering by conspecific cagemates [461–463] Mites [464,465] Trauma [466,467] Allergens, irritants [468,469]	Molecular, genetic assessments (for relevant genotypes) Microbial assessments, culture, PCR, parasitology Pathology, macroscopic and histopathology

[a]immunology, immunity may be a relevant research area whenever infections cause or contribute to phenotypes.
[b]MPD, Mouse Phenome Database http://phenome.jax.org/ accessed July 1, 2013.
[c]See embryo phenotyping, above, Chapter 16.
[d]MTB, Mouse Tumor Biology Database http://tumor.informatics.jax.org/mtbwi/index.do, accessed July 1, 2013.

References in Order of Citation in Table A.16.1

[1] C.F. Brayton, P.M. Treuting, J.M. Ward, Pathobiology of aging mice and GEM: background strains and experimental design, Vet. Pathol. 49 (1) (2012) 85−105.

[2] R.E. Brown, A.A. Wong, The influence of visual ability on learning and memory performance in 13 strains of mice, Learn. Mem. 14 (3) (2007) 134−144.

[3] C. Brayton, Spontaneous diseases in commonly used mouse strains, in: J.G. Fox, S.W. Barthold, et al. (Eds.), The Mouse in Biomedical Research, Elsevier (Academic Press), New York, NY, 2006, pp. 623−717.

[4] A.A. Wong, R.E. Brown, Visual detection, pattern discrimination and visual acuity in 14 strains of mice, Genes Brain Behav. 5 (5) (2006) 389−403.

[5] M.F.W. Festing, Inbred strains: index of Major Mouse Strains. (1998) April 9, 1998, July 1, 2013]. Available from: <http://www.informatics.jax.org/external/festing/mouse/STRAINS.shtml>.

[6] O. Puk, C. Dalke, J. Favor, M.G., J Hrabe de Angelis, Variations of eye size parameters among different strains of mice, Mamm. Genome 17 (8) (2006) 851−857.

[7] S.J. Clapcote, N.L. Lazar, A.R. Bechard, G.A. Wood, et al., NIH Swiss and Black Swiss mice have retinal degeneration and performance deficits in cognitive tests, Comp. Med. 55 (4) (2005) 310−316.

[8] S.J. Clapcote, N.L. Lazar, A.R. Bechard, J.C. Roder, Effects of the rd1 mutation and host strain on hippocampal learning in mice, Behav. Genet. 35 (5) (2005) 591−601.

[9] L.M. Serfilippi, D.R. Pallman, M.M. Gruebbel, T.J. Kern, et al., Assessment of retinal degeneration in outbred albino mice, Comp. Med. 54 (1) (2004) 69−76.

[10] S.J. Pittler, C.E. Keeler, R.L. Sidman, W. Baehr, PCR analysis of DNA from 70-year-old sections of rodless retina demonstrates identity with the mouse rd defect, Proc. Natl. Acad. Sci. USA 90 (20) (1993) 9616−9619.

[11] A.J. Jimenez, J.M. Garcia-Fernandez, B. Gonzalez, R.G. Foster, The spatio-temporal pattern of photoreceptor degeneration in the aged rd/rd mouse retina, Cell Tissue Res. 284 (2) (1996) 193−202.

[12] C.E. Keeler, The inheritance of a retinal abnormality in white mice, Proc. Natl. Acad. Sci. USA 10 (7) (1924) 329−333.

[13] R. Smith, J. Sundberg, Inbred C57 black mice: microphthalmia and ocular infections, JaxNotes 463 (Fall) (1995)<http://jaxmice.jax.org/jaxnotes/archive/463a.html> (accessed 01.07.13).

[14] R.S. Smith, T.H. Roderick, J.P. Sundberg, Microphthalmia and associated abnormalities in inbred black mice, Lab. Anim. Sci. 44 (6) (1994) 551−560.

[15] M.M. Simon, S. Greenaway, J.K. White, H. Fuchs, et al., A comparative phenotypic and genomic analysis of C57BL/6J and C57BL/6N mouse strains, Genome. Biol. 14 (7) (2013) R82.

[16] M.J. Mattapallil, E.F. Wawrousek, C.C. Chan, H. Zhao, et al., The Rd8 mutation of the Crb1 gene is present in vendor lines of C57BL/6N mice and embryonic stem cells, and confounds ocular induced mutant phenotypes, Invest. Ophthalmol. Vis. Sci. 53 (6) (2012) 2921−2927.

[17] R.A. Rachel, G. Dolen, N.L. Hayes, A. Lu, et al., Spatiotemporal features of early neuronogenesis differ in wild-type and albino mouse retina, J. Neurosci. 22 (11) (2002) 4249−4263.

[18] G.W. Balkema, U.C. Drager, Impaired visual thresholds in hypopigmented animals, Vis. Neurosci. 6 (6) (1991) 577−585.

[19] M.G. Anderson, K.S. Nair, L.A. Amonoo, A. Mehalow, et al., GpnmbR150X allele must be present in bone marrow derived cells to mediate DBA/2J glaucoma, BMC Genet. 9 (2008) 30.

[20] A.A. Wong, R.E. Brown, Age-related changes in visual acuity, learning and memory in C57BL/6J and DBA/2J mice, Neurobiol. Aging 28 (10) (2007) 1577−1593.

[21] B. Chang, R.S. Smith, N.L. Hawes, M.G. Anderson, et al., Interacting loci cause severe iris atrophy and glaucoma in DBA/2J mice, Nat. Genet. 21 (4) (1999) 405−409.

[22] S.W. John, R.S. Smith, O.V. Savinova, N.L. Hawes, et al., Essential iris atrophy, pigment dispersion, and glaucoma in DBA/2J mice, Invest. Ophthalmol. Vis. Sci. 39 (6) (1998) 951−962.

[23] W.G. Sheldon, A.R. Warbritton, T.J. Bucci, A. Turturro, Glaucoma in food-restricted and ad libitum-fed DBA/2NNia mice, Lab. Anim. Sci. 45 (5) (1995) 508−518.

[24] J. Graw, Mouse models of cataract, J. Genet. 88 (4) (2009) 469−486.

[25] W. Pendergrass, P. Penn, D. Possin, N. Wolf, Accumulation of DNA, nuclear and mitochondrial debris, and ROS at sites of age-related cortical cataract in mice, Invest. Ophthalmol. Vis. Sci. 46 (12) (2005) 4661−4670.

[26] M.F. Hubert, G. Gerin, G. Durand-Cavagna, Spontaneous ophthalmic lesions in young Swiss mice, Lab. Anim. Sci. 49 (3) (1999) 232−240.

[27] C. Taradach, P. Greaves, Spontaneous eye lesions in laboratory animals: incidence in relation to age, Crit. Rev. Toxicol. 12 (2) (1984) 121−147.

[28] T.J. Van Winkle, M.W Balk, Spontaneous corneal opacities in laboratory mice, Lab. Anim. Sci. 36 (3) (1986) 248−255.

[29] J. Yamate, M. Tajima, Y. Maruyama, S. Kudow, Observations on soft tissue calcification in DBA/2NCrj mice in comparison with CRJ:CD-1 mice, Lab. Anim. 21 (4) (1987) 289−298.

[30] C. Verhagen, T. Rowshani, B. Willekens, N.J. van Haeringen, Spontaneous development of corneal crystalline deposits in MRL/Mp mice, Invest. Ophthalmol. Vis. Sci. 36 (2) (1995) 454−461.

[31] D.L. Greenman, P. Bryant, R.L. Kodell, W. Sheldon, Influence of cage shelf level on retinal atrophy in mice, Lab. Anim. Sci. 32 (4) (1982) 353−356.

[32] M.A. Bermudez, A.F. Vicente, M.C. Romero, M.D. Arcos, et al., Time course of cold cataract development in anesthetized mice, Curr. Eye. Res. 36 (3) (2011) 278−284.

[33] A.C. Vieira, A.F. Vicente, R. Perez, F. Gonzalez, Chloral hydrate anesthesia and lens opacification in mice, Curr. Eye. Res. 34 (5) (2009) 355−359.

[34] W.H. Ridder, S. Nusinowitz, J.R. Heckenlively, Causes of cataract development in anesthetized mice, Exp. Eye. Res. 75 (3) (2002) 365.

[35] L. Calderone, P. Grimes, M. Shalev, Acute reversible cataract induced by xylazine and by ketamine-xylazine anesthesia in rats and mice, Exp. Eye. Res. 42 (4) (1986) 331−337.

[36] F.T. Fraunfelder, R.P. Burns, Acute reversible lens opacity: caused by drugs, cold, anoxia, asphyxia, stress, death and dehydration, Exp. Eye. Res. 10 (1) (1970) 19.

[37] J.D. West, G. Fisher, Further experience of the mouse dominant cataract mutation test from an experiment with ethylnitrosourea, Mutat. Res. 164 (2) (1986) 127−136.

[38] M.M. Jablonski, X. Wang, L. Lu, D.R. Miller, et al., The Tennessee Mouse Genome Consortium: identification of ocular mutants, Vis. Neurosci. 22 (5) (2005) 595−604.

[39] S.M. Jones, T.A. Jones, K.R. Johnson, H. Yu, et al., A comparison of vestibular and auditory phenotypes in inbred mouse strains, Brain Res. 1091 (1) (2006) 40−46.

[40] E.M. Keithley, C. Canto, Q.Y. Zheng, N. Fischel-Ghodsian, et al., Age-related hearing loss and the ahl locus in mice, Hear. Res. 188 (1-2) (2004) 21−28.

[41] J.F. Willott, L. Tanner, J. O'Steen, K.R. Johnson, et al., Acoustic startle and prepulse inhibition in 40 inbred strains of mice, Behav. Neurosci. 117 (4) (2003) 716−727.

[42] J.F. Willott, L.C. Erway, Genetics of age-related hearing loss in mice. IV. Cochlear pathology and hearing loss in 25 BXD recombinant inbred mouse strains, Hear. Res. 119 (1-2) (1998) 27−36.

[43] J.F. Willott, J.G. Turner, S. Carlson, D. Ding, et al., The BALB/c mouse as an animal model for progressive sensorineural hearing loss, Hear. Res. 115 (1-2) (1998) 162−174.

[44] A.P. Nagtegaal, S. Spijker, T.T. Crins, J.G. Borst, A novel QTL underlying early-onset, low-frequency hearing loss in BXD recombinant inbred strains, Genes Brain Behav. (2012).

[45] K.R. Johnson, C. Longo-Guess, L.H. Gagnon, H. Yu, et al., A locus on distal chromosome 11 (ahl8) and its interaction with Cdh23 ahl underlie the early onset, age-related hearing loss of DBA/2J mice, Genomics 92 (4) (2008) 219−225.

[46] R.S. Sellers, C.B. Clifford, P.M. Treuting, C. Brayton, Immunological variation between inbred laboratory mouse strains: points to consider in phenotyping genetically immunomodified mice, Vet. Pathol. 49 (1) (2012) 32−43.

[47] C.J. MacArthur, D.A. Pillers, J. Pang, J.M. Degagne, et al., Gram-negative pathogen *Klebsiella oxytoca* is associated with spontaneous chronic otitis media in Toll-like receptor 4-deficient C3H/HeJ mice, Acta Otolaryngol. 128 (2) (2008) 132−138.

[48] C.R. Mitchell, J.B. Kempton, B. Scott-Tyler, D.R. Trune, Otitis media incidence and impact on the auditory brain stem response in lipopolysaccharide-nonresponsive C3H/HeJ mice, Otolaryngol. Head. Neck. Surg. 117 (5) (1997) 459−464.

[49] P.M. Treuting, C.B. Clifford, R.S. Sellers, C.F. Brayton, Of mice and microflora: considerations for genetically engineered mice, Vet. Pathol. 49 (1) (2012) 44−63.

[50] H.M. Dietrich, D. Khaschabi, B. Albini, Isolation of *Enterococcus durans* and *Pseudomonas aeruginosa* in a scid mouse colony, Lab. Anim. 30 (2) (1996) 102−107.

[51] K.R. Short, D.A. Diavatopoulos, R. Thornton, J. Pedersen, et al., Influenza virus induces bacterial and nonbacterial otitis media, J. Infect. Dis. 204 (12) (2011) 1857−1865.

[52] M. Klein, C. Schmidt, S. Kastenbauer, R. Paul, et al., MyD88-dependent immune response contributes to hearing loss in experimental pneumococcal meningitis, J. Infect. Dis. 195 (8) (2007) 1189−1193.

[53] J.E. Prince, C.F. Brayton, M.C. Fossett, J.A. Durand, et al., The differential roles of LFA-1 and Mac-1 in host defense against systemic infection with *Streptococcus pneumoniae*, J. Immunol. 166 (12) (2001) 7362−7369.

[54] Y. Kataoka, M. Haritani, M. Mori, M. Kishima, et al., Experimental infections of mice and pigs with *Streptococcus suis* type 2, J. Vet. Med. Sci. 53 (6) (1991) 1043−1049.

[55] A. Bleich, P. Kirsch, H. Sahly, J. Fahey, et al., *Klebsiella oxytoca*: opportunistic infections in laboratory rodents, Lab. Anim. 42 (3) (2008) 369−375.

[56] M.F. Goelz, J.E. Thigpen, J. Mahler, W.P. Rogers, et al., Efficacy of various therapeutic regimens in eliminating *Pasteurella pneumotropica* from the mouse, Lab. Anim. Sci. 46 (3) (1996) 280−285.

[57] M.D. McGinn, D. Bean-Knudsen, R.W. Ermel, Incidence of otitis media in CBA/J and CBA/CaJ mice, Hear. Res. 59 (1) (1992) 1−6.

[58] L.E. Davis, Comparative experimental viral labyrinthitis, Am. J. Otolaryngol. 11 (6) (1990) 382−388.

[59] K. Shimokata, Y. Nishiyama, V. Ito, Y. Kimura, et al., Affinity of Sendai virus for the inner ear of mice, Infect. Immun. 16 (2) (1977) 706−708.

[60] C.H. White, J.D. Ohmen, S. Sheth, A.F. Zebboudj, et al., Genome-wide screening for genetic loci associated with noise-induced hearing loss, Mamm. Genome 20 (4) (2009) 207–213.

[61] J.G. Turner, J.L. Parrish, L.F. Hughes, L.A. Toth, et al., Hearing in laboratory animals: strain differences and nonauditory effects of noise, Comp. Med. 55 (1) (2005) 12–23.

[62] R.R. Davis, J.K. Newlander, X.-B. Ling, G.A. Cortopassi, et al., Genetic basis for susceptibility to noise-induced hearing loss in mice, Hear. Res. 155 (1-2) (2001) 82.

[63] J.E. Jeskey, J.F. Willott, Modulation of prepulse inhibition by an augmented acoustic environment in DBA/2J mice, Behav. Neurosci. 114 (5) (2000) 991–997.

[64] J.F. Willott, L.S. Bross, S.L. McFadden, Morphology of the cochlear nucleus in CBA/J mice with chronic, severe sensorineural cochlear pathology induced during adulthood, Hear. Res. 74 (1-2) (1994) 1–21.

[65] S. Someya, M. Tanokura, R. Weindruch, T.A. Prolla, et al., Effects of caloric restriction on age-related hearing loss in rodents and rhesus monkeys, Curr. Aging Sci. 3 (1) (2010) 20–25.

[66] J.F. Willott, L.C. Erway, J.R. Archer, D.E. Harrison, Genetics of age-related hearing loss in mice. II. Strain differences and effects of caloric restriction on cochlear pathology and evoked response thresholds, Hear. Res. 88 (1-2) (1995) 143–155.

[67] A.L. Poirrier, P. Van den Ackerveken, T.S. Kim, R. Vandenbosch, et al., Ototoxic drugs: difference in sensitivity between mice and guinea pigs, Toxicol. Lett. 193 (1) (2010) 41–49.

[68] W.J. Wu, S.H. Sha, J.D. McLaren, K. Kawamoto, et al., Aminoglycoside ototoxicity in adult CBA, C57BL and BALB mice and the Sprague–Dawley rat, Hear. Res. 158 (1-2) (2001) 165–178.

[69] M. Hultcrantz, M.L. Spangberg, Pathology of the cochlea following a spontaneous mutation in DBA/2 mice, Acta Otolaryngol. 117 (5) (1997) 689–695.

[70] D. Bloom, M. Hultcrantz, Vestibular morphology in relation to age and circling behavior, Acta Otolaryngol. 114 (4) (1994) 387–392.

[71] M. Hewicker, G. Trautwein, Sequential study of vasculitis in MRL mice, Lab. Anim. 21 (4) (1987) 335–341.

[72] W.M. Qu, T. Miyazaki, M. Terada, L.M. Lu, et al., Genetic dissection of vasculitis in MRL/lpr lupus mice: a novel susceptibility locus involving the CD72c allele, Eur. J. Immunol. 30 (7) (2000) 2027–2037.

[73] K. Maita, M. Hirano, T. Harada, K. Mitsumori, et al., Mortality, major cause of moribundity, and spontaneous tumors in CD-1 mice, Toxicol. Pathol. 16 (3) (1988) 340–349.

[74] T. Southard, C.F. Brayton, Spontaneous unilateral brainstem infarction in Swiss mice, Vet. Pathol. 48 (3) (2011) 726–729.

[75] D.F. Kohn, W.F. MacKenzie, Inner ear disease characterized by rolling in C3H mice, J. Am. Vet. Med. Assoc. 177 (9) (1980) 815–817.

[76] L.D. Olson, R.D. Ediger, Histopathologic study of the heads of circling mice infected with *Pseudomonas aeruginosa*, Lab. Anim. Sci. 22 (4) (1972) 522–527.

[77] P. Boadas-Vaello, E. Jover, S. Saldana-Ruiz, C. Soler-Martin, et al., Allylnitrile metabolism by CYP2E1 and other CYPs leads to distinct lethal and vestibulotoxic effects in the mouse, Toxicol. Sci. 107 (2) (2009) 461–472.

[78] C. Soler-Martin, N. Diez-Padrisa, P. Boadas-Vaello, J. Llorens, Behavioral disturbances and hair cell loss in the inner ear following nitrile exposure in mice, guinea pigs, and frogs, Toxicol. Sci. 96 (1) (2007) 123–132.

[79] S.M. Chou, T. Miike, W.M. Payne, G.J. Davis, Neuropathology of "spinning syndrome" induced by prenatal intoxication with a PCB in mice, Ann. N.Y. Acad. Sci. 320 (1979) 373–395.

[80] J.P. McLin, O. Steward, Comparison of seizure phenotype and neurodegeneration induced by systemic kainic acid in inbred, outbred, and hybrid mouse strains, Eur. J. Neurosci. 24 (8) (2006) 2191−2202.

[81] M.H. Mohajeri, R. Madani, K. Saini, H.P. Lipp, et al., The impact of genetic background on neurodegeneration and behavior in seizured mice, Genes Brain Behav. 3 (4) (2004) 228−239.

[82] I.G. M. McKhann, H.J. Wenzel, C.A. Robbins, A.A. Sosunov, et al., Mouse strain differences in kainic acid sensitivity, seizure behavior, mortality, and hippocampal pathology, Neuroscience 122 (2) (2003) 551−561.

[83] G.T. Golden, T.N. Ferraro, G.G. Smith, R.L. Snyder, et al., Acute cocaine-induced seizures: differential sensitivity of six inbred mouse strains, Neuropsychopharmacology. 24 (3) (2001) 291−299.

[84] T.N. Ferraro, G.T. Golden, R. Snyder, M. Laibinis, et al., Genetic influences on electrical seizure threshold, Brain Res. 813 (1) (1998) 207−210.

[85] A.C. Hom, I.E. Leppik, C.A. Rask, Effects of estradiol and progesterone on seizure sensitivity in oophorectomized DBA/2J mice and C57/EL hybrid mice, Neurology 43 (1) (1993) 198−204.

[86] A. Smolen, T.N. Smolen, Genetic influence on increased seizure susceptibility in pregnancy, Life Sci. 39 (17) (1986) 1525−1530.

[87] F.L. Engstrom, D.M. Woodbury, Seizure susceptibility in DBA and C57 mice: the effects of various convulsants, Epilepsia 29 (4) (1988) 389−395.

[88] M.F. Goelz, J. Mahler, J. Harry, P. Myers, et al., Neuropathologic findings associated with seizures in FVB mice, Lab. Anim. Sci. 48 (1) (1998) 34−37.

[89] W.B. Iturrian, G.B. Fink, Effect of noise in the animal house on seizure susceptibility and growth of mice, Lab. Anim. Care. 18 (5) (1968) 557−560.

[90] K.R. Henry, Audiogenic seizure susceptibility induced in C57BL-6J mice by prior auditory exposure, Science 158 (3803) (1967) 938−940.

[91] M.M. Niaussat, Experimentally induced otitis and audiogenic seizure in the mouse, Experientia 33 (4) (1977) 473−474.

[92] P.F. Fabene, A. Sbarbati, In vivo MRI in different models of experimental epilepsy, Curr. Drug Targets 5 (7) (2004) 629−636.

[93] M.M. Mirrione, S.E. Tsirka, Neuroimaging in animal seizure models with (18)FDG-PET, Epilepsy Res. Treat. 2011 (2011) 369295.

[94] M.G. Drage, G.L. Holmes, T.N. Seyfried, Hippocampal neurons and glia in epileptic EL mice, J. Neurocytol. 31 (8-9) (2002) 681−692.

[95] D. Wahlsten, A. Bachmanov, D.A. Finn, J.C. Crabbe, Stability of inbred mouse strain differences in behavior and brain size between laboratories and across decades, Proc. Natl. Acad. Sci. USA 103 (44) (2006) 16364−16369.

[96] C.C. Zygourakis, G.D. Rosen, Quantitative trait loci modulate ventricular size in the mouse brain, J. Comp. Neurol. 461 (3) (2003) 362−369.

[97] M. Tanaka, T. Marunouchi, Abnormality in the cerebellar folial pattern of C57BL/6J mice, Neurosci. Lett. 390 (3) (2005) 182.

[98] P.E. Neumann, J.D. Garretson, G.P. Skabardonis, G.G. Mueller, Genetic analysis of cerebellar folial pattern in crosses of C57BL/6J and DBA/2J inbred mice, Brain Res. 619 (1-2) (1993) 81−88.

[99] P.A. Cooper, R.H. Benno, M.E. Hahn, J.K. Hewitt, Genetic analysis of cerebellar foliation patterns in mice (Mus musculus), Behav. Genet. 21 (4) (1991) 405−419.

[100] E.S. Brodkin, BALB/c mice: low sociability and other phenotypes that may be relevant to autism, Behav. Brain Res. 176 (1) (2007) 53−65.

[101] D. Wahlsten, K.M. Bishop, H.S. Ozaki, Recombinant inbreeding in mice reveals thresholds in embryonic corpus callosum development, Genes Brain Behav. 5 (2) (2006) 170—188.

[102] S.A. Balogh, C.S. McDowell, A.J. Stavnezer, V.H. Denenberg, A behavioral and neuroanatomical assessment of an inbred substrain of 129 mice with behavioral comparisons to C57BL/6J mice, Brain Res. 836 (1-2) (1999) 38—48.

[103] D.J. Livy, D. Wahlsten, Tests of genetic allelism between four inbred mouse strains with absent corpus callosum, J. Hered. 82 (6) (1991) 459—464.

[104] C.C. Filgueiras, A.C. Manhães, Effects of callosal agenesis on rotational side preference of BALB/cCF mice in the free swimming test, Behav. Brain. Res. 155 (1) (2004) 13.

[105] C.A. Carton, J.H. Perry, A. Winter, V. Tennyson, Studies of hydrocephalus in C57 black mice, Trans. Am. Neurol. Assoc. 81 (1956) 147—149 (81st Meeting).

[106] V.H. Denenberg, G.F. Sherman, G.D. Rosen, L. Morrison, et al., A behavior profile of the MRL/Mp lpr/lpr mouse and its association with hydrocephalus, Brain Behav. Immun. 6 (1) (1992) 40—49.

[107] I. Yonekura, N. Kawahara, H. Nakatomi, K. Furuya, et al., A model of global cerebral ischemia in C57 BL/6 mice, J. Cereb. Blood Flow Metab. 24 (2) (2004) 151—158.

[108] G. Yang, K. Kitagawa, K. Matsushita, T. Mabuchi, et al., C57BL/6 strain is most susceptible to cerebral ischemia following bilateral common carotid occlusion among seven mouse strains: selective neuronal death in the murine transient forebrain ischemia, Brain Res. 752 (1-2) (1997) 209—218.

[109] N. Beckmann, High resolution magnetic resonance angiography non-invasively reveals mouse strain differences in the cerebrovascular anatomy in vivo, Magn. Reson. Med. 44 (2) (2000) 252—258.

[110] D. Wahlsten, Deficiency of corpus callosum varies with strain and supplier of the mice, Brain Res. 239 (2) (1982) 329—347.

[111] A. Holtz, G. Borman, C.P. Li, Hydrocephalus in mice infected with polyoma virus, Proc. Soc. Exp. Biol. Med. 121 (4) (1966) 1196—1200.

[112] K. Kristensson, J. Leestma, B. Lundh, E. Norrby, Sendai virus infection in the mouse brain: virus spread and long-term effects, Acta Neuropathol. 63 (2) (1984) 89—95.

[113] I. Tsunoda, I.J. McCright, L.Q. Kuang, A. Zurbriggen, et al., Hydrocephalus in mice infected with a Theiler's murine encephalomyelitis virus variant, J. Neuropathol. Exp. Neurol. 56 (12) (1997) 1302—1313.

[114] J. Lagace-Simard, J.P. Descoteaux, G. Lussier, Experimental pneumovirus infections: 1. Hydrocephalus of mice due to infection with pneumonia virus of mice (PVM), Am. J. Pathol. 101 (1) (1980) 31—40.

[115] M. Tardieu, A. Goffinet, G. Harmant-van Rijckevorsel, G. Lyon, Ependymitis, leukoencephalitis, hydrocephalus, and thrombotic vasculitis following chronic infection by mouse hepatitis virus 3 (MHV 3), Acta Neuropathol. 58 (3) (1982) 168—176.

[116] M. Hausler, B. Sellhaus, S. Scheithauer, M. Engler, et al., Murine gammaherpesvirus-68 infection of mice: a new model for human cerebral Epstein—Barr virus infection, Ann. Neurol. 57 (4) (2005) 600—603.

[117] G. Margolis, L. Kilham, Hydrocephalus in hamsters, ferrets, rats, and mice following inoculations with reovirus type I. II. Pathologic studies, Lab. Invest. 21 (3) (1969) 189—198.

[118] J.C. Ramirez, A. Fairen, J.M. Almendral, Parvovirus minute virus of mice strain i multiplication and pathogenesis in the newborn mouse brain are restricted to proliferative areas and to migratory cerebellar young neurons, J. Virol. 70 (11) (1996) 8109—8116.

[119] S.A. Ferguson, Neuroanatomical and functional alterations resulting from early postnatal cerebellar insults in rodents, Pharmacol. Biochem. Behav. 55 (4) (1996) 663–671.

[120] M. Vainzof, D. Ayub-Guerrieri, P. Onofre, P. Martins, et al., Animal models for genetic neuromuscular diseases, J. Mol. Neurosci. 34 (3) (2008) 241.

[121] K. Wenzel, C. Geier, F. Qadri, N. Hubner, et al., Dysfunction of dysferlin-deficient hearts, J. Mol. Med. 85 (11) (2007) 1203.

[122] M. Ho, C.M. Post, L.R. Donahue, H.G. Lidov, et al., Disruption of muscle membrane and phenotype divergence in two novel mouse models of dysferlin deficiency, Hum. Mol. Genet. 13 (18) (2004) 1999–2010.

[123] J.P. Lefaucheur, C. Pastoret, A. Sebille, Phenotype of dystrophinopathy in old mdx mice, Anat. Rec. 242 (1) (1995) 70–76.

[124] M.M. Zdanowicz, A.E. Slonim, I. Bilaniuk, M.M. O'Connor, et al., High protein diet has beneficial effects in murine muscular dystrophy, J. Nutr. 125 (5) (1995) 1150–1158.

[125] G. Jasmin, E. Bajusz, Myocardial lesions in strain 129 dystrophic mice, Nature 193 (1962) 181–182.

[126] A.M. Michelson, E.S. Russel, P.J. Harman, Dystrophia muscularis: a hereditary primary myopathy in the house mouse, Proc. Natl. Acad. Sci. USA 41 (1955) 1079–1084.

[127] A.V. Ceccarelli, N. Rozengurt, Outbreak of hind limb paralysis in young CFW Swiss Webster mice, Comp. Med. 52 (2) (2002) 171–175.

[128] R.E. Mitchel, J.S. Jackson, D.P. Morrison, S.M. Carlisle, Low doses of radiation increase the latency of spontaneous lymphomas and spinal osteosarcomas in cancer-prone, radiation-sensitive Trp53 heterozygous mice, Radiat. Res. 159 (3) (2003) 320–327.

[129] G.J. Krinke, W. Kaufmann, A.T. Mahrous, P. Schaetti, Morphologic characterization of spontaneous nervous system tumors in mice and rats, Toxicol. Pathol. 28 (1) (2000) 178–192.

[130] K.R. Bailey, N.R. Rustay, J.N. Crawley, Behavioral phenotyping of transgenic and knockout mice: practical concerns and potential pitfalls, ILAR J. 47 (2) (2006) 124–131.

[131] A.A. Dandekar, S. Perlman, Virus-induced demyelination in nude mice is mediated by gamma delta T cells, Am. J. Pathol. 161 (4) (2002) 1255–1263.

[132] S.E. Coley, E. Lavi, S.G. Sawicki, L. Fu, et al., Recombinant mouse hepatitis virus strain A59 from cloned, full-length cDNA replicates to high titers in vitro and is fully pathogenic in vivo, J. Virol. 79 (5) (2005) 3097–3106.

[133] E. Lavi, D.H. Gilden, M.K. Highkin, S.R. Weiss, The organ tropism of mouse hepatitis virus A59 in mice is dependent on dose and route of inoculation, Lab. Anim. Sci. 36 (2) (1986) 130–135.

[134] J.L. Woyciechowska, B.D. Trapp, D.H. Patrick, I.C. Shekarchi, et al., Acute and subacute demyelination induced by mouse hepatitis virus strain A59 in C3H mice, J. Exp. Pathol. 1 (4) (1984) 295–306.

[135] C. Deb, R.G. Lafrance-Corey, L. Zoecklein, L. Papke, et al., Demyelinated axons and motor function are protected by genetic deletion of perforin in a mouse model of multiple sclerosis, J. Neuropathol. Exp. Neurol. 68 (9) (2009) 1037–1048.

[136] N.P. Turrin, Central nervous system Toll-like receptor expression in response to Theiler's murine encephalomyelitis virus-induced demyelination disease in resistant and susceptible mouse strains, Virol. J. 5 (2008) 154.

[137] A.S. Kumar, H.V. Reddi, A.Y. Kung, M. Dal Canto, et al., Virus persistence in an animal model of multiple sclerosis requires virion attachment to sialic acid coreceptors, J. Virol. 78 (16) (2004) 8860–8867.

[138] A. Azoulay-Cayla, S. Syan, M. Brahic, J.F. Bureau, Roles of the H-2D(b) and H-K(b) genes in resistance to persistent Theiler's murine encephalomyelitis virus infection of the central nervous system, J. Gen. Virol. 82 (Pt 5) (2001) 1043–1047.

[139] I. Tsunoda, Y. Iwasaki, H. Terunuma, K. Sako, et al., A comparative study of acute and chronic diseases induced by two subgroups of Theiler's murine encephalomyelitis virus, Acta Neuropathol. 91 (6) (1996) 595–602.

[140] M. Girgenrath, M.L. Beermann, V.K. Vishnudas, S. Homma, et al., Pathology is alleviated by doxycycline in a laminin-alpha2-null model of congenital muscular dystrophy, Ann. Neurol. 65 (1) (2009) 47–56.

[141] C. Berthonneche, B. Peter, F. Schupfer, P. Hayoz, et al., Cardiovascular response to beta-adrenergic blockade or activation in 23 inbred mouse strains, PLoS One 4 (8) (2009) e6610.

[142] C.F. Deschepper, J.L. Olson, M. Otis, N. Gallo-Payet, Characterization of blood pressure and morphological traits in cardiovascular-related organs in 13 different inbred mouse strains, J. Appl. Physiol. 97 (1) (2004) 369–376.

[143] V.V. Colinayo, J.H. Qiao, X. Wang, K.L. Krass, et al., Genetic loci for diet-induced atherosclerotic lesions and plasma lipids in mice, Mamm. Genome. 14 (7) (2003) 464–471.

[144] B.D. Hoit, S. Kiatchoosakun, J. Restivo, D. Kirkpatrick, et al., Naturally occurring variation in cardiovascular traits among inbred mouse strains, Genomics 79 (5) (2002) 679.

[145] J.F. Van Vleet, V.J. Ferrans, Myocardial diseases of animals, Am. J. Pathol. 124 (1) (1986) 98–178.

[146] S. Fujihira, T. Yamamoto, M. Matsumoto, K. Yoshizawa, et al., The high incidence of atrial thrombosis in mice given doxorubicin, Toxicol. Pathol. 21 (4) (1993) 362–368.

[147] H. Meier, W.G. Hoag, Studies on left auricular thrombosis in mice, Exp. Med. Surg. 19 (1961) 317–322.

[148] H. Meng, I. Vera, N. Che, X. Wang, et al., Identification of Abcc6 as the major causal gene for dystrophic cardiac calcification in mice through integrative genomics, Proc. Natl. Acad. Sci. USA 104 (11) (2007) 4530–4535.

[149] J.I. Everitt, L.M. Olson, J.B. Mangum, W.J. Visek, High mortality with severe dystrophic cardiac calcinosis in C3H/OUJ mice fed high fat purified diets, Vet. Pathol. 25 (2) (1988) 113–118.

[150] G.J. Eaton, R.P. Custer, F.N. Johnson, K.T. Stabenow, Dystrophic cardiac calcinosis in mice: genetic, hormonal, and dietary influences, Am. J. Pathol. 90 (1) (1978) 173–186.

[151] C.A. Conover, L.K. Bale, J.R. Mader, M.A. Mason, et al., Longevity and age-related pathology of mice deficient in pregnancy-associated plasma protein-A, J. Gerontol. A Biol. Sci. Med. Sci. 65 (6) (2010) 590–599.

[152] P.M. Nishina, J.K. Naggert, J. Verstuyft, B. Paigen, Atherosclerosis in genetically obese mice: the mutants obese, diabetes, fat, tubby, and lethal yellow, Metabolism. 43 (5) (1994) 554–558.

[153] J. Hoover-Plow, A. Shchurin, E. Hart, J. Sha, et al., Genetic background determines response to hemostasis and thrombosis, BMC. Blood. Disord. 6 (2006) 6.

[154] H.M. Dansky, S.A. Charlton, J.L. Sikes, S.C. Heath, et al., Genetic background determines the extent of atherosclerosis in ApoE-deficient mice, Arterioscler. Thromb. Vasc. Biol. 19 (8) (1999) 1960–1968.

[155] M. Nose, M. Nishihara, J. Kamogawa, M. Terada, et al., Genetic basis of autoimmune disease in MRL/lpr mice: dissection of the complex pathological manifestations and their susceptibility loci, Rev. Immunogenet. 2 (1) (2000) 154–164.

[156] J.W. Mullink, G.T. Haneveld, Polyarteritis in mice due to spontaneous hypertension, J Comp Pathol. 89 (1) (1979) 99−106.

[157] K. Doi, Experimental encephalomyocarditis virus infection in small laboratory rodents, J. Comp. Pathol. 144 (1) (2011) 25−40.

[158] H.F. Krous, N.E. Langlois, Ljungan virus: a commentary on its association with fetal and infant morbidity and mortality in animals and humans, Birth. Defects. Res. A. Clin. Mol. Teratol. 88 (11) (2010) 947−952.

[159] H.L. Lipton, A.S.M. Kumar, S. Hertzler, Cardioviruses: encephalomyocarditis virus and Theiler's mouse encephalmyelitis virus, in: J.G. Fox, S.W. Barthold, et al. (Eds.), The Mouse in Biomedical Research, Elsevier (Academic Press), New York, NY, 2006, pp. 311−323.

[160] J.C. Lenzo, D. Fairweather, V. Cull, G.R. Shellam, et al., Characterisation of murine cytomegalovirus myocarditis: cellular infiltration of the heart and virus persistence, J. Mol. Cell. Cardiol. 34 (6) (2002) 629−640.

[161] B. Sherry, C.J. Baty, M.A. Blum, Reovirus-induced acute myocarditis in mice correlates with viral RNA synthesis rather than generation of infectious virus in cardiac myocytes, J. Virol. 70 (10) (1996) 6709−6715.

[162] A.L. Armstrong, S.W. Barthold, D.H. Persing, D.S. Beck, Carditis in Lyme disease susceptible and resistant strains of laboratory mice infected with Borrelia burgdorferi, Am. J. Trop. Med. Hyg. 47 (2) (1992) 249−258.

[163] P. Price, K.S. Eddy, J.M. Papadimitriou, D.L. Faulkner, et al., Genetic determination of cytomegalovirus-induced and age-related cardiopathy in inbred mice. Characterization of infiltrating cells, Am. J. Pathol. 138 (1) (1991) 59−67.

[164] N. Tomioka, C. Kishimoto, A. Matsumori, C. Kawai, Mural thrombus in experimental viral myocarditis in mice: relation between thrombosis and congestive heart failure, Cardiovasc. Res. 20 (9) (1986) 665−671.

[165] D.J. Chess, B. Lei, B.D. Hoit, A.M. Azimzadeh, et al., Effects of a high saturated fat diet on cardiac hypertrophy and dysfunction in response to pressure overload, J. Card. Fail. 14 (1) (2008) 82−88.

[166] J. Joven, A. Rull, N. Ferre, J.C. Escola-Gil, et al., The results in rodent models of atherosclerosis are not interchangeable: the influence of diet and strain, Atherosclerosis 195 (2) (2007) e85−e92.

[167] D. Teupser, A.D. Persky, J.L. Breslow, Induction of atherosclerosis by low-fat, semisynthetic diets in LDL receptor-deficient C57BL/6J and FVB/NJ mice: comparison of lesions of the aortic root, brachiocephalic artery, and whole aorta (en face measurement), Arterioscler. Thromb. Vasc. Biol. 23 (10) (2003) 1907−1913.

[168] J.I. Everitt, P.W. Ross, D.A. Neptun, J.B. Mangum, Effect of a purified diet on dystrophic cardiac calcinosis in mice, Lab. Anim. Sci. 38 (4) (1988) 426−429.

[169] L.M. Klevay, Atrial thrombosis, abnormal electrocardiograms and sudden death in mice due to copper deficiency, Atherosclerosis 54 (2) (1985) 213−224.

[170] C.R. Ball, B.R. Clower, W.L. Williams, Dietary-induced atrial thrombosis in mice, Arch. Pathol. 80 (4) (1965) 391−396.

[171] A. Phinikaridou, M.E. Andia, P. Saha, B. Modarai, et al., In vivo magnetization transfer and diffusion-weighted magnetic resonance imaging detects thrombus composition in a mouse model of deep vein thrombosis, Circ. Cardiovasc. Imaging 6 (3) (2013) 433−440.

[172] H. Fuchs, V. Gailus-Durner, S. Neschen, T. Adler, et al., Innovations in phenotyping of mouse models in the German Mouse Clinic, Mamm. Genome 23 (9-10) (2012) 611−622.

[173] D. Weinreb, J. Aguinaldo, J. Feig, E. Fisher, et al., Non-invasive MRI of mouse models of atherosclerosis, NMR Biomed. 20 (3) (2007) 256–264.

[174] B.D. Hoit, Echocardiographic characterization of the cardiovascular phenotype in rodent models, Toxicol. Pathol. 34 (1) (2006) 105–110.

[175] B.T. Kile, C.L. Mason-Garrison, M.J. Justice, Sex and strain-related differences in the peripheral blood cell values of inbred mouse strains, Mamm. Genome. 14 (1) (2003) 81–85.

[176] L.L. Peters, E.M. Cheever, H.R. Ellis, P.A. Magnani, et al., Large-scale, high-throughput screening for coagulation and hematologic phenotypes in mice, Physiol. Genomics. 11 (3) (2002) 185–193.

[177] D. Doeing, J. Borowicz, E. Crockett, Gender dimorphism in differential peripheral blood leukocyte counts in mice using cardiac, tail, foot, and saphenous vein puncture methods, BioMed Cent. Clin. Pathol. 3 (3) (2003)<http://www.pubmedcentral.nih.gov/articlerender.fcgi?artid = 201031> (accessed 01.07.13).

[178] L.M. Serfilippi, D.R. Stackhouse Pallman, B. Russell, C.B. Spainhour, Serum clinical chemistry and hematology reference values in outbred stocks of albino mice from three commonly used vendors and two inbred strains of albino mice, Contemp. Topics LAS 42 (3) (2003) 46–52.

[179] T. Itoh, H. Iwai, K. Ueda, Comparative lung pathology of inbred strain of mice resistant and susceptible to Sendai virus infection, J. Vet. Med. Sci. 53 (2) (1991) 275–279.

[180] A.Y. Simon, K. Moritoh, D. Torigoe, A. Asano, et al., Multigenic control of resistance to Sendai virus infection in mice, Infect. Genet. Evol. 9 (6) (2009) 1253–1259.

[181] A. Hirai, N. Ohtsuka, T. Ikeda, R. Taniguchi, et al., Role of mouse hepatitis virus (MHV) receptor murine CEACAM1 in the resistance of mice to MHV infection: studies of mice with chimeric mCEACAM1a and mCEACAM1b, J. Virol. 84 (13) (2010) 6654–6666.

[182] A.L. Yancey, H.L. Watson, S.C. Cartner, J.W. Simecka, Gender is a major factor in determining the severity of mycoplasma respiratory disease in mice, Infect. Immun. 69 (5) (2001) 2865–2871.

[183] S.C. Cartner, J.W. Simecka, D.E. Briles, G.H. Cassell, et al., Resistance to mycoplasmal lung disease in mice is a complex genetic trait, Infect. Immun. 64 (12) (1996) 5326–5331.

[184] M.T. Whary, J.G. Fox, Natural and experimental Helicobacter infections, Comp. Med. 54 (2) (2004) 128–158.

[185] A. Garcia, M.M. Ihrig, R.C. Fry, Y. Feng, et al., Genetic susceptibility to chronic hepatitis is inherited codominantly in Helicobacter hepaticus-infected AB6F1 and B6AF1 hybrid male mice, and progression to hepatocellular carcinoma is linked to hepatic expression of lipogenic genes and immune function-associated networks, Infect. Immun. 76 (5) (2008) 1866–1876.

[186] D.L. Rosenstreich, A.C. Weinblatt, A.D. O'Brien, Genetic control of resistance to infection in mice, Crit. Rev. Immunol. 3 (4) (1982) 263–330.

[187] J.R. Schurr, E. Young, P. Byrne, C. Steele, et al., Central role of toll-like receptor 4 signaling and host defense in experimental pneumonia caused by Gram-negative bacteria, Infect. Immun. 73 (1) (2005) 532–545.

[188] P.N. Paradkar, P.S. Blum, M.A. Berhow, H. Baumann, et al., Dietary isoflavones suppress endotoxin-induced inflammatory reaction in liver and intestine, Cancer Lett. 215 (1) (2004) 21–28.

[189] J. Shen, H. Ren, C. Tomiyama-Miyaji, M. Watanabe, et al., Resistance and augmentation of innate immunity in mice exposed to starvation, Cell. Immunol. 259 (1) (2009) 66−73.

[190] M. Watanabe, C. Tomiyama-Miyaji, E. Kainuma, M. Inoue, et al., Role of alpha-adrenergic stimulus in stress-induced modulation of body temperature, blood glucose and innate immunity, Immunol. Lett. 115 (1) (2008) 43−49.

[191] K.A. Jhaveri, R.A. Trammell, L.A. Toth, Effect of environmental temperature on sleep, locomotor activity, core body temperature and immune responses of C57BL/6J mice, Brain Behav. Immun. 21 (7) (2007) 975−987.

[192] X. Peng, C.M. Lang, C.K. Drozdowicz, B.M. Ohlsson-Wilhelm, Effect of cage population density on plasma corticosterone and peripheral lymphocyte populations of laboratory mice, Lab. Anim. 23 (4) (1989) 302−306.

[193] S.B. Pruett, R. Fan, Q. Zheng, C. Schwab, Patterns of immunotoxicity associated with chronic as compared with acute exposure to chemical or physical stressors and their relevance with regard to the role of stress and with regard to immunotoxicity testing, Toxicol. Sci. 109 (2) (2009) 265−275.

[194] K.L. Svenson, R. Von Smith, P.A. Magnani, H.R. Suetin, et al., Multiple trait measurements in 43 inbred mouse strains captures the phenotypic diversity characteristic of human populations, J. Appl. Physiol. 102 (6) (2007) 2369−2378.

[195] J.-L. Mu, J.K. Naggert, K.L. Svenson, G.B. Collin, et al., Quantitative trait loci analysis for the differences in susceptibility to atherosclerosis and diabetes between inbred mouse strains C57BL/6J and C57BLKS/J, J. Lipid. Res. 40 (7) (1999) 1328−1335.

[196] D. Teupser, M. Tan, A.D. Persky, J.L. Breslow, Atherosclerosis quantitative trait loci are sex- and lineage-dependent in an intercross of C57BL/6 and FVB/N low-density lipoprotein receptor −/− mice, Proc. Natl. Acad. Sci. USA 103 (1) (2006) 123−128.

[197] N. Ishimori, R. Li, P.M. Kelmenson, R. Korstanje, et al., Quantitative trait loci that determine plasma lipids and obesity in C57BL/6J and 129S1/SvImJ inbred mice, J. Lipid. Res. 45 (9) (2004) 1624−1632.

[198] J.W. Yoon, H.S. Jun, Viruses cause type 1 diabetes in animals, Ann. N.Y. Acad. Sci. 1079 (2006) 138−146.

[199] H.-T. Wu, C.K. Chang, C.-W. Tsao, Y.-J. Wen, et al., Insulin resistance without obesity induced by cotton pellet granuloma in mice, Lab. Invest. 89 (3) (2009) 362.

[200] M. Tous, V. Ribas, J.C. Escola-Gil, F. Blanco-Vaca, et al., Manipulation of inflammation modulates hyperlipidemia in apolipoprotein E-deficient mice: a possible role for interleukin-6, Cytokine 34 (3-4) (2006) 224−232.

[201] M. Tous, V. Ribas, N. Ferre, J.C. Escola-Gil, et al., Turpentine-induced inflammation reduces the hepatic expression of the multiple drug resistance gene, the plasma cholesterol concentration and the development of atherosclerosis in apolipoprotein E deficient mice, Biochim. Biophys. Acta 1733 (2-3) (2005) 192−198.

[202] K.J. Maurer, M.M. Ihrig, A.B. Rogers, V. Ng, et al., Identification of cholelithogenic enterohepatic helicobacter species and their role in murine cholesterol gallstone formation, Gastroenterology 128 (4) (2005) 1023−1033.

[203] R. Li, K.L. Svenson, L.R.B. Donahue, L.L. Peters, et al., Relationships of dietary fat, body composition, and bone mineral density in inbred mouse strain panels, Physiol. Genomics. 33 (1) (2008) 26−32.

[204] A.E. Petro, J. Cotter, D.A. Cooper, J.C. Peters, et al., Fat, carbohydrate, and calories in the development of diabetes and obesity in the C57BL/6J mouse, Metabolism 53 (4) (2004) 454−457.

[205] B. Martin, S. Ji, S. Maudsley, M.P. Mattson, "Control" laboratory rodents are metabolically morbid: why it matters, Proc. Natl. Acad. Sci. USA 107 (14) (2010) 6127–6133.

[206] B. Niklasson, A. Samsioe, M. Blixt, S. Sandler, et al., Prenatal viral exposure followed by adult stress produces glucose intolerance in a mouse model, Diabetologia 49 (9) (2006) 2192–2199.

[207] S.M. Clee, A.D. Attie, The genetic landscape of type 2 diabetes in mice, Endocr. Rev. 28 (1) (2007) 48–83.

[208] Z. Qi, H. Fujita, J. Jin, L.S. Davis, et al., Characterization of susceptibility of inbred mouse strains to diabetic nephropathy, Diabetes 54 (9) (2005) 2628–2637.

[209] M. Schwarz, D.L. Davis, B.R. Vick, D.W. Russell, Genetic analysis of intestinal cholesterol absorption in inbred mice, J. Lipid. Res. 42 (11) (2001) 1801–1811.

[210] M. Schwarz, D.L. Davis, B.R. Vick, D.W. Russell, Genetic analysis of cholesterol accumulation in inbred mice, J. Lipid. Res. 42 (11) (2001) 1812–1819.

[211] F. Zheng, G.E. Striker, C. Esposito, E. Lupia, et al., Strain differences rather than hyperglycemia determine the severity of glomerulosclerosis in mice, Kidney Int. 54 (6) (1998) 1999.

[212] R.C. Davis, E.E. Schadt, A.C.L. Cervino, M. Peterfy, et al., Ultrafine mapping of SNPs from mouse strains C57BL/6J, DBA/2J, and C57BLKS/J for loci contributing to diabetes and atherosclerosis susceptibility, Diabetes 54 (4) (2005) 1191–1199.

[213] D.R. Garris, B.L. Garris, Cytochemical analysis of pancreatic islet hypercytolipidemia following diabetes (db/db) and obese (ob/ob) mutation expression: influence of genomic background, Pathobiology 71 (5) (2004) 231–240.

[214] M. Haluzik, C. Colombo, O. Gavrilova, S. Chua, et al., Genetic background (C57BL/6J versus FVB/N) strongly influences the severity of diabetes and insulin resistance in ob/ob mice, Endocrinology 145 (7) (2004) 3258–3264.

[215] M. Rossmeisl, J.S. Rim, R.A. Koza, L.P. Kozak, Variation in type 2 diabetes-related traits in mouse strains susceptible to diet-induced obesity, Diabetes 52 (8) (2003) 1958–1966.

[216] S.A. Schreyer, D.L. Wilson, R.C. LeBoeuf, C57BL/6 mice fed high fat diets as models for diabetes-accelerated atherosclerosis, Atherosclerosis 136 (1) (1998) 17–24.

[217] E.H. Leiter, P.H. Le, D.L. Coleman, Susceptibility to db gene and streptozotocin-induced diabetes in C57BL mice: control by gender-associated, MHC-unlinked traits, Immunogenetics 26 (1-2) (1987) 6–13.

[218] A. Bleich, A.K. Hansen, Time to include the gut microbiota in the hygienic standardisation of laboratory rodents, Comp. Immunol. Microbiol. Infect. Dis. 35 (2) (2012) 81–92.

[219] L. Wen, R.E. Ley, P.Y. Volchkov, P.B. Stranges, et al., Innate immunity and intestinal microbiota in the development of type 1 diabetes, Nature 455 (7216) (2008) 1109–1113.

[220] N.L. Webster, C. Zufferey, J.A. Pane, B.S. Coulson, Alteration of the thymic T cell repertoire by rotavirus infection is associated with delayed type 1 diabetes development in non-obese diabetic mice, PLoS One 8 (3) (2013) e59182.

[221] J.A. Pane, N.L. Webster, K.L. Graham, G. Holloway, et al., Rotavirus acceleration of murine type 1 diabetes is associated with a T helper 1-dependent specific serum antibody response and virus effects in regional lymph nodes, Diabetologia 56 (3) (2013) 573–582.

[222] K.A. Saunders, T. Raine, A. Cooke, C.E. Lawrence, Inhibition of autoimmune type 1 diabetes by gastrointestinal helminth infection, Infect. Immun. 75 (1) (2007) 397–407.

[223] U. Christen, D. Benke, T. Wolfe, E. Rodrigo, et al., Cure of prediabetic mice by viral infections involves lymphocyte recruitment along an IP-10 gradient, J. Clin. Invest. 113 (1) (2004) 74–84.

[224] H.S. Jun, J.W. Yoon, The role of viruses in type I diabetes: two distinct cellular and molecular pathogenic mechanisms of virus-induced diabetes in animals, Diabetologia 44 (3) (2001) 271–285.

[225] I. Takei, Y. Asaba, T. Kasatani, T. Maruyama, et al., Suppression of development of diabetes in NOD mice by lactate dehydrogenase virus infection, J. Autoimmun. 5 (6) (1992) 665.

[226] S. Wilberz, H.J. Partke, F. Dagnaes-Hansen, L. Herberg, Persistent MHV (mouse hepatitis virus) infection reduces the incidence of diabetes mellitus in non-obese diabetic mice, Diabetologia 34 (1) (1991) 2.

[227] K. Doi, H. Matsuzaki, T. Tsuda, T. Onodera, Rapid development of renal lesions in diabetic DBA mice infected with the D-variant of encephalomyocarditis virus (EMC-D), Br. J. Exp. Pathol. 70 (3) (1989) 275–281.

[228] M.B. Oldstone, Prevention of type I diabetes in nonobese diabetic mice by virus infection, Science 239 (4839) (1988) 500–502.

[229] A. Tishon, M.B. Oldstone, Persistent virus infection associated with chemical manifestations of diabetes. II. Role of viral strain, environmental insult, and host genetics, Am. J. Pathol. 126 (1) (1987) 61–72.

[230] E.H. Leiter, The NOD mouse: a model for analyzing the interplay between heredity and environment in development of autoimmune disease, ILAR News 35 (1993) 4–14.

[231] E. Kounoue, S. Nagafuchi, M. Nakamura, S. Nakano, et al., Encephalomyocarditis (EMC) virus-induced diabetes mellitus prevented by *Corynebacterium parvum* in mice, Experientia 43 (4) (1987) 430–431.

[232] S.A. Huber, P.G. Babu, J.E. Craighead, Genetic influences on the immunologic pathogenesis of encephalomyocarditis (EMC) virus-induced diabetes mellitus, Diabetes 34 (11) (1985) 1186–1190.

[233] J.E. Craighead, Viral diabetes mellitus in man and experimental animals, Am. J. Med. 70 (1) (1981) 127–134.

[234] J.E. Craighead, M.F. McLane, Diabetes mellitus: induction in mice by encephalomyocarditis virus, Science 162 (856) (1968) 913–914.

[235] M.K. Montgomery, N.L. Hallahan, S.H. Brown, M. Liu, et al., Mouse strain-dependent variation in obesity and glucose homeostasis in response to high-fat feeding, Diabetologia 56 (5) (2013) 1129–1139.

[236] C. Podrini, E.L. Cambridge, C.J. Lelliott, D.M. Carragher, et al., High-fat feeding rapidly induces obesity and lipid derangements in C57BL/6N mice, Mamm. Genome 24 (5–6) (2013) 240–251.

[237] Y. Matsuda, M. Kobayashi, R. Yamauchi, M. Ojika, et al., Coffee and caffeine improve insulin sensitivity and glucose tolerance in C57BL/6J mice fed a high-fat diet, Biosci. Biotechnol. Biochem. 75 (12) (2011) 2309–2315.

[238] S.K. Panchal, L. Brown, Rodent models for metabolic syndrome research, J. Biomed. Biotechnol. 2011 (2011) 351982.

[239] J.J. Heindel, F.S vom Saal, Meeting report: batch-to-batch variability in estrogenic activity in commercial animal diets—importance and approaches for laboratory animal research, Environ. Health. Perspect. 116 (3) (2008) 389–393.

[240] R.L. Ruhlen, K.L. Howdeshell, J. Mao, J.A. Taylor, et al., Low phytoestrogen levels in feed increase fetal serum estradiol resulting in the "fetal estrogenization syndrome" and obesity in CD-1 mice, Environ. Health. Perspect. 116 (3) (2008) 322–328.

[241] P.E. Beales, R.B. Elliott, S. Flohe, J.P. Hill, et al., A multi-centre, blinded international trial of the effect of A(1) and A(2) beta-casein variants on diabetes incidence in two rodent models of spontaneous Type I diabetes, Diabetologia 45 (9) (2002) 1240–1246.

[242] R.S. Surwit, M.N. Feinglos, J. Rodin, A. Sutherland, et al., Differential effects of fat and sucrose on the development of obesity and diabetes in C57BL/6J and A/J mice, Metabolism 44 (5) (1995) 645.

[243] J.J. Heindel, F.S vom Saal, Role of nutrition and environmental endocrine disrupting chemicals during the perinatal period on the aetiology of obesity, Mol. Cell. Endocrinol. 304 (1-2) (2009) 90–96.

[244] P.C. Reifsnyder, G. Churchill, E.H. Leiter, Maternal environment and genotype interact to establish diabesity in mice, Genome Res. 10 (10) (2000) 1568–1578.

[245] A.J. Williams, J. Krug, E.F. Lampeter, K. Mansfield, et al., Raised temperature reduces the incidence of diabetes in the NOD mouse, Diabetologia 33 (10) (1990) 635–637.

[246] C. Pettan-Brewer, P.M. Treuting, Practical pathology of aging miceed, 2011 (2011).

[247] J.P. Sundberg, A. Berndt, B.A. Sundberg, K.A. Silva, et al., The mouse as a model for understanding chronic diseases of aging: the histopathologic basis of aging in inbred mice, Pathobiol. Aging Age Relat. Dis. 1 (2011) 10.

[248] T. Korenaga, X. Fu, Y. Xing, T. Matsusita, et al., Tissue distribution, biochemical properties, and transmission of mouse type A AApoAII amyloid fibrils, Am. J. Pathol. 164 (5) (2004) 1597–1606.

[249] K. Kitagawa, J. Wang, T. Mastushita, K. Kogishi, et al., Polymorphisms of mouse apolipoprotein A-II: seven alleles found among 41 inbred strains of mice, Amyloid 10 (4) (2003) 207–214.

[250] L. Wang, J.J. Liepnieks, M.D. Benson, B. Kluve-Beckerman, Expression of SAA and amyloidogenesis in congenic mice of CE/J and C57BL/6 strains, Amyloid 7 (1) (2000) 26–31.

[251] M.J. Hoenerhoff, M.F. Starost, J.M. Ward, Eosinophilic crystalline pneumonia as a major cause of death in 129S4/SvJae mice, Vet. Pathol. 43 (5) (2006) 682–688.

[252] D.C. Haines, S. Chattopadhyay, J.M. Ward, Pathology of aging B6;129 mice, Toxicol. Pathol. 29 (6) (2001) 653–661.

[253] W.C. Son, Factors contributory to early death of young CD-1 mice in carcinogenicity studies, Toxicol. Lett. 145 (1) (2003) 88–98.

[254] R.D. Lipman, E.T. Gaillard, D.E. Harrison, R.T. Bronson, Husbandry factors and the prevalence of age-related amyloidosis in mice, Lab. Anim. Sci. 43 (5) (1993) 439–444.

[255] H.Y. Rienhoff Jr., J.H. Huang, X.X. Li, W.S. Liao, Molecular and cellular biology of serum amyloid A, Mol. Biol. Med. 7 (3) (1990) 287–298.

[256] S. Sharmin, Y. Shimizu, M. Hagiwara, K. Hirayama, et al., Staphylococcus aureus antigens induce IgA-type glomerulonephritis in Balb/c mice, J. Nephrol. 17 (4) (2004) 504–511.

[257] J. Lule, F. Puginier-Carentz, J.P. Basile, I. Duga-Neulat, et al., The spontaneous development of immune complex type glomerular lesions in outbred mice is dependent on environmental factors and sex, J. Clin. Lab. Immunol. 29 (3) (1989) 111–118.

[258] E.C. Weir, D.G. Brownstein, A.L. Smith, E.A. Johnson, Respiratory disease and wasting in athymic mice infected with pneumonia virus of mice, Lab. Anim. Sci. 38 (2) (1988) 133–137.

[259] E.C. Weir, D.G. Brownstein, S.W. Barthold, Spontaneous wasting disease in nude mice associated with Pneumocystis carinii infection, Lab. Anim. Sci. 36 (2) (1986) 140–144.

[260] S.W. Barthold, A.L. Smith, M.L. Povar, Enterotropic mouse hepatitis virus infection in nude mice, Lab. Anim. Sci. 35 (6) (1985) 613–618.

[261] A. Sebesteny, R. Tilly, F. Balkwill, D. Trevan, Demyelination and wasting associated with polyomavirus infection in nude (nu/nu) mice, Lab. Anim. 14 (4) (1980) 337–345.

[262] G. Buchler, M.L. Wos-Oxley, A. Smoczek, N.H. Zschemisch, et al., Strain-specific colitis susceptibility in IL10-deficient mice depends on complex gut microbiota—host interactions, Inflamm. Bowel. Dis. 18 (5) (2012) 943—954.

[263] D.W. Rosenberg, C. Giardina, T. Tanaka, Mouse models for the study of colon carcinogenesis, Carcinogenesis 30 (2) (2009) 183—196.

[264] R. Suzuki, H. Kohno, S. Sugie, H. Nakagama, et al., Strain differences in the susceptibility to azoxymethane and dextran sodium sulfate-induced colon carcinogenesis in mice, Carcinogenesis 27 (1) (2006) 162—169.

[265] M. Mahler, I.J. Bristol, E.H. Leiter, A.E. Workman, et al., Differential susceptibility of inbred mouse strains to dextran sulfate sodium-induced colitis, Am. J. Physiol. Gastrointest Liver Physiol. 274 (3) (1998) G544—G551.

[266] A.B. Rogers, Gastric *Helicobacter* spp. in animal models: pathogenesis and modulation by extragastric coinfections, Methods Mol. Biol. 921 (2012) 175—188.

[267] J.G. Fox, Z. Ge, M.T. Whary, S.E. Erdman, et al., *Helicobacter hepaticus* infection in mice: models for understanding lower bowel inflammation and cancer, Mucosal Immunol. 4 (1) (2011) 22—30.

[268] M. Chichlowski, J.M. Sharp, D.A. Vanderford, M.H. Myles, et al., *Helicobacter typhlonius* and *Helicobacter rodentium* differentially affect the severity of colon inflammation and inflammation-associated neoplasia in IL10-deficient mice, Comp. Med. 58 (6) (2008) 534—541.

[269] S.R. Compton, L.J. Ball-Goodrich, L.K. Johnson, E.A. Johnson, et al., Pathogenesis of enterotropic mouse hepatitis virus in immunocompetent and immunodeficient mice, Comp. Med. 54 (6) (2004) 681—689.

[270] F.R. Homberger, L. Zhang, S.W. Barthold, Prevalence of enterotropic and polytropic mouse hepatitis virus in enzootically infected mouse colonies, Lab. Anim. Sci. 48 (1) (1998) 50—54.

[271] R. Boot, H. van Herck, J. van der Logt, Mutual viral and bacterial infections after housing rats of various breeders within an experimental unit, Lab. Anim. 30 (1) (1996) 42—45.

[272] R.S. Livingston, C.L. Franklin, C.L. Besch-Williford, R.R. Hook Jr., et al., A novel presentation of Clostridium piliforme infection (Tyzzer's disease) in nude mice, Lab. Anim. Sci. 46 (1) (1996) 21—25.

[273] A.S. Fries, O. Ladefoged, The influence of *Bacillus piliformis* (Tyzzer) infections on the reliability of pharmacokinetic experiments in mice, Lab. Anim. 13 (3) (1979) 257—261.

[274] A.D. Smith, S. Botero, T. Shea-Donohue, J.F. Urban Jr., The pathogenicity of an enteric *Citrobacter rodentium* infection is enhanced by deficiencies in the antioxidants selenium and vitamin E, Infect. Immun. 79 (4) (2011) 1471—1478.

[275] N.K. Petty, R. Bulgin, V.F. Crepin, A.M. Cerdeno-Tarraga, et al., The *Citrobacter rodentium* genome sequence reveals convergent evolution with human pathogenic *Escherichia coli*, J. Bacteriol. 192 (2) (2010) 525—538.

[276] B.A. Vallance, W. Deng, K. Jacobson, B.B. Finlay, Host susceptibility to the attaching and effacing bacterial pathogen *Citrobacter rodentium*, Infect. Immun. 71 (6) (2003) 3443—3453.

[277] L.M. Higgins, G. Frankel, G. Douce, G. Dougan, et al., *Citrobacter rodentium* infection in mice elicits a mucosal Th1 cytokine response and lesions similar to those in murine inflammatory bowel disease, Infect. Immun. 67 (6) (1999) 3031—3039.

[278] E.J. Theve, Y. Feng, K. Taghizadeh, K.S. Cormier, et al., Sex hormone influence on hepatitis in young male A/JCr mice infected with *Helicobacter hepaticus*, Infect. Immun. 76 (9) (2008) 4071—4078.

[279] M.D. Stout, G.E. Kissling, F.A. Suarez, D.E. Malarkey, et al., Influence of *Helicobacter hepaticus* infection on the chronic toxicity and carcinogenicity of triethanolamine in B6C3F1 mice, Toxicol. Pathol. 36 (6) (2008) 783−794.

[280] M. Ihrig, M.D. Schrenzel, J.G. Fox, Differential susceptibility to hepatic inflammation and proliferation in AXB recombinant inbred mice chronically infected with *Helicobacter hepaticus*, Am. J. Pathol. 155 (2) (1999) 571−582.

[281] P.S. Bhathal, N.R. Rose, I.R. Mackay, S. Whittingham, Strain differences in mice in carbon tetrachloride-induced liver injury, Br. J. Exp. Pathol. 64 (5) (1983) 524−533.

[282] V.P. Tryndyak, J.R. Latendresse, B. Montgomery, S.A. Ross, et al., Plasma microRNAs are sensitive indicators of inter-strain differences in the severity of liver injury induced in mice by a choline- and folate-deficient diet, Toxicol. Appl. Pharmacol. 262 (1) (2012) 52−59.

[283] I.P. Pogribny, A. Starlard-Davenport, V.P. Tryndyak, T. Han, et al., Difference in expression of hepatic microRNAs miR-29c, miR-34a, miR-155, and miR-200b is associated with strain-specific susceptibility to dietary nonalcoholic steatohepatitis in mice, Lab. Invest. 90 (10) (2010) 1437.

[284] A. Garcia, Y. Zeng, S. Muthupalani, Z. Ge, et al., *Helicobacter hepaticus*—induced liver tumor promotion is associated with increased serum bile acid and a persistent microbial-induced immune response, Cancer Res. 71 (7) (2011) 2529−2540.

[285] Y. Huang, X.F. Tian, X.G. Fan, C.Y. Fu, et al., The pathological effect of *Helicobacter pylori* infection on liver tissues in mice, Clin. Microbiol. Infect. 15 (9) (2009) 843−849.

[286] C.M. Nagamine, J.J. Sohn, B.H. Rickman, A.B. Rogers, et al., Helicobacter hepaticus infection promotes colon tumorigenesis in the BALB/c-Rag2($-/-$) Apc(Min/+) mouse, Infect. Immun. 76 (6) (2008) 2758−2766.

[287] J.M. Ward, J.G. Fox, M.R. Anver, D.C. Haines, et al., Chronic active hepatitis and associated liver tumors in mice caused by a persistent bacterial infection with a novel *Helicobacter* species, J. Natl. Cancer Inst. 86 (16) (1994) 1222−1227.

[288] S.R. Compton, L.J. Ball-Goodrich, C.J. Zeiss, L.K. Johnson, et al., Pathogenesis of mouse hepatitis virus infection in gamma interferon-deficient mice is modulated by co-infection with *Helicobacter hepaticus*, Comp. Med. 53 (2) (2003) 197−206.

[289] D.S. Huang, S.N. Emancipator, D.R. Fletcher, M.E. Lamm, et al., Hepatic pathology resulting from mouse hepatitis virus S infection in severe combined immunodeficiency mice, Lab. Anim. Sci. 46 (2) (1996) 167−173.

[290] A.H. Harrill, P.K. Ross, D.M. Gatti, D.W. Threadgill, et al., Population-based discovery of toxicogenomics biomarkers for hepatotoxicity using a laboratory strain diversity panel, Toxicol. Sci. 110 (1) (2009) 235−243.

[291] G.J. Moser, J. Foley, M. Burnett, T.L. Goldsworthy, et al., Furan-induced dose-response relationships for liver cytotoxicity, cell proliferation, and tumorigenicity (furan-induced liver tumorigenicity), Exp. Toxicol. Pathol. 61 (2) (2009) 101−111.

[292] J.H. Smith, K. Maita, S.D. Sleight, J.B. Hook, Mechanism of chloroform nephrotoxicity. I. Time course of chloroform toxicity in male and female mice, Toxicol. Appl. Pharmacol. 70 (3) (1983) 467−479.

[293] S.L. Byers, S.J. Payson, R.A. Taft, Performance of ten inbred mouse strains following assisted reproductive technologies (ARTs), Theriogenology 65 (9) (2006) 1716−1726.

[294] S.M. Yellon, L.T. Tran, Photoperiod, reproduction, and immunity in select strains of inbred mice, J. Biol. Rhythms. 17 (1) (2002) 65−75.

[295] L.M. Silver, Mouse Genetics: Concepts and Applicationsed, Oxford University Press, New York, Oxford, 1995 (Adapted for the web by Mouse Genome Informatics, the Jackson Laboratory, Bar Harbor, ME; Online at <http://www.informatics.jax.org/silverbook/>).

[296] M. Holt, T. Meuwissen, O. Vangen, Long-term responses, changes in genetic variances and inbreeding depression from 122 generations of selection on increased litter size in mice, J. Anim. Breed. Genet. 122 (3) (2005) 199−209.

[297] O. Suzuki, T. Asano, Y. Yamamoto, K. Takano, et al., Development in vitro of preimplantation embryos from 55 mouse strains, Reprod. Fertil. Dev. 8 (6) (1996) 975−980.

[298] S. Gearhart, J. Kalishman, H. Melikyan, C. Mason, et al., Increased incidence of vaginal septum in C57BL/6J mice since 1976, Comp. Med. 54 (4) (2004) 418−421.

[299] T.L. Cunliffe-Beamer, D.B. Feldman, Vaginal septa in mice: incidence, inheritance, and effect on reproductive, performance, Lab. Anim. Sci. 26 (6 Pt 1) (1976) 895−898.

[300] J.M. Sharp, D.A. Vanderford, M. Chichlowski, M.H. Myles, et al., Helicobacter infection decreases reproductive performance of IL10-deficient mice, Comp. Med. 58 (5) (2008) 447−453.

[301] K. Mikazuki, T. Hirasawa, H. Chiba, K. Takahashi, et al., Colonization pattern of Pasteurella pneumotropica in mice with latent pasteurellosis, Jikken. Dobutsu. 43 (3) (1994) 375−379.

[302] J.K. Davis, D.J. Gaertner, N.R. Cox, J.R. Lindsey, et al., The role of Klebsiella oxytoca in utero-ovarian infection of B6C3F1 mice, Lab. Anim. Sci. 37 (2) (1987) 159−166.

[303] L. Maggio-Price, K.L. Nicholson, K.M. Kline, T. Birkebak, et al., Diminished reproduction, failure to thrive, and altered immunologic function in a colony of T-cell receptor transgenic mice: possible role of Citrobacter rodentium, Lab. Anim. Sci. 48 (2) (1998) 145−155.

[304] A.K. Banerjee, A.F. Angulo, A.A. Polak-Vogelzang, A.M. Kershof, Naturally occurring genital mycoplasmosis in mice, Lab. Anim. 19 (4) (1985) 275−276.

[305] G.H. Cassell, J.R. Lindsey, J.K. Davis, Respiratory and genital mycoplasmosis of laboratory rodents: implications for biomedical research, Isr. J. Med. Sci. 17 (7) (1981) 548−554.

[306] J.I. Ackerman, J.G. Fox, Isolation of Pasteurella ureae from reproductive tracts of congenic mice, J. Clin. Microbiol. 13 (6) (1981) 1049−1053.

[307] G.E. Ward, R. Moffatt, E. Olfert, Abortion in mice associated with Pasteurella pneumotropica, J. Clin. Microbiol. 8 (2) (1978) 177−180.

[308] J. Whitaker, S.S. Moy, V. Godfrey, J. Nielsen, et al., Effects of cage size and enrichment on reproductive performance and behavior in C57BL/6Tac mice, Lab. Anim. (NY) 38 (1) (2009) 24−34.

[309] S. Rasmussen, G. Glickman, R. Norinsky, F.W. Quimby, et al., Construction noise decreases reproductive efficiency in mice, J. Am. Assoc. Lab. Anim. Sci. 48 (4) (2009) 363−370.

[310] J.M. Cline, A.A. Franke, T.C. Register, D.L. Golden, et al., Effects of dietary isoflavone aglycones on the reproductive tract of male and female mice, Toxicol. Pathol. 32 (1) (2004) 91−99.

[311] P.P. Tsai, D. Oppermann, H.D. Stelzer, M. Mahler, et al., The effects of different rack systems on the breeding performance of DBA/2 mice, Lab. Anim. 37 (1) (2003) 44−53.

[312] A.G. Peters, P.M. Bywater, M.F. Festing, The effect of daily disturbance on the breeding performance of mice, Lab. Anim. 36 (2) (2002) 188−192.

[313] C.K. Reeb-Whitaker, B. Paigen, W.G. Beamer, R.T. Bronson, et al., The impact of reduced frequency of cage changes on the health of mice housed in ventilated cages, Lab. Anim. 35 (1) (2001) 58−73.

[314] S. Eskola, E. Kaliste-Korhonen, Nesting material and number of females per cage: effects on mouse productivity in BALB/c, C57BL/6J, DBA/2 and NIH/S mice, Lab. Anim. 33 (2) (1999) 122−128.

[315] J.G. Vandenbergh, C.L. Huggett, The anogenital distance index, a predictor of the intrauterine position effects on reproduction in female house mice, Lab. Anim. Sci. 45 (5) (1995) 567−573.

[316] R.E. Chapin, D.K. Gulati, P.A. Fail, E. Hope, et al., The effects of feed restriction on reproductive function in Swiss CD-1 mice, Fundam. Appl. Toxicol. 20 (1) (1993) 15−22.

[317] JAX®Notes. Malocclusion in the laboratory mouse. 2003. #489.

[318] H. Petznek, R. Kappler, H. Scherthan, M. Muller, et al., Reduced body growth and excessive incisor length in insertional mutants mapping to mouse chromosome 13, Mamm. Genome. 13 (9) (2002) 504−509.

[319] P.G. Reeves, K.L. Rossow, J. Lindlauf, Development and testing of the AIN-93 puri-fied diets for rodents: results on growth, kidney calcification and bone mineralization in rats and mice, J. Nutr. 123 (11) (1993) 1923−1931.

[320] J.F. Bureau, S. Le Goff, D. Thomas, A.F. Parlow, et al., Disruption of differentiated func-tions during viral infection in vivo. V. Mapping of a locus involved in susceptibility of mice to growth hormone deficiency due to persistent lymphocytic choriomeningitis virus infection, Virology 281 (1) (2001) 61−66.

[321] S. Kring, C. King, K. Spindler, Susceptibility and signs associated with mouse adenovirus type 1 infection of adult outbred Swiss mice, J. Virol. 69 (12) (1995) 8084−8088.

[322] M.L. Walters, N.F. Stanley, R.L. Dawkins, M.P. Alpers, Immunological assessment of mice with chronic jaundice and runting induced by reovirus 3, Br. J. Exp. Pathol. 54 (3) (1973) 329−345.

[323] H.Y. Chang, W. Mitzner, Sex differences in mouse models of asthma, Can. J. Physiol. Pharmacol. 85 (12) (2007) 1226−1235.

[324] S.E. Soutiere, C.G. Tankersley, W. Mitzner, Differences in alveolar size in inbred mouse strains, Respir. Physiol. Neurobiol. 140 (3) (2004) 283−291.

[325] K. Huang, R. Rabold, B. Schofield, W. Mitzner, et al., Age-dependent changes of air-way and lung parenchyma in C57BL/6J mice, J. Appl. Physiol. 102 (1) (2007) 200−206.

[326] C.G. Tankersley, R. Rabold, W. Mitzner, Differential lung mechanics are genetically determined in inbred murine strains, J. Appl. Physiol. 86 (6) (1999) 1764−1769.

[327] H.Y. Chang, W. Mitzner, J. Watson, Variation in airway responsiveness of male C57BL/6 mice from 5 vendors, J. Am. Assoc. Lab. Anim. Sci. 51 (4) (2012) 401−406.

[328] J.M. Bishai, W. Mitzner, Effect of severe calorie restriction on the lung in two strains of mice, Am. J. Physiol. Lung. Cell. Mol. Physiol. 295 (2) (2008) L356−L362.

[329] S.K. Majeed, Survey on spontaneous systemic amyloidosis in aging mice, Arzneimittelforschung 43 (2) (1993) 170−178.

[330] K. Higuchi, H. Naiki, K. Kitagawa, M. Hosokawa, et al., Mouse senile amyloidosis. ASSAM amyloidosis in mice presents universally as a systemic age-associated amy-loidosis, Virchows Arch. B Cell Pathol. Incl. Mol. Pathol. 60 (4) (1991) 231−238.

[331] F. Neff, D. Flores-Dominguez, D.P. Ryan, M. Horsch, et al., Rapamycin extends murine lifespan but has limited effects on aging, J. Clin. Invest. 123 (8) (2013) 3272−3291.

[332] R.A. Miller, D.E. Harrison, C.M. Astle, J.A. Baur, et al., Rapamycin, but not resveratrol or simvastatin, extends life span of genetically heterogeneous mice, J. Gerontol. A Biol. Sci. Med. Sci. 66 (2) (2011) 191−201.

[333] V.N. Anisimov, M.A. Zabezhinski, I.G. Popovich, T.S. Piskunova, et al., Rapamycin increases lifespan and inhibits spontaneous tumorigenesis in inbred female mice, Cell Cycle 10 (24) (2011) 4230−4236.

[334] W.C. Son, C. Gopinath, Early occurrence of spontaneous tumors in CD-1 mice and Sprague−Dawley rats, Toxicol. Pathol. 32 (4) (2004) 371−374.

[335] G.S. Smith, R.L. Walford, M.R. Mickey, Lifespan and incidence of cancer and other diseases in selected long-lived inbred mice and their F 1 hybrids, J. Natl. Cancer Inst. 50 (5) (1973) 1195−1213.

[336] W.G. Hoag, Spontaneous cancer in mice, Ann. N.Y. Acad. Sci. 108 (1963) 805−831.

[337] P.M. Treuting, N.J. Linford, S.E. Knoblaugh, M.J. Emond, et al., Reduction of age-associated pathology in old mice by overexpression of catalase in mitochondria, J. Gerontol. A Biol. Sci. Med. Sci. 63 (8) (2008) 813−822.

[338] D. Brownstein, P.N. Bhatt, R.O. Jacoby, Mousepox in inbred mice innately resistant or susceptible to lethal infection with ectromelia virus. V. Genetics of resistance to the Moscow strain, Arch Virol. 107 (1-2) (1989) 35−41.

[339] G.N. Rao, W.W. Piegorsch, D.D. Crawford, J. Edmondson, et al., Influence of viral infections on body weight, survival, and tumor prevalence of B6C3F1 (C57BL/6N x C3H/HeN) mice in carcinogenicity studies, Fundam. Appl. Toxicol. 13 (1) (1989) 156−164.

[340] O. Foreman, A.M. Kavirayani, S.M. Griffey, R. Reader, et al., Opportunistic bacterial infections in breeding colonies of the NSG mouse strain, Vet. Pathol. 48 (2) (2010) 495−499.

[341] L.K. Gibbs, D.L. Hickman, A.D. Lewis, L.M. Colgin, Staphylococcus-induced urolithiasis in estrogen-treated ovariectomized nude mice, J. Am. Assoc. Lab. Anim. Sci. 46 (4) (2007) 61−65.

[342] J.D. Macy Jr., E.C. Weir, S.R. Compton, M.J. Shlomchik, et al., Dual infection with *Pneumocystis carinii* and *Pasteurella pneumotropica* in B cell-deficient mice: diagnosis and therapy, Comp. Med. 50 (1) (2000) 49−55.

[343] D.H. Percy, J.R. Barta, Spontaneous and experimental infections in scid and scid/beige mice, Lab. Anim. Sci. 43 (2) (1993) 127−132.

[344] J.F. Bradfield, J.E. Wagner, G.P. Boivin, E.K. Steffen, et al., Epizootic fatal dermatitis in athymic nude mice due to *Staphylococcus xylosus*, Lab. Anim. Sci. 43 (1) (1993) 111−113.

[345] F. Dagnaes-Hansen, M.R. Horsman, Experience with mouse hepatitis virus sanitation in three transplantable murine tumour lines, Lab. Anim. 39 (4) (2005) 394−399.

[346] K. Yagami, Y. Goto, J. Ishida, Y. Ueno, et al., Polymerase chain reaction for detection of rodent parvoviral contamination in cell lines and transplantable tumors, Lab. Anim. Sci. 45 (3) (1995) 326−328.

[347] B.W. Mahy, C. Dykewicz, S. Fisher-Hoch, S. Ostroff, et al., Virus zoonoses and their potential for contamination of cell cultures, Dev. Biol. Stand. 75 (1991) 183−189.

[348] N.C. Peterson, From bench to cageside: risk assessment for rodent pathogen contamination of cells and biologics, ILAR J. 49 (3) (2008) 310−315.

[349] K. Flurkey, C.M. Astle, D.E. Harrison, Life extension by diet restriction and *N*-acetyl-L-cysteine in genetically heterogeneous mice, J. Gerontol. A Biol. Sci. Med. Sci. 65 (12) (2010) 1275−1284.

[350] O. Lazarov, J. Robinson, Y.P. Tang, I.S. Hairston, et al., Environmental enrichment reduces Abeta levels and amyloid deposition in transgenic mice, Cell 120 (5) (2005) 701–713.

[351] G.N. Rao, P.W. Crockett, Effect of diet and housing on growth, body weight, survival and tumor incidences of B6C3F1 mice in chronic studies, Toxicol. Pathol. 31 (2) (2003) 243–250.

[352] K.P. Keenan, G.C. Ballam, K.A. Soper, P. Laroque, et al., Diet, caloric restriction, and the rodent bioassay, Toxicol. Sci. 52 (2 Suppl.) (1999) 24–34.

[353] A. Taylor, R.D. Lipman, J. Jahngen-Hodge, V. Palmer, et al., Dietary calorie restriction in the Emory mouse: effects on lifespan, eye lens cataract prevalence and progression, levels of ascorbate, glutathione, glucose, and glycohemoglobin, tail collagen breaktime, DNA and RNA oxidation, skin integrity, fecundity, and cancer, Mech. Ageing Dev. 79 (1) (1995) 33–57.

[354] M.J. Hoenerhoff, H.H. Hong, T.V. Ton, S.A. Lahousse, et al., A review of the molecular mechanisms of chemically induced neoplasia in rat and mouse models in National Toxicology Program bioassays and their relevance to human cancer, Toxicol. Pathol. 37 (7) (2009) 835–848.

[355] J. Kool, A. Berns, High-throughput insertional mutagenesis screens in mice to identify oncogenic networks, Nat. Rev. Cancer 9 (6) (2009) 389–399.

[356] L.P. Hale, D. Perera, M.R. Gottfried, L. Maggio-Price, et al., Neonatal co-infection with helicobacter species markedly accelerates the development of inflammation-associated colonic neoplasia in IL-10(−/−) mice, Helicobacter 12 (6) (2007) 598–604.

[357] S.J. Engle, I. Ormsby, S. Pawlowski, G.P. Boivin, et al., Elimination of colon cancer in germ-free transforming growth factor beta 1-deficient mice, Cancer Res. 62 (22) (2002) 6362–6366.

[358] B.A. Diwan, J.M. Ward, D. Ramljak, L.M. Anderson, Promotion by *Helicobacter hepaticus*-induced hepatitis of hepatic tumors initiated by *N*-nitrosodimethylamine in male A/JCr mice, Toxicol. Pathol. 25 (6) (1997) 597–605.

[359] J.R. Hailey, J.K. Haseman, J.R. Bucher, A.E. Radovsky, et al., Impact of *Helicobacter hepaticus* infection in B6C3F1 mice from twelve National Toxicology Program two-year carcinogenesis studies, Toxicol. Pathol. 26 (5) (1998) 602–611.

[360] J.J. Wirth, L.G. Martin, M.M. Fluck, Oncogenesis of mammary glands, skin, and bones by polyomavirus correlates with viral persistence and prolonged genome replication potential, J. Virol. 71 (2) (1997) 1072–1078.

[361] A.E. Lukacher, Y. Ma, J.P. Carroll, S.R. Abromson-Leeman, et al., Susceptibility to tumors induced by polyoma virus is conferred by an endogenous mouse mammary tumor virus superantigen, J. Exp. Med. 181 (5) (1995) 1683–1692.

[362] C.H. Frith, J.E. Heath, Tumours of the salivary gland, IARC Sci. Publ. 111 (1994) 115–139.

[363] A.S. Tischler, R. Freund, J. Carroll, A.L. Cahill, et al., Polyoma-induced neoplasms of the mouse adrenal medulla. Characterization of the tumors and establishment of cell lines, Lab. Invest. 68 (5) (1993) 541–549.

[364] R.P. Gollard, H.C. Slavkin, M.L. Snead, Polyoma virus-induced murine odontogenic tumors, Oral Surg. Oral Med. Oral Pathol. 74 (6) (1992) 761–767.

[365] R. Freund, T. Dubensky, R. Bronson, A. Sotnikov, et al., Polyoma tumorigenesis in mice: evidence for dominant resistance and dominant susceptibility genes of the host, Virology 191 (2) (1992) 724–731.

[366] J.W. Hartley, L.H. Evans, K.Y. Green, Z. Naghashfar, et al., Expression of infectious murine leukemia viruses by RAW264.7 cells, a potential complication for studies with a widely used mouse macrophage cell line, Retrovirology 5 (2008) 1.

[367] K. Weiser, B. Liu, G. Hansen, D. Skapura, et al., Retroviral insertions in the VISION database identify molecular pathways in mouse lymphoid leukemia and lymphoma, Mamm. Genome 18 (10) (2007) 709.

[368] J.P. Stoye, C. Moroni, J.M. Coffin, Virological events leading to spontaneous AKR thymomas, J Virol. 65 (3) (1991) 1273–1285.

[369] J.W. Hartley, S.K. Chattopadhyay, M.R. Lander, L. Taddesse-Heath, et al., Accelerated appearance of multiple B cell lymphoma types in NFS/N mice congenic for ecotropic murine leukemia viruses, Lab. Invest. 80 (2) (2000) 159–169.

[370] L. Taddesse-Heath, S.K. Chattopadhyay, D.L. Dillehay, M.R. Lander, et al., Lymphomas and high-level expression of murine leukemia viruses in CFW mice, J. Virol. 74 (15) (2000) 6832–6837.

[371] J.P. Stoye, J.M. Coffin, Polymorphism of murine endogenous proviruses revealed by using virus class-specific oligonucleotide probes, J. Virol. 62 (1) (1988) 168–175.

[372] T. Dupressoir, J.M. Vanacker, J.J. Cornelis, N. Duponchel, et al., Inhibition by parvovirus H-1 of the formation of tumors in nude mice and colonies in vitro by transformed human mammary epithelial cells, Cancer Res. 49 (12) (1989) 3203–3208.

[373] M. Malerba, L. Daeffler, J. Rommelaere, R.D. Iggo, Replicating parvoviruses that target colon cancer cells, J. Virol. 77 (12) (2003) 6683–6691.

[374] S.I. Lang, N.A. Giese, J. Rommelaere, C. Dinsart, et al., Humoral immune responses against minute virus of mice vectors, J. Gene. Med. 8 (9) (2006) 1141–1150.

[375] D.J. Marino, The effect of study type on body weight and tumor incidence in B6C3F1 mice fed the NTP-2000 diet, Toxicol. Mech. Methods 22 (6) (2012) 466–475.

[376] A.F. Chambers, Influence of diet on metastasis and tumor dormancy, Clin. Exp. Metastasis. 26 (1) (2009) 61–66.

[377] E.J. Masoro, Caloric restriction-induced life extension of rats and mice: a critique of proposed mechanisms, Biochim. Biophys. Acta 1790 (10) (2009) 1040–1048.

[378] A.C. Patel, N.P. Nunez, S.N. Perkins, J.C. Barrett, et al., Effects of energy balance on cancer in genetically altered mice, J. Nutr. 134 (12 Suppl) (2004) 3394S–3398S.

[379] V. Mai, L.H. Colbert, D. Berrigan, S.N. Perkins, et al., Calorie restriction and diet composition modulate spontaneous intestinal tumorigenesis in Apc(Min) mice through different mechanisms, Cancer Res. 63 (8) (2003) 1752–1755.

[380] R.T. Bronson, R.D. Lipman, Reduction in rate of occurrence of age related lesions in dietary restricted laboratory mice, Growth Dev. Aging 55 (3) (1991) 169–184.

[381] R. Weindruch, Dietary restriction, tumors, and aging in rodents, J. Gerontol. 44 (6) (1989) 67–71.

[382] C. Keenan, S. Elmore, S. Francke-Carroll, R. Kemp, et al., Best practices for use of historical control data of proliferative rodent lesions, Toxicol. Pathol. 37 (5) (2009) 679–693.

[383] J.K. Haseman, J.E. Huff, G.N. Rao, S.L. Eustis, Sources of variability in rodent carcinogenicity studies, Fundam. Appl. Toxicol. 12 (4) (1989) 793–804.

[384] J.R. Sabine, Exposure to an environment containing the aromatic red cedar, Juniperus virginiana: procarcinogenic, enzyme-inducing and insecticidal effects, Toxicology 5 (2) (1975) 221–235.

[385] C. Dasenbrock, T. Tillmann, H. Ernst, W. Behnke, et al., Maternal effects and cancer risk in the progeny of mice exposed to X-rays before conception, Exp. Toxicol. Pathol. 56 (6) (2005) 351–360.

[386] Y. Shimada, M. Nishimura, S. Kakinuma, T. Ogiu, et al., Genetic susceptibility to thymic lymphomas and K-ras gene mutation in mice after exposure to X-rays and *N*-ethyl-*N*-nitrosourea, Int. J. Radiat. Biol. 79 (6) (2003) 423–430.

[387] M. Utsuyama, K. Hirokawa, Radiation-induced-thymic lymphoma occurs in young, but not in old mice, Exp. Mol. Pathol. 74 (3) (2003) 319–325.

[388] H. Szymanska, M. Sitarz, E. Krysiak, J. Piskorowska, et al., Genetics of susceptibility to radiation-induced lymphomas, leukemias and lung tumors studied in recombinant congenic strains, Int. J. Cancer 83 (5) (1999) 674–678.

[389] W.P. Rowe, T. Pincus, Quantitative studies of naturally occurring murine leukemia virus infection of AKR mice, J. Exp. Med. 135 (2) (1972) 429–436.

[390] J.M. Ward, Lymphomas and leukemias in mice, Exp. Toxicol. Pathol. 57 (5-6) (2006) 377.

[391] H. Kobayashi, M. Potter, T.B. Dunn, Bone lesions produced by transplanted plasma-cell tumors in BALB/c mice, J. Natl. Cancer Inst. 28 (1962) 649–677.

[392] M. Potter, R.C. Maccardle, Histology of developing plasma cell neoplasia induced by mineral oil in Balb/C mice, J. Natl. Cancer Inst. 33 (1964) 497–515.

[393] M. Potter, J.S. Wax, E. Blankenhorn, BALB/c subline differences in susceptibility to plasmacytoma induction, Curr. Top. Microbiol. Immunol. 122 (1985) 234–241.

[394] C.F. Hendriksen, W de Leeuw, Production of monoclonal antibodies by the ascites method in laboratory animals, Res. Immunol. 149 (6) (1998) 535–542.

[395] M.L. Mucenski, H.G. Bedigian, M.M. Shull, N.G. Copeland, et al., Comparative molecular genetic analysis of lymphomas from six inbred mouse strains, J. Virol. 62 (3) (1988) 839–846.

[396] P.P.L. Chiu, E. Ivakine, S. Mortin-Toth, J.S. Danska, Susceptibility to lymphoid neoplasia in immunodeficient strains of nonobese diabetic mice, Cancer Res. 62 (20) (2002) 5828–5834.

[397] D.V. Serreze, E.H. Leiter, M.S. Hanson, S.W. Christianson, et al., Emv30null NOD-scid mice. An improved host for adoptive transfer of autoimmune diabetes and growth of human lymphohematopoietic cells, Diabetes 44 (12) (1995) 1392–1398.

[398] J.C. Tang, F.C. Ho, A.C. Chan, G. Srivastava, Clonality of lymphomas at multiple sites in SJL mice, Lab. Invest. 78 (2) (1998) 205–212.

[399] J.C. Tang, F.C. Ho, A.C. Chan, E.Y. Chow, et al., Progression of spontaneous lymphomas in SJL mice: monitoring *in vivo* clonal evolution with molecular markers in sequential splenic samples, Lab. Invest. 78 (11) (1998) 1459–1466.

[400] C.G. Crispens, Some characteristics of strain SJL-JDg mice, Lab. Anim. Sci. 23 (3) (1973) 408–413.

[401] M. Giknis, C.B. Clifford, Spontaneous neoplastic lesions in the Crl:CD-1 (ICR) mouse in control groups from 18 month to 2 year studies, 2010.

[402] J.P. Yun, J.W. Behan, N. Heisterkamp, A. Butturini, et al., Diet-induced obesity accelerates acute lymphoblastic leukemia progression in two murine models, Cancer Prev. Res. (Phila) 3 (10) (2010) 1259–1264.

[403] D. Dixon, R.A. Herbert, G.E. Kissling, A.E. Brix, et al., Summary of chemically induced pulmonary lesions in the National Toxicology Program (NTP) toxicology and carcinogenesis studies, Toxicol. Pathol. 36 (3) (2008) 428–439.

[404] B.N. Blackwell, T.J. Bucci, R.W. Hart, A. Turturro, Longevity, body weight, and neoplasia in ad libitum-fed and diet-restricted C57BL6 mice fed NIH-31 open formula diet, Toxicol. Pathol. 23 (5) (1995) 570–582.

[405] J.M. Ward, J.E. Rehg, H.C. Morse III, Differentiation of rodent immune and hematopoietic system reactive lesions from neoplasias, Toxicol. Pathol. (2012).

[406] J.E. Rehg, D. Bush, J.M. Ward, The utility of immunohistochemistry for the identification of hematopoietic and lymphoid cells in normal tissues and interpretation of proliferative and inflammatory lesions of mice and rats, Toxicol. Pathol. 40 (2) (2012) 345–374.

[407] H.C. Morse III, M.R. Anver, T.N. Fredrickson, D.C. Haines, et al., Bethesda proposals for classification of lymphoid neoplasms in mice, Blood 100 (1) (2002) 246–258.

[408] S.C. Kogan, J.M. Ward, M.R. Anver, J.J. Berman, et al., Bethesda proposals for classification of nonlymphoid hematopoietic neoplasms in mice, Blood 100 (1) (2002) 238–245.

[409] S. de Seranno, R. Meuwissen, Progress and applications of mouse models for human lung cancer, Eur. Respir. J. 35 (2) (2010) 426–443.

[410] N. Wakamatsu, T.R. Devereux, H.H. Hong, R.C. Sills, Models of human lung cancer, Toxicol. Pathol. 35 (1) (2007) 75–80.

[411] P.C. Zeidler-Erdely, M.L. Kashon, L.A. Battelli, S.H. Young, et al., Pulmonary inflammation and tumor induction in lung tumor susceptible A/J and resistant C57BL/6J mice exposed to welding fume, Part. Fibre. Toxicol. 5 (2008) 12.

[412] G. Manenti, A. Acevedo, F. Galbiati, R. Gianni Barrera, et al., Cancer modifier alleles inhibiting lung tumorigenesis are common in inbred mouse strains, Int. J. Cancer 99 (4) (2002) 555–559.

[413] G. Manenti, M. Gariboldi, A. Fiorino, A.I. Zedda, et al., Pas1 is a common lung cancer susceptibility locus in three mouse strains, Mamm. Genome. 8 (11) (1997) 801–804.

[414] J.F. Mahler, W. Stokes, P.C. Mann, M. Takaoka, et al., Spontaneous lesions in aging FVB/N mice, Toxicol. Pathol. 24 (6) (1996) 710–716.

[415] P. Huang, D.G. Duda, R.K. Jain, D. Fukumura, Histopathologic findings and establishment of novel tumor lines from spontaneous tumors in FVB/N mice, Comp. Med. 58 (3) (2008) 253–263.

[416] H.H. Hong, T.V. Ton, Y. Kim, N. Wakamatsu, et al., Genetic alterations in K-ras and p53 cancer genes in lung neoplasms from B6C3F1 mice exposed to cumene, Toxicol. Pathol. 36 (5) (2008) 720–726.

[417] Y. Yan, Q. Tan, Y. Wang, D. Wang, et al., Enhanced lung tumor development in tobacco smoke-exposed p53 transgenic and Kras2 heterozygous deficient mice, Inhal. Toxicol. 19 (Suppl. 1) (2007) 183–187.

[418] R. Renne, A. Brix, J. Harkema, R. Herbert, et al., Proliferative and nonproliferative lesions of the rat and mouse respiratory tract, Toxicol. Pathol. 37 (7 Suppl) (2009) 5S–73S.

[419] A.Y. Nikitin, A. Alcaraz, M.R. Anver, R.T. Bronson, et al., Classification of proliferative pulmonary lesions of the mouse: recommendations of the mouse models of human cancers consortium, Cancer Res. 64 (7) (2004) 2307–2316.

[420] J.J. Bittner, Mammary cancer in C3H mice of different sublines and their hybrids, J. Natl. Cancer Inst. 16 (5) (1956) 1263–1286.

[421] L. Kilham, Isolation in suckling mice of a virus from C3H mice harboring Bittner milk agent, Science 116 (3015) (1952) 391–392.

[422] J.J. Bittner, The milk-influence of breast tumors in mice, Science 95 (2470) (1942) 462–463.

[423] T. Dunn, Morphology of mammary tumors in mice, in: F. Homburger (Ed.), The Physiopathology of Cancer, Paul B. Hoeber, Inc., New York, NY, 1958, pp. 38–84.

[424] M.B. MacLennan, B.M. Anderson, D.W. Ma, Differential mammary gland development in FVB and C57Bl/6 mice: implications for breast cancer research, Nutrients 3 (11) (2011) 929–936.

[425] A.C. Whitmore, S.P. Whitmore, Subline divergence within L.C. Strong's C3H and CBA inbred mouse strains. A review, Immunogenetics 21 (5) (1985) 407–428.

[426] A. Raafat, L. Strizzi, K. Lashin, E. Ginsburg, et al., Effects of age and parity on mammary gland lesions and progenitor cells in the FVB/N-RC mice, PLoS One 7 (8) (2012) e43624.

[427] E. Radaelli, A. Arnold, A. Papanikolaou, R.A. Garcia-Fernandez, et al., Mammary tumor phenotypes in wild-type aging female FVB/N mice with pituitary prolactinomas, Vet. Pathol. 46 (4) (2009) 736–745.

[428] A.I. Nieto, G. Shyamala, J.J. Galvez, G. Thordarson, et al., Persistent mammary hyperplasia in FVB/N mice, Comp. Med. 53 (4) (2003) 433–438.

[429] R.D. Cardiff, N Kenney, Mouse mammary tumor biology: a short history, Adv. Cancer Res. 98 (2007) 53–116.

[430] H.C. Morse, Retroelements in the mouse, in: J.G. Fox, S.W. Barthold, et al. (Eds.), The Mouse in Biomedical Research, Elsevier, Inc. New York, NY, 2006, (Vol 2; Chapter 10) pp. 269–279.

[431] L.M. Hook, Y. Agafonova, S.R. Ross, S.J. Turner, et al., Genetics of mouse mammary tumor virus-induced mammary tumors: linkage of tumor induction to the gag gene, J. Virol. 74 (19) (2000) 8876–8883.

[432] K. Podsypanina, K. Politi, L.J. Beverly, H.E. Varmus, Oncogene cooperation in tumor maintenance and tumor recurrence in mouse mammary tumors induced by Myc and mutant Kras, Proc. Natl. Acad. Sci. USA 105 (13) (2008) 5242–5247.

[433] R. Callahan, MMTV-induced mutations in mouse mammary tumors: their potential relevance to human breast cancer, Breast Cancer Res. Treat. 39 (1) (1996) 33–44.

[434] J.J. Bittner, Breast cancer in mice as influenced by nursing, J. Natl. Cancer. Inst. 1 (1940) 155–168.

[435] V.P. Rao, T. Poutahidis, Z. Ge, P.R. Nambiar, et al., Innate immune inflammatory response against enteric bacteria Helicobacter hepaticus induces mammary adenocarcinoma in mice, Cancer Res. 66 (15) (2006) 7395–7400.

[436] R.R. Gordon, K.W. Hunter, P. Sorensen, D. Pomp, Genotype X diet interactions in mice predisposed to mammary cancer. I. Body weight and fat, Mamm. Genome. 19 (3) (2008) 163–178.

[437] M. Luijten, A. Verhoef, J.A. Dormans, R.B. Beems, et al., Modulation of mammary tumor development in Tg.NK (MMTV/c-neu) mice by dietary fatty acids and life stage-specific exposure to phytoestrogens, Reprod. Toxicol. 23 (3) (2007) 407–413.

[438] M. Luijten, A.R. Thomsen, J.A. van den Berg, P.W. Wester, et al., Effects of soy-derived isoflavones and a high-fat diet on spontaneous mammary tumor development in Tg.NK (MMTV/c-neu) mice, Nutr. Cancer 50 (1) (2004) 46–54.

[439] M.J. Dirx, M.P. Zeegers, P.C. Dagnelie, T. van den Bogaard, et al., Energy restriction and the risk of spontaneous mammary tumors in mice: a meta-analysis, Int. J. Cancer 106 (5) (2003) 766–770.

[440] G.N. Rao, E. Ney, R.A. Herbert, Influence of diet on mammary cancer in transgenic mice bearing an oncogene expressed in mammary tissue, Breast Cancer Res. Treat. 45 (2) (1997) 149–158.

[441] W.E. Heston, Testing for possible effects of cedar wood shavings and diet on occurrence of mammary gland tumors and hepatomas in C3H-A-vy and C3H-Avy-fB mice, J. Natl. Cancer Inst. 54 (4) (1975) 1011–1014.

[442] D. Rudmann, R. Cardiff, L. Chouinard, D. Goodman, et al., Proliferative and nonpro-
       liferative lesions of the rat and mouse mammary, Zymbal's, preputial, and clitoral
       glands, Toxicol. Pathol. 40 (6 Suppl.) (2012) 7S−39S.
[443] R.D Cardiff, The pathology of EMT in mouse mammary tumorigenesis, J. Mammary
       Gland Biol. Neoplasia 15 (2) (2010) 225−233.
[444] R.D. Cardiff, M.R. Anver, G.P. Boivin, M.W. Bosenberg, et al., Precancer in mice:
       animal models used to understand, prevent, and treat human precancers, Toxicol.
       Pathol. 34 (6) (2006) 699−707.
[445] E.G. Jeffrey, C. Robert, H. Lothar, W. Lalage, et al., Validation of transgenic mam-
       mary cancer models: goals of the NCI Mouse Models of Human Cancer Consortium
       and the mammary cancer CD-ROM, Transgenic. Res. V11 (6) (2002) 635.
[446] R.D. Cardiff, M.R. Anver, B.A. Gusterson, L. Hennighausen, et al., The mammary
       pathology of genetically engineered mice: the consensus report and recommendations
       from the Annapolis meeting, Oncogene 19 (8) (2000) 968−988.
[447] R.D. Cardiff, S.R Wellings, The comparative pathology of human and mouse
       mammary glands, J. Mammary Gland Biol. Neoplasia 4 (1) (1999) 105−122.
[448] A. Bilger, L.M. Bennett, R.A. Carabeo, T.A. Chiaverotti, et al., A potent modifier of
       liver cancer risk on distal mouse chromosome 1: linkage analysis and characterization
       of congenic lines, Genetics 167 (2) (2004) 859−866.
[449] H.B. Andervont, Studies on the occurrence of spontaneous hepatomas in mice of
       strains C3H and CBA, J. Natl. Cancer Inst. 11 (3) (1950) 581−592.
[450] T. Tillmann, K. Kamino, U. Mohr, Incidence and spectrum of spontaneous neoplasms
       in male and female CBA/J mice, Exp. Toxicol. Pathol. 52 (3) (2000) 221−225.
[451] F. Feo, M.R. De Miglio, M.M. Simile, M.R. Muroni, et al., Hepatocellular carcinoma
       as a complex polygenic disease. Interpretive analysis of recent developments on
       genetic predisposition, Biochim. Biophys. Acta 1765 (2) (2006) 126.
[452] R.R. Maronpot, T. Fox, D.E. Malarkey, T.L. Goldsworthy, Mutations in the ras
       proto-oncogene: clues to etiology and molecular pathogenesis of mouse liver tumors,
       Toxicology 101 (3) (1995) 125−156.
[453] R.R. Maronpot, J.K. Haseman, G.A. Boorman, S.E. Eustis, et al., Liver lesions in
       B6C3F1 mice: the National Toxicology Program, experience and position, Arch.
       Toxicol. Suppl. 10 (1987) 10−26.
[454] U. Bach, J.R. Hailey, G.D. Hill, W. Kaufmann, et al., Proceedings of the 2009 National
       Toxicology Program Satellite Symposium, Toxicol. Pathol. 38 (1) (2010) 9−36.
[455] A. Brix, J. Ward, B. Mahler, R. Maronpot, A digitized atlas of rat liver lesions.
       Laboratory of Experimental Pathology, National Toxicology Program. Distributed by
       NIEHS, NTP. Free of Charge: Research Triangle Park, NC, 2005.
[456] R.J. Kastenmayer, M.A. Fain, K.A. Perdue, A retrospective study of idiopathic ulcerative
       dermatitis in mice with a C57BL/6 background, J. Am. Assoc. Lab. Anim. Sci. 45 (6)
       (2006) 8−12.
[457] M.M. Slattum, S. Stein, W.L. Singleton, T. Decelle, Progressive necrosing dermatitis of
       the pinna in outbred mice: an institutional survey, Lab. Anim. Sci. 48 (1) (1998) 95−98.
[458] K.J. McElwee, P. Freyschmidt-Paul, M. Zoller, R. Hoffmann, Alopecia areata suscep-
       tibility in rodent models, J. Investig. Dermatol. Symp. Proc. 8 (2) (2003) 182−187.
[459] K.J. McElwee, K. Silva, D. Boggess, L. Bechtold, et al., Alopecia areata in C3H/HeJ
       mice involves leukocyte-mediated root sheath disruption in advance of overt hair loss,
       Vet. Pathol. 40 (6) (2003) 643−650.

[460] J. Sun, K.A. Silva, K.J. McElwee, L.E. King Jr., et al., The C3H/HeJ mouse and DEBR rat models for alopecia areata: review of preclinical drug screening approaches and results, Exp. Dermatol. 17 (10) (2008) 793–805.

[461] J.R. Sarna, R.H. Dyck, I.Q. Whishaw, The Dalila effect: C57BL6 mice barber whiskers by plucking, Behav. Brain. Res. 108 (1) (2000) 39–45.

[462] J.P. Garner, S.M. Weisker, B. Dufour, J.A. Mench, Barbering (fur and whisker trimming) by laboratory mice as a model of human trichotillomania and obsessive-compulsive spectrum disorders, Comp. Med. 54 (2) (2004) 216–224.

[463] A.V. Kalueff, A. Minasyan, T. Keisala, Z.H. Shah, et al., Hair barbering in mice: implications for neurobehavioural research, Behav. Processes. 71 (1) (2006) 8.

[464] P. Jungmann, J.L. Guenet, P.A. Cazenave, A. Coutinho, et al., Murine acariasis: I. Pathological and clinical evidence suggesting cutaneous allergy and wasting syndrome in BALB/c mouse, Res. Immunol. 147 (1) (1996) 27–38.

[465] P. Jungmann, A. Freitas, A. Bandeira, A. Nobrega, et al., Murine acariasis. II. Immunological dysfunction and evidence for chronic activation of Th-2 lymphocytes, Scand. J. Immunol. 43 (6) (1996) 604–612.

[466] P.L. Van Loo, J.A. Mol, J.M. Koolhaas, B.F. Van Zutphen, et al., Modulation of aggression in male mice: influence of group size and cage size, Physiol. Behav. 72 (5) (2001) 675–683.

[467] Y. Litvin, D.C. Blanchard, N.S. Pentkowski, R.J. Blanchard, A pinch or a lesion: a reconceptualization of biting consequences in mice, Aggress. Behav. 33 (6) (2007) 545–551.

[468] P. Johansen, Y. Wackerle-Men, G. Senti, T.M. Kundig, Nickel sensitisation in mice: a critical appraisal, J. Dermatol. Sci. 58 (3) (2010) 186–192.

[469] M. Yamamoto, T. Haruna, C. Ueda, Y. Asano, et al., Contribution of itch-associated scratch behavior to the development of skin lesions in Dermatophagoides farinae-induced dermatitis model in NC/Nga mice, Arch. Dermatol. Res. 301 (10) (2009) 739–746.

17 Design of Vectors for Optimizing Transgene Expression

Louis-Marie Houdebine

Institut National de la Recherche Agronomique, Jouy en Josas, France

I. Introduction

Animal trangenesis is used most frequently to: (i) create models for the study of gene function and of human diseases, (ii) modify organs and tissues (such as from swine) for xenotransplantation into human patients, (iii) prepare recombinant pharmaceutical proteins (principally in milk and egg white), and (iv) improve animal production characteristics (efficiency and value-added traits). This implies the use of different animal species for specific endpoints, requiring the implementation of various gene transfer techniques and a tight control of transgene expression. The techniques of gene transfer to generate transgenic animals are not the same for the different species and they are more or less appropriate to allow either random gene addition or gene targeting. Gene addition is still the most frequently used protocol. Gene targeting, which includes gene knockout, gene knock-in, and targeted mutation (insertional mutagenesis), is becoming more commonplace across disciplinary boundaries.

From the earliest gene transfer studies involving DNA microinjection, a variety of methods were developed to generate transgenic animals (Figure 17.1).

It soon appeared that expression of transgenes generated by DNA microinjection and randomly integrated into the nuclear genome was very variable between the different lines of transgenic animals. This fact was not totally surprising, as a similar observation was done years before in the different clones of cultured cells harboring integrated foreign genes. This phenomenon was particularly frequent with transgenes and attributable to position effects of chromatin. The transgenes integrated in genome regions devoid of genes such as centromeres and telomeres were generally silent. It was also observed that multicopies of the transgene even in tandem may be less efficiently transcribed than a single copy in the same genome site (Garrick et al., 1998).

A comparison of the expression of gene constructs in cultured cells and in transgenic mice revealed that a transferred gene may be active in *in vitro* cultures, but poorly expressed or nonfunctional when introduced into transgenic animals. The reverse was generally not true: an active transgene is usually also well expressed in cells. This meant that the test of a gene construct in cells did not uniformly predict

Transgenic Animal Technology. DOI: http://dx.doi.org/10.1016/B978-0-12-410490-7.00017-7

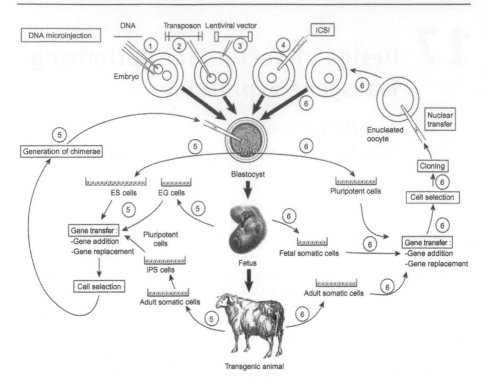

Figure 17.1 Different methods to generate transgenic animals: (1) DNA transfer via direct microinjection into pronucleus or cytoplasm of embryo; (2) DNA transfer via a transposon: the foreign gene is introduced in the transposon, which is injected into a pronucleus; (3) DNA transfer via a lentiviral vector: the gene of interest introduced in a lentiviral vector is injected between the zona pellucida and membrane of the oocyte or the embryo; (4) DNA transfer via sperm: sperm is incubated with the foreign gene and injected into the oocyte cytoplasm for fertilization by ICSI (intracytoplamic sperm injection); (5) DNA transfer via pluripotent or multipotent cells. The foreign gene is introduced into pluripotent cell lines (ES, embryonic stem cell lines established from early embryos, or iPS, induced pluripotent stem cells obtained after dedifferentiation of somatic cells) or into multipotent cell lines (EG, gonad cell lines established from primordial germ cells of fetal gonads). The pluripotent cells containing the foreign gene are injected into an early embryo to generate chimeric animals harboring the foreign gene DNA. The multipotent EG cells containing the foreign gene are injected into chicken embryos to generate gametes harboring the transgene. In both cases the transgene is transmitted to progeny; (6) DNA transfer via cloning: the foreign gene is transferred into a somatic cell, the nucleus of which is introduced into the cytoplasm of an enucleated oocyte to generate a transgenic clone. Methods 1, 2, 3, and 4 allow traditionally random gene addition, whereas methods 5 and 6 allow random gene addition and targeted gene integration via homologous recombination for gene addition or gene replacement including gene knockout and knock-in. The use of engineered endonucleases to cut both DNA strands makes it possible to target gene knock-in and knockout in one-cell embryos.

levels of expression in transgenic animals. This suggested that an integrated foreign gene was not submitted to the same events when it was added into cells and embryos.

It is now well established that gametes derived from gonad-derived somatic cells, which express a few thousand genes, whereas all the genes must be functional to allow the development of embryos. In somatic cells, inactive genes are methylated and embedded in compact chromatin (heterochromatin) in which histones are methylated rather than acetylated. On the contrary, active genes are poorly methylated and the associated histones are acetylated, generating sites hypersensitive to DNAse 1 (HS), indicating that the chromatin is locally in an open configuration (euchromatin) in which DNA is accessible to transcription factors and to RNA polymerase.

During gametogenesis, DNA is demethylated and this event is followed by a full remethylation before fertilization. During the early stage of embryo development until the blastocyst stage prior to implantation, the genome is again demethylated. The two successive demethylation events aim at erasing completely the program of gene expression of the gonad cells. At the time of implantation, a subtle and progressive methylation of the embryonic genome takes place so as to establish step by step the program of gene expression specific to the different cell types of the fetus, newborn, and adult. The mechanism of this programming is still poorly understood, but it is known that some signals in the DNA control the specific methylation of the genome. The conventional gene constructs have little chance to contain all the appropriate signals to be programmed as expected. In particular, the transgene must be protected against silencing mechanisms.

In the case of the β-globin cluster, it was observed that the human β-globin gene containing the transcribed sequence surrounded by the promoter and terminator regions was well expressed when transfected into red blood cells but not after having been transferred into mouse embryos. It was proved later that several essential regulatory regions controlling the expression of the four globin genes of the cluster are located far upstream and downstream of the transcribed region. The addition of some of these regulatory regions to the β-globin construct allowed the gene to be expressed in transgenic mice. Decades later, the functional mechanisms associated with these regulatory regions are only partly known. Thus the existence of remote regulatory elements was discovered via transgenesis, which took advantage of this phenomenon to successfully generate better and more effective gene constructs.

Transgenes must therefore be protected against incorrect or incomplete programming and silencing effectors. Ideally, transgenes should not alter the overall functioning of the host genome. Reaching these two goals is almost impossible, because the genes properly occupy about 2% of the genome in higher organisms and the genomes contain multiple regulators. Recently a vast population of long noncoding RNA (lncRNA) was shown to specifically control gene expression. The subsequent pattern of expression is to a large extent specific to the cell types. This pattern changes during cell differentiation, thus they could participate in the differentiation (Guttman et al., 2011; Lardenois et al., 2011; Sun et al., 2013).

Three major approaches are possible to improve expression vectors for transgenes: (i) adding known regulatory elements favoring transcription as well as mRNA stabilization and translation, (ii) using long genomic DNA fragments expected to contain most of the transcriptional regulatory elements, and (iii) targeting the gene of interest into genomic regions known to favor a reliable transgene expression.

II. Basic Vectors

A certain number of rules should be respected to obtain efficient expression vectors. These rules concern the different regions of the gene construct depicted in Figure 17.2.

A. Nucleotidic Composition of the Vectors

Integrated retroviral sequences and transposons are more or less inactivated by a cytosine methylation of the CpG motifs and the local formation of condensed chromatin (heterochromatin) in which histones are deacetylated and methylated in some specific sites. Transgenes seem to be silenced by similar mechanisms. Most of the vertebrate genes contain CpG islets in their regulatory regions that contribute to their expression. Some of the CpG motifs belong to the binding site of the transcription factor Sp1, which is present not only in the promoter region of the gene but also sometimes in the first introns. This is the case for the eF1-α gene, which is highly expressed in animals and in cultured cells but poorly as a transgene (Taboit-Dameron et al., 1999). Paradoxically, an exceedingly large number of CpG motifs in vectors on both side and in the vicinity of the promoter induce transgene silencing. The replacement of some of the GC regions by AT-rich regions improves transgene expression. The *E. coli* β-galactosidase gene is rich in CpG and is known

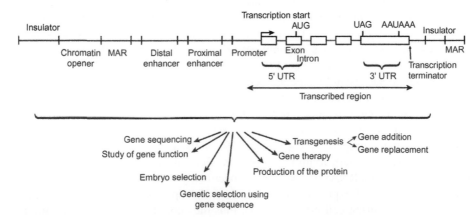

Figure 17.2 General animal gene structure and different uses of isolated genes. Transgenesis, which includes random and targeted gene addition as well as specific gene inactivation and replacement, is an essential tool for gene study and for biotechnological applications.

to be a strongly silenced transgene. This silencing proved to be markedly reduced when the number of CpG was diminished. The coding sequences of a transgene may thus be obtained by chemical synthesis to replace a part of the CpG-rich codons by others without modifying the sequence of the corresponding protein.

Matrix attached regions (MARs) are frequently found in the vicinity of genes, and they bind locally DNA to the nuclear matrix. MARs are generally AT rich and have been added into vectors to tentatively improve transgene expression. This approach has met with variable success.

B. Addition of Introns

A transgene must contain at least one intron, required to favor the transfer of mRNA to the cytoplasm. The first intron of many genes contains sites that bind transcription factors and may favor transgene expression. The intron splicing is dependent upon several signals comprising consensus sequences in both splicing sites (CAG GUA/GAGUA/UGGG in 5′ and CAG G...GAA/G...GAA/G...in 3′), a CU-rich region immediately upstream of the 3′ splicing site and a branched point sequence site, U/CNCUGAC, at about 30 nucleotides upstream of the 3′ splicing site and upon splicing enhancers (Mersch et al., 2008). The sequences involved in the splicing mechanism may be modified and optimized if needed by mutagenesis. This may significantly enhance transgene expression but also reduce possible alternative splicing based on the use of weak cryptic splicing sites leading to the synthesis of truncated or altered proteins.

The intron(s) must preferably be put before the coding region. If an intron is added after the translated region, the 5′ splicing site must be located not more than 50 nucleotides from the termination codon to avoid the activation of the nonsense-mediated decay that degrades the mRNA (Chang et al., 2007).

C. Optimization of the Transcribed Region

The optimization of vectors for the expression of transgenes has been focused initially on promoters and on transcription. It is now clear that the transcribed region of genes contains multiple signals that control mRNA translation and stability. Constructing a gene for transgenesis consists often of taking DNA fragments containing unknown as well as known signals and associating them with the risk of inactivating important mechanisms for transgene expression and generating new unknown signals. The transcribed region of the genes and transgenes is also submitted to these rules. To avoid problems, the following precautions can be taken.

The 5′ UTR (untranslated region) of the mRNA coded by the transgene must be as poor as possible in GC sequences, which can stabilize double-strand hairpin structures that do not favor ribosome migration to the initiation codon. The AUG initiation codon must preferably be in the Kozak consensus sequence GCCA/GCCAUGG to optimize translation initiation. The natural 5′ UTR of the gene of interest may contain sequences regulating translation. It may then be helpful not to keep this region and replace it by a short (preferably not less than 80 nucleotides)

AT-rich 5' UTR region from genes known to be efficiently translated in many cell types or in the targeted cells of the animals. Some mRNAs encode proteins that are not naturally secreted. Peptide signals may be added to their cDNA to allow secretion of the proteins.

The 3' UTR region of a number of mRNAs contains signals for mRNA translation and stability. A number of mRNAs have an AU-rich region, with the AUUUA motif in their 3' UTR. These mRNAs have a short half-life controlled by the cell cycle (Beelman and Parker, 1995). The fortuitous presence of such sequences must be searched and eliminated to prevent a poor transgene expression.

Some mRNAs contain translation regulators that bind to proteins favoring the recycling of ribosomes by binding to the 5' UTR. CU-rich regions in the 3' UTR enhance the stability of the mRNAs and may be added in the vectors downstream of the cDNAs. Stabilizing sequence can be taken in the 3' UTR of the human or bovine genes and of the α-globin gene, which also contain efficient transcription terminators (Chkheidze et al., 1999). Translation stimulators may also be added in the 3' UTR (Knirsch and Biadasz Clerch, 2000).

Some proteins are anchored to the plasma membrane by a GPI structure (glycophosphatidylinositol). A protein normally not anchored in this way acquires this property by inclusion in the 3' end of the cDNA the peptide allowing the addition of GPI (Malassagne et al., 2003).

MicroRNAs (miRNA), the role of which was recently discovered, inhibit specifically the translation of an mRNA after forming a hybrid with its 3' UTR. The presence of target sequences for a miRNA may unduly inhibit the expression of a transgene. In such cases, this target sequence should then be deleted.

A cDNA and other regions of the vectors may thus preferably be chemically synthesized. This allows reducing the number of CpG motifs, to choose the best codons adapted to the cell type in which the transgene is to be expressed and to eliminate cryptic 5' or 3' splicing sites and sequences known to prevent transcription or translation. A complete sequencing of the vectors and a test of its capacity to express the foreign gene *in vitro* may be achieved using transfected cells.

D. Addition of Insulators

The expression of most animal genes is controlled by remote regulatory sequences. These regulators, which are poorly known for most of the genes, may be enhancers, chromatin openers, or insulators. MARs are also found in the genome regions containing genes. In some cases, the remote regulators govern not only one gene but also a cluster of genes. They are then considered as locus control regions (LCRs). The β-globin locus is one example of a cluster forming an LCR.

MARs have been considered as essential elements to favor gene and transgene expression. It is now admitted that MARs have no direct impact but might contribute to favor the effects of genuine stimulators (Houdebine, 2010).

Insulators are known to protect a gene from the action of the regulators of neighbor genes. Insulators are also capable of preventing the heterochromatin from spreading in a gene region and can silence it. An insulator activity has been found in the 5′ HS4 region of the chicken β-globin locus. This element contains an insulator proper and a chromatin opener (Gaszner and Felsenfeld, 2006). The 5′ HS4 element can improve the expression of a number of unrelated transgenes in mammals when added into the vectors (Taboit-Dameron et al., 1999; Giraldo et al., 2003). However, the potency of the 5′ HS4 element remains generally insufficient to express transgenes in a fully satisfactory manner.

E. Coexpression of Two Cistrons from the Same Vector

It is sometimes necessary to express two or even three genes in the same transgenic animals. The coinjection of several independent vectors makes possible the generation of up to 80% of the animals harboring the two or three genes that are cointegrated at the same site. An alternative consists of using internal ribosome entry sites (IRES). Such sequences exist in the 5′ UTR of a number of mRNAs, the translation of which is controlled by these sequences that bind specific cellular inducible proteins. Such sequences may be added between two cistrons and allow their simultaneous translation from a single vector. The addition of the IRES's 80 nucleotides after the termination codon of the first cistron may contribute to favor the expression of the second cistron (Houdebine and Attal, 1999).

Another particularly efficient method is based on the addition of the peptide 2A between the two cistrons. The peptide is spontaneously cleaved in most cells separating the two proteins (Deng et al., 2011).

F. Control of Transgene Expression by Exogenous Inducers

The basic expression vectors are essentially dependent on host inducers. Several systems have been proposed to induce transgene expression by exogenous inducers not acting on host genes. The most popular of these systems is based on the use of an antibiotic, doxycycline or tetracycline, that may induce (Tet-On) or repress (Tet-Off) the transgene constructed accordingly. The effects of doxycycline are obtained even after oral administration and are readily reversible. This system and the others were depicted elsewhere (Malphettes and Fussenegger, 2006). A data bank listing available mouse (about 500) and rat lines with the Tet system functional in various cell types was also recently published (Schönig et al., 2013).

III. Vectors Using Long Genomic DNA Fragments

Most of the vectors used to express transgenes contain at least a promoter, one or several introns, the coding sequence (cDNA or genomic DNA) and a transcription terminator. These simple vectors are often sufficient to study the effects of a transgene, since only a limited amount of protein coded by the transgene is required in many cases. Moreover, in species like mice, the number of available transgenic lines is not limiting, allowing the selection of those expressing the transgene in a satisfactory manner. For some projects, such as the production of recombinant proteins in milk, a high expression level of the transgenes is necessary and the use of large farm animals is needed. Gene transfer in these species is less efficient than in mice. Reliable vectors for the expression of the transgenes are then needed.

Experiments carried out in the 1980s revealed that the human β-globin gene needed very remote regulatory sequences to be expressed in transgenic mice. The most remote upstream regulatory element was found at −62.5/−60.7 kb from the β-globin gene promoter within the cluster of the olfactory receptors. Then it was found that most of the animal genes are controlled by multiple regulatory regions located far upstream and downstream of their promoters. It remained to decipher the underlying mechanisms allowing remote regulators to control transcription. Experiments using the technique known as 3C (Chromatin Conformation Capture) showed that a looping process allows the formation of an active chromatin hub (ACH) in which the different transcription factors are in close vicinity to the transcription start site (de Laat et al., 2008). These factors are then able to participate in the formation of the transcription initiation complex (Figure 17.3).

These data support the idea that long genomic DNA fragments surrounding a gene contain all the regulatory elements of the gene and are thus likely to express efficiently the gene in transgenic animals. It was postulated that the transgenes present in a bacterial artificial chromosome (BAC) vector could be expressed in all transgenic lines independently of their integration sites and at a level proportional to the number of integrated copies. In other words, in an ideal situation, transgenes

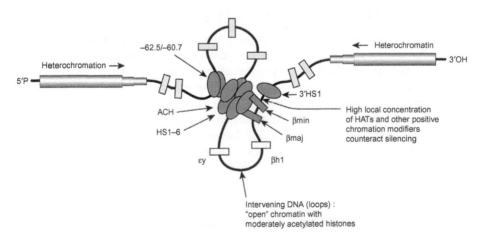

Figure 17.3 Mechanism of action of the remote regulatory elements of the β-globin gene. Red blood cells contain multiple factors binding to the regulatory sequences of the β-globin gene including HS (hypersensitive sites to DNase1) sites and the −62.5/−60.7 region. The HAT (histone acetyltransferase) factor acetylates histones leading to a local chromatin opening. Chromatin forms loops that bring the remote factors to the vicinity of the promoter, allowing the formation of the transcription initiation complex. This model is valid for most of the animal genes.

surrounded by long genomic fragments could be expressed autonomously as would be evident in a genomic context.

In a number of cases, it was observed that using large genomic DNA fragments (50—250 kb) cloned in BACs improved the efficiency of transgene expression (Long and Miano, 2007). An attractive approach may consist in using as vectors BACs harboring a long genomic DNA fragment (50—250 kb) known to favor the expression of the gene(s) it contains. The foreign DNA containing the coding sequence (cDNA or genomic DNA) must then be introduced in the BAC vector using homologous recombination in bacteria (Cobb and Zhao, 2012). The expression of the transgene added in the genomic DNA of the BAC is generally better or much better than that obtained with basic vectors. However, each BAC carrying genomic DNA shows specific, largely unpredictable properties. BACs often contain several genes that may interfere with the expression of the transgene or have side effects on animals.

An example is the case of the whey acidic protein (WAP) which expresses WAP gene poorly with a 30 kb genomic fragment of much better with 80 kb, weakly with 120 kb, and the best with 135 kb (Saidi et al., 2007). The luciferase gene introduced in the 135 kb WAP locus replacing the WAP coding sequence was much better expressed in transgenic mice than with a basic vector containing the WAP gene promoter (Jolivet et al., unpublished data). In no case was the expression level dependent on the number of integrated copies.

Other examples include the mouse HPRT and Rosa 26 sites, which are known to express nicely most, if not all, of the foreign genes targeted in these sites by homologous recombination, irrespectively of the promoter used (Palais et al., 2009). It remains unclear if a BAC containing these loci would express the foreign genes as well.

A BAC containing the DHFR locus proved to express foreign genes as a function of the integrated copy number. However, the vectors were transfected into cultured somatic cells and not in embryos (Bian and Belmont, 2010). It is thus not known if the capacity of the DHFR locus would be kept after the genome reprogramming that occurs during early embryo development.

It is thus important to note that the transgenes driven by BACs rarely work in an ideal fashion. Indeed, genes have been designed during evolution to work in their specific sites and not in any position in the genome. Long genomic DNA fragments are expected to suppress the position effects on transgenes, which is rarely the case, but it is clear that the variegated expression that characterizes the conventional transgenes is much less frequent in animals harboring BAC vectors. A higher proportion of animals expressing the transgenes is generally observed with BAC than with plasmid vectors. Some BACs may contain all the elements providing transgenes with a complete independence of the integration site. If not, a BAC vector may still contain enough regulatory elements that significantly improve transgene expression to justify its use. A more sophisticated approach could consist of using vectors containing not all the DNA sequence of BACs but only the major elements involved in the control of gene and transgene expression. This is generally not presently possible, as most of the active elements present in BACs are unknown.

Despite the advantage of BAC vectors to express transgenes, these vectors are not frequently used, essentially for technical reasons. The genomic DNA regions that can be used as vectors are not all known, and their validation may require a relatively great effort. The construction of BAC vectors is usually based on homologous recombination, which necessitates more complex construction requirements than those of basic vectors. Additionally, manipulation of the BAC vectors requires special methods to minimize fragmentation. Moreover, uncontrolled recombination leading to an alteration of the transgene is more frequent using BAC vectors rather than basic vectors (Le Saux et al., 2010).

IV. Gene Targeting

Targeting the integration of a foreign gene into chosen sites of a genome may be a way to avoid the position effects and to provide the transgene with all the genomic transcription regulators.

As noted in Chapters 4 and 5, it is possible to target the integration of the foreign gene using homologous recombination. This mechanism is based on the perfect recognition between a chosen genome sequence and the sequence of the exogenous DNA. This recognition leads to the formation of hybrids and finally to the targeted integration of the foreign DNA (Figure 17.4). Homologous recombination is naturally implemented to repair mutated genes. It is a rare event in animal cells, corresponding to about 0.1–1% of the heterologous recombination. Therefore, homologous recombination is traditionally not implemented directly in early embryos but in intermediate cells further used to generate transgenic animals.

Several applications of this approach are possible: (i) the precise integration of a functional foreign gene in a chosen genomic region (gene knock-in), (ii) the replacement of an allele by another allele (cisgenesis), or (iii) the replacement of a gene by a nonfunctional DNA sequence leading to the inactivation of the targeted gene (gene knockout) (Figure 17.4).

A. Use of Engineered Endonucleases

The frequency of homologous recombination is greatly increased when both DNA strands are specifically cleaved at the targeted site of the genome. To reach this goal, several endonucleases having a domain recognizing the targeted sites in the genome and another domain able to cleave DNA nonspecifically in the vicinity of the binding site may be used. These endonucleases may be meganucleases found in yeast. They cleave DNA at specific sites nonexisting in most species. Meganucleases must therefore be engineered to target the genomic DNA cleavage and to induce specific homologous recombination. An alternative consists of generating fully engineered fusion endonucleases, known as ZFNs, containing a zinc finger region recognizing specifically the chosen genome site and a common nonspecific endonuclease, Fok1. A third, particularly attractive possibility is based on the generation of fusion enzymes in which the Fok1 domain is associated to a

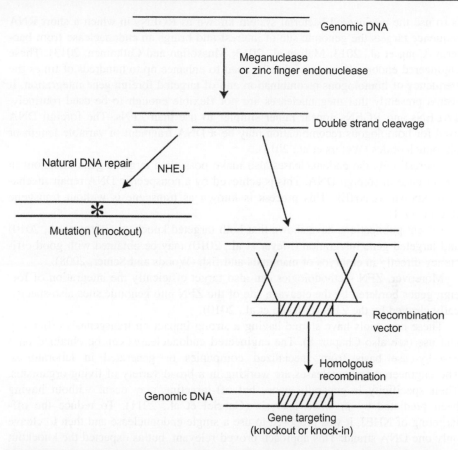

Figure 17.4 Gene targeting using homologous recombination. The introduction into a cell of a DNA fragment having part of its sequence similar to genomic DNA may lead to a replacement of the genomic sequence at a very low frequency. If the transferred DNA contains two sequences homologous to genomic DNA regions surrounding foreign DNA, the homologous sequences recombine (indicated by crosses) and the foreign DNA is integrated into the genome in a targeted manner. The targeted genomic gene is interrupted and thus inactivated (knocked out) by the foreign DNA. Alternatively, the foreign sequence may be a functional gene, the integration of which is precisely targeted (knocked in). The homologous recombination process is enhanced up to 1000-fold when both strands of genomic DNA are locally cleaved by targeted endonucleases (meganucleases, zinc finger nucleases (ZFNs), transcription activator-like endonucleases (TALENs), or RNA guided endonucleases (RGENs)). When the endonucleases are injected into the embryos without any recombinant vector, DNA break is repaired but often according to a random process known as NHEJ (nonhomologous end joining) generating a targeted mutation and thus a knockout.

domain recognizing specific DNA sequences. These second domains belong to the plant transcription activator-like effectors. The resulting endonucleases known as TALENs can be virtually engineered to target any genomic site (Li et al., 2011; Reyon et al., 2012; Nature Methods, 2012; Wefers et al., 2013). A fourth possibility is to use the engineered bacterial system known as RGENs in which a short RNA sequence targets the genomic site of interest and brings an endonuclease from bacteria (Cong et al., 2013; Mali et al., 2013; Mussolino and Cathomen, 2013). These engineered endonucleases may thus be used to enhance up to hundreds of times the frequency of homologous recombination and of targeted foreign gene integration. It seems presently that meganucleases are not flexible enough to be used routinely. TALENs and RGENs appear rather simpler to use than ZFNs. The foreign DNA used for homologous recombination may be a DNA fragment of variable length or oligonucleotides (Wefers et al., 2013).

Interestingly, the endonucleases also make possible a targeted gene knockout in the absence of foreign DNA. This is achieved by a nonspecific DNA repair mechanism known as NHEJ. This process is known as transgenesis without transgene (Figure 17.4).

Recent publications have shown that both targeted knockout (Rémy et al., 2010) and targeted gene integration (Meyer et al., 2010) may be obtained with good efficiency directly in embryos of mammals and fish (Woods and Schier, 2008).

Moreover, ZFN methodologies can also target efficiently the integration of foreign genes bordered by the cleavage site of the ZFN into genomic sites also specifically cleaved by the ZFN (Orlando et al., 2010).

These new tools have started having a strong impact on transgenesis efficiency and use (see also Chapter 8). The engineered endonucleases can be obtained on a case-by-case basis from specialized companies or generated in laboratories. The engineered endonucleases are working in a broad variety of living organisms. Their specificity is generally good, but off-targeting may occur without having been predicted by *in silico* analysis (Gabriel et al., 2011). To reduce the off-targeting of NHEJ, it was proposed to use a single endonuclease and then to cleave only one DNA strand. This approach proved relevant, but as expected the knockout frequency was reduced.

A genome-scale collection of TALENs for efficient and scalable gene targeting in human cells corresponding to 18,740 protein-coding genes has been obtained (Kim et al., 2013).

B. Use of the PhiC31 Integrase-Mediated System and AAV

In a number of species, cells contain a specific integrase, PhiC31, capable of recombining the DNA sequences *attB* and *attP* added to vectors with similar genomic sequences with a good efficiency. This recombination integrates the vector and it generates *attL* and *attR* sites that are not recognized by the PhiC31 integrase. The integration of the foreign DNA is thus irreversible. This system proved efficient to generate transgenic *Drosophila* (Bateman et al., 2006). This system is efficient in other species but its use is limited.

The adeno-associated virus (AAV) is known to integrate in host genomes via a homologous recombination process. AAV vectors have thus been designed to target gene integration (Larochelle et al., 2011). This system is efficient, but it can carry only a few kilobase DNA fragments; it is being used by a small number of laboratories to generate transgenic animals.

V. Specific Inhibition Gene Expression

Inhibiting specifically the expression of a host gene (or of a transgene) is essential to understand gene functions and to create models for the study of human diseases. Several methods of inhibition are possible: at the levels of DNA, mRNAs, or proteins.

A. Cre-LoxP and Flp-FRT Systems

Cre recombinase may be synthesized by the corresponding gene under the direction of a cell-specific promoter including promoters under the control of doxycycline (see Section 17.II.F). Another level of control can be obtained by using an engineered Cre recombinase which becomes reversibly active in the presence of an estrogen analogue, 4-hydroxy tamoxifen (Metzger and Chambon, 2001). This offers the advantage of having the active Cre recombinase for short periods of time. This prevents the nonspecific action of the Cre recombinase, which can recognize cryptic sites in the host genome and induce illegitimate recombination, damaging the host DNA. This tool is appropriate to delete genes for resistance to antibiotics (Figure 17.5) but mainly to allow conditional knockout, which may be induced in a chosen cell type and at a given time. The models obtained in this way have a better chance to mimic a physiological situation.

Cre-LoxP and Flp-FRT may also be used to target the integration of a foreign DNA fragment in a genome site. To reach this goal, a LoxP (or FRT) sequence must be added to the animal genome at the chosen site and another sequence in one end of the foreign DNA. The recombinases will then induce the targeted integration. However, the action mass law implies that the excision of the integrated foreign DNA is more efficient than its integration. These systems can be used as tools for targeted foreign gene integration only if the integration process generates DNA sequences unable to recombine and eject the integrated foreign DNA. This approach is known as recombinase-mediated cassette exchange (Baer and Bode, 2001). LoxP and FRT systems are more often used to delete a DNA region previously bordered by the LoxP or the FRT sequences (see Chapters 4 and 5 for further discussion).

B. Gene Inactivation

Gene knockout is a potent and irreversible means to inactivate a gene. The Cre-LoxP system is one possibility (see Section 17.V.A). The knockout can be achieved using conventional homologous recombination or with engineered endonucleases (see Section 17.IV). A gene knockout may also be obtained using the NHEJ after a

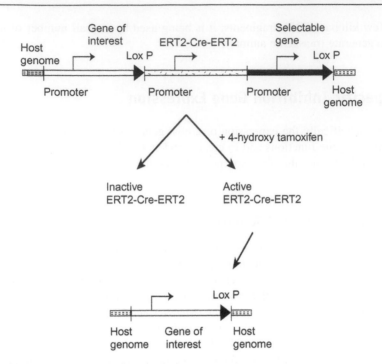

Figure 17.5 Activation of Cre recombinase and selectable gene elimination by 4-hydroxy tamoxifen. The Cre recombinase gene expression may be under the control of the Tet-On system, itself under the control of a cell-specific promoter. The fusion protein Cre recombinase-mutated estrogen receptor is active only in the presence of 4-hydroxy tamoxifen. The elimination of the DNA region bordered by LoxP sequences is thus sharply controlled. The selection gene and the Cre recombinase may thus be eliminated from transgenic animals at any stage of their lives.

double break of DNA at the chosen site using engineered endonucleases or the RGEN system (see Section 17.IV). The advantages of this approach are its high efficiency and the fact that the inactivation may be achieved in one-cell embryos.

C. Use of siRNAs

The use of engineered endonucleases, namely ZFNs, TALENs, and RGENs, has greatly facilitated gene knockout efficiency (see also Chapter 8). In comparison, the utilization of small interfering RNA (siRNA) appeared less easy and potent than using endonucleases. Yet siRNAs remain a valuable tool for rapid inhibition of specific genes in cultured cells. They also are appropriate to inhibit the expression of exogenous genes, namely viral genes, even if the inhibition of the targeted genes is usually not complete. siRNAs expressed in transgenic animals are expected to protect the animals against viral infection. The success of this approach in plants strongly supports the idea that the genetic vaccination by the use of siRNA transgenes will become a reality in laboratories and breeding.

Long double-strand RNAs present in cells are randomly cut into 19−21 nucleotide fragments known as siRNAs. One of the two strands of the siRNA is kept and targeted to an mRNA having a complementary sequence. This induces the degradation of the mRNA. In practice, a synthetic gene containing the targeted 19−21 nucleotide sequence follows a short random sequence and by the targeted sequence in the opposite orientation is linked to a promoter acting with RNA polymerase III (usually U6 or H1 gene promoters). The RNAs synthesized by such vectors form a 19−21 nucleotide double-strand RNA known as shRNA (short hairpin RNA), which is processed in cells to generate active siRNAs. An appropriate expression of siRNA genes in transgenic animals can be obtained when they are introduced into lentiviral vectors (Tiscornia et al., 2003).

Gene constructs containing the U6 or H1 promoters and the gene coding for the siRNAs are fully active in cultured cells but for unknown reasons not in transgenic mice. To circumvent this problem, the U6/H1-siRNA vectors were introduced at different positions and in both orientations in a vector dependent of RNA polymerase II (promoter of the human eF1-α gene) and expressed in all cell types. The U6/H1-siRNA vectors appear protected from silencing by the RNA polymerase II, since the siRNA's genes were expressed at high levels in cells and in transgenic mice. This protocol enabled the protection of the mice against the pseudorabies virus using siRNAs to target the immediate early gene of the virus (Daniel-Carlier et al., 2013).

In another experiment, 11 siRNA sequences embedded in the consensus part of a miRNA gene were chemically synthesized in tandem as a single DNA fragment that was introduced in the intron of a conventional vector containing the WAP gene promoter. The siRNAs were targeted to recognize the mRNAs of the four more important rabbit milk proteins. The milk of the transgenic rabbits contained a very low concentration of the four proteins, whereas the nontargeted proteins were at a normal or a higher level (Ripoll et al., 2010). This protocol thus appears efficient to inhibit gene expression at the mRNA level in a single cell type.

The recent discovery of the role of microRNAs has increased the possibility of using interfering RNAs. MicroRNAs are encoded by short genes expressed under the control of RNA polymerase II promoters. Their primary products are transformed into siRNAs. The mature miRNAs, which are fully complementary to the targeted mRNA, induce a degradation of this mRNA. The miRNAs that are only partially complementary to the targeted mRNA recognize a sequence located in the 3′ UTR of the mRNA and inhibit reversible translation of this mRNA without inducing its degradation.

The application of the siRNA approach raises specific problems in animals. Long double-strand RNAs induce interferons and some unspecific immune reactions (Sioud, 2006). On the other hand, as opposed to plants, siRNAs are not auto-amplified in higher animals and this reduces their potency. Vectors to express miRNA genes are available, but simple shRNA genes are also easily expressed in transgenic animals using conventional vectors. Moreover, siRNAs may off-target mRNAs and generate deleterious side effects.

Several programs based on empirical data indicate the putative optimal shRNA sequences use to allow the preferential use of the siRNA strand complementary to the mRNA (RNAi, 2009). A very important point is to choose a target region of the mRNA which is not in a double-strand structure and thus accessible to the siRNA. Banks of shRNA genes in lentiviral vectors are available for the mRNAs of different species. It remains that most of the siRNAs do not inhibit the targeted gene to more than 70–80% which may be insufficient to obtain some relevant animal models. It is tempting to use vectors expressing the shRNA genes at a relatively high level, but this may lead to no important increase of the inhibition and to a higher off-targeting, which may be detrimental for the animals or even lethal. In fact, it seems that a well-targeted siRNA can be highly active even at a low concentration. It appears therefore of paramount importance to select the shRNA capable of strongly inhibiting the targeted mRNA even at a low concentration in cell systems before generating transgenic animals.

D. Use of Decoys

Another possibility to inhibit the specific expression of a gene is to use decoys. This may include proteins, RNAs, or other molecules.

A *trans*-dominant negative mutant of insulin receptor overexpressed in transgenic mice and playing the role of decoy for the hormone led to the generation of a new model for diabetes study (Chang et al., 1994). Another example is the overexpression of the soluble part of the pseudorabies virus receptor in mice. The virus preferentially bound the decoy, and the mice were protected against the virus (Ono et al., 2004).

In transgenic chicken, the overexpression of a mutant RNA of the influenza virus acted as a decoy preventing the formation of viral particles and leading to the generation of chicken resistant to the virus (Lyall et al., 2011). The major difficulty with this approach is to design decoys having a potent and specific action.

VI.   Relations between Vectors and Gene Transfer Techniques

The use of a vector to express a transgene must be chosen, taking into account the specific techniques implemented to generate the transgenic animals (Figure 17.1).

A. DNA Microinjection

As noted in Chapters 2 and 9, DNA microinjection into mammalian pronuclei is primarily used to generate transgenic mice and rabbits. Indeed, this method has become less efficient than several other techniques in a number of species. DNA microinjection into cytoplasm of lower vertebrates and invertebrates is successful and largely used in some species and not in others. The reasons for this specificity are not known.

DNA microinjection is appropriate for the transfer of basic vectors and of BACs. In the latter case, special care must be taken to avoid mechanical fragmentation of DNA.

B. Use of Lentiviral Vectors

Lentiviral vectors are extensively used despite some specific limits. The efficiency of these vectors is high in various species. In chicken, DNA microinjection was totally inefficient. These vectors cannot harbor more than 8 kb of foreign DNA. The amount of vector used must be adjusted to give a good yield of transgenic animals with a limited number of independent integrated copies. If not, it is necessary to eliminate some of the copies and to obtain a constant number of copies in each mouse line. The integrated lentiviral vectors and the transgene are subjected to silencing. Each integrated copy is different from this point of view. The addition of the insulator $5'$ HS4 (see Section 17.II.D) reduces the silencing effect. The genes introduced in lentiviral vectors are generally under the dependency of the viral promoter. Adding specific promoters independent of the viral promoter is possible but laborious. The lentiviral vectors are appropriate to express genes for siRNAs (Tiscornia et al., 2003). Banks of these vectors harboring siRNA genes directed against human and mouse mRNAs are available on the market. Finally, special confinement may be needed to manipulate the lentiviral vectors.

C. Use of Transposons

Transposons are natural genomic DNA sequences that can replicate and integrate within the same genome. Several transposons are being used as vectors to transfer genes into cells and embryos. These vectors have been markedly improved by mutagenesis and mainly by enhancing the activity of the transposase (or integrase) that integrates the vector. The transposons are of various origins; most of them are active in a broad spectrum of cells and organisms. Transposons have been used extensively for more than two decades to generate transgenic insects.

Piggy Bac is particularly used to generate transgenic birds via the transfer of the foreign genes into primordial germ cells (PGCs) (Park and Han, 2012; Yang and Kim, 2012; Macdonald et al., 2012; Urschitz et al., 2010; Liu et al., 2013). Tol2 proved efficient in fish and mice, with the capacity to transfer up to 66 kb of foreign DNA (Suster et al., 2009; Sumiyama et al., 2010). Sleeping Beauty has been greatly improved and the SB100X mutant appears particularly attractive (Garrels et al., 2011; Mátés et al., 2009). Interestingly, Piggy Bac integrates preferentially in the vicinity of genes or within genes. piggy Bac may thus generate alteration of the host genome. Several piggy Bac-derived vectors containing the transposase gene or self-inactivating transposase are available (Urschitz et al., 2010; Marh et al., 2012). SB100X is not found in gene-rich regions of the genome and it is not expected to interfere with host gene expression. SB100X may be present in multiple independent copies in the host genome. All the copies appear equally active and the level of expression is proportional to the number of integrated copies. SB100X is thus likely integrated preferentially in euchromatin and protected from epigenetic mechanisms leading to silencing. SB100X is stable as its expression level in the descendants of the transgenic founders.

The transposons can transfer genes efficiently in embryos from various species, in ES cells, in PGCs, and in somatic cells to be used for SCNT. Their manipulation is not at risk and it requires no particular confinement.

D. Use of ICSI

Using sperm as DNA carrier to generate transgenic animals was shown for the first time about two decades ago (see Chapter 22). It soon appeared that this approach was poorly reproducible and thus not utilizable. The protocol was thus abandoned.

The idea was reappraised via ICSI, an *in vitro* fertilization technique that consists of injecting sperm into the cytoplasm of oocytes. This technique is currently used for *in vitro* fertilization in humans. To transfer genes, sperm from which plasma membrane has been damaged by freezing and thawing are incubated in the presence of the gene of interest and further used for fertilization by ICSI. This method has proved efficient in mice with random integration (Moreira et al., 2007; Shinohara et al., 2007) and pigs (Yong et al., 2006). Interestingly, the yield of transgenesis is often higher than with DNA microinjection and it works as well with short and long DNA fragments. This technique is expected to be extended to the other species in which ICSI is possible.

E. Use of Intermediate Cells

When the yield of genetic modification is too low in embryos, it appears possible to use intermediate cells. The genetic modifications are then achieved in the cells, which are further used to participate in the development of the embryos with a transmission of their genome to progeny. The methods to transfer genes are those used for cell transfections. One advantage of this approach is that the efficiency of the gene transfer is relatively high. Another advantage is that it is possible to analyze in detail the gene modifications in cells before using them to generate transgenic animals.

Pluripotent cells are those present in early embryos (morula and blastocysts). In the best conditions, the embryonic pluripotent cells can be cultured and keep their pluripotency. The resulting lines of ES cells can be genetically modified, selected, and transferred into recipient early embryos at the morula or blastocyst stages. These cells participate to the development of the embryo to give birth to chimeric transgenic animals (Figure 17.1).

As outlined in Chapters 4 and 5, the first ES cells implemented to genetically modify animals (mice) were used at the end of the 1980s (Capecchi, 1989; Bronson and Smithies, 1994). After about a decade of effort, genuine rat ES cell lines were obtained, allowing gene knockout and knock-in as had been the case in mice for more than two decades (Hamra, 2010).

Recent experiments have shown that the transfer of four and even only three of the genes normally expressed in pluripotent cells into mouse somatic cells can induce a relatively rapid dedifferentiation of these organ cells into pluripotent cells known as iPS cells. The pluripotency of these cells was ascertained by their capacity to participate in the development of chimeric mice with a transmission of their genome to progeny (Pera and Hasegawa, 2008). iPS cells might also be implemented for transgenesis, particularly in species in which ES cells are not available (Figure 17.1). The concept was validated in mice, but until now iPS cells have not been used to generate other transgenic animals.

Multipotent cells, organ stem cells able to give rise to mature gonads and gametes, have been identified in several species. These cells known as PGCs have been shown to be able to form chimera when transferred to early embryos. Foreign genes can be transferred into EG cells, which can be implanted into recipient embryos and participate to gonad development, thus transmitting their transgene to progeny with a high yield. This approach has greatly simplified the generation of transgenic chicken (van de Lavoir et al., 2006; Han, 2009). Recent studies indicated that gene transfer into PGCs is facilitated by the use of the transposon piggy Bac vector (Yang and Kim, 2012).

The birth of Dolly the sheep demonstrated that the genome of somatic cells can be reprogrammed after being introduced into enucleated oocytes. This generates pseudo-embryos capable, with a relatively low yield, of giving birth to clones of the cell donor. This technique, known as SCNT (somatic cell nuclear transfer), was initially designed to improve transgenesis efficiency in farm animals (Schnieke et al., 1997; Robl et al., 2007). The principle of this method is described in Figure 17.1.

VII. Conclusion

During the last decade, the techniques to generate transgenic animals and control the expression of transgenes have been markedly improved. The modification of animal genomes has become easier, cheaper, and more precise. This will improve the relevance of the models, particularly for human disease study. This may lead to a further increase in the number of experimental animals, but fewer animals per project should be needed. The new techniques should also contribute to develop relevant and safe projects for food production and for improving farm animal welfare.

References

Baer, A., Bode, J., 2001. Coping with kinetic and thermodynamic barriers: RMCE, an efficient strategy for the targeted integration of transgenes. Curr. Opin. Biotechnol. 12 (5), 473–480.

Bateman, J.R., Lee, A.M., Wu, C., 2006. Site-specific transformation of Drosophila via ϕC31 integrase-mediated cassette exchange. Genetics 173, 769−777.

Beelman, C.A., Parker, R., 1995. Degradation of mRNA in eukaryotes. Cell 81, 179−183.

Bian, Q., Belmont, A.S., 2010. BAC TG-EMBED: one-step method for high-level, copy-number-dependent, position-independent transgene expression. Nucleic Acids Res. 38 (11), e127. Available from: http://dx.doi.org/10.1093/nar/gkq178.

Bronson, S.K., Smithies, O., 1994. Altering mice by homologous recombination using embryonic stem cells. J. Biol. Chem. 26, 27155−27158.

Capecchi, M.R., 1989. The new mouse genetics: altering the genome by gene targeting. Trends Genet. 5, 70−76.

Chang, P.Y., Benecke, H., Le Marchand-Brustel, Y., Lawitts, J., Moller, D.E., 1994. Expression of a dominant-negative mutant human insulin receptor in the muscle of transgenic mice. J. Biol. Chem. 269, 16034−16040.

Chang, Y.F., Imam, J.S., Wilkinson, M., 2007. The nonsense-mediated decay RNA surveillance pathway. Ann. Rev. Biochem. 76, 51−74.

Chkheidze, A.N., Lyakhov, D.L., Makeyev, A.V., Morales, J., Kong, J., Liebhaber, S.A., 1999. Assembly of the α-globin mRNA stability complex reflects binary interaction between the pyrimidine-rich 3′ untranslated region determinant and poly(C) binding protein αCP. Mol. Cell. Biol. 19, 4572−4581.

Cobb, R.E., Zhao, H., 2012. Direct cloning of large genomic sequences. Nat. Biotechnol. 30 (5), 405−406. Available from: http://dx.doi.org/10.1038/nbt.2207.

Cong, L., Ann Ran, F., Cox, D., Lin, S., Barretto, R., Habib, N., et al., 2013. Multiplex genome engineering using CRISPR/Cas systems. Science 339 (6121), 819−823. Available from: http://dx.doi.org/10.1126/science.1231143.

Daniel-Carlier, N., Sawafta, A., Passet, B., Thépot, D., Leroux-Coyau, M., Lefèvre, F., et al., 2013. Viral infection resistance conferred on mice by siRNA transgenesis. Transgenic Res. 22, 489−500. Available from: http://dx.doi.org/10.1007/s11248-012-9649-4.

Deng, W., Yang, D., Zhao, B., Ouyang, Z., Song, J., Fan, N., et al., 2011. Use of the 2A peptide for generation of multi-transgenic pigs through a single round of nuclear transfer. PLoS ONE 6, e 19986. Available from: http://dx.doi.org/10.1371/journal.pone.0019986.

de Laat, W., Klous, P., Kooren, J., Noordermeer, D., Palstra, R.J., Simonis, M., et al., 2008. Three-dimensional organization of gene expression in erythroid cells. Curr. Top. Dev. Biol. 82, 117−139.

Gabriel, R., Lombardo, A., Arens, A., Miller, J.C., Genovese, P., Kaeppel, C., et al., 2011. An unbiased genome-wide analysis of zinc-finger nuclease specificity. Nat. Biotechnol. 29, 816−823. Available from: http://dx.doi.org/10.1038/nbt.1948.

Garrels, W., Mátés, L., Holler, S., Dalda, A., Taylor, U., Petersen, B., et al., 2011. Germline transgenic pigs by Sleeping Beauty transposition in porcine zygotes and targeted integration in the pig genome. PLoS ONE 6, e23573.

Garrick, D., Fiering, S, Martin, D.I., Whitelaw, E., 1998. Repeat-induced gene silencing in mammals. Nat. Genet. 18, 56−59.

Gaszner, M., Felsenfeld, G., 2006. Insulators: exploiting transcriptional and epigenetic mechanisms. Nat. Rev. Genet. 7, 703−713.

Giraldo, P., Rival-Gervier, S., Houdebine, L.M., Montoliu, L., 2003. The potential benefits of insulators on heterogonous constructs in transgenic. Transgenic Res. 12, 751−755.

Guttman, M., Donaghey, J., Carey, B.W., Garber, M., Grenier, J.K., Munson, G.G., et al., 2011. lincRNAs act in the circuitry controlling pluripotency and differentiation. Nature 477 (7364), 295−300. Available from: http://dx.doi.org/10.1038/nature10398.

Hamra, F.K., 2010. Enter the rat. Nature 467, 161−162.

Han, J.Y., 2009. Germ cells and transgenic chicken. Comp. Immunol. Microbiol. Infect. Dis. 32, 61–80.

Houdebine, L.M., 2010. Design of expression cassettes for the generation of transgenic animals (including insulators). In: Anegon, I. (Ed.), Rat Genomics: Methods in Molecular Biology, vol. 597. Humana Press, a part of Springer Science Business Media, LLC, pp. 55–69.

Houdebine, L.M., Attal, J., 1999. Internal ribosome entry sites (IRESs) : Reality and use. Transgenic Res. 8, 157–177.

Kim, Y., Kweon, J., Kim, A., Chon, J.K., Yoo, J.Y., Kim, H.J., et al., 2013. A library of TAL effector nucleases spanning the human genome. Nat. Biotechnol. 31 (3), 251–258. Available from: http://dx.doi.org/10.1038/nbt.2517.

Knirsch, L., Biadasz Clerch, L., 2000. A region in the 39 UTR of MnSOD RNA enhances translation of a heterologous RNA. Biochem. Biophys. Res. Commun. 272, 164–168. 10.1006/bbrc.2000.2754. Available online from: <http://www.idealibrary.common>.

Lardenois, A., Liu, Y., Walther, T., Chalmel, F., Evrard, B., Granovskaia, M., et al., 2011. Execution of the meiotic noncoding RNA expression program and the onset of gametogenesis in yeast require the conserved exosome subunit Rrp6. Proc. Natl. Acad. Sci. USA 108 (3), 1058–1063. Available from: http://dx.doi.org/10.1073/pnas.1016459108.

Larochelle, N., Stucka, R., Rieger, N., Schermelleh, L., Schiedner, G., Kochanek, S., et al., 2011. Genomic integration of adenoviral gene transfer vectors following transduction of fertilized mouse oocytes. Transgenic Res. 20, 123–135.

Le Saux, A., Houdebine, L.M., Jolivet, G., 2010. Chromosome integration of BAC (bacterial artificial chromosome): evidence of multiple rearrangements. Transgenic Res. 19, 923–931. Available from: http://dx.doi.org/10.1007/s11248-010-9368-7.

Li, T., Huang, S., Jiang, W.Z., Wright, D., Spalding, M.H., Weeks, D., et al., 2011. TAL nucleases (TALNs): hybrid proteins composed of TAL effectors and FokI DNA-cleavage domain. Nucleic Acids Res. 39 (1), 359–372. Available from: http://dx.doi.org/10.1093/nar/gkq704.

Liu, X., Li, N., Hu, X., Zhang, R., Li, Q., Cao, D., et al., 2013. Efficient production of transgenic chickens based on piggyBac. Transgenic Res. 22 (2), 417–423. Available from: http://dx.doi.org/10.1007/s11248-012-9642-y.

Long, X., Miano, J.M., 2007. Remote control of gene expression. J. Biol. Chem. 282, 15941–15945.

Lyall, J., Irvine, R.M., Sherman, A., McKinley, T.J., Núñez, A., Purdie, A., et al., 2011. Suppression of avian influenza transmission in genetically modified chickens. 331, 223–226.

Macdonald, J., Taylor, L., Sherman, A., Kawakami, K., Takahashi, Y., Sang, H.M., et al., 2012. Efficient genetic modification and germ-line transmission of primordial germ cells using piggyBac and Tol2 transposons. Proc. Nat. Acad. Sci. USA 109, E1466–1472.

Malassagne, B., Regimbeau, J.M., Taboit, F., Troalen, F., Chéreau, C., Moir, N., et al., 2003. Hypodermin A, a new inhibitor of human complement for the prevention of xenogeneic hyperacute rejection. Xenotransplantation 10, 267–277.

Mali, P., Yang, L., Esvelt, K.M., Aach, J., Guell, M., DiCarlo, J.E., et al., 2013. RNA-guided human genome engineering via Cas9. Science 339 (6121), 823–826. Available from: http://dx.doi.org/10.1126/science.1232033.

Malphettes, L., Fussenegger, M., 2006. Improved transgene expression fine-tuning in mammalian cells using a novel transcription-translation network. J. Biotechnol. 124, 732–746.

Marh, J., Stoytcheva, Z., Urschitz, J., Sugawara, A., Yamashiro, H., Owens, J.B., et al., 2012. Hyperactive self-inactivating piggyBac for transposase-enhanced pronuclear microinjection transgenesis. Proc. Natl. Acad. Sci. USA 109 (47), 19184–19189. Available from: http://dx.doi.org/10.1073/pnas.1216473109.

Mátés, L., Chuah, M.K., Belay, E., Jerchow, B., Manoj, N., Acosta-Sanchez, A., et al., 2009. Molecular evolution of a novel hyperactive Sleeping Beauty transposase enables robust stable gene transfer in vertebrates. Nat. Genet. 41, 753–761.

Mersch, B., Gepperth, A., Suhai, S., Hotz-Wagenblatt, A., 2008. Automatic detection of exonic splicing enhancers (ESEs) using SVMs. BMC Bioinformatics 9, 369. Available from: http://dx.doi.org/10.1186/1471-2105-9-369.

Metzger, D., Chambon, P., 2001. Site- and time-specific gene targeting in the mouse. Methods 24, 71–80.

Meyer, M., de Angelis, M.H., Wurst, W., Kühn, R., 2010. Gene targeting by homologous recombination in mouse zygotes mediated by zinc-finger nucleases. Proc. Natl. Acad. Sci. USA 107, 15022–15026. Available from: http://dx.doi.org/doi:10.1073/pnas.1009424107.

Moreira, P.N., Pozueta, J., Pérez-Crespo, M., Valdivieso, F., Gutiérrez-Adán, A., Montoliu, L., 2007. Improving the generation of genomic-type transgenic mice by ICSI. Transgenic Res. 16, 163–168.

Mussolino, C., Cathomen, T., 2013. RNA guides genome engineering. Nat. Biotechnol. 31 (3), 208–209. Available from: http://dx.doi.org/10.1038/nbt.2527.

Nature Methods, 2012. A special issue containing six reviews on ZFN and TALEN. Nature Methods 9, 28–34.

Ono, E., Amagai, K., Taharaguchi, S., Tomioka, Y., Yoshino, S., Watanabe, Y., et al., 2004. Transgenic mice expressing a soluble form of porcine nectin-1/herpesvirus entry mediator C as a model for pseudorabies-resistant livestock. Proc. Nat. Acad. Sci. USA 101, 16150–16155.

Orlando, S.J., Santiago, Y., DeKelver, R.C., Freyvert, Y., Boydston, E.A., Moehle, E.A., et al., 2010. Zinc-finger nuclease-driven targeted integration into mammalian genomes using donors with limited chromosomal homology. Nucleic Acids Res. 38, e152. Available from: http://dx.doi.org/10.1093/nar/gkq512.

Palais, G., Nguyen Dinh Cat, A., Friedman, H., Panek-Huet, N., Millet, A., Tronche, F., et al., 2009. Targeted transgenesis at the HPRT locus: an efficient strategy to achieve tightly controlled in vivo conditional expression with the tet system. Physiol. Genomics 37 (2), 140–146. Available from: http://dx.doi.org/10.1152/physiolgenomics.90328.2008.

Park, T.S., Han, J.Y., 2012. piggyBac transposition into primordial germ cells is an efficient tool for transgenesis in chickens. Proc. Natl. Acad. Sci. USA 109, 9337–9341.

Pera, M.F., Hasegawa, K., 2008. Simpler and safer cell reprogramming. Nat. Biotechnol. 26, 59–60.

RNAi, 2009. RNAi: Multi-author review. Nature 457 (7228).

Rémy, S., Tesson, L., Menoret, S., Usal, C., Scharenberg, A.M., Anegon, I., 2010. Zinc-finger nucleases: a powerful tool for genetic engineering of animals. Transgenic Res. 19, 363–371.

Reyon, D., Tsai, S.Q., Khayter, C., Foden, J.A., Sander, J.D., Joung, J.K., 2012. FLASH assembly of TALENs for high-throughput genome editing. Nat. Biotechnol. Available from: http://dx.doi.org/10.1038/nbt.2170.

Ripoll, P.J., Turquois, V., Lebouris, C., Delprat, J., Martin, K., Chabert, C., et al., 2010. Knockdown of milk protein genes in transgenic rabbits. UC Davis Transgenic Animal Research Conference VII. Tahoe City, USA. Transgenic Res. 19, 146.

Robl, J.M., Wang, Z., Kasinathan, P., Kuroiwa, Y., 2007. Transgenic animal production and animal biotechnology. Theriogenology 67, 127−133.

Saidi, S., Rival-Gervier, S., Daniel-Carlier, N., Thépot, D., Morgenthaler, C., Viglietta, C., et al., 2007. Distal control of the pig whey acidic protein (WAP) locus in transgenic mice. Gene 401, 97−107.

Schnieke, A.E., Kind, A.J., Ritchie, W.A., Mycock, K., Scott, A.R., Ritchie, M., 1997. Human factor IX transgenic sheep produced by transfer of nuclei from transfected fetal fibroblasts. Science 278, 2130−2133.

Schönig, K., Freundlieb, S., Gossen, M., 2013. Tet-transgenic rodents: a comprehensive, up-to date database. Transgenic Res. 22 (2), 251−254. Available from: http://dx.doi.org/10.1007/s11248-012-9660-9.

Shinohara, E.T., Kaminski, J.M., Segal, D.J., Pelczar, P., Kolhe, R., Ryan, T., 2007. Active integration: new strategies for transgenesis. Transgenic Res. 16, 333−339.

Sioud, M., 2006. Innate sensing of self and non-self RNAs by Toll-like receptors. Trends Mol. Med. 12, 167−176.

Sumiyama, K., Kawakami, K., Yagita, K., 2010. A simple and highly efficient transgenesis method in mice with the Tol2 transposon system and cytoplasmic microinjection. Genomics 95, 306−311.

Sun, L., Goff, L.A., Trapnell, C., Alexander, R., Lo, K.A., Hacisuleyman, E., et al., 2013. Long noncoding RNAs regulate adipogenesis. Proc. Natl. Acad. Sci. USA 110 (9), 3387−3392. Available from: http://dx.doi.org/10.1073/pnas.1222643110.

Suster, M.L., Sumiyama, K., Kawakami, K., 2009. Transposon-mediated BAC transgenesis in zebrafish and mice. BMC Genomics 10, 477. Available from: http://dx.doi.org/10.1186/1471-2164-10-477.

Taboit-Dameron, F., Malassagne, B., Viglietta, C., Puissant, C., Leroux-Coyau, M., Chéreau, C., et al., 1999. Association of the 5′ HS4 sequence of the chicken beta-globin locus control region with human EF1-alpha gene promoter induces ubiquitous and high expression of human CD55 and CD59 cDNAs in transgenic rabbits. Transgenic Res. 8, 223−235.

Tiscornia, G., Singer, O., Ikawa, M., Verma, I.M., 2003. A general method for gene knockdown in mice by using lentiviral vectors expressing small interfering RNA. Proc. Natl. Acad. Sci. USA 100, 1844−1848.

Urschitz, J., Kawasumi, M., Owens, J., Morozumi, K., Yamashiro, H., Stoytchev, I., et al., 2010. Helper-independent piggyBac plasmids for gene delivery approaches: strategies for avoiding potential genotoxic effects. Proc. Natl. Acad. Sci. USA 107, 8117−8122.

van de Lavoir, M.C., Diamond, J.H., Leighton, P.A., Mather-Love, C., Heyer, B.S., Bradshaw, R., et al., 2006. Germline transmission of genetically modified primordial germ cells. Nature 441 (7094), 766−769.

Wefers, B., Meyer, M., Ortiz, O., Hrabé de Angelis, M., Hansen, J., Wurst, W., et al., 2013. Direct production of mouse disease models by embryo microinjection of TALENs and oligodeoxynucleotides. Proc. Natl. Acad. Sci. USA 110 (10), 3782−3787. Available from: http://dx.doi.org/10.1073/pnas.1218721110.

Woods, I.G., Schier, A.F., 2008. Targeted mutagenesis in zebrafish. Nat. Biotechnol. 26, 650−651.

Yang, J.H., Kim, S., 2012. Establishment of an efficient and stable transgene expression system in chicken primordial germ cells. Bull. Korean Chem. Soc. 33 (5). Available from: http://dx.doi.org/10.5012/bkcs.2012.33.5.1536.

Yong, H.Y., Hao, Y., Lai, L., Li, R., Murphy, C.N., Rieke, A., et al., 2006. Production of a transgenic piglet by a sperm injection technique in which no chemical or physical treatments were used for oocytes or sperm. Mol. Reprod. Dev. 73, 595−599.

18 Analysis of Transgene Integration

Cristina Vicente-García[1,2], Almudena Fernández[1,2] and Lluís Montoliu[1,2]

[1]National Centre for Biotechnology (CNB-CSIC), Department of Molecular and Cellular Biology, Madrid, Spain, [2]CIBERER, ISCIII, Madrid, Spain

I. Introduction

The integration event of DNA transgenes has been a black box, an unexplained event that occurred shortly after microinjection and resulted in the successful insertion of the exogenous microinjected DNA molecule into the host genome of the recipient embryo used. However, upon systematic analysis of the structure of DNA transgenes integrated in many independent transgenic mouse lines, it has become apparent that those DNA molecules were inserted as tandem arrays, normally at single places (although multiple integrations were not infrequent). They were mostly arranged head-to-tail and eventually integrated through illegitimate recombination processes, not involving homologous recombination with the targeted endogenous genomic sites. This integration process was also observed upon co-transfecting DNA in cells in culture along with a selective marker, so it was postulated the associated mechanism is likely shared between these two different approaches to genetically alter an organism, whether a cell or a transgenic animal (Bishop and Smith, 1989). From early studies, it was postulated that direct repeat (head-to-tail) transgene tandem arrays are assembled extrachromosomally, using the several hundreds to thousands molecules normally injected with a standard plasmid-based DNA transgene. This assembly would occur prior to integration, by a process known as extrachromosomal nonconservative recombination between foreign DNA molecules, as observed previously in transfected cultured mammalian cells (Lin et al., 1984, 1987; Wake et al., 1985). The process has also been demonstrated for the integration of reconstituted DNA molecules from partially overlapping cotransfected DNA fragments in cells (Lin and Sternberg, 1984) and microinjected mouse embryos (Shimoda et al., 1991; Strouboulis et al., 1992). This assembly is thought to begin by an attack conducted by a DNA exonuclease at the $5'$ ends of the linear DNA molecules injected, followed by attempts at base-pairing

Transgenic Animal Technology. DOI: http://dx.doi.org/10.1016/B978-0-12-410490-7.00018-9

between the resulting exposed 3' ends, and including refilling, removal of unpaired nucleotides, and finally a ligation step to seal the newly joined DNA molecules (reviewed by Bishop, 1997).

However, the mechanism by which the linear directly arranged tandem array of DNA molecules is formed requires taking into account additional processes. First, some of the incoming/injected/transfected linear DNA molecules would be circularized, either by spontaneous end-joining or through illegitimate recombination at their open ends followed by a ligation. The self-circularization of single molecules, being a unimolecular event, would be a favored process over the end-joining of different molecules. The rapid end-joining of DNA molecules has been observed shortly after microinjection of linear DNA transgenes into the pronuclei of mouse fertilized oocytes (Burdon and Wall, 1992). Second, some of these newly created circular DNA molecules would be linearized randomly (as has been observed in transfected animal cells; Bishop and Smith, 1989) and will result in the generation of circularly permuted molecules. Eventually these molecules will recombine among themselves through a homologous recombination mechanism, since they share identical DNA sequences, and form these tandem arrays that eventually will integrate, also at random places (Bishop, 1997).

Therefore, the model of homologous recombination between circularly permuted DNA molecules is the currently accepted mechanism by which microinjected or transfected DNA transgenes are thought to integrate in the host genome (Bishop, 1997). This mechanism largely explains the effective integration of intact bacterial artificial chromosome (BAC) DNA molecules sustaining optimal expression levels, originating from DNA preparations that are microinjected as circular supercoiled DNA, as demonstrated by Thomas Saunders and collaborators in their large study of BAC transgenic mice (Van Keuren et al., 2009).

In summary, a number of extrachromosomal processes occur prior to integration, including the formation of circular molecules and their subsequent cleavage at random places to trigger the formation of tandem arrays. Hence, the microinjection of linear or circular DNA transgenes, as initially investigated by Brinster et al. (1985), favoring the use of linear molecules, is equally possible. In both cases, the generation of transgenic mice is achieved through the mechanisms reported in this chapter, which explain how foreign DNA integrates into the host genome.

II.   Chromosomal Position Effects

Most of the mammalian genomes are formed by repeated DNA elements and intergenic regions, accounting for about 98% of all DNA sequences. The remaining 2% correspond to DNA sequences coding for proteins, and altogether are currently referred to as the exome of a given organism, since it corresponds to a virtual assembly of all exons from all genes. Therefore, if the integration of DNA transgenes is largely a random process, it is highly possible that the insertion of exogenous DNA would normally occur at intergenic areas, where most of the heterochromatin (and hence, condensed unexposed chromatin) is found. This would generally result, in the worse case, in the silencing of the transgene expression, or, at the very least, in the alteration of the expression patterns of the transgene, including the appearance of variegated expression. Alternatively (and/or simultaneously), if the DNA transgene lands near a transcriptional regulatory element or near an active transcriptional locus, it will likely receive the influence of the endogenous expression pattern, which will be then partially or wholly adopted by the transgene. All these alterations in the expected transgene expression patterns are globally known as chromosomal position effects (Wilson et al., 1990; Sippel et al., 1997; Grosveld et al., 1998) and are the primary explanation of the often common but unexpected behavior of transgenes inserted at random places in the host genome (Giraldo and Montoliu, 2001; Montoliu, 2002; Montoliu et al., 2009).

What can be done to overcome these chromosomal position effects? To seriously take them into account, and given their unpredictable nature because the integration of microinjected DNA transgenes proceeds at random places, one must aim to produce multiple transgenic mouse lines with several founder individuals, each representing a different genomic location where the DNA transgene has been integrated. In this scenario, if a similar transgene expression pattern is seen in various independent transgenic mouse lines carrying the same DNA transgene, it can be inferred that the observed phenotype is derived from the exogenous DNA. Otherwise, if each of the transgenic mouse lines display different expression patterns, they are likely to reflect the endogenous loci next to which the transgene has integrated and would not display the associated phenotype to be expected from the microinjected DNA transgene. Therefore, the uncertain nature of the integration site where our favorite DNA transgene will land into the host genome mandates the analysis of several independent mouse lines (usually two to five lines) per microinjected DNA transgene.

Other than analyzing multiple mouse transgenic lines, overcoming the chromosomal position effects requires a focus on the construction of the DNA transgene itself (see also Chapter 17). A number of strategies have been evaluated and reported in the literature, including adding specific transcriptional regulatory elements, matrix-attachment regions, the use of genomic-type constructs, the use of chromosomal-type transgenes, and inserting the desired transgene under the control of the endogenous locus whose expression pattern is required (reviewed in Giraldo and Montoliu, 2001). The combination of several of these strategies is certainly possible. It is now theoretically feasible to build transgenes with optimal capacity for transgene expression, irrespective of the landing site in the host genome. A number of rules for designing transgene vectors for optimal expression have been reported (Molto et al., 2011; see also Chapter 17).

Chromosome-type transgenes, whether yeast artificial chromosomes (YACs), BACs, or P1-derived artificial chromosomes (PACs), have been regularly used since 1993 and are normally associated with optimal transgenic expression results. The sizes of their heterologous DNA inserts range from 100 kb (PACs) to 300 kb (BACs), up to over 1 Mb (YACs). Thus, they have enough room to accommodate entire expression domains, formed by coding and neighboring regions that contain the transcriptional regulatory elements that are essential for the locus to be expressed faithfully and can be located far away from the coding area of the gene. The discovery of the existence of relevant and powerful enhancers located away from the immediate promoter has been reported earlier; in various loci, some key but distant transcriptional regulatory elements were found to be essential for the proper expression of transgenes (i.e., Pinkert et al., 1987). The analysis of many transgenic mouse lines produced with BACs or YACs suggests that the large size of the genomic inserts they can transfer to the host genome ensures the inclusion of all or nearly all elements that are required by the expression domain for proper behavior and correct function (Forget, 1993; Montoliu et al., 1994; Jakobovits, 1994; Lamb and Gearhart, 1995; Peterson et al., 1997; Giraldo and Montoliu, 2001; Peterson, 2003; Yang and Gong, 2005; Van Keuren et al., 2009). Such evidence suggests that the integration of genomic-type constructs, whether PACs, BACs, or YACs, normally results in optimal transgene expression patterns. So a good way to overcome chromosomal position effects related to the host genome is to use large genomic-type vectors (Montoliu et al., 2009). In some cases, it has been possible to decipher and uncover the precise elements traveling into these large genomic regions responsible for the optimal expression of the transgenes included. This is, for example, the case of the mouse tyrosinase locus, where an essential element, a locus control region (LCR), was found 15 kb upstream of its promoter and shown to be enough to sustain copy-number-dependent and position-independent transgene expression, with a per copy value comparable to that of the corresponding endogenous locus (Montoliu et al., 1996; Giraldo et al., 1999; Giraldo and Montoliu, 2002). In its absence, the lack of this LCR element triggered a variegated expression pattern of the transgenes, hence triggering *de novo* the appearance of the no-longer-compensated chromosomal position effects (Giménez et al., 2001, 2005).

Alternatively, the use of genomic boundaries or insulators also ensures optimal transgene expression independent of the site of integration. Through innovative validation strategies using transgenic zebrafish as heterologous systems (Bessa et al., 2009), we have been able to discover new genomic boundaries associated to a subset of CTCF sites evolutionarily and functionally conserved (Martin et al., 2011), or associated to genes regulated by hypoxia (Tiana et al., 2012). Genomic boundaries have been tested ectopically, shielding transgenes from integration site effects. Lastly, if large genomic constructs, such as YACs or BACs, are not available for a given construct, the next recommendation to ideally build a transgene for achieving optimal expression is the use of genomic boundaries or insulators.

III. Promoting the Integration at Selected Genomic Regions

Since the large majority of transgenes integrate randomly and this triggers a chromosomal position effect, we have just seen what can be done at the level of the construct, of the heterologous DNA sequences used to build the transgenes.

However, perhaps we could try favoring the integration of transgenes at certain regions, either to expose them to a heterochromatic area (Dillon and Festenstein, 2002) full of condensed chromatin, or to be sure they land in genomic regions fully compatible with transcription. Rather than choosing integration by gene-targeting approaches through homologous recombination, an issue we will discuss in the following section, this approach promotes integration in certain types of genomic DNA, selected because of their features.

Several attempts have been reported with variable success, but one of the most successful examples was done in the laboratory of Neil Dillon, where a transgene was efficiently targeted to pericentric heterochromatin by flanking it at either side with several copies of the major mouse γ-satellite repeat (Lundgren et al., 2000).

The use of transposons to generate transgenic mice also follows this route. Several types of transposon systems have been used for transgenesis, including *sleeping beauty*, *piggyBac*, *marine*, *Tol2*, and *Tn5*, all derived from different organisms. In general, transposons integrate nonrandomly, since they prefer to jump to certain regions. Transposons generally display a tendency to jump into actively transcribed regions where genes are functional and loci are expressed, although the specific preferences for integration sites, the specific target sequences that transposons expect to find upon jumping, do vary a lot between different transposon systems (Largaespada, 2003; Clark et al., 2004; Ivics et al., 2009).

A. Using Preselected Docking Sites for Transgene Integration

The next step in controlling the integration of our transgenes is to decide *a priori* on an endogenous locus in which to insert all our different transgenes. These places in the genome are known as "docking sites" and have been used with great success in the short history of animal transgenesis. Many of them are related to commercial developments of biotechnology-based companies that offer the tools and strategies to systematically integrate any transgene of choice that needs to be tested into the same genomic region, primarily chosen to be neutral (safe) but also selected because whatever chromosomal position effects exist would equally apply for all different transgenes whose expression patterns must be compared (Rossant et al., 2011).

Several examples of docking sites useful in transgenic mouse experiments include the 3' UTR of the *ColA1* gene (Beard et al., 2006); the *Hprt* locus (Bronson et al., 1996), also used extensively at the Pleiades project (http://www.pleiades.org/) for the analysis of promoters (Yang et al., 2009); and the *Rosa26* locus, whose use has also been promoted by the site-directed PhiC31 integrase-mediated transgenesis approach (Tasic et al., 2011). Another example is the site-specific recombination transgenesis approach mediated by Cre/loxP (Ohtsuka et al., 2010), thus resulting in increased transgenesis efficiencies supporting robust transgene expression at predefined loci. The use of recombinase-mediated cassette exchange (RMCE) strategies has clearly evolved as a valid and robust protocol to integrate transgenes at preselected loci of choice, thus overcoming chromosomal position effects and allowing comparison of transgene behavior with different constructs, all inserted at the same genomic locus (Bouhassira et al., 1997; Feng et al., 1999). New versions of RMCE have been created that combine the use of several recombinases (i.e., Cre and Flipase) to achieve higher transgenic efficiencies in the absence of selectable markers (Lauth et al., 2002). New developments are expected that will use the almost infinite number of approaches based on site-directed recombinases, their specific targets, and intelligent solutions to integrate transgenes at preselected loci (Turan et al., 2013).

These neutral sites for the integration of transgenes have also been investigated systematically, at genome-wide level, through elegant search strategies, resulting in the initial characterization of numerous potential docking sites useful for this purpose (Wallace et al., 2000).

B. Determining the Transgene Site of Integration in the Host Genome

When the first transgenic mice were established, there were no easy methods or tools to determine the site of integration of transgenes into the host genome, so quite often this was not addressed experimentally. Also, it was recognized that some chromosomal rearrangements could be triggered by the integration of foreign DNA, including chromosomal translocation events (Mahon et al., 1988). Subsequently, it became possible to apply a fluorescence *in situ* hybridization (FISH) approach, in combination with transgene-specific fluorescent probes, to detect, on fixed metaphase chromosomes, where the transgene had integrated (Montoliu et al., 1994). However, the FISH technique was tedious and required specific skills in cytogenetics not available in all laboratories producing transgenic mice. Hence, it was not universally applied and very few transgenes were characterized by integration site, in spite of the publication of improved FISH protocols for the specific detection of transgene insertion sites (Matsui et al., 2002). The characterization of transgene insertion sites was sometimes combined with the confirmation of structural integrity of the DNA transgene inside the host genome. This is a most relevant feature to address when working with large chromosome-type transgenes such as YAC transgenic mice. The use of Southern blot analyses in combination with Alu-specific probes (for heterologous DNA of human origin) provided a suboptimal but practical way of determining the integrity of the inserted YAC constructs in embryonic stem (ES) cells (Lamb et al., 1993; Jakobovits et al., 1993). For BAC transgenic mice, the use of real-time quantitative polymerase chain reaction (PCR) approaches in combination with BAC-specific polymorphisms provided a general protocol to assess the copy number as well as the integrity of the inserted BAC transgene (Chandler et al., 2007). Recently, a more sophisticated method including the use of microarrays and a next-generation sequencing step was devised for the analysis of the integrity and the integration site of BAC transgenes in mice (Dubose et al., 2013).

Notably, in some cases, the characterization of the landing site was indirect, since the integration of the transgene at a given preexisting locus was detrimental, lethal, or generated an obvious phenotype when the transgenic mouse line was brought to the homozygous state. One of the best-known examples is the identification of a situs-inversus mutation, thanks to the locus interruption mediated by a transgene integration (Yokoyama et al., 1993). Because of this publication and similar reports that appeared subsequently, for some time most transgenic laboratories were systematically breeding their transgenes to homozygosis, hoping to come across interesting phenotypes due to the inactivation of the interrupted genomic region, particularly if this corresponded to a gene whose function was essential during embryo or early postnatal development. However, contrary to expectations, these mutations were rarely found. In general, it has been estimated that in about 5–7% of transgenic integrations the exogenous DNA lands and interrupts a previously existing locus. The insertional mutagenesis capacity of transgenes, upon integration in the host genomes, has been exploited in several large genome-wide studies, where the use of the so-called "gene-trap" vectors has allowed the accumulation of thousands of integration sites, covering a large proportion of endogenous genes and allowing the functional characterization of many of these loci (Skarnes et al., 1992; Hansen et al., 2003; Nord et al., 2006; Araki et al., 2009).

Perhaps the most comprehensive analysis of transgene integration sites was conducted by the systematic identification of 142 integration sites of an EGFP-reporter transgene, primarily using FISH and then molecular approaches to decipher the genomic area where all these transgenes had landed (Nakanishi et al., 2002). The large majority of these integration sites were single (>80%), although multiple integration sites within the same transgenic mouse line were also observed. Although it had always been assumed that the integration of DNA transgenes was a mostly random process (Dellaire and Chartrand, 1998), the investigation of more than 100 integration sites provided evidence for an uneven distribution of these sites within the genome. In particular, DNA genomic regions rich in AT and with a low density of genes were the hot integration spots found by this team (Nakanishi et al., 2002).

DNA transgenes can integrate as single copies or, more often, as tandem arrays. Accordingly, it becomes relevant to investigate the complexity of this array, the number of copies of the transgene integrated in the host genome. Several strategies have been developed in the field to estimate the number of copies, using initially Southern or dot-blot standard approaches (the latter relegated more to history) or, more commonly, using real-time quantitative PCR (qPCR) methods (Ballester et al., 2004; Mitrecić et al., 2005; Shepherd et al., 2009) or sometimes alternative pyrosequencing approaches (Liu et al., 2009).

We have already seen that FISH can be used to efficiently detect the integration site of the transgene of choice (nicely illustrated by the large study of Nakanishi et al., 2002). What alternative methods exist based on molecular biology approaches to detect the transgene insertion site in the host genome? Several methods have been developed in the field of animal transgenesis to identify transgene insertion sites at the molecular level. Next, we will review the most relevant methods currently used to clone and identify the integration site of transgenes. One of the simplest and most direct approaches requires (i) using a suitable enzyme to digest the genomic DNA of transgenic mice, (ii) cutting inside the transgene together with recircularization and ligation of the resulting molecules to enable the use of transgene-derived primers to extend by inverse PCR into the surrounding genomic DNA regions, (iii) crossing over the insertion site of the transgene, and finally (iv) sequencing the resulting amplified molecules in a standard way, and (v) identifying the integration site by homology comparison of the sequenced DNA molecules with the reference genome (Takemoto et al., 2006; Liang et al., 2008). This method can be very successful but also can suffer if the transgene integrates near stretches of repetitive DNA, resulting in the amplification of many artifacts that would be misleading regarding the actual integration site of the transgene.

A related method, also involving inverse PCR, adds some specific linkers to the digested genomic DNA of transgenic animals and uses them as anchored primers, in combination with internal, transgene-specific primers, to amplify the site of integration eventually resolved by sequencing and by comparison to a reference genome to localize the insertion site (Noguchi et al., 2004). Progressively more sophisticated methods have been deployed, including the use of several internal and external primers in relation to the transgene along with several rounds of nested PCR (Pillai et al., 2008), and the use of next-generation sequencing approaches to determine, by comparison to the reference genome of the species, the place where the transgene integrated (Zhang et al., 2012).

Most of these methods to determine the site of integration (Noguchi et al., 2004; Takemoto et al., 2006; Liang et al., 2008; Dubose et al., 2013), or slight variations of them (Tesson et al., 2010; McHugh et al., 2012), as well as new methods based on the use of multiplex ligation-dependent probe amplification (MLPA; Notini et al., 2009) can be also applied to the determination of zygosity, to assess whether transgenic animals are heterozygous/hemizygous or homozygous.

C. How to Detect Transgene Integration, Part 1: PCR

Perhaps the easiest and most commonly used method to analyze the potential integration of transgenes in the host genomes is using PCR approaches (Erlich, 1989). Using a pair of transgene-specific primers and tested robust conditions to amplify the transgenic DNA within the rest of the host genome, one can easily aim to detect single-copy integration events. To do so, careful attention should be devoted to optimizing PCR conditions to ensure that they can actually amplify the equivalent of a single-copy transgene in the 50–500 ng of genomic DNA generally used for analysis purposes (Figure 18.1).

Even simple methods can go wrong. The key points for successful, robust, reproducible PCR detection protocols for transgenes include (i) the use of appropriate positive and negative controls in every single PCR assay, (ii) the confirmation of genomic DNA with a quality compatible with PCR amplification, in case of unexpected negative samples, by amplifying a known endogenous locus, and (iii) the repetition of PCR assays using 1/10 and/or 1/50 diluted genomic DNA, in case these preparations were too concentrated and/or too dirty to allow adequate

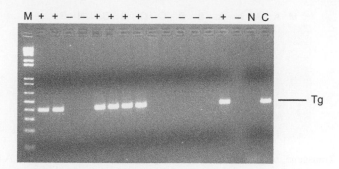

Figure 18.1 Transgene integration detection using the standard PCR approach. A transgene-specific DNA band (Tg) can be detected in seven lanes of this standard agarose 1% gel, along with DNA molecular size markers (M) and the most important and mandatory negative (no DNA, only PCR reagents) (N) and positive (C) control lanes, which must always be included to confirm that DNA bands indeed correspond to positive transgenic mice, and, likewise, that the absence of the transgene-specific DNA band corresponds to wild-type nontransgenic animals. Suitable protocols for the robust analysis of transgene integration by PCR can be obtained from Lluis Montoliu's laboratory web page (http://www.cnb. csic.es/~montoliu/prot.html).

transgene amplification. Suitable protocols for the robust analysis of transgene integration by PCR can be freely obtained from the Montoliu laboratory web page (http://www.cnb.csic.es/~montoliu/prot.html). Additionally, optimization parameters can also be found in Chapter 19.

D. How to Detect Transgene Integration, Part 2: Southern Blot Analysis

Standard PCR analysis is suitable for detecting the presence of transgenes using limited amounts of genomic DNA. However, these methods do not inform about the structure of the integrated transgene, whether the DNA construct has been inserted intact or with some unexpected insertions/deletions/rearrangements, nor the number of copies that may have eventually integrated. Southern blot strategies, although rightly considered to be somewhat more difficult to execute correctly compared to simple PCR protocols, are able to provide much more information than indicating whether or not an animal carries a given transgene. Southern blot analyses are regularly used to detect the intactness of the DNA transgene, the possible number of copies integrated, and whether the animal is heterozygous/hemizygous or homozygous for the transgene (Schedl et al., 1993; Montoliu et al., 1996) (Figure 18.2). If the DNA transgene used includes some partial sequences that are also present in an endogenous locus, normally present as a single copy in a mammalian genome, then one can devise strategies that would highlight both transgene-derived and endogenous locus-derived DNA sequences. Besides confirming the transgenicity of the individuals, the same experiment can be used to assess the number of copies inserted, simply comparing the intensity of the endogenous locus signal

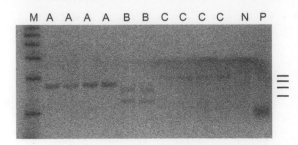

Figure 18.2 Transgene integration detection using the Southern blot approach. A transgene-specific DNA band (Tg), indicated by the black bars on the right side, can be detected at different positions, according to the DNA size marker lane (M), depending on the transgenic mouse line, depending on the integration. The transgenic mice in lanes indicated with A appear to carry one single insertion. Likewise, the DNA from transgenic animals included on the lanes indicated with C also suggests a single integration, with a single DNA band but at a different size, indicating that this is a different genomic integration event. In contrast, the genomic DNA from transgenic mice in lanes indicated with B appear to carry two independent integrations, also at different places as compared to lanes A and C. Control lanes include a wild-type and hence negative (N) control, and a diluted DNA transgene plasmid-derived (P) lane to monitor the hybridization success. In order to detect one DNA band per integration event, a unique restriction enzyme cutting inside the transgene must be selected, along with a DNA probe detecting one of the resulting fragments, whose final size will vary depending on the exact location of the nearest site for the same restriction enzyme in the landing spot of the host genome.

(equivalent to two alleles, the ones inherited from both progenitors) with that of the transgene signal. A simple ratio would then be used to estimate the number of trans-gene copies integrated in the host genome (Furlan-Magaril et al., 2011).

Confirming the integrity of the inserted transgenes is particularly relevant when working with chromosome-type transgenes, whose large size makes them prone to partially integrated events (Giraldo et al., 1999; Giraldo and Montoliu, 2001). The integrity of YAC/BAC-based transgenes, if the heterologous inserts are from human DNA origin, can be assessed by DNA fingerprinting, using *Alu*-derived, human-specific DNA probes, as reported previously (Jakobovits et al., 1993; Lamb et al., 1993).

Southern blot protocols can be obtained from a variety of sources and are also described in this manual. Suitable methods for Southern blot analysis are included in the collection of transgenic mouse protocols "Generation of Transgenic Mice" (1993) available from the Montoliu laboratory web page (http://www.cnb.csic.es/~montoliu/transgenic.html).

E. How to Detect Transgene Integration, Part 3: Slot-Blot Analysis

A particular case of Southern blot analysis is trapping and fixing the DNA into a small surface of a suitable membrane, usually positively charged (i.e., Hybond, GE

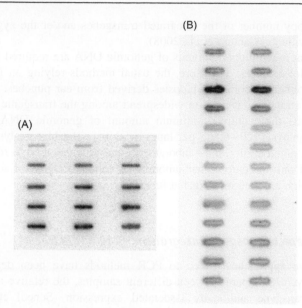

Figure 18.3 Transgene integration detection using the slot-blot approach.
A transgene-specific DNA band, derived from SV40-polyA signal, is used to hybridize
equivalent amounts of genomic DNA immobilized onto a membrane, concentrated in a
small area (Giménez et al., 2004; Lavado et al., 2006; Cantero et al., 2008). The strongest
positive signals indicate the presence of DNA transgene in this animal. (A) Triplicate
diluted amounts of DNA transgene, corresponding to (from top to bottom) 1×, 5×,
10×, 20×, and 50× copies of the transgene in 5 μg of genomic DNA, can be mixed
with wild-type DNA, in order to use the signals obtained to compare intensities with
those from (B) duplicate experimental mouse samples, and assess the number of
copies integrated or whether the individuals are hemizygous/heterozygous or
homozygous animals.

Healthcare). In this case, we lose structural information since there is no electrophoresis involved; simply a limited amount of genomic DNA from mice is bound
to small areas, shaped like dots (hence, dot-blot analysis) or like slots (hence, slot-
blot analysis). While this technology has for the most part been superseded by PCR
analyses, there are still instances where it can be very useful. To detect the integration of transgenes in these DNA-trapped samples, we must then use transgene-
specific probes that are not represented in the host genome. DNA probes that also
cross-detect endogenous loci cannot be used, since these signals will compete with
those from the integrated transgenes, and we would not be able to distinguish
between transgenic and wild-type nontransgenic individuals. This approach can be
robustly undertaken only with transgenes carrying suitable unique DNA elements
not existing in the host genome, such as the popular SV40-derived polyadenylation
signals (Giménez et al., 2004; Lavado et al., 2006) (Figure 18.3). In combination
with some reference DNA transgene amounts, this approach can also be employed

to assess the copy number of the integrated transgenes and/or the zygosity of the transgenic individuals (Cantero et al., 2008).

Be aware that much higher amounts of genomic DNA are required for Southern blot and slot-blot analyses. Therefore, the usual methods relying on DNA extraction protocols obtained from tiny biopsies derived from ear punches, tips of tails, or toe-clipping strategies, which are widespread among the transgenic community, will not produce the required minimum amount of genomic DNA needed for Southern blot analysis (20–60 µg per lane) or slot- or dot-blot analyses (5 µg per slot or dot). As noted earlier, most laboratories and facilities tend to focus on PCR approaches and only prepare higher amounts of genomic DNA for some selected individuals, which are then investigated further.

F. How to Detect Transgene Integration, Part 4: qPCR

Semiquantitative approaches linked to PCR methods have been developed that serve to explore, or compare between different samples, the relative amounts of a given DNA transgene and/or its associated expression (Schedl et al., 1993). However, proper real-time PCR, also called qPCR, has been widely used to robustly quantify gene expression and gene presence in a variety of scenarios (Giménez et al., 2003; Lavado et al., 2006).

Methods using qPCR strategies can also be applied to detect the integration of a DNA transgene, as well as to detect the zygosity of the positive individuals, whether hemi- or heterozygous, versus homozygous (Shitara et al., 2004; Haurogné et al., 2007; Sakurai et al., 2008). Figure 18.4 illustrates the use of qPCR approaches to determine transgene integration and the zygosity of transgenic individuals (Cantero et al., 2008), according to methods developed by Haurogné et al. (2007). In contrast to Southern and slot-blot analyses, this approach has the obvious advantage that only very small quantities of DNA are required for analysis (in the range of nanograms). Furthermore, these genomic DNA preparations must be precisely quantified and exactly the same amount of DNA must be applied to all individuals being analyzed. Otherwise, results will tend to be difficult to interpret or to cause technical artifacts. Moreover, the use of negative and positive controls is also mandatory when using qPCR methods, to ensure reproducibility and to trust the experimental values obtained.

These qPCR methods can be wonderful, and are perhaps the optimal and recommended approach to determine integration and zygosity, especially for laboratories and facilities that regularly work with the same transgenes using the same transgenic mouse lines. However, everything must be in place and double-checked before the experiment can be trusted and used routinely. Every parameter must be optimized, in order to rapidly and efficiently run and analyze any given mouse DNA sample. Methods involving qPCR are also clean, because they do not use radioactively labeled DNA probes. But, as indicated, they require dedicated time to set up the exact conditions, the precise primer concentrations where the amplification progresses in the linear range; otherwise, the quantification might be erroneous.

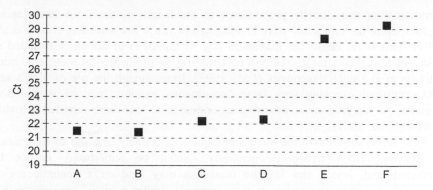

Figure 18.4 Transgene integration detection using qPCR approach. A qPCR approach was used to determine the zygosity of transgenic mice using qPCR equipment (ABI Prism 7000 Detection System), 2 ng of genomic DNA amplified with transgene-specific internal primers, and Power SYBR Green PCR Master Mix (ABI) (Cantero et al., 2008), according to Haurogné et al. (2007). The Ct values for six mice are shown (mean of duplicates). Mice A and B are homozygous individuals and transgenic mice C and D are hemizygous. Animals E and F are wild-type, nontransgenic, and hence display a much higher Ct value compared with DNA negative controls, because the primers fail to amplify any specific DNA product within the expected range of Ct values and only show up, eventually, as primer–dimers and other rearranged unspecific DNA molecules whose fluorescence is eventually detected. Please note that, as expected, doubling the amount of target DNA (from hemi- to homozygous individuals) decreases the Ct values one unit. This is of course a subtle variation that is required to be shown robustly; otherwise, hemi- and homozygous individuals can be easily mixed up.

IV. Discussion

Brinster's pioneering studies (1989) also clearly demonstrated that the integration of DNA transgenes into the host mouse genome was a random event. Integrations at the corresponding homologous DNA sequences in the genome, while theoretically possible, were an extremely rare event. In their experiment to assess this possibility they nonetheless microinjected more than 10,000 mouse embryos with a given DNA transgene and produced about 500 independent transgenic mouse lines, before coming across one founder transgenic mouse in which the integration appeared to have occurred at the endogenous homologous locus, thus confirming that integration of DNA transgenes mostly occurs at random places, and very rarely at the homologous loci (Brinster et al., 1989). These experiments were revisited years later, when the generation of transgenic mice with very large chromosome-type transgenes (YACs or BACs) was achieved (Schedl et al., 1993; Antoch et al., 1997). But soon it became obvious that even using these large genomic-type transgenes, which sustained optimal and robust transgene expression independent of the integration site in the host genome, the insertion also occurred regularly at random places, not at the homologous loci (Giraldo and Montoliu, 2001, 2002). The experiments carried out in mouse ES cells indicated that the efficiency of homologous recombination was directly proportional to the length of the DNA homologous regions, up to the limits that were investigated of just a few kilobases (Hasty et al., 1991). This is far below the maximum size of hundreds of kilobases included in YACs or BACs and transfected into ES cells for the production of transgenic mice (Strauss et al., 1993; Jakobovits et al., 1993). Even though the use of YACs and BACs ensured exposing very large homologous genomic regions to the host genome, no integrations at homologous endogenous loci were reported (Giraldo and Montoliu, 2001).

This observation has some direct consequences, namely, the site of integration of standard DNA transgenes generally cannot be anticipated, cannot be predicted, and, hence, our favorite transgene may land on a genomic region unsuitable for gene expression or may integrate within a given gene expression domain, thus resulting in the interruption, interference, and likely suppression of the gene function associated, and perhaps even generating a mutation at the targeted endogenous locus. These undesirable events, collectively known as chromosomal position effects, were discussed in Section 18.II and have been the subject of research in several laboratories, including ours, aiming to define optimal expression vectors with which to clone transgenes in order to ensure the best achievable expression level following successful integration at random places (Montoliu, 2002; Montoliu et al., 2009).

V. Conclusion and Perspectives

The standard technique of random integration of DNA transgenes microinjected into the pronuclei of fertilized oocytes has been used to populate all transgenic laboratories for more than 30 years. Its inherent simplicity has permitted its universal use. However, injecting a DNA transgene without being able to control where it will be integrated is troublesome, leading to problems and unexpected alterations of the pattern of transgene expression. To overcome these chromosomal position effects one could optimize the transgene design, plus one could decide to use one of the reported strategies to favor or target the integration of the transgene at preselected sites in the genome. The integration of the transgene at target sites through knock-in strategies, using ES cells, or through the use of site-directed recombinases has been possible but still remains intrinsically more technically challenging and demanding. Therefore, it has not become as universally widespread as pronuclear microinjection and maintains its limitations even today.

The scenario is likely to change dramatically, and the generation of transgenic animals using random methods is likely to vanish soon. The appearance of gene-editing nucleases allows one now to easily direct the integration of transgenes to anywhere in the genome, provided the target site is unique. Several types of gene-editing nucleases have been developed, namely: Zn-finger nucleases (ZFNs), TALENs, and CRISPR-Cas nucleases (Gaj et al., 2013; Wei et al., 2013; see Chapter 8). The use of gene-editing nucleases will have profound effects and will change the way we plan our experiments. We will forget the uncertainty of integration of DNA transgenes since they will no longer be inserted randomly but will be planned to land at preselected sites in the host genome, mediated by these gene-editing nucleases, at loci where their expression can best be analyzed.

Acknowledgments

This work is supported by MINECO project BIO2012-39980 to LM. CVG has been supported by a fellowship from the JAE-CSIC program. AF is supported by CIBERER-ISCIII. Eduardo Moltó and Marta Cantero are greatly acknowledged for the technical work illustrated in Figures 18.3 and 18.4.

References

Antoch, M.P., Song, E.J., Chang, A.M., Vitaterna, M.H., Zhao, Y., Wilsbacher, L.D., et al., 1997. Functional identification of the mouse circadian Clock gene by transgenic BAC rescue. Cell 89 (4), 655–667.

Araki, M., Araki, K., Yamamura, K., 2009. International Gene Trap Project: towards gene-driven saturation mutagenesis in mice. Curr. Pharm. Biotechnol. 10 (2), 221–229.

Ballester, M., Castelló, A., Ibáñez, E., Sánchez, A., Folch, J.M., 2004. Real-time quantitative PCR-based system for determining transgene copy number in transgenic animals. Biotechniques 37 (4), 610–613.

Beard, C., Hochedlinger, K., Plath, K., Wutz, A., Jaenisch, R., 2006. Efficient method to generate single-copy transgenic mice by site-specific integration in embryonic stem cells. Genesis 44 (1), 23–28.

Bessa, J., Tena, J.J., de la Calle-Mustienes, E., Fernández-Miñán, A., Naranjo, S., Fernández, A., et al., 2009. Zebrafish enhancer detection (ZED) vector: a new tool to facilitate transgenesis and the functional analysis of cis-regulatory regions in zebrafish. Dev. Dyn. 238 (9), 2409–2417.

Bishop, J.O., Smith, P., 1989. Mechanism of chromosomal integration of microinjected DNA. Mol. Biol. Med. 6 (4), 283–298.

Bishop, J.P., 1997. Chromosomal insertion of foreign DNA. In: Houdebine, L.M. (Ed.), Transgenic Animals, Generation and Use. Harwood Academic Publishers, Amsterdam, pp. 219–223.

Bouhassira, E.E., Westerman, K., Leboulch, P., 1997. Transcriptional behavior of LCR enhancer elements integrated at the same chromosomal locus by recombinase-mediated cassette exchange. Blood 90 (9), 3332–3344.

Brinster, R.L., Chen, H.Y., Trumbauer, M.E., Yagle, M.K., Palmiter, R.D., 1985. Factors affecting the efficiency of introducing foreign DNA into mice by microinjecting eggs. Proc. Natl. Acad. Sci. USA 82 (13), 4438–4442.

Brinster, R.L., Braun, R.E., Lo, D., Avarbock, M.R., Oram, F., Palmiter, R.D., 1989. Targeted correction of a major histocompatibility class II E alpha gene by DNA micro-injected into mouse eggs. Proc. Natl. Acad. Sci. USA 86 (18), 7087–7091.

Bronson, S.K., Plaehn, E.G., Kluckman, K.D., Hagaman, J.R., Maeda, N., Smithies, O., 1996. Single-copy transgenic mice with chosen-site integration. Proc. Natl. Acad. Sci. USA 93 (17), 9067–9072.

Burdon, T.G., Wall, R.J., 1992. Fate of microinjected genes in preimplantation mouse embryos. Mol. Reprod. Dev. 33 (4), 436−442.

Cantero, M., Molto, E, Montoliu, L., 2008. Distinguishing between homozygous and hemizygous transgenic mice. In: Program and Abstracts of the 8th Transgenic Technology Meeting (TT2008). Transgenic Res. 17, 993−1023.

Chandler, K.J., Chandler, R.L., Broeckelmann, E.M., Hou, Y., Southard-Smith, E.M., Mortlock, D.P., 2007. Relevance of BAC transgene copy number in mice: transgene copy number variation across multiple transgenic lines and correlations with transgene integrity and expression. Mamm. Genome 18 (10), 693−708.

Clark, K.J., Geurts, A.M., Bell, J.B., Hackett, P.B., 2004. Transposon vectors for gene-trap insertional mutagenesis in vertebrates. Genesis 39 (4), 225−233.

Dellaire, G., Chartrand, P., 1998. Direct evidence that transgene integration is random in murine cells, implying that naturally occurring double-strand breaks may be distributed similarly within the genome. Radiat. Res. 149 (4), 325−329.

Dillon, N., Festenstein, R., 2002. Unravelling heterochromatin: competition between positive and negative factors regulates accessibility. Trends Genet. 18 (5), 252−258.

Dubose, A.J., Lichtenstein, S.T., Narisu, N., Bonnycastle, L.L., Swift, A.J., Chines, P.S., et al., 2013. Use of microarray hybrid capture and next-generation sequencing to identify the anatomy of a transgene. Nucleic Acids Res. 41 (6), e70.

Erlich, H.A., 1989. PCR Technology. Principles and Applications for DNA Amplification. Stockton Press, New York, NY.

Feng, Y.Q., Seibler, J., Alami, R., Eisen, A., Westerman, K.A., Leboulch, P., et al., 1999. Site-specific chromosomal integration in mammalian cells: highly efficient CRE recombinase-mediated cassette exchange. J. Mol. Biol. 292 (4), 779−785.

Forget, B.G., 1993. YAC transgenes: bigger is probably better. Proc. Natl. Acad. Sci. USA 90 (17), 7909−7911.

Furlan-Magaril, M., Rebollar, E., Guerrero, G., Fernández, A., Moltó, E., González-Buendía, E., et al., 2011. An insulator embedded in the chicken α-globin locus regulates chromatin domain configuration and differential gene expression. Nucleic Acids Res. 39 (1), 89−103.

Gaj, T., Gersbach, C.A., Barbas III, C.F., 2013. ZFN, TALEN, and CRISPR/Cas-based methods for genome engineering. Trends Biotechnol. 31 (7), 397−405.

Giménez, E., Giraldo, P., Jeffery, G., Montoliu, L., 2001. Variegated expression and delayed retinal pigmentation during development in transgenic mice with a deletion in the locus control region of the tyrosinase gene. Genesis 30 (1), 21−25.

Giménez, E., Lavado, A., Giraldo, P., Montoliu, L., 2003. Tyrosinase gene expression is not detected in mouse brain outside the retinal pigment epithelium cells. Eur. J. Neurosci. 18 (9), 2673−2676.

Giménez, E., Lavado, A., Giraldo, P., Cozar, P., Jeffery, G., Montoliu, L., 2004. A transgenic mouse model with inducible Tyrosinase gene expression using the tetracycline (Tet-on) system allows regulated rescue of abnormal chiasmatic projections found in albinism. Pigment Cell Res. 17 (4), 363−370.

Giménez, E., Lavado, A., Jeffery, G., Montoliu, L., 2005. Regional abnormalities in retinal development are associated with local ocular hypopigmentation. J. Comp. Neurol. 485 (4), 338−347.

Giraldo, P., Montoliu, L., 2001. Size matters: use of YACs, BACs and PACs in transgenic animals. Transgenic Res. 10 (2), 83−103.

Giraldo, P., Montoliu, L., 2002. Artificial chromosome transgenesis in pigmentary research. Pigment Cell Res. 15 (4), 258−264.

Giraldo, P., Giménez, E., Montoliu, L., 1999. The use of yeast artificial chromosomes in transgenic animals: expression studies of the tyrosinase gene in transgenic mice. Genet. Anal. 15 (3–5), 175–178.

Grosveld, F., de Boer, E., Dillon, N., Gribnau, J., McMorrow, T., Milot, E., et al., 1998. The dynamics of globin gene expression and position effects. Novartis Found. Symp. 214, 67–79.

Hansen, J., Floss, T., Van Sloun, P., Füchtbauer, E.M., Vauti, F., Arnold, H.H, et al., 2003. A large-scale, gene-driven mutagenesis approach for the functional analysis of the mouse genome. Proc. Natl. Acad. Sci. USA 100 (17), 9918–9922.

Hasty, P., Rivera-Pérez, J., Bradley, A., 1991. The length of homology required for gene targeting in embryonic stem cells. Mol. Cell. Biol. 11 (11), 5586–5591.

Haurogné, K., Bach, J.M., Lieubeau, B., 2007. Easy and rapid method of zygosity determination in transgenic mice by SYBR Green real-time quantitative PCR with a simple data analysis. Transgenic Res. 16 (1), 127–131.

Ivics, Z., Li, M.A., Mátés, L., Boeke, J.D., Nagy, A., Bradley, A., et al., 2009. Transposon-mediated genome manipulation in vertebrates. Nat. Methods 6 (6), 415–422.

Jakobovits, A., 1994. YAC vectors: humanizing the mouse genome. Curr Biol. 4 (8), 761–763.

Jakobovits, A., Moore, A.L., Green, L.L., Vergara, G.J., Maynard-Currie, C.E., Austin, H.A., et al., 1993. Germ-line transmission and expression of a human-derived yeast artificial chromosome. Nature 362 (6417), 255–258.

Lamb, B.T., Gearhart, J.D., 1995. YAC transgenics and the study of genetics and human disease. Curr. Opin. Genet. Dev. 5 (3), 342–348.

Lamb, B.T., Sisodia, S.S., Lawler, A.M., Slunt, H.H., Kitt, C.A., Kearns, W.G., et al., 1993. Introduction and expression of the 400 kilobase amyloid precursor protein gene in transgenic mice. Nat. Genet. 5 (1), 22–30.

Largaespada, D.A., 2003. Generating and manipulating transgenic animals using transposable elements. Reprod. Biol. Endocrinol. 1, 80.

Lauth, M., Spreafico, F., Dethleffsen, K., Meyer, M., 2002. Stable and efficient cassette exchange under non-selectable conditions by combined use of two site-specific recombinases. Nucleic Acids Res. 30 (21), e115.

Lavado, A., Jeffery, G., Tovar, V., de la Villa, P., Montoliu, L., 2006. Ectopic expression of tyrosine hydroxylase in the pigmented epithelium rescues the retinal abnormalities and visual function common in albinos in the absence of melanin. J. Neurochem. 96 (4), 1201–1211.

Liang, Z., Breman, A.M., Grimes, B.R., Rosen, E.D., 2008. Identifying and genotyping transgene integration loci. Transgenic Res. 17 (5), 979–983.

Lin, F.L., Sternberg, N., 1984. Homologous recombination between overlapping thymidine kinase gene fragments stably inserted into a mouse cell genome. Mol. Cell. Biol. 4 (5), 852–861.

Lin, F.L., Sperle, K., Sternberg, N., 1984. Model for homologous recombination during transfer of DNA into mouse L cells: role for DNA ends in the recombination process. Mol. Cell. Biol. 4 (6), 1020–1034.

Lin, F.L., Sperle, K.M., Sternberg, N.L., 1987. Extrachromosomal recombination in mammalian cells as studied with single- and double-stranded DNA substrates. Mol. Cell. Biol. 7 (1), 129–140.

Liu, Z., Obenauf, A.C., Speicher, M.R., Kopan, R., 2009. Rapid identification of homologous recombinants and determination of gene copy number with reference/query pyrosequencing (RQPS). Genome Res. 19 (11), 2081–2089.

Lundgren, M., Chow, C.M., Sabbattini, P., Georgiou, A., Minaee, S., Dillon, N., 2000. Transcription factor dosage affects changes in higher order chromatin structure associated with activation of a heterochromatic gene. Cell 103 (5), 733−743.

Mahon, K.A., Overbeek, P.A., Westphal, H., 1988. Prenatal lethality in a transgenic mouse line is the result of a chromosomal translocation. Proc. Natl. Acad. Sci. USA 85 (4), 1165−1168.

Martin, D., Pantoja, C., Fernández-Miñán, A., Valdes-Quezada, C., Moltó, E., Matesanz, F., et al., 2011. Genome-wide CTCF distribution in vertebrates defines equivalent sites that aid the identification of disease-associated genes. Nat. Struct. Mol. Biol. 18 (6), 708−714.

Matsui, S., Sait, S., Jones, C.A., Nowak, N., Gross, K.W., 2002. Rapid localization of transgenes in mouse chromosomes with a combined Spectral Karyotyping/FISH technique. Mamm. Genome 13 (12), 680−685.

McHugh, D., O'Connor, T., Bremer, J., Aguzzi, A., 2012. ZyFISH: a simple, rapid and reliable zygosity assay for transgenic mice. PLoS One 7 (5), e37881.

Mitrecić, D., Huzak, M., Curlin, M., Gajović, S., 2005. An improved method for determination of gene copy numbers in transgenic mice by serial dilution curves obtained by real-time quantitative PCR assay. J. Biochem. Biophys. Methods 64 (2), 83−98.

Moltó, E., Vicente-García, C., Montoliu, L., 2011. Designing transgenes for optimal expression. In: Pease, S., Saunders, T. (Eds.), Advanced Protocols for Animal Transgenesis: and ISTT Manual. Springer Protocols Handbooks, Berlin, pp. 43−56.

Montoliu, L., 2002. Gene transfer strategies in animal transgenesis. Cloning Stem Cells 4 (1), 39−46.

Montoliu, L., Schedl, A., Kelsey, G., Zentgraf, H., Lichter, P., Schütz, G., 1994. Germ line transmission of yeast artificial chromosomes in transgenic mice. Reprod. Fertil. Dev. 6 (5), 577−584.

Montoliu, L., Umland, T., Schütz, G., 1996. A locus control region at -12 kb of the tyrosinase gene. EMBO J. 15 (22), 6026−6034.

Montoliu, L., Roy, R., Regales, L., García-Díaz, A., 2009. Design of vectors for transgene expression: the use of genomic comparative approaches. Comp. Immunol. Microbiol. Infect. Dis. 32 (2), 81−90.

Nakanishi, T., Kuroiwa, A., Yamada, S., Isotani, A., Yamashita, A., Tairaka, A., et al., 2002. FISH analysis of 142 EGFP transgene integration sites into the mouse genome. Genomics 80 (6), 564−574.

Noguchi, A., Takekawa, N., Einarsdottir, T., Koura, M., Noguchi, Y., Takano, K., et al., 2004. Chromosomal mapping and zygosity check of transgenes based on flanking genome sequences determined by genomic walking. Exp. Anim. 53 (2), 103−111.

Nord, A.S., Chang, P.J., Conklin, B.R., Cox, A.V., Harper, C.A., Hicks, G.G., et al., 2006. The International Gene Trap Consortium Website: a portal to all publicly available gene trap cell lines in mouse. Nucleic Acids Res. 34 (Database issue), D642−D648.

Notini, A.J., Li, R., Western, P.S., Sinclair, A.H., White, S.J., 2009. Rapid and reliable determination of transgene zygosity in mice by multiplex ligation-dependent probe amplification. Transgenic Res. 18 (6), 987−991.

Ohtsuka, M., Ogiwara, S., Miura, H., Mizutani, A., Warita, T., Sato, M.., et al., 2010. Pronuclear injection-based mouse targeted transgenesis for reproducible and highly efficient transgene expression. Nucleic Acids Res. 38 (22), e198.

Peterson, K.R., 2003. Transgenic mice carrying yeast artificial chromosomes. Expert Rev. Mol. Med. 5 (13), 1−25.

Peterson, K.R., Clegg, C.H., Li, Q., Stamatoyannopoulos, G., 1997. Production of transgenic mice with yeast artificial chromosomes. Trends Genet. 13 (2), 61−66.

Pillai, M.M., Venkataraman, G.M., Kosak, S., Torok-Storb, B., 2008. Integration site analysis in transgenic mice by thermal asymmetric interlaced (TAIL)-PCR: segregating multiple-integrant founder lines and determining zygosity. Transgenic Res. 17 (4), 749−754.

Pinkert, C.A., Ornitz, D.M., Brinster, R.L., Palmiter, R.D., 1987. An albumin enhancer located 10 kb upstream functions along with its promoter to direct efficient, liver-specific expression in transgenic mice. Genes Dev. 1 (3), 268−276.

Rossant, J., Nutter, L.M., Gertsenstein, M., 2011. Engineering the embryo. Proc. Natl. Acad. Sci. USA 108 (19), 7659−7660.

Sakurai, T., Kamiyoshi, A., Watanabe, S., Sato, M., Shindo, T., 2008. Rapid zygosity determination in mice by SYBR Green real-time genomic PCR of a crude DNA solution. Transgenic Res. 17 (1), 149−155.

Schedl, A., Montoliu, L., Kelsey, G., Schütz, G., 1993. A yeast artificial chromosome covering the tyrosinase gene confers copy number-dependent expression in transgenic mice. Nature 362 (6417), 258−261.

Shepherd, C.T., Moran Lauter, A.N., Scott, M.P., 2009. Determination of transgene copy number by real-time quantitative PCR. Methods Mol. Biol. 526, 129−134.

Shimoda, K., Cai, X., Kuhara, T., Maejima, K., 1991. Reconstruction of a large DNA fragment from coinjected small fragments by homologous recombination in fertilized mouse eggs. Nucleic Acids Res. 19 (23), 6654.

Shitara, H., Sato, A., Hayashi, J., Mizushima, N., Yonekawa, H., Taya, C., 2004. Simple method of zygosity identification in transgenic mice by real-time quantitative PCR. Transgenic Res. 13 (2), 191−194.

Sippel, A.E., Saueressig, H., Hubler, M.C., Faust, N., Bonifer, C., 1997. Insulation of transgenes from chromosomal position effects. In: Houdebine, L.M. (Ed.), Transgenic Animals, Generation and Use. Harwood Academic Publishers, Amsterdam, pp. 267−272.

Skarnes, W.C., Auerbach, B.A., Joyner, A.L., 1992. A gene trap approach in mouse embryonic stem cells: the lacZ reported is activated by splicing, reflects endogenous gene expression, and is mutagenic in mice. Genes Dev. 6 (6), 903−918.

Strauss, W.M., Dausman, J., Beard, C., Johnson, C., Lawrence, J.B., Jaenisch, R., 1993. Germ line transmission of a yeast artificial chromosome spanning the murine alpha 1(I) collagen locus. Science 259 (5103), 1904−1907.

Strouboulis, J., Dillon, N., Grosveld, F., 1992. Efficient joining of large DNA fragments for transgenesis. Nucleic Acids Res. 20 (22), 6109−6110.

Takemoto, Y., Keighren, M., Jackson, I.J., Yamamoto, H., 2006. Genomic localization of a Dct-LacZ transgene locus: a simple assay for transgene status. Pigment Cell Res. 19 (6), 644−645.

Tasic, B., Hippenmeyer, S., Wang, C., Gamboa, M., Zong, H., Chen-Tsai, Y., et al., 2011. Site-specific integrase-mediated transgenesis in mice via pronuclear injection. Proc. Natl. Acad. Sci. USA 108 (19), 7902−7907.

Tesson, L., Rémy, S., Ménoret, S., Usal, C., Anegon, I., 2010. Analysis by quantitative PCR of zygosity in genetically modified organisms. Methods Mol. Biol. 597, 277−285.

Tiana, M., Villar, D., Pérez-Guijarro, E., Gómez-Maldonado, L., Moltó, E., Fernández-Miñán, A., et al., 2012. A role for insulator elements in the regulation of gene expression response to hypoxia. Nucleic Acids Res. 40 (5), 1916−1927.

Turan, S., Zehe, C., Kuehle, J., Qiao, J., Bode, J., 2013. Recombinase-mediated cassette exchange (RMCE): a rapidly expanding toolbox for targeted genomic modifications. Gene 515 (1), 1–27.

Van Keuren, M.L., Gavrilina, G.B., Filipiak, W.E., Zeidler, M.G., Saunders, T.L., 2009. Generating transgenic mice from bacterial artificial chromosomes: transgenesis efficiency, integration and expression outcomes. Transgenic Res. 18, 769–785.

Wake, C.T., Vernaleone, F., Wilson, J.H., 1985. Topological requirements for homologous recombination among DNA molecules transfected into mammalian cells. Mol. Cell. Biol. 5 (8), 2080–2089.

Wallace, H., Ansell, R., Clark, J., McWhir, J., 2000. Pre-selection of integration sites imparts repeatable transgene expression. Nucleic Acids Res. 28 (6), 1455–1464.

Wei, C., Liu, J., Yu, Z., Zhang, B., Gao, G., Jiao, R., 2013. TALEN or Cas9: rapid, efficient and specific choices for genome modifications. J. Genet. Genomics 40 (6), 281–289.

Wilson, C., Bellen, H.J., Gehring, W.J., 1990. Position effects on eukaryotic gene expression. Annu. Rev. Cell Biol. 6, 679–714.

Yang, G.S., Banks, K.G., Bonaguro, R.J., Wilson, G., Dreolini, L., de Leeuw, C.N., et al., 2009. Next generation tools for high-throughput promoter and expression analysis employing single-copy knock-ins at the Hprt1 locus. Genomics 93, 196–204.

Yang, X.W., Gong, S., 2005. An overview on the generation of BAC transgenic mice for neuroscience research. Curr. Protoc. Neurosci. (Chapter 5:Unit 5.20), <http://www.ncbi.nlm.nih.gov/pubmed/18428622>.

Yokoyama, T., Copeland, N.G., Jenkins, N.A., Montgomery, C.A., Elder, F.F., Overbeek, P.A., 1993. Reversal of left-right asymmetry: a situs inversus mutation. Science 260 (5108), 679–682.

Zhang, R., Yin, Y., Zhang, Y., Li, K., Zhu, H., Gong, Q., et al., 2012. Molecular characterization of transgene integration by next-generation sequencing in transgenic cattle. PLoS One 7 (11), e50348.

19 PCR Optimization for Detection of Transgene Integration

Michael H. Irwin[1], Wendy K. Pogozelski[2] and Carl A. Pinkert[1,3]

[1]Department of Pathobiology, College of Veterinary Medicine, Auburn University, Auburn, AL, [2]Department of Chemistry, State University of New York at Geneseo, Geneseo, NY, [3]Department of Biological Sciences, College of Arts and Sciences, The University of Alabama, Tuscaloosa, AL

I. Introduction

This chapter is not intended to serve as a comprehensive review of the use of the polymerase chain reaction (PCR) for the identification of transgenic founder mice or their offspring. Numerous sources of information are available dealing with the theory, performance, and optimization of PCR analysis in a variety of applications (reviewed in Dawson et al., 1996; Nagy et al., 2002; Green and Sambrook, 2012). Rather, this brief overview is intended to address some of the general considerations for PCR analysis and some of the particular concerns when using this technique to detect transgene integration in experimental animals (especially laboratory mice).

Amplification of DNA by PCR, first described in a revolutionary paper by Mullis and Faloona (1987), has made *in vitro* amplification of specific DNA target sequences both rapid and easy to perform. The impact of this technology on diagnostic screening of genomes for specific target sequences would be difficult to overstate. The utility of this technique is now taken for granted in virtually all laboratories involved in molecular cloning and other molecular biological technologies used to study the genomes of experimental animals and humans.

II. Discussion

A. General

For the purpose of detecting the integration of transgene constructs into mouse (or other animal) genomes, as described in Chapter 18, PCR analyses have proven invaluable in providing accurate, cost-effective, and timely data. If the proper safeguards are employed, PCR analyses can be very reliable in the initial identification

Transgenic Animal Technology. DOI: http://dx.doi.org/10.1016/B978-0-12-410490-7.00019-0

of founder transgenic animals and in characterizing germ line transmission of genetic modifications transmitted to their offspring. This discussion will address various parameters and practical concerns of PCR analysis that relate to its use in the diagnostic testing of experimental animal genomic DNA samples for transgene integration.

Although PCR is indeed exceedingly useful for integration analysis, it does have limitations. For example, PCR detects integration of the transgene only, and, in its basic form, provides no information on expression of the transgene. Reverse transcriptase-PCR (RT-PCR) can be very useful in detecting and quantifying gene expression; however, RT-PCR is somewhat more complicated than conventional/ basic PCR and poses its own unique set of practical considerations (see Chapter 21). One should also bear in mind that the PCR in its basic form is not intended to be a quantitative test and that PCR results can be misleading when one attempts (wrongly) to extrapolate information regarding the number of copies of the transgene present in the genome of an experimental animal without more precise methodology (from quantitative real-time PCR to various hybridization analyses outlined in Chapter 21). Additionally, when quantitative information is desired, real-time PCR or competitive PCR methods may also be employed with specific advantages as described in Section II.C.

B. PCR Considerations

If the target template is not specific in relationship to the strain-specific genomic DNA, the PCR results will be of limited value. False positives resulting from a PCR of questionable specificity will invariably lead to significant expense related to maintaining and breeding animals that have no experimental value. Therefore, time spent optimizing and verifying the specificity of the PCR in the preliminary phases of an experiment will be critical. That said, optimization of any PCR analysis must initially focus on the specificity of the primers to the target sequence one wishes to amplify. Thus, primer design is the most fundamental consideration for the success of the PCR. Many software packages are available today that simplify the design of primers for use in the PCR, for example, GeneFisher (http://bibiserv.techfak.uni-bielefeld.de/genefisher/); OligoPerfect™ Designer (http://tools.lifetechnologies.com/content.cfm?pageid=9716); Primer 3 (http://bioinfo.ut.ee/primer3-0.4.0/primer3/); and OLIGO (http://www.oligo.net/). Although such programs no doubt greatly simplify primer design, one can easily devise suitable primers without such programs. All of the primers used in our labora-tory and national cores over the years were designed without the use of primer-design programs, and in some cases were shown to outperform primers designed using these programs. For most purposes, the length of primers should be at least 15 base pairs. If a nonrepetitive sequence is used, the probability of an exact duplication of a 15 bp sequence elsewhere in the genome is very small. This probability obvi-ously shrinks as the primer increases in length. However, a point of diminishing returns may be reached with larger primers, where the increased specificity is out-weighed by the loss of sensitivity, as larger primers hybridize less efficiently than smaller primers. As a rule of thumb, we routinely use primers in the range of 15−30 bp.

There are just a few practical considerations to keep in mind when designing primers for PCR without the use of specific software programs. First, the possibility of "primer−dimer" formation must be examined. These are small amplification products that result from hybridization of a given primer with itself or with the opposite direction primer. Although such species are rather easily identified on agarose gels due to their small size and are seldom confused with *bona fide* amplification products from the target sequence, they decrease the sensitivity of the PCR by diverting primers from the true target. Indeed, some reactions may show absolutely no amplification of the target sequence due to primer−dimer formation. By comparing the sequence of a given primer both to itself and to the opposite direction primer, one can easily identify stretches of the primer that may be prone to primer−dimer formation. In general, we avoid using primers that demonstrate complementarity of four or more bases, rejecting primers that show complementarity of three bases more than once. Furthermore, even a three-base match is unacceptable if it occurs at the 3′ end of any primer (from which extension proceeds). Another important consideration is the G + C ratio of each primer. A G + C ratio that is too high will result in more difficulty in strand separation during the denaturation step, while a low G + C ratio will necessitate a lower annealing temperature, which is equivalent to lower stringency, thus increasing the likelihood of nonspecific amplification products. Therefore, practically speaking, it is best to have primers with a G + C ratio in the range of 40−60%. In addition, although a primer may fit this criterion, stretches of continuous Gs and Cs within a region of the primer are also problematic and should be avoided. Finally, it is also mandatory that both forward and reverse primers have very similar G + C ratios so that optimal reaction conditions for one primer will also be optimal for the opposite direction primer.

The length of the target sequence (the stretch of bases amplified by PCR) is also a practical concern. Amplification products less than about 200 bp sometimes appear diffuse in agarose gel electrophoresis. Alternatively, long target sequences are amplified less efficiently. Again, as a general rule, we attempt to identify target sequences in the range of 200−1000 bp. It is a common mistake to attempt amplification of an entire (or nearly complete) transgene sequence, unless the transgene is relatively small. If one wished to ascertain the complete integration of a large transgene, two separate PCRs may be employed to amplify each end of the transgene. But for accurate verification of the integration of the complete transgene, Southern blotting along with DNA sequence determination provides much more reliable data.

There are several tests one must perform prior to utilizing PCR analysis for detection of transgenic founder animals. First, the reaction should be tested for efficiency of amplification of the transgene itself, with no other DNA present. This is a good time not only to verify that the proper product is amplified but also to perform adjustments in various parameters such as buffer composition (especially in relation to magnesium concentration), cycling temperatures and times, and number of cycles required for optimal amplification (Figure 19.1). Next, it is imperative

If PCR analysis is to be used to identify transgenic mice, appropriate oligonucleotide primers and specification of proven conditions for the reaction are recommended. Mouse DNA (strain-specific tissue samples; e.g., B6SJL F1 hybrid, C57BL/6, or FVB tissues, as appropriate) for evaluating PCR specificity. One should complete the information below and if desired, maintain archival photographs (originals or discernable copies) of the control reactions for troubleshooting protocols and results.

5′ primer name (≤12 characters)   _____

5′ primer length (bp), molar conc., total conc.   _____,_____,_____

3′ primer name (≤12 characters)   _____

3′ primer length (bp), molar conc., total conc.   _____,_____,_____

Length of PCR product (bp):   _____

Amount of each primer per reaction (μl):   5′ _____   3′ _____

Denaturation temp. (°C): _____   Denaturation time:____

Annealing temp. (°C):   _____   Annealing time: _____

Extension temp. (°C):   _____   Extension time: _____

Cycles: _____

Other reaction conditions: pH ____, MgCl₂ conc. ____, Taq conc.____, dNTPs ___, KCl conc. ___

Reaction conditions (for diagnostic evaluation, primer sets should be tested with the following control amplifications):
1. Normal mouse DNA (designated NM; use the specific mouse strain appropriate to the project)
2. NM + 0.1 to 0.5 gene copy/cell equivalent of the DNA construct
3. NM + 1 gene copy/cell equivalent of the DNA construct
4. NM + 5 (or 10) gene copies/cell equivalent of the DNA construct
5. 1 gene copy/cell equivalent of the DNA construct (no NM DNA)
6. 5 (or 10) gene copies/cell equivalent of the DNA construct (no NM DNA)
7. Appropriate marker DNA

(Genomic conversion for DNA copies per cell—6×10^9 base pairs per diploid genome.)

Marker DNA (lane 8 above): Type: _____ Concentration: _____ Volume (μL): _____

Reaction volume (μL): _____

Amount of mouse DNA per NM reaction (ng):_____

Amount of DNA used in 1 copy control (pg): _____

Figure 19.1 PCR analysis worksheet.

that the reaction be tested using normal genomic DNA from the animal species/ strain used in the study, with no transgene DNA added. This important step is included to verify that the reaction does not amplify any erroneous "target" sequences that might either affect the efficiency of the desired reaction or even lead to the generation of false positive signals. After the PCR has been optimized and tested to ensure no discernable reaction with normal genomic DNA, the sensitivity of the reaction must be assessed in a "real-world" situation. This means that the sensitivity and specificity of the PCR must be tested in the presence of normal

Calculation of copy number controls (amount of construct-specific DNA) is beneficial in characterizing the sensitivity and specificity of the PCR reaction. If whole vector (e.g., whole plasmid containing a given sequence) is used to spike normal mouse DNA (NMDNA), then one should use a ratio of construct to vector in the final calculation. Additionally, the use of strain-specific NMDNA should reflect the host genome.

The mouse diploid genome has a mass of approximately 6.42×10^{-12} g.

The amount of NMDNA used in the assay divided by this number gives the equivalent number of mouse diploid genomes. If, for example, a PCR assay was set up with 100 ng mouse DNA per sample,

$$100 \, \text{ng} = 1 \times 10^{-7} \, \text{g}$$

$$\frac{1 \times 10^{-7} \, \text{g}}{6.42 \times 10^{-12} \, \text{g/diploid genome}} = 15{,}576 \text{ diploid genomes}$$

The size of the construct in bp multiplied by 1.07×10^{-21} g/bp = the mass of the construct in grams.

The mass of the construct multiplied by the number of diploid genome equivalents = the single gene copy equivalent. For example, if the construct is 5000 or 10,000 bp and we are using 100 ng of DNA in our PCR, then the mass of the construct is

5000 bp $\times \ (1.07 \times 10^{-21} \, \text{g/bp}) = 5.35 \times 10^{-18} \, \text{g}$

10,000 bp $\times (1.07 \times 10^{-21} \, \text{g/bp}) = 10.70 \times 10^{-18} \, \text{g}$

Therefore, for 100 ng (or 1.5576×10^4 genome equivalents), the single gene copy equivalent is:

For 5000 bp: $(5.35 \times 10^{-18} \, \text{g}) \times (1.5576 \times 10^4) = 8.33 \times 10^{-14} \, \text{g}$ or 0.0833 pg

For 10,000 bp: $(10.7 \times 10^{-18} \, \text{g}) \times (1.5576 \times 10^4) = 16.67 \times 10^{-14} \, \text{g}$ or 0.1667 pg

Figure 19.2 Calculating gene copy/genome equivalent.

genomic DNA to which has been added known quantities (equivalent to 0.1 copy, 1 copy, 10 copies, and 100 copies) of the transgene construct (Figure 19.2). If a given PCR is unable to amplify the target sequence present at the level of one-copy equivalent, it will be useless in detecting experimental animals that have integrated only a single copy of the transgene. And although mosaics may be rare in transgenic animals produced by pronuclear microinjection, using a 0.1- to 0.5-copy control will increase confidence that the sensitivity of the reaction is sufficient to detect all of the founders thus produced. Controls of 10- and 100-copy equivalents are used in case the lower copy number controls do not amplify sufficiently, so that at least one has an idea of how much more optimization is needed to obtain the desired sensitivity.

It is wise to always include these controls, including a zero-copy (negative) control both with and without normal genomic DNA present, in every subsequent PCR so that one may have a high level of confidence in the results obtained.

One final concern related to the PCR of genomic DNA sequences involves the method of DNA extraction and purification utilized. Most techniques for preparation of genomic DNA involve the use of a protease (e.g., proteinase K) in the initial digestion of tissue samples before DNA extraction. Unfortunately, proteases are very capable of digesting the thermostable polymerase used in the amplification reaction. Therefore, extreme care should be exercised during the DNA extraction procedure to eliminate carryover of proteases. This also includes recognition of proteases normally present in the tissue used as the source of DNA. This problem is of particular concern when using a standard organic extraction procedure (e.g., phenol:chloroform extraction) where pipetting the liquid phase containing DNA can introduce proteins (including proteases) contained in the interphase. It is best when using such protocols to sacrifice a portion of the aqueous DNA-containing phase rather than attempt to recover it all and inadvertently carry over proteases from the interphase. Using one of several widely available DNA purification kits will eliminate this problem.

C.   Quantitative (Real-Time) PCR

Quantitative PCR (qPCR) methods are gaining considerable popularity in transgene analyses due to their power as quantitation tools. Although PCR techniques have readily replaced labor-intensive blotting procedures, qPCR has indeed become commonplace since the last edition of this chapter. What distinguishes qPCR from conventional methods is that the quantity of amplification product is measured as it is being made (after each extension), rather than at the completion of all the cycles (endpoint determination).

Two approaches can be used for quantifying PCR products via qPCR. The first is TaqMan™-PCR (Holland et al., 1991; Heid et al., 1996), which relies on a polymerase with exonuclease activity and a sequence-specific probe that anneals to a region between the forward or reverse primer site. The probe contains a fluorescent dye at the 5' end to serve as the reporter, and a quencher dye is tagged to the 3' end. The emission spectrum of the reporter overlaps the absorption spectrum of the quencher. As long as the reporter and quencher are in close proximity, or as long as the probe is intact, Förster energy transfer occurs and no fluorescence is detected. As the polymerase begins extension from the primer, it encounters the probe in its path and hydrolyzes the probe (hence, the analogy with the classic video game *Pac-Man*). Hydrolysis separates the reporter from the quencher, reducing energy transfer and enabling fluorescence to occur. Fluorescence increases proportionally each time the probe is hydrolyzed. Eventually, fluorescence will reach a detectable level called the threshold cycle or C_t. This value, representing the number of amplification cycles required for positive fluorescence, is directly related to the amount of target in the initial sample.

One requirement of the TaqMan™ approach is that the sequence between the forward and reverse primers must be known. When the sequence information is not available, or when a researcher wishes to prescreen and avoid the expense of labeled probes, the fluorescent dye SYBR Green can be used (Schneeberger et al., 1995). This dye binds to the minor groove of double-stranded DNA; therefore, the assay may not be specific for the target sequence. To counteract this problem, researchers often perform a melting point analysis to determine if the fluorescence data correspond to those from the target sequence.

Most real-time systems are purchased as a unit that contains a thermal cycler attached to a light source and fluorescence reader. The data are fed directly into a computer and analyzed with specialty software; however, this specialized equipment is not completely necessary. A researcher can obtain real-time data by removing aliquots from a standard thermal cycler at various times during amplification. Of course, lack of automation does risk producing data that are less reproducible, and the specialized systems are far more convenient to use.

qPCR systems can be purchased with either lasers or tungsten lamps as light sources. The laser systems are more expensive. Although the bandwidths are wider with the tungsten light sources, the data obtained with the tungsten lamps are, in our experience, equally acceptable to those produced by the laser systems.

Quantification assays require reproducibility in setting up the reactions. To minimize pipetting variations, researchers often use master mixes. These generally include nucleotides, the polymerase, cations, and a dye such as ROX (6-carboxy-X-rhodamine) for an internal reference. Labeled probes can be purchased from many manufacturers. Typical fluorescent reporter labels are TET (tetrachloro-6-carboxy-fluorescein, $\lambda_{max(emission)} = 538$ nm), FAM (6-carboxyfluorescein, $\lambda_{max} = 518$ nm), JOE ($\lambda_{max} = 554$ nm), and VIC ($\lambda_{max} = 554$ nm), while a typical quencher dye is TAMRA (6-carboxytetramethylrhodamine, $\lambda_{max} = 582$ nm). Probe design requires a set of considerations similar to that for primer design. Software programs such as Primer Express (Applied Biosystems, Foster City, CA, http://www.appliedbiosystems.com) can facilitate probe selection. A given probe must be specific and should avoid regions that dimerize or form secondary structures. There are several differences between primers and probes. The probe T_m should be about 10°C higher than that of the primers (assuming a primer T_m in the region of 58−60°C) and the sequence should avoid a G at the 5′ end. In addition, probes should be highly purified (e.g., high-performance liquid chromatography (HPLC) purification) to avoid carryover of residual contaminants including quencher dyes.

qPCR is amenable to multiplex analysis; that is, two or more products can be monitored simultaneously in the same sample by using different reporter dyes. The dyes must be selected with care, so that the spectra do not overlap. FAM and JOE or FAM and VIC are good choices in this respect. Multiplex analysis requires extra caution, however. If one target is initially present in a much greater amount, it will deplete reagents, thereby altering the amplification kinetics of the less abundant target. This situation can be avoided by prior optimization and by using conditions in which one set of primers is limiting.

Quantification can be either absolute or relative. For absolute quantification, the researcher generally constructs a set of standard curves for the target whose copy number is desired and for a control (either external or internal) whose copy number is known. Typical internal controls are the β-actin gene, the 18 S rRNA gene, and the glyceraldehyde 3-phosphate dehydrogenase gene. The threshold cycle for a dilution series, plotted against the log of the copy number, will yield a straight line, facilitating extrapolation and interpolation.

It is important that one takes extreme care to avoid contamination of reagents. The technique is so sensitive that trace amounts of template in a reagent can cause large problems. Assays should always include no-template controls to test for reagent contamination. If RT-PCR is used, the procedure should also include a no-amplification control to test for contamination of RNA by DNA.

The high sensitivity of qPCR makes it especially useful in transgene analysis. The technique can identify and quantify insertion events and can indicate the stability of an insert over time (Norris et al., 2000; Ingham et al., 2001). One important use is in determination of copy number and zygosity (whether an organism is heteroygotic/hemizygotic, or homozygotic for a targeted allele), particularly with low-copy integrations and differentiation from wild-type offspring when endogenous genomic sequences are introduced into the genome (Becker et al., 1999). qPCR can also be used to assess the amount of viral vector in virus-mediated gene transfer modeling (Kozlowski et al., 2001; Wang et al., 2001). Finally, as outlined in Chapter 21, another use is to combine quantitative methods with RT-PCR of mRNA products to monitor gene expression (Fairman et al., 1999).

III.   Summary

In routine nucleic acid analyses, DNA samples are qualitatively and/or quantitatively analyzed first by the PCR, and the results are subsequently verified by Southern hybridization and expression analyses. The PCR procedure can be completed within one day, while the more labor- and cost-intensive procedures may take 3−7 days to complete.

The invention of the PCR utilized a common sense approach that relied on some knowledge of how DNA behaves along with the curious existence of a class of DNA polymerases that maintain function at high temperature. Indeed, the basic concept is so straightforward that many of us were left wondering, "Why didn't I think of that?" It is therefore not surprising that common sense is the most important consideration in designing and optimizing PCRs useful in specific applications and in interpreting data obtained from this ubiquitous and powerful technique.

References

Becker, K., Pan, D., Whitley, C.B., 1999. Real-time quantitative polymerase chain reaction to assess gene transfer. Hum. Gene Ther. 10, 2559−2566.

Dawson, M.T., Powell, A., Gannon, F., 1996. Gene Technology. Bios Scientific Publishers, Oxford.

Fairman, J., Roche, L., Pieslak, I., Lay, M., Corson, S., Fox, E., et al., 1999. Quantitative RT-PCR to evaluate *in vivo* expression of multiple transgenes using a common intron. Biotechniques 27 (3), 566–570, 572–574.

Green, M.R., Sambrook, J., 2012. Molecular Cloning: A Laboratory Manual. fourth ed. Cold Spring Harbor Laboratory Press, Cold Spring Harbor, NY.

Heid, C.A., Stevens, J., Livak, K.J., Williams, P.M., 1996. Real time quantitative PCR. Genome Res. 6, 986–994.

Holland, P.M., Abramson, R.D., Watson, R., Gelfand, D.H., 1991. Detection of specific polymerase chain reaction product by utilizing the $5' \to 3'$- exonuclease activity of *Thermus aquaticus* DNA polymerase. Proc. Natl. Acad. Sci. USA 88, 7276–7280.

Ingham, D.J., Beer, S., Money, S., Hansen, G., 2001. Quantitative real-time PCR assay for determining transgene copy number in transformed plants. Biotechniques 31 (1), 132–134, 136–140.

Kozlowski, D.A., Bremer, E., Redmond Jr., D.E., George, D., Larson, B., Bohn, M.C., 2001. Quantitative analysis of transgene protein, mRNA, and vector DNA following injection of an adenoviral vector harboring glial cell line-derived neurotrophic factor into the primate caudate nucleus. Mol. Ther. 3, 256–261.

Maudru, T., Pden, K.W., 1998. Adaptation of the fluorogenic 5'-nuclease chemistry to a PCR-based reverse transcriptase assay. Biotechniques 25, 972–975.

Mullis, K.B., Faloona, F.A., 1987. Specific synthesis of DNA *in vitro* via a polymerase-catalyzed chain reaction. Meth. Enzymol. 155, 335–350.

Nagy, A., Gertsenstein, M., Vintersten, K., Behringer, R., 2002. Manipulating the Mouse Embryo: A Laboratory Manual. third ed. Cold Spring Harbor Press, Cold Spring Harbor, NY.

Norris, M.D., Burkhart, C.A., Marshall, G.M., Weiss, W.A., Haber, M., 2000. Expression of N-myc and MRP genes and their relationship to N-myc gene dosage and tumor formation in a murine neuroblastoma model. Med. Pediatr. Oncol. 35, 585–589.

Schneeberger, C., Speiser, P., Kury, F., Zeillinger, R., 1995. Quantitative detection of reverse transcriptase-PCR products by means of a novel and sensitive DNA stain. PCR Methods Appl. 4, 234–238.

Wang, K., Pesnicak, L., Guancial, E., Krause, P.R., Straus, S.E., 2001. The 2.2-kilobase latency-associated transcript of herpes simplex virus type 2 does not modulate viral replication, reactivation, or establishment of latency in transgenic mice. J. Virol. 75, 8166–8172.

20 Analysis of Transgene Expression

Michael H. Irwin[1] and Carl A. Pinkert[1,2]

[1]Department of Pathobiology, College of Veterinary Medicine, Auburn University, Auburn, AL, [2]Department of Biological Sciences, College of Arts and Sciences, The University of Alabama, Tuscaloosa, AL

I.   Introduction

This chapter will introduce the techniques available to assay transgene expression, critically evaluate each approach, and provide some examples of the results one might expect to achieve with these methods. Actual protocols for carrying out assays for gene expression are not given; most of these have been compiled in various formats or are available by web searching. Most of these techniques are extant in molecular biology laboratories. Many of the issues to be considered when assaying transgenic animals for gene expression are the same as for transfecting genes into cultured cells. For example, how does one distinguish between an endogenous transcript or protein and its transgene-derived counterpart? How does one compare the relative levels of transgene expression between cells? How does one recognize the use of alternate transcriptional start sites in different cell types? The complexities of measuring transgene expression in the whole animal often pose unique challenges, and we evaluate each approach in light of this fact.

A.   Design of the Transgene to Allow Expression Analysis

Before beginning a transgenic project (in fact, even before the actual construction of the transgene), it is wise to consider the potential problems of analyzing transgene expression. In this regard, there are two broad categories of transgenes: those that have an endogenous counterpart already present in the genome of the animal and those that do not. Those that do not are in general straightforward to assay for expression, as they will produce a novel transcript that can be detected by a unique nucleic acid probe, or they will produce a novel protein or enzyme activity that can be detected using antisera and/or histochemical methods. Those that do have an endogenous counterpart are more problematic, especially if the tissue distribution of endogenous and transgene expression overlap, in which case one must include features distinguishing the transgene-derived mRNA or protein from the endogenous. The most facile approach is to alter the nucleotide sequence of one or more exons in the transgene. For example, one can delete a portion of the 5' or 3'

Transgenic Animal Technology. DOI: http://dx.doi.org/10.1016/B978-0-12-410490-7.00020-7

noncoding region of the gene and distinguish between the two on the basis of size. One may also insert a piece of foreign DNA to serve as a molecular tag. The latter approach has the advantage of providing a unique sequence in the transgene mRNA that can be detected by a variety of methods. The extent and nature of each distinguishing feature designed into the transgene will be dictated by individual experimental circumstances and by the preference of the investigator in terms of how transgene expression will be detected. The most important consideration, though, is whether altering the transgenic mRNA will perturb gene expression, that is, will it alter the half-life of the transcript or change its translation efficiency? Because of the uncertainties those issues hold, the minimal change that allows one to distinguish between and yet detect both transcripts is generally preferred.

B.   Considerations in the Analysis of Transgene Expression

Characterizing the expression of a transgene is a crucial step in all transgenic projects. In some cases, it may represent the crux of the whole experiment—for example, if one is using transgenic animals to define the specificity of a putative cis-regulatory element. In other experiments, it may be only the first step in a long process of assessing a disease phenotype. Indeed, in the latter case, a careful definition of transgene expression may prompt the investigator to anticipate the phenotypic consequences of transgene expression. Having derived a set of transgenic founder animals, the first decision to be made with regard to analysis of transgene expression is whether to analyze the founder animals themselves or to carry out the analysis on the transgenic offspring. An important consideration in making this decision is whether it is possible to assess expression in a tissue that can readily be biopsied without endangering the health or reproductive capacity of the founder. For example, one can easily use a tail biopsy to look for expression in skin, muscle, and bone. One can also safely remove a full thickness of skin biopsy from other regions of the body. Expression in liver, spleen, kidney, adrenal gland, lymph node, testis, ovary, and pancreas can be assessed by partial or complete removal of these organs without any long-term deleterious effects on the health of the animal. Blood can also be safely removed by retro orbital or tail bleeding. Procedures for performing many of these surgeries in rats have been compiled (Waynforth, 1980), and most can be readily adapted to mice with minor modifications. With practice, they can be safely carried out and generally do not require specialized equipment.

Although it is generally desirable to avoid any invasive procedures that may impose a risk of losing a founder animal before deriving F1 offspring, there is still considerable merit to analyzing expression in the founder animals. Knowing ahead of time which founders to breed to derive an F1 generation can save a significant amount of time and resources that would otherwise go to analyzing offspring of nonexpressing lineages. Implicit in this discussion is the assumption that the offspring of a founder animal that does not express a transgene will themselves also be nonexpressors.

In general, analyses of transgene expression are carried out on material derived from Fl generation offspring of founder animals. Any of several methods may be used to determine which offspring have inherited the transgene, and these are described in detail in Chapter 18. However, before an accurate assessment of expression can be carried out, the pattern of transgene inheritance must be well defined. For this purpose, a Southern blot analysis will yield the most information with regard to transgene copy number and arrangement (Jacobsen et al., 2010). To complicate matters, integration at a single site or at multiple genomic sites may occur later than the one-cell stage of development, resulting in the generation of a genetically mosaic animal. It is important to be absolutely certain that expression is analyzed from a single integration site.

Once the number and arrangement of transgene integration sites have been well defined, one can consider whether it is appropriate to breed the transgene to homozygosity by establishing transgenic brother—sister breeding pairs. In general, the most efficient approach is to first determine which lines express the transgene, then breed a selected few to homozygosity if necessary, because the amount of time and resources involved in generating homozygotes is significant and success is not guaranteed. Although hemizygous mice are usually the starting point in expression analyses, there may be situations where having homozygotes is essential. For example, if the level of transgene expression is anticipated to be near the threshold for detection, then increasing the gene dosage by achieving homozygosity may be important. Alternatively, if a very small tissue is to be assayed and samples from many animals must be pooled, then deriving homozygotes early on may be useful.

II. Analyses of Steady-State Levels of Transgenic RNA Transcripts

Frequently, the initial analysis to evaluate transgene expression involves a broad survey of a variety of tissues at the mRNA level. The number of different samples taken from an individual mouse is limited primarily by the dissecting skill of the investigator but most often includes at least portions of all major organ systems: brain, salivary gland, thymus, heart, lung, liver, spleen, kidney, small intestine, large intestine, pancreas, testis or ovary, muscle, bone, lymph nodes, and skin.

A. Isolation of RNA from Tissues

Isolation of intact RNA, free of protein and DNA, is the critical first step in analyzing steady-state levels of transgene mRNA. The most effective method for inactivating endogenous nucleases and deproteinizing RNA is by rapid homogenization in the presence of guanidinium thiocyanate or guanidine hydrochloride (MacDonald et al., 1987). Rapid and thorough homogenization using a high-speed homogenizer is essential for isolating intact RNA, especially in the case of tissues with relatively high levels of endogenous RNase activity (e.g., pancreas and spleen). The use of specialized reagents such as TRIzol® Reagent (Ambion®, Life Technologies) and assorted tissue-specific RNA isolation kits has simplified and standardized methodologies. TRIzol® is a monophasic solution of phenol and guanidine isothiocyanate that is designed for the simultaneous isolation of RNA, DNA, and protein from a variety of biological samples of human, animal, plant, yeast, or bacterial origin.

B. Slot-Blot Hybridization Analysis

One approach to analyzing a large number of RNA samples in a single experiment is by slot-blot analysis (Green and Sambrook, 2012). In this method, denatured total or poly(A)-enriched RNA is applied to a solid support (typically a nitrocellulose membrane) with the aid of a vacuum applied through slots in a Plexiglas manifold. The bound RNA is then hybridized with a radiolabeled nucleic acid probe, and a signal is detected by autoradiography. A major disadvantage of the slot-blot method is that no information with regard to the size of the transgene-derived transcript or transcripts is obtained. This problem can be circumvented by performing a northern blot analysis, which is an equally effective method for measuring the steady-state level of mRNA (Selden, 1989). In this technique, total or poly(A)-enriched RNA is separated by electrophoresis in a denaturing agarose gel and transferred to a solid membrane support. The support is then hybridized to a radioactive probe to detect the transgenic mRNA (see also Chapter 18).

C. Northern Blot Hybridization Analysis

Although it is more laborious than the slot-blot method, northern analysis has the added benefit of allowing the investigator to determine the size of the transcript of interest and can reveal the presence of multiple transcripts that differ from one another by greater than 50–100 bases. If, however, multiple transcripts are detected, a northern analysis will usually not reveal the basis of the size difference (e.g., alternate transcription start sites, alternate splicing, or alternate use of poly(A) sites) unless a specific probe is available to address each possibility. Northern analysis has the advantages of high sample throughput and quantitation, and often allows one to measure the transgenic mRNA and its endogenous counterpart in the same sample. As an example of the results of a typical northern blot analysis, Figure 20.1A shows an autoradiogram of liver RNA extracted from three independent strains of mice carrying a metallothionein I (MT) major histocompatibility complex (MHC) D^d class I transgene. Liver RNA was extracted from mice that were either not treated or treated with heavy metals administered through the drinking water to induce transcription from the metallothionein promoter. The ratio of normal to induced transgenic mRNA levels, as quantitated by densitometry, is indicated. The probe used to detect the transgene was a ^{32}P end-labeled, 17-base oligonucleotide specific for the transgenic D^d class I gene. Figure 20.1B shows an autoradiogram of the same northern blot membrane that was stripped and rehybridized with an oligonucleotide probe to detect the endogenous K^b class I gene. The latter is not affected by treatment with Zn^{2+} and serves as an internal control.

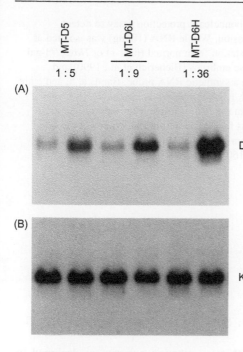

Figure 20.1 Northern blot analysis of RNA isolated from livers of transgenic mice. Poly(A)$^{+}$/RNA (5 μg) from three independent transgenic lines carrying an MT I-D^d transgene (see text) was separated on a denaturing formaldehyde agarose gel and transferred to nitrocellulose. Mice were maintained on either normal drinking water or water supplemented with zinc sulfate to induce MT promoter activity prior to RNA isolation. (A) Hybridization was carried out with a D^d-specific oligonucleotide probe. The ratios of uninduced to induced levels of expression are indicated for each line. (B) The same filter shown in (A) was stripped and hybridized again with a K^b-specific probe to detect expression of the endogenous K^b MHC class I gene.

D. Nuclease Protection Assay

Another approach to measuring the steady-state level of the transgenic mRNA is by the nuclease protection assay (Gilman, 1987; Green and Sambrook, 2012). In this procedure, a radiolabeled nucleic acid probe of defined length is incubated in solution with an RNA sample. The probe is typically in one of three forms: end-labeled DNA, uniformly labeled DNA, or uniformly labeled synthetic RNA. Following the hybridization step, a nuclease or combination of nucleases is used to digest away any part of the probe that does not hybridize to the specific mRNA. The sample is then analyzed by electrophoresis through a denaturing polyacrylamide gel, and the signal is detected autoradiographically.

Although the sample throughput in nuclease protection assays is somewhat lower than slot-blot or northern blot analyses owing to the number of manipulations involved, it offers several advantages over these methods. Because both the probe and the mRNA are in solution, hybridization follows second-order kinetics, compared to first-order kinetics for filter hybridization where the mRNA is immobilized. Rapid and quantitative hybridization can, therefore, be achieved at comparatively low probe concentrations. That translates into a higher signal-to-noise ratio for nuclease protection assays compared to northern or slot-blot analysis. Furthermore, specific information with regard to transcription initiation or termination can be obtained depending on what region of the mRNA is covered by the probe. This is a particularly important consideration in analyzing gene expression in transgenic mice. Because transgenes often consist of heterologous pieces of DNA juxtaposed by virtue of cloning rather than by evolutionary selection, cryptic initiation, splice,

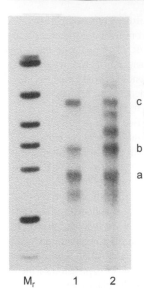

Figure 20.2 Ribonuclease protection assay to detect transgene expression. Whole RNA (10 mg) was isolated at 12.5 days of gestation from normal (lane 1) or *Hox3.1*/β-gal transgenic mouse embryos (Bieberich et al., 1990) and hybridized with a synthetic RNA probe. Bands a, b, and c are common to both RNA samples, whereas bands a′, b′, and c′ are found only in the RNA sample from the transgenic embryo (see text for explanation). The lengths of the size markers are, from top to bottom (in bases), 242, 238, 217, 201, 190, 180, 160, and 147.

or termination sites may come into play. Some of these may be tissue specific, whereas others may appear more globally.

Figure 20.2 shows the results of an RNase protection experiment designed to detect expression of a transgene consisting of a mouse homeobox (*Hox3.1*) promoter driving the β-galactosidase (β-*gal*) reporter gene. The probe covers the 5′ end of the *Hox3.1* transcription unit and detects three major transcriptional start sites evidenced by the protected fragments labeled a, b, and c. Lane 1 (Figure 20.2) contains RNA from a normal midgestation mouse embryo. The probe also contains 25 bases near its 3′ end that are present in the transgenic *Hox3.1*/β-*gal* transcript but not present in the endogenous *Hox3.1* mRNA. When RNA from a transgenic embryo (Figure 20.2, lane 2) is hybridized to the same probe, six protected fragments are observed. The three novel protected fragments, labeled a′, b′, and c′, are each 25 bases longer than their endogenous counterparts (a, b, and c) and indicate the presence of the transgene-derived mRNA species. Because the same probe is used to detect both sets of transcripts, one can directly compare the steady-state level of endogenous versus transgenic mRNA in the same sample. This example demonstrates that the transgene-derived mRNAs are initiated at the same three major start sites as the endogenous *Hox3.1* transcripts. However, the level of transgenic mRNA initiated at start site c is lower than that of the endogenous transcript, as evidenced by the relatively low signal observed for the c′ protected fragment.

E. Reverse Transcriptase-Polymerase Chain Reaction Analysis

The most sensitive method for detecting transgene expression is by the reverse transcriptase-polymerase chain reaction (RT-PCR) method (Shibata et al., 2002; Kawasaki, 1990). To carry out this assay, a specific "antisense" primer is allowed to anneal in solution to an RNA sample, and reverse transcriptase enzyme is used to convert a region of the mRNA of interest to cDNA. A second "sense" primer is then added to the sample along with a thermostable DNA polymerase, and a region of the newly formed cDNA is amplified as in a typical PCR. In most cases, the products of an RT-PCR amplification can be examined by agarose gel electrophoresis with ethidium bromide staining.

Many PCR amplifications result in the formation of multiple products, including the "correct" product. Sorting out the desired product from the background components by electrophoresis and ethidium bromide staining alone can be difficult. It is often desirable to increase the sensitivity of product detection by using a radiolabeled oligonucleotide probe. The probe can be used in a Southern blot analysis or by performing a solution hybridization experiment directly with the PCR product before analysis on a nondenaturing polyacrylamide gel (Yoshioka et al., 1991). Not only is sensitivity increased, but a confirmation of specificity can also be achieved by using a probe that does not overlap with the PCR primers.

The merits of using RT-PCR to study transgene expression are manifold. It is indisputably the most sensitive method now available to detect the presence of a specific transcript; a single mRNA molecule can be converted to cDNA and amplified to a detectable level. The procedure is technically straightforward and can accommodate many samples in a single experiment. Importantly, for transgenic work, it can be used to detect very small structural differences between endogenous and transgenic transcripts. In addition, RNA isolated from formalin-fixed and paraffin-embedded tissues can be analyzed by this method, allowing for retrospective studies of transgene expression (Rupp and Locker, 1988).

Along with those advantages come significant pitfalls. In most cases, only a small region of the mRNA is amplified, so no information on overall size of the transcript is obtained. Also, amplification of genomic DNA that virtually always copurifies to some degree with RNA can give rise to false-positive signals. This is of particular concern when analyzing tissues from a transgenic mouse with many copies of a transgene and hence many DNA-based target sites for the PCR primers. In some cases, it is possible to derive PCR primers that span an intron, thereby allowing one to distinguish between genomic DNA and cDNA-based products on the basis of size. In other instances, where transcripts are derived from intronless transgenes, more rigorous steps to eliminate trace amounts of DNA from the RNA sample may be required. For example, one can include a step to degrade the DNA enzymatically, or one can perform differential precipitation using lithium chloride to separate DNA from RNA (Cathala et al., 1983). An additional precaution one can take in the PCR experiment is to include a "no reverse transcriptase" control for each RNA sample by simply omitting the RT step. The great care that must be taken to avoid contamination of RNA samples with exogenous DNA, such as products of previous amplification reactions, also encumbers this technique.

Another significant shortfall with the RT-PCR method is the fact that it is difficult to use as a quantitative technique for measuring mRNA levels, due to the exponential nature of the amplification process. The amount of product (N) in a

PCR amplification is dependent on the number of input molecules (N_0), the number of rounds of amplification (n), and the efficiency (eff) of the reaction:

$$N = N_0(1 + \text{eff})^n$$

Clearly, small differences in the efficiency of each reaction brought about by, for example, differing levels of RNA purity can have a significant impact on the number of molecules produced. A 10% difference in the efficiency of a reaction between two samples with the same number of input molecules can lead to a five-fold difference in the amount of product produced after 30 rounds of amplification. Despite these difficulties, several semiquantitative and quantitative approaches for measuring mRNA levels using RT-PCR have been devised (Wang and Mark, 1990; Gilliland et al., 1990; Chelly et al., 1988; reviewed by Csakos, 2006; O'Connor and Glynn, 2010).

An example of a semiquantitative analysis of gene expression using RT-PCR is shown in Figure 20.3. In this experiment, RNA samples from various tissues of a transgenic mouse carrying a transgene in which the myelin basic protein (MBP) promoter derives the expression of the MHC K^b class I gene were analyzed

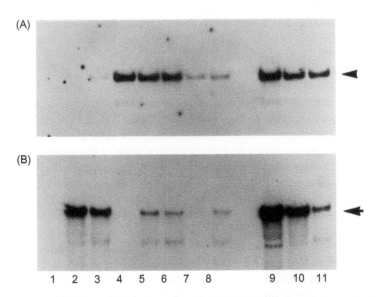

Figure 20.3 RT-PCR analysis of expression of an MBP−K^b transgene. (A) Whole RNA (100 ng) was analyzed by RT-PCR to detect expression of the endogenous K^b class I MHC gene (arrowhead) (Yoshioka et al., 1991). (B) A different aliquot of the same RNA samples was analyzed using primers to detect expression of the MBP−K^b transgene. Liver RNA was titrated in lane 9 (300 ng), lane 10 (100 ng), and lane 11 (33 ng), to demonstrate that the level of signal was dependent on the RNA concentration. Lanes 1, no RNA; lanes 2, brain; lanes 3, spinal cord; lanes 4, thymus; lanes 5, lung; lanes 6, liver; lanes 7, kidney; lanes 8, testis.

(Yoshioka et al., 1991). The analysis of transgene expression in the mice was complicated by the fact that the endogenous K^b class I gene is expressed in virtually all tissues. The transgenic mRNA differed only slightly from the endogenous mRNA; the first 12 nucleotides of the endogenous K^b transcript were replaced by 36 nucleotides immediately downstream of the MBP transcription start site. The results of RT-PCR with primers to detect either the endogenous K^b mRNA (Figure 20.3A) or the transgenic MBP–K^b mRNA (Figure 20.3B) are shown. After amplification, the reaction products were denatured and hybridized in solution to a ^{32}P end-labeled oligonucleotide probe common to both of the predicted PCR products. The hybridized products were then separated from free probe by electrophoresis in a nondenaturing polyacrylamide gel. The solution hybridization step serves three purposes: to increase the sensitivity of the assay by several orders of magnitude, to confirm the specificity of the amplified product, and to allow for quantitation across a broad range of signals. For each pair of primers, a titration of input RNA was carried out simultaneously (Figure 20.3, lanes 9–11) to demonstrate that the amount of product produced was dependent on the number of input target mRNA molecules. A marked difference in the pattern of expression of the endogenous K^b gene and the transgenic MBP–K^b gene is apparent. Most notably, expression of the endogenous gene is barely detectable in the brain and spinal cord, where the transgenic mRNA is abundant. In contrast, the endogenous mRNA is abundant in thymus, lung, liver, and kidney, whereas the transgenic mRNA was either low or below detection.

F. In Situ *Hybridization*

All of the approaches to determine transgene expression described thus far suffer from a common shortcoming: they all depend on homogenization of tissues to extract mRNA. As a result, no information with regard to cell type specificity of expression can be obtained without employing painstaking cell separation techniques prior to grinding up the sample. The signal that is observed is necessarily an average over the entire population of cells in the homogenate, and no information as to the number of cells expressing the transgene or their relative levels of expression is gained. To achieve a higher level of resolution of gene expression, one must turn to *in situ* methods.

In situ hybridization is an effective method for analyzing transgene expression at the cellular level (Awgulewitsch and Atset, 1991; Bandtlow et al., 1987); however, because of the technical difficulty of this technique, its use is increasingly rarer. In this technique, microtome sections of either frozen or paraffin-embedded tissues are fixed to a solid support, typically a glass microscope slide. Following a series of treatments to fix, dehydrate, and permeabilize the tissue sections, a radiolabeled nucleic acid probe is allowed to hybridize to the mRNA *in situ*. After washing to remove unbound probe, the slide is either exposed directly to X-ray film, exposed to a photographic emulsion-coated cover slip, or dipped directly into emulsion to detect the signal. Sulfur-35 is the radionuclide of choice for *in situ* analyses because its relatively low-energy β particle provides adequate resolution with reasonable exposure times. Recent advances in nonisotopic *in situ* hybridization have increased the achievable sensitivity to that approaching radioactive detection methods. Notable in this group is the RNAscope® (Advanced Cell Diagnostics) methodology, which claims single molecule sensitivity, far exceeding previous nonisotopic systems'. This novel multiplex (simultaneous detection of multiple targets) nucleic acid *in situ* hybridization technology uses a patented probe design and signal amplification methodology, that is, the fluorescent (or enzymatic) signal is multiplied manyfold by including a step for signal addition after hybridization. Probes are available for colormetric or fluorescent detection, using bright field or fluorescence microscopy. By using short, end-to-end nucleic acid hybridization probes, that are only recognized in pairs by the signal amplification reagent, this technique is able to distinguish genes with high homology while avoiding signal addition to unbound singlet nucleic acid probes.

The high degree of resolution of *in situ* hybridization comes at a price; the isotopic technique is time consuming and technically demanding, requires specialized equipment, and cannot accommodate a large number of samples in a single experiment. Although it is certainly the most difficult of the methods described thus far, it yields the most information with regard to localization of gene expression. Subsets of transgene-expressing cells within an organ can readily be identified. In fact, if fixation and hybridization conditions are optimal, the level of resolution can approach the single cell. The nonisotopic detection method claims potential single mRNA molecule resolution, but no transgene expression study has yet been published using this system. The "sensitivity" of *in situ* hybridization depends somewhat on the pattern of expression of the gene under study. For example, if the gene of interest is expressed in all cells of an organ or embryo and its mRNA accumulates to a relatively low steady-state level, then any filter or solution hybridization method of detection would likely yield a relatively strong signal. On the other hand, an *in situ* hybridization analysis would show a weak and diffuse signal everywhere that would be difficult to interpret. However, if the gene of interest is expressed in only a small subset of cells within a tissue or embryo, but its mRNA is relatively abundant within that small group of cells, then the *in situ* analysis would give an unequivocally positive signal on a few of the cells in the section, whereas other detection methods (with the exception of RT-PCR) might give a negative result. The combined use of *in situ* hybridization for transcript localization along with protein immunolocalization is a powerful method for gene expression analysis (Karunakaran et al., 2013).

III.   Analyses of Steady-State Levels of Transgenic Protein Products

Although the analysis of RNA is the usual first step for determining transgene expression, it is by no means given in all situations. For genes that are expressed at very low levels and whose gene products have high enzymatic activities that can be easily detected without significant background interference, analysis of the protein product may prove more sensitive and direct. Furthermore, detection of RNA does not necessarily imply that a protein product will be made and that it will be active. Efficient translation of certain mRNAs, such as the ferritin transcript (Koeller et al., 1989), requires specific *trans*-acting factors that may be restricted in tissue distribution. Alternatively, some gene products may require tissue-specific posttranslational modifications, such as proteolytic cleavage or glycosylation, for functionality. In each of these cases, detection of transgenic mRNA in a particular tissue does not imply the expression of a functional protein product. Equally important is the recognition that the site of synthesis of the product of a transgene need not be its site of action. The protein may be secreted into the circulation and may have one or more target tissues. Given these circumstances, analysis of the transgene protein may prove more informative and rewarding. The choice of whether to analyze the expression of the mRNA, the protein product, or both would depend on the specific transgene and its expected function.

Protein analyses frequently make use of antibodies with exquisite specificity. Each of the antibody-based protein detection methods has its strengths and weaknesses, and the technique of choice frequently depends on the characteristics of the particular antibody, the properties of the antigen, and the type of tissues to be analyzed.

A. Western Immunoblot Analysis of Tissue Extracts

The western immunoblot analysis approach involves separating total protein in different tissue extracts by sodium dodecyl sulfate (SDS)-polyacrylamide gel electrophoresis (PAGE) analysis, transferring the protein from the polyacrylamide gel to a membrane (typically nitrocellulose) with the aid of an electric field, and detecting the protein on the membrane with a specific antibody (Towbin et al., 1979). Typically, nonisotopic methods are used to detect the immune complex, such as subsequent incubation with a secondary peroxidase-conjugated antibody and subsequent reaction with any of several peroxidase substrates. The most commonly used method is the enhanced chemiluminescence (ECL) technique, in which a horseradish peroxidase (HRP)-conjugated secondary antibody catalyzes the conversion of the ECL substrate into a sensitized reagent. Upon oxidation by hydrogen peroxide, the sensitized ECL substrate emits light, exposing an X-ray film. ECL western blotting can also be visualized on an apparatus designed for chemiluminescence documentation and quantitation. ECL allows detection down to femtomole quantities, well below the detection limit for most assay systems. Western immunoblot analysis allows relative quantitation of the product of the transgene in virtually all tissues that can be dissected from an animal (Figure 20.4). Because it involves the analysis of tissue extracts, it does not provide information as to whether individual cells within the tissue express similar or differing levels of the protein, or whether only a subset of the cells within the tissue express the transgene. Furthermore, not all antibodies are suitable for this purpose. Crude antisera may give a high nonspecific background, and monoclonal antibodies that detect SDS-sensitive epitopes will not work. In general, the limitation of the method is dependent on how much of the protein extract can be loaded onto a gel for the analysis.

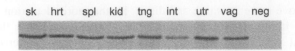

Figure 20.4 Western blotting of carcinoembryonic antigen-related cellular adhesion molecule 1 (CEACAM1) transgene expression in transgenic mice by western blotting. Total proteins extracted from tissues of CEACAM1 transgenic mice line 53 screened with anti-hCEACAM1 mAb. Total protein extract from mixed tissues of nontransgenic mice was used as negative control. sk = skin, hrt = heart, spl = spleen kid = kidney, tng = tongue, int = intestine, utr = uterus, vag = vagina, neg = negative control.
Source: From Li et al. (2011), used by permission.

B. Immunolabeling of Tissue Sections

Immunohistochemical staining of tissue sections allows single-cell analysis of transgenic protein products (Beltz and Burd, 1989). Frozen, fixed and frozen, or fixed and paraffin-embedded tissues, depending on the particular antibody to be used, are sectioned in a microtome and mounted onto glass microscope slides. Following a series of treatments to fix, dehydrate, and permeabilize the tissue sections, an appropriate primary antibody is added and allowed to bind to the target protein *in situ*. The immune complex is then visualized by a secondary antibody, commonly peroxidase-conjugated, and a suitable substrate. After immunochemical staining, the sections may be further treated with a light counterstain, such as hematoxylin and/or eosin, to facilitate histological identification of tissue architecture. An example of the use of this method is shown in Figure 20.5, where a fixed and paraffin-embedded section from the liver of a transgenic mouse carrying the *HBx* gene of the hepatitis B virus (Kim et al., 1991) has been immunostained with an antipeptide antibody specific to the HBx protein. The immunostaining is not detected in all cells in the liver, but rather is restricted to groups of cells surrounding the blood vessels. The staining is predominantly cytoplasmic. The counterstain by hematoxylin allows identification of the nuclei of both HBx-positive and HBx-negative cells. This procedure permits evaluation of the percentage of cells within the tissue expressing the transgenic protein product as well as the relative level of expression among the cells. Admittedly, these estimates are at best subjective and crude. With appropriate morphological evaluation, such as the use of different cytochemical stains or histochemical markers, one may also define the specific cell type involved. Success in this procedure is predicated on the availability of an adequate antibody with minimal background cross-reactivity. Antibodies that can be used on fixed and paraffin-embedded sections usually are more revealing than those that work only on frozen sections. At times, antibodies that detect different epitopes on the same protein may stain different subsets of cells within the same tissue section. Such information can only come from procedures that allow single-cell analysis.

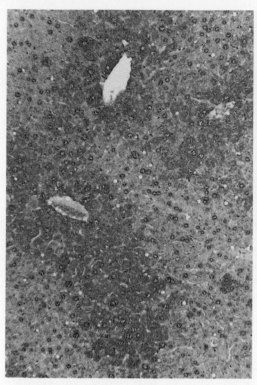

Figure 20.5 Immunohistochemical staining of a liver biopsy from an *HBx* transgenic mouse. A 10-μm fixed and paraffin-embedded section of liver from a 6-month-old mouse with early focal lesions of altered hepatocytes was deparaffinized and reacted with a rabbit antiserum against a C-terminal peptide (amino acids 139–154) of the HBx protein. The immune reaction was then visualized by the avidin–biotin complex (ABC) method, which involved sequential incubations with biotinylated goat anti-rabbit immunoglobulin G and a mixture of avidin and biotinylated HRP, followed by a further incubation with the peroxidase substrate. The slide was subsequently counterstained with hematoxylin.

Alternatively, fluorescence-labeled secondary antibodies are commonly used to label proteins in frozen tissue sections. With careful planning, a variety of proteins can be labeled within the same tissue section using primary antibodies from different species and secondary antibodies conjugated to different fluorescent molecules.

C.   Fluorescence Labeling of Individual Cells

For circulating cells and those that can be derived from tissues as single cells, direct or indirect immunostaining of live or fixed cells, either in suspension or as a monolayer, can be performed. Using fluorescent antibodies, including those that are conjugated with fluorescein or rhodamine, single cells may be treated and analyzed either under a fluorescence microscope or by fluorescence-activated cell sorting (FACS) and flow cytometry (McCoy, 2002). An example of the use of this method for the analysis of transgenic mice carrying the MT promoter driving the MHC D^d class I gene is shown in Figures 20.6 and 20.7. When live epithelial cells, removed by hyaluronidase treatment of the intestine from transgenic mice that were either untreated (Figure 20.6A) or zinc treated (Figure 20.6B), were incubated with anti-D^d antibody and subsequently incubated with a phycoerythrin-conjugated second antibody, specific cell surface immunofluorescence was detected using an

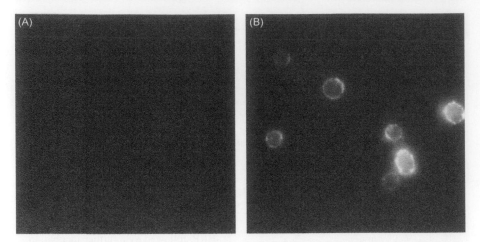

Figure 20.6 Fluorescence microscopic analysis of intestinal epithelial cells from the MT-D^d transgenic mice. Live cells obtained from the intestines of transgenic mice that were either untreated (A) or zinc treated (B) were stained with an anti-D^d monoclonal antibody and a rhodamine−phycoerythrin-conjugated second antibody. The stained cells were photographed under an epifluorescence microscope.

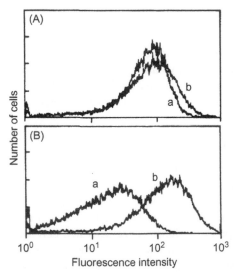

Figure 20.7 Flow microfluorometric analysis of intestinal epithelial cells from the MT-D^d transgenic mice. Intestinal epithelial cells stained with either an anti-K^bL^b or an anti-D^d monoclonal antibody, followed by a R-phycoerythrin-conjugated second antibody, were analyzed in an EPICS 752 flow cytometer. (A) Cells from the untreated (trace a) and zinc-treated (trace b) mice showed similar levels of fluorescence for the endogenous class I antigen when stained with the anti-K^bL^b monoclonal antibody. (B) Cells from the zinc-treated mouse (trace b) showed a much higher level of fluorescence than those from the untreated mouse (trace a) when stained for the transgenic class I antigen using the anti-D^d monoclonal antibody.

epifluorescence microscope only in cells from the treated mouse. The extent of fluorescence was variable among the individual cells, with some cells staining significantly more intensely than others. When the same cells after immunostaining were subjected to FACS analysis, a fluorescence intensity profile for each cell population was obtained (Figure 20.7). In this study, one can deduce not only the mean fluorescence of each population but also the heterogeneity among the individual cells within a population.

Although such methods allow direct quantitation and comparison, one has to assume that the metabolic state of the isolated cells was not too different from that within the tissues of an animal, and that the observed gene expression or the lack thereof was not an artifact resulting from an altered state of metabolism of the isolated cells. At times, this may be difficult to control. The main advantage of this procedure, however, lies in the fact that one can select for subsets of cells, such as those with a specific cell surface marker, for analysis. This greatly facilitates identification of cells that express the transgene.

In situations where antibodies are not available, one has to resort to specific properties of the gene product of interest to facilitate its detection. For example, knowledge of its isoelectric point, state of phosphorylation, or ability to bind specific cofactors may allow the detection of the gene product. In the case of enzymes, substrates can be designed that will permit direct detection and quantitation.

D. Nonimmunologic Reporter Gene Analysis

The transgenic mouse is the ultimate proving ground for measuring the regulatory potential of DNA elements that act in *cis* to direct gene expression. Potential *cis*-regulatory sequences are used to direct expression of a so-called reporter gene. Any transcription unit can be used as a reporter provided its product, transcriptional or translational, is innocuous and can be detected and distinguished from any endogenous counterpart. Detection of reporter gene expression can be based on the presence of unique target sequences for a probe within the transcript, altered transcript length, or the presence of unique epitopes on a protein product. However, the most commonly used reporter genes code for a fluorescent protein product such as green fluorescent protein or a functional enzyme which would confer the presence of a novel enzyme activity. The use of fluorescent proteins or enzymes as reporters is particularly attractive, because they provide an opportunity to study *cis*-acting elements that are either weakly active or are restricted to a very small subset of cells.

The luciferase gene isolated from the common North American firefly *Photinas pyralis* has also been employed as a reporter in transgenic mice (DiLella et al., 1988). The luciferase enzyme catalyzes the production of light in the presence of adenosine triphosphate (ATP), molecular oxygen, and a heterocyclic carboxylic acid substrate termed luciferin (de Wet et al., 1987). The emission of light can be detected with a luminometer. As a reporter gene, luciferase shares with chloramphenicol acetyltransferase (CAT) the advantages of sensitivity, quantitation, and low background. Luciferase has the added advantage of being detectable using nonisotopic methods.

Bioluminescence imaging utilizes light emission from one of several luminescent proteins incorporated into the DNA of transgenic mice. Bioluminescence is a very powerful, albeit low-resolution technique that can be very useful in transgenic animals to define the specificity of putative *cis*-regulatory elements (reviewed by de Almeida et al., 2011). A variety of *in vivo* imaging genetic constructs are available for assaying promoter/regulatory sequence function noninvasively in small animals such as laboratory mice. The most commonly used sources of luminescent proteins are firefly luciferase and other proteins derived from various marine organisms such as the sea pansy and bacteria such as *Photorhabdus luminescens*. For firefly luciferase, injection of the substrate molecule D-luciferin allows detection in tissues expressing the reporter enzyme and can be visualized *in vivo* in anesthetized whole animals using an ultrasensitive CCD camera to image bioluminescence externally. Bioluminescence is a convenient, noninvasive method for obtaining expression data in live animals and in whole organs (Zelko et al., 2013); however, the spatial resolution offered is considered to be poor in comparison to techniques such as *in situ* hybridization or various immunolocalization techniques.

The most versatile enzyme reporter for high-resolution screening is *Escherichia coli* β-galactosidase. This 100-kDa protein, encoded by the *lacZ* gene, catalyzes the cleavage of lactose into glucose and galactose. Importantly, several chromogenic and fluorogenic substrates have been developed to detect β-galactosidase activity. The two most widely used are the chromogenic molecule 5-bromo-4-chloro-3-indolyl-β-D-galactopyranoside (X-gal) and the fluorogenic substrate fluorescein di-β-D-galactopyranoside (FDG). The availability of several different substrates allows for considerable flexibility in terms of detecting β-galactosidase activity.

The use of FDG as a substrate in combination with flow cytometric analysis provides the most sensitive assay for β-galactosidase activity (Nolan et al., 1988). The FDG assay can also be carried out on live cells, providing the opportunity to enrich for cells expressing the reporter gene by FACS. Although this technique, called FACS−FDG, is easily applied to hematopoietic cells, its general utility is limited to tissues that can be readily dissociated into single-cell populations suitable for flow cytometric analyses.

X-gal is the only β-galactosidase substrate that has been routinely used to detect transgene activity *in situ*. When X-gal is cleaved by the β-galactosidase enzyme, a striking blue halogenated indolyl derivative is formed, pinpointing the site of reporter gene expression. X-gal is particularly useful in that it can be used to detect β-galactosidase activity in whole mounts of embryos as well as whole adult organs or razor blade sections of adult organs with minimal tissue preparation time (Sanes et al., 1986). Essentially, tissues or embryos are dissected, fixed, and incubated directly in the staining cocktail. In the presence of detergents, X-gal can readily penetrate through several millimeters of most tissues. A notable exception is skin, which can act as an effective barrier to penetration. After incubation with X-gal, the whole mounts can be processed by standard paraffin embedding techniques and counterstained to facilitate histological identification of expressing cells. However, one must beware that some counterstains (e.g., hematoxylin and Giemsa) can interfere with detection of weak X-gal staining because these stains also yield a bluish color. It may be advisable then, in some cases, to counterstain only alternate sections.

Some mouse tissues (e.g., salivary gland, kidney, and intestine) have considerable amounts of endogenous lysosomal β-galactosidase activity that can interfere with interpretation of results of reporter gene expression studies. To minimize background activity, it is important to perform the FDG and X-gal assays at a

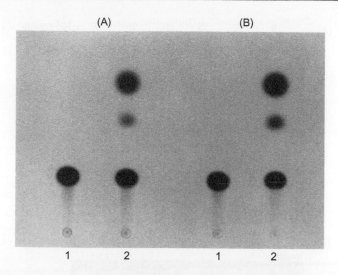

Figure 20.8 CAT assay to detect reporter transgene expression in two mice from independent lines, A and B. Protein extracted from the tails of normal (lanes 1) or transgenic (lanes 2) mice carrying the human T-lymphotropic virus (HTLV)-1 long terminal repeat (LTR) driving expression of the CAT reporter gene was analyzed for activity by thin-layer chromatography.

neutral pH to favor the activity of the bacterial β-galactosidase. Minimizing staining time is also critical for keeping background low (Figures 20.8 and 20.9).

As an alternative to whole-mount staining to detect β-galactosidase activity, tissues can be cryosectioned prior to incubation with X-gal. Although it is more time consuming, this approach alleviates any potential problems of substrate penetration. Tissues can also be fixed and cryoprotected to preserve morphology before sectioning without inhibiting β-galactosidase activity (Mucke et al., 1991).

The advantages of analyzing reporter gene activity *in situ* are obvious. One can rapidly determine not only which tissues show reporter gene activity but also whether all, or only a subpopulation of cells within a given tissue, are capable of supporting transgene expression. An X-gal staining analysis of three 10.5-day-gestation mouse embryos that carry a homeobox/*lacZ* transgene is shown in Figure 20.10. Figure 20.10A shows a darkfield micrograph of a whole mount after fixation and staining with X-gal. β-galactosidase activity is evident in a broad posterior region of the embryo, as well as in a small patch of skin overlying the hindbrain. Figure 20.10B shows a cross section through the neural tube of a 10.5-day-gestation embryo. Strong β-galactosidase activity is evident in two groups of cells that flank the neural tube. Weaker activity is detected in a central band of neuronal precursor cells within the neural tube. Figure 20.10C shows a parasagittal section through a third embryo. In this section, it is possible to distinguish single blue cells on a large background of cells that show no β-galactosidase activity. The photomicrographs in Figure 20.10B and C were produced using differential interference contrast optics without counterstaining on paraffin-embedded sections of embryos that had been stained with X-gal as whole mounts.

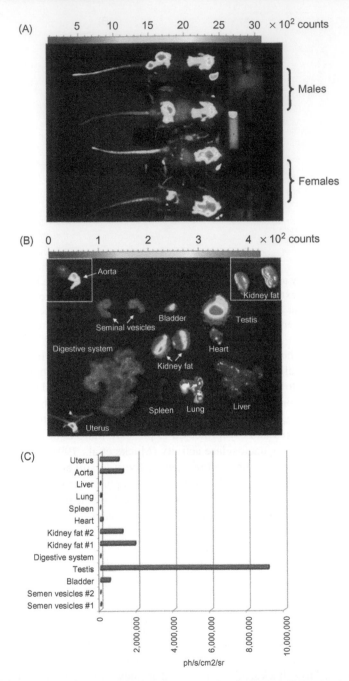

Figure 20.9 Analysis of the reporter activity distribution in tissues of transgenic mice. (A) The ventral view of two representative mice are depicted for each gender. The color scale at the top indicates the signal intensity at the surface of the animals in photons per second. (B) The representative bioluminescent imaging of promoter-directed luciferase

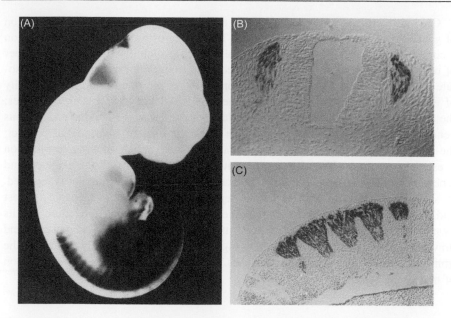

Figure 20.10 Analysis of β-galactosidase activity in transgenic embryos. Transgenic embryos carrying the *lacZ* gene under the control of a homeobox gene promoter were analyzed for β-galactosidase activity using the X-gal histochemical assay. (A) Whole-mount embryo shows a broad posterior domain of dark staining indicative of β-galactosidase activity. (B) Cross section through the neural tube of a 10.5-day β-galactosidase-expressing embryo reveals activity in the developing spinal ganglia flanking the neural tube. (C) Parasagittal section through a 10.5-day transgenic embryo also reveals strong β-galactosidase activity in spinal ganglia. Dorsal is toward the top in (B) and (C).

IV. Conclusion

Once a pattern of transgene expression has been determined by any of the techniques described here, it is critical to confirm the pattern in a second independent transgenic strain, given that the site of integration may strongly influence gene expression. In cases where two strains of mice are examined and their expression patterns do not corroborate one another, then more strains must be analyzed to reveal a consistent pattern.

◄ activity in the isolated mouse organs. Images were collected using photon imaging system (Biospace, France). The acquisition time to allow for sufficient collection of photons to generate color map overlay was 10 min. (C) The quantitative analysis of luciferase-driven photon emission detected in the tissues of transgenic mice. The collected luminescence emission signal was quantified as photons per second per centimeter squared per steridian (ph/s/cm^2/sr) (Zelko et al., 2013).

Obviously, no one particular method is any better than another for the evaluation of gene expression in transgenic animals. The choice of procedures will depend on a combination of factors, including the following: availability of reagents; number of tissues to be analyzed; sites of suspected gene expression; requirement for absolute quantitation of the level of expression; whether information on heterogeneity of expression among cells is required; the need to identify expressing cell types; the ability to distinguish between the products of the transgene and the endogenous gene; the need to confirm the size of the transcriptional or translational product; the ability to determine the transcriptional start sites, splice sites, or termination sites; the need to confirm the posttranslational modifications of the product; and whether information on the subcellular localization of the gene product is required. In short, the method of choice will be dependent on the specific gene of interest and the type of information to be obtained. The fact that each procedure has its own potentials and limitations is axiomatic. By considering the advantages and drawbacks of each method before deriving animals and by carefully designing the transgene to facilitate expression analyses, one can expedite the timely completion of many transgenic mouse experiments.

Acknowledgment

We are grateful to Dr. Gilbert Jay and colleagues for developing this chapter in the first two editions of *Transgenic Animal Technology* and allowing us to revise and update this edition.

References

Awgulewitsch, A., Atset, M.F., 1991. Detection of specific RNA sequences in tissue sections by *in situ* hybridization. In: Chao, L., Karam, J., Warr, G. (Eds.), Methods in Nucleic Acids Research. CRC Press, Boca Raton, FL, pp. 359–375.

Bandtlow, C.E., Heumann, R., Schwab, M.E., Thoenen, H., 1987. Cellular localization of nerve growth factor synthesis by *in situ* hybridization. EMBO J. 6, 891–899.

Beltz, B.S., Burd, G.D., 1989. Immunocytochemical Techniques: Principles and Practice. Blackwell, Cambridge, MA.

Bieberich, C.J., Utset, M.F., Awgulewitsch, A., Ruddle, F.H., 1990. Evidence for positive and negative regulation of the *Hox 3.1* gene. Proc. Natl. Acad. Sci. USA. 87, 8462–8466.

Cathala, G., Savouret, J.F., Mandez, B., West, B.L., Karin, M., Martial, J.A., et al., 1983. A method for isolation of intact, translationally active ribonucleic acid. DNA. 2, 329–335.

Chelly, J., Kaplan, J.C., Maire, P., Gautron, S., Kahn, A., 1988. Transcription of the dystrophin gene in human muscle and non-muscle tissues. Nature. 333, 858–860.

Chomczynski, P., Sacchi, N., 1987. Single-step method of RNA isolation by acid guanidinium thiocyanate-phenol-chloroform extraction. Anal. Biochem. 162, 156–159.

Csako, G., 2006. Present and future of rapid and/or high-throughput methods for nucleic acid testing. Clin. Chim. Acta. 363, 6–31.

de Almeida, P.E., van Rappard, J.R., Wu, J.C., 2011. *In vivo* bioluminescence for tracking cell fate and function. Am. J. Physiol. Heart Circ. Physiol. 301, H663—H671.

de Wet, J.R., Wood, K.V., DeLuca, M., Helinski, D.R., Subramani, S., 1987. Firefly luciferase gene: structure and expression in mammalian cells. Mol. Cell Biol. 7, 725—737.

DiLella, A.G., Hope, D.A., Chen, H., Trumbauer, M., Schwartz, R.J., Smith, R.G., 1988. Utility of firefly luciferase as a reporter gene for promoter activity in transgenic mice. Nucleic Acids Res. 16, 4159.

Gilliland, G., Perrin, S., Bunn, H.F., 1990. Competitive PCR for quantitation of mRNA. In: Innis, M.A., Gelfand, D.H., Sninsky, J.J., White, T.J. (Eds.), PCR Protocols: A Guide to Methods and Applications. Academic Press, San Diego, CA, pp. 60—69.

Gilman, M., 1987. Ribonuclease protection assay. In: Ausubel, F.M., Brent, R., Kingston, R.E., Moore, D.D., Seidman, J.G., Smith, J.A., et al., Current Protocols in Molecular Biology. Wiley, New York, NY, pp. 4.7.1—4.7.8.

Green, M.R., Sambrook, J., 2012. Molecular Cloning: A Laboratory Manual, fourth ed. Cold Spring Harbor Laboratory Press, Cold Spring Harbor, NY.

Jacobsen, J.C., Bawden, C.S., Rudiger, S.R., McLaughlan, C.J., Reid, S.J., Waldvogel, H.J., et al., 2010. An ovine transgenic Huntington's disease model. Hum. Mol. Genet. 19, 1873—1882.

Karunakaran, D.K., Congdon, S., Guerrette, T., Banday, A.R., Lemoine, C., Chhaya, N., et al., 2013. The expression analysis of Sfrs10 and Celf4 during mouse retinal development. Gene Expr. Patterns. 13, 425—436.

Kawasaki, E.S., 1990. Amplification of RNA. In: Innis, M.A., Gelfand, D.H., Sninsky, J.J., White, T.J. (Eds.), PCR Protocols: A Guide to Methods and Applications. Academic Press, San Diego, CA, pp. 21—27.

Kim, C.-M., Koike, K., Saito, I., Miyamura, T., Jay, G., 1991. HBx gene of hepatitis B virus induces liver cancer in transgenic mice. Nature. 351, 317—320.

Koeller, D.M., Casey, J.L., Hentze, M.W., Gerhardt, E.M., Chan, L.N.L., Klausner, R.D., et al., 1989. A cytosolic protein binds to structural elements within the iron regulatory region of the transferrin receptor mRNA. Proc. Natl. Acad. Sci. USA. 86, 3574—3578.

Kwon, S., 2013. Single-molecule fluorescence *in situ* hybridization: quantitative imaging of single RNA molecules. BMB Rep. 46, 65—72.

Li, G., Jiao, H., Yan, H., Wang, J., Wang, X., Ji, M., 2011. Establishment of a human CEACAM1 transgenic mouse model for the study of gonococcal infections. J. Microbiol. Methods. 87 (3), 350—354 (Western blot).

MacDonald, R.J., Swift, G.H., Przybyla, A.E., Chirgwin, J.M., 1987. Isolation of RNA using guanidinium salts. In: Berger, S.L., Kimmel, A.R. (Eds.), Methods in Enzymology, vol. 152. Academic Press, San Diego, CA, pp. 219—227.

McCoy Jr., J.P., 2002. Basic principles of flow cytometry. Hematol. Oncol. Clin. North Am. 2, 229—243.

Mucke, L., Oldstone, M.B.A., Morris, J.C., Nirenberg, M.I., 1991. Rapid activation of astrocyte-specific expression of GFAP-*lacZ* transgene by focal injury. New Biol. 3, 465—474.

Nolan, G.P., Fiering, S., Nicolas, J.-F., Herzenberg, L.A., 1988. Fluorescence-activated cell analysis and sorting of viable mammalian cells based on β-D-galactosidase activity after transduction of *Escherichia coli lacZ*. Proc. Natl. Acad. Sci. USA. 85, 2603—2607.

O'Connor, L., Glynn, B., 2010. Recent advances in the development of nucleic acid diagnostics. Expert Rev. Med. Devices. 7, 529—539.

Rupp, G.M., Locker, J., 1988. Purification and analysis of RNA from paraffin-embedded tissues. Biotechniques. 6, 56—60.

Sanes, J.R., Rubenstein, J.L.R., Nicolas, J.-F., 1986. Use of a recombinant retrovirus to study post-implantation cell lineage in mouse embryos. EMBO J. 5, 3133−3142.

Selden, R.F., 1989. Analysis of RNA by Northern hybridization. In: Ausubel, F.M., Brent, R., Kinston, R.E., Moore, D.D., Seidman, J.G., Smith, J.A., et al., Current Protocols in Molecular Biology. Wiley, New York, NY, pp. 4.9.1−4.9.8.

Shibata, N., Oda, H., Hirano, A., Kato, Y., Kawaguchi, M., Dal Canto, M.C., et al., 2002. Molecular biological approaches to neurological disorders including knockout and transgenic mouse models. Neuropathology. 22, 337−349.

Towbin, H., Staehelin, T., Gordon, J., 1979. Electrophoretic transfer of proteins from polyacrylamide gels to nitrocellulose sheets: procedure and some applications. Proc. Natl. Acad. Sci. USA. 76, 4350−4354.

Wang, A.M., Mark, D.F., 1990. Quantitative PCR. In: Innis, M.A., Gelfand, D.H., Sninsky, J.J., White, T.J. (Eds.), PCR Protocols: A Guide to Methods and Applications. Academic Press, San Diego, CA, pp. 70−75.

Waynforth, H.B., 1980. Experimental and Surgical Technique in the Rat. Academic Press, New York, NY.

Yoshioka, T., Feigenbaum, L., Jay, G., 1991. Transgenic mouse model for central nervous system demyelination. Mol. Cell. Biol. 11, 5479−5486.

Zelko, I.N., Stepp, M.W., Folz, R.J., 2013. A bioluminescent transgenic mouse model: real-time *in vivo* imaging of antioxidant EC-SOD gene expression and regulation by interferon gamma. Gene. 530, 75−82.

21 Compound Transgenics: Recombinase-Mediated Gene Stacking

Benjamin P. Beaton[1] and Kevin D. Wells[2]

[1]Division of Animal Sciences, University of Missouri, Columbia, MO,
[2]Division of Animal Sciences, National Swine Resource and
Research Center, University of Missouri, Columbia, MO

I. Introduction

Genetic engineering technologies have matured to a point where there are no technological barriers that differentiate mammalian models, from mice to pigs, in regard to the types or extent of genetic modifications that can be accomplished. Due to the physiological and anatomical similarities between pigs and humans, it is likely that pigs will continue to increase in importance as a biomedical model for both discovery and for clinical trials. However, due to timeline differences between mice and pigs associated with development and generation interval, genetic engineering in swine requires special consideration for the impact of genetic engineering strategies on future logistics and costs. This consideration is particularly important in regard to maintenance and propagation of animals with multiple genetic modifications.

For example, suppose that a project required four genetic modifications, and for breeding purposes, suppose that these modifications needed to be in a homozygous state. Further, suppose that for each transgene a particular range of transgene expression is acceptable. This scenario is likely to be required for xenotransplantation (Zhou et al., 2005) or for engineered biosynthetic pathways (Rees et al., 1990). One strategy could involve cointegration of all four transgenes through random concatenation and integration via pronuclear microinjection or transfection of cultured cells followed by somatic cell nuclear transfer (SCNT). However, this strategy relies on the assumption that a sufficient number of founder animals could be produced to identify an individual animal that has acceptable levels of expression of each of the four transgenes. Although this strategy is technically possible, it does not appear likely that success would be guaranteed within the scope of most

Transgenic Animal Technology. DOI: http://dx.doi.org/10.1016/B978-0-12-410490-7.00021-9

budgets. For those projects in which all four of the transgenes require expression in the same cell type, it may be possible to develop screening methods for expression in cultured cells prior to SCNT to enrich for founders that provide appropriate expression of the transgenes. However, when expression of the transgenes is required in separate cell types or when the target cell type cannot be cultured, this strategy is not feasible. Yet when this strategy is successful, all of the transgenes are genetically linked and will be inherited as a single *linkage group*: an animal that inherits one transgene inherits all transgenes.

A second strategy may be to produce founder animals in a stepwise progression, such that each transgene is added individually and a founder with acceptable expression is selected as the genetics for the next step. In the end, it is likely that a combination of transgene integrations could be obtained that provided an appropriate expression pattern and level for each transgene. However, upon propagation of this genotype, each transgene would segregate independently. In an outcross between a wild-type animal and a founder that is hemizygous for four transgenes, only 1/16 of the progeny would inherit all four transgenes based on Mendelian inheritance and independent assortment. Although this proportion of the progeny may be acceptable in mice for some projects, it will rarely be practical in swine. However, the scenario at hand supposes that each transgene needs to be rendered homozygous. Upon breeding of two animals that carry all four transgenes in a hemizygous state, only 1/256 of the progeny are expected to harbor all four transgenes in a homozygous state. It does not seem likely that independent integration would be a feasible approach for most goals. Similar issues arise if a viral-, retroviral-, or transposon-based strategy is selected.

A third approach, and the interest of this chapter, would be a stepwise strategy that placed each new transgene adjacent to the previous transgene(s) such that each modification becomes genetically linked to all previous modifications. At each step, the additional transgene could be evaluated for expression and impact on previously inserted transgenes. Each new transgene could be modified for increased or decreased expression as well as genetically insulated when transgenes were found to interact in an unwanted way. In the end, transmission of the desired genotype in a wild-type cross would be to one-half of the progeny. Upon breeding of two hemizygous animals, one-quarter of the progeny would be expected to be homozygous. These efficiencies are comparable to single gene modifications and are therefore well within the realm of feasibility.

Currently, there are two methods that can enable this third strategy of stacking each transgene within a previously selected locus. The first strategy would be through the use of homologous recombination. The second strategy would be through the use of a site-specific recombinase. This chapter describes a methodology for the use of bacterial recombinases for transgene stacking. In addition, the project described here combines gene stacking with gene targeting such that a series of transgenes will be placed within a disrupted locus, alpha-1,3-galactosyltransferase (*GGTA1*) (Figure 21.1).

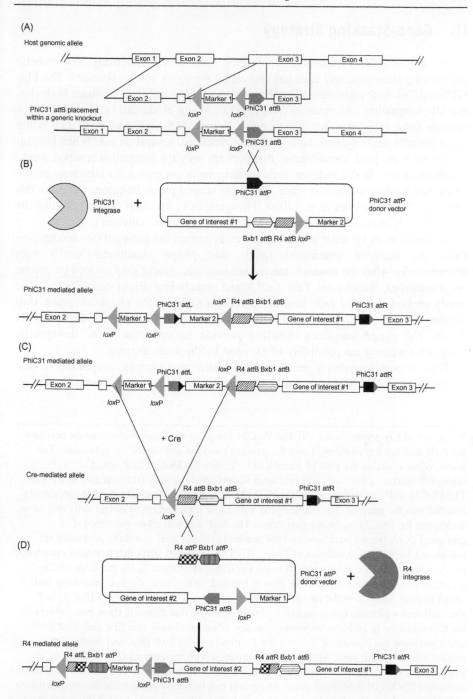

Figure 21.1 Strategy to sequentially stack transgenes in a site-specific manner through the use of a dual recombinase-mediated gene-stacking system. (A) A PhiC31 *att*B site is inserted into the host genome, represented by site-specific integration by homologous recombination. The marker gene is flanked by *lox*P sites and will be excised at a later time

II. Gene-Stacking Strategy

There are several recombination systems that work efficiently for genetic engineering strategies, and each has individual strengths and weaknesses. The Flp-FRT and Cre-*lox*P recombinase systems are reversible and thus facilitate both site-specific integration and excision (Nagy, 2000; Wang et al., 2011). However, these systems favor excision rather than insertion (Nagy, 2000; Wang et al., 2011). While some recombination systems have reversible reactions, several do not. When integration alone is the goal, recombinases that perform only the integration reaction would be advantageous. In this instance, once a transgene is integrated, the transgene cannot be excised by the recombinase alone. There are several phage integrases that have this property: PhiC31 (Groth et al., 2000; Thyagarajan et al., 2001), Bxb1 (Mediavilla et al., 2000; Kim et al., 2003), and R4 (Matsuura et al., 1996; Olivares et al., 2001).

Recombination by these phage integrases is carried out between two recognition sites, the bacterial attachment (*att*B) and phage attachment (*att*P) sites (Figure 21.2). After the reaction, the recombined attachment sites create two unique sequences/sites, attachment Left (*att*L) and attachment Right (*att*R) sites. The newly created *att*L and *att*R sites are unrecognizable by the phage integrase, thus creating an irreversible integration reaction (Thorpe and Smith, 1998; Thorpe et al., 2000). The phage integrases therefore provide an opportunity for site-specific integration without the possibility of excision by the same enzyme.

This chapter describes a gene-stacking system that uses an integrating recombinase system (PhiC31/*att*) for transgene insertion, and in conjunction, an

---

◄ by the use of Cre recombinase. (B) The PhiC31 integrase mediates recombination between the *att*B docking site (integrated into the genome) and the *att*P donor site (plasmid). The donor vector contains the gene of interest to be inserted, a PhiC31 *att*P site, a selectable marker, a *lox*P site, and two heterologous docking sites (R4 *att*B and Bxb1 *att*B). The PhiC31 *att*P site will recombine with the PhiC31 *att*B docking site that was previously inserted into the genome. The marker gene will allow for selection to isolate cells that have undergone the PhiC31 recombination event. The *lox*P site will allow the removal of unwanted DNA (vector backbone and the marker). The R4 *att*B and Bxb1 *att*B sites are introduced for the future addition of genes. (C) The unwanted DNA that has been introduced in (A) and (B) above can be removed using the Cre-*lox*P system. In the presence of Cre recombinase, sequence between *lox*P sites is excised. In this case, the two outermost *lox*P sites facilitate removal of the unwanted DNA (the two selectable markers, a PhiC31 *att*L site, and vector plasmid components). It should be noted that there are three possibilities for the Cre excision. In addition to the excision described previously (the first and third *lox*P sites), excision can also occur between the first and second *lox*P sites, and between the second and third *lox*P sites. Although recombination with the internal *lox*P will not be the predominant recombination, screening strategies must be designed to confirm removal of all unwanted DNA. (D) Additional genes of interest can be added through the use of alternating recombinase. An R4-mediated event is depicted. The components and mechanism are the same as in (B), with the exception that the recombination *att* sites originate from phage R4 and the PhiC31 *att*B site is reintroduced for future use. Since the various phage systems do not cross-react, the different integrases can be used in any order in subsequent steps.

(A)   *att*B: 5' - GGTGCCAGGGCGTGCCC *TTG* GGCTCCCCGGGCGCG - 3'

        *att*P: 5' - ccccaactggggtaacct *TTG* agttctctcagttgggg - 3'

        *att*L: 5' - GGTGCCAGGGCGTGCCC *TTG* agttctctcagttgggg - 3'

        *att*R: 5' - ccccaactggggtaacct *TTG* GGCTCCCCGGGCGCG - 3'

(B)

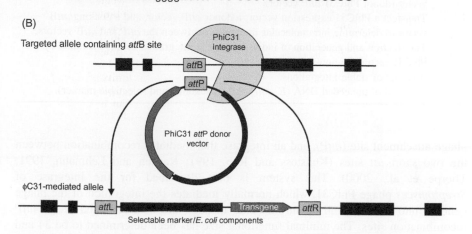

Figure 21.2 Site-specific integration of PhiC31 *att*P into preintegrated chromosomal *att*B. (A) The heterotypic *att*B (35 bp) and *att*P (39 bp) sequences for PhiC31 attachment sites. (B) Graphical representation of PhiC31 integrase mediating the integration of a transgene donor plasmid that contains an *att*P site into a previously chromosomally integrated *att*B site within a targeted location. Once integrated, the donor plasmid is flanked by *att*L and *att*R sequences.

excision-favoring recombinase system (Cre-*lox*P) to remove DNA that becomes superfluous after integration. Although the overall strategy is likely feasible in any eukaryotic cell, this protocol references porcine fibroblasts. Methods are described to: (i) generate cells that contain a prepositioned PhiC31 *att*B site (PhiC31 docking site); (ii) transfect a PhiC31 expression vector and a PhiC31 *att*P donor plasmid (PhiC31 donor site, with transgene); (iii) remove undesirable DNA (*E. coli* vector backbone and selectable marker) by the Cre-*lox*P system; and (iv) a strategy to stack multiple genes using a dual recombinase system (PhiC31 and R4 or Bxb1). See Figure 21.1 for a graphical outline of the methods listed. Table 21.1 provides an experimental outline.

III.   Materials and Methods

A.   *PhiC31 Vector Maps and Details*

Three components are required to achieve gene stacking mediated by a site-specific integration. These components include the bacterial attachment site (*att*B), the

Table 21.1 The Basic Steps Needed to Perform Irreversible Gene Stacking

Step	Action
1	Develop a strategy for stable integration of a PhiC31 *att*B docking site.
2	Generation of cell lines that contain docking site.
3	Develop and execute a strategy to ligate gene of interest into the pBB53 vector (donor PhiC31 *att*P vector).
4	Transfect a PhiC31 expression vector, a donor *att*P vector, and a docking *att*B vector to determine intramolecular integration between the *att*P and *att*B vectors.
5	Transfection and generation of individual stable integration events mediated by PhiC31 recombination.
6	Analysis of stable integrations.
7	Removal of unneeded DNA (*E. coli* vector backbone and selectable marker).

phage attachment site (*att*P), and an integrase that mediates recombination between the two short *att* sites (Kuhstoss and Rao, 1991; Rausch and Lehmann, 1991; Thorpe et al., 2000). This system is best described for the integrase of *Streptomyces* phage PhiC31, which normally mediates the integration of the phage genome into the bacterial chomromosome through the host *att*B and phage *att*P recombination sites. The minimal functional size has been determined to be 34 and 39 bp for the *att*B and *att*P sites, respectively (Groth et al., 2000). The two sequences share only a common three-bp central region (TTG), which is flanked by imperfect repeats. Recombination between the two attachment sites creates two hybrid sites, which are referenced as *att*L and *att*R (Thorpe et al., 2000; Smith et al., 2004). The *att*L and *att*R sequences are not substrates for the PhiC31 recombinase, thus making this a unidirectional tool. In our gene-stacking approach, we have chosen to follow in suit with the natural context of the *att*B located in the host genome, where it serves as a docking site (see Figures 21.1 and 21.2). The *att*P site is located in the incoming donor vector, to mimic the phage *att*P integration into the bacterial host *att*B. This selection is arbitrary, since the integration can be accomplished in either direction (Groth et al., 2000; Thyagarajan et al., 2001). Several mammalian expression vectors have been generated and proven functional in mammalian cells. We have been able to confirm the utility of two different vectors (Groth et al., 2000: Addgene plasmid 18935; Raymond and Soriano, 2007: Addgene plasmid 13794) in pig cells. The data presented in this chapter were generated with the integrase-expression vector, pCMVint (Groth et al., 2000: Addgene plasmid 18935).

a. PhiC31 gene-stacking vectors
    i. The PhiC31 expression vector, pCMVint, was previously described (Groth et al., 2000).
    ii. To generate the PhiC31 *att*B docking site, we chose to use a gene-targeting strategy via homologous recombination. We generated a vector to target the porcine *GGTA1* gene. The method to generate a *GGTA1* gene-targeting vector has previously been described (Beaton et al., 2013). We have since made a modification to the targeting

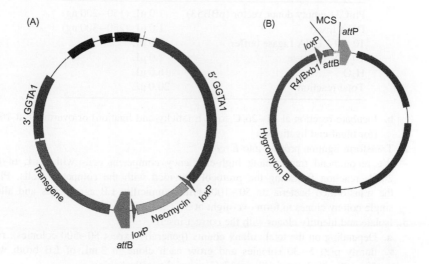

Figure 21.3 Graphical representation of PhiC31 *att*B and *att*P vectors. (A) This plasmid was used to introduce the PhiC31 *att*B docking site into the pig genome at the GGTA1 locus. This targeting vector consisted of sequence homologous to the porcine *GGTA1* gene, a mammalinized neomycin-resistant gene flanked by *lox*P sites, a PhiC31 *att*B docking site (red block), and a transgene. (B) The PhiC31 attP donor plasmid was created to contain a mammalianized Hygromycin B resistance gene, a *lox*P site, the PhiC31 donor *att*P site (orange block), an R4 *att*B site (light blue box), a Bxb1 *att*B site (pink box), and a series of multiple cloning sites (MCS). The multiple cloning sites were designed to receive the gene of interest.

    vector used in that experiment. The minimum PhiC31 *att*B docking site (Groth et al., 2000) was inserted into pBB7 to make pBB18 (Figure 21.3A) and used to create a heterozygous *GGTA1* targeted allele that harbors a PhiC31 *att*B. A similar strategy can be applied to any genomic target of interest.

  iii. To generate a universal PhiC31 *att*P donor plasmid, we modified an *att*P-containing plasmid, pDRAV-3 (Osterwalder et al., 2010: Addgene plasmid 26748). The generated plasmid (pBB53) was created with a strategically located multiple cloning region that is ready to receive a user's gene of interest (Figure 21.3B). The donor plasmid contains a selectable marker (Hygromycin B; Invitrogen) to select for integration.

b. Cloning and preparation of plasmids

  i. Clone the gene of interest (must be a self-contained functional unit) into the multiple cloning region of pBB53 using the appropriate restriction enzymes and buffers.

    1. Ligation of transgene into pBB53. (See Notes 1 and 2 in Section 21.V.)

      a. Set up 20 μL ligation reaction.

Ingredient	Volume
PhiC31 empty donor vector (pBB53)	1.0 μL (150–200 ng)
DNA insert	1.0 μL (250–300 ng)
10 × T4 DNA Ligase buffer	2.0 μL
T4 DNA Ligase	1.0 μL
H$_2$O	16.0 μL
Total reaction	20.0 μL

    b. Incubate reaction at 24–26°C for 1 h (sticky-end ligation) or overnight at 16°C (for blunt-end ligations).

2. Transform ligation product into *E. coli.*

We recommend transforming high-efficiency competent cells with 2 μL of the 20 μL reaction following the protocol provided with the competent cells. Plate the transformed bacteria on 50–100 μg/mL ampicillin LB agar plates and allow single colony clones to form overnight at 37°C.

3. Isolate and identify clones with the correct insert.

    a. Depending on the total colony counts (generally expect 50–500 colonies), randomly pick 5–30 colonies and grow each clone in 3 mL of LB broth with 50–100 μg/mL ampicillin at 37°C overnight, with shaking.

    b. The following day, purify the construct using a plasmid purification kit or alkaline lysis protocol. Verify insert using either polymerase chain reaction (PCR) or restriction enzyme diagnostics.

    c. Grow a positive clone containing the insert in an appropriate amount of LB ampicillin broth (500 mL culture to produce a large quantity of plasmid DNA), and purify the construct using a plasmid purification kit or alkaline lysis protocol. Verify clone with either PCR or restriction enzyme diagnostics.

B.   *Intramolecular Integration Assay*

We have observed that the efficiency of PhiC31-mediated recombination differs between donor plasmids (and likely integration sites). This efficiency can be estimated prior to experimental procedures by transient transfection. To verify that the precise recombination between the PhiC31 *att*B docking site and the PhiC31 *att*P donor vector has occurred, the following intramolecular integration assay can be performed on crude cell lysate from cells that have been transiently transfected. The cells must be transfected with three pieces of DNA: (i) a PhiC31 expression vector that will transiently express the PhiC31 integrase (i.e., pCMVint); (ii) the PhiC31 *att*B docking site vector (i.e., pBB18, the *GGTA1* targeting vector that contains the PhiC31 *att*B); and (iii) the PhiC31 *att*P donor site plasmid (plasmid pBB53 with inserted gene of interest). The goal of this assay is to determine if the reaction between the two *att* sites is being catalyzed by the recombinase in the cell type and cell culture conditions that are being used. A PCR assay is performed using primers that flank the newly created *att*L or *att*R site.

The proper conditions and ratios have been optimized for porcine fibroblasts (see Note 3 and Figure 21.4). The described protocol is per manufacturer

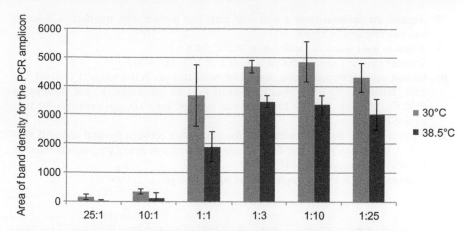

Figure 21.4 Determination of the culture temperature and the DNA ratios of PhiC31 integrase to *att*P donor vector. To calculate the occurrence of the recombination event, a PCR that flanks the junction of a PhiC31 *att*R was utilized. The yield of PCR product estimates recombination events that occurred in a pool of cells. Analysis was performed using ImageJ software (http://rsbweb.nih.gov/ij/) by calculating the density of the bands (arbitrary units). The data represent mean band density $\pm$ SD ($n = 4$ transfections). PhiC31 dependent integration is more efficient at 30°C as compared to 38.5°C.

recommendations, FuGENE HD (Promega; Madison, WI). We recommend using a 1:10 ratio of integrase plasmid to donor vector.

a. Culture porcine fibroblast cells in complete medium (Dulbecco's modified Eagle's medium (DMEM) (Cellgro) containing 2.77 mM glucose, 1.99 mM L-glutamine, and 0.5 mM sodium pyruvate, supplemented with 1 mM L-alanyl-L-glutamine (Cellgro) and 12% fetal bovine serum (FBS)). Porcine fibroblasts should be cultured at 38.5°C, with 5.5% CO_2, 5.0% O_2, and 100% humidity.

b. Passage cells and plate them at a density of 30,000 cells/well in a 24-well plate. The cells will be ready to be transfected the next day or when they are at 50–60% confluency.

c. Transfect the cells according to manufacturer instructions using a lipid-based transfection method (see Note 5). The protocol uses a reagent to DNA ratio of 3.5:1. When transfecting a single well of a 24-well plate, $\sim$ 600 ng DNA are to be transfected.

   1. Allow FuGENE HD Transfection Reagent to reach room temperature.
   2. Mix by inverting or vortexing briefly. No precipitate should be visible.
   3. Dilute a total of $\sim$ 600 ng plasmid DNA to a volume of 25 μL in water in a 1.5-mL tube (see below).

      Transfection is to be performed with the three DNA components at a 1:10:10 ratio (PhiC31 integrase:*att*P-containing plasmid:*att*B containing plasmid). Transfections are to be performed using supercoiled plasmid DNA for all three vectors. Transfect 30 ng PhiC31 vector, 300 ng *att*B vector, and 300 ng *att*P vector.

   4. Add 2 μL FuGENE HD to same tube. Mix by pipetting the entire 27 μL, eight times.
   5. Incubate for 5–10 min at room temperature.
   6. Add 500 μL fresh complete medium to the 27 μL of the DNA:lipid complex. Mix gently by pipetting.

7. Aspirate the medium from a well with cells and replace with transfection solution that was prepared in Step 6.
8. Return to gas controlled incubator at 38.5°C for 4 h.
9. Transfer plate to a gas controlled incubator at 30°C for ~44 h.
10. Harvest cells by trypsinization, resuspend in 40 μL lysis buffer (40 mM Tris, pH 8.9; 0.9% Triton X-100; 0.9% Nonidet P-40; 0.4 mg/mL proteinase K) and incubate at 65°C for 15 min to lyse the cells followed by 95°C for 10 min to inactivate the proteinase K.
11. Perform PCR to assay for intramolecular integration between the *att*P and *att*B vectors. A PCR assay should be designed using primers that detect the junction flanking the newly created *att*L or *att*R site.
12. Evaluate the PCR by electrophoresis on an agarose gel.

C. Culture and Transfection of Porcine Fibroblast Cells

To demonstrate the efficacy of this protocol, cells were generated that contain the PhiC31 *att*B site within the *GGTA1* locus as per Beaton et al. (2013). These cells serve as a recipient for the second piece of DNA, the donor vector that contains a PhiC31 *att*P site and a transgene. The third and final DNA is a mammalian expression vector that contains the coding sequence for the PhiC31 recombinase (pCMVint). The strategy for integration of the donor *att*P vector into the established *GGTA1*-located *att*B docking site is to co-transfect the two vectors, pCMVint (PhiC31 expression vector) and pBB53, which includes the gene of interest (PhiC31 *att*P donor vector).

The transfection method is similar to the previous method outlined in the intramolecular integration assay. The major differences are that after the transfection, the transfected cells are split onto 100-mm tissue culture plates and put under Hygromycin B selection.

a. Culture porcine fibroblast cells in complete medium, (DMEM containing 2.77 mM glucose, 1.99 mM L-glutamine, and 0.5 mM sodium pyruvate, supplemented with 1 mM L-alanyl-L-glutamine and 12% FBS). Porcine fibroblasts should be cultured at 38.5°C with 5.5% CO_2, 5.0% O_2, and 100% humidity.
b. Passage cells and plate them at a density of 30,000 cells/well in a 24-well plate. The cells will be ready to be transfected the next day when they are 50–60% confluent (see Note 4).
c. Transfect the cells according to manufacturer instructions using a lipid-based transfection method (see Note 5). The protocol uses a reagent to DNA ratio of 3.5:1. When transfecting a single well of a 24-well plate, a total of ~600 ng DNA is used.
    1. Allow FuGENE HD Transfection Reagent to reach room temperature.
    2. Mix by inverting or vortexing briefly. No precipitate should be visible.
    3. Dilute a total of ~600 ng plasmid DNA to a volume of 25 μL in water in a 1.5-mL tube (see below).
        Transfection is to be performed with the PhiC31 expression vector and the *att*P donor vector at a 1:10 ratio (PhiC31 integrase:*att*P-containing plasmid). Transfections are to be performed using supercoiled plasmid DNA for both vectors. Transfect 55 ng pCMVint and 550 ng *att*B donor vector.

4. Add 2 μL FuGENE HD to same tube. Mix entire 27 μL by pipetting, eight times.
5. Incubate for 5—10 min at room temperature.
6. Add 500 μL fresh, complete medium to the 27 μL of the DNA:lipid complex. Mix gently by pipetting.
7. Aspirate the medium from a well with cells and replace with transfection solution that was prepared in Step 6.
8. Incubate the cells at 38.5°C under normal porcine cell culture conditions for 4 h.
9. Transfer plate to 30°C for ~44 h.
10. Forty-eight hours post transfection, split cells onto ten 100-mm culture plates with 6 mL complete medium.
11. Twenty-four hours post Step 10, administer Hygromycin B drug selection. Add 1 mL of complete medium that has Hygromycin B (1750 mg/L), which is a 7× concentration. The 7× concentration allows the entire drug selection medium on the plates to be equivalent to 1× concentration.
12. Culture the cells for 13 days. A medium change can be performed after 7—9 days of culture using 1× concentration Hygromycin B (250 mg/L) in complete medium.
13. After a total of 13 days in Hygromycin medium, colonies should range in size from 5 to 15 mm.
14. Harvest the cells by trypsinization. We recommend using cloning cylinders as a method to harvest individual colonies. Two-thirds of cells harvested are transferred to one well of a 24-well tissue culture plate for expansion, and one-third of cells isolated are used for PCR analysis of integration (user-developed assay from the transient assays above).
   a. Pellet assay cells by centrifugation.
   b. Remove supernatant and resuspend cell pellet in 5 μL lysis buffer (40 mM Tris, pH 8.9; 0.9% Triton X-100; 0.9% Nonidet P-40; 0.4 mg/mL proteinase K) and incubate at 65°C for 15 min to lyse the cells followed by 95°C for 10 min to inactivate the proteinase K.
15. Perform PCR to assay for intramolecular integration between the attP and attB vectors. A PCR assay is performed using primers that flank the newly created attL or attR site.
16. Evaluate the PCR by electrophoresis on an agarose gel.
17. Expand and cryopreserve positive colonies. SCNT can then be performed to generate fetuses for fibroblast collection for additional genetic manipulations or live pigs with the PhiC31 stacked transgene.

D. Cre-loxP-Mediated Excision of Unwanted DNA by Transient Expression of Cre Recombinase

To ensure expression of the transgene, it is important to remove plasmid sequences (Shani, 1986). In addition, the presence of a selectable marker can also influence the expression of neighboring genes (Fiering et al., 1995). Further, removal of superfluous DNA presents the possibility of recycling the selectable marker, as demonstrated in mouse embryonic stem cells (Rossant and Nagy, 1995; Abuin and Bradley, 1996; Nagy and Rossant, 1996). In this method, removal of all unwanted sequences is facilitated by Cre recombinase (Nagy et al., 1998). Cre expression is provided by transient expression in cultured cells. The transfection method that has

Table 21.2 Results of Cre-*lox*P-Mediated Deletion of a Selectable Marker

Transfection Set	Number of Colonies Picked	Number of Cre-*lox*P-Deleted Colonies (%)
Transfection 1	24	21 (87.5%)
Transfection 2	23	19 (82.6%)
Transfection 3	10	7 (70.0%)
Transfection 4	21	17 (81.0%)
Overall	78	64 (82.1%)

been optimized is a modification of the electroporation methods described by Ross et al. (2010) and Beaton et al. (2013). We have been able to achieve an efficiency of 82.1% through the transient expression of a mammalianized Cre-expression vector (Table 21.2).

1. Culture porcine fibroblast cells in complete medium, (DMEM containing 2.77 mM glucose, 1.99 mM L-glutamine, and 0.5 mM sodium pyruvate, supplemented with 1 mM L-alanyl-L-glutamine and 12% FBS). Porcine fibroblasts should be cultured at 38.5°C, with 5.5% CO_2, 5.0% O_2, and 100% humidity.
2. Harvest fibroblasts by trypsinization and resuspend the cells at a concentration of 1×10^6 cells/mL in an electroporation medium (25% OptiMEM (Invitrogen) + 75% buffered KCl (135 mM KCl; 15 mM Tris; 11.3 mM N,N-Bis(2-hydroxyethyl)-2-aminoethanesulfonic acid (BES); 3.7 mM 1,4-Piperazinediethanesulfonic acid (PIPES), pH 7.3).
3. In a 4-mm cuvette, aliquot 400 μL of the cell suspension (400,000 cells) and add 7.5 μg of supercoiled pCAG-Cre:GFP, a mammalian Cre-expression vector (Matsuda and Cepko, 2007: Addgene plasmid 13776).
4. Using a square wave generator, electroporate the cells at 490 volts for three pulses at 1 ms per pulse.
5. After the electroporation, resuspend the cells in a total volume of 10 mL complete medium.
6. Aliquot 5 μL of the 10 mL suspension to 7 mL of complete medium in a 100-mm tissue culture plate. This dilution is required in order to have a clonal density of 1−10 colonies per culture plate. Perform this dilution onto 5−20 plates. Generally, fewer than five plates are required.
7. After 7 days, change medium with fresh complete medium (as in Step 1).
8. Allow cells to grow for an additional 6 days without any selection. After a total of 13 days in culture after the transfection, colonies should range in size from 5 to 15 mm.
9. Harvest the cells by trypsinization. We recommend using cloning cylinders as a method to harvest individual colonies. Two-thirds of cells harvested are transferred to one well of a 24-well tissue culture plate for expansion, and one-third of cells isolated are used for PCR analysis of Cre-*lox*P-mediated excision of unwanted DNA.
   a. Pellet assay cells by centrifugation.
   b. Remove supernatant and resuspend cell pellet in 5 μL lysis buffer (40 mM Tris, pH 8.9; 0.9% Triton X-100; 0.9% Nonidet P-40; 0.4 mg/mL proteinase K) and incubate at 65°C for 15 min to lyse the cells, followed by 95°C for 10 min to inactivate the proteinase K.

10. Screen the colonies by PCR to determine which colonies have undergone the Cre-mediated excision. A PCR screen needs to be designed to flank the unwanted DNA in order to confirm the excision.

11. Expand and cryopreserve positive colonies. SCNT can then be performed to generate fetuses for fibroblast collection for additional genetic manipulations or live pigs with the Cre-excised genotype.

IV. Summary

We would like to note that the given protocol demonstrates the addition of a targeted PhiC31 attachment site (attB), followed by the PhiC31-mediated integration of a PhiC31 attachment site (attP) donor vector. We have also given an approach to remove unwanted DNA by the Cre-loxP system using transient expression of Cre for excision. The final contribution is a gene-stacking strategy that has unlimited potential if a dual recombinase attachment system is employed, as described in Figure 21.1. The gene-stacking system presented is a straightforward method that is unidirectional, site-specific, and enables the sequential insertion of multiple transgenes.

V. Notes

1. If assistance with DNA ligations, *E. coli* transformations, restriction enzyme analysis, or DNA sequencing is needed, we recommend you consult product manuals or protocol literature such as *Molecular Cloning: A Laboratory Manual* (Sambrook et al., 1989) or *Current Protocols in Molecular Biology* (Ausubel et al., 2002).

2. For all restriction enzymes and ligation reactions required, we recommend following manufacturer directions; use the protocol given in this chapter as guidance only.

3. The transfection method has been optimized for the ratio of DNA concentration (PhiC31 expression vector:attP donor vector). In addition, we performed a temperature optimization that would allow for the most suitable reaction to occur. We chose to explore varying ratios due to the wide range in ratios that have previously been reported using PhiC31 in mammalian culture (Groth et al., 2000; Thyagarajan et al., 2001; Chalberg et al., 2006). Groman and Suzuki (1962) reported that phages' reproduction efficiency varies depending on temperature. In the reported data, phage reproduction is drastically reduced at temperatures above 37°C. Porcine cell culture is performed at 38.5°C. Therefore, PhiC31 expression may not be optimal under normal porcine cell culture conditions. The results from varying ratios and the temperature comparison of 30°C and 38.5°C are shown in Figure 21.4.

4. If transfection of larger plates is desired, the amount of each DNA and the transfection reagent should be scaled. We used the amounts of each DNA to ensure a low level of background, which allowed us to efficiently pick well-isolated colonies.

5. For the transfections performed in the experiments demonstrated, we determined that when using FuGENE HD (Promega), ~90% of porcine cells are transfected using reporter DNA (data not shown). Therefore, all experiments performed using PhiC31 have been performed using the transfection reagent FuGENE HD.

References

Abuin, A., Bradley, A., 1996. Recycling selectable markers in mouse embryonic stem cells. Mol. Cell. Biol. 16 (4), 1851−1856.

Ausubel, F.M., Brent, R., Kingston, R.E., Moore, D.D., Seidman, J.G., Smith, J.A., et al., 2002. Short Protocols in Molecular Biology: A Compendium of Methods from Current Protocols in Molecular Biology, vol. 2. Wiley, New York, NY.

Beaton, B.P., Mao, J., Murphy, C.N., Samuel, M.S., Prather, R.S., Wells, K.D., 2013. Use of single stranded targeting DNA or negative selection does not provide additional enrichment from a GGTA1 promoter trap. J. Mol. Cloning Genet. Recomb. 2 (1), 2.

Chalberg, T.W., Portlock, J.L., Olivares, E.C., Thyagarajan, B., Kirby, P.J., Hillman, R.T., et al., 2006. Integration specificity of phage φC31 integrase in the human genome. J. Mol. Biol. 357 (1), 28−48.

Fiering, S., Epner, E., Robinson, K., Zhuang, Y., Telling, A., Hu, M., et al., 1995. Targeted deletion of 5′HS2 of the murine beta-globin LCR reveals that it is not essential for proper regulation of the beta-globin locus. Genes Dev. 9 (18), 2203−2213.

Groman, N.B., Suzuki, G., 1962. Temperature and lambda phage reproduction. J. Bacteriol. 84 (3), 431−437.

Groth, A.C., Olivares, E.C., Thyagarajan, B., Calos, M.P., 2000. A phage integrase directs efficient site-specific integration in human cells. Proc. Natl. Acad. Sci. USA 97 (11), 5995−6000.

Kim, A.I., Ghosh, P., Aaron, M.A., Bibb, L.A., Jain, S., Hatfull, G.F., 2003. Mycobacteriophage Bxb1 integrates into the *Mycobacterium smegmatis* groEL1 gene. Mol. Microbiol. 50 (2), 463−473.

Kuhstoss, S., Rao, R.N., 1991. Analysis of the integration function of the streptomycete bacteriophage φC31. J. Mol. Biol. 222 (4), 897−908.

Matsuda, T., Cepko, C.L., 2007. Controlled expression of transgenes introduced by *in vivo* electroporation. Proc. Nat. Acad. Sci. USA 104 (3), 1027−1032.

Matsuura, M., Noguchi, T., Yamaguchi, D., Aida, T., Asayama, M., Takahashi, H., et al., 1996. The sre gene (ORF469) encodes a site-specific recombinase responsible for integration of the R4 phage genome. J. Bacteriol. 178 (11), 3374−3376.

Mediavilla, J., Jain, S., Kriakov, J., Ford, M.E., Duda, R.L., Jacobs, W.R., et al., 2000. Genome organization and characterization of mycobacteriophage Bxb1. Mol. Microbiol. 38 (5), 955−970.

Nagy, A., 2000. Cre recombinase: the universal reagent for genome tailoring. Genesis 26 (2), 99−109.

Nagy, A., Rossant, J., 1996. Targeted mutagenesis: analysis of phenotype without germ line transmission. J. Clin. Invest. 97 (6), 1360.

Nagy, A., Moens, C., Ivanyi, E., Pawling, J., Gertsenstein, M., Hadjantonakis, A.K. , et al., 1998. Dissecting the role of *N-myc* in development using a single targeting vector to generate a series of alleles. Curr. Biol. 8 (11), 661−666.

Olivares, E.C., Hollis, R.P., Calos, M.P., 2001. Phage R4 integrase mediates site-specific integration in human cells. Gene 278 (1), 167−176.

Osterwalder, M., Galli, A., Rosen, B., Skarnes, W.C., Zeller, R., Lopez-Rios, J., 2010. Dual RMCE for efficient re-engineering of mouse mutant alleles. Nat. Methods 7 (11), 893−895.

Rausch, H., Lehmann, M., 1991. Structural analysis of the actinophage ΦC31 attachment site. Nucleic Acids Res. 19 (19), 5187−5189.

Raymond, C.S., Soriano, P., 2007. High-efficiency FLP and φC31 site-specific recombination in mammalian cells. PLoS One 2 (1), e162.

Rees, W.D., Flint, H.J., Fuller, M.F., 1990. A molecular biological approach to reducing dietary amino acid needs. Nat. Biotechnol. 8 (7), 629−633.

Ross, J.W., Whyte, J.J., Zhao, J., Samuel, M., Wells, K.D., Prather, R.S., 2010. Optimization of square-wave electroporation for transfection of porcine fetal fibroblasts. Transgenic Res. 19 (4), 611−620.

Rossant, J., Nagy, A., 1995. Genome engineering: the new mouse genetics. Nat. Med. 1 (6), 592−594.

Sambrook, J., Fritsch, E.F., Maniatis, T., 1989. Molecular Cloning, vol. 2. Cold Spring Harbor Laboratory Press, New York, NY, pp. 14−19.

Shani, M., 1986. Tissue-specific and developmentally regulated expression of a chimeric actin-globin gene in transgenic mice. Mol. Cell. Biol. 6 (7), 2624−2631.

Smith, M.C., Till, R., Brady, K., Soultanas, P., Thorpe, H., Smith, M.C., 2004. Synapsis and DNA cleavage in ϕC31 integrase-mediated site-specific recombination. Nucleic Acids Res. 32 (8), 2607−2617.

Thorpe, H.M., Smith, M.C., 1998. *In vitro* site-specific integration of bacteriophage DNA catalyzed by a recombinase of the resolvase/invertase family. Proc. Natl. Acad. Sci. USA 95 (10), 5505−5510.

Thorpe, H.M., Wilson, S.E., Smith, M., 2000. Control of directionality in the site-specific recombination system of the *Streptomyces* phage φC31. Mol. Microbiol. 38 (2), 232−241.

Thyagarajan, B., Olivares, E.C., Hollis, R.P., Ginsburg, D.S., Calos, M.P., 2001. Site-specific genomic integration in mammalian cells mediated by phage φC31 integrase. Mol. Cell. Biol. 21 (12), 3926−3934.

Wang, Y., Yau, Y.Y., Perkins-Balding, D., Thomson, J.G., 2011. Recombinase technology: applications and possibilities. Plant Cell Rep. 30 (3), 267−285.

Zhou, C.Y., McInnes, E., Copeman, L., Langford, G., Parsons, N., Lancaster, R., et al., 2005. Transgenic pigs expressing human CD59, in combination with human membrane cofactor protein and human decay-accelerating factor. Xenotransplantation 12 (2), 142−148.

22 Assisted Reproductive Technologies and Embryo Culture Methods for Farm Animals

Robert A. Godke[1], Marina Sansinena[1] and Curtis R. Youngs[2]

[1]Embryo Biotechnology Laboratory, Department of Animal Science, Louisiana State University, Baton Rouge, LA, [2]Department of Animal Science, Iowa State University, Ames, IA

I. Introduction

It has been said that males of an animal species tend to have an advantage over females in the propagation of their genes for future generations. For example, in wild populations it is common for males of most of the hoofstock species to mate with multiple females in seasonal breeding groups, as long as the male is strong enough to ward off subordinate males. Also, males constantly renew their supply of gametes so they may pass on their genes to subsequent offspring. Throughout the male's life the testes can produce spermatozoa in almost unlimited quantities.

In contrast, the female farm animal is born with all the oocytes she will ever produce in her lifetime already stored in her ovaries. A beef heifer, for example, may have 200,000 or more primordial follicles (each containing one oocyte) in her ovaries at the time of birth. That female may produce only 10 calves in her lifetime, so what happens to the remaining 199,990 oocytes of the ovarian pool? Firstly, cycling heifers ovulate at 20- to 22-day intervals throughout the year, removing one ovum from this oocyte pool at ovulation during each estrous cycle. Secondly, a significantly greater number of oocytes are lost from the ovarian pool through atresia. Bovine follicles develop in waves of 4−12 follicles at a time, and a cow usually has 2−3 waves of these growing follicles during each estrous cycle. For a heifer with three follicular waves per estrous cycle, 12−36 follicles develop from the cohort and most of them undergo atresia (only one follicle will ovulate per estrous cycle). Follicular waves also occur during the early part of pregnancy, further removing oocytes from the ovarian pool. Assuming a beef female produces 10 calves in her lifetime and has three follicular waves per estrous cycle, she would have activated and used less than 4200 oocytes from her ovarian follicle pool.

Transgenic Animal Technology. DOI: http://dx.doi.org/10.1016/B978-0-12-410490-7.00022-0

In other farm animal females (e.g., sheep, goats, and pigs) it has been estimated that the number of primordial follicles in the ovaries at the time of birth ranges from 40,000 to 500,000.

It has been estimated that the newborn horse filly usually has more than 60,000 primordial follicles in the ovaries at the time of birth, but the young mare takes longer to reach puberty than a beef heifer. The horse has a 20- to 22-day estrous cycle, most often producing one potentially fertile ovum per estrous cycle. Unlike the cow, however, the mare is a long-day seasonal breeder that exhibits estrous cycles only from mid-spring through mid-fall (in the northern hemisphere), and she typically has only one or two follicular waves per estrous cycle. The mare is also unique among the domestic farm animals because she exhibits follicular waves (with follicles that ovulate) during the first one-third of pregnancy. If a mare lives to age 20 and produces 12 foals during her reproductive lifetime, she would have used less than 8600 primordial follicles from her ovarian follicle pool.

This difference in male and female gamete production presents an interesting point for an academic discussion; however, livestock producers who wish to maximize production of live offspring from their most genetically superior seedstock females can take advantage of this "underutilized" pool of ovarian follicles. For example, oocytes can be harvested for *in vitro* fertilization (IVF) or for use as nuclear transfer (NT) recipient oocytes. In the case of the mare, it is possible to aspirate *in vivo* matured oocytes directly from the ovarian follicles of cyclic mares at a frequency greater than the once every 21 days she would normally ovulate. Assuming aspiration of 12 preovulatory follicles per breeding season from each of 10 mares, the potential exists to produce large numbers of offspring. Approximately 15% of oocyte aspiration attempts would likely be unsuccessful due to ovulation occurring earlier than anticipated, thus leaving approximately 102 preovulatory follicles available for aspiration. Approximately 85% of those follicles should respond as expected to exogenous human chorionic gonadotropin (hCG), thus enabling aspiration of ~ 86 preovulatory follicles containing an *in vivo* matured oocyte. With an oocyte recovery rate of 75%, this would provide ~ 65 *in vivo* matured oocytes for IVF per 10 mares per year. If ovarian superstimulation with exogenous hormonal agents (such as equine follicle-stimulating hormone) were to become more successful in the mare, this would drastically increase the number of *in vivo* matured oocytes harvested, as well as the number of offspring that could be produced.

II. Assisted Reproductive Technologies

A. Artificial Insemination and Sexed Semen

Assisted reproductive technologies allow livestock producers to make continued genetic advances in their seedstock. One widely known success story is artificial insemination (AI); the use of AI technology in the United States' dairy cattle industry, beginning in the 1950s and continuing though the present time, has led to an increase of milk production per cow of more than 300%. During that same time period the number of cows in the United States was reduced by more than 50%. This improvement in productivity occurred because researchers developed the technology, extension professionals transferred the technology to the field, and progressive producers incorporated this technology to remain competitive in the global animal industry. Enhanced genetic selection for milk production, made possible through use of frozen semen for AI and improved herd management, dramatically changed the dairy industry in North America. A recent study reported the carbon footprint per billion kilograms of milk produced in 2007 was 37% of equivalent milk production in 1944 (Capper et al., 2009).

In the 1980s, there was a major breakthrough that subsequently had a substantial impact on the bovine AI industry. Researchers at the US Department of Agriculture developed and patented a flow cytometric method for segregating X- from Y-chromosome-bearing spermatozoa. The underlying basis for this method is to separate spermatozoa based on the DNA content difference between the larger X and the smaller Y chromosome (spermatozoa possessing a Y chromosome possess, depending on species, approximately 2.5−4.0% less DNA than spermatozoa possessing an X chromosome). Following early success with rabbits (Johnson et al., 1989), this technology was subsequently adapted for use in a variety of livestock species including cattle, horses, sheep, and swine. (Garner, 2011). Depending on the sorting parameters, the efficiency of sorting typically ranges between 85% and 92%. The commercial bovine AI companies have heartily embraced this technology, and most bull studs market sexed semen. In calendar year 2009, more than six million units of sexed semen were marketed globally, and this number has undoubtedly increased since that time. Two potential drawbacks of this technology are that the price per unit of sexed semen is greater than that for unsexed semen, and conception rate typically averages no more than 70−75% of that obtained with semen from the same bull but not subjected to flow cytometric sorting (DeJarnette et al., 2009). Application of this technology to nondomestic species (Bahr et al., 2009), as well as other domestic species, is presently being developed.

B. In Vivo *Embryo Production*

A second assisted reproductive technology that received tremendous interest from cattle producers beginning in the late 1970s and early 1980s was embryo transfer (ET). Although the first ET producing a live calf was reported in the early 1950s (Willet et al., 1951), it was not until 1976 that nonsurgical collection and transfer procedures were developed for cattle. Once ET transitioned from a surgical to a nonsurgical procedure, commercial adoption of this technology in the livestock industry exploded. Even though AI enabled genetically superior males to produce large numbers of offspring, ET was the first practical procedure that allowed livestock producers to tap into the "underutilized" pool of ovarian follicles of their genetically outstanding females. With the standard ET procedure, donor females are exposed to gonadotropic hormones (such as follicle-stimulating hormone—the same hormone she produces herself) to produce multiple ovulations for *in vivo* fertilization. Nonsurgical embryo collection and nonsurgical ET procedures are now considered routine in the cattle industry, with more than 100,000 embryo recoveries and more than 850,000 ETs per year.

Although ET technology has been used successfully to increase the number of offspring produced from selected females, there are still drawbacks associated with this approach. For example, some females simply do not respond (or stop responding) to the exogenous stimulatory agents, or they develop physiological conditions that make it difficult to retrieve the embryos. Some estimates are that as many as 20% of all superovulation and embryo recovery attempts do not produce a transferrable quality embryo. Another potential drawback is that genetically elite cows are most often kept "open" so that they may be repeatedly used for embryo collection. These superovulated donor cows often take 60 days or longer to conceive after embryo collection has ceased. Some producers have observed that some of their donor cows also develop cystic ovarian follicles during the process. Although high beef cattle prices, industry promotion, and producers' keen interest enhanced the use of ET technology in the late 1970s and early 1980s, ET in North America today is more often used by dairymen than by beef cattle producers.

The mare has presented a unique problem to those scientists involved in equine-assisted reproductive technology research. Whether it is due to the unique anatomical structure of the equine ovary and/or other unknown factors, attempts to super-ovulate donor horses generally produce poor results. This factor, coupled with restrictions imposed by certain horse breed associations on the registration of off-spring produced via ET, has made ET less acceptable to the horse industry than to other livestock industries. Fortunately, equine follicle-stimulating hormone is showing great promise to achieve mild superovulation, and some breed associations are relaxing their restrictions of registration of ET offspring due to the availability of DNA testing for parentage verification. Nonsurgical embryo collection and transfer procedures are considered relatively simple in the mare and can produce acceptable pregnancy rates when using good quality recipient females.

Application of ET technologies to the commercial sheep, goat, and swine industries has been limited for two major reasons: (i) the economics of animal production in those species historically has not justified the expense of ET and (ii) ET techniques typically are performed surgically—limiting the number of times a donor female can be used and greatly increasing the costs compared with the non-surgical methods used in cattle and horses. Nonetheless, activity in swine ET technologies has blossomed in the biomedical research community, and recent advances in nonsurgical ET in swine (Martinez et al., 2013) may lead to more widespread adoption of this technology in the commercial swine industry, especially in central nucleus group breeding schemes.

Over the past decade, there has been a major change in the commercial ET industries due to the commercialization of IVF procedures for farm animals. These techniques, originally developed in the 1980s, had previously been used to make laboratory-derived embryos available for research on subjects such as NT, transgenesis, and embryo reconstruction for study of developmental biology. Due to advances in equipment (e.g., ultrasound machines and incubators) and scientific understanding of oocyte maturation, sperm capacitation, fertilization, and culture of *in vitro* produced zygotes, success rates have climbed to a commercially acceptable level. Approximately one-third of all bovine embryos transferred globally each year are now produced *in vitro*.

C. In Vitro *Embryo Production*

IVF of mammalian oocytes has been of interest to scientists for over a century. This *in vitro* technique was first successfully used to produce offspring in rabbits (Chang, 1959). A major breakthrough occurred when the first human IVF baby was born in England in 1978 (Steptoe and Edwards, 1978); this breakthrough was worthy of the 2010 Nobel Prize in Physiology or Medicine. The first farm animal produced by IVF was a healthy bull calf reported by Brackett et al. (1982), and this birth was followed by IVF offspring in sheep (Cheng et al., 1996), swine (Cheng et al., 1996), goats (Hanada, 1985), and horses (Palmer et al., 1991).

Initial *in vitro* embryo production (IVP) utilized oocytes collected from slaughterhouse ovaries. This source of oocytes worked well for experimentation requiring large numbers of immature oocytes and was necessary to develop and refine *in vitro* laboratory procedures (and to train technical personnel). In the 1980s, it was proposed that the application of IVP in animals would likely be used with genetically valuable farm animal seedstock and possibly with endangered exotic animals.

Oocyte collection from live donors, coupled with laboratory-based IVF procedures, became commercially available to dairy and beef cattle producers in the United States in the early 1990s. With thousands of bovine offspring now being produced worldwide each year (Stroud, 2012), IVF has gone from a risky technique to a standard genetic improvement tool. With the advent of temperature-regulated portable shippers, oocytes can now be collected on the farm, placed into oocyte maturation medium in an appropriate gaseous incubation environment, and shipped via overnight courier service to the laboratory where matured oocytes will be fertilized and zygotes cultured before shipment back to the field for ET.

IVF is a multi-step process that requires a well-equipped laboratory and a skilled technician. The IVF procedure involves harvesting the oocytes from the female's ovaries, maturing them *in vitro*, and fertilizing them *in vitro* with *in vitro* capacitated sperm. The resulting embryos are held in an incubator for 7 or 8 days and then transferred nonsurgically to recipient females at an appropriate stage of the estrous cycle. With improvements in oocyte maturation and sperm capacitation methods, IVF rates of bovine oocytes are expected to be >85% (Zhang et al., 1992 and others). The pregnancy success rate for excellent quality IVF-derived bovine embryos is expected to range from 50% to 65%. However, one of the current limitations of this technology is the relative inability to successful cryopreserve IVP embryos.

IVF in the mare has not developed to the same level for in-field use as is present for cattle. Although several offspring were reported in France in the early 1990s using IVF in the mare, repeatable IVF protocols are presently not available. Attempts by many other research groups to produce IVF foals have not been successful, making it clear that more research is needed before this technology can be commercialized in horses. The reasons for the low success rate of equine IVF remain unclear. Equine oocytes have a thick zona pellucida (ZP) compared with other species, and *in vitro* maturation (IVM) of these oocytes takes longer than other domestic farm animal species (Hinrichs et al., 1993). The thick (and sometime hardened) ZP of *in vitro* matured equine oocytes may act as a barrier to sperm cells prepared *in vitro* (Li et al., 1995). The potentially altered ZP found in *in vitro* matured oocytes (Cohen et al., 1990), in addition to less than adequate sperm cell preparation, likely contributes to poorer than expected IVF rates in the horse. Although the first IVF offspring were reported in the mid to late 1980s for sheep, goats, and swine, IVF has not been accepted by the commercial livestock industry primarily due to cost of production.

Because *in vitro* oocyte maturation and *in vitro* penetration of the ZP by *in vitro* capacitated spermatozoa are fundamental problems hindering IVF development in farm animals, other assisted reproductive technologies, such as zona drilling, zona renting, subzonal sperm injection, and intracytoplasmic sperm injection (ICSI), are now under investigation for use in the farm animals.

D. Microfertilization Methods

1. Zona Renting

Zona drilling, zona renting (partial zona dissection, PZD), and subzonal insemination techniques have been developed to facilitate passage of sperm through the ZP into the ovum for fertilization (Figure 22.1). Live offspring were first produced by

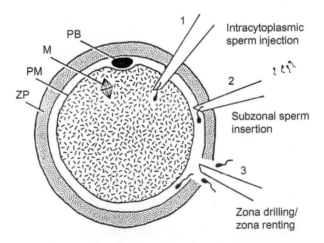

Figure 22.1 Three methods of ova microfertilization evaluated for mammalian embryos. (1) ICSI using a single spermatozoon, (2) zona renting with subzonal sperm insertion (SUZI) using a single or multiple sperms, and (3) zona drilling/zona renting allowing multiple sperm to enter through the opening in the ZP. Abbreviations: PB, polar body; M, metaphase plate; PM, perivitelline membrane; and Z, zona pellucida.

zona drilling in laboratory mice in the mid-1980s (Gordon and Talansky, 1986), and live offspring were subsequently produced in humans using a similar zona drilling procedure (Cohen et al., 1991, 1992a,b). Pregnancies and live births have also resulted from PZD and insemination of human oocytes (Malter and Cohen, 1989a,b; Cohen et al., 1989a).

Microinjecting sperm directly under the ZP with resultant full-term conceptus development and live offspring was reported in laboratory mice (Mann, 1988). Also, pregnancies have resulted from the injection of multiple, acrosome-reacted sperm cells into the perivitelline space of human ova (Ng et al., 1990). Using a subzonal insemination procedure (SUZI) to transfer multiple sperm into human oocytes, six healthy IVF babies were reported (Antinori et al., 1990). Attempts have been made to produce blastocysts using SUZI with both bovine and porcine oocytes, but only limited success with blastocyst production was attained (Clement-Sengewald et al., 1996; Nagashima et al., 1995). Parthenogenetic embryos were common in the sham control treatment groups in the swine oocyte studies. Perforating the ZP to enhance fertilization was first used on equine oocytes by Choi et al. (1994), but the size of the opening, the concentration of sperm, and the time of sperm exposure to the oocytes were found to alter expected IVF rates in the horse. Subsequently, the SUZI procedure was used successfully to produce a live offspring in the domestic cat in the late 1990s (Pope et al., 1995).

In an effort to improve zona renting technology, a microbeam from a dye tuned, compact nitrogen laser was used to make a small opening in the ZP of murine and bovine oocytes (Godke et al., 1990). Although the new approach was successful at a wavelength of 440 nm, the acridine orange stain used for activating the laser beam for microsurgery at the zona surface was detrimental to subsequent *in vitro* development of the oocytes. Further studies have since evaluated the feasibility of this approach for IVF.

2. Intracytoplasmic Sperm Injection

Studies by research groups in many countries targeted techniques to produce offspring from microinjection of sperm cells into unfertilized ova (Uehara and Yanagimachi, 1976; Thadani, 1980; Markert, 1983, and others). Using this technique, the premise was that the ovum of the female would be activated by a microinjected sperm cell (Figure 22.1). Direct injection of sperm into the ooplasm has been used extensively to study fertilization in sea urchin egg (Hiramoto, 1962) and *Xenopus laevis* egg (Brun, 1974) models. Although there had been many prior unsuccessful attempts, the ICSI procedure has been successfully used to produce live offspring from zebrafish (*Danio rerio*) eggs (Poleo et al., 2001). The first ICSI experiments in mammals were conducted in rodent species and yielded highly variable results. Formation of the male pronuclei and subsequent early cleavage was first achieved in the hamster (Uehara and Yanagimachi, 1976). Because the hamster oocyte allows for pronuclear formation with sperm from a multitude of different species, it has been a useful research tool for the development of sperm injection techniques for other animals.

The first live ICSI offspring was produced by microinjection of sperm cells into the ooplasm of rabbit ova (Hosoi et al., 1988). Subsequently, a slightly modified ICSI method created much interest after producing viable pregnancies and healthy babies in humans (Palermo et al., 1992a,b). The advent of ICSI has increased the potential of assisted reproductive technologies to propagate mammalian species, with one of the best examples being *Homo sapiens*. In humans, ICSI was first used as a method for treating male infertility (Palermo et al., 1992b; Ng et al., 1993).

IVF and normal cleavage of *in vitro* matured oocytes was first reported for cattle following ICSI with *in vitro* capacitated sperm (Younis et al., 1989). The first live ICSI offspring in farm animals, however, was reported by Goto et al. (1990), who microinjected a bovine sperm cell into bovine oocytes and produced live calves. Although others had attempted to produce ICSI-derived calves (Keefer et al., 1990; Chen and Seidel, 1997, and others); surprisingly, no additional offspring were born until Hamano et al. (1999) reported the birth of two heifer calves and eight bull calves resulting from ICSI with gender-sorted sperm.

When SUZI and ICSI procedures were evaluated on oocytes from donor mares, cleavage rates tended to be higher with ICSI (Li et al., 1995). Shortly thereafter, pregnancies produced with ICSI using oocytes from abattoir ovaries (Squires et al., 1996), nonpregnant mares (Meintjes et al., 1996; McKinnon et al., 1998; Cochran et al., 1999b), and pregnant mares (Cochran et al., 1998a,b, 2000) were reported. Multiple ICSI embryos were transferred to the oviducts of surrogate females in an effort to conserve recipient mares, and one of those females resulted in a twin-ICSI pregnancy (Cochran et al., 1999b). Sperm injection of oocytes has resulted in the birth of live offspring from mice (Roknabadi et al., 1994; Kimura and Yanagimachi, 1995) and sheep (Catt et al., 1996). Offspring from other mammals have been produced by ICSI with cat (Pope et al., 1998; Gomez et al., 2000), pig (Kolbe et al., 2000; Martin, 2000), and rhesus monkey (Nusser et al., 2001) oocytes.

SUZI and ICSI procedures were evaluated in human infertility units in an attempt to increase fertilization rates in humans with unexplained infertility. However, because of the increased efficiency and eliminating polyspermic fertilization, ICSI became preferred over the SUZI procedure (Palermo et al., 1993). Other methods of human sperm insertion have also been abandoned in favor of ICSI, which has increased from 11% of the IVF procedures in 1995 to 40% of the cases in 1998, and to >85% in some human infertility units today.

The possibility exists that freeze-dried sperm cells could be used to fertilize farm animal oocytes with microinjection techniques in the future. Although details are lacking, at least one calf was born following fertilization with previously dried bovine sperm cells (Meryman, 1960; Graham et al., 1974). It is interesting to note that live offspring were reported in mice from freeze-dried sperm cells (Wakayama and Yanagimachi, 1998). Certainly the potential exists for incorporating freeze-dried sperm cells or possibly DNA extracted from cryopreserved tissue of extinct animals to create embryos for transfer into a recipient female.

In addition, efforts are now under way to reevaluate the use of the ICSI procedure to direct foreign DNA by the sperm into the ovum to incorporate into the genome of the zygote at the time of or shortly after fertilization (see Gandolfi, 2000;

Wall, 2002; and Chapter 11, for additional insight on the potential for gene incorporation in swine with ICSI).

E.   Transvaginal Ultrasound-Guided Oocyte Aspiration for IVF

ET in cattle has become an acceptable management practice to obtain multiple offspring from valuable donor animals. Time and cost of animal preparation, limited repeatability of superovulation procedures, individual variation in response, and the interruption of the donor's estrous cycle at the expense of establishing a pregnancy, however, have made IVM and IVF of *in vivo*-derived oocytes an attractive alternative for the production of live offspring from genetically elite seedstock. Relatively simple and consistent procedures have been developed to produce IVM/IVF embryos, and these are now routinely used not only in research laboratories but also in progressive commercial ET companies in North and South America and Asia. A major barrier remaining in IVP, however, is obtaining a sufficient number of high-quality oocytes.

Early attempts at retrieving oocytes from donor cattle included surgical and the somewhat less invasive laparoscopic procedures; however, there was a limit to the number of procedures that could be performed safely without causing injury to the donor animal. In the mid-1980s, ultrasound-guided aspiration of preovulatory follicles through the vagina was developed to assist in the treatment of human infertility cases (Dellenbach et al., 1985). With this approach, oocytes can be retrieved using ultrasonography to visualize the ovary and to guide a needle transvaginally into the ovarian follicle. Fresh *in vivo*-derived oocytes aspirated from individual follicles could be subjected to IVM, IVF, and short-term *in vitro* culture prior to transfer.

A transvaginal ultrasound-guided oocyte aspiration (TUGA) method was developed for use in cattle by researchers in the Netherlands (Pieterse et al., 1988; Kruip et al., 1991) and refined by others (Looney et al., 1994; Hasler et al., 1995; Meintjes et al., 1995b). To retrieve oocytes for IVF, a trained professional inserts an ultrasound-guided stainless steel needle through the wall of the vagina near the cervix to extract the oocytes from the ovarian follicles. The ultrasonographic procedure may be especially attractive to veterinary practitioners and others already using ultrasound equipment for breeding/reproductive management and for pregnancy testing. TUGA is a viable alternative to the conventional superovulation and embryo flushing procedures for valuable donor cattle. The use of follicle aspiration and IVF has proven to aid in diagnosing the cause of infertility and possibly for the treatment of problem-breeder females (Looney et al., 1994). Currently, TUGA, also known as ovum pick-up (OPU) in human medicine (see Wikland et al., 1993), is being utilized not only in cattle, but also in mares, goats, pigs, sheep, and exotic hoofstock species.

1.   Cattle

The oocyte donor female is restrained in a suitable holding chute and administered an epidural block. A convex 5-MHz transducer is fitted into the distal end of a

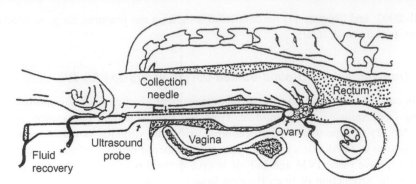

Figure 22.2 A diagram of the TUGA procedure used to harvest oocytes from cattle. Collection is guided by ultrasound with the probe fitted on the distal end of an extended handle. The ovary with the follicles to be punctured is adequately positioned on the distal end of the transducer by the hand in the rectum.
Source: Drawing by E. Meintjes.

specially designed 500-mm long plastic handle to visualize the ovaries on the ultrasound monitor. The handle (with a latex protective covering) is inserted into the vaginal canal, and then the ovary is grasped *per rectum* and placed against the vaginal wall adjacent to the transducer (Figure 22.2). An 18-gauge, 53- to 60-cm long needle is inserted through the needle guide in the plastic handle. (Alternatively, there is also equipment that utilizes disposable hypodermic injection needles.) This needle is connected to a suction pump by means of polyethylene tubing, passing through an embryo filter or into a 50-mL conical-shaped test tube for collection of the follicular fluid containing the oocytes. The oocyte recovery medium used is typically phosphate-buffered saline (PBS) with 10% bovine serum, antibiotics, and heparin. Using this aspiration method, 60–70% of the medium- to large-size follicles punctured on both ovaries result in oocytes recovered, with an average of 3–10 oocytes recovered per nonstimulated donor female.

The frequency of follicle aspiration varies. Oocyte retrievals have been performed twice a week for up to 3 months in cows with no adverse effects reported for the donor females (Broadbent et al., 1997), and other investigations have also demonstrated that oocytes may be collected multiple times per week for a period of several weeks (Aetrs et al., 2005; Chaubal et al., 2006). Most commercial facilities in North America perform aspirations 2–4 times per month per donor. Donors receiving mild ovarian stimulation will produce an average of 17 (dairy) to 22 (beef) oocytes per retrieval session, resulting in the production of 4.2 (dairy) to 5.4 (beef) transferrable quality embryos (AETA, 2012).

Oocytes can be harvested from donor cows at any time of the estrous cycle, including the early growth phase of the first follicular wave (Paul et al., 1995) and during the early postpartum (<30 days) period (Perez et al., 2000, 2001). This latter approach provides the opportunity to produce additional calves from a cow before she is mated to establish a term pregnancy.

Figure 22.3 The first calf in the world produced from an IVF embryo where the oocyte was harvested from a live pregnant cow. The Brangus donor female remained pregnant and produced her own biological calf from AI after a normal-length gestation period. The IVF calf pictured was carried by a surrogate female and was born a month later than the AI calf born to the biological mother.
Source: Photograph by M. Meintjes.

Offspring also can be produced from oocytes collected using TUGA in pregnant cows. Successful oocyte aspiration has been reported from first trimester pregnant cows (Ryan et al., 1993; Meintjes et al., 1995b), and this approach allows females who ordinarily cannot produce additional offspring during pregnancy to do so. Oocyte collection during early stages of pregnancy in cattle has not been associated with compromising an ongoing pregnancy (Ryan et al., 1993; Meintjes et al., 1995b). The first offspring produced from oocytes collected by TUGA from pregnant donors were from cattle (Figure 22.3) (Meintjes et al., 1995b). Live offspring are now routinely produced from IVF-derived embryos from both cyclic and pregnant donor cattle throughout the world.

Using a smaller transvaginal probe, a concerted effort has been made to harvest oocytes by TUGA on prepubertal dairy and beef heifers (Looney et al., 1995; Adams et al., 1996; Brogliatti and Adams, 1996; Damiani et al., 1996; Fry et al., 1996: Presicce et al., 1997). Although oocytes have been recovered from these young females, the IVF offspring born to date have been fewer than expected (Damiani et al., 1996). TUGA is now being used to harvest valuable oocytes from domestic females representing rare maternal bloodlines, reproductively senescent females, and clinically infertile cows.

2. Horses

The primary factor limiting development of new reproductive technologies in the mare is the lack of sufficient numbers of excellent quality oocytes. Due primarily to the unique anatomical structure of the horse ovary, typically only one oocyte is ovulated per estrous cycle. Coupled with the relative inability to use exogenous hormones to achieve growth of multiple ovarian follicles, one of the greatest challenges to widespread adoption of IVF in horses is the limited number of oocytes available. Also, for reason(s) unknown, typical IVF procedures have not been effective in the horse.

Mares have one or two waves of multiple ovarian follicles growing and developing during an estrous cycle (Ginther and Pierson, 1989). If this developing ovarian follicle population were available for oocyte collection using the TUGA procedure, this would help in efforts to develop a more efficient IVF procedure for the mare. Thus, researchers began to evaluate TUGA procedures on horse and pony mares to collect oocytes for IVF research (Brück et al., 1992; Cook et al., 1992; Bracher et al., 1993; Meintjes et al., 1995a).

The basic oocyte aspiration method is similar to that used in cattle, but with minor modifications. Briefly, mares require sedation instead of an epidural block, and most often a 12-gauge needle is used to puncture the ovarian follicles. Extra rinsing/scraping is necessary in the horse because the oocyte is usually well embedded in the granulosa cell layer of the follicle wall. The follicular fluid is collected into a 500-mL glass bottle, and then later passed through a standard bovine embryo filter. Using this approach, oocytes have been successfully recovered from cyclic mares (Meintjes et al., 1995a), altrenogest-treated mares (Cochran et al., 1999b), and free-ranging zebras in South Africa (Meintjes et al., 1997a).

One of the limiting factors when using the TUGA technique in mares is the operator's ability to recover appropriate-age matured oocytes from the ovarian follicles. Exogenous hCG is administered to enhance oocyte maturation, with oocyte aspiration occurring ∼24 h post-hCG administration. Oocyte recovery rate from hCG-treated mares usually ranges between 35% and 70% of follicles punctured per mare.

Various research groups began to develop ICSI for horses, where individual sperm cells are injected into the oocytes to trigger the fertilization process. Live foals have been produced using abattoir-derived oocytes (Squires et al., 1996; Meintjes et al., 1996; and others). The first pregnancies and live foals produced by ICSI with oocytes aspirated from live mares were reported the same year (Cochran et al., 1998a,b; McKinnon et al., 1998). Shortly thereafter, a twin-ICSI pregnancy was produced from the transfer of two sperm-injected equine oocytes harvested via TUGA from altrenogest-treated mares (Cochran et al., 1999b). One of the fetuses was manually eliminated, and the remaining ICSI conceptus produced a healthy live foal. After ICSI, embryos are most often surgically transferred at the two- to four-cell stage into the oviducts of suitable recipients (Cochran et al., 2000), since *in vitro* culture has still not been perfected for IVF-derived equine embryos.

The use of TUGA and IVF procedures does offer an alternative to horse owners who have genetically valuable mares that for whatever reason are unable to produce viable embryos through standard embryo collection/transfer procedures. This technology can be used on oocytes harvested from older ovulating or nonovulating mares, females with physical injuries (e.g., leg injury) and problem mares having uterine anomalies. Oocytes may also be harvested oocytes from early postpartum mares (<15 days postfoaling), allowing production of extra ICSI-derived embryos for transfer to recipient females.

One impediment to *in vivo* embryo production from the larger farm animal species (e.g., cow, mare) is that their gestation periods are lengthy in comparison with those of smaller domestic mammals and that conventional *in vivo* embryo production during gestation is not possible. However, mares exhibit ovarian follicular wave development during early to mid-gestation, and it is possible to take advantage of those developing follicles by harvesting oocytes using TUGA during early pregnancy (Meintjes et al., 1994, 1995a,b). Initially, the main concern was whether the oocyte aspiration procedure would affect the oocyte donors. Successful oocyte aspiration has been reported from first trimester pregnant mares (Meintjes et al., 1994, 1995a, 1996). This oocyte aspiration approach proved not to be a problem, and pregnant donors were found to consistently produce more oocytes per collection than similar nonpregnant, cyclic females. With this novel transvaginal approach to oocyte collection, no apparent detrimental effects on donor estrous cyclicity or evidence of ovarian adhesions were reported from repeated use of this procedure. In addition, oocyte collection during early stages of pregnancy in horses has not been associated with compromising an ongoing pregnancy (Meintjes et al., 1997b). The first offspring produced from oocytes collected by TUGA from pregnant donor mares was reported in 1998 (Cochran et al., 1998a). Live offspring are now routinely produced from IVF-derived embryos from both cyclic and pregnant donor horses throughout the world.

When using pregnant mares as oocyte donors, often there is a lack of cyclic females available to use as embryo recipients. It should not be overlooked, however, that hormone-supplemented noncyclic mares have been used successfully as embryo recipients. Various studies established that noncyclic mares could be used successfully for oocyte transfer and gamete intrafallopian transfer (Carnevale et al., 1999; Hinrichs et al., 2000). Daily administration of equine somatotropin (eST) has been used on cyclic mares to increase the number of small- to medium-size follicles for TUGA procedures (Cochran et al., 1999a). The administration of eST to seasonally anovulatory mares increased the development of small- to medium-size follicle in anestrous mares (Cochran et al., 1999c). When seasonally anestrous mares were administered daily eST followed by administration of a gonadotropin releasing hormone (GnRH) agonist, the mares developed small to medium-size follicles of which at least one follicle progressed to an ovulatory size. Based on more recent studies at Louisiana State University, equine oocytes harvested by TUGA from both medium- and ovulatory-size follicles of cyclic and pregnant mares are potentially viable for IVP.

3. Goats

Goats are another farm animal species in which IVP has proven successful. Transvaginal aspirations have also been performed on cyclic and noncyclic adult goats with good success (Graff et al., 1999). Although the methodology for puncturing the follicle is similar to that of the cow and the horse, the goat is sedated, placed under anesthesia, and placed in dorsal recumbency for the procedure.

Manual pressure is applied to the doe's abdomen in an effort to stabilize the ovaries for aspiration. The ultrasound probe, which is correspondingly smaller (similar to that used for humans), has a convex transducer at the distal end of the handle, which is inserted into the vagina. The aspiration does not require extra rinsing of the follicles to recover the oocytes as is needed for the mare.

Although the oocyte recovery rates usually range from 60% to 80% of the follicles punctured per donor female, there are some problems with aspiration of ovarian follicles from does using the ultrasound-guided transvaginal method. Firstly, the ovaries cannot be grasped *per rectum* for optimum visualization with ultrasonography. Secondly, because the ovaries cannot be easily grasped, it is more difficult to puncture follicles and aspirate the oocytes. Oocyte recovery is usually a little slower than desired because not all follicles can be visualized, and not all follicles visualized can be adequately punctured. Goat offspring have been produced using the transvaginal aspiration procedure together with IVP methods (Han et al., 2001). Although this noninvasive procedure takes expertise and patience, it is an important technology needed to reduce the risk of ovarian adhesions (and/or death loss) from using the standard surgical method to harvest oocytes from valuable donors (e.g., transgenic founder does). Efforts are now underway to modify the caprine TUGA procedure for use in mature ewes as well.

4. Swine

The TUGA technique was used successfully to harvest oocytes from mature adult sows (Bellow and Davis, 1999, unpublished data; Bellow et al., 2001). In the latter report, the 15 sows that yielded oocytes had a mean of 4.5 oocytes per female. This procedure was effective if the sow was properly restrained and if manipulation of the ovaries *per rectum* was possible for the donor female. More research is needed to refine the methodology and validate this transvaginal approach to oocyte collection in the sow.

F. Embryo Micromanipulation

Application of assisted reproductive technologies over the years generated a great deal of interest among researchers and commercial livestock producers alike. In the 1970s, it was thought that, unlike laboratory animal embryos, embryos of farm animals would not be able to survive and to produce live young following embryo micromanipulation procedures. However, studies in the early 1980s demonstrated that morulae and blastocysts of farm animal species were able to survive following microsurgery and could develop normally *in utero* resulting in the production of viable offspring. Embryo microsurgical procedures such as embryo bisection, inner cell mass (ICM) transfer, assisted hatching, and embryo sexing are now available for use in animal reproduction management schemes for farm animals.

Early studies in mammalian embryo micromanipulation evaluated the developmental potential of single blastomeres isolated from early-stage embryos of the rabbit (Pincus, 1936; Seidel, 1952; Daniel and Takahashi, 1965), mouse (Tarkowski, 1959), and rat (Nicholas and Hall, 1942). These studies provided evidence that individual blastomeres of early-stage mammalian embryos (eight-cell stage) were totipotent, and the potential existed for the production of multiple offspring from a single embryo. A major obstacle to the production of multiple offspring from early-stage embryos was the inability of precompaction-stage embryos to survive *in vivo* without an intact or near-intact ZP. The obstacle was later overcome by the development of an agar-embedding technique for micromanipulated mammalian embryos (Willadsen, 1979, 1982).

A variation on this embryo manipulation technique involved the separation of blastomeres of early-stage embryos by microsurgery and the transfer of one or more blastomeres from one embryo to one or more evacuated ZP. The separated blastomeres contained within surrogate ZP were then embedded in agar chips (see Willadsen, 1982). The agar-embedded blastomeres were transferred to the ligated oviducts of either diestrous or anestrous ewes to allow subsequent development *in vivo*. After development to the morula or blastocyst stage, the micromanipulated embryos were recovered from the intermediate host, carefully removed from the agar, and transferred to recipients. This procedure was used to produce the first sets of identical twin lambs from eight-cell embryos (Willadsen, 1979). This approach was subsequently used to produce the first identical twin calves and one set of triplet calves (Willadsen and Polge, 1981), twin foals (see Willadsen, 1982), and twin calves from day-5 to day-6 bovine embryos collected nonsurgically (Willadsen et al., 1981).

1. Embryo Bisection

Although the agar-embedding technique is an excellent method for the micromanipulation and culture of early-stage embryos, this procedure is generally considered too impractical to be used by commercial ET companies. From a practical standpoint, procedures for bisecting nonsurgically collected day-6 to day-8 embryos would allow for efficient bisection and immediate transfer of split embryos into recipient animals. In 1982, three independent research groups reported micromanipulation techniques for use on later-stage bovine embryos collected nonsurgically collected (Ozil et al., 1982; Williams et al., 1982; Lambeth et al., 1982).

Ozil et al. (1982) described a technique to produce split embryos in which two glass microneedles were used to open the ZP and a glass transfer pipette was used to remove the embryonic cell mass from the ZP. The ZP-free embryo was bisected with a microscalpel, and the resultant halves were replaced into separate evacuated ZPs with a transfer pipette. Using this micromanipulation procedure, 14 early blastocysts were bisected and transferred nonsurgically to 14 recipients, resulting in a pregnancy rate of 64% and a twinning rate of 66%.

A somewhat similar embryo bisection procedure for nonsurgically collected bovine embryos was described that same year (Lambeth et al., 1982). With this method, a fine glass needle was used to make a rent in the ZP. The embryonic cell mass was removed with the same flexible glass needle, and the embryo was bisected on a vertical plane while it rested on the bottom of the Petri dish. A glass transfer pipette was used to transfer each "half" embryo (demi-embryo) to separate evacuated ZPs. In this study, eight demi-embryo pairs and six individual demi-embryos were transferred nonsurgically to 14 beef recipients. The twin demi-embryo transfers resulted in a pregnancy rate of 63%, and the single demi-embryo transfers resulted in a pregnancy rate of 17%.

Also the same year, Williams et al. (1982) reported a procedure for splitting the bovine embryo while remaining within the ZP. This method used a razor blade chip mounted with superglue on a small glass capillary pipette to bisect the intact embryo. A fine glass pipette was then used to transfer each of the demi-embryos to an evacuated ZP. In this study, 20 good quality bovine embryos (morula to early blastocyst stage) were bisected and transferred as demi-embryo pairs either surgically or nonsurgically to 20 recipient cattle. Fourteen surgical and six nonsurgical demi-embryo transfers resulted in pregnancy rates of 64% and 17%, respectively.

Over the years, the more simplified razor blade chip method has become more widely adapted for use by commercial ET firms than the other procedures. However, in a comparative study, Mertes and Bondioli (1985) suggested there may be an advantage to the glass needle bisection procedure (Lambeth et al., 1983), because it caused less cell damage. With practice and patience, all three microsurgery procedures can be used effectively to bisect farm animal embryos and produce viable split-embryo offspring.

A simplified method for bisecting morula to hatched-blastocyst stage ovine embryos was subsequently reported by Willadsen and Godke (1984). This procedure used a fine glass needle to penetrate and to bisect ZP-intact morulae or blastocysts and ZP-free hatched blastocysts. The overall transfer pregnancy rate in sheep using this procedure was 89%. This procedure was readily adapted for use with swine, goat, and cattle embryos.

Soon thereafter, a simplified technique for bisecting later-stage farm animal embryos was reported (Rorie et al., 1985). This technique allowed for rapid bisection of intact embryos without the use of a commercial micromanipulator unit. With this method, an intact embryo was placed in a small microdrop of holding medium on a microscope slide, and a razor blade held by a hemostat was used to bisect the embryo. In that study, 98% of the intact bovine embryos were successfully bisected using this razor blade method. Pregnancy rates obtained with demi-embryos produced by the hand-held razor blade procedure were not different than those obtained with the glass needle bisection procedure (Lambeth et al., 1983). The new razor blade procedure is simple to execute, inexpensive, and easy to learn.

After efficient micromanipulation procedures were developed, research efforts were directed toward identifying factors that contribute to optimal pregnancy rates from bisected embryos. Experience of skilled operators revealed that excellent and good quality embryos were suitable for embryo bisection. The transfer of demi-embryos produced from either excellent or good quality bovine embryos resulted in pregnancy rates similar to that obtained with intact embryos of the same embryo quality grades (Voelkel et al., 1984b,c; Baker et al., 1984). Pregnancy rates expected from single demi-embryos nonsurgically transplanted to individual bovine recipient females ranged from 45% to 65%.

Studies using bovine embryos have shown that later-stage ZP-free demi-embryos survive *in vivo* as well as do ZP-enclosed demi-embryos (Voelkel et al., 1984b; Warfield et al., 1986). The highest pregnancy and twinning rates have been obtained when demi-embryo pairs rather than a single demi-embryo were transferred to recipient females (Ozil, 1983; Lambeth et al., 1983; Baker and Shea, 1985). Pregnancy rates were similar when demi-embryo pairs were transferred to the same uterine horn or when one demi-embryo was transferred to each uterine horn of recipient cattle (Baker and Shea, 1985).

Initially, it was suggested that a single quarter embryo was less likely to produce a sufficient luteotrophic signal *in utero* needed to prevent luteal regression in recipient females (Willadsen and Polge, 1981). Correspondingly, the reduced size (mass) of the developing blastocyst from a quarter embryo may lack sufficient embryonic cells in the ICM to form a viable conceptus. However, Voelkel et al. (1985b) produced live multiple embryo offspring from quartering a bovine embryo. Each quarter embryo was cultured and placed into a single recipient female. Voelkel et al. (1986) bisected cultured blastocysts that had been originally derived from bovine demi-embryos. In this case, only 17.6% of the quarter embryos derived from cultured demi-embryos produced a second blastocyst during *in vitro* culture.

2. ICM Transfer

In terms of saving endangered species, one important ET procedure is the transfer of embryos between closely related donor and recipient species. During the 1980s, some success was reported from the transfer of embryos from one exotic species to another more abundant surrogate species (Stover et al., 1981; Dresser et al., 1984). This technique had the potential for rapidly expanding small animal populations by using the rare animals only as embryo donors, while animals from a more plentiful species could serve as recipients. In cases where a closely related species would not carry the donor's embryo to term due to differences in gestation length, placental morphology, and/or maternal-conceptus immunocompatibility problems, a microsurgical embryo reconstruction technique known as ICM transfer was hypothesized as a viable option.

The ICM-transfer method involves removal of ICM cells from a blastocyst-stage embryo of the donor species and placing the isolated ICM cells into a blastocyst-stage embryo of the recipient species whose ICM had previously been removed. Because the trophoblastic cells develop into outer fetal placental tissue layers, this reconstructed ICM-transfer embryo would have placental tissue characteristics that matched those of the recipient animal's uterus. The fetus developing within the amnion will be of the donor species, because it originates from the transplanted ICM.

A good example of this technology would be to use the sheep and goat (two different but related animal species) as embryo donor females. When a sheep embryo is transferred to a female goat recipient or a goat embryo is transferred to a female sheep recipient, the transplanted embryos do not develop to term. If the ICM of the sheep embryo was microsurgically removed and transferred into the goat embryo with the ICM removed, the composite embryo could then be transferred to a goat recipient. The chance for offspring to be born from this composite embryo is enhanced, because the placental membranes that develop from this embryo would be derived from the original goat embryo. In this procedure, the goat ICM could then be placed in the remaining trophectoderm cells of the sheep embryo and this composite embryo could be transferred to a female sheep recipient. The end result would be an interspecies transfer of two embryos, with the goat giving birth to a lamb and the ewe giving birth to a goat offspring. Using this approach, sheep and goat embryos were used to produce unique chimeric and ICM-transfer offspring (Butler et al., 1987; Polzin et al., 1987). Using a new simplified micromanipulation procedure (Figure 22.4), Rorie et al. (1994) produced viable lambs from a goat recipient using ICM transfer from the ovine embryo to the caprine embryo without ICM.

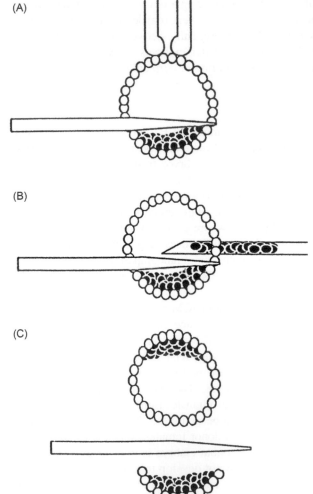

(A)

(B)

(C)

Figure 22.4 A simplified method used for the ICM transfer procedure between blastocysts. After the ZP is removed, the foreign ICM is placed in the blastocoel cavity of the host blastocyst. The host ICM is then microsurgically removed with a flexible glass needle (Rorie et al., 1994).

The use of ICM-transfer techniques could be used to save some endangered species, as the more abundant recipient species could carry the embryos of endangered species to term. Examples include Indian Gaur cattle embryos transferred into domestic cattle and Przewalski's horse embryos placed in domestic mare recipients.

3. Assisted Hatching

Embryo hatching is a critical step in embryo development. In most mammalian species, there are three primary factors involved in the hatching process. These factors include a physical pressure from blastocyst expansion (Nieder and Caprio, 1990), hormone-dependent uterine events (McLaren, 1970), and the effect of embryo-derived enzymes (McLaren, 1968; Schiewe et al., 1995a).

The *in vitro* production of farm animal embryos requires culturing embryos *in vitro* to support development to a morphological stage that is suitable for optimal transfer pregnancy rates. Zona hardening has been reported to occur during extended *in vitro* culture, and this likely impairs the normal hatching process (Alikani and Cohen, 1992). Previous results show that human embryos derived from IVF and *in vitro* culture may lack the ability to produce ZP lysin, even though the embryos appear to be developing normally *in vitro* (Doklas et al., 1991). Such embryos fail to implant properly in the uterine wall and are often lost because they fail to hatch from their zonae pellucidae (Doklas et al., 1991).

Zona opening (renting) techniques have been used in an effort to assist embryos during the hatching process. Opening the ZP has developed into an "assisted zona hatching" (AZH), and this technique has been used to increase embryo implantation rates (Matler et al., 1989b; Cohen, 1990,1991,1992a,b; Obruca et al. 1994). To perform assisted hatching, an opening is made in the ZP to help the embryo hatch from the opening. This opening (rent) is made using either physical (Cohen, 1991) or chemical methods (Gordon and Dapunt, 1993). Additionally, the laser has been used to open the zona to assist hatching (Godke et al., 1990, Laufer et al., 1993; Reshef et al., 1993; Obruca et al., 1993; Nccv ct al., 1995; Schiewe et al., 1995b).

Research has shown that AZH not only significantly improves the hatching rate of mouse embryos (Alikani and Cohen 1992; Obruca et al., 1994), but also improves the implantation rate after ET (Gordon and Dapunt, 1993). In our experience, frozen-thawed mouse embryos were used to evaluate the effects of laser-assisted hatching on the *in vitro* hatching process of the embryos when they were either cocultured with a somatic cell monolayer or cultured in medium only. The results showed that, although the laser treatment and the coculture with somatic cells had no significant effect on the blastocyst hatching of murine embryos, both treatments increased the *in vitro* hatching rate of the embryos.

Clinical trials have suggested that the embryonic implantation rates of human embryos with thick zonae (>15 μm), and with less than five blastomeres on the morning of day 3, or with over 20% fragmentation rate within the zona, can be significantly increased by assisted hatching (Cohen et al., 1992a,b). Assisted hatching is considered to enhance embryo implantation not only by mechanically facilitating the hatching process but also by allowing earlier embryo-endometrial contact, which likely optimizes synchronization between the embryo and the endometrium (Liu et al., 1993).

Kruger et al. (1996) has reported that PZD improved the *in vitro* hatching rate of IVF *in vitro* cultured bovine blastocysts over that of the *in vitro* control group. Using a compact nitrogen laser, Gao et al. (1996) also was able to enhance zona hatching in later stage, frozen-thawed cattle embryos.

G. Embryo Sex Determination and Preimplantation Genetic Diagnosis

One simple approach to gender determination was to bisect the embryo and identify the genetic sex of one half of the embryo. Once the genetic sex was established, then the remaining half of the embryo was transferred to a recipient female (Nakagowa et al., 1985). In an early study, White et al. (1987) bisected bovine embryos and sexed one demi-embryo of the pair using an H-Y antibody procedure. Each demi-embryo of the pair was transferred to a different recipient animal. The success rate for embryo sexing was 90%, and there was no significant difference found in pregnancy rates between the sexed demi-embryos and control demi-embryos (47% versus 44%).

The application of polymerase chain reaction (PCR) DNA amplification technology to sex determination in biopsies from farm animal embryos proved successful (Peura et al., 1991; Kirkpatrick and Monson, 1993) and did not result in a significant reduction in postbiopsy ET pregnancy rates. In addition to the amplification of Y-chromosome-specific DNA, amplification of X-chromosome or autosomal DNA (as a control for the PCR technique) is necessary. The removal of a few cells (or perhaps only one cell) from the trophoblast of the blastocyst-stage embryo enables a very rapid ($\sim$ 1 h) and accurate ($\sim$98%) determination of genetic sex. A single cell of a bovine embryo may also be used for multiple genotype analysis (Chrenek et al., 2001).

There is great interest, particularly in the dairy cattle industry, in the production of offspring with a predetermined genetic sex. Although the equipment and the supplies needed to sex bovine embryos are commercially available and the PCR-based embryo sexing technique is rapid and reliable, the advent of sexed semen in the cattle industry has led to a significant drop in demand for embryo sexing.

At present, research efforts with embryo biopsy are being directed toward the harvest of cells for subsequent preimplantation genetic diagnosis (PGD). Preimplantation embryos at risk for carrying a deleterious genetic condition can be screened for the presence of the defect, and either be discarded or potentially "fixed" using the TALENS (Liu et al., 2014) or CRISPR/Cas9 (Niu et al., 2014) gene editing technique. As the technology for single nucleotide polymorphism (SNP) chip technology advances, it may soon be possible to simultaneously screen for thousands of genes from a single biopsy. It is likely that breeding studs, as well as elite seedstock producers, will use embryo biopsy in conjunction with PGD to identify genetically superior animals to enhance food and fiber production to feed and clothe the growing global human population.

III. Embryo Culture Systems

A. Background

Experimental evidence had indicated that ET success markedly improves once embryos of large animal species reach the morula and blastocyst developmental stages prior to transfer to recipients. If simple and effective *in vitro* culture systems were available, the chances of successful pregnancy would likely increase if micromanipulated embryos (e.g., NT, gene insertion) were cultured to morulae and blastocysts before transfer to surrogate females. Furthermore, embryo culture prior to transfer may also serve as a valuable screening tool to identity embryos that appear to be of marginal quality yet still possess developmental potential. Embryo culture methodology may also have potential application for use with poor-quality embryos resulting from IVF procedures.

Mammalian embryo culture has been an active area of biological research for over a half of a century. Although a multitude of research papers have been published in the scientific literature, comparatively little is understood about the biochemical components necessary to maintain normal development of mammalian embryos in an *in vitro* environment (Bavister, 1987, 1988; Whittingham and Biggers (1968)). Research studies have established that effective *in vitro* culture systems require proper temperature, pH, buffering capacity, nutrients, and the presence of still undefined biological components for embryonic development to proceed *in vitro* (Kane, 1987).

Some of the earliest efforts in maintaining embryo development outside of the female's reproductive tract were conducted using rabbit embryos. In an early study, Brachet (1912) reported development of primitive groove and rudimentary placental structures when culturing rabbit blastocysts within small plasma clots in glass dishes. However, it was not until some years later that Lewis and Gregory (1929) reported using blood plasma as the medium for the development of one-cell rabbit embryos to the eight-cell stage during 48 h of *in vitro* culture.

Efforts to develop culture media for embryos have subsequently centered around the addition of diluted blood plasma or serum and other biological fluids to enhance embryo growth and to increase buffering capacity of the *in vitro* culture system. Carrel (1913) first noted that extracts from chick embryos increased the growth of mammalian cells maintained *in vitro*. Subsequently, Pincus (1930), using hanging-drop cultures containing various mixtures of rabbit plasma, chick plasma, rabbit embryo, and chick embryo extracts (CEEs), observed *in vitro* development of two- and four-cell rabbit embryos to the morula stage. Waddington and Waterman (1933) showed that when CEEs and chicken plasma were incorporated into the culture medium for later-stage rabbit embryos, embryonic cell differentiation reached the primitive streak developmental stage *in vitro*. One of the first successful cultures of mouse embryos, incubation in a physiological saline, required supplementation with egg white and yolk from the domestic chicken egg (Hammond, 1949). However, when an attempt was made to culture bovine embryos in physiological saline supported with egg white or yolk, this approach did not support successful embryo development (Dowling, 1949).

Chang (1949) demonstrated that heat-inactivated serum could be used successfully to supplement the culture medium for two-cell rabbit embryos. This success with rabbit embryos stimulated attempts to develop culture systems for farm animal embryos using heat-inactivated bovine serum as a supplement to the culture medium. In retrospect, this was likely due to use of fresh rather than heat-inactivated serum in the culture medium. Onuma and Foote (1969) first reported success culturing one-cell bovine embryos when heat-treated bovine and rabbit sera were added in the culture medium. In this study, 45% cleavage rate was obtained from 184 fertilized ova incubated *in vitro*.

Gordon (1975) was one of the first to use PBS supplemented with heat-treated bovine serum for short-term culture of bovine embryos, and 60% of the embryos developed *in vitro* in this study. Wright et al. (1976) used a bicarbonate-buffered medium (Ham's F-10) with 10% heat-treated fetal bovine serum in a 5% CO_2 atmosphere to culture later-stage embryos, and they reported very good success in developing ovine embryos with this culture medium. These early culture studies and others (see reviews by Renard et al., 1976; Wright and Bondioli, 1981) using PBS, Ham's F-10 medium, Tissue Culture Medium-199, and heat-treated fetal bovine serum have formed the basis of the embryo culture media recipes to culture prehatched farm animal embryos in commercial ET companies over the last 20–30 years.

B. In Vitro *Block to Development*

One common finding in the early attempts at culturing early-stage mammalian embryos was the apparent *in vitro* developmental block that occurred at various morphological stages in the animal species evaluated. The *in vitro* "developmental block" was first described for mouse embryos and subsequently documented for embryos of other laboratory and farm animal species (Table 22.1). Early studies

Table 22.1 *In Vitro* Developmental Block Stage for Common Mammalian Species

Mammal	Developmental Stage	Scientific Reference
Hamster	2- to 4-cell	Whittingham and Bavister (1974)
Rat	4- to 8-cell	Markert (1984)
		Bavister (1988)
Rabbit	Morula	Kane and Foote (1970)
Cat	Morula	Swanson (1990)[a]
Cow	8- to 16-cell	Thibault (1966)
Sheep	8- to 16-cell	Wintenberger et al. (1953)
Goat	8- to 16-cell	Betteridge (1977)
Swine	4- to 8-cell	Davis and Day (1978)
Horse	?	N/A[b]
Human	No observed block	Edwards et al. (1981)

[a]Unpublished data (1990).
[b]Data not available; little success has been identified for culture of precapsulated equine embryos.

indicated that the one-cell mouse embryos could easily develop to the two-cell stage *in vitro*; however, these two-cell embryos failed to undergo further cleavage and characteristically degenerated in the culture medium (Cole and Paul, 1965; Whitten, 1971). Subsequent studies later identified medium components suitable for culture of mouse zygotes through the blastocyst stage of development *in vivo* (Brinster, 1972).

In early experiments involving bovine embryos, development of early-stage embryos (one to four cells) were observed to proceed to the 8- to 16-cell stage *in vitro* before they ceased in development (Thibault, 1966). It was noted, however, that corresponding bovine embryos collected from donors at the 8- to 16-cell stages would readily develop to the morula and blastocysts stages *in vitro* using a simple culture medium. Eyestone and First (1986) showed that bovine embryos "blocked" at the 8- to 12-cell stage *in vitro* usually could not be rescued, even if the embryos were returned to an *in vivo* environment. These culture observations with embryos of laboratory and farm animals suggest an inadequacy in the *in vitro* culture systems at this developmental stage, thus resulting in the characteristic *in vitro* block to development.

Eyestone and First (1986) proposed that the embryonic cells were alive during the initial stage of the block but that they were not capable of dividing after being arrested in development. Although *in vitro* developmental blocks have been shown to occur at different developmental stages in laboratory and farm animal species, this *in vitro* block has not been reported for the human zygote (Edwards et al., 1981).

Evidence indicates that the transition from maternal to embryonic messenger RNA (mRNA) dependence in bovine embryos occurs at the same time as the 8- to 16-cell bovine *in vitro* developmental block. In an important study, Frei et al. (1989) cultured both fertilized oocytes and embryos of cattle with radiolabeled methionine and noted that a progressive decrease in protein synthesis occurred from the oocyte to the eight-cell stage. This was in contrast to an increase in protein synthesis from the eight-cell to the blastocyst stage of development. King et al. (1989) have reported that bovine embryos begin to synthesize ribosomal RNA (rRNA) at the time of maternal to embryo transition, starting at the eight-cell stage, to further substantiate this hypothesis.

C. In Vivo *Oviduct Culture Systems*

In an effort to overcome apparent inadequacies when *in vitro* culture of embryos is conducted in medium alone, alternative *in vivo* embryo culture methods were developed and evaluated. One of the early successful methods for enhancing early-stage mammalian embryo development was to use *in vivo* culture in the oviducts of an intermediate host animal (e.g., rabbit). In a classic study, Averill et al. (1955) transferred early-stage 2- to 12-cell ovine embryos to the ligated oviducts of pseudopregnant rabbits for 4 or 5 days, resulting in morula- and blastocyst-stage embryos developing for transfer to recipient ewes. This method was used by Hunter et al. (1962) to transport sheep embryos collected in Cambridge, England, to South Africa for subsequent transfer to recipient ewes. The storage interval for embryos in rabbit oviducts ranged from 101 to 128 h in this study; four lambs were born following the transfer of 16 transported embryos.

In a more extensive study, 456 early-stage ovine embryos (2- to 12-cell stage) were transferred to the ligated oviducts of pseudopregnant rabbits (Lawson et al., 1972a). Following *in vivo* culture, 87% of the embryos were recovered from the intermediate host with 93% of these developing to later-stage embryos while in the oviducts. Survival rates of up to 69% were recorded when embryos were subsequently transferred to synchronized recipient ewes. The same year, Lawson et al. (1972b) transferred 48 one- to eight-cell bovine embryos to the ligated oviducts of pseudopregnant rabbits. Of the embryos recovered 2–4 days later, 83% advanced to later stages. Fifteen embryos were transferred and 73% resulted in live calves. This finding substantiated the initial studies of successful development of bovine (Sreenan and Scanlon, 1968; Sreenan et al., 1969) and caprine embryos (Agrawal et al., 1983) in the oviducts of rabbits. The feasibility of using rabbit oviducts for viability screening of bovine embryos prior to transfer was critically reviewed (see Boland, 1984).

Culture of agarose-embedded embryos in the ligated oviducts of sheep became an important *in vivo* method when IVF and/or micromanipulated (e.g., NT) farm animal embryos required culturing prior to transfer. This *in vivo* culture procedure was developed by Willadsen (1979) to facilitate the *in vitro* development of micromanipulated ovine reconstituted embryos. In this classic study, blastomeres of two-cell embryos were embedded in agarose and transferred to the ligated oviducts of ewes for 3–4.5 days. Of the 40 embryos surgically recovered from the ligated oviducts (65% of those embedded), 88% developed in the intermediate hosts to late morulae and early blastocysts. This same approach using the ligated oviducts of sheep was also successful for micromanipulated embryos of cattle, swine, and horses (Willadsen, 1982). Although the rabbit oviduct was initially used for *in vivo* culture of farm animal embryos, the sheep oviduct culture system became the method of choice for culturing IVF (Gordon and Lu 1990) and NT–derived zygotes (Westhusin et al., 1989; Marek et al., 1990; Bondioli et al., 1990). As the expertise level increased with this *in vivo* culture method, embryo recovery rates ranged from 93% to 97% (Bondioli et al., 1990).

In spite of the effectiveness of this *in vivo* culture technique over that of media-based *in vitro* culture systems, it has not been readily accepted for general use by most commercial animal ET companies. The primary disadvantages of using ligated sheep oviducts for intermediate embryo culture is the time required, amount of effort needed and difficulty involved in performing multiple surgical procedures on the sheep intermediate host. In addition, a flock of ewes must be maintained as host animals, and this often proves to be costly for the ET companies.

In an effort to reduce the cost of *in vivo* embryo culture, Ebert and Papaioannou (1989) evaluated mouse oviducts for culturing porcine embryos. In the initial study, porcine embryos were transferred to the oviducts of immature mice. When four- to six-cell porcine embryos were cultured in mouse oviducts, 77% reached the blastocyst stage of development compared with only 10% in the *in vitro* cultured group. Correspondingly, when early blastocysts were cultured in mouse oviducts for

2 days, the embryos recovered had twice as many blastomeres compared with that of control embryos cultured in medium alone. Although this method of *in vivo* culture shows promise, placement and recovery of embryos in mouse oviducts requires a great deal of time, skill, and patience.

IV.   Embryo Coculture Systems

A.   Uterine Cell Coculture

In a classical cell culture study, Cole and Paul (1965) used irradiated HeLa cells to coculture two-cell murine embryos through the hatching blastocyst stage. They noted marked improvement in *in vitro* developmental rates of cocultured embryos when compared with that of embryos cultured in medium alone.

Many different coculture systems have been developed to enhance embryonic growth and development *in vitro*. Initial sources of cells for coculture were of reproductive origin, used in hopes of mimicking *in vitro* the *in vivo* environment of the uterus or the oviduct. In the early 1980s, coculture systems incorporated monolayers of fibroblast and fibroblast-like cells derived from different animal sources. The fibroblast monolayers for embryo coculture were most often prepared following three to five passages of cells after their initial outgrowth from uterine endometrial explants (Kuzan and Wright, 1982a,b; Voelkel et al., 1984a; Voelkel et al., 1985a). In one of the earliest studies coculturing cattle embryos with these "helper" cells, Kuzan and Wright (1982a) cultured morula-stage embryos and reported a higher percentage of hatched blastocysts with the use of either uterine or testicular fibroblast monolayers (40% and 41%) compared with morulae cultured in medium alone (4%).

Allen and Wright (1984) subsequently showed that early-stage porcine embryos (four-cell to morula) had a greater developmental rate when cocultured on passaged cells derived from porcine endometrial tissue compared with embryos cultured in medium alone. In addition, the bovine uterine fibroblast coculture system was found to enhance *in vitro* development of porcine embryos (Kuzan and Wright, 1981, 1982b). Furthermore, porcine embryos cocultured on a bovine fibroblast monolayer resulted in a greater number of embryos that hatched and subsequently attached to the substratum during coculture. Kuzan and Wright (1982a) proposed that the fibroblasts of the monolayer released embryo growth factor(s) into the culture medium that enhanced embryo development, or alternatively that the fibroblasts removed toxic substances from the medium, thus resulting in enhanced *in vitro* development. Allen and Wright (1984) further suggested that cell contact between embryos and the fibroblast monolayer was needed to enhance embryonic development in a fibroblast monolayer culture system.

Several years later, a similar uterine fibroblast monolayer system was evaluated using bovine demi-embryos (Voelkel et al., 1985a). Improved viability of bovine demi-embryos was evident on uterine fibroblast monolayers after 72 h of coculture, compared with corresponding demi-embryos cultured with Ham's F-10 medium alone. Furthermore, a beneficial effect of the fibroblast monolayer on these micromanipulated demi-embryos resulted during their *in vitro* cellular repair process. Shortly thereafter, Wiemer et al. (1987a) developed a fetal uterine cell monolayer culture system that incorporated uterine fibroblasts derived from near-term ($\sim$270 days) bovine fetuses. This fetal bovine uterine fibroblast coculture system enhanced *in vitro* development of mouse (Wiemer and Cohen, 1988, unpublished data), bovine (Wiemer et al., 1987a,b), equine (Wiemer et al., 1988a,b, 1989a), and human (Wiemer et al., 1989b,c; Cohen et al., 1989b) embryos over those cultured in medium alone.

A majority of the studies have reported the culture of embryos on monolayer systems using primarily fibroblasts originating from reproductive tissue. In subsequent embryo culture studies, endometrial cells (e.g., epithelial or epithelial-like cells) were isolated from the uterine lining of the cow (Rodriguez et al., 1990; Thibodeaux et al., 1991b, c) and the goat (Blakewood et al., 1990a; Prichard et al., 1992). Prichard et al. (1992) cocultured two- to four-cell caprine embryos on caprine oviductal and uterine epithelial monolayers. Adequate hatching rates were noted on uterine cell monolayers (63%); however, oviductal cell coculture had a greater number of embryos hatching (87%) compared with the uterine cell monolayers.

In an effort to maximize embryotrophic capabilities within a single culture system, Pool et al. (1988b) evaluated the use of both the endometrial fibroblast monolayer and trophoblastic vesicles in one culture system for developing bovine embryos *in vitro*. Quality grade 1 morulae were randomly and individually placed into single wells of culture and either incubated with control medium with a bovine trophoblastic vesicle (bTV), with a fetal bovine endometrial fibroblast monolayer, or with a combination of a bTV and a fetal fibroblast monolayer. The number of viable embryos at 72 h in culture in both the control and the combination culture treatments was significantly less than those cocultured with either a bTV alone or with a fetal fibroblast monolayer. The reason for the poor showing when the two different helper cell types were used in a combination treatment remains unclear.

In the early 1990s, uterine epithelial cells became the helper cell of choice for coculture preimplantation farm animal embryos. It appears that the source of cells and the day of the cycle on which cells are harvested for culture influence the functional capability of the uterine cell coculture systems. Thibodeaux et al. (1991a,b) reported that changes in cell function *in vitro* were associated with a changing hormonal profile occurring across different days of the bovine estrous cycle. Both uterine and oviductal epithelial cells had the lowest percentage of cell attachment during primary culture when isolated from animals between days 8 to 10 and days 14 to 16 of the estrous cycle. In contrast, the highest percentage of cell attachment was noted for cells isolated between days 4 and 6 of the cycle.

B. Oviduct Cell Coculture

Although positive results were reported with the use of uterine cells as a coculture system, subsequently oviduct cells have been shown to more effectively enhance the development of early-stage embryos *in vitro*. It was proposed that oviduct epithelial cell coculture systems provide an *in vitro* environment similar to embryonic development *in vivo* (Bavister, 1988). It has also been shown that oviduct cells of various animal species are capable of stimulating embryonic development in other animal species. Earlier studies have demonstrated the capabilities of the rabbit oviduct to enhance or maintain embryonic development in mouse, sheep, cow, pig, goat, and horse embryos (see Bavister, 1988; Boland, 1984).

Oviduct cells used for coculture are generally obtained by scraping the luminal surface of excised oviducts (Rexroad and Powell, 1986, 1988) or by collecting cells during embryo recoveries (Gandolfi and Moor, 1987). An effective method of flushing the excised oviducts of farm animals to harvest epithelial cells was reported soon thereafter (Ouhibi et al., 1989; Thibodeaux et al., 1991b,c). Unlike earlier studies using explanted oviducts for culture, with this approach epithelial cells and epithelial-like cells from the lumen of the oviduct are harvested to prepare growing populations of cells *in vitro* for embryo coculture.

Rexroad and Powell (1986) were among the first to report the use of oviductal epithelial cell monolayers for short-term culture of early-stage ovine embryos. A number of laboratories confirmed that oviductal cell cultures were capable of supporting development of early-stage ovine embryos when compared with that of control medium alone (Bunch et al., 1987; Maciulis et al., 1987; Rexroad and Powell, 1988).

Gandolfi and Moor (1987) evaluated the developmental potential of pronuclear-stage ovine embryos cocultured on ovine oviductal epithelial and uterine fibroblast monolayers. In this study, both monolayer systems supported embryo development to the blastocyst stage. However, after 6 days of incubation, 42% of embryos developed into expanded blastocysts using oviductal cell monolayer coculture compared with only 5% cocultured on uterine fibroblast monolayers. The transfer of embryos cocultured on oviduct cell monolayers resulted in higher pregnancy rates when compared with those cultured on fibroblast monolayers.

Oviductal epithelial cell monolayers were successfully used for *in vitro* coculture of embryos from cattle (Rodriguez et al., 1990; Eyestone et al., 1987; Eyestone and First, 1988, 1989; Ellington et al., 1990a,b,c), sheep (Maciulis et al., 1987), and goats (Prichard et al., 1992). Eyestone and First (1989) further evaluated the effects of oviductal cell coculture on embryonic growth by comparing this culture system with that of oviductal-cell-conditioned medium. A higher rate of development was obtained when one- to eight-cell bovine embryos were cocultured on oviduct cell monolayers compared with that of culture in medium alone. When oviduct cell coculture was compared with that of conditioned medium, *in vitro* developmental rates were similar.

Oviductal cell coculture systems provide adequate developmental rates for early-stage mammalian embryos. These same coculture systems have been used in development of IVF-derived bovine embryos (Eyestone and First, 1988; Fukui, 1989; Rexroad, 1989; Kitiyanant et al., 1989) and IVF-derived chimeric embryos (Kinis et al., 1990). In each case, oviduct cell monolayers were more effective than culturing embryos in medium alone. There were no differences in blastocyst development among oviduct cells, trophoblastic vesicles, and amniotic sac cells, with developmental rates of 39–51% (Aoyagi et al., 1990). These culture systems were also more effective in stimulating blastocyst development than cumulus cell coculture, culture in rabbit oviducts, or culture in medium alone.

A bilayered coculture system of bovine oviductal cells and bovine granulosa cells was developed to culture morulae-stage bovine embryos (Rodriguez et al., 1991). Costar7 transwell culture plates with a collagen treated, microporous membrane (0.4 μm pore size) were used to separate the two-cell populations during *in vitro* culture. After 72 h of *in vitro* culture, a greater percentage of hatched embryos resulted when they were cocultured in a bilayered system with oviduct and granulosa cells (75%) than when embryos were cultured in either oviductal cells (60%) or granulosa cells (50%) alone. This finding suggests a positive embryotrophic interaction when bovine oviduct cells and granulosa cells were maintained in the same coculture system.

An ongoing hypothesis suggests that oviductal epithelial cell feeder layers enhance the development of preimplantation embryos by one or more mechanisms. Firstly, it is likely that oviductal cells secrete specific growth factors (e.g., peptides) that are required by early-stage embryos to maintain normal rate of development (Gandolfi and Moor, 1987). Another possibility is that these cell feeder layers also have the ability to remove toxic components that are detrimental to the embryo from the surrounding medium during coculture (Bavister, 1988). In addition to secreting embryotrophic components and/or removing embryotoxic substances from the medium, Bavister (1988) has proposed that the oviductal cells may lower oxygen tension in the culture medium to enhance embryo development. It was suggested that the major difference between uterine fibroblast and oviductal epithelial cell coculture systems is that oviduct epithelial cell feeder layers are capable of reducing the oxygen tension, thereby improving embryo development over that of culture medium alone. Fibroblasts, however, are thought to be able to secrete only embryotrophic substances.

Results thus far suggest that oviductal cell monolayers are the most effective coculture system for early-stage farm animal embryos, whereas fibroblast monolayers are adequate for development of later-stage (morula to blastocyst) embryos (Fukui, 1989). Correspondingly, Gandolfi and Moor (1987) have shown that pronuclear-stage sheep embryos cultured to later stages on fibroblast monolayers generally produce adequate pregnancy rates.

C. Granulosa Cell Coculture

Evidence has accumulated to show that bovine granulosa/cumulus cells are important for the IVM of bovine oocytes obtained from abattoir ovaries. In fact, granulosa/cumulus cells not only are beneficial for IVM of oocytes but also appear to play a role in subsequent development of IVF-derived embryos to the morula and blastocyst stages *in vitro* (Critser and First, 1986). Faundez et al. (1988) reported that IVF rates for bovine oocytes were higher in the presence of granulosa cells than when similar oocytes were exposed to IVF procedures without the aid of granulosa cells. The highest fertilization rates were achieved when cumulus-intact bovine oocytes were incubated on granulosa cell monolayers prior to and during the IVF procedure.

In addition, Goto et al. (1988a) reported that ET pregnancies could be obtained following coculture of IVF-derived bovine embryos with bovine cumulus cells for 6 or 7 days *in vitro*. In another study using granulosa cell coculture, Goto et al. (1988b) reported that 25% of *in vitro* matured, *in vitro* fertilized (IVM/IVF) bovine oocytes reached the eight-cell stage after 3–4 days of incubation, and 21% reached the morula and blastocyst stages. Berg and Brem (1990) noted a significantly higher rate of development to morulae and blastocysts (32%) with granulosa cell coculture compared with coculture on a monolayer of oviductal epithelial cells (17%). These findings were similar to those IVM/IVF bovine oocytes cocultured with bovine cumulus cells (Zhang et al., 1990, 1991a,b,c, 1992). Zhang et al. (1991a) reported that 54% of the IVF bovine oocytes cleaved and 41% reached the morula stage of development *in vitro* with a simple bovine cumulus cell coculture system. These findings were similar to those of Younis and Brackett (1990), who used a bovine cumulus cell coculture system for IVF-derived bovine embryos.

The major advantage of the granulosa cell coculture system is that the cells are readily obtained from the ovarian follicles at the time of oocyte collection. This makes the procedure simple to implement at little or no additional cost when compared with obtaining trophoblastic, uterine, and/or oviductal tissues for preparation of the other *in vitro* culture systems.

One interesting note is that the embryotrophic effect of granulosa cells exists across animal species. Baird et al. (1990) reported good developmental success when culturing mouse embryos with granulosa cells from hamsters, showing that an embryotrophic response likely exists across these two species. The bovine cumulus cell coculture system has been effective in supporting development of porcine IVF-derived embryos (Zhang et al., 1991b). Possibly in the future, commercially frozen granulosa/cumulus cells from laboratory animal species will be used to coculture farm animal embryos.

D. Trophoblastic Vesicle Coculture

In a classic study, Heyman et al. (1984) microsurgically sectioned trophoblastic cell layers of day-12 to day-14 elongating conceptuses (cattle and sheep) and noted that these pieces formed spherical vesicles after several days of *in vitro* incubation at 37°C. These newly formed spherical vesicles (termed *trophoblastic vesicles*) were thought to secrete cell-active luteotropic and embryo growth factors, and it was recommended that these small vesicles be used for *in vitro* embryo culture.

Camous et al. (1984) overcame the 8- to 16-cell *in vitro* block to bovine embryo development by coculturing embryos with vesicles prepared from trophoblastic layers of day-13 or day-14 elongated bovine blastocysts. In this study, 46% of the one- to eight-cell embryos developed to the morula stage when cultured with trophoblastic vesicles for 3–4 days compared with <20% reaching the morula stage when embryos were cultured in the control medium alone. Similar studies were conducted by Heyman et al. (1987a), where one- to eight-cell bovine embryos were cocultured with trophoblastic vesicles prepared from day-14 elongated bovine blastocysts. A total of 46% reached the morula stage compared with only 18% reaching the morula stage in medium alone. These findings suggest that the day-12 to day-14 developing conceptus of sheep and cattle produces embryotrophic factors that are capable of enhancing development of embryos from the one-cell stage to the morula stage *in vitro*. Camous et al. (1984) and Heyman and Menezo (1987) suggested that trophoblastic vesicles may provide important metabolic component (s) (such as lipids), normally present in uterine tract, that are required for embryo cleavage *in utero*. Also, it was proposed that these embryotrophic factors are secreted directly into the culture medium, and that there was no need for direct contact between the developing embryo and the trophoblastic cells during coculture. The latter hypothesis was verified when one- to two-cell bovine embryos developed to the 16-cell stage by simply culturing bovine embryos in conditioned medium at a rate similar to corresponding embryos cocultured with bTVs (Heyman and Menezo, 1987).

In an attempt to improve the efficiency of the trophoblastic vesicle coculture system for farm animals, Pool et al. (1988a) used microsurgery to individually place early-stage bovine morulae into bTVs prior to culture and evaluated subsequent embryo development *in vitro*. Unexpectedly, embryos placed in the lumen of the trophoblastic vesicles had a lower percentage developing to quality grades 1 and 2 blastocysts (36%) than corresponding embryos cocultured outside of the trophoblastic vesicles (69%) after 60 h of incubation. This pattern of embryo development was also evident after 96 h of culture. Why embryo development is slowed when placed inside the trophoblastic vesicle during culture is not clear. Possibly, the level of embryotrophic factors produced by the trophoblastic cells becomes too concentrated inside the vesicle during incubation and/or the metabolic by-products of the confined embryo accumulate to a toxic level in the lumen of the vesicle during coculture. Thus, placement of the bovine embryos inside trophoblastic vesicles was not as effective as a simple coculture method for bovine embryos.

Based on the results with trophoblastic vesicle cocultures reported to date, it appears that trophoblastic vesicles should be prepared after day 10 and before day 15 of development in the cow, sheep, and goat for maximum embryotrophic activity during *in vitro* embryo coculture. For a review of the biochemical involvement of trophoblast cells and the developing embryo, see Heyman and Menezo (1987). If trophoblastic vesicles are to be used effectively in commercial ET companies for *in vitro* culture, these vesicles must be stored and ready for use when the need arises. In a preliminary study (Rorie et al., 1987), it was found that bovine and caprine trophoblastic vesicles could be frozen and stored for months in liquid nitrogen for subsequent use in embryo coculture. After thawing, 30–70% of the trophoblastic vesicles were recovered during incubation and gave evidence of having embryotrophic properties. Although both fresh and frozen-thawed trophoblastic vesicles can appear viable for weeks in culture, their embryotrophic properties are most evident during the first week after preparation (or the first few days after thawing).

E. Chick Embryo Coculture

The avian egg represents a complete *in situ* environment for the development of the avian embryo from the blastoderm to the stage of hatching. The egg albumen and yolk supply virtually all the nutrient requirements of the developing embryo during the early stages of embryonic development. Based on early observations describing the attributes of the fertile chicken egg as a near "perfect" biological unit, the chick embryo amnion was evaluated as a culture system for *in vitro* development of early-stage mammalian embryos (Blakewood et al., 1988).

CEE was among the first factors used to stimulate the growth of mammalian cells in culture. Carrel (1913) prepared CEE by grinding tissues of 6- to 20-day-old chicks in Ringer's solution, added the extracts to *in vitro* cultures of canine connective tissue, and noted a 3- to 30-fold increase in the rate of tissue growth. In a later study, Willmer and Jacoby (1936) prepared CEE from 7-day-old chick embryos and found that this extract stimulated the development of avian cells that had previously ceased to grow in culture. The rate of cell proliferation was proportional to the concentration of CEE added to the culture medium.

New and Stine (1964) first reported the use of CEE in mammalian embryo culture. Postimplantation of mouse and rat embryos (days 7–10) were placed in plasma clots that contained 15 drops of fowl plasma and 5 drops of CEE prepared from day-13 chick embryos. After 36 h of culture, 50% of the mouse embryos at the one to seven somite stages developed blood circulation, and some developed tail and posterior limb buds. This finding further suggested that the developing fertile chicken egg produces growth-promoting factors that could be valuable for supplementation of mammalian embryo culture systems.

The first report using the amniotic cavity of developing chick embryos to culture mammalian embryos involved pronuclear mouse embryos embedded in agarose and injected into the amniotic cavity of a 96-hour-old chick embryo and incubated at 37°C (Blakewood and Godke, 1989; Blakewood et al., 1989a). In initial experiments, mouse zygotes from two different lines were placed in the amniotic cavity of chick embryos for 72–96 h of incubation. Following incubation, significantly more embryos had developed into hatching blastocysts when placed in the chick embryo culture system compared with those from the control culture medium alone.

Initial success using the chick embryo coculture system with precompaction-stage bovine morulae and two- to eight-cell caprine embryos was also reported. More expanded (80%) and hatching blastocysts (35%) resulted from precompaction-stage bovine morulae following culture in the chick embryo amnion than following culture in Ham's F-10 medium (15% and 0% expanding and hatching blastocysts, respectively).

Blakewood et al. (1989c) subsequently demonstrated that the culture of early-morula stage bovine embryos with either a uterine cell monolayer or in the chick embryo culture resulted in significantly greater postthaw embryo development than when embryos were cultured in the Ham's F-12 control medium alone prior to freezing. This approach of culturing prior to freezing suggests that this may be a way to improve postthaw embryo survival in farm animals.

The ability of the chick embryo coculture system to enhance development of two- to eight-cell goat embryos through the *in vitro* developmental block stage was characterized (Blakewood et al., 1990a). In the first experiment, the use of the fetal bovine uterine fibroblast monolayer and chick embryo coculture for 72 h were both able to produce higher blastocyst yields compared to culture only in the Ham's F-10 medium. The results were even more dramatic in the second experiment, where the two- to eight-cell caprine embryos were placed in the amniotic cavity of the chick embryo for 96 h. In this case, no early-stage embryos developed to the expanded or the hatched-blastocyst stages with the fetal uterine monolayer or with the control medium; however, 86% of the embryos cocultured in the chick embryo amnion reached the expanded blastocyst stage and 82% developed to the hatched-blastocyst stage. Following culture, four recipient goats received transplanted chick embryo coculture morulae, and two maintained pregnancies to term, with a total of six live transplant offspring (50% of all embryos transferred) born (Blakewood et al., 1990b). These findings further indicate that the developing chick embryo has potent embryotrophic properties and these properties are evident with mammalian embryos across species.

Chick embryo coculture is being used with early-stage NT cattle and goat embryos at Louisiana State University. The success with culturing NT embryos in various media for bovine and caprine embryos has been less than anticipated using standard culture media.

F. Nonreproductive Cells for Coculture

Cole and Paul (1965) first reported that a high percentage of mouse blastocysts hatched from the ZP when cultured with a feeder layer of irradiated HeLa cells. More than a decade later, Glass et al. (1979) evaluated several different types of helper cells (L-cells, liver, JLS-V11, and teratocarcinoma cells) for the culture of mouse embryos to the hatching stage and observed no differences in culture efficiency among these cell types.

Overskei and Cincotta (1987) noted *in vitro* developmental success (83%) when two-cell mouse embryos were cultured to blastocysts on a monolayer of hamster hepatocytes. Similarly, Hu et al. (1989) reported use of a commercially available cell line (Buffalo rat liver cells) to coculture two-cell mouse embryos *in vitro*. It was interesting to note that embryos cultured on Buffalo rat liver cells exhibited better *in vitro* embryonic development than did embryos incubated on mouse oviduct cells. In subsequent studies, we confirmed that Buffalo rat liver cells cocultured with early stage and IVF-derived bovine embryos produced excellent results.

Precompaction-stage mouse embryos have been cocultured with either bovine fetal spleen (BFS) cell monolayers or chicken embryo fibroblast (CEF) monolayers (Kim et al., 1989a). Although the CEF monolayer offered no marked advantage over that of control medium alone, more murine embryos hatched during coculture when BFS cells were incorporated into the culture system. A similar study was conducted using BFS and CEF monolayers to coculture bovine morulae *in vitro* (Kim et al., 1989b); the BFS cell and the CEF coculture systems produced more hatched embryos (75% and 83%, respectively) than the control culture medium (45%). In a subsequent study, these two monolayer coculture systems were compared with a bovine cumulus cell coculture system for culture of IVF-derived bovine embryos (Kim et al., 1991). Better *in vitro* development and hatching resulted when using bovine cumulus cells compared with either the BFS cell or the CEF monolayers.

Ouhibi et al. (1990) evaluated development of one-cell mouse embryos cocultured on reproductive tract cells (oviduct epithelium) or on cells established from established cell lines that were not derived from reproductive origin (e.g., kidney cells). The coculture treatments consisted of mouse oviduct organ cultures, mouse oviduct cells, bovine oviductal monolayers consisting of both polarized and unpolarized cells, and polarized and unpolarized kidney cell (Vero and MDBK) monolayers. The highest percentage of morulae and blastocysts was obtained from embryos placed in mouse oviductal organ cultures (77%), and the percentages of combined morulae and blastocysts resulting from mouse oviductal, bovine unpolarized, and bovine polarized monolayers were 67%, 48% and 14%, respectively. In addition, Vero cell monolayers had lower development rates (8% and 4%, respectively) compared with MDBK unpolarized and polarized monolayers (74% and 21%, respectively). From this study, it seems that the Vero cell line was not suitable for development of mouse embryos.

Collectively, these studies indicate that various cell lines and cell monolayers that were developed from cells that did not originate from adult and fetal reproductive tissue are capable of supporting development of early-stage embryos for up to 72 h. Certainly, the ability to purchase a pathogen-free cell line that possesses embryotrophic properties, particularly one that could be cryopreserved without loss of function, would make embryo coculture an acceptable procedure for commercial ET units.

V. The Use of Embryo Culture Systems

During early mouse embryo culture experiments, Whitten (1956) investigated the use of bicarbonate-buffered Kreb's medium for mouse embryos in lieu of the standard physiological saline culture medium as a means to stabilize the pH of the culture environment. No development of eight-cell mouse embryos was noted in the Kreb's medium alone; however, supplementation of the medium with 1% egg white resulted in development of these embryos to blastocysts. More importantly, Whitten showed that crystalline bovine serum albumin (BSA) could be substituted for egg whites in the culture system to produce similar *in vitro* embryonic development. As research continued to develop a somewhat more chemically defined culture medium, it was discovered that fertilized oocytes from some inbred strains of mice could develop from the pronuclear stage to the blastocyst stage in a defined medium, without the addition of BSA to the medium (Whitten and Biggers, 1968).

Following this discovery that various strains of mouse embryos could undergo development to the blastocyst stage in a defined medium, efforts were made to culture embryos from domestic animal species in defined media. Restall and Wales (1966) were among the first to report success in culturing embryos using a defined medium based on the components of sheep oviductal fluid. Tervit et al. (1972) cultured bovine and ovine embryos using synthetic oviductal fluid (SOF), and they reported *in vitro* development not only from the 1-cell stage to the 16-cell stage but also from the eight-cell to the blastocyst stage. In a subsequent study (Shea et al., 1974), Brinster's modified ova culture medium (BMOC-3) was compared with SOF for *in vitro* culturing of 8- to 16-cell bovine embryos, resulting in 26% and 57% of the embryos developing to the morula stage, respectively. Over the years, the farm animal embryo culture studies using SOF, modified SOF, CR1aa, and other partially defined media have continued to produce variable results.

A more complete understanding of factors influencing embryonic growth might be possible if embryos could be cultured in medium with completely defined components. Although BSA is a component of virtually all "defined" culture media, individual batches of commercially supplied BSA are poorly characterized and are often found to have different growth-promoting effects on mammalian embryo development *in vitro* (Kane, 1983). Unfortunately, attempts at *in vitro* culture of mammalian embryos from the one-cell to the blastocyst stage in a completely defined medium have been consistently successful only with specific strains of mouse embryos (Whitten and Biggers, 1968). Supplementation of culture medium with complex, undefined biological fluids (e.g., fetal bovine serum) apparently are still required to advance the embryos through the *in vitro* developmental block stage of domestic mammalian species (see review by Wright and Bondioli, 1981). The exception in this case may be the early-stage human embryo, which apparently does not need serum supplementation during early embryo culture (Menezo et al., 1984). However, the human embryo has been reported not to exhibit the characteristic *in vitro* developmental block during culture (Edwards et al., 1981).

Repeatable techniques for producing IVF-derived farm animal embryos (Goto et al., 1988a; Zhang et al., 1992; and others) have made the early-stage embryo more plentiful for experimentation and developmental study. This accessibility to laboratory-derived embryos has illustrated the need for culture systems that have the ability to promote normal *in vitro* development of early-stage farm animal embryos. Although the first live calf produced from IVF oocyte was produced more than 30 years ago (Brackett et al., 1982), the current lack of a chemically defined medium for consistent and repeatable *in vitro* development of early-stage farm animal embryos will likely make *in vitro* embryo culture one of the major research areas of reproductive physiologists and embryologists during the decades ahead.

The potential availability of a large number of oocytes has stimulated increased efforts to develop gene transfer and NT techniques for farm animals (Baguisi et al., 1999; Bondioli et al., 2001; Keefer et al., 2001; and others). The ability to produce larger numbers of early-stage embryos in the laboratory would offer little benefit without developing an *in vitro* culture system that would allow the embryos to progress through the *in vitro* block stage. This is particularly true because early-stage embryos of these animals (e.g., cattle) often are not considered to be good candidates for transfer to recipient females, since acceptable pregnancy rates are not usually achieved until embryos reach the morula or blastocyst stage of development. If early stage, IVF-derived embryos could be cultured *in vitro* to later morphological stages to improve ET success rates, marked gains could then be made with a wide range of new reproductive technologies in the commercial livestock industry.

Prior to the 1980s, the production of offspring following IVF or embryo micromanipulation techniques was considerably restricted in most mammals (with the exception of some strains of mice) by the absence of suitable culture systems (Whittingham, 1977; Kane, 1987) capable of effectively supporting *in vitro* development of early-stage embryos to a stage suitable for transfer (i.e., to morula or blastocyst). The limitations imposed by inadequate culture media for farm animal embryos were markedly reduced with procedures for the temporary culture of early-stage embryos in the oviducts of rabbits (Boland, 1984) or sheep (Willadsen, 1979, 1982). This *in vivo* culture approach, however, is labor intensive, time consuming, and has not been well accepted by commercial embryo transplant companies.

With advances made in the last decade in embryo culture systems for farm animal embryos, it appears that *in vitro* incubation problems have been partially alleviated with the development of helper cell coculture systems that permit embryonic growth and development through the *in vitro* block stage to morulae and blastocysts while in culture. Uterine and oviduct epithelial and epithelial-like cells have an advantage over corresponding fibroblasts (originating from these same two-cell types) for culture of farm animal embryos. Oviduct cells (although more difficult to maintain in culture compared with fibroblasts) have become the first choice for embryo coculture by researchers producing laboratory-derived embryos for experimentation. Others are now using the new chick embryo amnion culture system (Blakewood and Godke, 1989) and/or cumulus cell coculture system (Goto et al., 1989; Zhang et al., 1992) for embryo culture, with results similar to those using oviductal cell cocultures for IVF-derived farm animal embryos. Efforts have been made to develop a completely defined medium for embryo culture (Ellington et al., 1990c), but progress thus far has been limited. Many research efforts have been directed toward using multicell bilayered or three-dimensional coculture systems, and efforts in this area will likely continue in years to come.

Perhaps one of the most interesting uses of culture systems for embryos in the future will be to nurture (or "rescue") poor quality or questionable-quality embryos *in vitro* prior to transfer to recipient females. By screening embryos with poor developmental potential, it may be possible to significantly reduce the number of females used as recipients. In addition, developing efficient embryo culture methods would aid in evaluating the viability of transgenic and/or NT embryos prior to transfer.

The need for a chemically defined medium capable of producing consistent and repeatable results is strong, particularly with respect to *in vitro* produced embryos. From the standpoint of biosecurity issues, it will be important to develop a chemically defined medium devoid of any components of animal origin. A number of studies have focused on the inclusion of compounds with direct or indirect antioxidant properties (e.g., β-mercaptoethanol, cysteamine) as a means to reduce cell damage caused by oxygen-free radicals (Caamaño et al., 1996, 1998; Geshi et al., 1999). Commercial IVF companies are making significant headway in this area; however, most of their research is proprietary and has not been made publicly available.

VI. Additional Assisted Reproductive Technologies

A. Oocyte Reconstruction

Cytoplasmic oocyte maturation involves the synthesis of proteins from nuclear and mitochondrial transcripts used by the oocyte during early embryonic development. Oocytes collected for IVM are normally obtained at the germinal vesicle stage, a stage prior to the time when cytoplasmic maturation has occurred. Protein synthesis *in vitro* may differ from that which occurs *in vivo*, thus potentially affecting fertilization and subsequent embryonic development. Therefore, it seems reasonable to assume that differences in oocyte competence may affect embryonic development after sperm injection or NT procedures. Some studies have addressed the issue of oocyte competence using both *in vivo* and *in vitro* matured oocytes in sperm injection studies with various farm animal species, including the cow (Heuwieser et al., 1992) and the pig (Kolbe and Holtz, 2000).

Key events such as meiotic maturation of the oocyte, fertilization, and the activation of the embryonic genome involve both nuclear and cytoplasmic processes. In humans, women suffering from idiopathic infertility continually experience poor embryonic development, high rates of embryo fragmentation, and implantation failure after their IVF treatment procedure. Ooplasm (or cytoplasm) donation techniques have been developed to help solve difficult human infertility cases (Cohen et al., 1997). With this approach, a portion of the cytoplasm from the donor oocyte (5–15%) is microinjected into a recipient oocyte either prior to sperm injection or along with the sperm during the ICSI procedure (Cohen et al., 1998).

The mechanisms by which ooplasm transplantation can restore growth and viability of developmentally compromised oocytes remain unclear. Cytoplasmic components that might benefit development include the pool of stored mRNA and proteins that regulate oocyte function and embryo development, as well as mitochondria and other cell organelles. Cytoplasmic transfer has become the subject of debate because of the possibility of heteroplasmy of mitochondrial DNA. Barritt et al. (2001) demonstrated that active mitochondria were transferred from human donor oocytes to recipient oocytes during cytoplasmic transfer. Recipient oocytes contained a heteroplasmic population of mitochondria after cytoplasmic transfer, and these mitochondria persisted in the resultant preimplantation embryos. Mitochondrial heteroplasmy is of concern because of the growing list of genetic diseases associated with mutations, inversions, and deletions of the mitochondrial genome.

Further development of novel-assisted reproductive technologies such as insertion of somatic cell nuclei into enucleated oocytes (Wakayama et al., 1998) will continue to provide a tool for research into cell cycle control and the mechanisms involved in sperm-induced oocyte activation. The potential application of ICSI, germinal vesicle transfer, and ooplasm transfer in farm animal species still remains to be explored, but it may be an alternative to circumvent *in vivo* fertilization and/ or implantation failures.

B. Nuclear Transfer

It has long been a goal of scientists to be able to produce genetically identical animals from differentiated mammalian cells obtained from genetically superior animals. Although research efforts with tadpoles in the 1950s and 1960s were encouraging to developmental biologists, the progress made replicating these experiments on other species was limited at best Willadsen (1986) until recently (see Chapters 7 and 14).

Following the birth of the first adult somatic cell NT farm animal in 1996 (Wilmut et al., 1997), the curiosity and interest in assisted reproductive technologies grew considerably. It was originally proposed that NT techniques would be used to efficiently generate NT transgenic farm animal offspring capable of producing valuable proteins, and that this would likely have a marked impact on the pharmaceutical industry (Echelard, 1996). One of the limitations of somatic cell NT in two farm animal species (sheep and cattle) was the very low success rate, due in part to a high proportion of fetal loss (Campbell et al., 1996a; Wilmut et al., 1997; Schnieke et al., 1997; Wells et al., 1997; Zakhartchenko et al., 1999), as well as an increase in perinatal morbidity/mortality (Cibelli et al., 1998a,b; Kato et al., 1998; Vignion et al., 1998; Shiga et al., 1999; Hill et al., 1999). Reported causes of fetal death and/or wastage include abnormal liver development (Wilmut et al., 1997), insufficient placentation leading to spontaneous abortions or prolonged gestation (Cibelli et al., 1998a,b; Wells et al., 1999; Hill et al., 1999; Bertolini and Anderson, 2002) and oversized fetuses (Vignion et al., 1998). Common causes of perinatal death also included metabolic and cardiopulmonary abnormalities (Wells et al., 1999; Zakhartchenko et al., 1999; Hill et al., 1999; 2000).

In contrast to other species, a high incidence of perinatal loss was not reported in somatic cell NT in goats (Baguisi et al., 1999). The reasons for this difference were not clear but may be due to either the relatively low number of NT goat offspring produced to date or the minimal *in vitro* culture period (reconstructed NT embryos were transferred at the two- to four-cell stage). However, Yong and Yuqiang (1998) produced 45 NT goats from the transfer of 141 serially reconstructed embryos into 29 recipients. In this study, blastomeres were used as karyoplast donors and embryos were cultured to the morula stage prior to transfer. Thus, it may be that manipulated goat embryos are less sensitive to *in vitro* culture conditions than cow and sheep embryos (Reggio et al., 2001).

Although there are a variety of potential applications for somatic cell NT technology, the most common application of NT technology in the North American livestock industry is to re-create show-winning steers as bulls so that semen can be collected, processed, and marketed. Of course, there is also substantial use of somatic cell NT technology in the biomedical research community Swanson et al. (1992).

The combination of somatic cell NT methodology with existing transgenic animal production techniques has the potential to increase the efficiency of transgenic animal production (Ziomek, 1998). Transfecting cell lines with genes of interest, coupled with somatic cell NT, has become a viable method for introducing foreign DNA into the genome of animals (Schnieke et al., 1997; Cibelli et al., 1998a; McCreath et al., 2000). Supplementing the foreign DNA with promoters that direct gene expression to the mammary gland (Ebert et al., 1994) enables straightforward recovery of the protein of interest (Young et al., 1997). Although milk yields are higher in dairy cows, dairy goats offer a shorter generation interval and thus are more attractive for use in transgenic programs (Ziomek, 1998). Moreover, transgenic female goats can be hormonally induced to lactate at 2 months of age (Cammuso et al., 2000) to provide an initial sample for product testing (e.g., expression levels and biological activity). Larger quantities of milk can then be produced following a normal gestation initiated either through natural mating or AI of the transgenic female.

Several examples of improving animal production and/or the production of animal-derived products have been identified, and these include wool keratin genes to increase the quality of the wool (Ward et al., 1991; Ward and Nancarrow, 1992), growth-promoting genes to enhance growth in sheep (Rexroad et al., 1989), genes to improve the quality of milk (Maga and Murray, 1995), and a gene (alpha lactalbumin) to enhance piglet growth rates (Wheeler et al., 2003).

Transgenic cloned farm animals have already been produced that have resulted in human pharmaceutical production (Meade et al.,1998; Wright et al., 1991; Ziomek, 1998). For example, cloned transgenic female goats have been produced that carry and express a human gene for ATIII in their milk (Reggio et al., 2001). Schnieke et al. (1997) have reported the production of transgenic cloned sheep females that produced human clotting factor IX in their milk. Kuroiwa et al. (2002) reported cloned transchromosomic calves that produced human immunoglobulins in their blood. Being able to extract these human protein products from the milk or blood of transgenic farm animals has been termed "biopharming". This new technology may provide a more cost-effective production system and will likely be used by the pharmaceutical companies to develop and market human biomedical products in the years to come (see recent reviews by Maksimenko et al., 2013; Wang et al., 2013).

A potential limitation of using adult transfected fibroblast cells as nuclear donors is the prolonged culture period required to propagate, transfect, select, and expand the cell line. Each of these steps requires repeatedly passaging the cells and an extended culture period that may induce perturbations in the donor cells (such as chromosomal damage due to physical manipulation during transfection), thereby decreasing the efficiency of NT (Forsberg et al., 2001; Zakhartchenko et al., 2001). Also, there is a limit to the number of passages a primary cell line can be subjected to before a state of senescence is reached (Hayflick and Moorhead, 1961; Pignalo et al., 1992). Moreover, adult cells are capable of fewer population doublings before reaching senescence than are fetal cells (Cristofalo et al., 1998; Lanza et al., 2000).

The ability to generate live offspring using proliferating donor cells may be advantageous in transgenic NT programs that must rely on extended *in vitro* culture periods of the cell line to allow incorporation of the gene of interest. Bondioli et al. (2001) produced transgenic live piglets after NT using proliferating transfected fibroblasts, a noteworthy accomplishment as it has been reported that porcine fibroblast cells undergo extensive DNA fragmentation during just 3 days of serum deprivation (Kues et al., 2000). Hence, culture systems and conditions will remain a critical concern in the characterization of efficient and cost-effective NT programs.

VII. Summary and Future Direction

The list of assisted reproductive technologies has been expanding each year and will continue to do so in the future (Hansel and Godke, 1992; Nowak-Imialek and Niemann, 2012; Bähr and Wolf (2012); Ponsart et al., 2014). The transgenic cow, pig, and goat have become a prominent driving force within the biotechnology industry—particularly with respect to biomedical research. Many scientists celebrated when our colleagues developed ET and freezing technologies for farm animals. As new technologies such as TUGA, IVF, sperm sorting, embryo sexing, genetic testing, and NT become more available to the livestock industry, livestock producers will use basic nonsurgical ET procedures pioneered in the mid-1970s with greater frequency.

With IVF the potential exists for more embryos to be produced in a shorter period of time, because the TUGA procedures can be repeated on the same cow 3—4 times or more a month. The use of TUGA and IVF procedures offers an alternative for cattle and horse producers who have genetically valuable females that are unable to produce viable embryos through standard ET procedures. This gamete technology can be used on oocytes harvested from older acyclic or nonovulating animals, females with physical injuries (e.g., broken leg) and problem females (e.g., cows) exhibiting an abnormal cervix. Good success has been reported using IVF procedures on oocytes obtained from subordinate follicles of cows with cystic ovarian disease.

Currently, TUGA is being used to harvest oocytes from domestic females representing rare bloodlines and reproductively senescent mares and cows. Research continues to find applications for this technology, including harvesting oocytes from prepubertal heifers, fillies, and ewe lambs for IVP. We are harvesting oocytes from early postpartum (<40 days) beef and dairy cattle, before the female resumes her reproductive cyclicity. The approach provides an opportunity to produce one or more extra calves from the cow before she is mated to establish a natural pregnancy.

More oocytes, often of higher developmental potential, can be harvested from early pregnant cows and mares than from cyclic females. This approach would allow the pregnant female to produce her own natural-birth offspring each year in addition to allowing her to produce extra ET offspring from IVF procedures. Although only a small number of foals have been produced using *in vitro* methods, the ICSI procedure appears to be the method of choice to produce IVF-derived horse embryos at the present time.

More than 25 years ago, embryo-splitting trials excited many about the potential of implementing this technology to improve the reproductive efficiency of seed-stock production. The bisection methods have been successfully adapted for use by the commercial ET companies throughout the world. The method of splitting embryos with a hand-held razor blade is practical and shows promise for application under field conditions. Research has shown that excellent to good quality, late morula- to blastocyst-stage bovine embryos can be successfully bisected to produce split-embryo offspring. Under optimal conditions, bisected embryos can produce pregnancy rates comparable with those achieved with intact sibling embryos. It should not be overlooked as a tool to create identical twins for research. The use of IVF, embryo bisection, and ICM cell transfer might produce *in vitro*-derived embryos to aid in germplasm preservation of rare and endangered exotic species.

Significant progress has been made in the *in vitro* culture of large animal embryos during the last 20−30 years. This progress has been due primarily to developing new media recipes and the incorporation of "helper" cells into embryo culture procedures. These improvements in culture methodology have led to reproductive technology innovations never before considered in the livestock industry. The potential may exist for improving pregnancy rates from IVF, bisected, reconstructed, and NT embryos by culturing the embryos for a short period of time on either uterine, oviduct, or granulosa cells, or Buffalo rat liver cells prior to transfer.

Research efforts are being directed toward minimally invasive biopsy procedures to enable harvest of embryonic cells for use in PGD, including embryonic sex determination. Partial genomic selection using SNP chip testing of embryo biopsies will likely become commonplace and will gradually reduce emphasis on phenotypic selection of breeding stock.

Until recently, the interest in producing embryonic NT offspring was low. The high cost, the low success rate, and the existence of "large calf" syndrome were the primary reasons this NT technology was not immediately adapted by the commercial sector. In the late 1990s, the first somatic cell NT cattle were born, and now there are hundreds of somatic cell NT calves throughout the world. In one case, adult somatic cell NT has reported an 80% success rate from transferring NT cattle embryos (Kato et al., 1998). Also, NT has been used in an attempt to save an endangered cattle bloodline (Wells et al., 1998). Somatic cell NT technology is clearly developing a scientific and commercial presence, and it will likely change many of our breeding strategies in the livestock industry in the years to come.

Although initial progress was relatively slow, transgenic farm animals were first produced in the mid-1980s. In the early 1990s, the production of transgenic goats that were capable of producing specific human proteins in their milk was reported (Ebert et al., 1994). Since then, several commercial entities have begun to use transgenic animals to produce a number of human biopharmaceuticals. The production of recombinant proteins in the milk of transgenic animals now appears to have economic potential far beyond what was initially envisioned, and these transgenic technologies may also be useful to enhance animal performance.

There is still much to be studied and learned in the use of assisted reproductive technologies to maximize reproductive potential in genetically valuable animals. Now that repeatable oocyte retrieval methods have been fine-tuned, the next obstacle to overcome will be development of a chemically defined embryo culture system that supports high *in vivo* development rates following transfer of fresh and frozen-thawed (or vitrified) IVP embryos. It will be exciting to follow new developments and new applications of assisted reproductive technology in the decades ahead.

References

Adams, G.P., Brogliatti, G.M., Salamone, D.F., Mapletoft, R.J., 1996. Supestimulatory response and oocyte collection in calves. Theriogenology 45, 281 (abstr.).

Aerts, J.M., Oste, J.M., Bols, P.E.J., 2005. Development and practical applications of a method for repeated transvaginal, ultrasound-guided biopsy collection of the bovine ovary. Theriogenology 64, 947–957.

Agrawal, K.P., Mongha, I.V., Bhattacharyya, N.K., 1983. Survival of goat embryos in rabbit oviducts. Vet. Rec. 112, 200.

Alikani, M., Cohen, J., 1992. Micromanipulation of cleaved embryos cultured in protein-free medium: a mouse model for assisted hatching. J. Exp. Zool. 263, 458–463.

Allen, R.L., Wright, R.W., 1984. *In vitro* development of porcine embryos in coculture with endometrial cell monolayers or culture supernatants. J. Anim. Sci. 59, 1657–1661.

American Embryo Transfer Association, 2012. AETA survey results of embryo transfer businesses. <http://www.aeta.org/survey.asp> (accessed 17.02.14).

Antinori, S., Fischel, S., Versaci, C., Chairiell, F., Lisi, F., 1990. Evaluating 6 pregnancies established by subzonal insemination, "SUZI". In: Proceedings of the Seventh World Congress Human Reproduction, June 26–July 1, Helsinki, Finland. No. 262 (abstr.).

Aoyagi, Y., Fukui, Y., Iwazumi, Y., Urakawa, M., Ono, H., 1990. Effects of culture systems on development of *in vitro* fertilized bovine ova into blastocysts. Theriogenology 34, 749–759.

Averill, R.L.W., Adams, C.E., Rowson, L.E.A., 1955. Transfer of mammalian ova between species. Nature (London) 176, 167–168.

Baguisi, A., Behboodi, E., Melican, D.T., Pollock, J.S., Destrempes, M.M., Cammuso, C., et al., 1999. Production of goats by somatic cell nuclear transfer. Nat. Biotech. 17, 456–461.

Bahr, B., Rath, D., Mueller, P., Hildbrandt, T.B., Goeritz, F., Braun, B.C., et al., 2009. Feasibility of sex-sorting White and Black rhinoceros (*Ceratotherm simum, Diceros bicornis*). Theriogenology 73, 353–364.

Bähr, A., Wolf, E., 2012. Domestic animal models for biomedical research. Reprod. Domest. Anim. 47 (Suppl. 4), 59–71. Available from: http://dx.doi.org/doi:10.1111/j.1439-0531.2012.02056.x.

Baird, J.W.C., Johnson, C.A., Williams, S.R., Godke, R.A., Jenkins, C.L., Schmidt, G., 1990. Increased blastocyst formation in the mouse following culture of hamster cumulus cell monolayers. In: Proceedings of the American Fertility Society. p. S9 (abstr).

Baker, R.D., Shea, B.F., 1985. Commercial splitting of bovine embryos. Theriogenology 22, 3–12.

Baker, R.D., Eberhard, B.E., Leffel, B.E., Rhoade, R.E., Henschen, T.J., 1984. Pregnancy rates following surgical transfer of bovine demi-embryos, Proceedings of the Tenth International Congress of Animal Reproduction and Artificial Insemination, vol. 2. University of Illinois, Urbana, p. 220.

Barritt, J.A., Brenner, C.A., Malter, H., Cohen, J., 2001. Mitochondria in human offspring derived from ooplasmic transplantation. Hum. Reprod. 16, 513—516.

Bavister, B.D., 1987. Studies on the developmental block in cultured hamster embryos. In: Bavister, B.D. (Ed.), The Mammalian Embryo: Regulation of Growth and Differentiation *In Vitro*. Plenum Publishing Corporation, New York, NY, pp. 61—78.

Bavister, B.D., 1988. Role of oviductal secretions in embryonic growth *in vitro* and *in vivo*. Theriogenology 29, 143—154.

Bellow, M.S., Didion, B.A., Davis, D.L., Murphy, C., Ferguson, C.E., Godke, R.A., et al., 2001. A procedure for nonsurgical transvaginal ultrasound-guided oocyte aspiration in the sow. Theriogenology 55, 528 (abstr.).

Berg, U., Brem, G., 1990. Developmental rates of *in vitro* produced IVM—IVF bovine oocytes in different cell culture systems. Theriogenology 33, 195 (abstr).

Bertolini, M., Anderson, G.B., 2002. The placenta as a contribution to production of large calves. Theriogenology 57, 181—187.

Betteridge, K.J., 1977. Embryo transfer in farm animals. Agricultural Monograph No. 16, Ottawa, Canada, p. 39.

Blakewood, E.G., Godke, R.A., 1989. A method using the chick embryo amnion for mammalian embryo culture. J. Tissue Cult. Meth. 12, 73—76.

Blakewood, E.G., Jaynes, J.M., Godke, R.A., 1988. Culture of pronuclear mammalian embryos using domestic chicken eggs. Theriogenology 29, 226 (abstr.).

Blakewood, E.G., Jaynes, J.M., Johnson, W.A., Godke, R.A., 1989a. Using the amniotic cavity of the developing chick embryo for the *in vivo* culture of early stage mammalian embryos. Poultry Sci. 68, 1695—1702.

Blakewood, E.G., Wiemer, K.E., Godke, R.A., 1989c. Post-thaw viability of bovine embryos cultured in domestic chicken eggs or on epithelial monolayers prior to freezing in liquid nitrogen (LN_2). Theriogenology 31, 177 (abstr.).

Blakewood, E.G., Pool, S.H., Prichard, J.F., Godke, R.A., 1990a. Culturing two- to eight-cell caprine embryos using domestic chicken eggs. Mol. Reprod. Dev. 27, 288—294.

Blakewood, E.G., Prichard, J.F., Pool, S.H., Godke, R.A., 1990b. Live births following transfer of caprine embryos cultured for 72 hours in domestic chicken eggs. Theriogenology 33, 197 (abstr.).

Boland, M.P., 1984. Use of the oviduct as a screening tool for the viability of mammalian eggs. Theriogenology 21, 126—137.

Bondioli, K., Ramsoondar, J., Williams, B., Costa, C., Fodor, W., 2001. Cloned pigs generated from cultured skin fibroblasts derived from an H-transferase transgenic boar. Mol. Reprod. Dev. 60, 189—195.

Bondioli, K.R., Westhusin, M.E., Looney, C.R., 1990. Production of identical bovine offspring by nuclear transfer. Theriogenology 33, 165—174.

Bracher, V., Parlevliet, J., Fazelli, A.R., Pieterse, M.C., Vos, P.L., Dielman, S.J., et al., 1993. Repeated transvaginal ultrasound-guided follicle aspiration in the mare. Equine Vet. J. 15 (Suppl.), 75—78.

Brachet, A., 1912. Development *in vitro* de blastodermes et de jeunes embryons de mammiferes. C.R. Hebd. Seanc. Acad. Sci. (Paris) 55, 1191—1193.

Brackett, B.G., Bousquet, D., Boice, M.L., Donawick, W.J., Evans, J.F., Dressel, M.A., 1982. Normal development following *in vitro* fertilization in the cow. Biol. Reprod. 27, 147 (abstr.).

Brinster, R.L., 1972. Cultivation of the mammalian embryo. In: Rothblat, G., Cristofalo, V (Eds.), Growth, Nutrition and Metabolism of Cells in Culture, Vol II. Academic Press, Inc., New York, pp. 251–286.

Broadbent, P.J., Dolman, D.F., Watt, R.G., Smith, A.K., Franklin, M.F., 1997. Effect of the frequency of follicle aspiration on oocyte yield and subsequent superovulatory response in cattle. Theriogenology 47, 1027–1040.

Brogliatti, G.M., Adams, G.P., 1996. Ultrasound-guided transvaginal oocyte collection in prepubertal calves. Theriogenology 45, 1163–1176.

Brun, B., 1974. Studies on fertilization in *Xenopus laevis*. Biol. Reprod. 11, 513–518.

Brück, I., Raum, K., Synnestvedt, B., Greve, T., 1992. Follicle aspiration in the mare using a transvaginal ultrasound-guided technique. Equine Vet. J. 24, 58–59.

Bunch, T.D., Foote, W.C., Call, J.W., Wright Jr, R.W., Selgrath, J.P., Foote, W.D., 1987. Long term culture of two to eight-cell ovine embryos in various co-culture systems. Encyclia 64, 66–72.

Butler, J.E., Anderson, G.B., BonDurrant, R.H., Pashen, R.L., Penedo, M.C.T., 1987. Production of ovine chimeras by inner cell mass transplantation. J. Anim. Sci. 65, 317–324.

Caamaño, J.N., Ryoo, Z.Y., Thomas, J.A., Youngs, C.R., 1996. β-mercaptoethanol enhances blastocyst formation rate of bovine *in vitro* matured/*in vitro* fertilized embryos. Biol. Reprod. 55, 1179–1184.

Caamaño, J.N., Ryoo, Z.Y., Youngs, C.R., 1998. Promotion of development of bovine embryos produced *in vitro* by addition of cysteine and β-mercaptoethanol to a chemically defined culture system. J. Dairy Sci. 81, 369–374.

Cammuso, C., Porter, C., Nims, S., Gaucher, D., Melican, D., Bombard, S., et al., 2000. Hormonal induced lactation in transgenic goats. Anim. Biotech. 11, 1–17.

Camous, S., Heyman, Y., Meziou, W., Menezo, Y., 1984. Cleavage beyond the block stage and survival after transfer of early bovine embryos cultured with trophoblastic vesicles. J. Reprod. Fertil. 72, 479–485.

Campbell, K.H.S., McWhir, J., Ritchie, W.A., Wilmut, I., 1996a. Sheep cloned by nuclear transfer from a cultured cell line. Nature 380, 64–66.

Capper, J.L., Cady, R.A., Bauman, D.E., 2009. The environmental impact of dairy production: 1944 compared with 2007. J. Anim. Sci. 86 (6), 2160–2167. Available from: http://dx.doi.org/10.2527/jas.2009-1781.

Carnevale, E.M., Alveranga, M.A., Squires, E.L., 1999. Use of noncycling mares as recipients for oocyte transfer and GIFT. Society Theriogenology, September 22–24, Nashville, TN. p. 44.

Carrel, A., 1913. Artificial activation of the growth *in vitro* of connective tissue. J. Exp. Med. 17, 14–19.

Catt, S.L., Catt, J.W., Evans, G., Maxwell, W.M.C., 1996. Birth of a male lamb derived from an *in vitro* matured oocyte fertilised by intracytoplasmic sperm injection of a single presumptive male sperm. Vet. Rec. 139, 494–495.

Chang, M.C., 1949. Effects of heterologous sera on fertilized rabbit ova. J. Gen. Physiol. 32, 291–300.

Chang, M.C., 1959. Fertilization of rabbit ova *in vitro*. Nature 184, 466–467.

Chaubal, S.A., Molina, J.A., Ohlrichs, C.L., Ferre, L.B., Faber, D.C., Bols, P.E.J., et al., 2006. Comparison of different transvaginal ovum pick-up protocols to optimize oocyte retrieval and embryo production over a 10-week period in cows. Theriogenology 65, 1631–1648.

Chen, S.H., Seidel Jr., G.E., 1997. Effects of oocyte activation and treatment of spermatozoa on embryonic development following intracytoplasmic sperm injection in cattle. Theriogenology 48, 1265–1273.

Cheng, W.T.K., Moor, R.M., Polge, C., 1996. *In vitro* fertilization of pig and sheep oocytes matured *in vivo* and *in vitro*. Theriogenology 25, 146 (abstr.).

Choi, Y.H., Okada, Y., Hochi, S., Braun, H., Sato, K., Oguri, N., 1994. *In vitro* fertilization rates of horse oocytes with partially removed zonae. Theriogenology 42, 795–802.

Chrenek, P., Boulanger, L., Heyman, Y., Uhrin, P., Laurincik, J., Bulla, J., et al., 2001. Sexing and multiple genotype analysis from a single cell of bovine embryo. Theriogenology 55, 1071–1081.

Cibelli, J.B., Stice, S.L., Golueke, P.J., Kane, J.J., Jerry, J., Blackwell, C., et al., 1998a. Cloned transgenic calves produced from nonquiescent fetal fibroblasts. Science 280, 1256–1258.

Cibelli, J.B., Stice, S.L., Golueke, P.J., Kane, J.J., Jerry, J., Blackwell, C., et al., 1998b. Transgenic bovine chimeric offspring produced from somatic cell-derived stem-like cells. Nat. Biotech. 16, 642–646.

Clement-Sengewald, A., Schütze, K., Ashkin, A., Palma, G.A., Kerlen, G., Brem, G., 1996. Fertilization of bovine oocytes induced solely with combined laser microbeam and optical tweezers. J. Assist. Reprod. Genet. 13 (3), 259–265.

Cochran, R., Meintjes, M., Reggio, B., Hylan, D., Carter, J., Pinto, C., et al., 1998a. Live foals produced from sperm-injected oocytes derived from pregnant mares. J. Equine Vet. Sci. 18, 736–740.

Cochran, R., Meintjes, M., Reggio, B., Hylan, D., Carter, J., Pinto, C., et al., 1998b. *In vitro* development and transfer of *in vitro*-derived embryos produced from sperm-injected oocytes harvested from pregnant mares. In: Proceedings of the Seventh International Symposium Equine Reproductive. pp. 135–136.

Cochran, R., Reggio, B., Carter, J., Hylan, D., Paccamonti, D., Pinto, C., et al., 1999b. Twin pregnancies resulting from the transfer of sperm-injected equine oocytes harvested from Altrenogest-treated mares. Theriogenology 51, 281 (abstr.).

Cochran, R., Meintjes, M., Reggio, B., Hylan, D., Carter, J., Pinto, C., et al., 2000. Production of live foals from sperm-injected oocytes harvested from pregnant mares. J. Reprod. Fertil. Suppl. 56, 503–512.

Cochran, R.A., Leonardi-Cattolica, A.A., Sullivan, M.R., Kincaid, L.A., Leise, B.S., Thompson Jr., D.L., et al., 1999a. The effects of equine somatotropin (eST) on follicular development and circulating plasma hormone profiles in cyclic mares treated during different stages of the estrous cycle. Domestic Anim. Endocrinol. 16, 57–67.

Cochran, R.A., Wyble, A., Hylan, D.A., Carter, J.A., Johnson, H., Thompson, D.L., Jr, et al., 1999c. Effects of administration of exogenous eST to seasonally anovulatory mares. In: Proceedings of the 16th Equine Nutritional Physiology Symposium. vol. 16, pp. 83–84.

Cohen, J., 1991. Assisted hatching of human embryos. J. In Vitro Fertil. Embryo Transf. 8, 79–190.

Cohen, J., Malter, H., Wright, G., Kort, H., Massey, J., Mitchell, D., 1989a. Partial zona dissection of human oocytes when failure of zona pellucida penetration is anticipated. Hum. Reprod. 4, 435–442.

Cohen, J., Wiemer, K., Wiker, S., Malter, H., Kort, H., Massey, J., et al., 1989b. Co-culture of human zygotes on fetal bovine uterine fibroblasts. In: Proceedings of the Sixth World Congress *In Vitro* Fertility and Alternate Assisted Reproduction, Jerusalem, Israel. p. 41 (abstr.).

Cohen, J., Elsner, C., Kort, H., Malter, H., Massey, J., Mayer, M.P., et al., 1990. Impairment of the hatching process following IVF in the human and improvement of implantation by assisted hatching using micromanipulation. Hum. Reprod. 5, 7–13.

Cohen, J., Talansky, B.E., Malter, H., Alikani, M., Adler, A., Reing, A., et al., 1991. Microsurgical fertilization and teratospermia. Hum. Reprod. 6, 118–123.

Cohen, J., Talansky, B.E., Adler, A., Alikani, M., Rosenwaks, Z., 1992a. Controversies and opinions in clinical microsurgical fertilization. J. Assist. Reprod. Genet. 9, 94–96.

Cohen, J., Alikani, M., Trowbridge, J., Rosenwaks, Z., 1992b. Implantation enhancement by selective assisted hatching using zona drilling of human embryos with poor prognosis. Hum. Reprod. 5, 685–691.

Cohen, J., Scott, R., Schimmel, T., 1997. Birth of an infant after transfer of anucleate donor oocyte cytoplasm into recipient eggs. Lancet 350, 186–187.

Cohen, J., Scott, R., Aikani, M., 1998. Ooplasmic transfer in mature human oocytes. Mol. Hum. Reprod. 4, 269–280.

Cole, R.J., Paul, J., 1965. Properties of cultured preimplantation mouse and rabbit embryos and cell strains developed from them. In: Wolstenhoume, G.E.W., O'Conner, M. (Eds.), Preimplantation Stages of Pregnancy. Little, Brown & Co., Boston, MA, pp. 82–155.

Cook, N.L., Squires, E.L., Ray, B.S., Cook, V.M., Jesko, D.J., 1992. Transvaginal ultrasonically guided follicular aspiration of equine oocytes: preliminary results. J. Equine Vet. Sci. 12, 204–207.

Cristofalo, V.J., Allen, R.G., Pignalo, R.J., Martin, B.G., Beck, J.C., 1998. Relationship between donor age and replicative lifespan of human cells in culture: a reevaluation. Proc. Natl. Acad. Sci. USA 95, 10614–10619.

Critser, E.S., First, N.L., 1986. Use of a fluorescent stain for visualization of nuclear material in living oocytes and early embryos. Stain Technol. 61, 1–5.

Damiani, P., Bellow, M.S., Walstra, M., Looney, C.R., 1996. Repeatable transvaginal ultrasound-guided aspirations in prepuberal calves. Proc. Intl. Congr. Anim. Reprod. (Sidney, Australia) 3 (18), 12.

Daniel, J.C., Takahashi, K., 1965. Selective laser destruction of rabbit blastomeres and continued cleavage of survivors in vitro. Exp. Cell Res. 39, 475–482.

Davis, D.L., Day, B.N., 1978. Cleavage and blastocyst formation by pig eggs in vitro. J. Anim. Sci. 46, 1043–1053.

Dellenbach, P., Nissand, L., Moreau, L., Feger, B., Plumere, C., Gerlinger, P., 1985. Transvaginal sonographically controlled follicle puncture for oocyte retrieval. Fertil. Steril. 44, 656–662.

DeJarnette, J.M., Nebel, R.L., Marshall, C.E., 2009. Evaluating the success of sex-sorted semen in US dairy herds from on farm records. Theriogenology 71, 49–58.

Doklas, A., Sargent, I., Ross, C., 1991. The human blastocyst: its morphology and hCG secretion in vitro. Hum. Reprod. 6, 1143–1151.

Dowling, D.F., 1949. Problems of the transplantation of fertilized ova. J. Agric. Sci. (Cambridge) 39, 374–396.

Dresser, B.L., Pope, C.E., Kramer, L., Kuehn, G., Dahlhausen, R.D., Thomas, W.D., 1984. Superovulation of Bongo antelope (Tragelaphus euryceros) and interspecies embryo transfer to African eland (Tragelaphus oryx). Theriogenology 21, 232 (abstr.).

Ebert, K.M., Papaioannou, V.E., 1989. In vivo culture of embryos in the immature mouse oviduct. Theriogenology 31, 299–308.

Ebert, K.M., DiTullio, P., Barry, C.A., 1994. Induction of human tissue plasminogen activator in the mammary gland of transgenic goats. Biotechnology 12, 699–702.

Echelard, Y., 1996. Recombinant protein production in transgenic animals. Curr. Opin. Biotechnol. 7, 536–540.

Edwards, R.G., Purty, J.M., Steptoe, D.C., Walters, D.E., 1981. The growth of human preimplantation embryos in vitro. Am. J. Obstet. Gynecol. 141, 408.

Ellington, J.E., Farrell, P.B., Foote, R.H., 1990a. Comparison of six-day bovine embryo development in uterine tube (oviduct) epithelial cell co-culture versus in vivo development in the cow. Theriogenology 34, 837–844.

Ellington, J.E., Farrell, P.B., Simkin, M.E., Goldman, E.E., Foote, R.H., 1990b. Bovine zygote development in oviduct cell co-culture versus rabbit oviduct. Theriogenology 33, 224 (abstr.).

Ellington, J.E., Farrell, P.B., Simkin, M.E., Foote, R.H., Goldman, E.E., McGrath, A.B., 1990c. Development and survival after transfer of cow embryos cultured from 1−2-cells to morulae or blastocyst in rabbit oviducts or in a simple medium with bovine oviduct epithelial cells. J. Reprod. Fertil. 89, 293−299.

Eyestone, W.H., First, N.L., 1986. A study of the 8 to 16-cell developmental block in bovine embryos cultured in vitro. Theriogenology 23, 152 (abstr.).

Eyestone, W.H., First, N.L., 1988. Co-culture of bovine embryos with oviductal tissue. In: Proceedings of the 11th International Congress of Animal Reproduction and Artificial Insemination. vol. 4, p. 471 (abstr.).

Eyestone, W.H., First, N.L., 1989. Co-culture of early cattle embryos to the blastocyst stage with oviductal tissue or in conditioned medium. J. Reprod. Fertil. 86, 715−720.

Eyestone, W.H., Vignier, J., First, N.L., 1987. Co-culture of early bovine embryos with oviductal epithelium. Theriogenology 27, 228 (abstr.).

Faundez, R., Spohr, I., Boryczko, Z., 1988. Effect of follicle cells on maturation and in vitro fertilization of cattle oocytes. In: Proceedings of the 11th International Congress of Animal Reproduction and Artificial Insemination. vol. 5, p. 325 (abstr.).

Forsberg, E.L., Betthauser, J., Strelchenko, N., Golueke, P., Childs, L., Jurgella, G., et al., 2001. Cloning non-transgenic and transgenic cattle. Theriogenology 55, 269 (abstr.).

Frei, R.E., Schultz, G.A., Church, R.B., 1989. Qualitative and quantitative changes in protein synthesis occur at the 8−16-cell stage of embryogenesis in the cow. J. Reprod. Fertil. 86, 637−641.

Fry, R.C., Zuelke, K.A., Butler, K., Squires, T.L., 1996. Frequency of aspirations and oocyte recovery from 5-month old calves. Proc. Intl. Congr. Anim. Reprod. (Sidney, Australia) 3 (18), 14.

Fukui, Y., 1989. Effects of sera and steroid hormones on development of bovine oocytes matured and fertilized in vitro and co-cultured with bovine oviduct epithelial cells. J. Anim. Sci. 67, 1318−1323.

Gandolfi, F., 2000. Sperm-mediated transgenesis. Theriogenology 53, 127−137.

Gandolfi, F., Moor, R.M., 1987. Stimulation of early embryonic development in the sheep by co-culture with oviduct epithelial cells. J. Reprod. Fertil. 81, 23−28.

Gao, C., Eilts, B.E., Han, Y., Carter, J.A., Godke, R.A., 1996. Using a nitrogen pulse laser for assisted hatching of in vitro-cultured bovine embryos. Proc.13th Intl. Conf. Anim. Reprod. (Sidney, Australia) 3 (22), 13.

Garner, D.L., 2011. Mammalian sperm sexing. Encyclopedia Biotechnol. Agri. Food. Available from: http://dx.doi.org/10.1081/E-EBAF-120042263.

Geshi, M., Youngs, C.R., Nagai, T., 1999. Addition of cysteamine to a serum-free maturation medium enhances in vitro development of IVM−IVF bovine oocytes. J. Mamm. Ova Res. 16, 135−140.

Ginther, O.J., Pierson, R.A., 1989. Regular and irregular characteristics of ovulation and the interovulatory interval in mares. J. Equine Vet. Sci. 90, 9−12.

Glass, R.H., Spindle, A.I., Pederson, R.A., 1979. Mouse embryo attachment to substratum and interaction of trophoblast with cultured cells. J. Exp. Zool. 208, 327−335.

Godke, R.A., Beetem, D.D., Burleigh, D.W., 1990. A method for zona pellucida drilling using a compact nitrogen laser. In: Proceedings of the Seventh World Congress Human Reproduction, June 26−July 1, Helsinki, Finland. No. 258 (abstr.).

Gomez, M.C., Pope, C.E., Harris, R., Davis, A., Mikota, S., Dresser, B.L., 2000. Birth of kittens produced by intracytoplasmic sperm injection of domestic cat oocytes matured in vitro. Reprod. Fertil. Dev. 12, 423–433.

Gordon, I., 1975. Cattle twinning by the egg transfer approach. In: Rowson, L.E.A. (Ed.), Egg Transfer in Cattle. Commission of European Communities, Luxembourg, pp. 305–319.

Gordon, I., Lu, K.H., 1990. Production of embryos in vitro and its impact on livestock production. Theriogenology 33, 77–87.

Gordon, J.W., Dapunt, U., 1993. A new mouse model for embryos with a hatching deficiency and its use to elucidate the mechanism of blastocyst hatching. Fertil. Steril. 59, 1302–1307.

Gordon, J.W., Talansky, B.E., 1986. Assisted fertilization by zona drilling: a mouse model for correction of oligospermia. J. Exp. Zool. 239, 347–354.

Goto, K., Kajihara, Y., Kosaka, S., Koba, M., Nakanishi, Y., Ogawa, K., 1988a. Pregnancies after co-culture of cumulus cells with bovine embryos derived from in vitro fertilization of in vitro matured follicular oocytes. J. Reprod. Fertil. 83, 753–758.

Goto, K., Kajihara, Y., Kosaka, S., Koba, M., Nakanishi, Y., Ogawa, K., 1988b. Pregnancies after in vitro fertilization of cow follicular oocytes, incubation in vitro an their transfer to the cow uterus. Theriogenology 29, 251 (abstr.).

Goto, K., Koba, M., Takuma, Y., Nakanishi, Y., Ogawa, K., 1989. Co-culture of bovine embryos with cumulus cells. AJAS 2, 595–598.

Goto, K., Hinoshita, A., Takuma, Y., Ogawa, K., 1990. Fertilization of bovine oocytes by the injection of immobilised, killed spermatozoa. Vet. Rec. 127, 517–520.

Graff, K.J., Meintjes, M., Dyer, V.W., Paul, J.B., Denniston, R.S., Ziomek, C., et al., 1999. Transvaginal ultrasound-guided oocyte retrieval following FSH stimulation of domestic goats. Theriogenology 51, 1099–1119.

Graham, E.F., Larson, E.V., Crabo, B.G., 1974. Freezing and freeze-drying bovine spermatozoa. In: Proceedings of the Fifth Technology Conference of Artificial Insemination Reproduction. National Association of Animal Breeders, Columbia, Missouri, 14–20.

Hamano, K., Li, X., Qian, L., Funauchi, X., Furudate, K., Minato, Y., 1999. Gender preselection in cattle with intracytoplasmic injected, flow cytometrically sorted sperm heards. Biol. Reprod. 60, 1194–1197.

Hammond Jr, J., 1949. Recovery and culture of tubal mouse ova. Nature (London) 163, 28–29.

Han, Y., Meintjes, M., Graff, K., Denniston, R., Zhang, L., Ziomek, C., et al., 2001. Production of fresh and frozen-thawed transplant offspring from latter stage IVF-derived caprine embryos. Vet. Rec. 149, 714–716.

Hanada, A., 1985. In vitro fertilization in goat. J. Anim. Reprod. 31, 21–26.

Hansel, W., Godke, R.A., 1992. Future prospectives on animal biotechnology. Anim. Biotechnol. 3, 111–137.

Hasler, J.F., Henderson, W.B., Hurtgen, P.J., Jin, Z.Q., McCauley, A.D., Mower, S.A., et al., 1995. Production, freezing and transfer of bovine IVF embryos and subsequent calving results. Theriogenology 43, 141–151.

Hayflick, l., Moorhead, P.S., 1961. The serial cultivation of human diploid cell strains. Exp. Cell Res. 25, 585–621.

Heuwieser, W., Yang, X., Jiang, S., Foote, R.H., 1992. Fertilization of bovine oocytes after microsurgical injection of spermatozoa. Theriogenology 38, 1–9.

Heyman, Y., Menezo, Y., 1987. Interaction of trophoblastic vesicles with bovine embryos developing in vitro. In: Bavister, B.D. (Ed.), The Mammalian Preimplantation Embryo. Plenum Press, New York, NY, pp. 175–191.

Heyman, Y., Camous, S., Fevre, J., Meziou, W., Martal, J., 1984. Maintenance of the corpus luteum after uterine transfer of trophoblastic vesicles to cyclic cows and sheep. J. Reprod. Fertil. 70, 533–540.

Heyman, Y., Chesne, P., Chupin, D., Menezo, Y., 1987a. Improvement of survival rate of frozen cattle blastocysts after transfer with trophoblastic vesicles. Theriogenology 27, 477–484.

Hill, J., Winger, Q., Jones, K., Keller, D., King, W.A., Westhusin, M., 1999. The effect of donor cell serum starvation and oocyte activation compounds on the development of somatic cell cloned embryos. Cloning 1, 201–208.

Hill, J.R., Winger, Q.A., Long, C.R., Looney, C.R., Thompson, J.A., Westhusin, M.E., 2000. Development rates of male bovine nuclear transfer embryos derived from adult and fetal cells. Biol. Reprod. 62, 1135–1140.

Hinrichs, K., Schmidt, A.L., Friedman, P.P., Selgrath, J.P., Martin, M.G., 1993. In vitro maturation of horse oocytes: characterization of chromatin configuration using flouresenic microscopy. Biol. Reprod. 48, 363–370.

Hinrichs, K., Provost, P.J., Torello, E.M., 2000. Treatments resulting in pregnancy in nonovulating hormone-treated oocyte recipient mares. Theriogenology 54, 1285–1293.

Hiramoto, Y., 1962. Microinjection of live spermatozoa into sea urchin eggs. Exp. Cell Res. 27, 416–426.

Hosoi, Y., Miyake, M., Utsumi, K., Iritani, A., 1988. Development of rabbit oocytes after microinjection of spermatozoon. In: Proceedings of the 11th International Congress of Animal Reproduction and Artificial Insemination. vol. 3, pp. 331–333.

Hu, Y.X., Voelkel, S.A., Godke, R.A., 1989. One-cell murine embryos cultured on rat liver cell monolayers and mouse oviduct cells. Proceedings of the First AASS Conference Cornell University, Ithaca, NY. p. 83 (abstr.).

Hunter, G.I., Bishop, G.P., Adams, J.C.E., Rowson, L.E.A., 1962. Successful long-distance aerial transport of fertilized sheep ova. J. Reprod. Fertil. 3, 33–40.

Johnson, L.A., Flook, J.P., Hawk, H.W., 1989. Sex preselection in rabbits: live births from X- and Y-sperm separated by DNA and cell sorting. Biol. Reprod. 41, 199–203.

Kane, M.T., 1983. Variability in different lots of commercial bovine serum albumin affects cell multiplication and hatching of rabbit blastocyst in culture. J. Reprod. Fertil. 69, 555–558.

Kane, M.T., 1987. Culture media and culture of early embryos. Theriogenology 27, 49–57.

Kane, M.T., Foote, R.H., 1970. Culture of two- and four-cell rabbit embryos to expanding blastocyst stage in synthetic media. Proc. Soc. Exp. Biol. Med. 133, 921–925.

Kato, Y., Tani, T., Sotomaru, Y., Kurokawa, K., Kato, J., Doguchi, H., et al., 1998. Eight calves cloned from somatic cells of a single adult. Science 282, 2095–2098.

Keefer, C.L., Younis, A.I., Brackett, B.G., 1990. Cleavage development of bovine oocytes fertilized by sperm injection. Mol. Reprod. Dev. 25, 281–285.

Keefer, C.L., Baldassarre, H., Keystone, R., Wang, B., Bhatia, B., Bilodeau, A.S., et al., 2001. Generation of dwarf goat (Capra hircus) clones following nuclear transfer with transfected and nontransfected fetal fibroblasts and in vitro-matured oocytes. Biol. Reprod. 64, 849–856.

Kim, H.N., Hu, Y.X., Roussel, J.D., Godke, R.A., 1989a. Culturing murine embryos on bovine fetal spleen cell fibroblast and chick embryo fibroblast monolayers. Theriogneology 21, 211 (abstr.).

Kim, H.N., Roussel, J.D., Amborski, G.F., Hu, Y.X., Godke, R.A., 1989b. Monolayers of bovine fetal spleen cells and chick embryo fibroblasts for co-culture of bovine embryos. Theriogenology 31, 212 (abstr.).

Kim, H.N., Zhang, L., Roussel, J.D., Godke, R.A., 1991. Development of *in vitro* fertilized (IVF)-bovine oocytes using fetal spleen cell and chick embryo fibroblast monolayers. In: Proceedings of the Society Study Reproduction (abstr.).

Kimura, Y., Yanagimachi, R., 1995. Intracytoplasmic sperm injection. Biol. Reprod. 52, 709−720.

King, W.A., Chartrain, I., Kopeony, V., Betteridge, K.J., Bergeron, H., 1989. Nucleolus organizer regions and nucleoli in mammalian embryos. J. Reprod. Fertil. (Suppl.) 38, 63−71.

Kinis, A., Vergos, E., Gordon, I., Gordon, A., Gallagher, M., 1990. Studies in the production of chimeric cattle embryos by aggregation of blastomeres from embryos derived from oocytes matured and fertilized *in vitro*. Theriogenology 33, 268 (abstr.).

Kirkpatrick, B.W., Monson, R.L., 1993. Sensitive sex determination assay applicable to bovine embryos derived from IVM and IVF. J. Reprod. Fertil. 98, 336−340.

Kitiyanant, Y., Thonabulsombat, C., Tocharaus, C., Sanituongse, B., Pavasuthipaisit, K., 1989. Co-culture of bovine embryos from oocytes matured and fertilized *in vitro* to the blastocyst stage with oviductal tissue. J. Sci. Soc. Thailand 15, 251−260.

Kolbe, T., Holtz, W., 2000. Birth of a piglet derived from an oocyte fertilized by intracytoplasmic sperm injection (ICSI). Anim. Reprod. Sci. 64, 97−101.

Kruger, E., Schmoll, F., Schernthaner, W., Brem, G., Schellander, K., 1996. Increased *in vitro* hatching rate following partial zona dissection of *in vitro* produced bovine embryos. Proc.13th Intl. Conf. Anim. Reprod. (Sidney, Australia) 3 (21), 13.

Kruip, Th.A.M., Pieterse, M.C., Van Beneden, T.h.H., Vos, P.L.A.M., Wurth, Y.A., Taverne, M.A.M., 1991. A new method for bovine embryo production: a potential alternative to superovulation. Vet. Rec. 128, 208−210.

Kues, W.A., Anger, M., Carnwath, J.W., Paul, D., Motlik, J., Niemann, H., 2000. Cell cycle synchronization of porcine fetal fibroblasts: effects of serum deprivation and reversible cell cycle inhibitors. Biol. Reprod. 62, 412−419.

Kuroiwa, Y., Kasinathan, P., Choi, Y.J., Naeen, R., Tomizuka, K., Sullivan, E.J., et al., 2002. Cloned transchromosomic calves producing human immunoglobulin. Nat. Biotech. 20, 889−894.

Kuzan, F.B., Wright Jr, R.W., 1981. Attachment of porcine blastocyst to fibroblast monolayers *in vitro*. Theriogenology 16, 651−658.

Kuzan, F.B., Wright Jr, R.W., 1982a. Observations on the development of bovine morulae on various cellular and noncellular substrata. J. Anim. Sci. 54, 811−816.

Kuzan, F.B., Wright, R.W., 1982b. Blastocyst expansion, hatching, and attachment of porcine embryos co-cultured with bovine fibroblasts *in vitro*. Anim. Reprod. Sci. 5, 57−63.

Lambeth, V.A., Looney, C.R., Voelkel, S.A., Hill, K.G., Jackson, D.A., Godke R.A., 1982. Micromanipulation of bovine morulae to produce identical twin offspring. In: Proceedings of the Second World Congress Embryo Transfer Mammalian, September 20−22, Annecy, France. p. 55. (abstr.).

Lambeth, V.A., Looney, C.R., Voelkel, S.A., Hill, K.G., Jackson, D.A., Godke, R.A., 1993. Microsurgery on bovine embryos at the morula stage to produce monozygotic twin calves. Theriogenology 20, 85−95.

Lanza, R.P., Cibelli, J.B., Blackwell, C., Cristofalo, V.J., Francis, M.K., Baerlocher, G.M., et al., 2000. Extension of cell life-span and telomere length in animals cloned from senescent somatic cells. Science 288, 665−669.

Laufer, N., Palanker, D., Shufaro, W., Sarfan, A., Simon, A., Lewis, A., 1993. The efficiency and safety of zona pellucida drilling by a 193 nm excimer laser. Fertil. Steril. 59, 889−895.

Lawson, R.A.S., Rowson, L.E.A., Adams, C.E., 1972a. The development of sheep eggs in the rabbit oviduct and their viability after re-transfer to ewes. J. Reprod. Fertil. 28, 105−116.

Lawson, R.A.S., Rowson, L.E.A., Adams, C.E., 1972b. The development of cow eggs in the rabbit oviduct and their viability after re-transfer to heifers. J. Reprod. Fertil. 28, 313−315.

Lewis, W.H., Gregory, P.W., 1929. Chinematographs of living developing rabbit eggs. Science 69, 226−229.

Li, L.Y., Meintjes, M., Graff, K.J., Paul, J.B., Denniston, R.S., Godke., R.A., 1995. *In vitro* fertilization and development of *in-vitro* matured oocytes aspirated from pregnant mares. Biol. Reprod. Mongr. 1, 613−622.

Liu, H.C., Cohen, J., Alikani, N., 1993. Assisted hatching facilitates earlier implantation. Fertil. Steril. 60, 871−875.

Liu, H., Chen, Y, Niu, Y., Zhang, K., Kang, Y., Ge, W., et al., 2014. TALEN-mediated gene mutagenesis in Rhesus and Cynomolgus monkeys. Cell Stem Cell 14 (3), 323−328. Available from http://dx.doi.org/10.1016/j.stem.2014.01.018

Looney, C.R., Lindsey, B.R., Gonseth, C.L., Johnson, D.L., 1994. Commercial aspects of oocyte retrieval and *in vitro* fertilization (IVF) for embryo production in problem cows. Theriogenology 41, 67−72.

Looney, C.R., Damiani, P., Lindsey, B.R., Long, C.R., 1995. The use of prepubertal heifers as oocyte donors for IVF: effect of age and gonadotropin treatment. Theriogenology 43, 269 (abstr.).

Maciulis, A., Bunch, T.D., Foote, W.C., Call, J.W., 1987. The influence of oviductal and embryo cell co-cultures on the development of one to two-cell ovine embryos. Encyclia 64, 73−78.

Maga, E.A., Murray, J.D., 1995. Mammary gland expression of transgenes and the potential for altering properties of milk. Biotechnology 132, 1452−1457.

Maksimenko, O.G., Deykin, A.V., Khodarovich, Y.M., Georgie, P.G., 2013. The use of transgenic animals in biotechnology: problems and prospects. Acta Nat. 5 (1), 33−46.

Malter, H., Cohen, J., 1989a. Partial zona dissection of the human oocyte: a nontraumatic method using micromanipulation to assist zona pellucida penetration. Fertil. Steril. 51, 139−148.

Malter, H., Cohen, J., 1989b. Blastocyst formation and hatching *in vitro* following zona drilling of mouse and human embryos. Gamete Res. 24, 67−80.

Mann, J., 1988. Full term development of mouse eggs fertilized by a spermatozoon microinjected under the zona pellucida. Biol. Reprod. 38, 1077−1083.

Marek, D.E., Pryor, J.H., Whitesell, T.H., Looney, C.R., 1990. Nuclear transplantation in the bovine: effect of donor embryo age on subsequent embryo production. Theriogenology 33, 283. (abstr.).

Markert, C.L., 1983. Fertilization of mammalian eggs by sperm injection. J. Exp. Zool. 228, 195−201.

Markert, C.L., 1984. Genetic manipulation of mammalian embryos: current techniques and their potential usefulness in livestock improvement. Proc. 10th Intl. Congr. Anim. Reprod. Artif. Insem. 2, 13−19.

Martin, M.J., 2000. Development of *in vitro*-matured porcine oocytes following intracytoplasmic sperm injection. Biol. Reprod. 63, 109−112.

Martinez, E.A., Cuello, C., Parrilla, I., Rodriguez-Martinez, H., Roca, C., Vazquez, J.L., et al., 2013. Design, development, and application of a non-surgical deep uterine embryo transfer technique in pigs. Anim. Front. 3 (4), 40−47. Available from: http://dx.doi.org/10.2527/af.2013-0032.

McCreath, K.J., Howcroft, J., Campbell, K.H.S., Colman, A., Schnieke, A.E., Kind, A.J., 2000. Production of gene-targeted sheep by nuclear transfer from cultured somatic cells. Nature 405, 1066–1069.

McKinnon, A.O., Lacham-Kaplan, O., Trounson, A.O., 1998. Pregnancies produced from fertile and infertile stallions by intracytoplasmic sperm injection (ICSI) of a single frozen/thawed spermatozoa into *in vivo* matured mare oocytes. Proceedings of the Seventh International Symposium Equine Reproduction. University of Pretoria, South Africa, p. 137. (abstr.).

McLaren, A., 1968. A study of blastocysts during delay and subsequent implantation in lactating mice. J. Endocrinol. 42, 453–463.

McLaren, A., 1970. The fate of the zona pellucida in mice. J. Embryol. Exp. 23, 1–197.

Meade, H.M., Echelard, Y., Ziomek, C.A., Young, M.W., Harvey, M., Cole, E.S., et al., 1998. Expression of recombinant proteins in the milk of transgenic animals. In: Fernandez, J.M., Hoeffler, J.P. (Eds.), Gene Expression Systems: Using Nature for the Art of Expression. Academic Press, San Diego, CA, pp. 399–427.

Meintjes, M., Bellow, M.S., Broussard, J.R., Paccamonti, D., Eilts, B.E., Godke, R.A., 1994. Repeated transvaginal ultrasound-guided oocyte retrieval from pregnant mares. Theriogenology 41, 255 (abstr.).

Meintjes, M., Bellow, M.S., Paul, J.B., Broussard, J.R., Li, L.Y., Paccamonti, D., et al., 1995a. Transvaginal ultrasound-guided oocyte retrieval from cyclic and pregnant horse and pony mares for *in vitro* fertilization. Biol. Reprod. Monogr. 1, 281–292.

Meintjes, M., Bellow, M.S., Broussard, J.R., Paul, J.B., Godke, R.A., 1995b. Trans vaginal aspiration of oocytes from hormone treated pregnant beef cattle for *in vitro* fertilization. J. Anim. Sci. 73, 967–974.

Meintjes, M., Graff, K.J., Paccamonti, D., Eilts, B.E., Cochran, R., Sullivan, M., et al., 1996. *In vitro* development and embryo transfer of sperm-injected oocytes derived from pregnant mares. Theriogenology 45, 304 (abstr.).

Meintjes, M., Bezuidenhout, C., Bartels, P., Visser, D.S., Meintjes, J., Loskutoff, N.M., et al., 1997a. *In vitro* maturation and fertilization of oocytes recovered from free ranging Burchell's zebra (*Equus burchelli*) and Hartmann's zebra (*Equus zebra hartmannae*). J. Zoo Wildl. Med. 28, 251–259.

Meintjes, M., Graff, K.J., Paccamonti, D., Eilts, B.E., Paul, J.B., Thompson Jr, D.L., et al., 1997b. Effect of follicular aspiration and flushing, and genotype type of the fetus on circulating progesterone levels during pregnancy in the mare. Equine Vet. J. Suppl. 25, 25–32.

Menezo, Y., Testart, J., Perrone, D., 1984. Serum is not necessary for human *in vitro* fertilization, early embryo culture, and transfer. Fertil. Steril. 42, 750–755.

Mertes, P.C., Bondioli, K.R., 1985. Effect of splitting technique on pregnancy rate from half embryos. Theriogenology 23, 209. (abstr.).

Meryman, H.T., 1960. Drying of living mammalian cells. Ann. N. Y. Acad. Sci. 85, 729.

Nakagawa, A., Takahashi, Y., Kanagawa, H., 1985. Sexing of bisected mouse embryos by chromosomal analysis. J. Mam. Ova Res. 2, 79.

Nagashima, H., Grupen, C.G., Ashman, R.J., Nottle, M.B., 1995. Developmental competence of *in vivo* and *in vitro* matured porcine oocytes after subzonal injection. Proc. Aust. Soc. Reprod. Biol. 26, 53 (Abstr.).

Neev, J., Schiewe, M.C., Sung, W.V., Kang, D., Hezeleger, N., Berns, M.W., et al., 1995. Assisted hatching in mouse embryos using a no contact Ho: YSGG laser system. J. Assist. Reprod. Genet. 12, 288–293.

New, C.A.T., Stein, K.F., 1964. Cultivation of post-implantation mouse and rat embryos on plasma clots. J. Embryo Exp. Morphol. 12, 101–111.

Ng, S., Bongso, A., Sathananthan, H., Ratnam, S., 1990. Micromanipulation: its relevance to human *in vitro* fertilization. Fertil. Steril. 53, 203—219.

Ng, S.C., Liow, S.L., Sathananthan, H., Bongso, A., Ratnam, S.S., 1993. Review: microinjection of human sperm directly into human oocytes. J. Assist. Reprod. Genet. 10, 337—352.

Nicholas, J.S., Hall, B.V., 1942. Experiments on developing rats. II. The development of isolated blastomeres and fused eggs. J. Exp. Zool. 90, 441—449.

Nieder, G., Caprio, T.L., 1990. Early development in the Siberian hamster (*Phodopus sungorus*). Mol. Reprod. Dev. 27, 224—229.

Niu, Y., Shen, B., Cui, Y., Chen, Y., Wang, J., Wang, L., et al., 2014. Generation of gene-modified cynomolgus monkey via Cas9/RNA-mediated gene targeting in one-cell embryos. Cell 156 (4), 836—843. Available from: http://dx.doi.org/10.1016/j.cell.2014.01.027.

Nowak-Imialek, M., Niemann, H., 2012. Pluripotent cells in farm animals: state of the art and future perspectives. Reprod. Fertil. Dev. 25, 103—128.

Nusser, K.D., Mitalipov, S., Widmann, A., Gerami-Naini, B., Yeoman, R.R., 2001. Developmental competence of oocytes after ICSI in the rhesus monkey. Hum. Reprod. 16, 130—137.

Obruca, A., Strohmer, H., Sakkas, D., Menezo, Y., Kogosowski, A., Barak, Y., et al., 1993. Assisted hatching and micro-fertilization. Contracept. Fertil. Sex. 22 (5), 303—305.

Obruca, A., Strohmer, H., Sakkas, D., Menezo, Y., Kogosowski, S., Barak, Y., et al., 1994. Use of lasers in assisted fertilization and hatching. Hum. Reprod. 9, 1723—1726.

Onuma, H, Foote, RH, 1969. In vitro development of ova from prebuberal cattle. J. Dairy Sci. 52 (7), 1085—1087.

Ouhibi, N., Menezo, Y., Benet, G., Nicollet, B., 1989. Culture of epithelial cells derived from the oviduct of different species. Hum. Reprod. 4, 229—235.

Ouhibi, N., Hamidi, J., Guillaud, J., Menezo, Y., 1990. Co-culture of 1-cell mouse embryos on different cell supports. Hum. Reprod. 5, 737—743.

Overskei, T.L., Cincotta, A.H., 1987. A new approach to embryo co-culture. Theriogenology 27, 266. (abstr.).

Ozil, J.P., 1983. Production of identical twins by bisection of blastocysts in the cow. J. Reprod. Fertil. 69, 463—468.

Ozil, J.P., Heyman, Y., Renard, J.P., 1982. Production of monozygotic twins by micromanipulation and cervical transfer in the cow. Vet. Rec. 110, 126—127.

Palermo, G., Joris, H., Devroey, H., Van Steirteghem, A.C., 1992a. Pregnancies after intracytoplasmic sperm injection of single spermatozoon into an oocyte. Lancet 340, 17—18.

Palermo, G., Cohen, C., Rosenwaks, Z., 1992b. Intracytoplasmic sperm injection: a powerful tool to over come fertilization failure. Fertil. Steril. 65, 899—908.

Palermo, G., Joris, H., Derde, M.P., Camus, M., Devroey, M., Van Steirteghem, A.C., 1993. Sperm characteristics and outcome of human assisted fertilization by subzonal insemination and intracytoplasmic sperm injection. Fertil. Steril. 59, 826—835.

Palmer, E., Bezard, J., Magistrini, M., Duchamp, G., 1991. *In vitro* fertilization in the horse: a retrospective study. J. Reprod. Fertil. (Suppl.) 44, 375—384.

Paul, J.B., Looney, C.R., Lindsey, B.R., Godke, R.A., 1995. Gonadotropin stimulation of cattle donors at estrus for transvaginal oocyte collection. Theriogenology 43, 294. (abstr.).

Perez, O., Richard Iii, R., Green, H.L., Youngs, C.R., Godke, R.A., 2000. Ultrasound-guided transvaginal oocyte recovery from FSH-treated postpartum beef cows. Theriogenology 53, 364. (abstr.).

Perez, O., Boediono, A., Ferguson, E., Airhart, C., Richards III, R., Godke, R.A., 2001. Oocyte and embryo production from FSH-treated postpartum beef cows shortly after calving. Theriogenology 55, 516. (abstr.).

Peura, T., Hyttinen, J.-M., Turunen, M., Jänne, J., 1991. A reliable sex determination assay for bovine preimplantation embryos using the polymerase chain reaction. Theriogenology 35, 547−555.

Pieterse, M.C., Kappen, K.A., Kruip, Th.A.M., Taverne, M.A.M., 1988. Aspiration of bovine oocytes during transvaginal ultrasound scanning of the ovaries. Theriogenology 30, 751−762.

Pignalo, R.J., Masoro, E.J., Nicholas, W.W., Bradt, C.I., Cristofalo, V.J., 1992. Skin fibroblasts from Fischer 344 rats undergo similar changes in replicative lifespan but not immortilization with caloric restriction of donors. Exp. Cell Res. 201, 16−22.

Pincus, G., 1930. Observation of the living eggs of the rabbit. Proc. Royal Soc. (London) Ser. B 107, 132−167.

Pincus, G. (Ed.), 1936. The Eggs of Mammals. Macmillan Co., New York, NY.

Poleo, G.A., Denniston, R.S., Reggio, B.C., Godke, R.A., Tiersch, T.R., 2001. Fertilization of eggs of zebrafish (Danio rerio) by intracytoplasmic sperm injection. Biol. Reprod. 65, 961−966.

Polzin, V.J., Anderson, G.B., BonDurrant, R.H., Butler, J.E., Pashens, R.L., Penedo, M.C.T., et al., 1987. Production of sheep−goat chimeras by inner mass cell transplantation. J. Anim. Sci. 65, 325−330.

Ponsart, C., Le Bourhis, D., Knijn, H., Fritz, S., Guyader-Joly, C., Otter, T., et al., 2014. Reproductive technologies and genomic selection in dairy cattle. Reprod. Fertil. Dev. 26, 12−21. Available from: http://dx.doi.org/doi:10.1071/RD13328.

Pool, S.H., Rorie, R.W., Pendleton, R.J., Menino, A.R., Godke, R.A., 1988a. Culture of early-stage bovine embryos inside day-13 and day-14 precultured trophoblastic vesicles. Ann. N.Y. Acad. Sci. 541, 407−418.

Pool, S.H., Wiemer, K.E., Rorie, R.W., Godke, R.A., 1988b. The use of trophoblastic vesicles and fetal uterine fibroblast cells for the culture of precompaction-stage bovine embryos. Proc. 11th Intl. Congr. Anim. Reprod. Artif. Insem. 4, 479.

Pope, C.E., Johnson, C.A., McRae, M.A., Keller, G.L., Dresser, B.L., 1995. In vitro and In vivo development of domestic cat oocytes following intracytoplasmic sperm injection or subzonal insemination. Theriogenology 43, 302. (abstr.).

Pope, C.E., Johnson, C.A., McRae, M.A., Keller, G.L., Dresser, B.L., 1998. Development of embryos produced by intracytoplasmic sperm injection of cat oocytes. Anim. Reprod. Sci. 53, 221−236.

Presicce, G.A., Jiang, S., Simkin, M., Zhang, L., Looney, C.R., Godke, R.A., et al., 1997. Age and hormonal dependence on acquisition of oocyte competence for embryogenesis in prepubertal calves. Biol. Reprod. 56, 386−392.

Prichard, J.F., Thibodeaux, J.K., Pool, S.H., Blakewood, E.G., Menezo, Y., Godke, R.A., 1992. In vitro co-culture of early-stage caprine embryos with oviduct and uterine epithelial cells. Hum. Reprod. 7, 553−557.

Reggio, B.C., James, A.N., Green, H.L., Gavin, W.G., Behboodi, E., Echelard, Y., et al., 2001. Cloned transgenic offspring resulting from somatic cell nuclear transfer in the goat: oocytes derived from both follicle-stimulating hormone-stimulated and nonstimulated abattoir-derived ovaries. Biol. Reprod. 65, 1528−1533. Available from: http://dx.doi.org/10.1095/biolreprod65.5.1528

Renard, J.P., du Mesnil du Buisson, F., Winterberger-Torres, S., Menezo, Y., 1976. *In vitro* culture of cow embryos from day 6 and day 7. In: Rowson, L.E.A. (Ed.), Egg Transfer in Cattle. Commission of European Communities, Luxembourg, pp. 154–159.

Reshef, E., Haaksma, C.J., Bettinger, T.L., Haas, G.G., Sehafer, S.A., Zavy, M.T., 1993. Gamete and embryo micromanipulation using the Holmium:YAG laser. In: Proceedings of the 49th American Fertility Society, Montreal, Canada, October 11–14, Fertil. Steril. Suppl., p. 016, S88.

Restall, B.J., Wales, R.G., 1966. The fallopian tube of the sheep. III. The chemical composition of the fluid from the fallopian tube. Aust. J. Biol. Sci. 19, 687–698.

Rexroad Jr, C.E., 1989. Co-culture of domestic animal embryos. Theriogenology 31, 105–114.

Rexroad Jr., C.E., Powell, A.M., 1986. Co-culture of sheep ova and cells from sheep oviduct. Theriogenology 37, 859–866.

Rexroad Jr., C.E., Powell, A.M., 1988. Co-culture of ovine eggs with oviductal cells in medium 199. J. Anim. Sci. 66, 947–953.

Rexroad Jr., C.E., Hammer, R.E., Bolt, D.J., Mayo, K.M., Frohman, L.A., Palmiter, R.D., et al., 1989. Production of transgenic sheep with growth regulating genes. Mol. Reprod. Dev. 1, 164–169.

Rodriguez, H.F., Wiemer, K.W., Denniston, R.S., Godke, R.A., 1990. A bilayered fetal-cell co-culture system for culturing bovine embryos. Theriogenology 33, 309. (abstr.).

Rodriguez, H.F., Denniston, R.S., Godke, R.A., 1991. Co-culture of bovine embryos using a bilayer of bovine oviductal and granulosa cells. Theriogenology 35, 264. (abstr.).

Roknabadi, G.A., Ng, S.C., Liow, S.L., Bongso, A., Ratnam, S.S., 1994. Intracytoplasmic sperm injection in the mouse. In: Proceedings of the 13th Annals Meeting Fertility Society Australia. MP 116 (abstr.).

Rorie, R.W., McFarland, C.W., Overskei, T.L., Voelkel, S.A., Godke, R.A., 1985. A new method of splitting embryos without the use of a commercial micromanipulation unit. Theriogenology 23, 224. (abstr.).

Rorie, R.W., Pendleton, R.J., Pool, S.H., White, K.L., Godke, R.A., 1987. Cryopreservation of bovine trophoblastic vesicles. Theriogenology 27, 272. (abstr.).

Rorie, R.W., Pool, S.H., Prichard, J.F., Betteridge, K.J., Godke, R.A., 1994. A simplified procedure for interspecific and intergeneric embryo transfer. Vet. Rec. 115, 186–187.

Ryan, D.P., Blakewood, E.G., Swanson, W.F., Rodrigues, H., Godke, R.A., 1993. Using hormone-treated pregnant cows as potential source of oocytes for *in vitro* fertilization. Theriogenology 40, 1039–1055.

Schiewe, M., Hazeleger, N., Sclimenti, C., Balmaceda, J., 1995a. Physiological characterization of blastocyst hatching mechanisms by use of a mouse antihatching model. Fertil. Steril. 63 (2), 288–294.

Schiewe, M., Neev, J., Hazeleger, N.L., Balmaceda, J.P., Bern, M.W., Tadir, Y., 1995b. Developmental competence of mouse embryos following zona drilling using a non-contact holmium: yttrium scandian gallium garnet laser system. Hum. Reprod. 10 (7), 1821–1824.

Schnieke, A.E., Kind, A.J., Ritchie, W.A., Mycock, K., Scott, A.R., Ritchie, M., et al., 1997. Human Factor IX transgenic sheep produced by transfer of nuclei from transfected fetal fibroblasts. Science 278, 2130–2133.

Seidel, F., 1952. Die entwicklungspotenzen einer isolierten blastomere des zweizellenstadiums in säugetierei. Naturwissenschaften 39 (15), 355–356.

Shea, B.F., Hines, D.J., Lightfoot, D.E., Ollis, G.W., Olsa, S.M., 1974. In: Rowson, L.E.A. (Ed.), Egg Transfer in Cattle. Commission of European Communities, Luxembourg, pp. 145–152.

Shiga, K., Fujita, T., Hirose, K., Sasae, Y., Nagai, T., 1999. Production of calves by transfer of nuclei from cultured somatic cells obtained from Japanese black bulls. Theriogenology 52, 527–535.

Squires, E.L., Wilson, J.M., Kato, H., Blaszczyk, A., 1996. A pregnancy after intracytoplasmic sperm injection into equine oocytes matured in vitro. Theriogenology 45, 306. (abstr.).

Sreenan, J., Scanlon, P.F., 1968. Continued cleavage of fertilized bovine ova in the rabbit. Nature 217, 867.

Sreenan, J., Scanlon, P.F., Gordon, I., 1969. Culture of fertilized cattle eggs. J. Agric. Sci. (Cambridge) 70, 183–185.

Steptoe, P.C., Edwards, R.G., 1978. Birth after implantation of a human embryo. Lancet 2, 366.

Stover, J., Evans, J., Dolensek, E.P., 1981. Interspecies embryo transfer from the Gaur to domestic Holstein. In: Proceedings of the American Association of Zoo Veterinarians, Seattle, Washington, DC. pp. 122–124.

Stroud, B., 2012. The year 2011 worldwide statistics of embryo transfer in domestic farm animals. Embryo. Transf. Newsl. 30 (4), 16–26.

Swanson, M.E., Martin, M.J., O'Donell, J.K., Hoover, K., Lago, W., Huntress, F., et al., 1992. Production of functional human hemoglobin in transgenic swine. Biotechnology 10, 557–559.

Tarkowski, A.K., 1959. Experiments on the development of isolated blastomeres of mouse eggs. Nature 184, 1286–1287.

Tervit, H.R., Whittingham, D.G., Rowson, L.E.A., 1972. Successful culture in vitro of sheep and cattle ova. J. Reprod. Fertil. 30, 495–497.

Thadani, V.M., 1980. A study of heterospecific sperm-egg interactions in the rat, mouse and deer mouse using in vitro fertilization and sperm injection. J. Exp. Zool. 212, 435–453.

Thibault, C., 1966. La culture in vitro de l'oeuf de vache. Ann. Biol. Anim. Biochem. Biophys. 15, 159–164.

Thibodeaux, J.K., Goodeaux, L.L., Roussel, J.D., Amborski, G.F., Moreau, J.D., Godke, R.A., 1991a. Stage of the bovine estrous cycle and in vitro characteristics of uterine and oviductal epithelial cells. In: Proceedings of the Southern Security American Dairy Science p. 3. (abstr.).

Thibodeaux, J.K., Goodeaux, L.L., Roussel, J.D., Menezo, Y., Amborski, G.F., Moreau, J.D., et al., 1991b. Effects of stage of the bovine estrous cycle on in vitro characteristics of uterine and oviductal epithelial cells. Hum. Reprod. 6, 751–760.

Thibodeaux, J.K., Roussel, J.D., Menezo, Y., Godke, R.A., Goodeaux, L.L., 1991c. A method for in vitro cell culture of superficial bovine uterine endometrial epithelium. J. Tissue Cult. Meth. 13, 247–252.

Uehara, T., Yanagamachi, R., 1976. Microsurgical injection of spermatozoa into hamster eggs with subsequent transformation of sperm nuclei into male pronuclei. Biol. Reprod. 15, 467–470.

Vignion, X., Chesne, P., Le Bourhis, D., Flechion, J.E., Heyman, Y., Renard, J.P., 1998. Development potential of bovine embryos reconstructed from enucleated matured oocytes fused with cultured somatic cells. Comp. Rend. Acad. Sci. 321, 735–745.

Voelkel, S.A., Amborski, G.F., Hill, K.G., Godke, R.A., 1984a. Use of a uterine-cell monolayer culture system for micromanipulated bovine embryos. Theriogenology 21, 271. (abstr.).

Voelkel, S.A., Humes, P.E., Godke, R.A., 1984b. Pregnancy rates resulting from non-surgical transfer of micromanipulated bovine embryos, Proceedings of the Tenth International Congress of Animal Reproduction and Artificial Insemination (June 10–14), vol. 2. University of Illinois, Urbana, p. 251.

Voelkel, S.A., Viker, S.D., Humes, P.E., Godke, R.A., 1984c. Micromanipulation and non-surgical transfer of bovine embryos. J. Anim. Sci. 59 (Suppl. 1), 393 (abstr.).

Voelkel, S.A., Amborski, G.F., Hill, K.G., Godke, R.A., 1985a. Use of uterine-cell monolayer culture system for micromanipulated bovine embryos. Theriogenology 24, 271–281.

Voelkel, S.A., Viker, S.D., Johnson, C.A., Hill, K.G., Humes, P.E., Godke, R.A., 1985b. Multiple embryo transplant offspring produced from quartering a bovine morulae. Vet. Rec. 117, 528–530.

Voelkel, S.A., Rorie, R.W., McFarland, C.W., Godke, R.A., 1986. An attempt to produce quarter embryos from non-surgically recovered bovine blastocysts. Theriogenology 25, 207 (abstr.).

Waddington, C.H., Waterman, A.J., 1933. The development *in vitro* of young rabbit embryos. J. Anat. 67, 356–370.

Wakayama, T., Yanagimachi, R., 1998. Development of normal mice from oocytes injected with freeze-dried spermatozoa. Nat. Biotechnol. 16, 639–641.

Wakayama, T., Perry, A.C.F., Zuccotti, M., Johnson, K.R., Yanagimachi, R., 1998. Full-term development of mice from enucleated oocytes injected with cumulus cell nuclei. Nature 394, 369–374.

Wall, R.J., 2002. New gene transfer method. Theriogenology 57, 189–201.

Wang, Y., Zhao, S., Bai, L., Fan, J., Li, E., 2013. Expression systems and animal species used for transgenic animal bioreactors. BioMed. Res. Int. 1–9, article 580463. <http://dx.doi.org/10.1155/2013/580463>.

Ward, K.A., Byrne, C.R., Wilson, B.W., Leish, Z., Rigby, N.W., Townrow, C.R., et al., 1991. Regulation of wool growth in transgenic animals. Adv. Dermatol. 1, 70–76.

Ward, K.W., Nancarrow, C.D., 1992. The production of transgenic sheep for the improved wool production. In: Speedy, A.W. (Ed.), Progress in Sheep and Goat Research. CAB International, Wallingford, Oxon, UK, pp. 257–273. (Chapter 12).

Warfield, S.J., Siedel Jr., G.E., Elsden, R.P., 1986. Transfer of bovine demi-embryos with and without a zona pellucida. Theriogenology 25, 212. (abstr.).

Wells, D.N., Misica, P.M., Day, A.M., Tervit, H.R., 1997. Production of cloned lambs from an established embryonic cell line: a comparison between *in vivo-* and *in vitro*-matured cytoplasts. Biol. Reprod. 57, 385–393.

Wells, D.N., Misica, P.M., Tervit, H.R., Vivanco, W.H., 1998. Adult somatic cell nuclear transfer is used to preserve the last surviving cow of the Enderby Island cattle breed. Reprod. Fertil. Dev. 10, 369–378.

Wells, D.N., Misica, P.V., Tervit, R., 1999. Production of cloned calves following nuclear transfer with cultured adult mural granulosa cells. Biol. Reprod. 60, 996–1005.

Westhusin, M.E., Slapak, J.R., Fuller, D.T., Kraemer, D.C., 1989. Culture of agar-embedded one and two cell bovine embryos and embryos produced by nuclear transfer in the sheep and rabbit oviduct. Theriogenology 31, 371. (abstr.).

Wheeler, M.B., Walters, E.M., Clark, S.G., 2003. Transgenic animals in biomedicine and agriculture: outlook for the future. Anim. Reprod. Sci. 79, 265–289.

White, K.L., Anderson, G.B., BonDurrant, R.H., Donahue, S., Pashen, R.L., 1987. Viability of bisected bovine embryos after detection of H-Y antigen. Theriogenology 27, 293. (abstr.).

Whitten, W.K., 1956. Culture of tubal ova. Nature (London) 177, 96.

Whitten, W.K., 1971. Nutrient requirements for the culture of preimplantation embryos *in vitro*. Adv. Biol. Sci. 6, 129.

Whitten, W.K., Biggers, J.D., 1968. Complete development *in vitro* of the preimplantation stages of the mouse in a simple chemically defined medium. J. Reprod. Fertil. 17, 399–401.

Whittingham, D.G., 1977. Culture of mouse ova. J. Reprod. Fertil. (Suppl.) 14, 7–21.

Whittingham, D.G., Bavister, B.D., 1974. Development of hamster eggs fertilized *in vitro* or *in vivo*. J. Reprod. Fertil. 38, 489–492.

Whittingham, D.G., Biggers, J.D., 1968. Fallopian tube and early cleavage in the mouse. Nature (London) 213, 942–943.

Wiemer, K.E., Denniston, R.S., Amborski, G.F., White, K.L., Godke, R.A., 1987a. A fetal fibroblast monolayer system of *in vitro* culture of bovine embryos. J. Anim. Sci. 65 (Suppl. 1), 122 (abstr.).

Wiemer, K.E., Amborski, G.F., Denniston, R.S., White, K.L., Godke, R.A., 1987b. Use of a hormone-treated fetal uterine fibroblast monolayer system for *in vitro* culture of bovine embryos. Theriogenology 27, 294 (abstr.).

Wiemer, K.E., Casey, P.L., DeVore, D., Godke, R.A., 1988a. The culture of equine embryos using a new fetal uterine monolayer culture system. Theriogenology 29, 327. (abstr.).

Wiemer, K.E., Casey, P.L., Godke, R.A., 1988b. Short term storage of equine embryos on a fetal bovine uterine monolayer followed by transfer to recipients. In: Proceedings of the 11th International Congress of Animal Reproduction and Artificial Insemination. vol. 2. p. 198 (abstr.).

Wiemer, K.E., Casey, P.L., Mitchell, P.S., Godke, R.A., 1989a. Pregnancies following 24-hour co-culture of equine embryos on foetal bovine uterine monolayer cells. Equine Vet. J. 8 (Suppl.), 117–122.

Wiemer, K.E., Cohen, J., Amborski, G.F., Wiker, S., Wright, G., Munyakazi, L., et al., 1989b. *In vitro* development and implantation of human embryos following culture on fetal bovine uterine fibroblast cells. Hum. Reprod. 45, 595–600.

Wiemer, K.E., Cohen, J., Wiker, S.R., Malter, H.E., Wright, G., Godke, R.A., 1989c. Coculture of human zygotes on fetal bovine uterine fibroblast: embryonic morphology and implantation. Fertil. Steril. 52, 503–508.

Wikland, M., Nilsson, L., Hansson, R., Hamberger, L., Olof Johnson, P., 1993. Collection of human oocytes by the use of sonography. Fertil. Steril. 39, 603–608.

Willadsen, S.M., 1979. A method for culture of micromanipulated sheep embryos and its use to produce monozygotic twins. Nature (London) 227, 298–300.

Willadsen, S.M., 1982. Micromanipulation of embryos of the large domestic species. In: Adams, C.E. (Ed.), Mammalian Egg Transfer. CRC Press, Boca Rotan, FL, pp. 185–210.

Willadsen, S.M., 1986. Nuclear transplantation in sheep embryos. Nature (London) 320, 63–65.

Willadsen, S.M., Godke, R.A., 1984. A simplified procedure for the production of identical sheep twins. Vet. Rec. 114, 240–243.

Willadsen, S.M., Polge, C., 1981. Attempts to produce monozygotic quadruplets in cattle by blastomere separation. Vet. Rec. 108, 211–213.

Willadsen, S.M., Lehn-Jensen, H., Fehilly, C.B., Newcomb, R., 1981. The production of monozygotic twins of preselected parentage by micromanipulation of non-surgically collected cow embryos. Theriogenology 14, 23–29.

Willet, E.L., Black, W.G., Casida, L.E., Stone, W.B., Buckner, P.J., 1951. Successful transplant of a fertilized ovum. Science 113, 247.

Williams, T.J., Elsden, R.P., Seidel Jr., G.E., 1982. Identical twin bovine pregnancies derived from bisected embryos. Theriogenology 17, 114. (abstr.).

Willmer, E.N., Jacoby, F., 1936. Studies on the growth of tissues *in vitro*. J. Exp. Biol. 13, 237–248.

Wilmut, I., Schnieke, A.E., McWhir, J., Kind, A.J., Campbell, K.H.S., 1997. Viable offspring derived from fetal and adult mammalian cells. Nature 385, 810–813.

Wintenberger, S., Dauzier, L., Thibault, C., 1953. Le development *in vitro* de l'oeuf de la brebis et de celui de la chevre. C.R. Seanc. Soc. Biol. 147, 1971.

Wright, G., Carver, A., Cottom, D., Reeves, D., Scott, A., Simons, P., et al., 1991. High level expression of active human alpha-1-antitrypsin in the milk of transgenic sheep. Biotechnology 9, 830—834.

Wright Jr., R.W., Bondioli, K.R., 1981. Aspects of *in vitro* fertilization and embryo culture in domestic animals. J. Anim. Sci. 53, 702—729.

Wright Jr., R.W., Anderson, G.B., Cupps, P.T., Drost, M., Bradford, G.E., 1976. *In vitro* culture of embryos from adult and prepuberal ewes. J. Anim. Sci. 42, 912—917.

Yong, M.W., Yuqiang, M., 1998. Production of biopharmaceutical proteins in the milk of transgenic dairy animals. BioPharm 10, 34—38.

Young, M.W., Okita, W.B., Brown, E.M., Curling, J.M., 1997. Production of biopharmaceutical proteins in the milk of transgenic dairy animals. BioPharm 10, 34—38.

Younis, A.I., Brackett, B.G., 1990. *In vitro* development of bovine oocytes into morulae and blastocysts. Theriogenology 33, 355. (abstr.).

Younis, A.I., Keefer, C.L., Brackett, B.G., 1989. Fertilization of bovine oocytes by sperm microinjection. Theriogenology 31, 276.

Zakhartchenko, V., Durcova-Hills, G., Stojkovic, M., Schernthaner, W., Prelle, K., Steinborn, R., et al., 1999. Effects of serum starvation and re-cloning on the efficiency of nuclear transfer using bovine fetal fibroblasts. J. Reprod. Fertil. 115, 325—331.

Zakhartchenko, V., Mueller, S., Alberio, R., Scheernthaner, W., Stojkovic, M., Wenigerkind, H., et al., 2001. Nuclear transfer in cattle with non-transfected and transfected fetal or cloned transgenic fetal and postnatal fibroblasts. Mol. Reprod. Dev. 60, 362—369.

Zhang, L., Blakewood, E.G., Denniston, R.S., Godke, R.A., 1990. The effect of ovary temperature on oocyte maturation, *in vitro* fertilization and embryo development in cattle. In: Proceedings of the Southern Security American Dairy Science. p. 16. (abstr.).

Zhang, L., Blakewood, E.G., Denniston, R.S., Godke, R.A., 1991a. The effect of insulin on maturation and development of *in vitro*-fertilized bovine oocytes. Theriogenology 35, 30. (abstr.).

Zhang, L., Denniston, R.S., Godke, R.A., 1991b. The effect of insulin on the development of *in vitro*-fertilized porcine oocytes. In: Proceedings of the Southern Security American Dairy Science p. 49. (abstr.).

Zhang, L., Denniston, R.S., Bunch, T.D., Godke, R.A., 1991c. Cows versus heifers for the production of *in vitro*-matured, *in vitro*-fertilized (IVF) embryos. In: Proceedings of the Southern Security American Dairy Science p. 49. (abstr.).

Zhang, L., Denniston, R.S., Godke, R.A., 1992. A simple method for *in vitro* maturation, *in vitro* fertilization, and co-culture of bovine oocytes. J. Tissue Cult. Meth. 14, 107—112.

Ziomek, C.A., 1998. Commercialization of proteins produced in the mammary gland. Theriogenology 49, 139—144.

23 Modifying Mitochondrial Genetics

Carl A. Pinkert[1,2], Michael H. Irwin[2], Kumiko Takeda[3] and Ian A. Trounce[4]

[1]Department of Biological Sciences, College of Arts and Sciences, The University of Alabama, Tuscaloosa, AL, [2]Department of Pathobiology, College of Veterinary Medicine, Auburn University, Auburn, AL, [3]National Agricultural and Food Research Organization, NARO Institute of Livestock and Grassland Science, Tsukuba, Japan, [4]Center for Eye Research Australia, University of Melbourne, Department of Ophthalmology, Royal Victorian Eye and Ear Hospital, East Melbourne, Australia

I. Introduction

The field of mitochondrial medicine is now just emerging. The first published accounts of diseases caused by mutations of mitochondrial DNA (mtDNA) were reported in 1988; there are now scores of point mutations and rearrangements of the mitochondrial genome known to be the underlying causes of various degenerative disorders (Clayton, 1991, 1992; Wallace, 1992a, 1992b, 2001; Larsson and Clayton, 1995; Luft, 1995; Graff et al., 1999; see MITOMAP: http://infinity.gen.emory.edu/mitomap.html). Such mutations affect mainly tissues with high cellular energy requirements from the central nervous system, as well as cardiac and skeletal muscle and various endocrine organs (Wallace, 2001). Exclusively maternal inheritance of mitochondria is observed in mammals, and it has been further postulated that a developmental bottleneck exists in female germ cell development, whereby a small number of mitochondria give rise to $\sim 10^5$ mitochondria that populate the mature ovum (Marchington et al., 1997; Cummins, 1998). Through 1996, *in vivo* animal models of mtDNA mutation and human disease were nonexistent, with none yet reflecting specific human mtDNA mutation-based disease. Yet, as we have learned in relation to nuclear-encoded genes, physiologically relevant models are of critical importance in studying mitochondrial dynamics and mtDNA-based disease pathogenesis.

As described in a number of reviews (see Pinkert and Trounce, 2002; Trounce and Pinkert, 2007; Pinkert et al., 2010; Dunn et al., 2012), nuclear-encoded genes and knockout modeling continue to be informative in unraveling mitochondrial

Transgenic Animal Technology. DOI: http://dx.doi.org/10.1016/B978-0-12-410490-7.00023-2

disease pathogenesis as well as critical pathways associated with mitochondrial function. With initial characterization of nuclear gene encoded models (e.g., adenine nucleotide translocator-1 (Ant1), manganese superoxide dismutase (MnSOD), mitochondrial transcription factor A (Tfam), glutathione peroxidase (GPx1), and uncoupling proteins (UCP1, 2, and 3), it became abundantly clear that models for mitochondria and mitochondrial gene transfer would be crucial for *in vivo* experimentation.

A. Manipulation of Mitochondria in Cultured Cells

Manipulation of the mitochondrial genome *in vitro* was technically feasible by the mid-1980s. Human and murine mitochondrial genomes were cloned early on (see Clayton, 1991). However, full-length cloning of human mtDNA in bacteria was not initially feasible due to "poison" sequences in the D-loop region of the mitochondrial genome (Biggers et al., 2000). Subsequently, mtDNA was sequenced and available as constituents of plasmid *constructs*. Differences between nuclear and mitochondrial codon assignments were identified, and the general features of amino acid presequences responsible for mitochondrial targeting of nuclear-encoded mitochondrial proteins were described (Bibb et al., 1981; Tapper et al., 1983; Pfanner and Neupert, 1990a,b).

Employing genetically engineered mitochondria or mitochondrial genes for *in vivo* gene transfer or for gene therapy requires the ability to introduce foreign or altered mitochondrial genomes into somatic cells or germ cells or to perform site-directed mutagenesis in a practical fashion. In this regard, pioneering work in the development of cybrid technology, first using mtDNA mutants resistant to chloramphenicol (Wallace et al., 1975) and then using mtDNA-less (ρ^0) cells (King and Attardi, 1989), made the establishment of transmitochondrial mouse models inevitable. Cybrids can harbor more than one type of mtDNA derived from mitochondria of distinct parental cells. This coexistence of more than one genotype of mtDNA within a single cell, or within cells comprising an individual organism, is referred to as *heteroplasmy*. Results of these experiments have shown not only that more than one mitochondrial genotype can be maintained within cells but also that selective segregation and amplification of one of these genotypes often occurs. Recent work with mitoTALENs (transcription activator-like effector nucleases) further demonstrated a methodology with the potential to selectively remove mutated mtDNA (Bacman et al., 2013); interestingly, this technology could also be used to enrich the heteroplasmic balance for production of ES cell cybrids harboring almost exclusively mutant mtDNA.

B. First Transmitochondrial Mice

Segregation of mtDNA was investigated in maternal lineages of heteroplasmic mice created by cytoplast fusion (Jenuth et al., 1996, 1997) and by embryonic karyoplast transplantation (Meirelles and Smith, 1997). In contrast to these techniques, our efforts to devise a direct mitochondria transfer technique offered certain advantages (Figure 23.1). Principally, the ability to use isolated mitochondria for

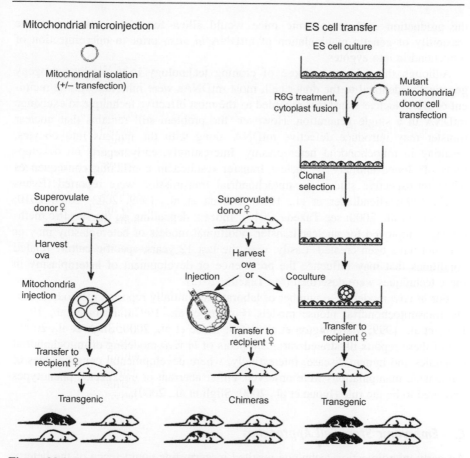

Figure 23.1 Production of transmitochondrial mice. Similar to techniques for nuclear gene transfer, mitochondrial injection and ES cell transfer technologies are the two most practiced methods for producing transmitochondrial mice at this time. A principal difference in the two techniques is the *in vitro* culture step, allowing propagation of targeted clones using immortalized ES cells (right). On the left, for microinjection of intact mitochondria, the mitochondrial preparation is injected into the cytoplasm of the pronuclear unicellular fertilized ovum (zygote) causing expansion of the vitelline membrane/boundary, occasionally with the appearance of the extranuclear vacuole as depicted. In general, using mitochondria injection, transgenic mice (or more specifically transmitochondrial mice), represented by the all-black mouse, possess heteroplasmic cells (harboring mutant and wild-type mitochondrial genomes). On the right, after clonal selection of transfected ES cells, one of two additional techniques is used for ES cell transfer. Ova are harvested between the eight-cell and blastocyst stages. R6G-treated and transfected ES cells are either injected directly into a host blastocyst (injection) or cocultured with eight-cell to morula stage ova, so that transfected ES cells are preferentially incorporated into the inner cell mass of the developing embryo (coculture). As noted in Chapter 2, with blastocyst injection transmitochondrial offspring are generally chimeric, because some of their cells are derived from the host blastocyst and some from transfected ES cells (denoted by white mice with black patches). Using coculture and tetraploid embryos, one can obtain founder mice derived completely from the transfected ES cells (denoted as all-black mice).
Source: From Pinkert and Trounce (2002), with permission.

the production of heteroplasmic mice would allow for investigations into the feasibility of genetic manipulation of mtDNA *in vitro* prior to microinjection of mitochondria into zygotes.

Although the primary purpose of cloning technology is to produce progeny genetically identical to the donor cell, most mtDNAs were inherited via the recipient oocyte. Nuclear transfer is regarded as the most effective technique to exchange mtDNA in a single generation. However, the problem still remains that nuclear transfer may introduce defective mtDNA along with the nucleus into oocytes, resulting in mitochondrial heteroplasmy. Interestingly, early reports on development of cloned animals by nuclear transfer resulted in conflicting consequences when retrospective studies on mitochondrial transmission were reported (Evans et al., 1999; Hiendleder et al., 1999; Takeda et al., 1999, 2002, 2005, 2010; Steinborn et al., 2000; see Takeda, 2013). Indeed, depending on the specific methodology employed for nuclear transfer, additional models of heteroplasmy may or may not have been created. Lastly, over the last 12 years, specific culture-related conditions that may influence the prevalence or development of heteroplasmy in these techniques were described (see Takeda, 2013).

For *in vivo* modeling, a number of laboratories initially reported methods to create transmitochondrial mouse models (Pinkert et al., 1997a; Irwin et al., 1999; Levy et al., 1999; Marchington et al., 1999; Inoue et al., 2000; Sligh et al., 2000). All of these reports illustrated various aspects of *in vivo* modeling of mitochondrial dynamics and human disease. Interestingly, where developmental consequences of the genetic manipulations were observed, either aberrant or unexpected phenotypes appeared to be the rule (Inoue et al., 2000; Sligh et al., 2000).

C. Embryonic Stem Cell Approaches

An early microinjection technique resulted in germ line competence of the heteroplasmic state (Irwin et al., 1999). Then, in support of our modeling direction but using an original stem cell approach, Levy et al. (1999) and Marchington et al. (1999) independently demonstrated that it was possible to fuse cytoplasts prepared from a chloramphenicol-resistant (CAP^R) cell line with embryonic stem (ES) cells and subsequently introduce the mutant stem cells into mouse blastocysts. The CAP^R mutation was expected to impart a respiratory deficiency in resultant animals (Levy et al. 1999). In initial reports, germ line-competent chimeras were not identified. Later, further characterizations were performed on mouse lineages in which the CAP^R mitochondria were transmitted to second-generation offspring following initial chimerism of <5−50% with varying levels of tissue-specific heteroplasmy (Sligh et al., 2000). Both homoplasmic and heteroplasmic offspring were obtained, all showing severe growth retardation and perinatal or *in utero* lethality. An important innovation was in the use of rhodamine-6G (R6G) to limit the transmission of endogenous ES cell mtDNA, something that had been previously demonstrated only for somatic cells. The advantage of this approach was illustrated by the failure of Marchington et al. (1999) to obtain homoplasmic ES cell cybrids or mice using chloramphenicol selection alone. These initial studies illustrate that such modeling would be feasible as a critical component in developing targeted animal models of human disease.

D.   Introduction of Mutant mtDNA into Mitochondria

A primary focus for our work included production of engineered mutations of mtDNA genomes and their subsequent transfection into isolated mitochondria *in vitro* prior to microinjection into zygotes—a key objective in creation of transmitochondrial mouse models of human diseases. At present, besides recapitulating a mitochondrial genomic deletion mutant (Irwin et al., 2001), methods for the experimental manipulation of mtDNA genomes in a precise, directed fashion have yet to be described. The transfection of DNA into mitochondria presents some very formidable challenges. Both the outer and inner mitochondrial membranes must be traversed, and their protein and lipid compositions are very different than the plasma membrane. Although *in vitro* fusion of inner mitochondrial membrane vesicles was reported, there were no reports of fusion of outer membrane vesicles or of whole mitochondria (Hackenbrock and Chazotte, 1986). Also, the basic mechanisms for mitochondrial membrane lipid addition are not well understood (i.e., how newly synthesized lipid is targeted to mitochondrial membranes). Recent advances in understanding the molecular mechanisms of mitochondrial fission and fusion might be very useful in attempts to design liposome-mediated DNA delivery systems (van der Bliek et al., 2013).

Using mitochondrial targeting presequence peptides covalently attached to DNA molecules, Seibel et al. (1995) demonstrated internalization of up to 322 bp DNA into intact mitochondria via the protein import pathway. However, delivery of mitochondrial-specific DNA sequences and subsequent integration into the host mitochondrial genome have yet to be demonstrated and may not even be possible if mechanisms for DNA recombination within mitochondria are lacking.

E.   Use of Transfected-ρ° Cells as Intermediate Mitochondrial Carriers in the Production of Mouse Models

By convention, ρ° cells are devoid of any mtDNA and require uridine and pyruvate in culture medium for survival (King and Attardi, 1989). Thus, immortalized ρ° cells provide an excellent vehicle for propagation of mtDNA deletion constructs. With transfer of transfected mitochondria (either by microinjection or cytoplast fusion), selection pressure could be exerted by removal of uridine and pyruvate from the culture medium. Such a system would then allow for enhanced survival rates of transfected mitochondria, enhance concentration of specifically modified mitochondria, act as a culture system to propagate mitochondria, and allow for more overall controlled experimental conditions to facilitate mitochondrial engineering.

An important advantage of this intermediate culture system is that heteroplasmic constructs can be produced at will by simply mixing cells of interest in the enucleation step, then fusing the mixed cytoplasts, in different ratios if desired, and genotyping many cybrid clones to select heteroplasmic clones of interest. These clones are then used in fusion with the R6G-treated ES cells as described below.

F. Allotopic Expression (AE)

Throughout mitochondrial evolution, gene transfer from the mitochondrial compartment to the nucleus is a continuing process (see Timmis et al., 2004). Through AE, a term first coined in respect to mtDNA gene expression in 1986 (Gearing and Nagley, 1986; Grasso et al., 1991), this phenomenon has led to speculation, experimentation, and subsequent mouse models (Dunn et al., 2012). AE as a potential gene therapy was postulated as a strategy for either circumventing or correcting diseases involving mtDNA mutation (DiMauro and Mancuso, 2007; Kyriakouli et al., 2008) and as a means to overcome the absence of animal models reflecting mtDNA mutations in human disease (see Dunn and Pinkert, 2012; Dunn et al., 2012). Accordingly, AE was initially demonstrated in cultured cells (Gearing and Nagley, 1986; Manfredi et al., 2002; Zullo et al., 2005) and in somatic tissues following delivery via viral vector (Qi et al., 2007; Guy et al., 2009). Our efforts resulted in the first germ line-competent transgenic mouse models of AE (Dunn and Pinkert, 2012).

G. Liposome-Mediated Mitochondrial Gene Transfer

Commercially available liposome formulations are routinely used for transfer of nucleic acids or proteins into cultured cells. Transfer of mitochondria into cultured cells represents a novel use for liposomes that was recently shown by our laboratory (Shi et al., 2012). Interspecies transfer was performed where mitochondria from *M. terricolor* were transferred into NIH 3T3 fibroblasts containing mtDNA of the normal laboratory species of mouse, *M. m. domesticus*. Using a species-specific polymerase chain reaction (PCR) assay, liposome-mediated transfer of *M. terricolor* mtDNA into 3T3 fibroblasts was demonstrated after 3 weeks in culture posttransfer, suggesting low efficiency of transfer or that the transferred mitochondria were selected against in favor of endogenous mitochondria. Efforts to improve transfer efficiency are ongoing.

H. Xenomitochondrial Cybrid Modeling of Oxidative Phosphorylation (OXPHOS) Defects

Another approach that circumvents the current lack of mtDNA mutagenesis techniques has been to create cybrid cells with mtDNAs from different species to the nuclear host (Trounce et al., 1994). Such xenomitochondrial cybrids were first produced using primate cells (Kenyon and Moraes, 1997). We extended this by producing a panel of mouse xenocybrids using increasingly divergent mtDNAs from various murid species (McKenzie and Trounce, 2000; McKenzie et al., 2003). The mismatching of the mtDNA and nuclear-encoded OXPHOS subunits create increasingly dysfunctional OXPHOS complexes I, III, IV, and V as the evolutionary

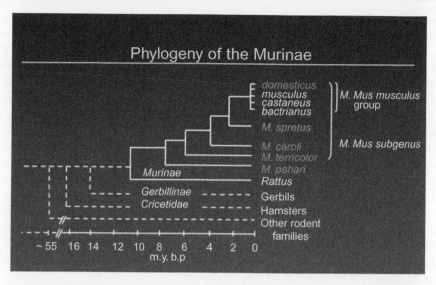

Figure 23.2 Brief Murinae phylogeny. The approximate divergence times in million years before present (m.y.b.p.) are shown for some Murinae species of interest in construction of xenomitochondrial cybrids (McKenzie and Trounce, 2000).

divergence between the *Mus musculus* host cell and mtDNA donor species increases (Figure 23.2, McKenzie et al., 2003). We have also produced the first xenomitochondrial mice by first creating cybrid ES cells using the R6G method outlined in Section II.D (McKenzie et al., 2004). After extensive backcrossing onto C57Bl/6 mice and aging colonies we are now beginning to uncover age-related pathologies that result from the partial OXPHOS deficiency in the model.

II. Methods

The majority of methods that follow were first outlined in Pinkert and Trounce (2002).

A. Preparation of Mitochondria for Microinjection into Zygotes

Mitochondria are prepared by differential centrifugation, using the protocol of Johnson and Lardy (1967) as modified by Berdanier and Kim (1993). All buffers and solutions are sterile-filtered and autoclaved. During the isolation procedure, mitochondria preparations are maintained at 4°C. *M. spretus* livers are exsanguinated and minced in 5 mM MOPS [3-(N-morpholino)propanesulfonic acid], pH 7.4, 250 mM sucrose, and 0.5 mM EDTA (Ethylenediaminetetraacetic acid). Minced tissues are rinsed twice and homogenized with a glass/Teflon homogenizer at 100 mg tissue/mL buffer. The crude homogenate is centrifuged twice for 10 min at $750 \times g$ and the pellets discarded. The $750 \times g$ supernatant is centrifuged for

15 min at $9800 \times g$ and the resulting supernatant is removed by vacuum aspiration with a Pasteur pipette. The light-colored material (microsomes) at the margins of the pellet, as well as the very dark material (blood pigments, etc.) in the center of the pellet are carefully aspirated leaving a homogeneous chocolate-brown-colored mitochondrial pellet. Pellets are slowly resuspended by dropwise addition of TES buffer (1 mM Tris—HCl, pH 7.4, 250 mM sucrose, and 1 mM EDTA (TES) Ebert et al., 1989) while stirring with a glass rod. Suspensions are rehomogenized with a glass/Teflon homogenizer until no visible aggregates remain, then centrifuged for 10 min at $9800 \times g$. The suspension and re-homogenization steps are repeated, followed by a final centrifugation at $9800 \times g$ for 10 min. The final pellet is resuspended in 200 μL of 150 mM KCl, 10 mM sodium phosphate buffer, pH 7.0. Large aggregates of mitochondria are disrupted by repeated gentle aspiration. Purified preparations of mitochondria are kept on ice until used for microinjection, generally within 30—90 min. Protein concentrations of final aliquots are determined by the Lowry method (Lowry et al., 1951). The total protein concentration of the mitochondria suspension used for microinjection is generally between 10 and 13 mg/mL prior to dilution, if necessitated for microinjection.

B. Mitochondrial Injection into Zygotes

Transfected mitochondria (enriched to minimize transfer of lysed or degraded mitochondria) are transferred into the cytoplasm of zygotes by microinjection as previously described (Pinkert et al., 1997a; Irwin et al., 1999; Ingraham and Pinkert, 2003). Surviving zygotes are initially cultured *in vitro* to evaluate utility of the transfection and microinjection procedure. Introduced mitochondria can be detected after culture to the blastocyst stage of development (d4.5 of culture), as well as in the development of live-born founders (Irwin et al., 1999). Tail biopsies from live-born mice are obtained at weaning to identify mice heteroplasmic for the introduced mitochondria as determined by analysis of mtDNA sequences.

Mitochondrial preparations too dense for microinjection are diluted by approximately 30% with 150 mM KCl, 10 mM sodium phosphate, pH 7.0. In early experiments, mitochondria from *M. spretus* donors were introduced by injection of approximately 3—5 μL of the mitochondrial suspension directly into the egg cytoplasm. Injections were performed using glass thin-wall capillary tubes (TW100F-4 with filaments; WPI, Sarasota, FL) drawn for injection on a P-87 pipette puller (Sutter, Novato, CA). Following microinjection, zygotes are washed twice in BMOC-3, then maintained for 4 days at 37°C in microdrops (15—20 μL) of modified BMOC-3 medium overlaid with silicone oil in 35 mm tissue culture dishes (Corning, Corning, NY) in a controlled and humidified atmosphere (5% CO_2, 5% O_2, and 90% N_2).

C. Collection and Treatment of Fertilized Ova

Procedures used for the collection and manipulation of ova mirror those described for pronuclear microinjection (Brinster et al., 1985; Pinkert et al., 1997a,b), and are as outlined in Chapter 2.

The use of embryos from B6SJLF1 hybrid mice was identified in preliminary studies as an optimal *M. m. domesticus* host strain for microinjection studies due to several significant experimental advantages, including minimal sequence divergence in the mtDNA between the two parental strain backgrounds (Ferris et al., 1982). Advantages of hybrid vigor, and increased reproductive and developmental performance, provide for a significant gain in efficiency over other inbred or outbred strains. Furthermore, there were no data to suggest that C57BL/6 or SJL mice (or hybrid derivatives) would have different alleles at loci that affect the ability to accept transferred mitochondria, nor were such found in our *M. spretus: M. m. domesticus* trials (again, representing two different murine species). In the end, a reasonable efficiency of producing heteroplasmic mice using the hybrid model was demonstrated; this hybrid background strain was then used in subsequent mitochondrial injection studies to generate heteroplasmic mice. With current egg survival rates (again, most cytoplasmic and nuclear microinjection techniques were originally optimized in hybrid models) and overall enhanced efficiencies in embryo survival, litter size, live-birth weights, growth performance, onset of puberty, and reproductive efficiency, the hybrid model is still deemed to be most appropriate and cost-effective for direct injection studies. In contrast, use of inbred strains for ES cell based efforts was another consideration. Discussion and use of hybrid and inbred mouse models for microinjection and stem cell studies are outlined in detail in Chapters 2, 3 and 4.

D. Production of Transmitochondrial ES Cells Using Rhodamine-6G Pretreatment

A female ES cell clone is pretreated with R6G to prevent the ongoing transmission of endogenous mtDNAs (Trounce and Wallace, 1996; Sligh et al., 2000). Cells are grown in ES-cell-tested Dulbecco's Modified Eagle Medium (DMEM) with 10% ES-cell-tested fetal bovine serum, 0.1 mM β-mercaptoethanol, Non-Essential Amino Acids (NEAA), and 1000 U/mL Leukemia Inhibitory Factor (LIF), and frozen in the same medium with the exception that the serum concentration was increased to 20% and dimethyl sulfoxide (DMSO) was added to a final concentration of 10%. Aliquots of cells were frozen at 10^7 cells/mL and test thawed directly into culture medium without centrifugation.

For R6G treatment, ES cells are plated at 10^6 cells/60 mm dish onto mitomycin-C inactivated STO (mouse embryonic fibroblast cell line) feeders. After 24 h, the cells are exposed to R6G at 1.0 μg/mL for 72 h, then trypsinized and washed in drug-free medium, and immediately fused with enucleated mitochondrial donor cells. Cybrids containing a given mitochondrial haplotype are enucleated by centrifugation in an isopycnic Percoll gradient in the presence of cytochalasin B (Trounce et al., 1996, 2000). Washed cytoplasts and R6G-treated ES cells are then mixed, centrifuged, and fused, either by electrofusion as previously described (Trounce et al., 1996) or by exposure to polyethylene glycol (PEG) for 1 min, gently diluted and resuspended in complete medium, and plated onto 60-mm tissue culture dishes containing STO feeders. We use 10^6 ES cells combined with 5×10^6

enucleated donor cells in fusions, and plate a range of densities between 10^5 and 5×10^5 ES cells/dish. After 24 h, the medium is replaced with sodium hypoxanthine, aminopterin and thymidine (HAT)-containing medium to select against surviving TK-mitochondrial donor cells. After 5−8 days, colonies of ES cells become visible and can be cloned using pipettes or cloning cylinders.

Treatment with R6G should be determined empirically for each new cell line used. The objective is to treat the cells just long enough to ensure that few cells ($<10^{-5}$) will grow to form colonies after treatment. Typical cybrid frequencies obtained from either electrofusion or PEG fusion are around 10^{-3} to 10^{-4}. If R6G treatment is not sufficient, "breakthrough" or noncybrid clones will overgrow any cybrids produced. Some treated ES cells should always be plated as controls alongside fusion cells to establish that this breakthrough level is sufficiently low.

E. Electroporation

Freshly isolated mitochondria (12 mg/mL mitochondrial protein) in 80 μL respiration buffer and 50 ng of the reconstituted mitochondrial genome in 20 μL water are added to a 0.1 cm electroporation cuvette and electroporated at 12 kV/cm field strength with 25 μF capacitance in a Bio-Rad Gene Pulser (Bio-Rad, Hercules, CA). Respirometric measurements of mitochondria electroporated at 8, 9, 10, 11, 12, 13, and 14 kV/cm were obtained in pilot experiments to confirm the substantial decrease in mitochondrial coupling at field strengths above 12 kV/cm as reported by Collombet et al. (1997) (Irwin et al., 2001).

1. Electroporation Optimization

The effect of electroporation on mitochondrial viability can be assessed by polarography. Mitochondria, isolated from mouse liver or ES cells, are subjected to a graded series of electroporation voltages/power settings (in the absence of exogenous DNA), and respirometric measurements are performed to determine the respiratory control ratio (RCR), a measure of coupling. Comparisons to RCR of control (nonelectroporated) mitochondria can be used to determine the maximum electroporation voltages that do not result in uncoupled mitochondria. Respirometric measurements of control, nonelectroporated mitochondria will also allow us to assess our isolation technique, by comparisons to lysed (uncoupled) mitochondria. The maximum electroporation voltage that still results in coupled mitochondria will be used as a starting point maximum to assess internalization of the 6.5 kb mtDNA deletion construct. Following electroporation of DNA at this maximum voltage, PCR will be used to ascertain the presence of electroporated DNA in the mitochondrial matrix as described by Collombet et al. (1997).

F. Analysis of Mice

Mice are maintained in a specific pathogen-free barrier facility. Procedures used for the collection and manipulation of ova (from zygotes to blastocysts) are routine and are described in detail in Chapters 2 and 3. For analysis of live-born mice, tail biopsies are routinely used. Mitochondria and mtDNA are isolated, as described previously, from tail and/or liver biopsies obtained from offspring. For early characterization of heteroplasmy, aliquots of mtDNA from founder mice and offspring are either subjected to a quantitative PCR or nested-PCR protocol, in the presence and absence of mtDNA-specific forward primers and common reverse, as described (Gyllensten et al., 1991; Pinkert et al., 1997a; Pinkert and Trounce, 2002). Additional negative controls consist of concurrently run reactions lacking target mtDNA. Normally, embryos and/or mice are analyzed for the presence and concentrations of mutant and wild-type mtDNA in lines generated (including founders and subsequent offspring/generations) in relation to: (i) tissue distribution, (ii) developmental specificities, (iii) maternal versus paternal derived lineages, and (iv) stability/instability (potential genetic drift).

G. Mating of Heteroplasmic Mice

Founder heteroplasmic founder females (G_0) are mated to control males (nonheteroplasmic) to generate heteroplasmic offspring (G_1). As heteroplasmic offspring vary in relative degrees of heteroplasmy, quantitative integration analyses (DNA slot-blot or Southern blot analyses) are generally warranted, not only for quantification but also for the study of mitochondrial population dynamics. Interbreeding of mice from different heteroplasmic lineages would not normally be anticipated, since this would confound data interpretation. Again, in order to identify individual heteroplasmic mice, quantitative DNA hybridization analyses (e.g., slot-blot or Southern blot) will be routinely performed. As a cautionary note, insertional mutagenesis and/or rearranged/fusion-mutants were reported (Chen et al., 1995; Meirelles and Smith, 1997), and may be problematic in individual mice or while establishing mouse lineages. As reviewed by Cummins, such a finding may further support the rearrangement hypothesis regarding paternal mitochondrial inheritance and some of the conflicting data regarding paternal inheritance of mitochondria in mouse lineages as reported by others (Gyllensten et al., 1991). Therefore, sequencing of mitochondrial genomes can be advantageous, independent of whether a phenotype is observed in mice. Detectable phenotypes may also be absent due to the specific deletion mutation causing embryonic or perinatal lethality. Cases may also exist that may not be directly related to the state of heteroplasmy. Fetal analyses from various lineages, even if germ line heteroplasmy is not observed, should therefore be used as a diagnostic tool in evaluating models.

H. Allotopic Expression

Our first effort at generating a nuclear-encoded version of a mitochondrial gene involved an *Atp6* sequence (Dunn and Pinkert, 2012). The following procedures recap this initial endeavor.

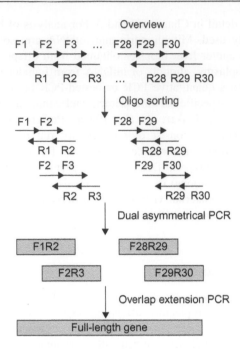

Figure 23.3 Cloning strategy for allotopic expression of a nuclear-encoded mitochondrial transgene. There were three PCR steps, dual-asymmetrical PCR (2× gene coverage), overlap-extension PCR, followed by full-length PCR. The resultant sequence is cloned into vector containing N-terminal mitochondrial transport signal, then linearized and sequence-verified prior to microinjection.
Source: Dunn (2010); adapted from Young and Dong (2004).

1. Generation of Transgenes

Two forms of the nuclear-encoded mitochondrial *Atp6* gene were synthesized *de novo*. Gene synthesis was performed using a three-step PCR-based technique (Young and Dong, 2004; Figure 23.3). *Atp6* genes were synthesized, coding for both murine wild-type and mutated amino acid sequences (L156R) using nuclear codons. Oligonucleotides/primers (25 nt) spanned forward and reverse sequences to be synthesized.

2. Plasmid Cloning/Transgenic Mouse Production

Synthesized *Atp6* DNAs were cloned in-frame into a pEF/*myc*/mito plasmid (Invitrogen). Elements in this expression system included the promoter for human EF-1α (Rizzuto et al., 1992), the mitochondrial targeting sequence from the human cytochrome c oxidase subunit VIII gene (Kim et al., 1990), and an in-frame 3′ myc

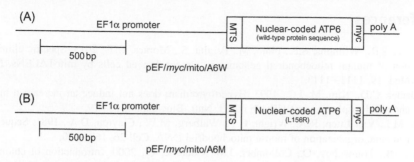

Figure 23.4 Schematic representation of AE DNA constructs. High-level transcription is driven by the human EF1α promoter. Protein coding elements include the N-terminal mitochondrial transport signal of human cytochrome oxidase VIII, nuclear-coded wild-type (A) or mutant (B) mouse *Atp6* gene sequence, and a C-terminal myc epitope tag.

epitope tag (Evan et al., 1985) (Figure 23.4). This gave rise to two plasmids, pEF/*myc*/mito/*A6W* and pEF/*myc*/mito/*A6M*. Procedures for generating transgenic mice were those in Chapter 2.

I. Liposome-Mediated Mitochondrial Gene Transfer

Isolated *M. terricolor* mitochondria were resuspended with a glass rod in 250 μL serum-free DMEM and 100 μL were added to 100 μL Lipofectamine® solution (20 μL liposome stock solution plus 80 μL medium, preincubated 30 min at room temperature). The mitochondria/Lipofectamine® suspension was incubated for 30 min at room temperature with gentle end-over-end mixing. One hour before transfer of mitochondria, complete medium was removed from cultured cells and replaced with serum-free DMEM. The liposome-encapsulated *M. terricolor* mitochondria were added to NIH 3T3 *M. m. domesticus* mouse fibroblasts in 10 mL of serum-free DMEM and incubated overnight (~16 h) at 37°C in 5% CO_2 in air. Control cells received either no treatment, mitochondria without liposomes, or liposomes without mitochondria. The next day, the medium was replaced with complete medium. Cells were passaged every 3−4 days (80−90% confluence) with aliquots of cells removed and used for DNA extraction and PCR analysis (Figures 23.3 and 23.4).

Acknowledgments

Our work over the years has been funded in part by the National Health and Medical Research Council of Australia, and the US organizations National Cancer Institute, National Institutes of Health (NCRR, NICHD, NIDCR, and NIEHS), National Science Foundation, USDA, the Alabama Agricultural Experiment Station, and the MitoCure Foundation. We thank Kosta Steliou, Mikhail Alexeyev, Matthew Cannon, David Dunn, Matthew McKenzie, and Kodeeswaran Parameshwaran for their assistance and support.

References

Bacman, S.R., Williams, S.L., Pinto, M., Peralta, S., Moraes, C.T., 2013. Specific elimination of mutant mitochondrial genomes in patient-derived cells by mitoTALENs. Nat. Med. 19, 1111−1113.

Berdanier, C.D., Kim, M.-J.C., 1993. Hyperthyroidism does not induce an increase n mitochondrial respiration in BHE/cdb rats. J. Nutr. Biochem. 4, 10−19.

Bibb, M.J., Van Etten, R.A., Wright, C.T., Walberg, M.W., Clayton, D.A., 1981. Sequence and gene organization of mouse mitochondrial DNA. Cell 26, 167−180.

Bigger, B., Tolmachov, O., Collombet, J.M., Coutelle, C., 2000. Introduction of chloramphenicol resistance into the modified mouse mitochondrial genome: cloning of unstable sequences by passage through yeast. Anal. Biochem. 277, 236−242.

Brinster, R.L., Chen, H.Y., Trumbauer, M.E., Yagle, M.K., Palmiter, R.D., 1985. Factors affecting the efficiency of introducing foreign DNA into mice by microinjecting eggs. Proc. Natl. Acad. Sci. USA 82, 4438−4442.

Chen, X., Prosser, R., Simonetti, S., Sadlock, J., Jagiello, G., Schon, E., 1995. Rearranged mitochondrial genomes are present in human oocytes. Am. J. Hum. Genet. 57, 239−247.

Clayton, D.A., 1991. Replication and transcription of vertebrate mitochondrial DNA. Annu. Rev. Cell Biol. 7, 453−478.

Clayton, D.A., 1992. Structure and function of the mitochondrial genome. J. Inher. Metab. Dis. 15, 439−447.

Collombet, J.M., Wheeler, V.C., Vogel, F., Coutelle, C., 1997. Introduction of plasmid DNA into isolated mitochondria by electroporation. A novel approach toward gene correction for mitochondrial disorders. J. Biol. Chem. 272, 5342−5347.

Cummins, J., 1998. Mitochondrial DNA in mammalian reproduction. Rev. Reprod. 3, 72−82.

DiMauro, S., Mancuso, M., 2007. Mitochondrial diseases: therapeutic approaches. Biosci. Rep. 27, 125−137.

Dunn, D.A., 2010. Allotopic Expression of ATP6 in the Mouse as a Model of Targeted Mitochondrial DNA Mutation. University of Rochester, 545 pages.

Dunn, D.A., Pinkert, C.A., 2012. Nuclear expression of a mitochondrial DNA gene: mitochondrial targeting of allotopically expressed mutant ATP6 in transgenic mice. J. Biomed. Biotechnol. 2012, 541245.

Dunn, D.A., Cannon, M.V., Irwin, M.H., Pinkert, C.A., 2012. Animal models of human mitochondrial DNA mutations. Biochim. Biophys. Acta 1820, 601−607.

Ebert, K.M, Alcivar, A., Liem, B., Goggins, R., Hecht, N.B., 1989. Mouse zygotes injected with mitochondria develop normally but the exogenous mitochondria are not detectable in the progeny. Mol. Reprod. Dev. 1, 156−163.

Evan, G.I., Lewis, G.K., Ramsay, G., Bishop, J.M., 1985. Isolation of monoclonal antibodies specific for human c-myc proto-oncogene product. Mol. Cell. Biol. 5, 3610−3616.

Evans, M.J., Gurer, C., Loike, J.D., Wilmut, I., Schnieke, A.E., Schon, E.A., 1999. Mitochondrial DNA genotypes in nuclear transfer-derived cloned sheep. Nat. Genet. 23, 90−93.

Ferris, S.D., Sage, R.D., Wilson, A.C., 1982. Evidence from mtDNA sequences that common laboratory strains of inbred mice are descended from a single female. Nature 295, 163−165.

Gearing, D.P., Nagley, P., 1986. Yeast mitochondrial ATPase subunit 8, normally a mitochondrial gene product, expressed in vitro and imported back into the organelle. EMBO J. 5, 3651−3655.

Graff, C., Clayton, D.A., Larsson, N.-G., 1999. Mitochondrial medicine—recent advances. J. Intern. Med. 246, 11−23.

Grasso, D.G., Nero, D., Law, R.H., Devenish, R.J., Nagley, P., 1991. The C-terminal positively charged region of subunit 8 of yeast mitochondrial ATP synthase is required for efficient assembly of this subunit into the membrane F0 sector. Eur. J. Biochem. 199, 203−209.

Guy, J., Qi, X., Koilkonda, R.D., Arguello, T., Chou, T.H., Ruggeri, M., et al., 2009. Efficiency and safety of AAV-mediated gene delivery of the human ND4 complex I subunit in the mouse visual system. Invest. Ophthalmol. Vis. Sci. 50, 4205−4214.

Gyllensten, U., Wharton, D., Josefsson, A., Wilson, A.C., 1991. Paternal inheritance of mitochondrial DNA in mice. Nature 352, 255−257.

Hackenbrock, C.R., Chazotte, B., 1986. Lipid enrichment and fusion of mitochondrial inner membranes. Meth. Enzymol. 125, 35−45.

Hiendleder, S., Schmutz, S.M., Erhardt, G., Green, R.D., Plante, Y., 1999. Transmitochondrial differences and varying levels of heteroplasmy in nuclear transfer cloned cattle. Mol. Reprod. Dev. 54, 24−31.

Ingraham, C.A., Pinkert, C.A., 2003. Developmental fate of mitochondria microinjected into murine zygotes. Mitochondrion 3, 39−46.

Inoue, K., Nakada, K., Ogura, A., Isobe, K., Goto, Y.-I., Nonaka, I., et al., 2000. Generation of mice with mitochondrial dysfunction by introducing mouse mtDNA carrying a deletion into zygotes. Nat. Genet. 26, 176−181.

Irwin, M.H., Johnson, L.W., Pinkert, C.A., 1999. Isolation and microinjection of somatic cell-derived mitochondria and germline heteroplasmy in transmitochondrial mice. Transgenic Res. 8, 119−123.

Irwin, M.H., Parrino, V., Pinkert, C.A., 2001. Construction of a mutated mtDNA genome and transfection into isolated mitochondria by electroporation. Adv. Reprod. 5, 59−66.

Jenuth, J.P., Peterson, A.C., Fu, K., Shoubridge, E.A., 1996. Random genetic drift in the female germline explains the rapid segregation of mammalian mitochondrial DNA. Nat. Genet. 14, 146−151.

Jenuth, J.P., Peterson, A.C., Shoubridge, E.A., 1997. Tissue-specific selection for different mtDNA genotypes in heteroplasmic mice. Nat. Genet. 16, 93−95.

Johnson, D., Lardy, H., 1967. Isolation of liver and kidney mitochondria. Meth. Enzymol. 10, 94−96.

Kenyon, L., Moraes, C.T., 1997. Expanding the functional human mitochondrial DNA database by the establishment of primate xenomitochondrial cybrids. Proc. Natl. Acad. Sci. USA 94, 9131−9135.

Kim, D.W., Uetsuki, T., Kaziro, Y., Yamaguchi, N., Sugano, S., 1990. Use of the human elongation factor 1 alpha promoter as a versatile and efficient expression system. Gene 91, 217−223.

King, M.P., Attardi, G., 1989. Human cells lacking mtDNA: repopulation with exogenous mitochondria by complementation. Science 246, 500−503.

Kyriakouli, D.S., Boesch, P., Taylor, R.W., Lightowlers, R.N., 2008. Progress and prospects: gene therapy for mitochondrial DNA disease. Gene. Ther. 15, 1017−1023.

Larsson, N.G., Clayton, D.A., 1995. Molecular genetic aspects of human mitochondrial disorders. Annu. Rev. Genet. 29, 151−178.

Levy, S.E., Waymire, K.G., Kim, Y.L., MacGregor, G.R., Wallace, D.C., 1999. Transfer of chloramphenicol-resistant mitochondrial DNA into the chimeric mouse. Transgenic Res. 8, 137−145.

Lowry, O.H., Rosebrough, N.J., Farr, A.L., Randall, R.J., 1951. Protein measurement with the Folin phenol reagent. J. Biol. Chem. 193, 265−275.

Luft, R, 1995. The development of mitochondrial medicine. Biochim. Biophys. Acta 1271, 1–6.

Manfredi, G., Fu, J., Ojaimi, J., Sadlock, J.E., Kwong, J.Q., Guy, J., et al., 2002. Rescue of a deficiency in ATP synthesis by transfer of MTATP6, a mitochondrial DNA-encoded gene, to the nucleus. Nat. Genet. 30, 394–399.

Marchington, D.R., Hartshorne, G.M., Barlow, D., Poulton, J., 1997. Homopolymeric tract heteroplasmy in mtDNA from tissues and single oocytes: support for a genetic bottleneck. Am. J. Hum. Genet. 60, 408–416.

Marchington, D.R., Barlow, D., Poulton, J., 1999. Transmitochondrial mice carrying resistance to chloramphenicol on mitochondrial DNA: developing the first mouse model of mitochondrial DNA disease. Nat. Med. 5, 957–960.

McKenzie, M., Trounce, I., 2000. Expression of *Rattus norvegicus* mtDNA in *Mus musculus* cells results in multiple respiratory chain defects. J. Biol. Chem. 275, 31514–31519.

McKenzie, M., Chiotis, M., Pinkert, C.A., Trounce, I.A., 2003. Functional respiratory chain analyses in murid xenomitochondrial cybrids expose coevolutionary constraints of cytochrome *b* and nuclear subunits of complex III. Mol. Biol. Evol. 20, 1117–1124.

McKenzie, M., Trounce, I.A., Cassar, C.A., Pinkert, C.A., 2004. Production of homoplasmic xenomitochondrial mice. Proc. Natl. Acad. Sci. USA 101, 1685–1690.

Meirelles, F.V., Smith, L.C., 1997. Mitochondrial genotype segregation in a mouse heteroplasmic lineage produced by embryonic karyoplast transplantation. Genetics 145, 445–451.

Pfanner, N., Neupert, W., 1990a. A mitochondrial machinery for membrane translocation of precursor proteins. Biochem. Soc. Trans. 18, 513–515.

Pfanner, N., Neupert, W., 1990b. The mitochondrial protein import apparatus. Annu. Rev. Biochem. 59, 331–353.

Pinkert, C.A., Irwin, M.H., Johnson, L.W., Moffatt, R.J., 1997a. Mitochondria transfer into mouse ova by microinjection. Transgenic Res. 6, 379–383.

Pinkert, C.A., Irwin, M.H., Moffatt, R.J., 1997b. Transgenic animal modeling. In: Meyers, R.A. (Ed.), Encyclopedia of Molecular Biology and Molecular Medicine, vol. 6. VCH, New York, NY, pp. 63–74.

Pinkert, C.A., Trounce, I.A., 2002. Production of transmitochondrial mice. Methods 26, 348–357.

Pinkert, C.A., Smith, L.C., Trounce, I.A., 2010. Transgenic animals: Mitochondrial genome modification. In: Ullrey, D.E., Baer, C.K., Pond, W.G. (Eds.), Encyclopedia of Animal Science, second ed. Dekkar, Taylor & Francis, New York, NY, pp. 1044–1046.

Polites, H.G., Pinkert, C.A., 2002. DNA microinjection and transgenic animal production. In: Pinkert, C.A. (Ed.), Transgenic Animal Technology: A Laboratory Handbook, second ed. Academic Press, San Diego, CA, pp. 15–70.

Qi, X., Sun, L., Lewin, A.S., Hauswirth, W.W., Guy, J., 2007. The mutant human ND4 subunit of complex I induces optic neuropathy in the mouse. Invest. Ophthalmol. Vis. Sci. 48, 1–10.

Rizzuto, R., Simpson, A.W., Brini, M., Pozzan, T., 1992. Rapid changes of mitochondrial Ca2+ revealed by specifically targeted recombinant aequorin. Nature 358, 325–327.

Seibel, P., Trappe, J., Villani, G., Klopstock, T., Papa, S., Reichmann, H., 1995. Transfection of mitochondria: strategy towards a gene therapy of mitochondrial DNA diseases. Nucleic Acids Res. 23, 10–17.

Shi, J., Irwin, M.H., Pinkert, C.A., 2012. Liposome-mediated transfer of mitochondria harboring foreign mitochondrial DNA into cultured fibroblasts. Auburn Univ. J. Undergrad. Scholarship 1, 8–11.

Sligh, J.E., Levy, S.E., Waymire, K.G., Allard, P., Dillehay, D.L., Heckenlively, J.R., et al., 2000. Maternal germ-line transmission of mutant mtDNAs from embryonic stem cell-derived chimeric mice. Proc. Natl. Acad. Sci. USA 97, 14461−14466.

Steinborn, R., Schinogl, P., Zakhartchenko, V., Achmann, R., Schernthaner, W., Stojkovic, M., et al., 2000. Mitochondrial DNA heteroplasmy in cloned cattle produced by fetal and adult cell cloning. Nat. Genet. 25, 255−257.

Takeda, K., 2013. Mitochondrial DAN transmission and confounding mitochondrial influences in cloned cattle and pigs. Reprod. Med. Biol. 12, 47−55.

Takeda, K., Takahashi, S., Onishi, A., Goto, Y., Miyazawa, A., Imai, H., 1999. Dominant distribution of mitochondrial DNA from recipient oocytes in bovine embryos and offspring after nuclear transfer. J. Reprod. Fertil. 116, 253−259.

Takeda, K., Akagi, S., Takahashi, S., Onishi, A., Hanada, H., Pinkert, C.A., 2002. Mitochondrial activity in response to serum starvation in bovine (Bos taurus) cell culture. Cloning Stem Cells 4, 223−229.

Takeda, K., Tasai, M., Iwamoto, M., Onishi, A., Tagami, T., Nirasawa, K., et al., 2005. Microinjection of cytoplasm or mitochondria derived from somatic cells affects parthenogenetic development of murine oocytes. Biol. Reprod. 72, 1397−1404.

Takeda, K., Tasai, M., Akagi, S., Matsukawa, K., Takahashi, S., Iwamoto, M., et al., 2010. Microinjection of serum-starved mitochondria derived from somatic cells affects parthenogenetic development of bovine and murine oocytes. Mitochondrion 10, 137−142.

Tapper, D.P., Van Etten, R.A., Clayton, D.A., 1983. Isolation of the mammalian mitochondrial DNA and RNA and cloning of the mitochondrial genome. Meth. Enzymol. 97, 426−434.

Timmis, J.N., Ayliffe, M.A., Huang, C.Y., Martin, W., 2004. Endosymbiotic gene transfer: organelle genomes forge eukaryotic chromosomes. Nat. Rev. Genet. 5, 123−135.

Trounce, I., Wallace, D.C., 1996. Production of transmitochondrial mouse cell lines by cybrid rescue of rhodamine-6G pretreated L cells. Somat. Cell Mol. Genet. 22, 81−85.

Trounce, I., Neill, S., Wallace, D.C., 1994. Cytoplasmic transfer of the mitochondrial DNA 8993T→G (ATP6) point mutation associated with Leigh syndrome into mtDNA-less cells demonstrates co-segregation of decreased state III respiration and ADP/O ratio. Proc. Natl. Acad. Sci. USA 91, 8334−8338.

Trounce, I., Kim, Y.L., Jun, A.S., Wallace, D.C., 1996. Assessment of mitochondrial oxidative phosphorylation in patient muscle, biopsies, lymphoblasts and transmitochondrial cell lines. Meth. Enzymol. 264, 484−509.

Trounce, I., Schmiedel, J., Yen, H.C., Hosseini, S., Brown, M.D., Olson, J.J., et al., 2000. Cloning of neuronal mtDNA variants in cultured cells by synaptosome fusion with mtDNA-less cells. Nucleic Acids Res. 28, 2164−2170.

Trounce, I.A., Pinkert, C.A., 2007. Cybrid models of mtDNA disease and transmission, from cells to mice. Curr. Top. Dev. Biol. 77, 157−183.

van der Bliek, A.M., Shen, Q., Kawajiri, S., 2013. Mechanisms of mitochondrial fission and fusion. Cold Spring Harb. Perspect. Biol. 5. Available from: http://dx.doi.org/10.1101/cshperspect.a011072.

Wallace, D.C., 1992a. Diseases of the mitochondrial DNA. Annu. Rev. Biochem. 61, 1175−1212.

Wallace, D.C., 1992b. Mitochondria genetics: a paradigm for aging and degenerative diseases? Science 256, 628−632

Wallace, D.C., 2001. Mouse models for mitochondrial disease. Am. J. Med. Genet. 106, 71−93.

Wallace, D.C., Bunn, C.L., Eisenstadt, J.M., 1975. Cytoplasmic transfer of chloramphenicol resistance in human tissue culture cells. J. Cell Biol. 67, 174−188.

Young, L., Dong, Q., 2004. Two-step total gene synthesis method. Nucleic Acids Res. 32, e59.

Zullo, S.J., Parks, W.T., Chloupkova, M., Wei, B., Weiner, H., Fenton, W.A., et al., 2005. Stable transformation of CHO Cells and human NARP cybrids confers oligomycin resistance (oli(r)) following transfer of a mitochondrial DNA-encoded oli(r) ATPase6 gene to the nuclear genome: a model system for mtDNA gene therapy. Rejuvenation. Res. 8, 18−28.

24 Databases, Internet Resources, and Genetic Nomenclature

Anna V. Anagnostopoulos and Janan T. Eppig

Mouse Genome Informatics, The Jackson Laboratory, Bar Harbor, ME

I. Introduction

Genetically engineered animals have revolutionized biomedical research. Although a variety of genetically tractable model systems are currently emerging, the mouse is still the only vertebrate for which highly versatile and systematic phenotype- and gene-driven mutagenesis strategies are well established (see Chapters 2–8). Genome manipulation tools include traditional, BAC/YAC, and lentiviral transgenesis, gene targeting by homologous, site- and/or time-specific recombination, gene trapping, chemical- and transposon-induced mutagenesis, nuclease-based engineering, and RNA-mediated interference (reviewed in Bockamp et al., 2008; Doyle et al., 2012; Gama Sosa et al., 2010; Miller, 2011). Such tools enable the creation of spatially or temporally restricted or induced mutations, ranging from single nucleotide changes to complex genomic rearrangements, that can provide unparalleled insight into individual gene function. These advances have compelled the development of sophisticated bioinformatics resources that support genome-wide manipulation and sequencing, and facilitate systematic phenotyping, archiving, and distribution of biomedically valuable mouse models (Bucan et al., 2012). In this regard, robust bioinformatics tools and databases to capture, annotate, analyze, integrate, publicize, and share large-scale data sets in a timely and intuitive manner are critical to the successful delivery of a comprehensive encyclopedia of mammalian gene function (Brown and Moore, 2012).

This chapter provides a current overview of major databases and online information resources relevant to transgenic and gene targeting mutagenesis in mice. Further, it seeks to raise awareness of the abundance, diversity, and collective impact of new functional genomics tools and genome-wide phenotyping initiatives on the potential application of genetically altered mice in human health and biotechnology. It is not intended to provide an exhaustive listing of all mouse resources related to any specific facet of transgenic technology. The chapter does not include websites pertaining to the production of transgenic animal species reviewed in Chapters 9–15.

Section 24.II. provides categorical lists of selected transgenic, targeted, and gene trap mutation databases and online resources in tabular form. The objective,

Transgenic Animal Technology. DOI: http://dx.doi.org/10.1016/B978-0-12-410490-7.00024-4

functionality, and relevance of a subset of these resources are highlighted, with emphasis on authoritative mouse resources that (i) grant free, unrestricted access to integrated information on the design, generation, characterization, and distribution of genetically engineered mouse (GEM) models and associated tool strains; (ii) include supportive bibliographical references; (iii) apply systematized nomenclature and semantic standards (e.g., controlled vocabularies and/or ontologies) as data annotation, integration, and analytical tools; (iv) host links to other accredited genetic and genomic resources for human or nonmouse model organisms; (v) offer a range of user-friendly querying and data mining tools to drive knowledge discovery; (vi) encourage resource interoperability and collaboration; (vii) deploy regular enhancements and updates that reflect and advance current research goals; (viii) provide mechanisms for electronic submission and/or automated transfer of relevant data sets; (ix) offer user support through online help documentation, FAQs, tutorials, and timely response to individual user queries; and (x) maintain electronic mailing lists to foster effective communication within the transgenic and broader biomedical community.

Section 24.III. addresses issues related to genetic nomenclature. It outlines the rules and guidelines set by the International Committee on Standardized Genetic Nomenclature for Mice (http://www.informatics.jax.org/nomen/inc.shtml) for naming transgenes, targeted and gene trap mutations, and strains, stocks, or embryonic stem (ES) cell lines carrying genetically manipulated entities in mouse or rat, and presents examples of the basic naming principles.

Finally, a table in the appendix lists alphabetically additional resource categories including: Anatomy, Development, and Imaging; E-mail List Services and Discussion Forums; Gene Expression; Gene Function and Pathways; Genetic Variation (SNPs); Ontologies and Vocabularies; Sequence, Genes, and Genome Browsers; and Welfare, Husbandry, and Colony Management. Readers are encouraged to peruse the International Society for Transgenic Technologies website (ISTT, http://www.transtechsociety.org/) to keep abreast of upcoming events and research activities or relay any new websites they may encounter for inclusion in the ISTT electronic resource collection.

II. Transgenic, Targeted, and Gene Trap Mutation Databases and Online Resources

A selection of worldwide transgenic, targeted, and gene trap mutation mouse databases and online resources, broadly categorized by their scope, utility, or relevance to a specific topic, are listed alphabetically in Table 24.1. Each listing includes the full resource name (and *acronym* in parentheses) and the current web address (URL). The categories are Cre and Other Recombinases, Gene Traps, International Societies, Mutagenesis Programs and Consortia, Phenotypes and Disease Models, Phenotyping Centers and Phenomics Networks, and Tet Expression Systems. Most of these represent well established, freely accessible community databases, such as

Table 24.1 Online Resources for Transgenic, Targeted, and Gene Trap Mouse Mutations

Resource Name (Acronym)	Web Address (URL)[a]
Cre and Other Recombinases	
Allen Brain Institute Mouse Connectivity Atlas (ABI Cre)	http://connectivity.brain-map.org/transgenic/
CANadian European Union Cre (CanEuCre) Project	http://www.caneucre.org/
Coordination of resources for conditional expression of mutated mouse alleles (CREATE) Portal	http://www.creline.org/search_cre_mice
Cre-X-Mice[b]	http://nagy.mshri.on.ca/cre_new/index.php
CreZOO Database	http://bioit.fleming.gr/crezoo/
EUCOMMTOOLS	http://www.mousephenotype.org/martsearch_ikmc_project/about/eucommtools
FaceBase Mouse Resource for Orofacial Clefting Research	http://www.jax.org/facebase/index.html
Gene Expression Nervous System Atlas (GENSAT) Cre Mice	http://www.gensat.org/cre.jsp
Genito-Urinary Developmental Anatomy Project2 (GUDMAP2) Reporter Strains	http://www.gudmap.org/Resources/MouseStrains/index.html
Institut Clinique de la Souris (ICS) MouseCre and CreER^T2 Zoo	http://www.ics-mci.fr/mousecre
International Mouse Strain Resource (IMSR) Cre Strains	http://www.findmice.org/summary?mutations=recombinase%28cre/flp%29
JAX Cre Repository	http://cre.jax.org/
JAX FLP-FRT System	http://jaxmice.jax.org/research/flp_frt.html
NIH Neuroscience Blueprint Cre-Driver Network	www.credrivermice.org
NorCOMM2LS Cre Drivers	https://www.norcomm2.org/norcomm2LS/index.php
Pleiades Promoter Project	http://pleiades.org/
Recombinase (cre) Portal (CrePortal)	http://www.creportal.org/
RIKEN BioResource Center (RBRC) Cre/Flp Strains	http://www.brc.riken.jp/lab/animal/catalogue/Cre_loxPsystem.html
	http://www.brc.riken.jp/lab/animal/catalogue/FLP_frtsystem.html
Vanderbilt University Renal specific transgenic database	http://www.mc.vanderbilt.edu/mkpdc/resources/mice/

(Continued)

Table 24.1 (Continued)

Resource Name (Acronym)	Web Address (URL)[a]

Gene Traps

Centre for Modeling Human Disease (CMHD) Gene Trap Resource	http://www.cmhd.ca/genetrap/index.html
European Conditional Mouse Mutagenesis (EUCOMM) Program	http://www.mousephenotype.org/martsearch_ikmc_project/about/eucomm
Exchangeable Gene Trap Clones (EGTC)	http://egtc.jp/action/main/index
German Gene Trap Consortium (GGTC)	http://www.genetrap.de/
International Gene Trap Consortium (IGTC)	http://www.genetrap.org/
RIKEN BioResource Center (RBRC) Gene Traps	http://www2.brc.riken.jp/lab/animal/search.php http://www2.brc.riken.jp/lab/mouse_es/ishida_es.html
Sanger Institute Gene Trap Resource (SIGTR)	http://www.sanger.ac.uk/resources/mouse/sigtr/
Soriano Lab Gene Trap Database	http://research.mssm.edu/soriano/lab/gene_trap.html
Telethon Institute of Genetics and Medicine (TIGEM) Gene Trap Project	http://genetrap.tigem.it/public
Texas A&M Institute for Genomic Medicine (TIGM) Mouse Knockout Database	http://www.tigm.org/database/
Trans Genic Inc. TG Resource Bank® Exchangeable Gene Traps	http://www.transgenic.co.jp/en/products/database/tgrb/
UniTrap Gene Trap Browser	http://unitrap.crg.es/index.php

International Societies[b]

International Embryo Transfer Society (IETS)	http://www.iets.org/index.asp
International Mammalian Genome Society (IMGS)	http://imgs.org/
International Society for Stem Cell Research (ISSCR)	http://www.isscr.org/
International Society for Transgenic Technologies (ISTT)	http://www.transtechsociety.org/

Mutagenesis Programs and Consortia

Asian Mouse Mutagenesis and Resource Association (AMMRA)	http://www.ammra.info/

(Continued)

Table 24.1 (Continued)

Resource Name (Acronym)	Web Address (URL)[a]
European Conditional Mouse Mutagenesis (EuCOMM) Program	http://www.mousephenotype.org/martsearch_ikmc_project/about/eucomm
EuCOMMTOOLS Project	http://www.mousephenotype.org/martsearch_ikmc_project/about/eucommtools
Germline Mutagenesis Database—Sleeping Beauty (SB) transposon	http://variation.osu.edu/germline/SB_transposon/index.html
International Knockout Mouse Consortium (IKMC) Project	http://www.mousephenotype.org/martsearch_ikmc_project/
Knockout Mouse Project (KOMP)	http://www.mousephenotype.org/martsearch_ikmc_project/aboutkomp
MicroRNA Knockout Project (Sanger mirKO ES Cells)	http://www.mmrrc.org/catalog/StrainCatalogSearchForm.jsp?jboEvent=Search&SourceCollection=Sanger+MirKO
Mutagenic Insertion and Chromosome Engineering Resource (MICER)	http://www.sanger.ac.uk/resources/mouse/micer/
North American Conditional Mouse Mutagenesis (NorCOMM) Project	http://www.norcomm.org/index.htm
PiggyBac Mutagenesis Information CEnter (PBmice)[b]	http://idm.fudan.edu.cn/PBmice/
Texas A&M Institute for Genomic Medicine (TIGM)	http://www.tigm.org/
Transposase and Recombinase-Associated Chromosomal Engineering Resource (TRACER) database	http://www.ebi.ac.uk/panda-srv/tracer/index.php

Phenotypes and Disease Models

Deltagen and Lexicon Knockout Mice	http://www.informatics.jax.org/external/ko/
Diabetic Complications (DiaComp) Consortium Mouse Models[b]	http://www.diacomp.org/shared/modelsPhenotype.aspx
eMice: National Cancer Institute (NCI) Mouse Cancer Models	http://emice.nci.nih.gov/aam/mouse
European Mutant Mouse Pathology Database (Pathbase)	http://eulep.pdn.cam.ac.uk/
EuroPhenome	http://www.europhenome.org/index.html

(Continued)

Table 24.1 (Continued)

Resource Name (Acronym)	Web Address (URL)[a]
FaceBase Mouse Resource for Orofacial Clefting Research	http://www.jax.org/facebase/index.html
GenitoUrinary Development Molecular Anatomy Project (GUDMAP)	http://www.gudmap.org/Resources/MouseStrains/index.html
Hereditary Hearing Impairment in Mice (HHIM)	http://hearingimpairment.jax.org/index.html
Human−Mouse Comparative Phenomics Server (PhenoHM)	http://phenome.cchmc.org/phenoBrowser/Phenome
JAX Disease Mouse Model Resources	http://research.jax.org/grs/disease-specific.html
JAX Rare disease mouse models	http://research.jax.org/rodc/rare-disease-models.html
Mouse Genome Informatics (MGI) Phenotypes, Alleles & Disease Models	http://www.informatics.jax.org/phenotypes.shtml
MouseMine	http://www.mousemine.org/mousemine/begin.do
Mouse Models of Human Cancers Consortium (MMHCC)	http://www.nih.gov/science/models/mouse/resources/hcc.html
Mouse Tumor Biology (MTB) Database	http://tumor.informatics.jax.org/mtbwi/index.do
MUGEN Mouse Database (MMdb): Models of Human Immunological Disease	http://bioit.fleming.gr/mugen/mde.jsp
Online Mendelian Inheritance in Man (OMIM®)	http://www.omim.org/
Pain Genes Database (PainGenesdb)	http://www.jbldesign.com/jmogil/enter.html
PhenomicDB	http://www.phenomicdb.de/
PHENOtype comparisons for DIsease and Gene Models (PhenoDigm)	http://www.sanger.ac.uk/resources/databases/phenodigm/
Phenotype Semantic Information with Terminology of Experiments (PhenoSite)	http://www.brc.riken.jp/lab/bpmp/index.html
Swiss NCCR Transgenic Mouse/Rat Database	http://www.nccrdatabase.ethz.ch/?qs=transgenic
The MUGEN Network of Excellence	http://www.mugen-noe.org/
The Visible Mouse	http://tvmouse.ucdavis.edu/

(Continued)

Table 24.1 (Continued)

Resource Name (Acronym)	Web Address (URL)[a]
Phenotyping Centers and Phenomics Networks	
Asian Mouse Phenotying Consortium (AMPC)	http://ampc.asia/
Australian Phenomics Network (APN)	http://www.australianphenomics.org.au/
Canadian Mouse Consortium (CMC)	http://www.mousecanada.ca/
Centre for Modeling Human Disease (CMHD) Phenotyping Services	http://www.phenogenomics.ca/services/phenotyping.html
European Mouse Disease Clinic (EuMODIC)	http://www.eumodic.org/
German Mouse Clinic (GMC)	http://www.mouseclinic.de/
Infrafrontier Project	http://www.infrafrontier.eu/
Institut Clinique de la Souris (ICS) Phenotyping Service	http://www.ics-mci.fr/dep_phenotyping_services.html
International Mouse Phenotyping Consortium (IMPC)	http://www.mousephenotype.org/
Japan Mouse Clinic (JMC)	http://www.brc.riken.jp/lab/jmc/mouse_clinic/en/index.html
Knockout Mouse Phenotyping Program (KOMP2)	http://commonfund.nih.gov/KOMP2/
MouseBook™	http://www.mousebook.org/
Mouse Metabolic Phenotyping Centers (MMPC)	http://www.mmpc.org/
Mouse Phenome Database (MPD)	http://phenome.jax.org/
Mouse Research Pathology	http://medicine.yale.edu/compmed/mrp/services/phenotyping.aspx
MRC Harwell Mouse Clinic	http://www.har.mrc.ac.uk/services/multi-systems-phenotyping-harwell
Neuro-Bsik Mouse Phenomics[b]	http://www.neurobsik.nl/index.html
Nordic Infrastructure for Mouse Models (NorIMM) Network	http://www.norimm.org/
PhenomeNet—Cross Species Phenotype Network	http://phenomebrowser.net/index.html
RIKEN BioResource Center (RBRC) Mouse Phenome Database (RMPD)	http://www.brc.riken.jp/rmpd/mouse_phenome_top.html
Sanger Mouse Resources Portal	http://www.sanger.ac.uk/mouseportal/
The Mouse Brain Library (MBL)	http://www.mbl.org/

(*Continued*)

Table 24.1 (Continued)

Resource Name (Acronym)	Web Address (URL)[a]
Taiwan Mouse Clinic	http://tmc.sinica.edu.tw/index.html
Toronto Centre for Phenogenomics (TCP)	http://www.phenogenomics.ca/
UC Davis Mouse Phenotyping Network	http://mouse.ucdavis.edu/iwant/pheno_mice.php

Tet Expression Systems

EMMA Tet Expression Systems	http://www.emmanet.org/mutant_types.php?keyword=tet_expression_system
JAX® Mice Tet Expression Systems	http://jaxmice.jax.org/research/tet.html
Mutant Mouse Regional Resource Centers (MMRRC) Tet Strains	http://www.mmrrc.org/catalog/StrainCatalogSearchForm.php?search_query=tTA%2C + rtTA%2C + tetO
RIKEN BioResource Center (RBRC) Tet System	http://www.brc.riken.jp/lab/animal/catalogue/Tetsystem.html
TET Systems GmbH—Tet-Transgenic Rodents Database	http://www.tetsystems.com/fileadmin/tettransgenicrodents.pdf

[a]All web addresses cited in this table were valid as of April 08, 2014, unless otherwise noted.
[b]User registration or subscription is required for access.

Mouse Genome Informatics (MGI), Mouse Phenome Database (MPD), and Online Mendelian Inheritance in Man (OMIM). Those requiring user registration or subscription are noted. Several websites are intended to ease the coordination and prioritization of work within a designated consortium, and serve as unified portals for worldwide access to critical mouse resources, such as CREATE, International Gene Trap Consortium (IGTC), and International Mouse Phenotyping Consortium (IMPC). Others focus on highly specialized research fields or innovative technologies. Whereas the great majority provide a wide range of data querying and mining tools, a few sites are limited to browsing.

Table 24.2 lists web resources for major public repositories and worldwide distributors of GEM strains, mutant ES (mES) cell lines, and/or targeting vectors and provides links to online catalogs of transgenic mouse/ES cell core facilities operated by United States and global academic institutions and nonprofit organizations. Auxiliary information on public mouse repositories and an expanded overview of their common and specialized services can be found in Donahue et al. (2012). Importantly, the majority of GEM strains and models held by centralized mouse repositories are accessible through the unified International Mouse Strain Resource (IMSR) portal described in Section 24II.A. Finally, Table 24.3 compiles a selection of global commercial distributors and animal production services, indicating the geographical location of their headquarters and main web address.

Table 24.2 Online Resources for Public Repositories and Transgenic Mouse and ES Cell Core Facilities

Resource Name (Acronym)	Web Address (URL)[a]	Location
Public Repositories and Distribution Centers		
Australian PhenomeBank (APB)	http://pb.apf.edu.au/phenbank/findstrains.html	Australia
Biological Resource Centre (BRC)	http://www.brc.a-star.edu.sg/index.php?sectionID=12#contract_breed	Singapore
Canadian Mouse Mutant Repository (CMMR)	http://www.cmmr.ca/index.html	Canada
Center for Animal Resources and Development (CARD)	http://cardb.cc.kumamoto-u.ac.jp/transgenic/index.jsp	Japan
European Mouse Mutant Archive (EMMA)	https://www.infrafrontier.eu/	Europe
European Mouse Mutant Cell Repository (EuMMCR)	http://www.eummcr.org/	Germany
Harwell Frozen Embryo and Sperm Archive (FESA)	http://www.har.mrc.ac.uk/services/harwell-frozen-embryo-and-sperm-archive	UK
Institute of Laboratory Animal Science, Chinese Academy of Medical Sciences (CAMS) & Peking Union Medical College (PUMC)	http://www.cnilas.org/html/en/	China
International Mouse Strain Resource (IMSR)—"Find Mice"	http://www.findmice.org/	USA
Japan Mouse/Rat Strain Resources (JMSR) Database	http://www.shigen.nig.ac.jp/mouse/jmsr/top.jsp	Japan
JAX® Mice (JAX)	http://jaxmice.jax.org/index.html	USA
Knockout Mouse Project (KOMP) Repository	https://www.komp.org/	USA
MRC Harwell MouseBook™	http://www.mousebook.org/catalog.php?catalog=stock	UK
Mutant Mouse Regional Resource Centers (MMRRC)	http://www.mmrrc.org/index.php	USA
National Resource Center for Mutant Mice — Model Animal Research Centre (NRCMM — MARC)	http://www.en.nrcmm.cn/mice/index.asp	China
National Cancer Institute Mouse Repository (NCIMR)	http://mouse.ncifcrf.gov/	USA
National Institute of Genetics (NIG) Mouse Genetic Resources	http://www.shigen.nig.ac.jp/mouse/nig/	Japan
OrientalBioService (OBS)	http://www.myv.ne.jp/obs/index.files/serviceguide_eng.htm	Japan
RIKEN BioResource Center (RBRC)	http://www.brc.riken.jp/lab/animal/en/	Japan

(Continued)

Table 24.2 (Continued)

Resource Name (Acronym)	Web Address (URL)[a]	Location
Rodent Model Resource Center— National Laboratory Animal Center (RMRC—NLAC)	http://www.nlac.org.tw/RMRC/ index_e.aspx	Taiwan
Texas A&M Institute for Genomic Medicine (TIGM)	http://www.tigm.org/repository/	USA

Transgenic Mouse and ES Cell Core Facilities[b]

International Society for Transgenic Technologies (ISTT)— List of Worldwide Transgenic and ES Cell Core Facilities	http://www.transtechsociety.org/ linkstg.php	Global
Stanford School of Medicine—List of US Transgenic and ES Cell Core Facilities	http://med.stanford.edu/transgenic/ links.html	USA

[a]All web addresses listed in this table were valid as of April 08, 2014.
[b]Transgenic mouse and ES cell core facilities listed in this table represent primarily academic institutions rather than commercial entities. For a list of selected commercial distributors and animal production services, see Table 24.3.

A selection of the most extensive online resources listed in Tables 24.1 and 24.2 are outlined in the following section.

A. Scope, Data Content, and Functionality of Selected Mutation Databases and Online Resources

1. The CREATE Portal

The CREATE (coordination of resources for conditional expression of mutated mouse alleles) consortium has developed a strategy for the creation, integration, and distribution of new Cre driver mouse strains for potential use in modeling complex human diseases (Smedley et al., 2011). The CREATE portal (http://www.creline.org/) includes key database fields, such as transgene and knock-in allele designations, MGI allele ID, driver, anatomical site of expression, PubMed ID, IMSR strain name and availability, and inducibility information downloaded from disparate Cre database sites. These include MGI's CrePortal (see the next resource in this list), the CreZoo database (http://bioit.fleming.gr/crezoo/), the Cre-X database (http://nagy.mshri.on.ca/cre_new/index.php), and the Institut Clinique de la Souris (ICS) Mouse Cre and CreER^{T2} zoo (http://www.ics-mci.fr/mousecre), thus covering a large fraction of all known Cre lines. CREATE also has conducted online surveys and fostered targeted workshops and discussion forums to assess new technologies and mutagenesis strategies and to prioritize the production of new conditional and inducible Cre driver mouse lines.

The CREATE portal is presented in a BioMart interface that can be queried by driver name or by anatomical site of Cre expression using terms from the Adult

Table 24.3 Online Resources for Commercial Distributors and Animal Production Services[a]

Resource Name (Acronym)	Web Address (URL)[b]	Main Location
Applied StemCell Inc.	http://www.appliedstemcell.com/services/	Menlo Park, CA, USA
B&K Universal Ltd.	http://www.bku.com/biocare.html	Grimston, Hull, UK
BGI Ark Biotechnology (BAB) Co., Ltd.	http://www.bab-genomics.com/list.aspx?catid=191	Shenzhen, China
Central Laboratory for Experimental Animals (CLEA) Japan, Inc.	http://www.clea-japan.com/en/index.html	Tokyo, Japan
Cellectis bioresearch Inc.	http://www.cellectis-bioresearch.com/genome-customization	Paris, France
Charles River Laboratories International, Inc.	http://www.criver.com/en-US/Pages/home.aspx	Wilmington, MA, USA
Cyagen Biosciences Inc.	http://www.cyagen.com/service.html	Sunnyvale, CA, USA
genOwayProducts & Services	http://www.genoway.com/	Lyon, France
Hamilton Thorne, Inc.	http://www.hamiltonthorne.com/index.php/transgenic-animal-production	Beverly, MA, USA
Harlan Laboratories, Inc.	http://www.harlan.com/products_and_services/	Indianapolis, IN, USA
HD Biosciences Co., Ltd.	http://www.hdbiosciences.com/EngTransgenicServices.htm	Shanghai, China
inGenious Targeting Laboratory, Inc.	http://www.genetargeting.com/	Ronkonkoma, NY, USA
Macrogen, Inc.	http://www.macrogen.com/eng/business/mouse_overview.html	Seoul, Korea
MuriGenics, Inc.	http://www.murigenics.com/index.html	Vallejo, CA, USA
Ozgene, Pty, Ltd.	http://www.ozgene.com/	Bentley, DC, Australia
PolyGene Transgenetics	http://www.polygene.ch/	Rumlang, Switzerland
Primogenix, Inc.	http://www.primogenix.com/	Laurie, MO, USA
Scanbur Research & Consumables	http://nova-scb.com/	Sollentuna, Sweden
Shanghai Laboratory Animal Center (SLAC) Laboratory Animal Co., Ltd.	http://english.sibs.cas.cn/rs/fs/ShanghaiLaboratoryAnimalCenterCAS/	Shanghai, China
Sigma-Aldrich®—CompoZr Zinc Finger Nuclease Technology	http://www.sigmaaldrich.com/life-science/zinc-finger-nuclease-technology.html	Saint Louis, MO, USA
Sigma Advanced Genetic Engineering Labs (SAGE®Labs)	http://www.sageresearchmodels.com/home	Saint Louis, MO, USA
Taconic Farms, Inc.	http://www.taconic.com/wmspage.cfm?parm1=26	Hudson, NY, USA
TransGenic, Inc.	http://www.transgenic.co.jp/en/products/	Kumamoto, Japan
Transposagen Biopharmaceuticals, Inc.	http://transposagenbio.com/place-your-order/	Lexington, KY, USA

[a]This table lists selected commercial vendors and animal production services. For selected online lists of transgenic mouse and ES cell core facilities operated by academic institutions and nonprofit organizations, see Table 24.2.
[b]All web addresses cited in this table were valid as of April 08, 2014.

Mouse Anatomy (AMA) dictionary or the Embryonic Mouse Atlas Project (EMAP) Ontology (see appendix). The BioMart allows users to apply filters and define the result output by selecting attributes for customized data display (see http://www.biomart.org/ for additional information on BioMart services). Search results link to each respective database source for additional details and to IMSR for strain availability information. The Ontology View enables users to quickly visualize whether Cre lines already exist for a particular organ, tissue, or cell lineage and navigate to the line of interest.

The CREATE website has facilitated the selection of more than 500 genes/promoters to be used to construct new Cre driver lines within the EUCOMMTOOLS project, and monitors progress in the EUCOMMTOOLS production tracker.

2. The crePortal

The Recombinase (cre) Activity Portal (crePortal, http://www.creportal.org/), hosted by MGI at The Jackson Laboratory, provides free access to expression and activity data for more than 2100 recombinase-containing transgenes and knock-in alleles created in mice. This portal enables users to search by anatomical structure or tissue where Cre is active, by specific promoter/driver element, or by symbol, synonym, or ID of the Cre-bearing transgene or knock-in allele of interest (Blake et al., 2014). Any recombinase alleles that match the selected criteria are returned in a tabular display providing, for each recombinase allele, the driver/promoter, the allele symbol and name (hyperlinked to the MGI transgene/allele detail page), a list of anatomical systems where recombinase activity is detected or not detected, whether the recombinase activity is inducible, a link to IMSR for Cre-containing strain availability, and links to all associated references. Each listed anatomical system is hyperlinked to the relevant Recombinase Activity Detail page, which summarizes the activity data for the selected allele and tissue combination and presents a molecular description of the Cre recombinase allele, a gallery of annotated images depicting Cre activity in specific anatomical structures, and the assays used to determine the activity level and pattern in each structure at a specified age. Access to phenotype data obtained from conditional genotypes using each Cre construct is provided via integration with other existing phenotype annotations on the MGI allele detail page. Users also can retrieve a full list of recombinase-bearing transgenes and knock-ins, including lines expressing non-Cre recombinases (e.g., Flp, Dre, or phiC31).

The MGI transgene/allele detail page (linked from the allele symbol) offers a matrix view of recombinase activity by age to provide users an overall view of the tissue distribution for Cre activity at given ages. Users can submit new recombinase activity data via a web-based submission form or data upload, and submit comments regarding usage of specific Cre alleles. The latter feature provides a unique opportunity for Cre line users to share information related to unreported or off-target Cre activity, mosaicism, inconsistent excision activity, parent-of-origin effects, or other confounding factors to be considered when selecting a Cre driver strain for conditional mutagenesis.

Efforts are currently underway to systematically characterize individual Cre driver mouse strains held at the JAX Cre Repository (http://cre.jax.org/) and the Allen Institute for Brain Science (http://connectivity.brain-map.org/transgenic). These data will be integrated into the crePortal and updated as work progresses, to complement a growing catalog of large-scale data sets from Cre-related projects, including GENSAT (http://www.gensat.org/cre.jsp), Pleiades (www.pleiades.org), the NIH Neuro Blueprint project (www.credrivermice.org), CanEuCre (www.caneucre.org), the ICS MouseCre (www.ics-mci.fr/mousecre/), and the new EUCOMMTOOLS and NorCOMM2LS projects (Murray et al., 2012).

3. The EuroPhenome Mouse Phenotyping Resource

The EuroPhenome database (Morgan et al., 2010) was established to capture, analyze, and disseminate phenotyping data arising from the EUropean MOuse DIsease Clinic (EUMODIC, http://www.eumodic.org/), the first EU-funded large-scale phenotyping pilot program, comprising MRC Harwell (http://www.har.mrc.ac.uk/), the Wellcome Trust Sanger Institute (WTSI, http://www.sanger.ac.uk/), the German Mouse Clinic (GMC, http://www.mouseclinic.de/), and Institut Clinique de la Souris (ICS, http://www.ics-mci.fr/). The EUMODIC consortium has completed primary phenotypic assessment of 500 mouse knockout lines (Gates et al., 2011). EuroPhenome includes EUMODIC data, as well as data from the Centre for Modeling Human Disease (CMHD, http://www.cmhd.ca/) pipeline in Canada. EuroPhenome performs automated statistical comparisons to identify significant phenotypic deviants, i.e., mutant lines that exhibit an aberrant phenotype relative to the control strain, and automatically assigns mouse phenotypes using the widely adopted hierarchical Mammalian Phenotype (MP) ontology (see appendix; Smith and Eppig, 2012).

The EuroPhenome portal (http://www.europhenome.org) provides a graphical data display, statistical analysis, web services, and a BioMart interface. The web interface allows users to access data from a gene- or phenotype-centric viewpoint. Querying the portal via a gene search returns a summary page displaying key information about the gene and allele, and a visual "heatmap" overview of statistically significant phenovariants. Mousing over any test with a positive result presents further details about that result, and a link to a graphical representation of the data. An advanced search option allows users to filter their gene query by procedure, pipeline, zygosity, or gender.

Querying the portal via a selected MP ontology term returns a list of all alleles that have a significant phenotypic deviant hit for that term or any of its descendants, as well as the specific parameters that generated these annotations. Finally, the OMIM Phenotype Mapper allows users to select a human gene or disorder and mine the underlying phenotype data to identify a mouse line of interest. Human disorders are retrieved from OMIM (see number 10 in this list of resources), and the resultant human genes are linked to mouse orthologs via the Ensembl Biomart (http://www.ensembl.org/Mus_musculus/Info/Index/).

As of January 7, 2013, the EuroPhenome portal contains data and annotations for 538 mutant strains and 45 inbred strains. The collected phenotyping results also are accessible through the MGI resource (number 9 in this list), while the data generated by WTSI are additionally reported through the Sanger Mouse Resources Portal (http://www.sanger.ac.uk/mouseportal/). All relevant mouse lines are available via EMMA (https://www.infrafrontier.eu/; Hagn et al., 2007), the KOMP Repository (https://www.komp.org/), and IMSR. Eventually, EuroPhenome will be superseded by the IMPC (number 7 in this list of resources).

4. The Infrafrontier Website

This website serves as the central information hub for the Infrafrontier project (http://www.infrafrontier.eu/), which aims to establish a sustainably financed, pan-European research infrastructure that provides open access to scientific platforms, services, and data related to the systemic phenotyping, archiving, and distribution of mouse disease models (Raess and Hrabé de Angelis, 2009). The Infrafrontier consortium, representing prominent mouse clinics, distribution nodes, and administrative partners from Europe and Canada, is largely based on joint European efforts such as EUMODIC and EMMA. The new EU-funded INFRAFRONTIER-I3 project joins the EMMA network and mouse clinics at the Helmholtz Zentrum München, ICS, MRC Harwell, and WTSI in a concerted effort to deliver up to 1215 new strains to the EMMA repository by 2016. Importantly, a website section devoted to a related project, InfraCoMP (https://www.infrafrontier.eu/infrafrontier-research-infrastructure/international-collaborations-and-projects/infracomp-0), will host a series of workshops and meeting reports targeted at coordinating Infrafrontier and IMPC activities, to help resolve redundancies, maximize mutual benefits, and leverage the infrastructural capacity of physical and data resources.

5. The IGTC Website

The IGTC was launched to create the first international resource of gene trap cell lines, publicize information through a centralized open access website, and integrate an exploding array of gene trap data into primary genome browsers and other informatics sites (Nord et al., 2006). Thousands of mES cell lines harboring insertional mutations in known and novel genes were generated by IGTC members, including CMHD, EUCOMM, the Sanger Institute Gene Trap Resource (SIGTR, http://www.sanger.ac.uk/resources/mouse/sigtr/), the German Gene Trap Consortium (GGTC, http://www.genetrap.de/), and the Texas A&M Institute for Genomic Medicine (TIGM, see below). As of April 08, 2014, this collaboration has yielded more than 125,034 characterized IGTC ES cell lines, representing nearly 40% of known mouse genes, all available to the scientific community. The IGTC website (http://www.genetrap.org/) provides access to gene trap cell lines generated by IGTC members and offers a suite of analytical tools to maximize the utility of public gene trap resources. The website maintains a list of IGTC partner sites, provides help documentation and online tutorials, and enables users to request IGTC cell lines of interest through links to provider sites.

Researchers can access the IGTC data via a gene-centric view that includes gene name and symbol and identifiers linking to gene records at Ensembl and Entrez Gene (http://www.ncbi.nlm.nih.gov/gene). Each IGTC cell line matching a given gene is listed by identification status, source, and vector name, and hyperlinked to its respective cell line page. Alignment images are available as graphical displays of all gene trap cell line sequence tags aligned to the trapped gene. When available, additional information includes the MGI gene symbol, a gene description, and the chromosomal position. The genomic localization of each cell line is available through the Ensembl and University of California Santa Cruz (UCSC) genome browsers, with IGTC cell lines depicted as an annotation track. The cell line pages offer imported annotation data including Gene Ontology (GO) classes (see appendix), protein domain, structure and function, as well as phenotype, homology, and orthology data.

Users can query the IGTC database by keyword, accession ID, or chromosomal location and retrieve a list of gene symbols or cell line IDs linked to the corresponding gene or cell line page. Users also can BLAST sequences against gene trap sequence tags, browse the database by MGI gene name/symbol or chromosome location, and view trapped genes within the context of biological pathways or functional GO categories. Finally, users can look for traps in genes with a defined expression profile by selecting a tissue of interest and/or the expression level of the gene relative to the median tissue expression. IGTC cell lines are currently accessible through major external informatics sites, including the UCSC, Ensembl, and MGI genome browsers, and IMSR.

6. The IKMC Project

The International Knockout Mouse Consortium (IKMC) was launched in 2007 as a global initiative to mutate (knockout) each of the more than 20,000 protein-coding genes of the mouse genome using a combination of gene targeting and gene trapping strategies in ES cells and render the generated resources accessible to the scientific community (Collins et al., 2007; International Mouse Knockout Consortium et al., 2007). Toward this end, four major multiinstitutional programs founded in Europe and North America, viz., EUCOMM (http://www.mousephenotype.org/martsearch_ikmc_project/about/eucomm; Friedel et al., 2007), KOMP (http://www.mousephenotype.org/martsearch_ikmc_project/aboutkomp), NorCOMM (http://www.norcomm.org/index.htm), and TIGM (http://www.tigm.org/), joined forces to coordinate programs, share expertise and technologies, maximize output, and reduce duplication of effort.

To date, IKMC members have collectively generated more than 18,022 mES cell clones, 16,370 vectors, and over 2434 mutant mouse strains, most of them conditional, providing >90% coverage of protein-coding genes. The Sanger microRNA Knockout (MirKO) program has also joined the IKMC and added more than 220 MirKO ES cell lines (Prosser et al., 2011) to this effort. The IKMC mES cell resources were developed primarily in a C57BL/6N genetic background. Most ES cells have been targeted using a "knockout-first, conditional ready" allele design, which enables the derivation of mice carrying a null allele or, alternatively, a "conditional ready" allele (Skarnes et al., 2011). The null configuration is the preferred allele for the generation and phenotyping of mice by members of the IMPC (to be discussed next). KOMP targeting vectors, ES cells, and mice are currently distributed by the KOMP Repository (https://www.komp.org/); EUCOMM vectors and ES cell lines by EuMMCR (www.eummcr.org); EUCOMM mice by EMMA (https://www.infrafrontier.eu/); NorCOMM targeting vectors, ES cells, and mice by the Canadian Mouse Mutant Repository (CMMR, http://www.cmmr.ca/index.html); TIGM gene trap ES cells and mice by the TIGM repository (http://www.tigm.org/repository/); and mirKO ES cells from either of two participating repositories, MMRRC (http://www.mmrrc.org/) and EuMMCR.

Although extensive, the IKMC resource is not yet complete. EUCOMMTOOLS (http://www.mousephenotype.org/martsearch_ikmc_project/about/eucommtools), a successor of EUCOMM, will pursue a binary objective to (i) create 3500 novel conditional mutant alleles for single-exon protein-coding mouse genes and (ii) produce a new Cre driver resource for each of the ~ 220 mouse adult tissues. In parallel, NorCOMM2LS (https://www.norcomm2.org/norcomm2LS/index.php), a successor of NorCOMM, will employ a novel transposon-based enhancer trap strategy termed "Cre-ping" to create up to 75 new embryonic tissue and stem-cell-specific Cre driver mouse lines.

Key data from the public IKMC web portal (Ringwald et al., 2011), which has been retired, have been incorporated into the new IMPC portal discussed below.

7. The IMPC Web Portal

Building on the success of pilot phenotyping programs like EUMODIC, the Sanger Mouse Genetics Project (www.sanger.ac.uk/resources/mouse), GMC, and the KOMP312 study (www.kompphenotype.org), the IMPC was launched in 2011 as a multicenter global consortium of leading mouse clinics and funding bodies tasked to complete the generation and broad based, primary phenotyping of all 20,000 IKMC mouse mutant lines by 2021 (Brown and Moore, 2012). In parallel, a distributed network of specialist centers with in-depth expertise in specific phenotyping domains will perform more sophisticated, secondary screens on selected mouse lines with notable primary phenotypes. An endeavor of this scale will ultimately uncover key aspects of organogenesis and embryo morphogenesis, help dissect the functional pleiotropy of individual genes, and lead to increasingly precise models of human disease and therapeutic interventions.

The initial IMPC phase aims to produce and phenotype up to 5000 mutants by 2016. Mice will be largely created using the "knockout-first, conditional ready" allele (tm1a) generated in ES cells which contains a reporter (lacZ) trapping cassette and a floxed promoter-driven neo cassette inserted into the intron of a gene, thus disrupting gene function (Skarnes et al, 2011). Breeding of mice carrying the tm1a allele to a suitable Cre driver line ablates a key early exon along with the selection cassette, and creates a lacZ-tagged tm1b null allele. All mice will be generated on an isogenic C57BL/6N background, and frozen sperm will be archived for both the tm1a and the tm1b allele, thus presenting an unprecedented resource of mouse mutants to the research community. Developmentally competent, viable mice homozygous for the tm1b null allele will enter the adult phenotyping pipeline. In the cases of homozygous lethality, the IMPC will embark on phenotyping embryos and heterozygous adults.

IMPC members have agreed on a core adult phenotyping pipeline of 20 phenotyping platforms, engaging a broad spectrum of in-life and terminal procedures on cohorts of seven males and seven females. It is expected that individual IMPC centers will add phenotyping modalities and/or challenge models of local interest. Of note, the design of the embryonic phenotyping pipeline is currently under final review by IMPC experts.

The informatics component of the IMPC is central to the success of the project. To this end, the NIH-funded KOMP2 Mouse Phenotyping Informatics Infrastructure (MPI2) consortium is tasked to develop and deploy a centralized data coordination center, database, and web portal to capture, annotate, integrate, and disseminate the phenotyping data from KOMP2 (http://commonfund.nih.gov/KOMP2/) and other IMPC programs to the scientific community (Mallon et al., 2012). The proposed solution envisions a central database that adheres to uniform data standards, captures raw data from local laboratory information management systems at participating production and phenotyping centers, and applies various quality control checks and validation procedures prior to deposition into a central data archive. The MPI2 program will deliver a statistical analysis pipeline, which will automatically assign MP terms to significant phenotypic outliers and summarize data for each mutant and assay for presentation to end users. An "international microinjection tracking system" is already in use to assign genes to individual IMPC centers, and audit the progress of transition from gene selection to ES cell injections, chimeras, mouse production, and phenotyping. In addition, an associated database, IMPReSS (http://www.mousephenotype.org/impress), has been released to manage and track the standardized phenotyping protocols implemented in IMPC.

The new IMPC portal (http://www.mousephenotype.org/) serves as a single point of access to all IMPC data and offers robust annotation, analytical, and integration tools to view and search IMPC data sets. The IMPC portal maintains a current member list and welcomes input and potential synergy with industrial partners. Current functionality enables users to participate in online forums focused on phenotyping protocols, express interest in a gene and receive updates on knockout strain production, perform gene queries to track the progress of mouse production and phenotyping, and view phenotype procedures and parameters held in IMPReSS (Koscielny et al., 2014).

8. The IMSR Web Portal

The IMSR web portal (http://www.findmice.org/) provides a searchable online database of globally available inbred and mutant mouse strains and stocks, including GEM strains and mES cell lines. The primary objective of IMSR is to ensure open access to unified information on mouse resource holdings and aid the international research community in locating or ordering mouse resources of interest (Eppig and Strivens, 1999; Strivens and Eppig, 2004). Researchers can search IMSR by strain type (e.g., inbred, mutant, congenic, coisogenic), strain status (e.g., live mice, cryopreserved embryos/germplasm, or mES cells), mutation (e.g., transgenic, targeted, gene trap, Cre/Flp recombinase), and repository site. Search results include the strain holder, links to detailed information on specific strains and alleles, and links to the repository site for further inquiries and order placement. As of April 04, 2014, IMSR contains data on more than 27,482 strains and 210,332 mES cell lines from various participating repositories, including JAX, MMRRC, RBRC, EMMA, CMMR, TIGM, the KOMP Repository, and Taconic. A list of contributing repositories can be found at http://www.findmice.org/repository. IMSR strongly encourages additional global providers to list their unique mouse stocks in the IMSR online catalog. A prominent link to the IMSR ("Find Mice") homepage is available on the MGI navigation bar. In addition, MGI allele detail pages provide a link that automatically returns strains and mES cell lines containing the featured allele.

9. The MGI Phenotypes, Alleles, and Disease Models Portal

The MGI resource (www.informatics.jax.org) represents a consortium of several bioinformatics programs working in concert to facilitate the use of the mouse as a premier mammalian surrogate for modeling human biology and disease (Blake et al., 2014; Bult et al., 2013). MGI serves as the authoritative source for official mouse gene, allele, and strain nomenclature, catalogs all mouse genes, genome features, and mutant alleles (available in mice or ES cells), and grants free access to current, integrated biological knowledge spanning from sequence and variation to phenotype and disease model data. Importantly, MGI adheres to semantic standards and applies a variety of bio-ontologies to ensure consistent annotation, retrieval, and analysis of gene function, phenotype, disease model, and anatomy-based gene expression information.

The MGI homepage features icon links to several content-specific mini portals, each encapsulating a different MGI bio-domain along with specific access instructions. Heterogeneous biological data are integrated from multiple sources, ranging from major data providers and mutagenesis consortia (e.g., GenBank, NCBI, Ensembl, VEGA, IKMC) to individual investigator laboratories and the biomedical literature, using a combination of automated processes, quality control checks, and expert manual curation. MGI also supports direct electronic data contributions from individual researchers via http://www.informatics.jax.org/submit.shtml. Data are released weekly, and public access is achieved through interactive web interfaces, FTP reports, web services, and the MGI BioMart. In addition, MGI maintains a moderated and active e-mail bulletin board (mgi-list; see appendix), serving more than 2000 subscribers.

MGI curates mouse phenotypes in the context of mutations (spontaneous, induced, or genetically engineered), strain variations, QTL, and complex traits that may serve as plausible models of human disorders, incorporating phenotype images and video clips, as available. Moreover, MGI offers versatile phenotype viewing options and supports customized retrieval of complex phenogenomic datasets, and disease model mining from a gene, allele, phenotype, or disease perspective. Use of human disease (OMIM) terms serves to associate aberrant phenotypic mouse features and human gene mutations or disease syndromes. To support data from large-scale mouse mutagenesis projects, including gene trap and knockout projects (Smith and Eppig, 2012), MGI integrates phenotype data from high-throughput phenotyping centers, including data sets provided by WTSI and EuroPhenome, and will incorporate data from the new IMPC as these become available. All newly acquired allele and phenotype data are integrated with data from individual laboratories and the biomedical literature, as well as other genomic, expression, function, tumor, and pathway data stored in MGI, to support comparative analyses and correlative discoveries. As of March 24, 2014, more than 753,786 alleles have been cataloged in MGI, representing >21,760 mouse genes. Over 38,808 alleles are propagated in live mice or available as cryopreserved embryos or sperm; the rest exist only in ES cell lines. At least 51,186 unique mouse genotypes have been curated with nearly 260,440 MP annotations. Finally, more than 4300 unique mouse genotypes have been annotated as models of nearly 1300 hereditary diseases or syndromes.

Available as part of the MGI resource, the Phenotypes, Alleles, and Disease Models portal (http://www.informatics.jax.org/phenotypes.shtml) offers a suite of data access tools, including a Quick Search tool, a Batch Query tool, web-based vocabulary browsers, and an advanced Phenotypes Query form. The Quick Search tool, found on all MGI pages, allows entry of any keywords, phrases, or IDs and quickly locates genome features (e.g., genes or alleles), vocabulary terms (e.g., GO, AMA, MP, OMIM), and other results by ID (e.g., sequences, orthologies, references). The Batch Query tool enables users to input a list of gene symbols or IDs and customize the result output to include any one of the following: phenotypic alleles; GO, MP, or OMIM terms; or GXD expression results. The Mammalian Phenotype Browser enables consistent retrieval of mouse phenotypes at the level of known data resolution, offering the ability to query with an MP term or ID and retrieve all relevant mouse genotypes annotated to that term or its descendants. The Human Disease (OMIM) Vocabulary Browser allows users to click on an OMIM disease of interest and retrieve the Human Disease and Mouse Model Detail page, which offers access to both the full OMIM entry and a summary table listing all mouse genotypes modeling the human disease, linked to their associated references and phenotype data.

Advanced MGI query forms enable users to formulate multiparametric, genome-scale queries and harness the full power of data. To illustrate, the Phenotypes, Alleles, and Disease Models Query form allows users to search for mouse mutations by phenotype, disease, allele, and gene nomenclature, map position, or allele category. Any combination of these parameters can be used to expand or limit the scope of the desired data output. Users can identify mutant alleles by defining a specific combination of phenotype and/or disease terms or IDs of interest. Any phenotypic alleles that meet the selected criteria are listed in a summary table and hyperlinked to their corresponding detailed records. Data displayed in the Phenotypic Allele Detail page include allele and gene nomenclature, details on the molecular mutation, the ES cell line IDs associated with the allele (where applicable), phenotype annotations and associated images, as well as established models of human disease, strain/ES cell availability via IMSR, and supporting references. Of note, phenotype data can be viewed in a matrix summary format or by genotype. The matrix view organizes affected anatomical systems by color-coded abbreviations of genotype states, delineated by sex and curation source. Presence of an abnormal phenotype is indicated and anatomical systems can be expanded to reveal annotations to more specific MP terms. Additionally, each abbreviated genotype is a hyperlink that expands to reveal the full phenotypic details for that genotype, including representative images and disease model associations.

10. Online Mendelian Inheritance in Man

OMIM is a freely accessible, comprehensive, authoritative knowledgebase of human genes and genetic phenotypes curated at the Johns Hopkins University School of Medicine. The OMIM database provides searchable, full text, referenced overviews of all known Mendelian disorders and over 12,000 genes, placing emphasis on the molecular relationship between genetic variation and phenotypic expression. OMIM provides links to numerous genetic, genomic, and model organism databases, as well as to variation, coding, clinical trials, and other research resources. Importantly, OMIM includes an Animal Model section that provides timely phenotypic descriptions of transgenic and knockout mouse models of human genetic disorders reported in the biomedical literature. A new website (http://www.omim.org/) recently was launched to facilitate a more structured view of OMIM contents and enhance interconnectivity with complementary clinical and genetics resources (Amberger et al., 2011). Current OMIM entry statistics, sample searches, FAQs, and an online tutorial are available at the homepage.

III. Standardized Nomenclature for Transgenic, Targeted, and Gene Trap Mutation Animals

The unique and systematized nomenclature of transgenic and targeted mutations is critical to their unambiguous identification in research and in the scientific literature. The MGI database is the authoritative source of official names for mouse genes, alleles, mutations, and strains (Bult et al., 2013). Nomenclature follows the rules and guidelines established by the International Committee on Standardized Genetic Nomenclature for Mice, a body elected by the mouse research community. Nomenclature guidelines are updated regularly to accommodate naming schemes that reflect new genome manipulation technologies and revised nomenclature policies. The MGI Mouse Nomenclature Home Page (http://www.informatics.jax.org/nomen) provides the current, unified, and complete rules and guidelines for naming genes,

Table 24.4 Online Guidelines on Standardized Genetic Nomenclature for Human, Mouse, and Rat

Resource Name (Acronym)	Web Address (URL)[a]
Human	
HUGO Gene Nomenclature Committee (HGNC)	http://www.genenames.org/
Mouse and Rat	
Guidelines for Nomenclature of Genes, Genetic Markers, Alleles, and Mutations in Mouse and Rat	http://www.informatics.jax.org/nomen/gene.shtml
Guidelines for Nomenclature of Mouse and Rat Strains	http://www.informatics.jax.org/nomen/strains.shtml
International ImMunoGeneTics Information System® (IMGT®) Nomenclature	http://www.imgt.org/IMGTindex/nomenclature.html
International Laboratory Code Registry (ILAR Lab Codes)	http://dels.nas.edu/global/ilar/Lab-Codes
Mouse ES Cell Line Names and Strains of Origin	ftp://ftp.informatics.jax.org/pub/reports/ES_CellLine.rpt
Nomenclature for Mouse Strains	http://jaxmice.jax.org/support/nomenclature/index.html

[a]All web addresses cited in this table were valid as of April 08, 2014.

alleles, mutations, and strains in mice and rats, along with up-to-date abbreviated guides, references to former published nomenclature guidelines, and previous archived online versions. The most recent print copy of mouse nomenclature guidelines can be found in Eppig (2006). However, the MGI online version represents the latest nomenclature policies and supersedes all previously published versions. This official nomenclature is widely propagated through regular data exchange and curation of shared links between MGI and other bioinformatics resources. The MGI nomenclature group collaborates with nomenclature experts for human (http://www.genenames.org/) and rat (Rat Genome Database, RGD; http://rgd.mcw.edu) to provide consistent nomenclature for mammalian species, and teams up with scientific journal editors to endorse compliance to nomenclature standards in publications. Relevant online nomenclature resources are listed in Table 24.4.

This section outlines the basic rules for naming transgenes, targeted and gene trap mutations, and strains or stocks carrying genetically manipulated entities, covering those nomenclature forms that are most frequently applied and encountered. Examples illustrating key naming principles are provided for each category. Readers should, however, consult the MGI Nomenclature website to ensure current usage, and to access additional or complex nomenclatures not explicitly covered in the rules and examples listed below.

Importantly, researchers can register new mutant mouse or rat strains using, respectively, the online MGI Mutant Alleles, Strains, and Phenotypes Submission

Form (http://www.informatics.jax.org/mgihome/submissions/amsp_submission.cgi) or the RGD Strain Registration Form (http://rgd.mcw.edu/tools/strains/strainRegistrationIndex.cgi). A key principle to keep in mind when symbolizing transgenes and targeted mutations is that the purpose of the symbol is to provide a unique identifier or name and not to convey all the information known about the genetic element being symbolized. For further assistance with nomenclature, contact the MGI nomenclature group via e-mail at nomen@jax.org.

A. Laboratory Registry Codes

Laboratory registry codes constitute an integral part of the authoritative nomenclature for transgenes, genetically engineered mutations, DNA loci, and chromosomal aberrations (see Table 24.4). Substrains, congenics, and other strains where several otherwise distinguishable forms exist should also be identified by laboratory codes. A laboratory registry code (also, "lab code" or "ILAR code") is a one-to-five letter designation that uniquely identifies an investigator, laboratory, or institution that produced and/or maintains a particular mouse or rat strain. Only the first letter of the lab code is capitalized. Unique lab codes are assigned from the International Laboratory Code Registry (http://dels.nas.edu/global/ilar/Lab-Codes) maintained by the Institute for Laboratory Animal Research (ILAR, http://dels.nas.edu/global/ilar/About-Us). Interested parties should consult the current alphabetical listing of approved lab codes or search the Laboratory Code Registry prior to registering a new code. Applicants can submit a new lab code directly to ILAR at http://ilarlabcode.nas.edu/register_code_nodep.php or via the MGI nomenclature coordinator at nomen@jax.org. Finally, investigators should use the same lab code for transgenes and targeted mutations that they create, and for strain sublines that they maintain for more than 10 generations. Examples of ILAR codes include J (The Jackson Laboratory), Mrt (Gail R. Martin), Lex (Lexicon Genetics, Inc.), and Unc (University of North Carolina).

B. Rules for Naming Transgenes

Any DNA that has been stably introduced into the germ line of mice or rats constitutes a transgene. This section addresses nomenclature rules for randomly inserted transgenes, where foreign DNA material is typically inserted into the genome by microinjection. Transgenes generated by homologous recombination as targeted events at particular loci warrant their own distinct nomenclature, described below. As transgenes do not comprise part of the native mouse genome, neither transgene symbols nor the gene symbols they contain are italicized. It is not necessary, or even pragmatic, to assign standardized symbols to all experimentally created transgenes. When several transgenic lines are reported in a publication, but not all are subsequently maintained or archived, only those that are studied and maintained require standardized nomenclature.

A transgene symbol consists of four parts, taking the form Tg(YYY)###Zzz, where Tg denotes transgene, (YYY) describes the inserted material, ### is a number assigned by the laboratory that created it, and Zzz is the ILAR registry code of the originating lab.

The insert designation or parenthetical element (YYY) contains the official gene symbol of the inserted DNA. The following basic rules apply:

1. The gene symbol should follow the nomenclature conventions of the species of origin and should not be italicized. For example, an inserted human gene would be in all uppercase letters, while the symbol for an inserted mouse gene would begin with an uppercase letter but the remaining letters would be lowercase. Short symbols, six characters or less, are preferred.
2. A series of different transgenic constructs containing the same gene or different transgenic animals with the same construct should use the same insert designation in parentheses; they are distinguished from each other by the serial number and lab code. Additional information about the nature of the construct inserted should be given in associated publications and database entries.
3. Where multiple genes are coinjected, each gene may be specified in a comma-separated list.
4. Where gene expression is driven by a nonendogenous promoter, the promoter can be specified, appearing first, followed by a hyphen and the expressed gene.
5. For transgenes that use reporter constructs or recombinases (e.g., lacZ, GFP, Cre), the promoter must be specified as the first part of the gene insertion designation, separated by a hyphen from the reporter or recombinase designation.
6. In the case of fusion gene inserts, where roughly equal parts of two or more genes compose the construct, a forward slash separates each of the component genes in parentheses.
7. For BAC transgenics, the insert designation is the BAC clone and follows the naming convention provided by the NCBI Clone Database (http://www.ncbi.nlm.nih.gov/clone).
8. When random insertion of a transgene occurs in or near an endogenous gene, a new mutant allele may be produced. The new allele of this gene should be designated by superscripting the transgene symbol or its abbreviation, if it is unique, to the gene symbol.
9. For transgenes containing RNAi constructs, the insert can be designated minimally as RNAi:geneX, where geneX is the gene knocked down.

Once a full transgene symbol is given in a paper, the symbol can be abbreviated by omitting the parenthetical element. It is important to note that while there is the option to include significant information on vectors, promoters, etc., within the parentheses of a transgene symbol, this should be minimized for brevity and clarity. The function of a symbol is to provide a unique designation to a gene, locus, or mutation. The fine molecular details of these loci and mutations should reside in databases such as MGI and RGD.

The laboratory-assigned number ### is a unique number assigned by an individual laboratory to each stable transgenic insertion once transmission is confirmed. It may be the laboratory's line, or a sequential number in a series for that laboratory, or the founder designation. No two lines generated within one laboratory should be given the same number. Numbers should be limited to five characters or less. Examples illustrating the basic scheme for naming transgenic entities include:

- Tg(Alox15)41FChed—A transgene containing the mouse *Alox15* gene in line 41F developed by Catherine C. Hedrick (Ched).
- Tg(LRRK2)66Mjff—A BAC transgene containing the entire human *LRRK2* gene, the 66th transgene insertion produced by The Michael J. Fox Foundation (Mjff).

- Tg(HLA-B*2705,B2M)33-3Trg—A double transgene in rat containing the human *HLA-B*2705* and *B2M* genes, which were coinjected, giving rise to line 33-3 by Joel D. Taurog (Trg). In this example, an asterisk (*) is used to denote an expressed sequence altered by mutation.
- Tg(TCF3/HLF)1Mlc—A transgene in which the human *TCF3* and human *HLF* genes were inserted as a fusion chimeric cDNA, the first transgenic mouse line produced by Michael L. Cleary's laboratory (Mlc).
- Tg(RP23-291P1)1Flp—A BAC transgene where the inserted BAC is from the RP23 BAC library, plate 291, row P, column 1, the first produced from the laboratories of Figge, Lammert, and Paigen (Flp).
- Tg(Wnt1-lacZ)206Amc—A lacZ transgene with a *Wnt1* promoter, from mouse line 206 in the laboratory of Andrew P. McMahon (Amc).
- Tg(Zp3-cre)3Mrt—A Cre transgene with a *Zp3* promoter, the third mouse line from the laboratory of Gail R. Martin (Mrt).
- Tg(RNU6-RNAi:Mpp5)13Wij—A RNAi transgene where the human *RNU6* promoter drives expression of a short hairpin RNA targeting the mouse *Mpp5* gene, the 13th transgene insertion produced by Jan Wijnholds (Wij).
- $Bmp7^{Tg(BCL2)114Cro}$—A transgenic allele where a transgene expressing the human *BCL2* gene has randomly inserted (as opposed to targeting) into the mouse *Bmp7* gene, resulting in a phenotypic mutation. This is the 114th transgene insertion produced in the laboratory of Carlo M. Croce (Cro).

C. Rules for Naming Targeted and Gene Trap Mutations

1. Knockout, Knock-in, Conditional, and Other Targeted Mutations

Mutations that are the result of gene targeting by homologous recombination in ES cells are symbolized by superscripting an allele symbol to the symbol for the targeted gene. The allele superscript symbol takes the form *tm#Zzz*, where *tm* denotes a targeted mutation, # is a serial number assigned by the laboratory of origin, and *Zzz* is the ILAR lab code identifying where the mutation was produced. This nomenclature covers knockouts, knock-ins, and other forms of genetic alteration targeted to specific genes. Allele symbols, like gene symbols, are italicized. For example:

- $Apoe^{tm1Unc}$—The first targeted (knockout) mutation of the *Apoe* gene produced at the University of North Carolina (Unc).

Knock-in mutations, in which foreign DNA sequence is inserted in an endogenous gene, should be given a tm symbol and the particular details of the knock-in should be reported with the symbol in publications or database entries. When the inserted gene segment results in expression of a foreign gene, or if it significantly changes the character of the expressed endogenous gene product, the symbol of the inserted gene may be used parenthetically in the allele symbol. However, reporter gene knock-ins are usually not indicated in allele symbols. Examples include:

- $Ntf3^{tm1(Bdnf)Tes}$—A knock-in mutation, where the coding region of the *Ntf3* gene was replaced by the *Bdnf* gene, the first originating from the laboratory of Lino Tessarollo (Tes).

- $Cd19^{tm1(cre)Cgn}$—A knock-in allele where a Cre recombinase was inserted in-frame into exon 1 of the $Cd19$ gene, the first targeted mutation made by the University of Cologne (Cgn).
- $Apoe^{tm1(APOE^*2)Mae}$—A DNA fragment containing exons 2–4 of the human $APOE$ gene (the $APOE2$ isoform) replaced the equivalent region of the mouse $Apoe$ gene, the first targeted mutation created by Nobuyo Maeda (Mae). The human protein is expressed from this allele and the endogenous mouse protein is not detectable.
- $Hprt^{tm2(CAG-Myof)Isrd}$—A transgene containing the mouse $Myof$ cDNA under the control of the ubiquitous CAG promoter was inserted into the $Hprt$ locus, the second targeted mutation made by Isabelle Richard (Isrd).

When a targeting vector is used to generate multiple germ-line transmissible alleles from the original insertion site, such as in the Cre-Lox system, the original knock-in is symbolized by the regular tm designation rules, and subsequent heritable alleles generated after mating with a Cre transgenic mouse would retain the parental designation followed by a decimal point and serial number. For example:

- $Adar^{tm1Knk}$—A targeted mutation where loxP sites were inserted into the $Adar$ gene, the first created by Kazuko Nishikura (Knk).
- $Adar^{tm1.1Knk}$—A distinct germ-line transmissible allele generated after mating an $Adar^{tm1Knk}$ bearing mouse with a Cre transgenic mouse. Note: somatic events generated in offspring from an $Adar^{tm1Knk}$ mouse and a Cre transgenic mouse that cause disruption of $Adar$ in selective tissues would not be assigned nomenclature.

Other more complex forms of gene replacement, such as partial knock-in, hit-and-run, double replacements, and loxP-mediated integrations are not conveniently abbreviated and should be given a conventional $tm\#Labcode$ superscript. Details of the targeted locus should be given in associated publications and database entries.

Large-scale projects that systematically produce a large number of alleles (>1000) may include a project abbreviation in parentheses as part of the allele designation. These should retain the accepted nomenclature features of other alleles of that class. For example:

- $Apof^{tm1(KOMP)Vlcg}$—The first targeted allele of the $Apof$ gene created by Velocigene (Regeneron Pharmaceuticals) for the KOMP knockout project.
- $Btk^{tm1a(EUCOMM)Hmgu}$—The initial "knockout-first" (tm1a) allele of the Btk gene created by Helmholtz Zentrum Muenchen GmbH (Hmgu) for the EUCOMM project.

Once fully designated in a publication, the allele can be abbreviated by omitting the portion of the allele designation in parentheses (e.g., $Apof^{tm1Vlcg}$), provided the symbol remains unique.

2. Gene Trap Mutations

Gene trap mutations are symbolized similarly to targeted mutations if the trapped gene is known. The superscripted portion of the gene-trapped allele symbol begins with the prefix Gt (for gene trap), followed by a vector designation in parentheses, a serial number assigned by the laboratory characterizing the locus, and the lab code. If the gene trap insertion site is unknown or not within a gene then only the gene trap symbol Gt(vector)#Labcode is utilized. This is the preferred symbol designation for individual gene trap alleles generated from private laboratories.

For high-throughput systematic gene trap pipelines, the mES cell line designation can be used in parentheses instead of the vector designation, and the serial number following the parentheses may be omitted. Examples include:

- $Gt(OST75692)Lex$—A gene trap insertion at an undefined locus in mES cell line OST75692, made by Lexicon Genetics, Inc.
- $Pfdn1^{Gt(GTR1.3)1Rul}$—A gene-trapped allele of the $Pfdn1$ gene generated by the insertion of gene trap vector pGTR1.3, the first made by H. Earl Ruley (Rul).
- $Aim2^{Gt(CSG445)Byg}$—A gene-trapped allele of the $Aim2$ gene, created by BayGenomics (Byg) using mES cell line CSG445.

D. Rules for Naming Strains or Stocks Carrying Genetically Manipulated Entities

The mouse or rat strain on which a transgene, targeted, or gene trap mutation is maintained should be named by giving the strain name for the genetic background, based on the Guidelines for Nomenclature of Mouse and Rat Strains (see Table 24.4), followed by a hyphen and the corresponding mutation designation. Examples for inbred strains, mixed background strains, coisogenic, and congenic strains are given below. The $Trp53$ targeted mutation strains held at JAX are good examples of how to designate strains in which the same mutation is on several different genetic backgrounds. Examples include:

- FVB/N-Tg(Zp3-cre)3Mrt/J—This transgenic insertion was made and continues to be maintained on the FVB/N inbred background.
- 129S2/SvPas-$Trp53^{tm1Tyj}$/J—The first targeted mutation of the $Trp53$ gene by Tyler Jacks (Tyj), made using D3 ES cells derived from the 129S2/SvPas substrain; chimeric founder mice were mated back to mice of the same strain as the ES cell line (129S2/SvPas), followed by maintenance on this background.
- B6;129S2-$Trp53^{tm1Tyj}$/J—This strain background is a mixture of C57BL/6J (B6) and 129S2/SvPas (129S2) obtained by mating the chimera to B6 mice and then mating within the stock.
- B6.129S2-$Trp53^{tm1Tyj}$/J—This is a congenic strain in which the same $Trp53^{tm1Tyj}$ mutation was backcrossed for at least five generations onto the B6 inbred background.
- C.129S2(B6)-$Trp53^{tm1Tyj}$/J—This is a congenic strain in which the same $Trp53^{tm1Tyj}$ mutation was backcrossed for at least five generations onto the BALB/c (C) inbred background after being crossed once or more times to B6 mice.
- B6.Cg-Tg(Gcg-cre)1Herr/Mmnc—This is a congenic strain in which a Cre transgene with a Gcg promoter, the first mouse line from Pedro L. Herrera (Herr), has been transferred from a mixed genetic background to the B6 strain. This strain is held at the MMRRC-UNC repository (Mmnc). Note that the Cg abbreviation can be used to

indicate that the donor strain is either mixed (as in this case), of complex origin, or unknown.

- B6C3-Tg(HD82Gln)81Dbo/J—This is a stock that is maintained at JAX by repeated backcrossing to B6C3F1 hybrid mice. The female parent of the F1 hybrid is given first and the male parent second. Note that these mice are not true F1s. The stock will always be segregating for genes that differ between the two parental strains but may be either heterozygous or homozygous for either parental allele.
- STOCK Tg(Mx1-cre)1Cgn/J—The STOCK designation is used for stocks or inbred strains that have derived from several different genetic backgrounds, are maintained by sibling matings, and have (or will) become an inbred strain unto themselves.

Acknowledgments

The authors thank Drs. Muriel T. Davisson and Susan M. Bello for helpful suggestions and critical review of the manuscript. This work is supported by NIH grant HG000330.

References

Amberger, J., Bocchini, C., Hamosh, A., 2011. A new face and new challenges for Online Mendelian Inheritance in Man (OMIM®). Hum. Mutat. 32 (5), 564–567. Available from: http://dx.doi.org/10.1002/humu.21466.

Blake, J.A., Bult, C.J., Eppig, J.T., Kadin, J.A., Richardson, J.E., The Mouse Genome Database Group, 2014. The Mouse Genome Database: integration of and access to knowledge about the laboratory mouse. Nucleic Acids Res. 42 (D1), D810–D817. Available from: http://dx.doi.org/doi:10.1093/nar/gkt1225.

Bockamp, E., Sprengel, R., Eshkind, L., Lehmann, T., Braun, J.M., Emmrich, F., et al., 2008. Conditional transgenic mouse models: from the basics to genome-wide sets of knockouts and current studies of tissue regeneration. Regen. Med. 3 (2), 217–235. Available from: http://dx.doi.org/10.2217/17460751.3.2.217.

Brown, S.D., Moore, M.W., 2012. The International Mouse Phenotyping Consortium: past and future perspectives on mouse phenotyping. Mamm. Genome 23 (9–10), 632–640. Available from: http://dx.doi.org/10.1007/s00335-012-9427-x.

Bucan, M., Eppig, J.T., Brown, S., 2012. Mouse genomics programs and resources. Mamm. Genome 23 (9–10), 479–489. Available from: http://dx.doi.org/10.1007/s00335-012-9429-8.

Bult, C.J., Eppig, J.T., Blake, J.A., Kadin, J.A., Richardson, J.E., The Mouse Genome Database Group, 2013. The mouse genome database: genotypes, phenotypes, and models of human disease. Nucleic Acids Res. 41 (D1), D885–D891. Available from: http://dx.doi.org/10.1093/nar/gks1115.

Collins, F.S., Finnell, R.H., Rossant, J., Wurst, W., 2007. A new partner for the international knockout mouse consortium. Cell 129 (2), 235.

Donahue, L.R., Hrabe de Angelis, M., Hagn, M., Franklin, C., Lloyd, K.C.K., Magnuson, T., et al., 2012. Centralized mouse repositories. Mamm. Genome 23 (9–10), 559–571. Available from: http://dx.doi.org/10.1007/s00335-012-9420-4.

Doyle, A., McGarry, M.P., Lee, N.A., Lee, J.J., 2012. The construction of transgenic and gene knockout/knockin mouse models of human disease. Transgenic Res. 21 (2), 327—349. Available from: http://dx.doi.org/10.1007/s11248-011-9537-3.

Eppig, J.T., 2006. Mouse strain and genetic nomenclature: an abbreviated guide. In: Fox, J., Barthold, S., Davisson, M.T., Newcomer, C., Quimby, F., Smith, A. (Eds.), The Mouse in Biomedical Research, vol. 1, second ed., Academic Press, Burlington, MA: Elsevier, pp. 79—98.

Eppig, J.T., Strivens, M., 1999. Finding a mouse: the International Mouse Strain Resource (IMSR). Trends Genet. 15 (2), 81—82.

Friedel, R.H., Seisenberger, C., Kaloff, C., Wurst, W., 2007. EUCOMM—the European conditional mouse mutagenesis program. Brief. Funct. Genomic Proteomic 6 (3), 180—185.

Gama Sosa, M.A., De Gasperi, R., Elder, G.A., 2010. Animal transgenesis: an overview. Brain Struct. Funct. 214 (2—3), 91—109. Available from: http://dx.doi.org/10.1007/s00429-009-0230-8.

Gates, H., Mallon, A.M., Brown, S.D., EUMODIC Consortium, E., 2011. High-throughput mouse phenotyping. Methods 53 (4), 394—404. Available from: http://dx.doi.org/10.1016/j.ymeth.2010.12.017.

Hagn, M., Marschall, S., Hrabè de Angelis, M., 2007. EMMA—the European mouse mutant archive. Brief. Funct. Genomic Proteomic 6 (3), 186—192.

International Mouse Knockout Consortium, Collins, F.S., Rossant, J., Wurst, W., 2007. A mouse for all reasons. Cell 128 (1), 9—13.

Koscielny, G., Yaikhom, G., Iyer, V., Meehan, T.F., Morgan, H., Atienza-Herrero, J., et al., 2014. The International Mouse Phenotyping Consortium Web Portal, a unified point of access for knockout mice and related phenotyping data. Nucleic Acids Res. 42 (D1), D802—D809. Available from: http://dx.doi.org/doi:10.1093/nar/gkt977.

Mallon, A.M., Iyer, V., Melvin, D., Morgan, H., Parkinson, H., Brown, S.D., et al., 2012. Accessing data from the International Mouse Phenotyping Consortium: state of the art and future plans. Mamm. Genome 23 (9—10), 641—652. Available from: http://dx.doi.org/10.1007/s00335-012-9428-9.

Miller, R.L., 2011. Transgenic mice: beyond the knockout. Am. J. Physiol. Renal Physiol. 300 (2), F291—F300. Available from: http://dx.doi.org/10.1152/ajprenal.00082.2010.

Morgan, H., Beck, T., Blake, A., Gates, H., Adams, N., Debouzy, G., et al., 2010. EuroPhenome: a repository for high-throughput mouse phenotyping data. Nucleic Acids Res. 38 (Suppl. 1), D577—D585. Available from: http://dx.doi.org/10.1093/nar/gkp1007.

Murray, S.A., Eppig, J.T., Smedley, D., Simpson, E.M., Rosenthal, N., 2012. Beyond knockouts: CRE resources for conditional mutagenesis. Mamm. Genome 23 (9—10), 587—599. Available from: http://dx.doi.org/10.1007/s00335-012-9430-2.

Nord, A.S., Chang, P.J., Conklin, B.R., Cox, A.V., Harper, C.A., Hicks, G.G., et al., 2006. The International Gene Trap Consortium Website: a portal to all publicly available gene trap cell lines in mouse. Nucleic Acids Res. 34 (Suppl. 1), D642—D648. Available from: http://dx.doi.org/10.1093/nar/gkj097.

Prosser, H.M., Koike-Yusa, H., Cooper, J.D., Law, F.C., Bradley, A., 2011. A resource of vectors and ES cells for targeted deletion of microRNAs in mice. Nat. Biotechnol. 29 (9), 840—845. Available from: http://dx.doi.org/10.1038/nbt.1929.

Raess, M., Hrabé de Angelis, M., 2009. Infrafrontier—mouse models and phenotyping data for the European biomedical research community. EMBnet.news 15 (2), 16—19, Retrieved from: <http://journal.embnet.org/index.php/embnetnews/article/view/7/12>.

Ringwald, M., Iyer, V., Mason, J.C., Stone, K.R., Tadepally, H.D., Kadin, J.A., et al., 2011. The IKMC web portal: a central point of entry to data and resources from the International Knockout Mouse Consortium. Nucleic Acids Res. 39 (Suppl. 1), D849–D855. Available from: http://dx.doi.org/10.1093/nar/gkq879.

Skarnes, W.C., Rosen, B., West, A.P., Koutsourakis, M., Bushell, W., Iyer, V., et al., 2011. A conditional knockout resource for the genome-wide study of mouse gene function. Nature 474 (7351), 337–342. Available from: http://dx.doi.org/10.1038/nature10163.

Smedley, D., Salimova, E., Rosenthal, N., 2011. Cre recombinase resources for conditional mouse mutagenesis. Methods 53 (4), 411–416. Available from: http://dx.doi.org/10.1016/j.ymeth.2010.12.027.

Smith, C.L., Eppig, J.T., 2012. The Mammalian Phenotype Ontology as a unifying standard for experimental and high-throughput phenotyping data. Mamm. Genome 23 (9–10), 653–668. Available from: http://dx.doi.org/10.1007/s00335-012-9421-3.

Strivens, M., Eppig, J.T., 2004. Visualizing the laboratory mouse: capturing phenotypic information. Genetica 122 (1), 89–97.

Appendix

Additional Mouse-Related Databases and Online Resources

Resource Name (Acronym)	Web Address (URL)[a]
Anatomy, Development, and Imaging	
Allen Developing Mouse Brain Atlas	http://developingmouse.brain-map.org/
DeltaBase™ Histology Mouse Atlas	http://www.deltagen.com/target/ histologyatlas/HistologyAtlas.html
e-Mouse Anatomy (EMA) Atlas of Embryo Development	http://www.emouseatlas.org/emap/ema/ home.html
Embryo Images—Normal and Abnormal Mammalian Development Tutorial	https://syllabus.med.unc.edu/courseware/ embryo_images/
GenitoUrinary Development Molecular Anatomy Project (GUDMAP) Tutorials	http://www.gudmap.org/About/Tutorial/ index.html
High Resolution Mouse Brain Atlas	http://www.hms.harvard.edu/research/ brain/index.html
Histology Atlas of the Mouse Mammary Gland	http://mammary.nih.gov/atlas/index.html
MICe Technologies—Mouse Atlas	http://www.mouseimaging.ca/ technologies/mouse_atlas.html
Mouse BIRN Atlasing Toolkit (MBAT)	http://mbat.loni.usc.edu/
Mouse Cochlea Database (MCD)	http://mousecochlea.umn.edu/index.php
Mouse Virtual Necropsy	http://tvmouse.ucdavis.edu/ virtualNecropsy/
The Mouse Brain Library (MBL)	http://www.mbl.org/
PerkinElmer Inc.	http://www.perkinelmer.com/pages/020/ imaging/invivouniversity.xhtml
E-Mail List Services and Discussion Forums[b]	
Embryo Mail	http://embryomail.org/
IMPReSS Phenotype Procedures and Parameters Mailing List	http://www.mousephenotype.org/my-impc/ newsletters
Mouse Genome Informatics (MGI) E-Mail List (mgi-list)	http://www.informatics.jax.org/mgihome/ lists/lists.shtml
MGI Technical E-Mail List (mgi-technical-list)	http://www.informatics.jax.org/mgihome/ lists/lists.shtml
RIKEN BRC E-MAIL News	http://www2.brc.riken.jp/lab/info/ mailnews1.php
Transgenic List (tg-l)	http://www.transtechsociety.org/ transgeniclist.php
Open Biological and Biomedical Ontologies (OBO)-Phenotype Mail List	https://lists.sourceforge.net/lists/listinfo/ obo-phenotype

Resource Name (Acronym)	Web Address (URL)[a]
Gene Expression	
Allen Brain Atlas Data Portal	http://www.brain-map.org/
ArrayExpress Archive	http://www.ebi.ac.uk/arrayexpress/
Brain Gene Expression Map (BGEM)	http://www.stjudebgem.org/web/ mainPage/mainPage.php
Cerebellar Development Transcriptome Database (CDT-DB)	http://www.cdtdb.neuroinf.jp/CDT/Top.jsp
e-Mouse Atlas of Gene Expression (EMAGE)	http://www.emouseatlas.org/emage/home. php
Embryonic Gene Expression Database as a Biomedical Research Source (EMBRYS)	http://embrys.jp/embrys/html/MainMenu. html
EURExpress Transcriptome Atlas	http://www.eurexpress.org/ee/
European Renal Genome (EuReGene) Expression Database	http://www.euregene.org/portal
Expression Database in 4D (4DXpress)	http://4dx.embl.de/4DXpress/welcome.do
Gene Expression Database (GXD)	http://www.informatics.jax.org/expression. shtml
Gene Expression Nervous System Atlas (GENSAT)	http://www.gensat.org/index.html
Gene Expression Notebook (GEN)	http://www.informatics.jax.org/mgihome/ GXD/GEN/
Gene Expression Omnibus (GEO)	http://www.ncbi.nlm.nih.gov/geo/
GenePaint	http://www.genepaint.org/Frameset.html
GenitoUrinary Development Molecular Anatomy Project (GUDMAP)	http://www.gudmap.org/Menu_Index/ Gene_Expression.html
Mouse Atlas of Gene Expression Project	http://www.mouseatlas.org
Molecular Anatomy of the Mouse Embryo Project (MAMEP)	http://mamep.molgen.mpg.de/index.php
Gene Function and Pathways	
Gene Ontology (GO@MGI)	http://www.informatics.jax.org/function. shtml
Gene Weaver	http://geneweaver.org/
Kyoto Encyclopedia of Genes and Genomes (KEGG)	http://www.genome.jp/kegg-bin/ show_organism?org=mmu
MGI Biochemical Pathways (MouseCyc)	http://www.informatics.jax.org/pathways. shtml
MouseMine	http://www.mousemine.org/mousemine/ begin.do
Reactome	http://www.reactome.org/ReactomeGWT/ entrypoint.html
Genetic Variation (SNPs)	
Center for Genome Dynamics (CGD) Mouse SNP Database	http://cgd.jax.org/cgdsnpdb/

Resource Name (Acronym)	Web Address (URL)[a]
Database of Short Genetic Variants (dbSNP)—Mouse SNPs	http://www.ncbi.nlm.nih.gov/projects/ SNP/MouseSNP.cgi
Database of genomic structural Variation (dbVar)	http://www.ncbi.nlm.nih.gov/dbvar/
Database of Genomic Variants archive (DGVa)	http://www.ebi.ac.uk/dgva/
MGI SNP Query Form	http://www.informatics.jax.org/javawi2/ servlet/WIFetch?page=snpQF
Mouse Phenome Database (MPD) Mouse SNPs	http://phenome.jax.org/db/q?rtn=snp/ret1
Sanger Mouse Genomes Project: SNPs, indels, and structural variants	http://www.sanger.ac.uk/cgi-bin/ modelorgs/mousegenomes/snps.pl

Ontologies and Vocabularies

Adult Mouse Anatomical (AMA) Dictionary Browser	http://www.informatics.jax.org/searches/ AMA_form.shtml
BioPortal	http://bioportal.bioontology.org/ontologies
e-Mouse Atlas Project (EMAP) Ontology Browser	http://www.emouseatlas.org/emap/ema/ DAOAnatomyJSP/abstract.html
Gene Ontology (GO) Browser	http://www.informatics.jax.org/searches/ GO_form.shtml
Human Disease (OMIM) Vocabulary Browser	http://www.informatics.jax.org/javawi2/ servlet/WIFetch?page=omimVocab &subset=A
Mammalian Phenotype (MP) Ontology Browser	http://www.informatics.jax.org/searches/ MP_form.shtml
Mouse UniProt Gene Ontology Annotations (GOA) Database	http://www.ebi.ac.uk/GOA/mouse_release
Ontology Lookup Service (OLS)	http://www.ebi.ac.uk/ontology-lookup/
Phenote	http://www.phenote.org/about.shtml
Phenotypic Quality Ontology (PATO)	http://obofoundry.org/wiki/index.php/ PATO:Main_Page
The Open Biological and Biomedical Ontologies (OBO) Foundry	http://www.obofoundry.org/

Sequence, Genes, and Genome Browsers

DNA Data Bank of Japan (DDBJ)	http://www.ddbj.nig.ac.jp/
Ensembl Mouse Genome Browser	http://www.ensembl.org/Mus_musculus/ Info/Index
European Nucleotide Archive (ENA)	http://www.ebi.ac.uk/ena/home
GenBank	http://www.ncbi.nlm.nih.gov/genbank/
Genome Reference Consortium (GRC) Mouse Genome Assembly	http://www.ncbi.nlm.nih.gov/projects/ genome/assembly/grc/mouse/
HomoloGene	http://www.ncbi.nlm.nih.gov/homologene
MGI Genome Browser	http://gbrowse.informatics.jax.org/cgi-bin/ gb2/gbrowse/mousebuild38/

Resource Name (Acronym)	Web Address (URL)[a]
National Center for Biotechnology Information (NCBI) Map Viewer	http://www.ncbi.nlm.nih.gov/projects/mapview/map_search.cgi?taxid=10090
NCBI Mouse Genome Resources	http://www.ncbi.nlm.nih.gov/genome?term=mus%20musculus
NCBI Sequence Read Archive (SRA)	http://www.ncbi.nlm.nih.gov/sra/
NCBI RefSeq	http://www.ncbi.nlm.nih.gov/RefSeq/
University of California Santa Cruz (UCSC) Genome Browser	http://genome.ucsc.edu/cgi-bin/hgGateway?org=mouse
Vertebrate Genome Annotation (VEGA) Genome Browser	http://vega.sanger.ac.uk/Mus_musculus/Info/Index
Sanger Mouse Genomes Project (newly sequenced 17 inbred mouse strains)	http://www.sanger.ac.uk/resources/mouse/genomes/

Welfare, Husbandry, and Colony Management

Association for Assessment and Accreditation of Laboratory Animal Care International (AAALAC)	http://www.aaalac.org/
BigBench Mouse (BBMouse™)	http://www.bigbenchsoftware.com/
eMice—Animal Husbandry for Mice	http://emice.nci.nih.gov/animal-husbandry/mice-1
Geneoz Pty Ltd.—Vivarium Management Solution (VMS)	http://www.geneoz.com/
International Council for Laboratory Animal Science (ICLAS)	http://www.iclas.org/
Iseehear Inc. SoftMouse DB™—Online Mouse Colony Management Database	http://www.softmouse.net/
JAX Colony Management System (JCMS)	http://colonymanagement.jax.org/index.html
JAX® Mice—Mouse Care & Husbandry	http://jaxmice.jax.org/support/husbandry/index.html
Locus Technology, Inc.—Animal Management Software	http://www.locustechnology.com/products.html
MausDB	http://vm-jupiter.helmholtz-muenchen.de/index.html
Python based Relational Animal Tracking (PyRAT)	http://www.scionics.de/pyrat
University of California, Irvine—Mouse Husbandry, Breeding and Development	http://www.research.uci.edu/tmf/husbandry.htm
University of Minnesota—Research Animal Resources (RAR)	http://www.ahc.umn.edu/rar/links.html
Virtual Chemistry, Inc.—Mosaic Vivarium	http://www.virtualchemistry.com/VCI/products/mosaicvivarium.aspx

[a]All web addresses cited in the appendix were valid as of April 08, 2014, unless otherwise noted.
[b]User subscription is required for access.

Glossary

Artificial chromosome (AC) Heterologous DNA molecule capable of replicating in a given host, bacterium, yeast, or mammalian cell, that contains all functional elements required for auto-replicating and segregating normally upon cell division.

Bacterial artificial chromosome (BAC) A circular episomal DNA molecule present in bacteria which may harbor up to 300 kb of DNA. A BAC also contains functional elements for self-replicating and segregating in bacteria.

Bicistronic mRNA mRNA containing two independent coding sequences separated by an IRES (internal ribosome entry site) allowing ribosomes to translate the second cistron or by a peptide sequence specifically cleaved by an endogenous protease.

Blastocyst An embryo of approximately 64−128 cells, where the inner cell mass (primordial embryo) and trophoblast layers (primordial placenta) are differentiated and a blastocoel cavity has formed.

Boundary See Insulator.

Chimera An organism carrying cell populations derived from two or more different zygotes or embryonic stem (ES) cell lineages of the same or a different species. Chimeras include animals in which only some cells contain an engineered gene, may be recipients of tissue grafts from other individuals, or may contain an organ that has been purposely damaged and then reconstituted with cells from another species.

Chromatin Chromosomal DNA associated with proteins and RNA, condensed (heterochromatin) or open (euchromatin), normally associated with silencing/repressing gene expression.

Chromatin looping A mechanism allowing remote regulatory elements to participate in the formation of a transcription initiation complex.

Chromosomal position effects Alteration in transgene expression patterns due to the integration site in the host genome. Position effects are caused by the interference of neighboring genomic sequences associated with the transgene and its expression.

Cis-acting elements DNA sequences in proximity to a given gene (often 5′). It is not the chromosomal environment that determines if a gene is accessible by a transactivator at a given place or time, it is cis-acting element(s) that determine the state of chromosomal accessibility and consequently the tissue distribution and developmental timing of expression.

Cloning (also nuclear transfer) A clone is a population of cells produced from a common ancestor. There is gene cloning—done in the laboratory as a common procedure today, and then the sometimes controversial cloning of animals; here specifically a technique called "nuclear transfer" where a new cell nucleus is transferred into an enucleated egg; an egg in which the original nucleus is removed. At this time, cloning or nuclear transfer cannot be done in animals in the same fashion as plants where roots or limbs are cut off and repotted; a host embryo is generally needed for success.

Clustered regularly interspaced short palindromic repeats (CRISPR) DNA elements containing short palindromic elements and internal "spacer" DNA. In genome editing use, delivery of CRISPRs with spacer DNA that is homologous to a targeted genomic locus, together with a Cas9 protein, can result in double-strand DNA breaks.

Concatemer A continuous DNA sequence of considerable length comprised of multiple copies of the same DNA sequence.

Conditional transgenic An animal in which the engineered gene is subject to control by an element that regulates when (at what development stage) and/or where (in what cell or organ) the gene will be expressed.

Congenic A strain of animals (generally rodents) in which continual backcrossing for 10 or more generations produces a consistent and homogenous genetic constitution at all major gene loci.

Construct (DNA construct) An assembly of DNA sequences borne on a bacterial plasmid or similar vector that is used to insert engineered genetic material into a cell.

CpG motifs Sequences associated in clusters which may influence transgene expression including transgene silencing.

Cre-lox Site-specific recombination system isolated from the bacteriophage P1 which utilizes a recombinase protein (i.e., Cre) and its 34 bp DNA recognition site (i.e., lox).

CRISPR/Cas systems A strategy for genome editing by which CRISPR DNA or RNA is coinjected with Cas9 protein to create double-strand breaks in targeted genomic loci.

Cybrid Terminology for "cytoplasmic hybrid" cells, which are created by introducing one or more mitochondrial DNAs (mtDNAs) of interest into cells that have been depleted of their endogenous mtDNA. Cybrids are cultured cells manipulated to contain introduced mtDNA.

DNA methylation and demethylation Mechanisms responsible for inactivation and activation of genes, respectively.

DNA microinjection A gene transfer technique utilizing micromanipulation to inject DNA constructs (transgenes) directly into pronuclei or nuclei of cells. The foreign DNA integrates into the host cell genome at a random location and usually in concatemers. DNA microinjection was previously considered the most commonly used gene transfer technique for creating transgenic mammalian species; however techniques involving the use of site-specific nucleases hold promise to replace more primitive random transgenesis techniques.

Dominant negative A mutation whose product overrides the function of the normal wild-type gene product within the same cell.

Dot-blot (and slot-blot) analysis A variation of Southern blot analysis where genomic DNA is not resolved through agarose gel electrophoresis, but rather concentrated and immobilized in an area of a transfer membrane, with a dot-shaped (or slot-shaped) well connected to a vacuum device.

Double-strand break (DSB) A lesion in which both strands of a DNA molecule are severed. These are commonly produced in a specified location as an initial step in genome editing.

Egg A female gamete, ovum, or a female reproductive cell at any stage before fertilization and/or its derivatives after fertilization.

Embryo In animals, the stage of development, following division of a fertilized ovum, up until birth. This is the period of most rapid development.

Embryonic germ cells (EG cells) Multipotent cells able to participate in the development of gonads and used to generate transgenic avian species.

Embryonic stem cells (ES cells) Pluripotent cells, derived from early embryos, used for the generation of transgenic animals.

Embryonic stem (ES) cell transfer The transfer of genetically manipulable pluripotent ES cells into a developing embryo by microinjection or coculture of techniques. Resultant transgenic founders may include chimeric animals possessing a proportion of cells descended from the ES cell lineage. In mammalian species, this gene transfer technique has only been successful in producing germ line transmission of ES cell lineages in mice.

Engineered nuclease A DNA endonuclease where the sequence specificity of the DNA binding and cleavage sites is customized to a defined genomic locus.

Episome Any extrachromosomal DNA molecule not physically associated with the host genome.

FLP−FRT (flippase−flippase recognition target) A site-specific recombination system isolated from *Saccharomyces cerevisiae* which utilizes a recombinase protein (i.e., FLP) and its 48 base pair DNA recognition site (i.e., FRT).

FLPe A thermostable variant of FLP developed by cycling mutagenesis to work efficiently in mammalian cells.

FLPo A codon-optimized FLPe designed to yield higher protein levels in mammalian cells.

Fluorescence *in situ* hybridization (FISH) Method used to reveal the presence of unique DNA sequences in fixed metaphase chromosomes. Also possible with interphase nuclei.

Gamete A reproductive cell (spermatozoon for males, ovum for females) that contains the haploid set of chromosomes (i.e., only one copy of each gene).

Gene stacking A genetic engineering strategy for transgene linkage where the addition of a transgene is sequentially added (stacked) next to a previously introduced transgene.

Gene targeting Use of homologous recombination to knock-out or knock-in genes or to replace a given allele. The genetic engineering technique whereby a given gene is replaced by homologous recombination.

Gene therapy Therapeutic delivery of engineered genetic material in a chemical solution or by means of a viral vector (often to an adult animal) meant to correct a genetic mutation or deficiency.

Gene transfer One of a set of techniques involving the introduction of foreign DNA sequences (genes) into living cells.

Genetic engineering Intentional modification of a genome. Modifications include gene addition through random integration, homologous recombination (knockout or knock-in), or enzyme-mediated integration or disruption. It also includes gene disruption through insertion or deletion of DNA (knockout or knock-in) or gene modifications such as point mutations or domain swaps.

Genome editing A range of technologies wherein gene knockout/knock-in experiments can be carried out in a wide range of species and cell types via creation of double-strand DNA breaks in specified genomic loci. Subsequent DNA repair often introduces experimentally introduced knock-in constructs or random mutations.

Genome engineering See Genome editing

Genotype The genetic makeup of an animal.

Haploinsufficiency The presence of a single functional gene in a cell, which cannot alone produce enough gene product to sustain full function.

Haplotype A classification of a group of gene sequences at adjacent positions on a DNA strand that will be transferred together during genome replication.

Hemizygous animal An animal into which one (or more copies) of a nonendogenous gene was inserted, with the genetic modification at only one of the two sister loci.

Heterozygous animal An animal with differing alleles at the same gene locus of paired chromosomes.

Histone posttranslational modification Mechanisms affecting DNA packaging molecules that are responsible for inactivation or activation of genes.

Homing endonuclease Naturally occurring endonucleases that recognize large (20−30 bp) DNA sequences.

Homologous recombination A chromosomal process by which homologous genomic sequences can pair and promote the exchange of DNA using the endogenous nucleus recombination machinery. This represents a mechanism of DNA repair in which a length of DNA is replaced by another DNA molecule with sequence homology.

Homozygous animal An animal with identical alleles at a gene locus.

Hypomorphic mutation A mutation in which the gene product functions at a reduced level.

Induced pluripotential stem cells (iPS cells) Pluripotent stem cells derived from somatic cells that can be used for the generation of transgenic animals.

Insulator Genomic DNA sequence usually found flanking gene expression domains and hence preventing any interaction from neighboring DNA sequences from taking place. Insulators can be used ectopically to shield transgenes and protect them from chromosomal position effects.

Interfering RNA (RNAi) Small RNA, potentially encoded by transgenes, able to knock down mRNA stability or translation.

Internal ribosome entry site (IRES) A nucleotide sequence that allows translation initiation within a messenger RNA sequence affecting protein synthesis.

Knockdown (gene knockdown, knockdown) An animal in which a gene's function has been reduced but not eliminated. Also, downregulation of gene expression, at the mRNA level, by RNAi.

Knock-in (gene knock-in) A gene replacement strategy using homologous recombination to switch an endogenous gene with a reporter or structural gene. Essentially, the endogenous gene is knocked out and the replacement gene is under the transcriptional control of the endogenous gene promoter.

Knockout (gene knockout) An animal in which an endogenous gene was replaced by a nonfunctional engineered gene.

Lentiviral transgenesis See Pseudotyped lentiviruses.

Linkage Physical juxtaposition of two genes such that inheritance of one gene essentially guarantees inheritance of the other gene.

Long noncoding RNA (lncRNA) RNA encoded by a genomic region containing no genes that participate in the specific control of gene expression.

Meganuclease See Homing endonuclease.

Morula An embryo consisting of a compact ball of around 16–32 cells.

Mosaicism Genetically different cell populations in an individual (generally one that developed from a single zygote) which are distinct at the level of the genotype or karyotype (i.e., at the chromosomal level). Mosaicism results from such events as a gene mutation or chromosomal nondisjunction during early embryogenesis. The extent of the mosaic state depends on the cleavage stage at which the genetic event occurred.

Nonhomologous end joining (NHEJ) A mechanism of DNA repair where double-strand breaks are ligated together. The process often results in short insertion or deletion (indel) mutations.

Nonsense-mediated decay Mechanism degrading mRNAs in which a 5′ splicing site is located less than 50 nucleotides from the termination codon.

Nuclear transfer See Cloning.

Oocyte A developing egg cell (ovum) at any developmental stage before or after ovulation but prior to fertilization.

Ootid A ripe ovum; one of four cells derived from the two consecutive divisions of the primary oocyte, and corresponding to the spermatids derived from division of the primary spermatocyte. In mammals, the second maturation division is not completed unless fertilization occurs; hence, the ootid has male as well as female pronuclear (haploid) elements.

Ova Plural of ovum.

Ovule The ovum within the Graafian follicle or any small, egg-like structure.

Ovum A female gamete synonymous with egg. Ovum is also used to designate any early stage of a conceptus.

P1-derived artificial chromosome (PAC) Circular episomal DNA molecule derived from P1 bacteriophage which can accommodate up to 100 kb of heterologous DNA and which also contains functional elements for self-replicating and segregating in bacteria.

Perivitelline space Region between the zona pellucida and cytoplasmic envelope of an oocyte/embryo.

Phenotype The sum of all functional and structural changes produced by the activity of an engineered gene.

Plasmid A self-replicating fragment of DNA that can be used to harbor a (superfluous) DNA fragment such as a transgene. Plasmids can serve to replicate fragments of DNA in a host species (usually bacteria) to facilitate construction and production of transgenes for use in genetic engineering projects. Plasmids are a type of DNA vector.

Polymerase chain reaction (PCR) A technique used to amplify minute amounts of target nucleic acids/sequences using specific oligonucleotides and a temperature-resistant polymerase, through repetitive cycles of denaturation, annealing, and extension.

Positive—negative selection A drug-based strategy to enrich for homologous recombination events in embryonic stem cells. This strategy uses a positive selectable marker (e.g., neomycin) that is incorporated into the genome, and a negative selectable marker (e.g., thymidine kinase) that is not introduced into the genome.

Primordial germ cells (PGCs) See Embryonic germ cells (EG cells).

Pseudotyped lentivirus A lentivirus packaged with an envelope protein from a different strain of virus, such as vesicular stomatitis virus envelope glycoprotein-G (VSV-G), to expand the host range by membrane fusion instead of presenting viral epitopes that are recognized by specific cell receptors. Such lentiviruses are used in the generation of lentiviral vectors for animal transgenesis.

Quantitative PCR (qPCR) Variant of PCR technology where the amplification is carefully monitored and the accumulation of the target DNA product can be efficiently and robustly quantified, using a variety of methods and approaches.

Real-time PCR see Quantitative PCR (qPCR).

Recombinase A protein that recognizes specific DNA sequences and promotes recombination between them, resulting in deletions, translocations, or inversions. An example of a recombinase is the Cre recombinase, which recognizes loxP DNA sequences.

Recombinase-mediated cassette exchange (RMCE) A strategy that uses the Cre-lox system to exchange a DNA sequence with another DNA sequence. Generally, a DNA sequence flanked by incompatible lox sites is inserted into the genome by random integration or homologous recombination. The newly placed sequence serves as a target for replacement by additional DNA sequences also flanked by incompatible lox sites.

Site-specific recombinase An enzyme that recognizes a specific sequence and facilitates insertions, deletions, and/or inversions between the recognized sequences. The recognition sequences and method of recombination are unique to each recombinase.

Slot-blot analysis See Dot-blot analysis.

Somatic cell nuclear transfer A technique used in the production of cloned animals where the nucleus of a somatic cell (or an intact somatic cell) is transferred into the cytoplasm of an enucleated oocyte and, which is then activated to create a diploid embryo.

Southern blot analysis A common analytical technique in molecular biology described by E.M. Southern in 1975, where DNA fragments resolved using horizontal agarose gel electrophoresis are transferred by means of capillary action onto suitable membranes where they are immobilized and subsequently used for hybridization with labeled DNA probes.

Stem cell A cell that upon division replaces itself and also gives rise to cells which differentiate further into one or more specialized types. Embryonic stem (ES) cells are cells obtained from an embryo generally at the blastocyst stage of preimplantation development.

Trans-acting elements DNA sequences that interact with genes in open domains and stimulate transcription.

Transcription activator-like effector nuclease (TALEN) A type of engineered endonuclease where the DNA-binding domain of the fusion protein is based on TALE proteins secreted by *Xanthomonas* bacteria.

Transgene Segment of DNA containing sequence from one or more organisms that possesses all appropriate elements critical for gene expression.

Transgene tandem arrays (or tandem arrays) Groups of various transgene DNA molecules inserted at a given site of the host genome, with different internal organizations, although mostly arranged as head-to-tail concatemers.

Transgenic animal A genetically engineered animal either harboring new/transferred DNA sequence in its genome following gene transfer or resulting from various manipulations of its own (endogenous) genomic DNA. Such manipulations will generally effect either loss or gain of specific genes or gene function.

Transgenic founder The first-generation transgenic organism directly produced as a result of gene transfer/germplasm manipulation procedures. The term does not imply that the transferred gene (transgene) is heritable to future generations.

Transgenic line A direct familial lineage of organisms derived from one or more transgenic founders, characterized by the passing of the transgene(s) to successive generations as a stable genetic element. The line includes the founder and any subsequent offspring inheriting the specific germ line manipulation.

Transmitochondrial animals Animals generated following introduction or modification of endogenous mitochondria populations, or after introduction of foreign mitochondrial cybrids into animal cells or embryos.

Transposon vector A vector containing a recombinant transposon allowing efficient foreign DNA integration and reliable transgene expression.

Xenomitochondrial mice Resultant mice following introduction of interspecies/interspecific mitochondria or xenomitochondrial cybrids into mouse models.

Xenotransplantation The transplant of cells, tissues, or organs between species.

Yeast artificial chromosome (YAC) Linear extrachromosomal DNA molecule present in yeast cells (the 17th chromosome) which can accommodate up to 2000 kb of heterologous DNA, containing functional elements for self-replicating and segregating in yeast cells.

Zinc finger nuclease (ZFN) A type of engineered endonuclease where the DNA-binding domain of the fusion protein is based on the zinc finger domain of naturally occurring DNA-binding proteins.

Zygote A fertilized ovum (single-cell, fertilized embryo).

*Some glossary terms were derived from earlier reviews including:

Dunn, D.A., Kooyman, D.L., Pinkert, C.A., 2005. Transgenic animals and their impact on the drug discovery industry. Drug Discov. Today 10, 757–767.

Galbreath, E.J., Pinkert, C.A., Bolon, B., Morton, D., 2013. Genetically engineered animals in product discovery and development. In: Haschek, W.M., Rousseaux, C.G., Wallig, M.A. (Eds.) Haschek and Rousseaux's Handbook of Toxicologic Pathology, third ed., pp. 405–460, Elsevier Inc., Academic Press.

Pinkert, C.A., Irwin, M.H., Moffatt, R.J., 1995. Transgenic animal modeling. In: Meyers, R.A. (Ed.) Molecular Biology and Biotechnology. pp. 901–907, VCH, New York.

Printed and bound by CPI Group (UK) Ltd, Croydon, CR0 4YY

03/10/2024

01040422-0017